17th Personal Armour Systems Symposium (PASS 2025)

Personal Armour

Bruges, Belgium
22-26 September 2025

ISBN: 979-8-3313-2614-2

Printed from e-media with permission by:

Curran Associates, Inc.
57 Morehouse Lane
Red Hook, NY 12571

Some format issues inherent in the e-media version may also appear in this print version.

Printed with permission by Curran Associates, Inc. (2025)

For permission requests, please contact Royal Military Academy (Belgium) at the address below.

Royal Military Academy (Belgium)
Department of Weapon Systems
Renaissance Avenue 30
1000 Brussels
Belgium

Additional copies of this publication are available from:

Curran Associates, Inc.
57 Morehouse Lane
Red Hook, NY 12571 USA
Phone: 845-758-0400
Fax: 845-758-2633
Email: curran@proceedings.com
Web: www.proceedings.com

TABLE OF CONTENTS

ARMOUR MATERIALS

BEHIND ARMOUR BLUNT TRAUMA

BEHIND HELMET BLUNT TRAUMA

BLAST INJURY AND MITIGATION

CASUALTY REDUCTION AND OPERATIONAL ANALYSIS

HUMAN FACTORS

HUMAN VULNERABILITY AND INJURY CRITERIA AND/OR CASUALTY REDUCTION AND OPERATIONAL ANALYSIS

PERSONAL ARMOUR END-ITEMS

TEST STANDARDS AND METHODS OF ASSESSMENTS

THREATS (INCLUDING LASER)

Cutting edge technology: an investigation into knife penetration performance of chain mail with differing dimensions

C. S. W. Partridge[1] and A. H. Jones[2]

[1]*Ministry of Defence, Defence Equipment & Support, MoD Abbey Wood, Bristol, BS34 8JH, UK. callum.partridge101@mod.gov.uk*
[2]*Materials and Engineering Research Institute, Sheffield Hallam University, Howard St, Sheffield, S1 1WB, UK. a.h.jones@shu.ac.uk*

Abstract. The prevalence of knife attacks makes knife protection essential for law enforcement, with 46,950 offences recorded in 2021 by UK police. Many armour systems use chain mail to defeat knife threats. However, no open literature investigates the influence of chain mail parameters (ring diameter, chain wire diameter and weld characteristics) on performance. These factors are likely to influence mass, cost and performance of armour systems. Five commercially available chain mail samples with differing wire diameters and ring diameters were tested to evaluate the correlation between chain mail parameters and penetration resistance against the UK Home Office's P1/B knife, the British Standards' EN-412 knife and two 'street' knives (one cook's knife and one carving knife). Using a bespoke test rig, parent material (PM) and welded regions were subjected to quasi-static knife penetration testing. Samples were also subjected to X-ray florescence, optical microscopy and hardness testing to evaluate the PM and weld properties. Results show welded regions typically account for 3% of the ring circumference. Lack of root penetration in the welds led to the consistent presence of V-notches, typically resulting in 41% lower penetration force compared to the PM when the knife was incident on the weld. This appeared to be due to lower wire diameter material present in the weld, rather than the welded material being inherently weaker. This was supported by hardness profiles across the weld showing no significant reduction across the welded region. Knife penetration force correlated well with ring cross sectional area, with a linear correlation for all knives when the blade was incident to the weld and PM. However, no correlation between penetration forces and knife characteristics (secondary bevel angle, spine thickness and tip angle) were observed, with linear correlation factors below 0.3.

1. INTRODUCTION

Knife protection continues to be an essential capability required by both civilian and military users due to the life threating potential that knife attacks pose [1, 2]. The UK Office of National Statistics indicates that, whilst knife attacks within the UK have decreased over recent years, they are still prevalent, with 46,950 offences recorded in 2021 [3], 22% of which were recorded in the London metropolitan area [1].

In general, the threat of knife attacks is less prevalent to military users due to the stand-off nature of small arms and fragmentation threats [4]. However, knife attacks may be the most prevalent threat to some specialist military user groups such as Military Police or close protection roles.

Whilst chain mail has been used as knife/sword protection for centuries [5], and remains an integral part of armour design, no open literature investigates the influence of chain mail parameters (ring diameter, chain wire diameter and weld characteristics) on stab resistance.

The Metropolitan Police Service (MPS) provided five commercially available chain mail samples with differing ring diameters and wire diameters (Table 1 and Figure 1) that were subjected to a variety of characterisation tests. A bespoke test rig was designed and built and samples subjected to quasi-static knife penetration testing using four knife types[1]. The blade edge was incident on the weld and 180° from the weld in the parent material (PM). Chain mail samples were also subjected to X-ray florescence (XRF) to determine steel composition, optical microscopy and hardness testing to evaluate and compare the chain mail's PM and weld properties.

Furthermore, knife characteristics (inclusive secondary bevel angle, spine thickness and tip angle) were measured to assess any correlation between chain mail penetration performance and knife characteristics.

The aim of this study was to evaluate if knife characteristics or the location of the cutting edge on individual chain mail rings had an effect on knife penetration forces.

[1] UK Home Office's P1/B knife, British Standards' EN-412 knife, store purchased cook's knife and store purchased carving knife

https://doi.org/10.52202/081621-0001

Table 1. Chain mail ring dimensions of samples provided by the MPS.

Wire Diameter (mm)	Ring Inner Diameter (mm)	Ring Outer Diameter (mm)
0.55	2.90	4.00
0.63	5.64	6.90
0.72	5.56	7.00
0.80	5.40	7.00
0.90	5.20	7.00

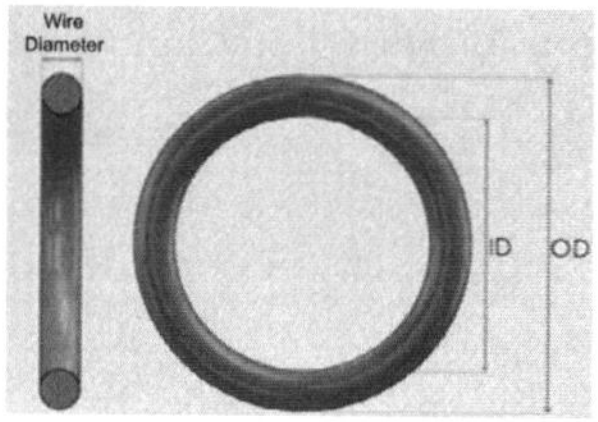

Figure 1. Chain mail ring dimensions indicating wire diameter, inner diameter (ID) and outer diameter (OD)

2. ELEMENTAL ANALYSIS

The MPS indicated the chain mail samples were stainless-steel, but no further information was known. Additional information was requested from the manufacturer but no response was received. XRF was conducted to provide assurance that samples were consistent, to determine the grade of stainless-steel and to enable estimates of mechanical properties, enabling appropriate testing equipment and procedures to be identified.

XRF was conducted using mobile and static XRF systems to provide additional confidence in the material grades. Mobile XRF provided quicker, but less accurate results, whilst static XRF provided slower but more accurate results. Furthermore, the mobile XRF provided a prediction of the material and grade from its internal library of materials.

2.1 EXPERIMENTAL PROCEDURE AND METHODOLOGY

2.1.1 Mobile XRF

Mobile XRF was conducted using a 3mm spot size ThermoFisher Scientific Niton XL3t Ultra Analyzer mounted on a ThermoFisher Scientific Mobile Test Stand in conjunction with Niton Data Transfer software.

2.1.2 Static XRF

Static XRF was conducted using a Rigaku ZSX Primus IV instrument in conjunction with ZSX Guidance Expert System XRF and EZ Analysis software.

2.2 LIMITATIONS

Due to XRF functionality, it was not possible to detect lighter elements that could influence the material's grade and structure, e.g., nitrogen or carbon content. Values are also possibly affected by contamination from the environment and handling.

2.3 RESULTS

Table 2 shows the average mass percentage of elements from all 5 chain mail samples. Mobile XRF predicted that all samples were made from super-duplex stainless-steel, grade 2205. Comparing the XRF data with literature [6], all samples are within the mass percentage limits associated with super-duplex stainless-steel 2205.

Table 2. XRF data indicating the average element mass % of all 5 chain mail samples

XRF method	Average element measured (Mass %)												Predicted material
	Al	Si	S	Cr	Mn	Fe	Ni	Cu	Mo	Na	Zn	V	
Mobile	0.00	0.24	0.00	22.30	1.50	66.78	5.77	0.11	3.25	0.00	0.01	0.02	SS 2205
Static	0.02	0.50	0.02	23.43	1.49	65.41	5.47	0.16	3.41	0.09	0.00	0.00	N/A

3. OPTICAL MICROSCOPY

All chain mail samples were subject to optical microscopy, evaluating welded and PM regions, to provide microstructural information such as; phase composition, grain orientation, inclusions, defects etc.

The software used allowed measurements to be taken of the weld, allowing the assessment of chain mail weld quality.

3.1 SAMPLE PREPARATION FOR MICROSTRUCTURAL ASSESSMENT

Individual chain mail rings were mounted in Struers conductive PolyFast Bakelite using a Buehler SimpliMet XPS1 mounting press for 150 seconds at 290 bar of pressure and 180°C. A subjective assessment of available space within the mounting determined the quantity of rings mounted for each sample (Figure 2).

Samples were ground using MetPrep abrasive paper and polished with MetPrep polycrystalline diamond suspension using a Buehler Alpha 2 speed grinder-polisher using a force of 20 Newtons at a speed of 300rpm in the contraflow rotation setting [7] until an appropriate surface finish was achieved. Samples were electro-etched using a Buehler PoliMat 2 for 25 seconds at 4.5 volts and 0.02 amps whilst submerged in oxalic acid (10% oxalic acid in aqueous solution) [8]. Samples were washed using methylated spirits (1% methanol in ethanol).
Due to the small surface area of the sample rings held in the Bakelite, three rings were ejected during the grinding and polishing procedure (One 0.55 mm ring, One 0.63 mm ring, and One 0.80 mm ring).

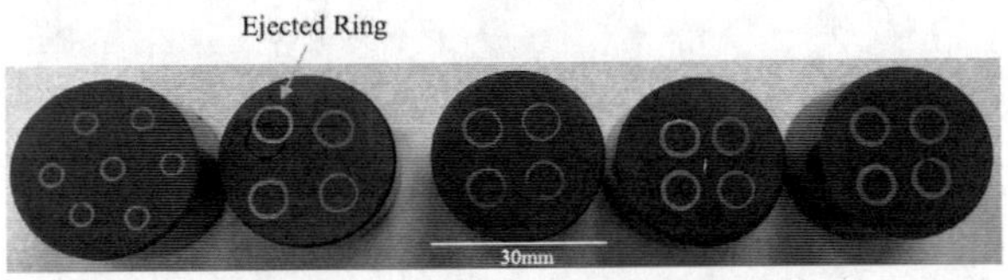

Figure 2. Mounted rings; smallest diameter (0.55 mm) on left increasing to largest diameter (0.90 mm) on right

3.2 OPTICAL MICROSCOPY PROCEDURE AND METHODOLOGY

A Leica DM2500M microscope was used to capture microscopy images in conjunction with Omnimet 10.1 software. Weld measurements were taken of the sample wire diameter, weld depth, cap width and effective weld depth (Figure 3) using Omnimet weld measuring software.

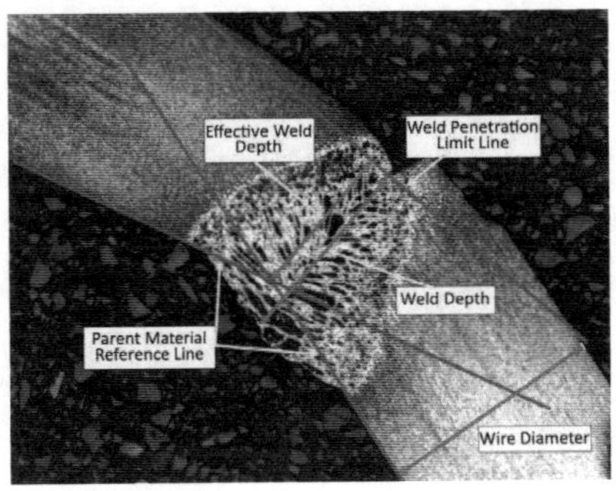

Figure 3. Weld measurement illustration

3.3 RESULTS

3.3.1 Optical microscopy Images

Optical microscopy images of all samples (n=20) identified 3 samples with PM defects consisting of; inconsistent ring wire diameter (Figure 4, a and b), and even one sample exhibiting a total lack of fusion in the weld (Figure 4, c).

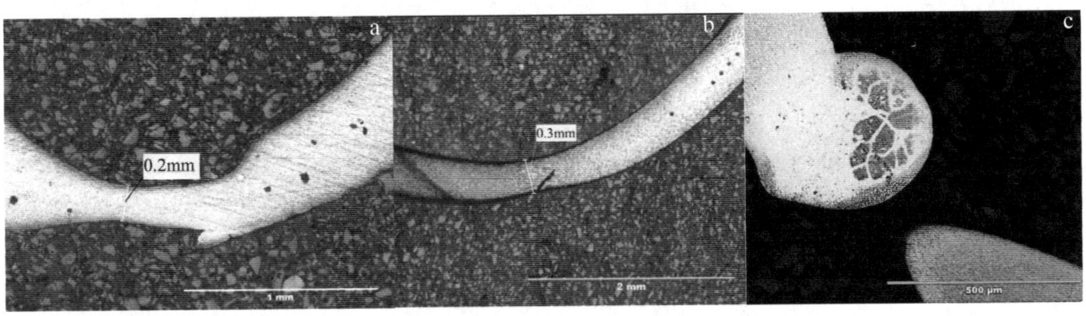

Figure 4. PM defects; (a) x5 magnification 0.55 mm diameter sample (b) x2.5 magnification 0.63 mm diameter sample (c) x10 magnification 0.90 mm diameter sample.

https://doi.org/10.52202/081621-0001

Images of the weld and Heat Affected Zone (HAZ) indicate changes in the microstructure compared to the PM for all samples, but to varying degrees (Figure 5). Unlike the PM, intergranular and intragranular black phases are observed within the weld and HAZ. An increase in size and quantity of lighter grains is also observed in the weld and HAZ compared to the PM, with all samples exhibiting a V-notches to varying extents on the weld root, indicating incomplete fusion.

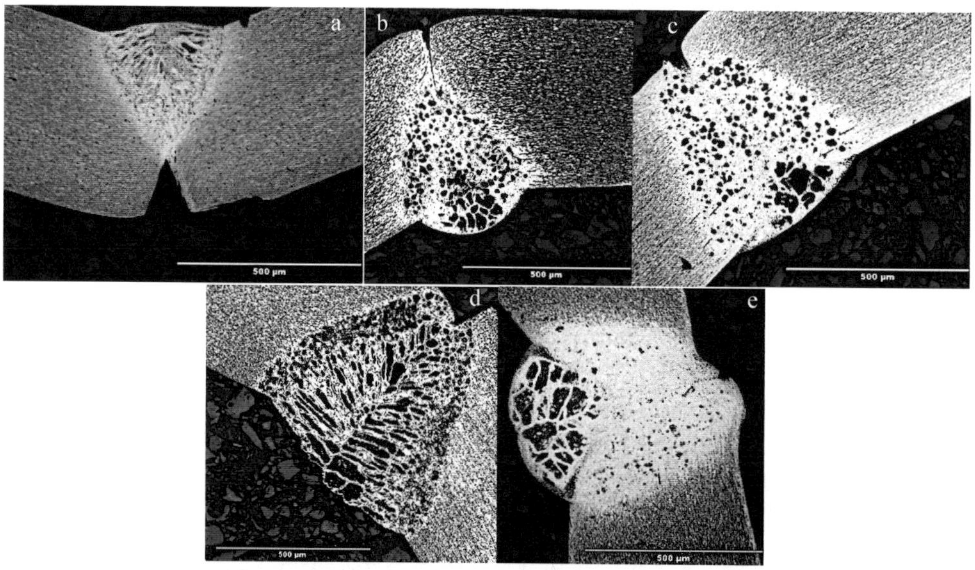

Figure 5. Typical x10 magnification optical microscopy images; sample diameter: (a) 0.55 mm (b) 0.63 mm (c) 0.72 mm (d) 0.80 mm (e) 0.90 mm.

3.3.2 Weld measurements

Table 3 indicates the width of welded region (cap width) as a proportion of the internal circumference, estimating the probability that a knife will be incident to the welded region when rings are randomly orientated.

Table 3. Average cap width as a percentage of ring internal diameter in accordance with Figure 3

Sample (nominal wire diameter, mm)	Nominal internal circumference (µm)	Average measured cap width (µm)	Cap width as a percentage of ring internal diameter (%)
0.55	10838	297	2.7
0.63	20012	441	2.2
0.72	19729	403	2.0
0.80	19478	690	3.5
0.90	19164	369	1.9

Variations in grinding and polishing times resulted in differences in material removal/depth, leading to variations in measured wire diameters compared with nominal diameters. Assuming consistent wire cross-section profiles, wire diameters and effective weld depths were measured to account for this variability, allowing the reduction in wire diameters to be calculated and compared. Results indicate a reduction in cross sectional wire diameter at the weld line across all samples (Table 4) due to the presence of V-notches.

Table 4. Average weld and PM cross sectional wire diameter measurements in accordance with Figure 3

Sample (nominal wire diameter, mm)	Average measured wire diameter (µm)	Average effective weld depth (µm)	Wire diameter reduction at weld (%)
0.55	483	273	43.5
0.63	576	431	25.2
0.72	562	373	33.7
0.80	694	417	39.9

https://doi.org/10.52202/081621-0001

0.90	667	465	30.2

5. HARDNESS TESTING

The same samples subjected to optical microscopy were subjected to hardness testing in the PM, the weld and the HAZ to assess any variation in hardness and therefore any differences in mechanical properties which may influence the knife penetration force if the blade edge was incident on the weld.

5.1 PROCEDURE AND METHODOLOGY

A Buehler Wilson VH3300 automatic hardness tester was used in conjunction with DiaMet software. The narrow ring wire diameter necessitated a force of HV0.2 to provide multiple valid readings.

Indentation distances were selected based on BS EN ISO 6507-1:2023 [9]. The PM, weld and HAZ were selected for testing (Figure 6) with six test locations selected in the PM and 12 across the HAZ (6 across the cap and 6 across the root). Quantity of tests along the weld varied based on the ring cross-sectional width.

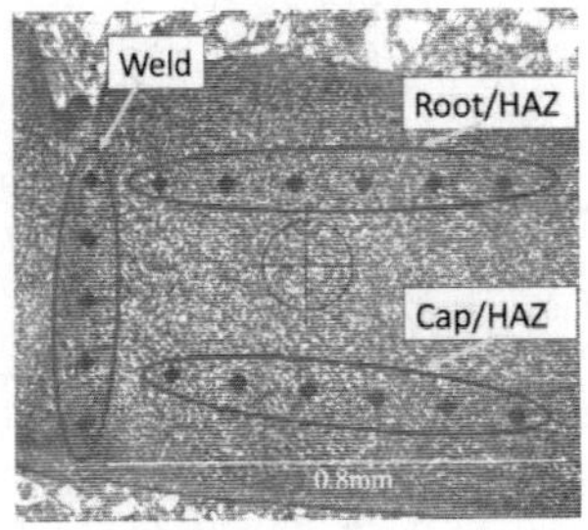

Figure 6. Example of hardness testing along the weld and across the HAZ

5.2 LIMITATIONS

Hardness testing results may not necessarily match the bulk stainless-steel due to variations caused by small wire dimensions and influence of the wire drawing process. Time constraints meant that electro-etched samples prepared for optical microscopy were not repolished prior to hardness testing, likely resulting in lower hardness values and increased variability compared to unetched samples, due to topographical relief introduced during the etching process.

5.3 HARDNESS RESULTS

Whilst there appears to be a small statistically significant difference in hardness in various locations and between some samples, the data indicates this isn't a large or consistent difference, suggesting consistent hardness across all regions of the chain mail ring (Figure 7). These results are consistent with similar studies [10].

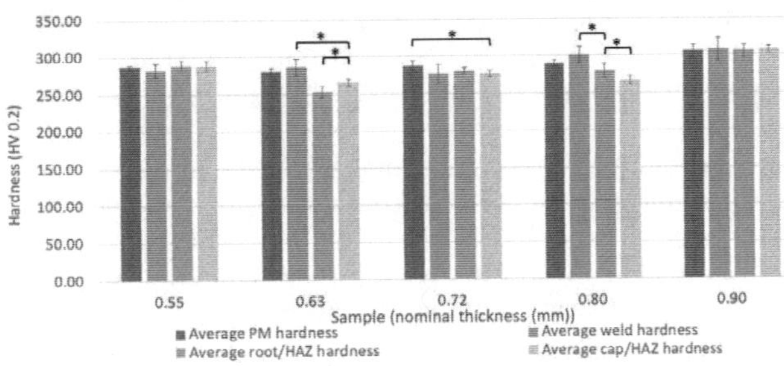

Figure 7. Graph comparing average sample micro-hardness. Error bars indicate 95% confidence. "*" indicates statistically significant ($p < 0.05$)

6. QUASI-STATIC STAB TESTING

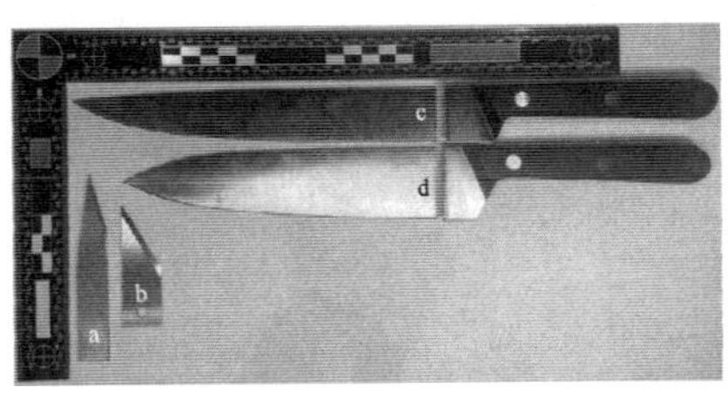

https://doi.org/10.52202/081621-0001

Quasi-static testing was conducted to assess the correlation between chain mail parameters and penetration force using the UK Home Office's P1/B knife [11], the British Standards' EN-412 [12] knife, one cook's knife and one carving knife (Figure 8).

Two quasi-static testing regimes were conducted; one to provide a performance comparison between chain mail with different dimensions, using P1/B and EN-412 knives only and the other to provide a performance comparison between knife characteristics, using all four knifes on the 0.72 mm chain mail only.

Microscopy images were taken of the penetrated chain mail rings post testing to compare the failure mechanisms exhibited.

Figure 8. Knives (a) P1/B (b) EN-412 (c) 23cm John Lewis carving knife (d) 20cm John Lewis cook's knife

6.1 EQUIPMENT DEVELOPMENT, PROCEDURE AND METHODOLOGY

Quasi-static testing was conducted using an Instron 3369 tensile tester. Knives were clamped in the upper driven crosshead. A bespoke chain mail retention rig (Figure 9) was designed to integrate with the stationary lower cross head.

The retention rig incorporates two sets of fixed pins, one small (Figure 9, b) for the 0.55 mm diameter chain mail and one large (Figure 9, c) for the remaining samples, with a central clearance hole for the knife tip. The retention rig pin distances were selected to support a square array[2] with the central ring being tested.

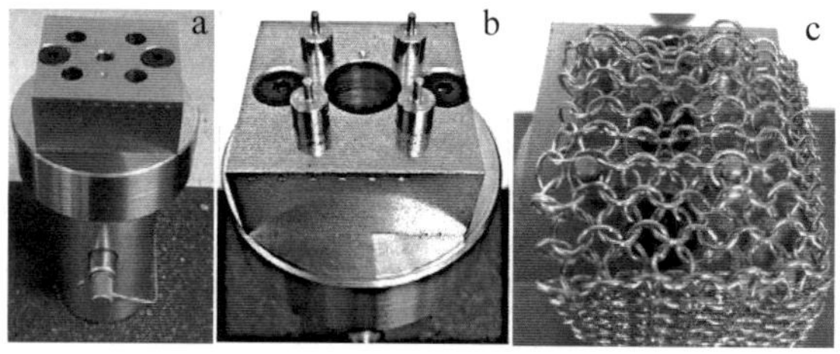

Figure 9. Chain mail retention rig; (a) pins removed (b) small pins (c) large pins with chain main set up

A 2 mm/min compression speed was selected for testing based on time limitations and similar existing tensile standards [13]. To negate chain mail 'jumping' and to standardise tension during initial loading, settings were adjusted to start recording readings at 3 Newtons of force.

Chain mail samples were positioned and orientated consistently, noting the chain mail exhibits a directionality (warp and weft). Central rings subjected to testing were rotated by hand to the desired position.

Due to limited numbers of P1/B knives, six preliminary tests were conducted using the same knife for multiple tests to assess the failure force or displacement variation. Results did not indicate significant variation from the first to the last test, informing the decision to use the same P1/B knife for subsequent tests where necessary. Preliminary testing was not necessary for the EN-412 knives as there were sufficient quantities to conduct one test per knife.

For chain mail comparison testing, 0.55 mm, 0.72 mm and 0.90 mm wire diameter samples (lower, middle and upper sample wire diameters respectively) were tested 6 times on the weld and 6 times on the PM using one P1/B knife every two tests to minimise variability. The 0.63 mm and 0.80 mm wire diameter samples were tested 6 times on the weld and 6 times on the PM using the 'used' knives, utilising two more tests per knife.

Microscopy images were taken of chain mail rings following chain mail comparison testing using a Leica CLS 150X microscope in conjunction with Omnimet software.

For knife characteristic comparison testing, the 0.72 mm wire diameter sample was tested using all four knives (Figure 8). The 0.72 mm wire diameter sample was selected as this was the middle wire diameter of chain mail samples. The cook's knife and carving knife were subjected to 6 tests on the weld and 6 on the PM due to only having one of each available. Two P1/B and two EN-412 knives were subject to 6 tests each (knife one; 6 tests on the weld, knife 2; 6 on the PM). This was a pragmatic compromise to minimise variability using the same knife for multiple tests whilst also providing a comparison with the store purchased knifes which were subjected to 12 tests each.

P1/B and EN-412 knife dimensions were identified using publicly accessible drawings [11, 12]. A Mk1 CATRA portable laser goniometer, unbranded vernier callipers and Trend DAR/300 Digital Angle Measurer were used to measure the store purchased knife's inclusive secondary bevel angle, spine thickness and tip angle, respectively, prior to testing (Table 7). These measurements were taken at the approximate location where the knives would engage the chain mail ring during testing.

6.2 RESULTS

6.2.1 Quasi-static stab testing (chain mail comparison)

[2] 8x6 ring array for 0.55mm diameter chain mail sample. 5x4 ring array for the remaining chain mail samples

Figure 10 and Figure 11 demonstrate that for all chain mail sizes, higher forces were required to cut the chain mail ring in the PM (green lines) than the welded region (red lines), and is summarised in Figure 12. Displacement and peak force results for both the weld and PM suggests a higher variability for wire diameter of 0.80mm and 0.90 mm. This may be due to the variability of the welding process with larger diameter wires, with more scope for a lack of fusion and larger notches being formed. However, variation on the parent metal results may also indicate a source of variation inherent to the experimental set-up, possibly related to the stiffness of the chain-mail fixture system and the higher forces required for failure on larger diameter wire. This would need to be explored further and improved upon in any extension of this work.

Table 5 indicates that the difference in failure force of the PM vs the weld is not consistent across both knives, with the 0.80 mm sample exhibiting the largest difference using the P1/B knife, and the 0.72 sample exhibiting the largest difference using the EN-412 knife.

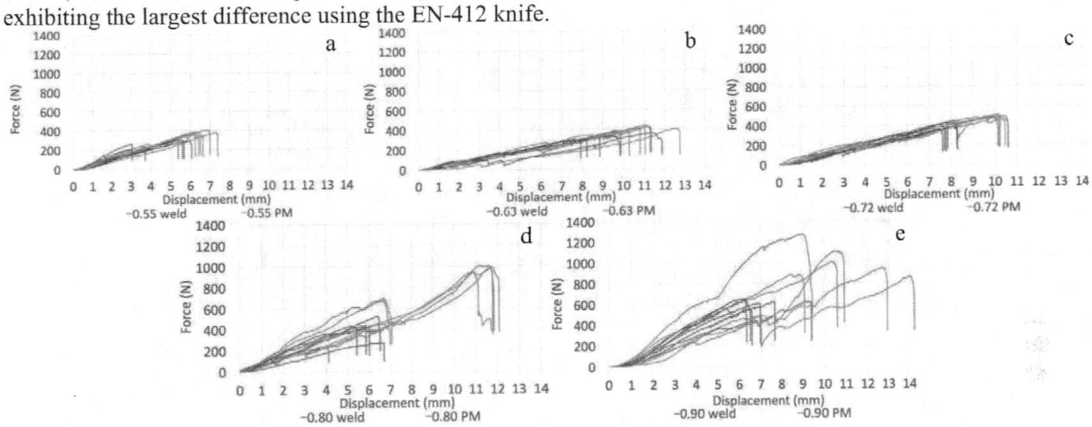

Figure 10. Chain mail force/displacement graph using P1/B knife. Sample diameter: (a) 0.55 mm (b) 0.63 mm (c) 0.72 mm (d) 0.80 mm (e) 0.90 mm.

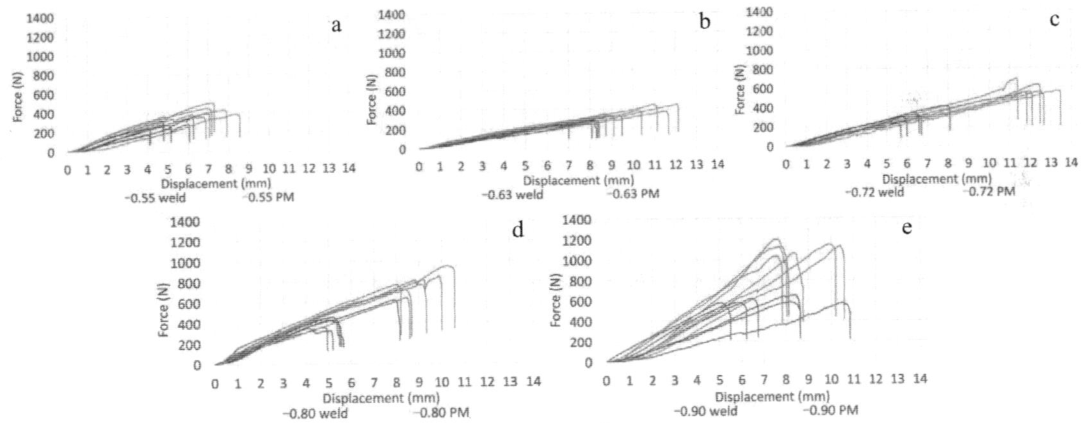

Figure 11. Chain mail force/displacement graph using EN-412 knife. Sample diameter: (a) 0.55 mm (b) 0.63 mm (c) 0.72 mm (d) 0.80 mm (e) 0.90 mm.

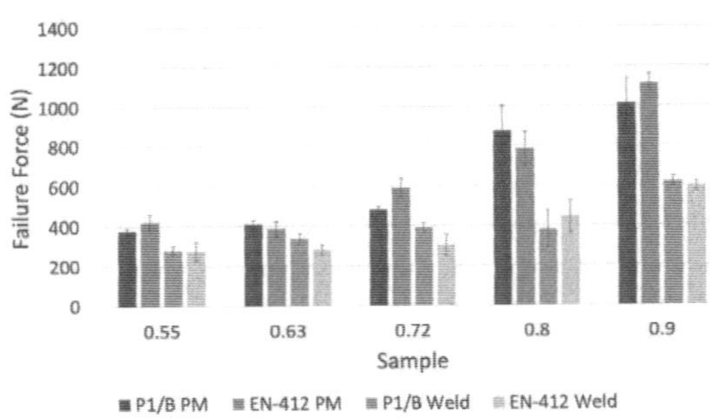

Table 5. Percentage difference in PM failure force vs weld failure force

Knife	Sample				
	0.55	0.63	0.72	0.80	0.90
P1/B	26%	18%	19%	57%	39%
EN-412	35%	27%	49%	44%	47%

https://doi.org/10.52202/081621-0001

Figure 12. Summary of average failure force using P1/B & EN-412 knives. Error bars indicate 95% confidence interval.

6.2.2 Post quasi-static stab testing (chain mail dimension comparison) microscopy

Figure 13 indicates different failure mechanisms observed when the blade edge was incident on the weld and PM. Penetrations on the weld indicate failure along the HAZ, rather than the weld itself, leaving the weld slug attached to one side, exhibiting a coarse surface finish. Cuts on the PM however demonstrate smoother fracture surface textures with a shear section from knife contact and shear lips from tensile failure, suggesting multiple failure mechanisms.

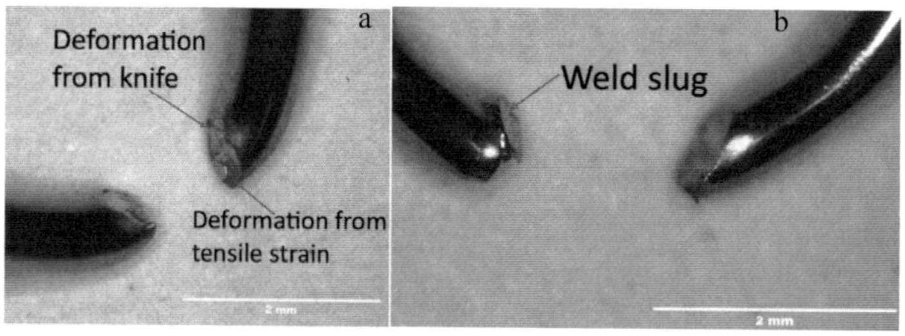

Figure 13. Typical post quasi-static stab testing microscopy images; (a) 0.72 mm diameter sample cut on PM, (b) 0.63 mm diameter sample cut on weld

6.2.3 Quasi-static stab testing (knife characteristic comparison)

Figure 14 demonstrates that for all knives the PM failure force is higher than that of the welded region and is summarised in Figure 15 and Table 6. Data indicates that the carving knife penetrates the chain mail with the least amount of force. This is consistent with previous studies where carving knives had some of the lowest forces to penetrate skin simulates [14]. However, there appears to be little to no correlation between isolated knife dimensions (Table 7) and chain mail failure force, with linear correlation factors below 0.3.

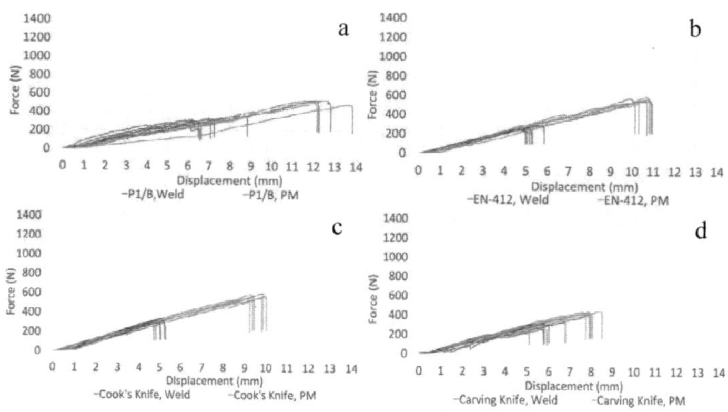

Figure 14. Diameter 0.72 mm chain mail force/displacement graph. (a) P1/B (b) EN-412 (c) cook's knife (d) carving knife.

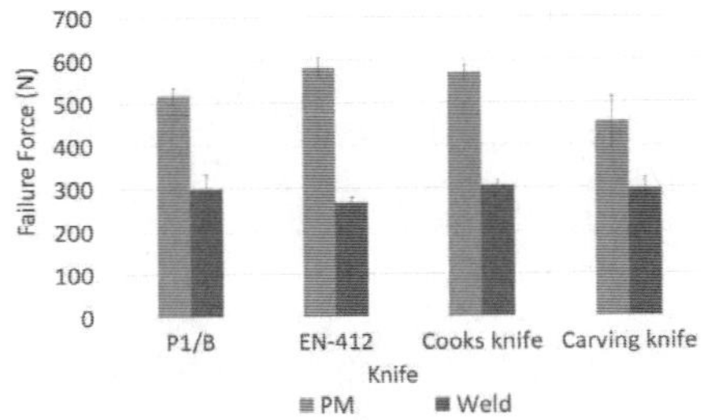

Figure 15. Summary graph of average failure force when testing diameter 0.72 mm samples against all knives. Error bars indicate 95% confidence interval.

Table 6. Percentage difference in PM failure force vs weld failure force

Sample	Knife			
	P1/B	EN-412	Cooks knife	Carving knife
0.72	42%	54%	46%	35%

Table 7. Knife characteristics measurements

Knife	tip angle (°)	Inclusive secondary bevel angle (°)	Spine thickness (mm)
P1/B	23	60	2.0
EN-412	30	30	1.5
Cook's knife	50	40	1.1
Carving Knife	30	40	1.2

7. DISCUSSION

Static and mobile XRF, corroborated by literature [6], identified all chain mail samples as duplex stainless-steel grade 2205. 2205 consists of 50% ferrite and 50% austenite [15], as illustrated by the light and dark grey areas in the parent material (PM) (Figure 5). Based on environments and forces associated with chain mail use, it is theorised 2205 has been selected based on its superior combination of strength, hardness, ductility, toughness and corrosion resistance when compared to other stainless steels such as, for example, 304 or 316 [16, 17]. Furthermore, duplex stainless steel's high specific strength allows manufacturers to provide thinner wire diameters for a desired strength [17], providing wearers with an armour system that is lighter and less bulky for the same strength, compared to other steels.

Despite variations in microstructure, hardness testing indicated few statistically significant differences between the PM, HAZ or weld. This is likely due to variations in phase volume fraction, size and distribution [18]. Whilst the weld itself had large volume fractions of precipitates that are typically harder than steel, their large size and heterogeneous dispersion would likely reduce its hardness. Conversely, the PM displays little to no precipitates, but has a smaller grain structure with a homogenous dispersion of ferrite and austenite grains.

Based on the weld cap width in relation to the ring circumference (Table 3), there is on average a 3% chance of a randomised impact resulting in the blade edge being incident on the weld. The chance of hitting the HAZ is slightly higher than this. Using the UK Home Office's stab protection standard for unformed armour [11] and cumulative probability of detecting a usability problem [19], there is a 31% probability that none of the 39 tests would impacting the weld. To achieve a probability of detection of 90%, 75 tests would be required. However, the relationship between knife penetration depths and dynamic forces associated with the UK Home Office standard is not known, and would be influenced by other system materials (cover, soft armour filler etc.). This would require further work to determine.

Figure 12 and Figure 15 show that there is a statistically significant decrease in weld failure forces at the weld compared to failure forces in the PM, for all samples and knives. This is likely the result of a combination the higher volume fraction of the weaker ferrite within the HAZ [20] and the observed V-notches (lack of fusion) resulting in lower cross-sectional area in the welded region which results in a lower force being needed for knife penetration.

When comparing the measured failure force (N) with the nominal cross-sectional area (mm^2) a good linear correlation is seen (Figure 16, a), with the gradient indicating the ring's "cut resistance" in N.mm^{-2}. This linear variation of force versus cross-sectional area is similar to that of tensile strength. Using this analysis, the cut resistance for the samples tested on the weld was 41% *lower* than those tested on the parent material. However, when using the measured weld depth, accounting for the presence of the V-notch (Table 4), to adjust for the actual reduced cross-sectional area at the weld (Figure 16, b) then the weld's cut resistance becomes 30% *higher* than the PM. This suggests that the performance limitation of the chain mail is the weld's cross-sectional area and not the weld's microstructure or inherent strength. However, it is noted that this does not necessarily correlate to the materials dynamic toughness or strength which would be more representative of threats.

a b

https://doi.org/10.52202/081621-0001

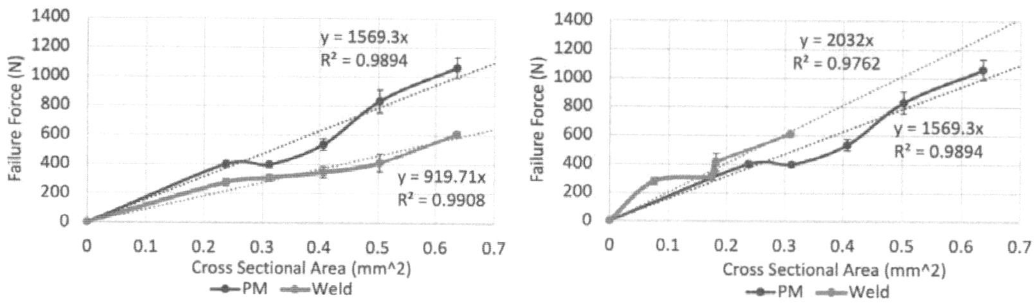

Figure 16. Average failure force vs cross-sectional area using P1/B & EN-412 knives (a) nominal cross-sectional area (b) measured cross-sectional area. Error bars indicate 95% confidence interval.

Measurement of the cut resistance of 0.72 mm diameter rings using knives with different geometries suggest there is little to no correlation between the knife characteristics and failure force, with linear correlation factors below 0.3.

8. CONCLUSION

Microscopy images show large variability in weld structure, including precipitates, phase quantity and grain size within all welds and HAZs. Furthermore, the weld's penetration depth in all samples is seen to be less than 100%, resulting in the formation of V-notches and reduced cross-sectional area. This combination of factors is the likely reason weld regions fail on average at 41% lower cut resistance during quasi-static testing compared to the parent material (PM). However, taking into account the reduction in cross-sectional area, welds have a 30% higher cut resistance. Based on the HOSDB stab testing standard [11], there is a 37% probability all 39 impacts will avoid impacting the weld. To achieve 90% confidence that welded regions are tested, 75 tests would be required, without taking into account other defects (e.g., thin sections) in the rings.

Although differences in weld and PM hardness were not statistically significant, suggesting similar properties to the PM throughout the weld, the literature [18] indicates phase transformation and precipitates in the weld may have an impact on properties. In particular, elongation to failure may be reduced.

No correlation was observed between the various knife characteristics and chain mail failure force for the knives tested in this work, which is a similar conclusion to Jones et al. [14] where no correlation between penetration force and knife geometry was found for a wide range of knife types, including the P1/B. A combination of factors may together influence the failure force but investigation of this would require further work.

Acknowledgments

The authors would like to acknowledge the contributions of Jane Barnes-Warden of the Metropolitan Police Service in providing support, expertise and facilitation of this research.

References

[1] J. Barnes-Warden, "Bladed Weapon Assaults and Human Vulnerability," in *PASS*, Dresden, 2023.
[2] E. Lewis and D. Carr, "Personal armor," in *Military and Law-Enforcement Applications*, Woodhead Publishing Series in Composites Science and Engineering, 2016, p. 217.
[3] Office of National Statistics, "Crime in England and Wales: year ending December 2021.," 28 April 2022.
[4] T. Atkins, "Punching Holes: Piercing and Perforating, Arms and Armour," in *The Science and Engineering of Cutting*, Butterworth-Heinemann, 2009, pp. 189-220.
[5] J. L. J. &. C. P. Gillingham, The Hutchinson Dictionary of Ancient and Medieval Warfare, Oxon: Taylor & Francis., 2016, p. 200.
[6] F. Cverna, "Worldwide Guide to Equivalent Irons and Steels," in *Worldwide Guide to Equivalent Irons and Steels*, ASM International, 2006, p. 956.
[7] Buehler, "SumMet: The Sum Of Our Experiences: A Guide to Materials Preparation & Analysis," in *The Sum Of Our Experiences: A Guide to Materials Preparation & Analysis*, Lake Bluff, Buehler, 2018, p. 45.
[8] J. C. Lippold, "Welding Metallurgy and Weldability," in *Welding Metallurgy and Weldability*, 2014.
[9] British Standards Institution, *Metallic materials- Vickers hardness test,* British Standards On-Line, 2023.
[10] G. S. Da Fonseca and et al., "Microstructural, Mechanical, and Electrochemical Analysis of Duplex and Superduplex Stainless Steels Welded with the Autogenous TIG Process Using Different Heat Input," *Metals,* vol. 7, no. 12, p. 538, 2017.
[11] Home Office Scientific Development Branch, *Body Armour Standard: CAST Publication number: 012/17.*
[12] British Standards Institution, *Protective clothing. Protective clothing – Protective aprons for use with hand knives; BS EN 412:1993,* British Standards On-Line, 1993.
[13] British Standards Institution, *Determination of tensile properties,* British Standards On-Line, 2023.
[14] A. H. Jones, I. Elomari and J. Barnes-Warden, "Characterising "Street Knives": A Study of the Tip Sharpness and Penetration Forces for Common Bladed Weapons," in *PASS*, Copenhagen, 2020.
[15] R. N. Gunn, "Duplex Stainless Steels," in *Duplex Stainless Steels*, Abington, 1997.
[16] R. A. Lula, "Effect of Nickel," in *Duplex Stainless Steels*, Ohio, 1983.
[17] International Molybdenum Association (IMOA), "Practical Guidelines for the Fabrication of Duplex Stainless Steel," in *Practical Guidelines for the Fabrication of Duplex Stainless Steel*, London, 2009.
[18] J. W. Martin, Precipitation Hardening, Oxform: Butterworth-Heinemann, 1998.
[19] British Standards Institute, "Medical Devices Part 2: Guidance on the application of usability engineering to medical devices," BSOL, 2016.
[20] M. Maslak, M. Stankiewicz and B. Slazak, "Duplex steels used in building structures and their resistance to chloride corrosion," *Microstructure and Corrosion Behavior of Advanced Alloys,* vol. 14, p. 5666, 2021.

https://doi.org/10.52202/081621-0001

Understanding Damage Propagation in UHMWPE Composite Armour Plate on Successive Impacts by 7.62 mm Projectiles

Kartikeya Kartikeya[1], Mohit Garg[1], Rajesh Punia[1] Khushi Ram[1], and **Naresh Bhatnagar**[1]

[1]*Department of Mechanical Engineering, Indian Institute of Technology Delhi, Hauz Khas, New Delhi – 110016, India*
nareshb@mech.iitd.ac.in

Abstract. The UHMWPE composite armour plates are extensively used to defeat 7.62 mm mild steel and lead core projectiles. However, after the first impact, the armour is weakened for subsequent impacts due to plate delamination associated with each impact. In this work, two identical UHMWPE composite armour plates were subjected to eight high-velocity impacts of 7.62 × 39 mm mild steel core (MSC) and 7.62 × 51 mm lead core (LC) projectiles. The delamination damage is quantified using ultrasonic C-Scans and Back Face Signature (BFS). It was noticed that both MSC and LC projectiles (8 shots) were defeated by UHMWPE composite armour. However, the damage inflicted by LC projectiles on an armour plate was considerably higher than by MSC projectiles. The cores of LC projectiles were destroyed during impact but the cores of MSC projectiles were retrieved and studied for steel deformation.

1. INTRODUCTION

Modern personal body armour is a lifesaver for security personnel. Personal body armour is typically made of multiple materials, given the diverse nature of threats. Hard armour plates (HAP) defeat high-velocity ammunition. These HAPs are typically fabricated from ultra-high molecular weight (UHMWPE) fibre-reinforced polymer composites and are added with either a hardened steel or ceramic strike face to defeat hardened core projectiles [1]. The UHMWPE composite armours are lightweight and provide excellent ballistic penetration resistance [2–4]. The HAPs from UHMWPE composite are generally fabricated by compression moulding [5–7].

Lee et al. [8] conducted one of the first impact studies on UHMWPE composite, Spectra® fibre-reinforced laminates. The authors noted that unidirectional fibre-reinforced composites performed better than woven fabric-reinforced composites. Taylor and Carr [9] observed that non-perforated UHMWPE composite panels showed extensive delamination whereas perforated specimens exhibited plugging and deformation. Karthikeyan et al. [10] conducted a study to understand the deformation and failure of UHMWPE fibre-reinforced laminates in comparison with carbon fibre-reinforced composite plastic (CFRP) laminates using the Moiré shadow technique. The UHMWPE laminates showed progressive delamination and failure and the CFRP laminates exhibited elastic behaviour till failure. The authors noted that UHMWPE laminate with the lowest shear strength matrix had the highest ballistic limit.

Heisserer et al.[11] observed that the depth of penetration in the UHMWPE composites had a linear relationship with the kinetic energy of the projectiles. They also observed that the amount of energy absorbed from perforated projectiles was similar in all cases showing that UHMWPE composite can absorb only a certain amount of energy. Nguyen et al. [12] compared V_{50} of thick UHMWPE composites with armour steel, aluminium and other fibre-reinforced composites using 12.7 mm and 20 mm FSPs. It was noticed that the UHMWPE composites were less space efficient than steels and glass fibre-reinforced composites, however, the UHMWPE composites were much more efficient than other tested materials. For thicker targets, the authors showed that the failure is a two-step process i.e., shear failure on the front surface up to a certain depth and membrane stretching on the back surface. Lassig et al. [13] showed that laminates fabricated at higher pressure had significantly improved ballistic limits. The plate planar impact tests were also carried out to determine the shock equation of state (EOS). The effect of consolidation pressure on shock EOS was evident below particle velocity of 1100 m s^{-1}.

Reddy et al. [14] subjected various 20 mm thick UHMWPE composites to impact by 7.62 × 39 mm MSC projectiles. The laminates had different hybridization using Dyneema® HB 50 and Tensylon® HSBD 30A. They showed that a laminate with equal weight per cent of Tensylon® and Dyneema® had the lowest back face signature when compared to other configurations. Cline et al. [15] showed that a compliant matrix of Dyneema® HB212 allowed greater energy dissipation through fibre rotation resulting in higher ballistic limit velocity. Zhang et al. [16] fabricated multilayered UHMWPE composite laminates with no bonding between layers. It was realized that as the number of layers increased, the

https://doi.org/10.52202/081621-0002

energy absorption capacity of laminate increased. The projectile was also able to draw more material inside thereby delaying high tensile stresses on the back face of the laminate.

Few authors studied 7.62 mm projectiles with deformable cores. Carbajal et al. [17] determined the mechanical properties of a 7.62 × 39 mm mild steel core (MSC) projectile. They performed compression tests on all projectile components at various strain rates. The mechanical tests and an iterative numerical model generated a reliable set of mechanical properties data for numerical simulations. Crouch et al. [18] studied the impact of 7.62 × 39 mm MSC projectiles on boron carbide targets. They found that the jacket and filler increase the penetrating capacity of the projectile by damaging ceramic before the arrival of the core. The authors determined that the penetration event was a two-step process, i.e., mushrooming of the core near the ceramic surface and erosion of the core as it penetrates [18]. However, this understanding is more relevant to imbibe for hardened steel cores. Recently, Giglio et al. determined the mechanical properties of 7.62 × 51 mm lead core (LC) projectiles for numerical modelling, which provided acceptable results for the projectile's impact on an aluminium target [19,20].

There are many studies on the understanding of UHMWPE composite armour. However, a study on post-test analysis of UHMWPE composite after HVI with 7.62 mm deformable core projectiles was found missing in the open literature and thus it is performed in this study at IIT Delhi.

HVI tests were conducted on in-house fabricated armour panels using 7.62 × 51 mm LC and 7.62 × 39 mm MSC projectiles. Subsequently, an investigation was done to understand the damage to the panels using an ultrasonic scan. The armour was cut using a waterjet machine to visualize the projectile's travel in the armour. Also, recovered projectiles were studied on their nature of deformation. The insights from the experiments will be beneficial in the numerical and analytical modelling of the penetration process of these studied projectiles and UHWMPE composite armour.

2. MATERIALS AND METHODS

This section provides manufacturing details of armour plates along with details of test set-up.

2.1. UHMWPE Armour Plates

The UHMWPE armour plates were fabricated in-house from Honeywell Spectra Shield® (SS6472). A single ply of SS6472 is around 120 μm thick and contains four laminae. The fibres in SS6472 are arranged in $(0°/90°)_2$ configuration in a polyolefin-based matrix.

First, the required number of plies were cut and stacked. The stack was then placed in a compression moulding machine where mould temperature was maintained in a range of 128-132 °C. A pressure was then applied on the stack for 20 mins. Four similar panels were fabricated with an areal density (AD) of 14 kg m^{-2}. The panels had a planar area of more than 1000 cm^2. The panel had nominal dimensions of 300 mm × 350 mm.

2.2. Projectiles

The projectiles used are shown in Figure 1 and their details are listed in Table 1. Both 7.62 × 51 mm LC and 7.62 × 39 mm MSC projectiles contain deformable cores unlike hardened steel cores in armour-piercing (AP) projectiles [21,22]. Table 1 also shows the threat level of these projectiles according to IS 17051:2018 i.e. Indian personal body armour testing standard [23]. The initial kinetic energy of an LC projectile is 1.7 times of an MSC projectile. The core length of the MSC projectile is 20.8 mm.

https://doi.org/10.52202/081621-0002

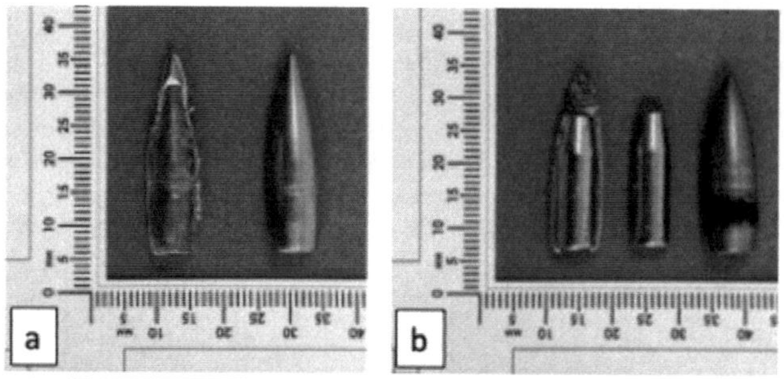

Figure 1. Projectiles used in this study (a) 7.62 × 51 mm LC and (b) 7.62 × 39 mm MSC

Table 1. Details of projectiles used in this study

Projectile	Weight (g)	Ordnance Velocity (m s^{-1})	Energy (J)	Core	IS 17051 *Threat Level*
7.62 × 39 mm MSC	7.45-8.05	710 ± 15	1953.39	Mild Steel	Level 2
7.62 × 51 mm LC	9.4-9.6	840 ± 15	3351.6	Lead	Level 3

2.3. High-velocity Impact Tests

The HVI tests were conducted at Terminal Ballistics Research Lab in Chandigarh, India as per IS 17051:2018 [23]. A schematic of the test set-up is shown in Figure 2. A weapon system was able to fire the required projectile is mounted 'a' distance from the target which in this case is 10 m. Also, between the weapon system and target, a velocity measurement system is present. The velocity measurement system is 'b' distance from the target which in this case is 2.5 m. The target is backed with Roma Plastilina No. 1 clay conditioned as per standard. Back-face signatures (BFS) are recorded on backing clay. There was at least a 51 mm distance on the panel between the shots as per IS 17051:2018.

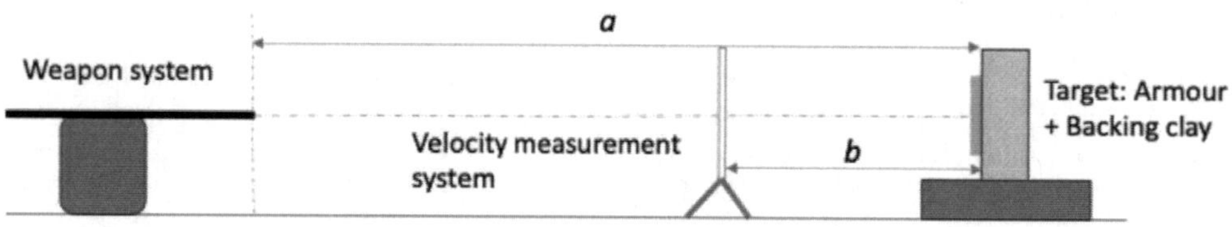

Figure 2. HVI test set-up

2.4. Post-Test Analysis

The BFS from each test was recorded on Roma Plastilina No. 1 clay after each test. The acceptable limit for BFS according to IS 17051:2018 is 25 mm for these projectiles. The ultrasonic sectional scan was performed using Olympus X3 equipment with a linear array transducer-receiver, glycerine was used as a medium for ultrasound wave propagation. Further, the armour panels were cut using a high-power waterjet machine. For UHMWPE panels subjected to the HVI test with 7.62 × 39 mm MSC projectile, reduced core length (RdCL) due to impact was also measured.

3. RESULTS AND DISCUSSIONS

3.1. BFS and Delamination

Results from HVI tests are presented in Table 2 and Table 3 for MSC and LC projectiles, respectively. It is quite apparent from these results that the BFS generated is considerably higher for LC projectiles than for MSC projectiles. Higher BFS will result in greater trauma to the user of armour. This signifies that with this armour design of UHMWPE composite, the user will endure greater trauma against LC projectiles. The high BFS values of LC projectiles can be attributed to the higher mass and velocity of these projectiles leading to almost double the kinetic energy of impact. It can be argued that BFS values of each shot is affected by prior shots on the clay. As this study followed IS 17051:2018 where although measurements of BFS are taken immediately after a shot but clay is not reset to default shape for the next shot.

Table 2. Results from HVI test for MSC projectile

MSC	HAP #1			HAP #2	
Shot No.	Velocity (m s^{-1})	BFS (mm)	RdCL (mm)	Velocity (m s^{-1})	BFS (mm)
1	711.3	11.9	13.6	714.7	10.2
2	715.8	7.4	14.2	715.3	8.5
3	714.9	4.0	15.1	712.5	6.5
4	715.7	4.6	16.4	713.5	5.2
5	714.5	4.6	15.3	715.9	5.0
6	708.2	1.5	14.1	714.9	5.2
7	722.8	3.3	14.6	715.8	4.0
8	715.1	2.7	14.9	715.1	6.1

Figure 3 also shows damage from both projectiles on similar panels. It is quite evident that, although both projectiles were defeated by UHMWPE composite panels, the delamination suffered in case of impact from the LC projectile was considerably more. Both panels delaminated into sub-laminates however, the number of sub-laminates was also higher in the case of LC projectiles.

Table 3. Results from HVI tests for LC projectile

LC	HAP #3			HAP #4	
Shot No.	Velocity (m s^{-1})	BFS (mm)	% Plies Perforated	Velocity (m s^{-1})	BFS (mm)
1	851.9	17.2	35.7	832.5	13.4
2	850.1	19.4	38.5	845.5	11.6
3	847.7	12.7	29.6	854.7	24.4
4	841.0	12.5	73.3	833.8	8.9
5	851.8	7.6	32.6	839.2	11.6
6	847.5	8.3	32.6	855.7	10.0
7	849.4	11.5	80.1	846.7	12.9
8	841.2	14.0	36.5	843.2	13.2

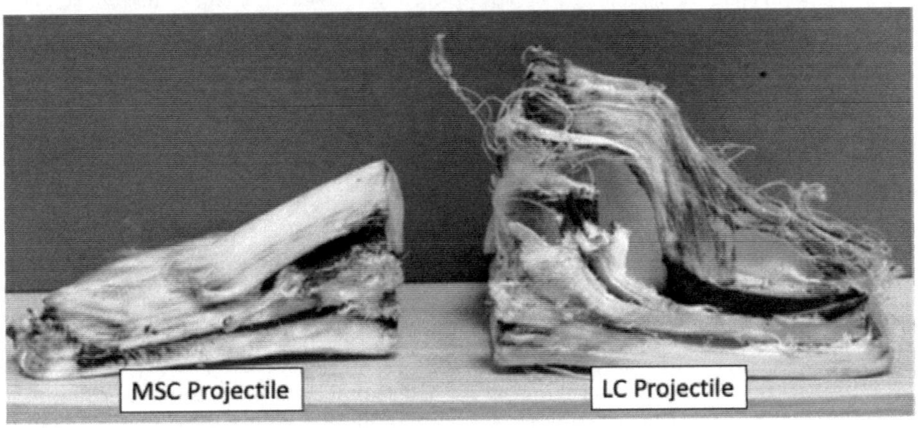

Figure 3. Delamination in UHMWPE composite panels due to impact from 7.62 mm projectiles with deformable cores

3.2. Post-Test Analysis

https://doi.org/10.52202/081621-0002

A typical LC projectile after the HVI test is shown in Figure 4. The LC projectile's lead deforms considerably around the copper jacket which also folds onto itself. On the other hand, the MSC projectile's mild steel deforms from the nose as shown in Figure 5. The nose is flattened and the final diameter at the nose side is much higher i.e. 8.6 ± 1.1 mm than the diameter at the shank (5.6 mm). The original diameter at the nose is 3.5 mm. Table 2 shows reduced core length RdCL for each shot for one panel only as the panel needs to be destroyed for it. The RdCL does not follow any trend with BFS and velocity. Also, the range of variation of RdCL values is limited between 13.6 and 16.4 mm. In contrast, core length reduction is quite a significant parameter in the defeat of hardened core projectiles [24]. Thus, the reduction in core length of the MSC projectile is insignificant in understanding the defeat of this projectile in UHMWPE composite panels at this velocity range.

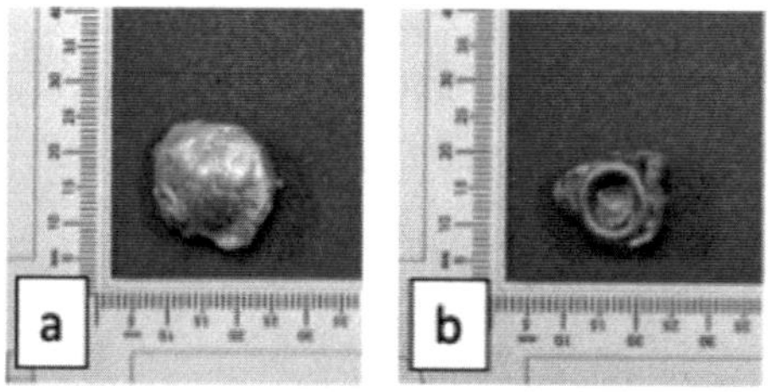

Figure 4. Typical state of 7.62 × 51 mm LC projectile

Figure 5. MSC cores from HAP #1 after the HVI test, cores are arranged in ascending shot order with the first one being the unused MSC core

Table 3 also shows the percentage of plies perforated with each shot in the case of LC projectile for one panel only as the panel needs to be destroyed for it. It was observed that LC projectiles were able to perforate the panel up to certain number of plies for each shot. The percentage of perforated plies was calculated by destructively opening the panel and taking a ratio of the weight of perforated plies to the weight of total panel. These values also do not seem to follow any trend with velocity and BFS and thus may be attributed to shot placement.

3.3. Ultrasonic Scan

A sectional ultrasonic scan was done to understand delamination from each shot. Figure 6 shows an ultrasonic scan of UHMWPE composite panels after HVI tests with 7.62 × 51 mm LC projectiles. In Figure 6 shot numbers are marked on the panel in black and the scanned region is marked in yellow along with arrows showing a relationship between them. Delamination after impact is quite evident in Figure 6. The colour from blue to red represents the intensity of the signal received. The faint blue contour at the bottom shows reflection from the back of the panel. The red contours show various places where delamination has occurred.

From Figure 3, it is apparent that the number of sub-laminates visible in the ultrasonic scan of Figure 6 is less. Since the signal gets dispersed due to such a high number of sub-laminates, only a few sub-laminates at the surface get registered in the ultrasonic scan. However, it is quite clear that near impact location a panel has undergone extensive delamination and will not provide effective ballistic resistance to the next shot.

Similarly, Figure 7 shows an ultrasonic scan of UHMWPE composite panels after HVI tests with 7.62 × 39 mm MSC projectiles. In the case of the MSC projectile, delamination in scans is like delamination visible in the sectional view shown in Figure 3. Also, it is apparent from Figure 7 that a lot

https://doi.org/10.52202/081621-0002

more signals are captured shown in various shades of blue which can be attributed to the closeness of delaminated sub-laminates. It is quite clear from ultrasonic scans that although a panel may look intact after the HVI test, it will not have original stiffness after delamination, which weakens it against the next shot.

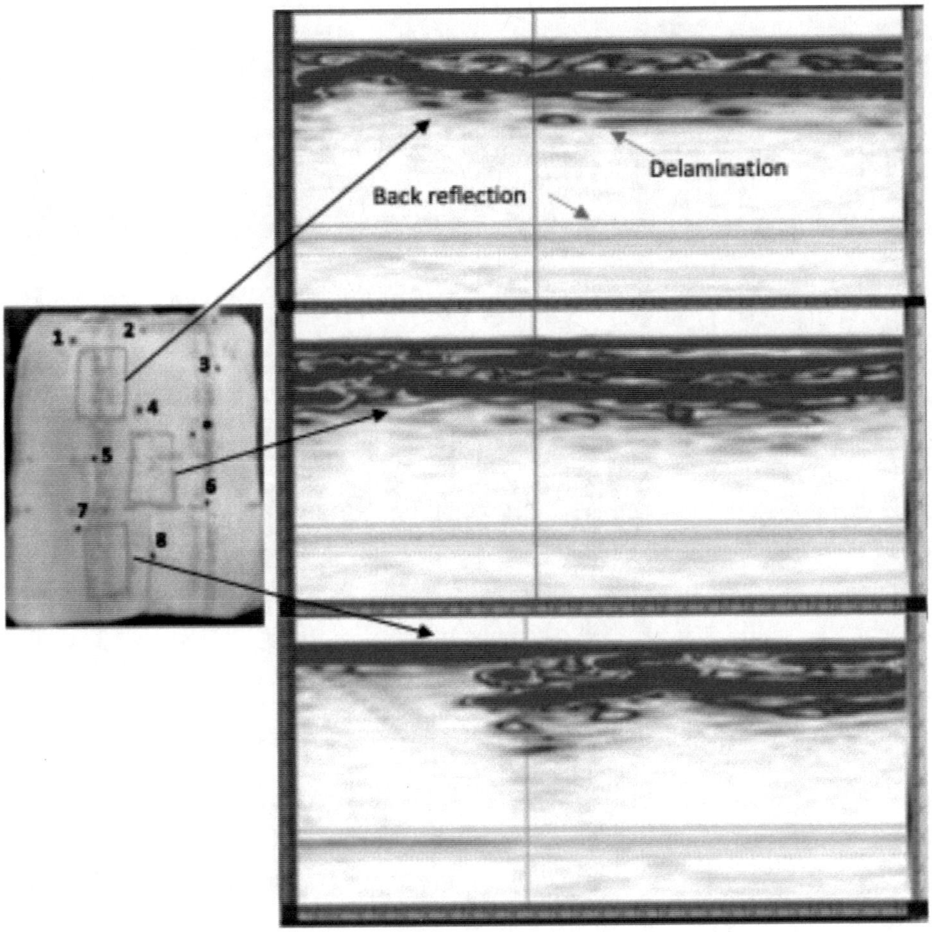

Figure 6. Ultrasonic scan of UHMWPE composite panels after HVI tests with 7.62 × 51 mm LC projectiles

https://doi.org/10.52202/081621-0002

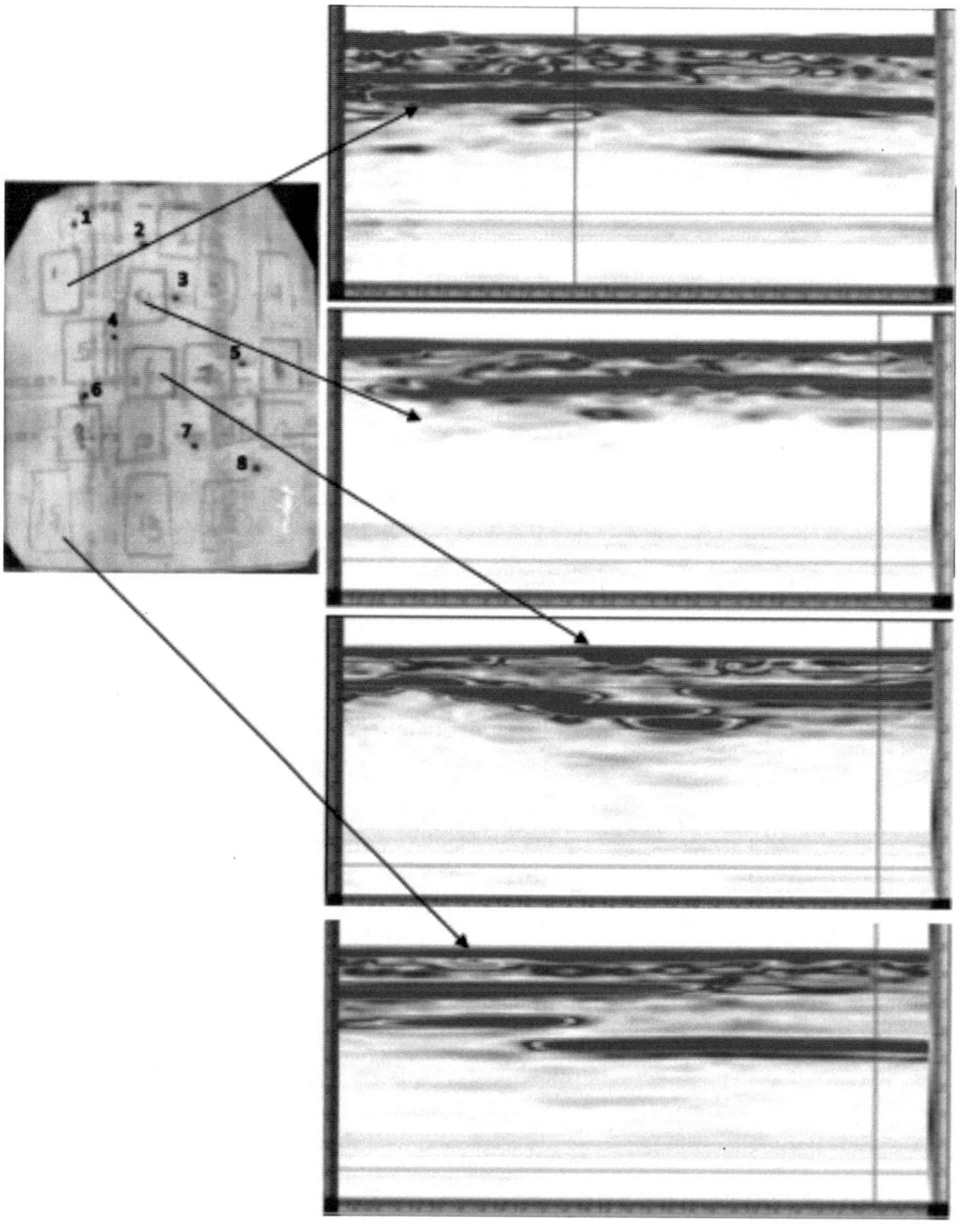

Figure 7. Ultrasonic scan of UHMWPE composite panels after HVI tests with 7.62 × 39 mm MSC projectiles

4. CONCLUSIONS

In this study, UHMWPE composite panels were subjected to HVI tests using 7.62 7.62 × 51 mm LC and 7.62 × 39 mm MSC projectiles. Subsequently, an investigation was done to understand the damage to the panels. The following conclusions can be drawn.

1. The BFS is higher for LC projectiles than for MSC projectiles which could be attributed to the higher initial energy of LC projectiles. Also, the number of delamination suffered by a UHMWPE composite panel is higher in the case of LC projectiles.
2. The reduced core length of the MSC projectile is not a significant parameter in determining the UHMWPE composite panel's ballistic resistance as is generally in the case of hardened core projectiles.

3. Similarly, number of plies perforated, in the case of LC projectiles, is dependent on shot placement.
4. Ultrasonic scan effectively shows delaminated areas in the case of both studied projectiles. However, an ultrasonic scan does not provide complete sectional detail which can only be studied after cutting the panel.
5. Future work can use inferences from this work like the number of sub-laminates in a section, delaminated areas or core deformation, for robust numerical analysis.

Acknowledgements

The authors acknowledge the use of manufacturing facilities at the Centre of Excellence in Personal Body Armour at the Indian Institute of Technology Delhi and testing facilities at Terminal Ballistics Research Lab DRDO Chandigarh.

References

[1] Bhatnagar A. Lightweight Ballistic Composites: Military and Law-Enforcement Applications: Second Edition. 2016. https://doi.org/10.1016/C2014-0-03657-X.

[2] Crouch IG. Body armour – New materials, new systems. Defence Technology 2019;15:241–53. https://doi.org/10.1016/j.dt.2019.02.002.

[3] O'Masta MR, Crayton DH, Deshpande VS, Wadley HNG. Mechanisms of penetration in polyethylene reinforced cross-ply laminates. Int J Impact Eng 2015;86:249–64. https://doi.org/10.1016/j.ijimpeng.2015.08.012.

[4] Nguyen LH, Ryan S, Cimpoeru SJ, Mouritz AP, Orifici AC. The effect of target thickness on the ballistic performance of ultra-high molecular weight polyethylene composite. Int J Impact Eng 2015;75:174–83. https://doi.org/10.1016/j.ijimpeng.2014.07.008.

[5] Greenhalgh ES, Bloodworth VM, Iannucci L, Pope D. Fractographic observations on Dyneema® composites under ballistic impact. Compos Part A Appl Sci Manuf 2013;44:51–62. https://doi.org/10.1016/j.compositesa.2012.08.012.

[6] O'Masta MR, Deshpande VS, Wadley HNG. Defect controlled transverse compressive strength of polyethylene fibre laminates. Int J Solids Struct 2015;52:130–49. https://doi.org/10.1016/j.ijsolstr.2014.09.023.

[7] Lässig TR, May M, Heisserer U, Riedel W, Bagusat F, van der Werff H, et al. Effect of consolidation pressure on the impact behaviour of UHMWPE composites. Compos B Eng 2018;147:47–55. https://doi.org/10.1016/j.compositesb.2018.04.030.

[8] Lee BL, Song JW, Ward JE. Failure of Spectra®* Polyethylene Fiber-Reinforced Composites under Ballistic Impact Loading. J Compos Mater 1994;28:1202–26. https://doi.org/10.1177/002199839402801302.

[9] Taylor SA, Carr DJ. Post-failure analysis of 0°/90°ultra high molecular weight polyethylene composite after ballistic testing. J Microsc 1999;196:249–56. https://doi.org/10.1046/j.1365-2818.1999.00632.x.

[10] Karthikeyan K, Russell BP, Fleck NA, Wadley HNG, Deshpande VS. The effect of shear strength on the ballistic response of laminated composite plates. European Journal of Mechanics - A/Solids 2013;42:35–53. https://doi.org/10.1016/j.euromechsol.2013.04.002.

[11] Heisserer U, Van Der Werff H, Hendrix J. Ballistic depth of penetration studies in Dyneema® composites. Proceedings - 27th International Symposium on Ballistics, BALLISTICS 2013 2013;2:1936–43.

[12] Nguyen LH, Ryan S, Cimpoeru SJ, Mouritz AP, Orifici AC. The effect of target thickness on the ballistic performance of ultra-high molecular weight polyethylene composite. Int J Impact Eng 2015;75:174–83. https://doi.org/10.1016/j.ijimpeng.2014.07.008.

[13] Lässig TR, May M, Heisserer U, Riedel W, Bagusat F, van der Werff H, et al. Effect of consolidation pressure on the impact behaviour of UHMWPE composites. Compos B Eng 2018;147:47–55. https://doi.org/10.1016/j.compositesb.2018.04.030.

[14] Reddy PRS, Reddy TS, Srikanth I, Kushwaha J, Madhu V. Development of cost effective personnel armour through structural hybridization. Defence Technology 2020;16:1089–97. https://doi.org/10.1016/j.dt.2019.12.004.

[15] Cline J, Love B. The effect of in-plane shear properties on the ballistic performance of polyethylene composites. Int J Impact Eng 2020;143. https://doi.org/10.1016/j.ijimpeng.2020.103592.

 https://doi.org/10.52202/081621-0002

[16] Zhang R, Han B, Zhong JY, Qiang LS, Ni CY, Zhang Q, et al. Enhanced ballistic resistance of multilayered cross-ply UHMWPE laminated plates. Int J Impact Eng 2022;159:104035. https://doi.org/10.1016/j.ijimpeng.2021.104035.
[17] Carbajal L, Jovicic J, Kuhlmann H. Assault riffle bullet-experimental characterization and computer (FE) modelling. Conference Proceedings of the Society for Experimental Mechanics Series 2011;6:647–64. https://doi.org/10.1007/978-1-4614-0222-0_77.
[18] Crouch IG, Appleby-Thomas G, Hazell PJ. A study of the penetration behaviour of mild-steel-cored ammunition against boron carbide ceramic armours. Int J Impact Eng 2015;80:203–11. https://doi.org/10.1016/j.ijimpeng.2015.03.002.
[19] Peroni L, Scapin M, Fichera C, Manes A, Giglio M. Mechanical properties at high strain-rate of lead core and brass jacket of a NATO 7.62 mm ball bullet. EPJ Web Conf, vol. 26, 2012. https://doi.org/10.1051/epjconf/20122601060.
[20] Giglio M, Gilioli A, Manes A, Peroni L, Scapin M. Investigation about the influence of the mechanical properties of lead core and brass jacket of a NATO 7.62 mm ball bullet in numerical simulations of ballistic impacts. EPJ Web Conf, vol. 26, 2012. https://doi.org/10.1051/epjconf/20122604010.
[21] Carbajal L, Jovicic J, Kuhlmann H. Assault riffle bullet-experimental characterization and computer (FE) modelling. Conference Proceedings of the Society for Experimental Mechanics Series 2011;6:647–64. https://doi.org/10.1007/978-1-4614-0222-0_77.
[22] Kiliç N, Ekici B. Ballistic resistance of high hardness armour steels against 7.62 mm armour piercing ammunition. Mater Des 2013;44:35–48. https://doi.org/10.1016/j.matdes.2012.07.045.
[23] Bureau of Indian Standards. Textiles — Bullet Resistant Jackets — Performance Requirements. IS 17051 : 2018 2018.
[24] Kartikeya K, Chouhan H, Ram K, Prasad S, Bhatnagar N. Ballistic evaluation of steel/UHMWPE composite armour system against hardened steel core projectiles. Int J Impact Eng 2022;164:104211. https://doi.org/10.1016/j.ijimpeng.2022.104211.

Effect of consolidation pressure on the impact behaviour of AK47 MSC on UHMWPE composite panels

H. van der Werff[1], U. Heisserer[1], J. van Elburg[1], W. Riedel[2]

[1]*Avient Protective Materials (APM), P.O. Box 1163, 6160 BD Geleen, The Netherlands, harm.werffvander@avient.com*
[2]*Fraunhofer Institute for High Speed Dynamics, Ernst-Mach-Institute (EMI), Zermelo Str. 4, D-79104, Freiburg, Germany.*

Abstract. Higher consolidation pressures during hot-pressing of UHMWPE composite panels are known to strongly improve the V_{50} performance, but the exact mechanism of this effect is not fully understood. In this study, the impact behaviour of the AK47 Mild Steel Core (MSC) projectile on four UHMWPE composites, mainly differing in resin type, processed at different consolidation pressures has been investigated in detail by the determination of V_{50}, and by depth-of-penetration and final core deformation measurements over a wide range of impact velocities. The V_{50} increased with higher consolidation pressures for all composite types. At projectile impacts below V_{50} with partial penetrations, increased consolidation pressures reduce the depth-of-penetration and result in higher deformation of the projectile steel core. All four UHMWPE composite types show different characteristic behaviours for V_{50}, depth-of-penetration and core deformation. Comparison of the results with shock wave properties from plate impact testing, suggests that higher consolidation pressures result in reduced porosity, a more homogeneous impact response and higher contact pressures between composite panels and projectile.

1. INTRODUCTION

Composites of high strength ultra-high molecular weight polyethylene fibres (UHWMPE) can be hot-pressed into panels offering high ballistic limit V_{50}'s for various threats at low weights [1]. It is known in practice that higher consolidation pressures during hot-pressing of the panels can lead to strong increases in V_{50} [2], but the exact physical mechanisms causing this behaviour are not fully known. This paper reports experimental results on the effect of panel consolidation pressure on 7.62x39mm AK47 MSC V_{50}, depth-of-penetration and deformation of the mild steel core, to provide more insight into the mechanism of the consolidation pressure effect.

2. EXPERIMENTAL

Experiments were carried out using four different UHMWPE composite types manufactured by Avient Protective Materials (APM) under the brand name Dyneema®: HB212, HB210, HB311 and HB480. All these four UHMWPE composites use the same UHMWPE fibre, but have different resin types. All the sheets of the composite types consist out of four uni-directional fibre layers stacked in a 0°/90° manner. Because of the different resin types, different consolidation temperatures have to be used for the different composites. Details are given in Table 1.

Table 1. Prepreg sheet areal densities of the UHMWPE composites and consolidation temperatures.

UHMWPE composites :	HB212	HB210	HB311	HB480
Areal density (g/m^2)	136	136	129	109
Consolidation temperature (°C)	125	130	130	145

HB212, HB210 and HB311 are conventional UHMWPE fibre composites obtained by impregnating the same fibre with three different resin systems. The resin stiffness increases from HB212 to HB210 and further to HB311. For AK47 MSC impact, higher resin stiffness in conventional UHMWPE fibre composites typically results in lower backface deformation but also slightly lower V_{50}. HB480 is not a conventional UHMWPE fibre composite. It is made using a proprietary manufacturing process (which we refer to as UD Film) and results into very stiff panels (so low backface deformation) but also higher V_{50} when impacted by AK47 MSC and Fragment Simulating Projectiles (FSP).

https://doi.org/10.52202/081621-0003

All panels were produced using an oil-heated hydraulic press with flat platens. 20x20 cm panels were prepared of each composite type varying the consolidation pressures from 20 bar, 165 bar to 300 bar while maintaining a constant areal density.
The thickness of the pressed panels was measured with a digital caliper on nine different locations in the panel.

The 7.62 mm AK47 MSC projectiles (124 gr / 8 g) were manufactured by Sellier & Bellot and fired in a range set-up shown schematically shown in Figure 1. The average Vickers Hardness measured on various locations of two cores was 177 HV10.

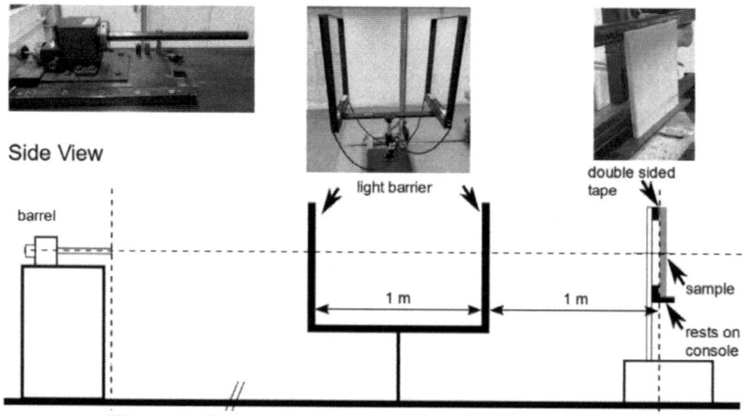

Figure 1. Schematical set-up of the shooting range.

The projectiles were shot from a barrel, adjusting the intended projectile velocities by the amount of powder in the cartridge. The projectile velocity was measured using two light barriers at 1 m distance, placed 1 m in front the target panel. The panel was mounted behind an upper and lower steel bar, maintaining its position by pressing the panel against double-side sticking tape on the bars. The distance from barrel to target was 12 m.

Each panel was shot only once with a projectile aimed at the center. Typically, over 25 samples were tested for each combination of composite type and consolidation pressure over a large range of impact velocities.
In case of a "stop" (i.e. a partial penetration), the panel was sawn open with a band-saw in order to retrieve the mild steel core. The length reduction from initially 20.0 mm was determined with a caliper. In case of a stop, the depth-of-penetration was determined as shown in Figure 2.

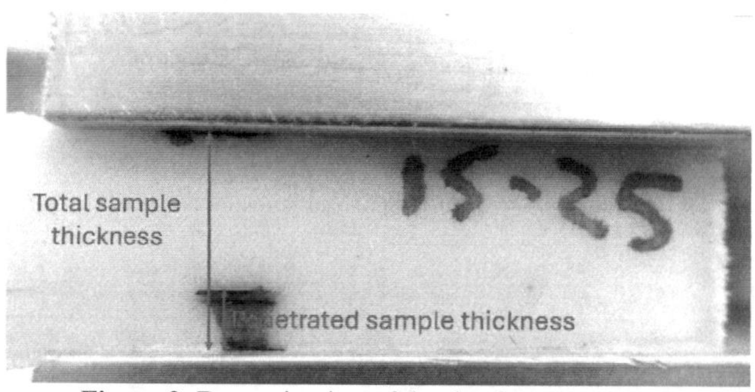

Figure 2. Determination of depth-of-penetration.

Deriving the depth-of-penetration by simply measuring the remaining sample thickness at the point of impact is inaccurate as the extensive shearing of the fibres leads to a thickening of the material. Therefore, the end of the penetrated sample thickness is taken at the point of impact, and this interlayer was tracked with a thin screw-driver towards and marked at the edge. The marked sample was then firmly clamped in an anvil, and the depth-of-penetration was determined as the penetrated fraction of the total sample thickness, as shown in Figure 2.

3. RESULTS

3.1 V_{50} and panel thickness

All individual shots were classified as stops (partial penetrations) or perforations (complete penetrations). The normalised results of the individual shots are shown in Figure 3. The V_{50} was calculated using a Probit analysis (based on the cumulative normal standard deviation).

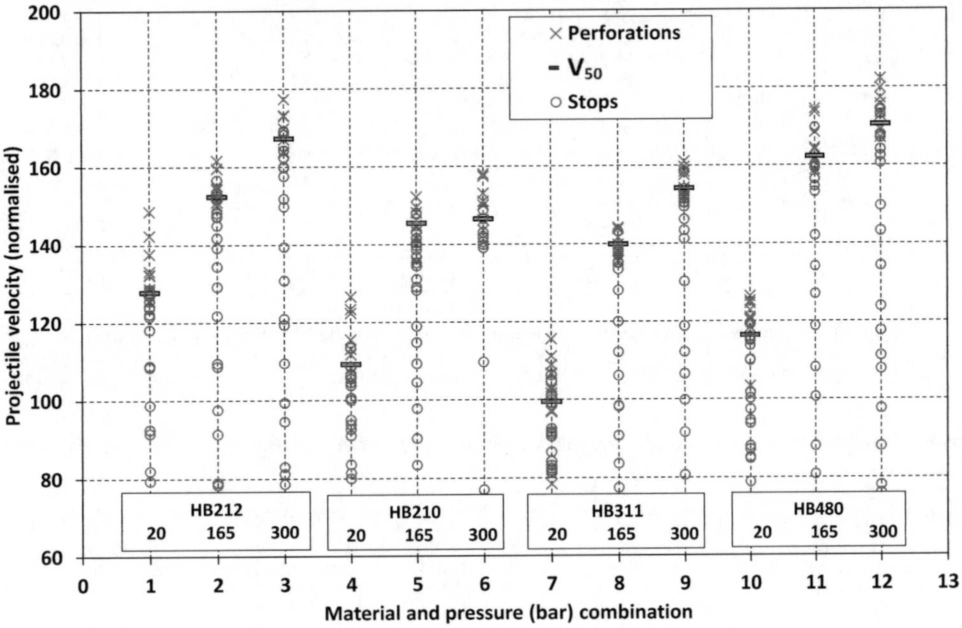

Figure 3. Individual shot result (normalised velocities) for all UHMWPE composite types pressed at the different consolidation pressures (20, 165 and 300 bar).

Figure 3 shows for almost all composites a very small region with mixed results. For the HB311, a low velocity penetration was observed at a normalised projectile velocity of 79, which is noticeably below the V_{50} of about 100. Low velocity penetrations of the AK47 MSC are known in the standards as the so-called shatter gap [3]. The stress that a projectile experiences upon impact is directly related to the impact velocity and the impedances of both projectile and target. If the impact velocity is too low the projectile core can not be deformed and the penetration probability increases.

The V_{50}'s calculated from the data in Figure 3 are plotted versus the consolidation pressure in Figure 4. The dashed lines are guides to the eye. For every composite type, the V_{50} increases very strongly from 20 to 165 bar. From 165 to 300 bar, the further enhancement is much smaller but still present. The results shown in Figure 4 also corroborate that AK47 MSC V_{50} for the conventional fibre UD composites decreases with higher resin stiffness (from HB212 to HB210 to HB311), and that the UD Film composite outperforms the conventional fibre UD composites.

Higher consolidation pressures generally result into significantly lower panel thicknesses. HB311 pressed at 300 bar is the only exception, as it has similar thickness to the panels pressed at 165 bar. This means that indeed voids are present in a panel after consolidation and that the total volume of voids is reduced by increasing consolidation pressures. In [4], defects (such as voids, missing fibres) in pressed UHWMPE panels (pressed only at one consolidation pressure of 206 bar) were clearly identified with micro X-ray computed tomography.
Figure 5 shows that there is no unique relation between panel thickness and V50. At a normalised panel thickness of 98, HB480 achieves a much higher V50 than HB210. The results indicate that the resin properties do significantly contribute to the V_{50} of a consolidated panel.

https://doi.org/10.52202/081621-0003

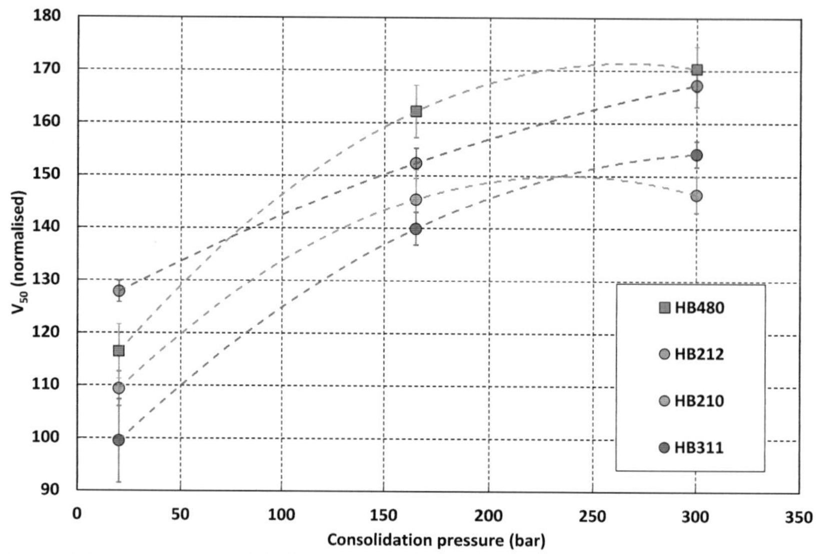

Figure 4. V_{50} versus consolidation pressure (error bars are the 95% confidence intervals).

The V_{50} is plotted versus the sample thickness in Figure 5. The consolidation pressures are given at the data points.

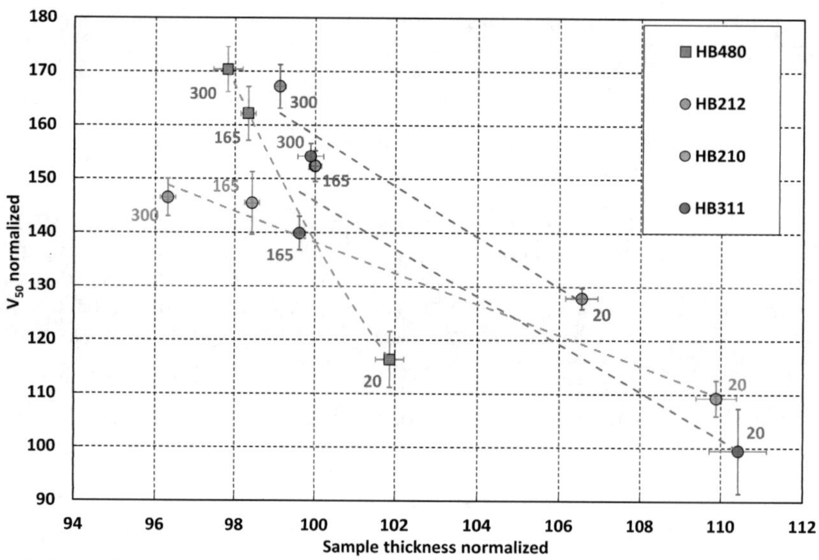

Figure 5. Normalised V_{50} versus normalised sample thickness (error bars are the 95% confidence intervals). The data labels provide the consolidation pressures (bar).

3.2 Core deformation and dept-of-penetration

Of the stopped projectiles, the depth-of-penetration and the residual length of the mild steel core were measured. After impact, the core is typically shortened and has a mushroom-like top [5]. The latter will result into a larger frontal area of the projectile and thus to a reduction of the mass per area of the projectile. It is known that the V_{50} in general scales with the ratio of the areal density of the target over the areal density of the projectile [6]. A higher ratio results in a higher V_{50}.

In Figure 6, the frontal surface area (= $\pi/4$ * diameter^2) of a selection of retrieved cores is plotted versus the length of the core.

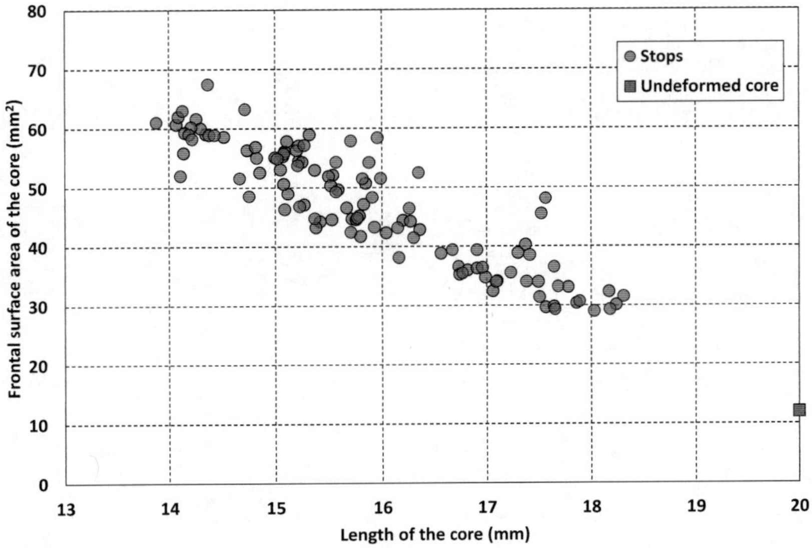

Figure 6. Frontal surface area of retrieved cores (from stops) versus the length of the core.

In Figure 6, the initial frontal surface area and core length of the undeformed core is also plotted. The residual core length clearly relates inversely to the final frontal surface area of the core. During core deformation, the frontal surface area of the cores increases very strongly. Ultimately, at a core length of 14 mm, the frontal surface area of the core has increased five times. Deformation of the core must thus be an important process in stopping the AK47 MSC projectile.

Figure 7 plots for all different UHMWPE composites the depth-of-penetration (left column) and the length of the cores versus the impact velocity (right column). The results demonstrate that an increase in consolidation pressure does lead to a decrease in depth-of-penetration at a given projectile velocity. For example, HB212 pressed at 20 bar, has a depth-of-penetration of almost 60% at a normalised projectile velocity of 100. HB212 pressed at 300 bar has a depth-of-penetration of around only 20% at the same projectile velocity. This is a drastically different response caused by the variation in consolidation pressure. Similar behaviour can be observed for the other composite types, although for HB212 the difference between 165 and 300 bar is most prominent, as can be seen in the left column of Figure 7.
To a lesser extent, the consolidation pressure affects the reduction of the core length. Going from 20 to 165 bar, there is more shortening of the core for all composite types, but less so going from 165 bar to 300 bar.

The data shown in Figure 7 have been rearranged into four graphs of length of the core versus depth-of-penetration per composite type (Figure 8). These graphs show different behaviour of the four composite types. All the graphs in Figure 7 and Figure 8 demonstrate that, at a consolidation pressure of 20 bar, the response of the target, in terms of depth-of-penetration and length of the core, does show more scatter than at the higher consolidation pressures. It indicates that the targets do become more homogeneous at higher consolidation pressures, which can be related to the distribution of resin around the fibre and/or reduction of void content.

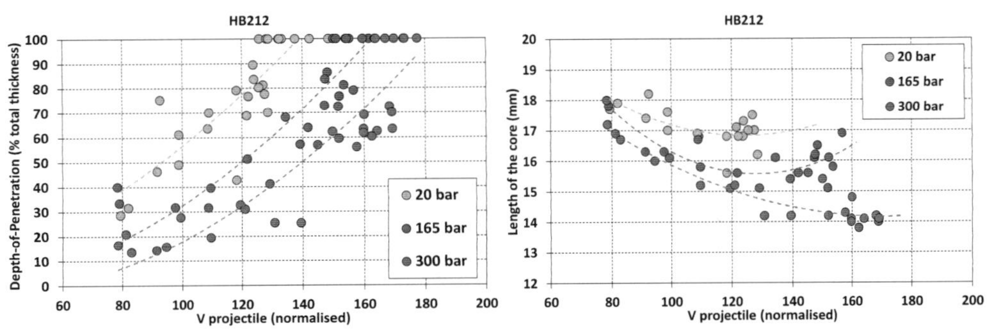

https://doi.org/10.52202/081621-0003

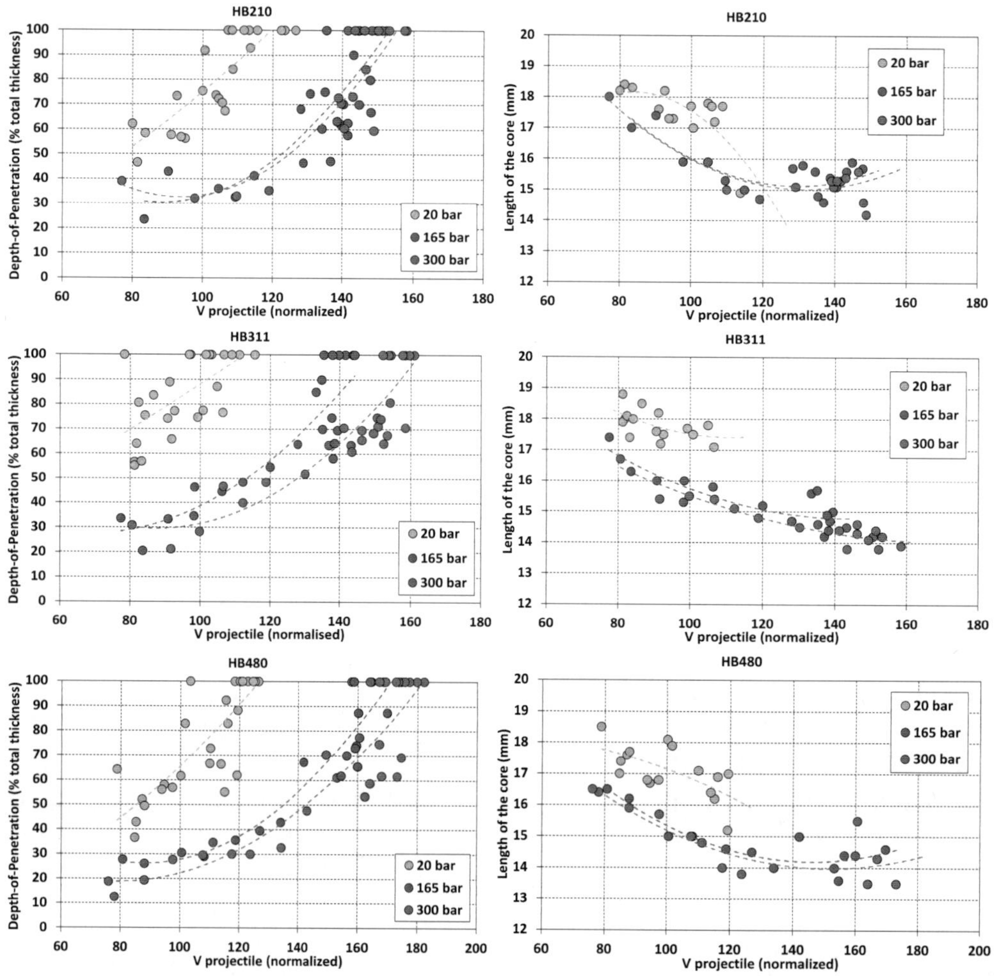

Figure 7. Depth-of-penetration versus projectile velocity (left column) and length of the retrieved mild steel core versus the projectile velocity (right column) for all different UHMWPE composite types.

The above results agree with a previous study [7], where the impact behaviour of an undeformable steel sphere was investigated on Dyneema® HB26 panels consolidated at 20, 165 and 300 bar. The V_{50} increased strongly only going from 20 to 165 bar. Also depth-of-penetration, at a given impact velocity, decreased going from 20 to 165 bar.

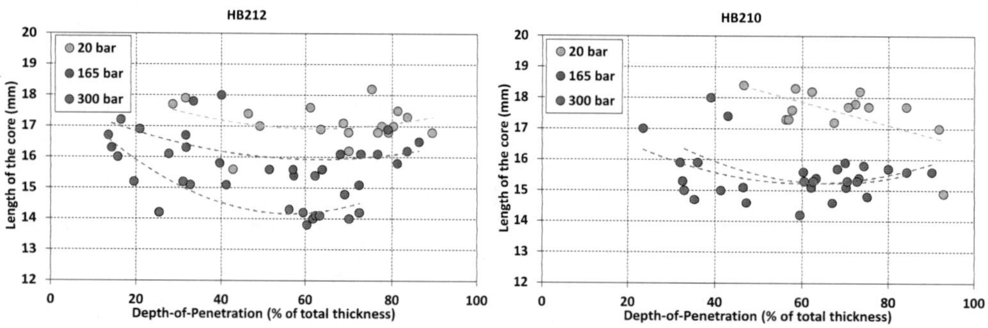

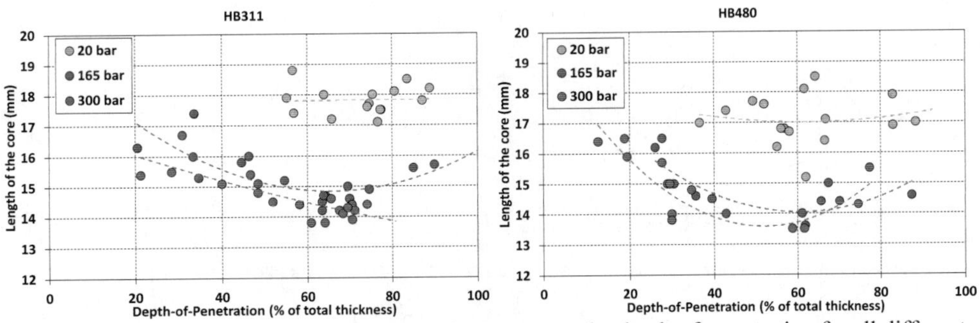

Figure 8. Length of the retrieved mild steel core versus the depth-of-penetration for all different UHMWPE composite types.

3.3 Shock properties

In addition to impact tests with complex three-dimensional stress and strain fields, the shock properties of panels were investigated using planar plate impact experiments as published in [7]. In this test, a UHMWPE sample plate, backed by a 5 mm C45 steel plate, is impacted onto a 5 mm C45 steel target plate leading in the center to a state plane strain. The target surface velocities are recorded with a VISAR laser system during the impact with a resolution of 2 ns. The HB26 data from [8] is given in Figure 9, together with new results measured on an UD Film prototype comparable to the HB480 composite used in this study.
When a projectile impacts on a pressed plate, the resulting pressure P depends directly on the shock properties, as derived from momentum conservation [8]:

$$P = \rho_0 \, U_S \, u_p \tag{1}$$

with the initial density ρ_0, the through-thickness shock wave speed U_S and material (or "particle") velocity u_p.

Figure 9 shows for HB26 below a particle velocity of 1100 m/s systematically lower shock velocities for a consolidation pressure of 20 bar compared to 165 bar. This difference can be attributed to pores in the panel consolidated at 20 bar. Similar behaviour has been identified for porous concrete and aluminium materials [9]. Going to 300 bar, the shock velocity again increases, at same particle velocities. Figure 9 also shows that the shock properties of a UD Film composite consolidated at 165 bar are even slightly higher that the shock property of HB26 consolidated at 300 bar.
Higher shock velocities in the composite and thus increased pressures on the projectile during impact and deep penetration result in stronger projectile deceleration and core deformation. Increased mushrooming will additionally lead to a larger frontal area with a higher deceleration force on the core. Both processes contribute to and explain the decreased depth-of-penetration, more core deformation and thus higher V_{50} with increasing consolidation pressures, as observed in Figures 7 and 8.

 https://doi.org/10.52202/081621-0003

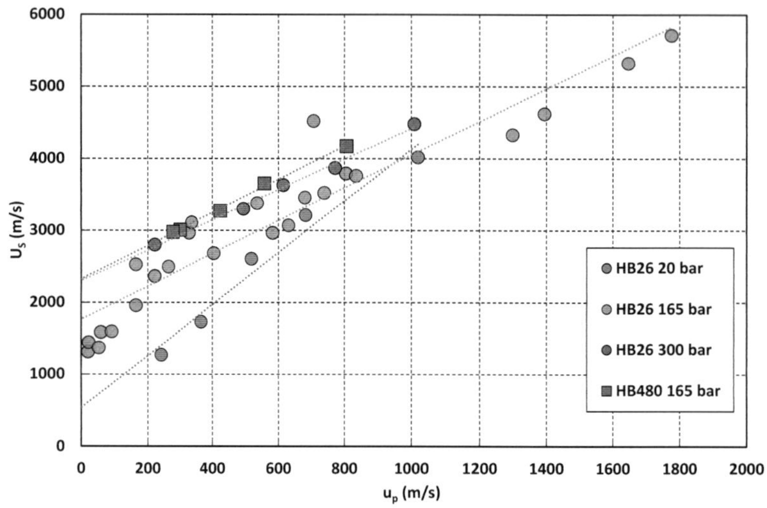

Figure 9. Shock properties under planar impact for HB26 consolidated at 20, 165 and 300 bar [7], and for HB480 consolidated at 165 bar.

3.4 Statistical relations

The obtained data on projectile velocities, depth-of-penetrations, core lengths and V_{50}'s were analyzed in statistical software Minitab® to find good statistical models to predict V_{50}. In Figure 10, the experimental V_{50} is plotted versus the predicted depth-of-penetration at an impact velocity of 100. The latter data were obtained by using a linear model for each composite, at each consolidation pressure. Figure 10 shows that the V_{50} of a composite can be estimated by shooting the projectile at a reference velocity below V_{50} and subsequent measurement of the depth-of-penetration. This can be very interesting if one, in practice, is confronted with an insufficient number of samples for V_{50} determinations. Figure 10 show remarkably that this estimation of V_{50} even works nicely for samples with a V_{50} much higher than the reference velocity of 100. For example, HB480 and HB212 pressed at 300 bars have V_{50}'s close to 170, but nevertheless fall on the general linear fit.
Most strikingly, the simple linear relation in Figure 10 describes very similarly the four different UHMPWE composites with different resins and different fibre compactions, at three different consolidation pressures.

Figure 11 plots the statistically predicted depth-of-penetration versus predicted length of the core, using linear models for both, at an impact velocity of 100.
Figure 11 does show that the different resin systems in conventional UHMWPE fibre composites can have different characteristics. HB311, with the stiffest resin, leads to much more penetration, at a given level of the length of the core, than HB212, with the softest resin. HB480 and HB210 are between of the former two resin systems.

The quality of the linear fits (expressed by the R^2 values) for depth-of-penetration and length of the core with projectile impact velocity are given in Table 2.
Table 2 shows that the quality of the linear fits increases strongly going from 20 to 300 bar consolidation pressures, especially for the depth-of-penetration. This indicates again more defined target response with increasing consolidation pressures. This improved homogeneity can originate from void removal and/or better distribution of the resins around the fibres in the stronger pressed composites.

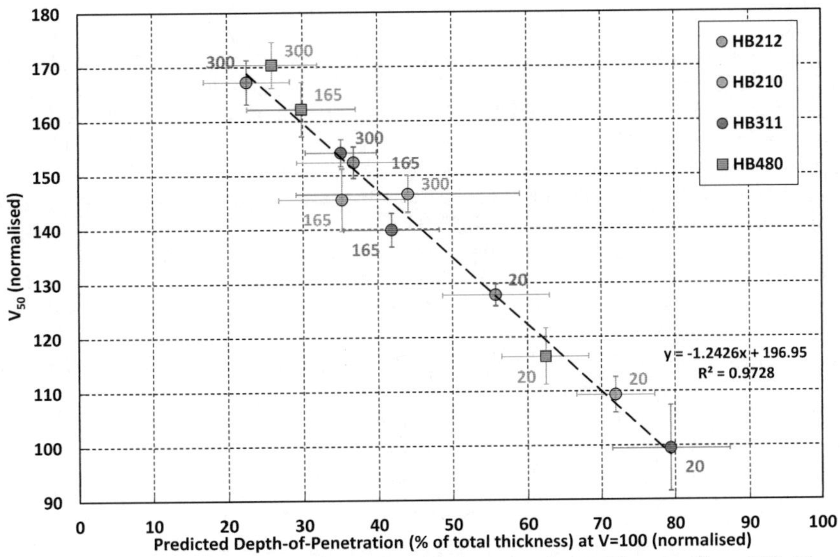

Figure 10. Experimental V_{50} versus predicted depth-of-penetration at V projectile = 100. The error bars are the 95% confidence intervals.

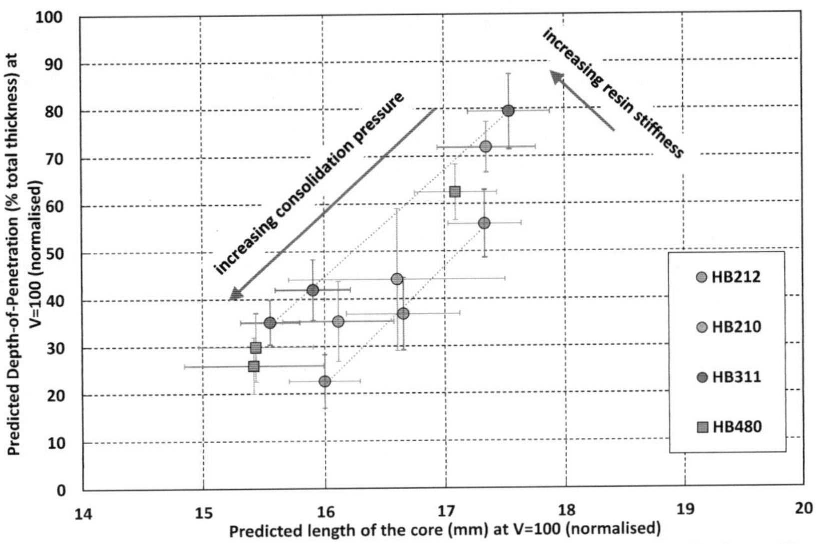

Figure 11. Predicted depth-of-penetration versus predicted length of the core, both at a V projectile = 100. The error bars are the 95% confidence intervals.

Table 2. R^2 of the linear fits per material and consolidation pressure

		R^2 (%) of linear fit with V projectile	
Material	Consolidation pressure (bar)	Depth-of-penetration	Length of the core
HB212	20	61	36
HB212	165	76	34
HB212	300	86	86
HB210	20	56	46
HB210	165	68	40
HB210	300	62	65
HB311	20	30	30
HB311	165	76	70
HB311	300	90	87
HB480	20	46	41
HB480	165	86	63
HB480	300	85	58

https://doi.org/10.52202/081621-0003

4. CONCLUSIONS

The effect of panel consolidation pressures of 20, 165 and 300 bar during hot-pressing on the impact resistance to AK47 MSC projectile has been investigated for four UHMWPE composites.
Higher consolidation pressures result into higher V_{50}'s for all investigated composites, and in decreased thicknesses of the pressed panels. For all composites, the depth-of-penetration at a given projectile velocity decreases strongly with higher consolidation pressure, with the strongest effect between 20 and 165 bar. In addition, increasing consolidation pressures lead consistently to stronger shortening and mushrooming of the mild steel core. In general, both relations show slightly different behaviour in all the different composites, which indicates that the resin type also influences the impact resistance.
The measuring of shock properties indicates that higher consolidation pressures lead to higher shock velocities, mainly below impact velocities of 1100 m/s. The underlying effect is probably reduced porosity around the fibres. The increased shock impedance of the composite leads to higher contact pressures, more core deformation and stronger mushrooming, causing the substantially increased ballistic resistance of the resulting UHMWPE targets.
A very important finding of statistical analyses is the fact that experimental V_{50}'s can be very well predicted from depth-of-penetration measurements carried out at a single projectile velocity, even much below the V_{50}. In all different analysis types, scatter decreases and quality of fits improve with increasing consolidation pressures.

References

[1] van der Werff H., Heisserer U., High-performance ballistic fibres: ultra-high molecular weight polyethylene (UHMWPE), in : Chen X. (Ed), Advanced Fibrous Composite Materials for Ballistic Protection (Woodhead Publishing, 2016), pp. 71-107.
[2] Zheng J.Q., Walsh S.M., Lightweight Ballistic Composites (Elsevier, 2016), pp. 393-437.
[3] NATO/PfP, AEP-55, Volume 1 (Edition 1), Procedures For Evaluating the Protection Level of Logistic and Light Armoured Vehicles (February 2005), Annex D.
[4] O'Masta M.R., Deshpande V.S., Wadley H.N.G., Defect controlled transverse compressive strength of polyethylene fibre laminates, Int. J. of Solid and Structures, 52 (2015), 130-149.
[5] Bolz K., Hensellek S., Acher V., Veehmayer M., V_{50} Study of Light-Weight Body Armour Inserts under Angle Shot, PASS 2023: Proceedings, pp. 11-18.
[6] Singletary J. Carbajal L., Boogh L., Fragment Simulating Projectile V_{50} Scaling rules, PASS 2008: Proceedings, pp. 287-296.
[7] Laessig T.R., May M., Heisserer U., Riedel W., Bagusat F., van der Werff H, Hiermaier S.J., Effect of consolidation pressure on the impact behaviour of UHMWPE composites, Composite Part B, 147 (2018), 47-55.
[8] Meyers M.A., Dynamic Behavior of Materials (John Wiley & Sons, New York, 1994), pp. 103-109.
[9] Riedel W., Wicklein M., Thoma K., Shock properties of conventional and high strength concrete: Experimental and mesomechanical analysis, Int. J. of Impact Eng., 35 (2008), pp 155-171.

DISCLAIMER

Skeletal Injuries Due to Behind Armour Blunt Trauma to the Spine

N. Hahne[1], E. Butler[1], KA. Carr[1], C. Howes[1], A. Rohrer[2], M. Clark[3], **A. Iwaskiw[1]**

[1] *The Johns Hopkins University Applied Physics Laboratory (JHU/APL), 11100 Johns Hopkins Road, Laurel, MD 20723, alexander.iwaskiw@jhuapl.edu*
[2] *Armed Forces Medical Examiner System, Defense Health Agency, Dover Air Force Base, DE*
[3] *U.S. Special Operations Command, 10 General Greene Avenue, Natick, MA 01760*

Abstract. Non-penetrating ballistic events inducing backface deformation in personal ballistic armour can cause blunt trauma to the human body. These deformation and injury patterns, known as behind armour blunt trauma (BABT), remain poorly understood. It is hypothesised that BABT directly over the spine could result in serious injury. In this study, live ammunition was used to strike ultra-high molecular weight polyethylene armour plates inducing BABT loading on instrumented postmortem human surrogates (PMHS). Nine tests were conducted using male PMHS specimens, with one ballistic plate per test. Impacts were delivered at near muzzle velocity. The 5th thoracic vertebral body was selected as the impact site, as it aligns with typical plate positioning under standard wear conditions. X-ray and computed tomography imaging were acquired before and after testing. Injury assessments were conducted through anatomical dissections by a board-certified forensic pathologist. Observed injuries ranged from superficial skin abrasions and lacerations to posterior element fractures, rib fractures, spinal cord impingement, and anterior vertebral body fractures. The PMHS model is limited in its ability to capture the functional outcome of these injuries, therefore it is of interest to compare these results with other biomechanical models that include living responses to these injuries. These findings highlight a potential risk of spinal injury from BABT and underscores the need to consider spinal trauma in future medical capability planning and ballistic armour design.

1. INTRODUCTION

The integration of advanced materials in contemporary personal armour design has significantly improved protection against ballistic penetration [1]. These designs strive to balance protection with weight reduction. While hard armour plates effectively stop ballistic projectiles, they can undergo significant backface deformation when stopping these threats. This rapid, high energy loading transfers energy through the plate to the body of the wearer potentially leading to behind armour blunt trauma (BABT), and resulting injuries can range from minor contusions to severe organ damage [2], [3]. In the interest of protecting military personnel, ongoing research aims to understand injury mechanisms, assess risks to the wearer, and evaluate the impact on operational readiness.

Relevant thresholds for maximum backface deformation (BFD) of ballistic armour plates is associated with 44 mm of deformation left in clay backing material, termed backface signature (BFS). This threshold was based on models that may not reflect the mechanisms of BABT on the human body accurately [4]. These standards were developed using in vivo animal surrogates and clay backed deformation studies to extrapolate human injury risk, however no direct relationship to human injury was established in these studies [1]. It is well documented that such animal models may not be representative of human injury risks for hard armour and clay does not accurately represent kinetic interactions between the armour and the body during impact, thus the risks of non-penetrating impacts to the wearer remain unclear [3], [5].

Typically, the same armour plate design is used to protect both the rear and front thoracic body regions. While this simplifies manufacturing considerations, this assumes that injury risk is equivalent for both locations. This assumption raises the question of whether the BFD standard for anterior impacts should be the same for posterior impacts. Understanding differences in vulnerability between the anterior and posterior thorax to blunt force loading is crucial for optimizing armour design [3], [5]. BABT to the anterior thorax has been modeled, and injury risk assessments indicate that such impacts can cause severe harm to the wearer [5], [6]. The potential impact of BABT over the spine is of particular concern given the risks to the spinal column and spinal cord. Computational modelling studies suggest that tissues beneath the armour experience pressure waves and shear deformation during BABT events, while animal studies indicate a risk of incapacitating injuries from such impacts [7], [8].

Epidemiological research on spinal injuries in both civilian and military populations has led to the development of several classification systems that help inform clinical implications and guide potential interventions [9], [10]. However, injuries associated with BABT involve unique mechanisms that remain

https://doi.org/10.52202/081621-0004

poorly understood in clinical practice and generally fall outside existing classification paradigms. This study introduces an experimental protocol utilizing instrumented post mortem human surrogates (PMHS), live ammunition, light weight hard ultra-high molecular weight polyethylene (UHMWPE) armour plates to induce BABT impacts to the thoracic spine. While numerous studies have examined the complex injury mechanisms of BABT in the thoracic region, this is the largest study to specifically investigate and characterise its effects on the thoracic spine in PMHS.

2. METHODS

2.1 Projectile and armour

All tests conducted in this study utilised the same projectile and armour pairing. Each plate was shot near muzzle velocity with a ballistic round relevant to current operational threats. The armour system investigated was designed to be standalone, therefore no backing material was used between the armour and test subject during impact. The armour was an UHMWPE composite multi-curved plate with a Small Arms Protective Insert (SAPI) shape. It was purposely developed to allow for large clay BFS under specific ballistic test conditions. Each plate was shot with the 7.62 x 51 mm M80 ball projectile to induce large backface deformation. The ballistic conditions selected for these tests were intended to induce associated BFS magnitudes around 60 mm. Each plate was tested once and shot 100 mm below the top edge along the mid-sagittal plane in line with the spinous process of the fifth thoracic vertebrae. Air gaps between the backside of the armour and the human anatomy at the point of impact ranged from 3.2-15.75 mm, and flesh thickness ranged from 12.4-39.8 mm. It is hypothesised that these local factors can affect loading to the skeleton.

2.2 PMHS demographics

Nine male, fresh frozen, full-body PMHS specimen were utilised for this study. All PMHS were acquired under institutional review board exemption approval through Johns Hopkins Medical Institution. PMHS were acquired with bilateral disarticulations at the distal femur. Specimens were selected using a specified range of metrics for age, stature, weight, body mass index (BMI) and lumbar bone mineral density (BMD) based on Dual Energy X-ray Absorptiometry (DEXA) imaging, Table 1. Specimen documented with prior trauma or severe thoracic disease were restricted from this study. Upon acquisition, specimens were imaged using computed tomography (CT) with slice thickness of 0.625 mm at the Johns Hopkins University Applied Physics Laboratory (JHU/APL) to ensure specimen had no prior trauma, medical interventions or thoracic conditions that could interfere with biomechanical testing or influence injury interpretation. To ensure that no specimens exhibited hyperkyphosis, the junctions between T1-T2 and T12-L1 were identified to measure the angle between these reference points using imaging software (3D Slicer 5.6.2 release). Specimens exceeding a kyphotic curvature of >40 degrees were classified as hyperkyphotic and excluded from the study. Specimen were stored at −20°C until thawed at 4°C six days prior to testing for instrumentation and test preparation.

Table 1. Specimen demographics

APL Identifier	Age	Cause of Death	Stature (cm)	Weight (kg)	BMI (kg/m^3)	DEXA (T-score)	Kyphosis Angle (degrees)
SPINE 01	56	Metoprolol/Amlodipine intoxication	188	108	30.7	2.4	34.8
SPINE 02	44	Brain tumour	191	100	27.5	0.5	38.8
SPINE 03	54	Hepatorenal syndrome	178	91	28.7	0.8	31.6
SPINE 04	63	Subdural hematoma	191	75	20.8	1.1	22.5
SPINE 05	49	Myocardial infarction	185	117	34.2	-1.2	32.0
SPINE 06	42	Cardiogenic shock, respiratory failure	165	65	23.8	-0.2	33.0
SPINE 07	42	Asphyxia	180	94	28.9	0.1	36.7
SPINE 08	45	Hepatic encephalopathy	178	106	33.4	-0.2	25.4
SPINE 09	33	Methamphetamine intoxication	175	94	30.6	2.4	33.1

2.3 PMHS instrumentation

While the focus of this study was on injury characterization rather than biomechanical analysis, each specimen was instrumented with biomechanical sensors to understand the mechanical and dynamic conditions during the ballistic event for use in future analyses. The techniques employed to affix sensors were minimally invasive to avoid affecting injury outcomes resulting from BFD. Sensors were placed near anatomical regions proximal to the impact area and in organs of clinical interest shown in Table 2. The instrumentation employed included sternal motion, skeletal strain, skeletal acoustic emissions, organ pressure and surface contact.

Table 2. Specimen instrumentation

Sensor Description	Location	Type	Sensor	#
Rib strain	Posteriolateral ribs 4-9	Single axis strain gauge	Kyowa KFWB-2-350	12
Rib acoustic emission	Posteriolateral ribs 4-9	Acoustic emission sensor	Mistras Nano30	12
Lung pressure	Left and right bronchi	Pressure transducer	Kulite XCL-100- 100A	4
Vascular pressure	Aortic arch	Pressure transducer	Kulite XCL-100- 100A	1
Vascular pressure	Descending aorta (T11 vertebra)	Pressure transducer	Kulite XCL-100- 100A	1
Sternum acceleration	Superior manubrium and xiphoid	Single-axis accelerometer	Endevco 727	1
Surface force	On skin under armour	Contact sensor	TekScan flexiforce 301	5

Fluid was introduced into the cardiovascular system to simulate intravascular pressure. Foley catheters were guided through the carotid arteries, one catheter terminated at the aortic valve and a second terminated at the aortic junction. A third catheter was placed in the descending aorta at the level of the 11th thoracic vertebra superior to renal artery branches via femoral artery. Deployed foley catheter balloons allowed fluid to remain in the isolated vascular regions and contrast fluid was introduced to ensure proper sensor placement verified by CT. A tracheotomy was performed through which lung pressure sensors were introduced into the left and right primary bronchi and a tracheostomy tube was placed enabling lung insufflation with approximately 7 kPa of positive pressure during the test. CT was used to verify lungs were fully expanded and contacting the interior surface of the anterior plural wall. Post-instrumentation CT imaging was performed to document final sensor locations and lung insufflation.

2.4 PMHS experimental setup

After instrumentation, the PMHS were transported to a National Institute of Justice certified commercial ballistic test lab. The specimens were placed on a custom mounting rig allowing for the horizontally placed specimen to be rotated up right simulating a standing position. The specimens were strapped to the rig using nylon straps to support the head, arms and legs. The PMHS were supported by a straddle seat under the pubis, and foam material was used against the head and abdomen to offset the anterior thorax from the rig. The impact location was positioned along the mid-sagittal plane, over the spinous process of the fifth thoracic vertebra. Each specimen was placed upright and symmetrically on the rig and the rig was adjusted such that the impact location aligned with the ballistic target to simulate a rear shot over the targeted location on the spinal column (Figure 1).

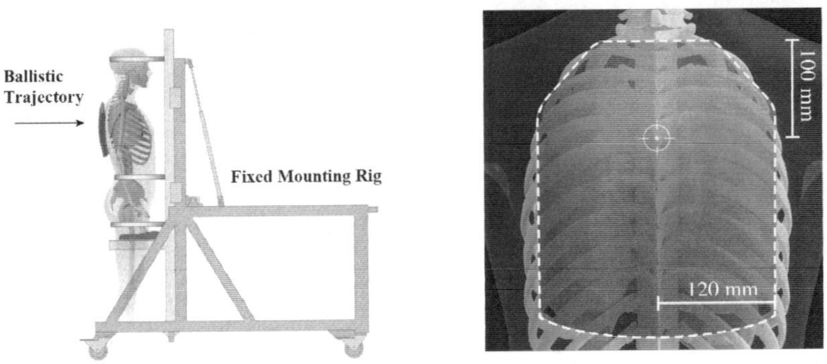

Figure 1. (Left) PMHS experimental setup. (Right) Schematic of armour alignment on specimen

The armour plate was aligned with the shot location on the specimen using a custom armour carrier with cutouts for the impact location to ensure no carrier material interferes with the BABT loading. The armour was donned to provide a "normal fit", where the armour carrier was secured snugly to the specimen. Due to anatomical variability in the curvature of the upper rear torso, an airgap was present between the external surface of the skin and the backface of the plate. This backface standoff of the plate from the external surface of the skin at the anticipated point of impact was measured using a FARO arm (FARO Technologies, Lake Mary, Florida) by comparing the impact location on the PMHS before donning the armour to the impact location on the impact face of the armour after donning and

https://doi.org/10.52202/081621-0004

subtracting the armour thickness. The armour standoff was measured and ranged from 3-16 mm and obliquity spanned from -3° below to 16° relative to the horizontal plane. Pretest x-rays and photos were taken of the perfused and insufflated PMHS directly before the ballistic impact to ensure proper anatomical targeting and to measure sensor locations in situ (Figure 2).

BABT loading was subsequently induced by a single live-fire ballistic projectile. Velocity measurements were obtained using two sets of infrared screens (Oehler Research model No. 57) with universal counter chronographs (Hewlett-Packard model No. 53131A). Velocity was measured 4.6 m from the armour plate, and then velocity loss equations were applied specific to the projectile to calculate striking velocity.

While biomechanical characterization was not the focus of this study, high-rate data was acquired for use in future analysis. These metrics included acoustic emissions and strain from rib cage, vascular pressure and surface loading behind armour backface. Additionally, two high-speed camera systems (ACS-1-M60, NAC Image Technology) recorded the ballistic event from both a lateral and oblique view of the specimen at a rate of 100k and 50k frames per second respectively. All systems were simultaneously triggered with the signal disruption of a frangible break screen by the incoming round. All instrumentation signals, except acoustic emission data, were subject to a 300-kHz anti-alias filter.

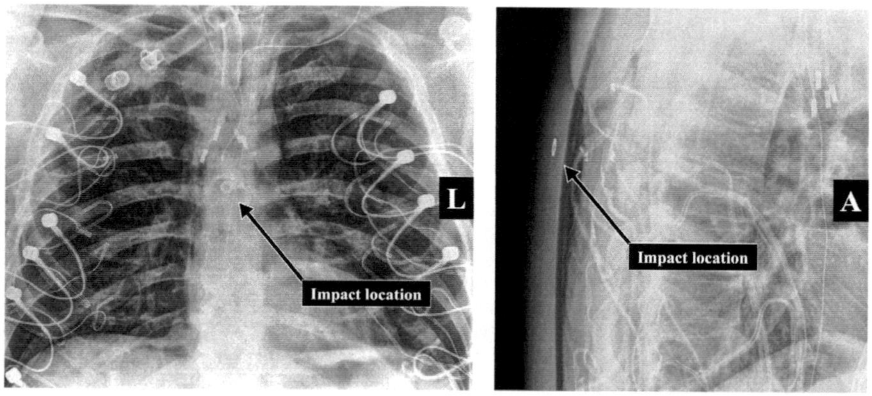

Figure 2. Characteristic specimen x-ray views of the thorax anteroposterior (left) and laterally (right) prior to impact (SPINE 03)

2.5 Post-test evaluation and forensic analysis

Immediately after each BABT loading event, post-test x-rays and photographs were taken without disruption of armour (Figure 3). Specimens were returned to a prone orientation and transported back to JHU/APL for post-test clinical CT imaging. Each specimen was scanned while in prone position with insufflated lungs with impacted armour in place and then scanned again with armour removed. The clinical CT imaging was reviewed by a board-certified forensic pathologist who subsequently led anatomical dissection of the PMHS. Dissection was conducted to document injury characteristics and evaluate injury severity. During this procedure, tissue layers were progressively reflected and organs removed sequentially for further analysis. Lungs were excised, inflated and inspected along with the chest wall for evidence of injury.

To visualise damage to the structural elements of the vertebral column, micro-computed tomography (μCT) was utilised (Nikon M2 Large Format). After the forensic analysis was completed, the spinal column from the second to the ninth thoracic vertebrae was disarticulated *en bloc* and μCT scanned at JHU/APL where each slice thickness was 75 μm providing six-fold increase in resolution compared the clinical CT modality. Finally, the vertebral bodies (anterior spinal column) were removed from the spinal column to visually examine the spinal cord and dural sheath (Cases SPINE 01-04).[1] Dissection observations were integrated into injury severity analysis using the Abbreviated Injury Scale (AIS). The AIS is a 7-digit, injury coding system that incorporates injury location, affected tissue and injury severity to calculate the New Injury Severity Score (NISS) [11]. The NISS is used to assess injury severity in the case where injury mechanisms induce polytrauma, and represents the sum of the square of the three most severe injuries occurring in a patient to be classified as mild (1-15), moderate (16-25),

---

[1] Removing the vertebral bodies was not routinely performed. This was a measured decision due to logistical considerations of obtaining initial micro CT imaging, the lack of significant spinal column intrusion by displaced bone, and understanding inherent limitations of identifying spinal cord injury in a PMHS.

severe (26-50) or critical (51-75) [12]. To quantify the most consequential injury, maximum abbreviated injury scale (MAIS) was used to compare cases.

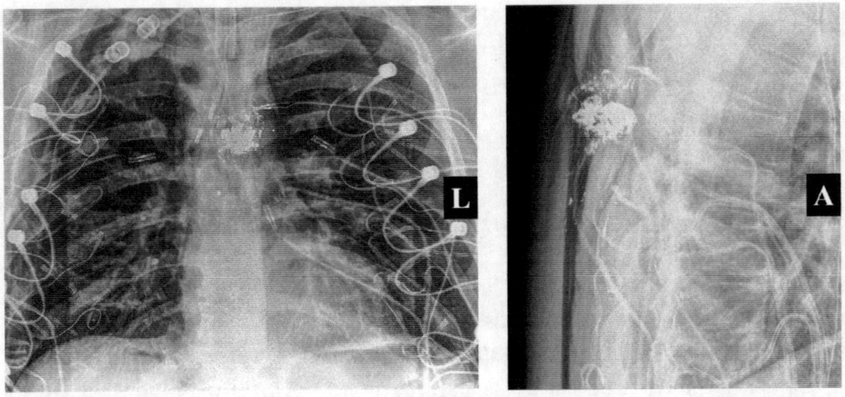

Figure 3. Characteristic specimen (SPINE 03) planar x-ray views of the thorax anteroposterior (left) and laterally (right) post impact visualizing the embedded projectile.

3. RESULTS

3.1 PMHS injuries

Injuries were observed in post-test medical imaging and during anatomical dissection. Representative images in Figure 4 and Figure 5 summarise extent of severe injury cases at T5. X-ray and clinical CT imaging demonstrate the residual armour deformation and resultant anatomical defects. μCT aided in visualizing vertebral fracture patterns for characterization. Associated injuries ranged from skin damage (AIS ≤ 1), pleural laceration (AIS 2), to multiple skeletal fractures (AIS ≤ 3). These injury outcomes can be found with corresponding injury codes in Appendix 1.

3.2 PMHS injury characteristics

The injuries observed at this impact location generally involved skin lacerations, ligament lacerations and skeletal fractures. All cases resulted in full thickness lacerations of the skin, exposing underlying tissue. These wounds ranged from approximately 6 -10 cm in width and 2 cm in depth. According to the NISS metric, these injuries are classified as 1, representing a minor injury severity. The second class of injury involved specific lacerations to the supraspinous ligament(s) and were broadly associated with

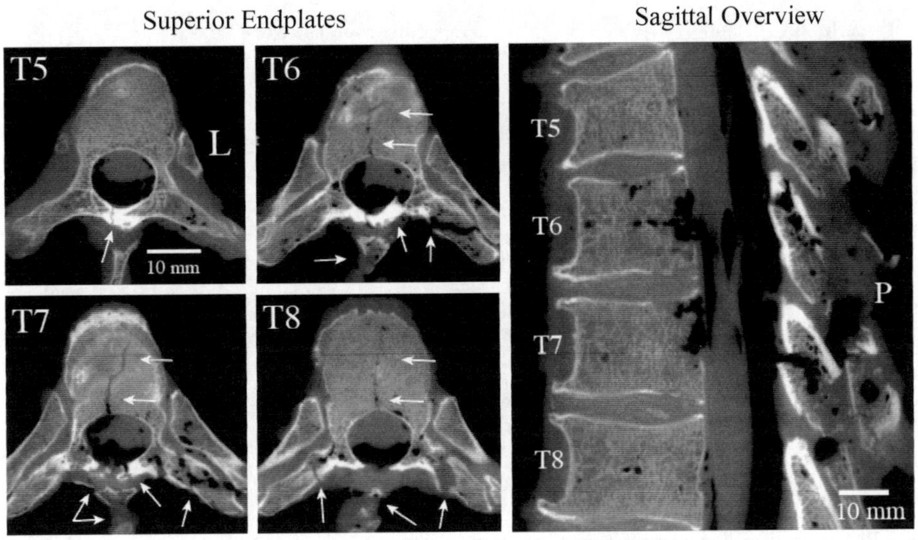

Figure 4. Resultant vertebral fractures from SPINE 04 post test

https://doi.org/10.52202/081621-0004

vertebral fractures. According to the NISS metric, these ligament injuries are classified as 1, also representing a minor injury severity.

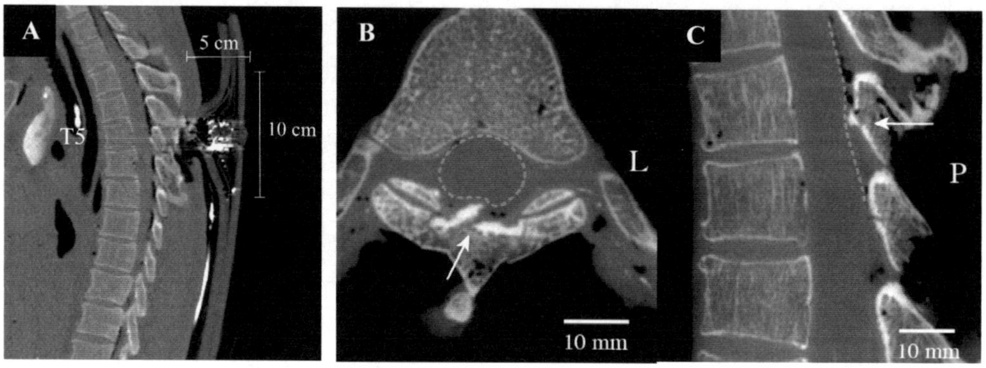

Figure 5. (A) Clinical CT scan showing lateral view of armour BFD spine intrusion. **(B)** and **(C)** Axial and lateral view (SPINE 06) of displaced fragments of vertebral fracture subsequent invasion into the vertebral foramen deforming the spinal cord (dotted line).

The third type of injury that was observed at this impact location were skeletal fractures involving the rib cage and vertebral segments. Rib cage fractures were identified in cases SPINE 02-04, 06-08 ranging from rib 5 to rib 8 and involved 2-3 ribs per case. Fractures generally occurred posteriorly with the exception of case SPINE 03 where anterior rib fractures were observed. These fractures included simple-nondisplaced, displaced and comminuted fracture patterns with simple-nondisplaced fractures occurring most frequently. AIS scoring for rib injury scales the severity according to the number of ribs involved. Injury involving 2 ribs is scored as 2 (moderate) and injuries involving 3 or more ribs are scored as 3 (serious).

The observed vertebral fractures involved the third through eighth thoracic vertebrae (T3-T8). In cases SPINE 01-04, 06-09, multiple vertebrae were fractured. These fracture patterns could be further characterised by the number of structural elements fractured and their location: (1) single fracture to a posterior vertebral element, (2) multiple fractures to posterior vertebral elements, (3) multiple fractures to posterior vertebral elements and vertebral body fracture (Figure 6). In order of prevalence, these included fractures of the spinous process, lamina, transverse process, vertebral body, superior articular facet, inferior articular facet and pedicle (Figure 7). Of the cases involving fractures, the most commonly fractured vertebrae were T6 and T7, with T6 sustaining the most fractures per case. T5, T6 and T7 were observed to have the largest variety of fracture types. The most common fractures involved the posterior elements of the vertebrae, namely the spinous process and lamina. A complete list of vertebral fractures can be found in Appendix 2.

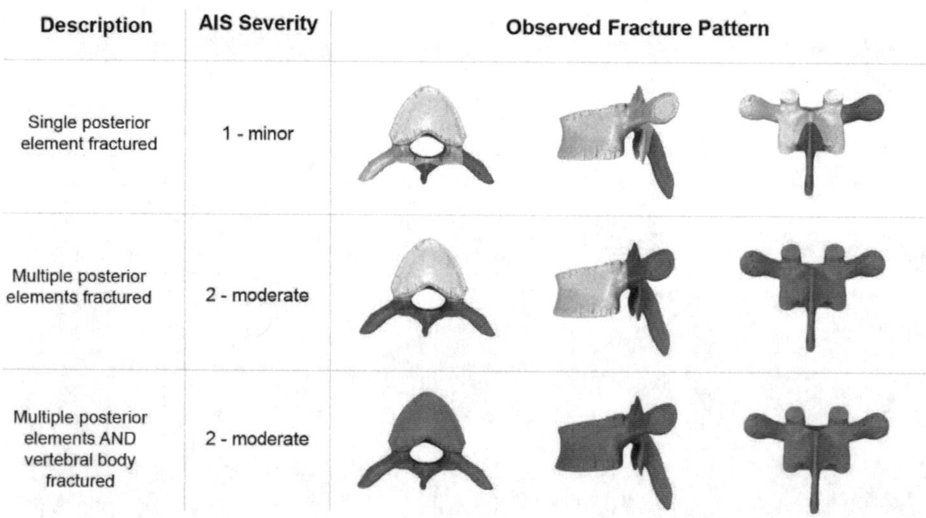

Figure 6. Illustration depicting the fracture patterns observed across all cases. Red areas indicate fractured regions.

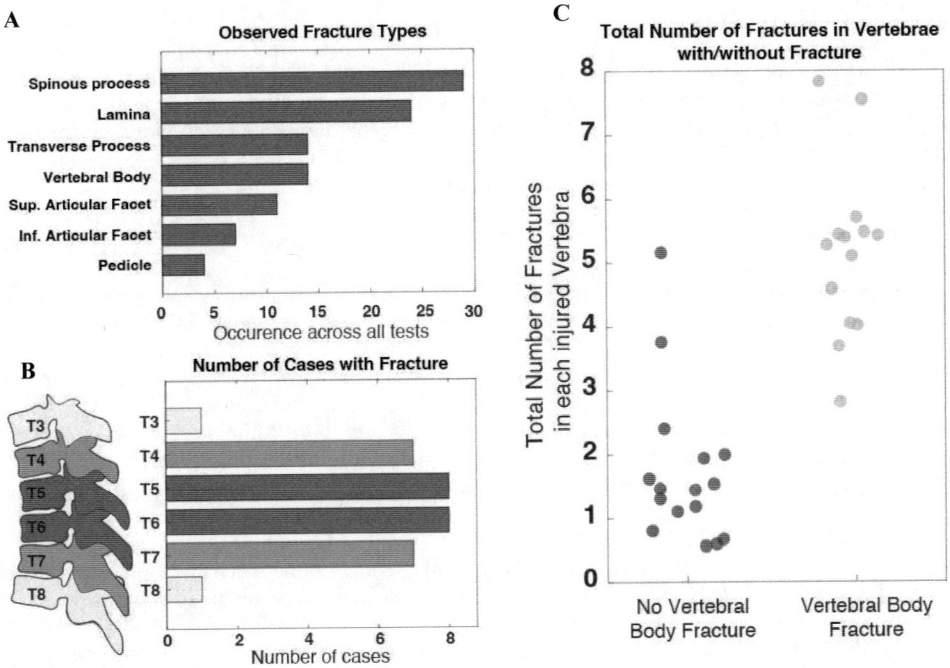

Figure 7. (A) Tabulation of all fracture types observed across all tests. **(B)** Distribution of fractures observed in injured vertebrae across all tests. **(C)** Fractures occurring in vertebrae with/without vertebral body fracture.

4. DISCUSSION

This ongoing study aims to characterise the breadth and severity of injuries resulting from BABT over the spine to inform clinical considerations and armour design policy. The clay BFS values in this study exceeds the 44 mm standard [1], the observed injuries represent a high-severity threshold for posterior BABT and approaches the ballistic penetration limit for this plate. We expect that providing data at these higher deformations provides critical insight into these injury mechanisms.

Fractures of the thoracic vertebrae are uncommon, accounting for only a small proportion of traumatic spinal injuries in both the general and military populations [10], [13]. The rib cage provides greater stability and rigidity to the thoracic vertebrae compared to other segments of the spinal column and are thus far more resistant to injury than the cervical and lumbar regions, regardless of impact mechanism [14]. For military personnel, spinal fractures are typically caused by falls or motor vehicle impacts. However, recent efforts have sought to better characterise combat-related spinal injury mechanisms [10], [15] . Isolated fractures of the posterior spinal elements are not typically associated with blunt trauma, whereas vertebral body fractures are commonly linked to compressive axial loading [12]. As an emerging threat with characteristically high velocity mechanics, the injuries observed in this study were unique and generally unobserved by the medical community. Notably, the third fracture pattern observed in this study involving destruction of vertebral posterior elements and vertebral body fracture was not represented in available datasets nor classification systems such as the widely adopted AO Spine Thoracolumbar spine injury system [16]. SPINE 03-04, 06-09 involved fractures of the vertebral body. This fracture pattern was associated with proximity to the impact location and correlated to a higher number of fractures per injured vertebrae (Figure 7). Since the anterior face of the vertebral bodies remained intact and no focal deformation was observed, it is possible that the posterior face of the vertebral body failed under tensile forces, appearing to be pulled apart.

To assess the severity of injury to the spinal column, this study utilised the AIS coding system. According to AIS conventions, each injured vertebral segment is assigned a specific code describing the fracture type and severity. Coding rules stipulate that the three most severe scores are used to tabulate a NISS score indicating overall injury severity. For multiple fractured vertebrae, the maximum severity score is 2 (moderate). Higher severity codes, up to 3 (serious), are available for vertebral injuries, but they specifically describe compression fractures resulting from axial loading. These classifications are inappropriate for this study, as BABT injuries involve posterior loading rather than axial compression.

 https://doi.org/10.52202/081621-0004

Additionally, the NISS only accounts for the three most severe injuries and does not scale injury severity beyond three affected vertebral segments. Under this metric, the severity outcome for an individual remains the same if injury involves 3 or more than 3 vertebral segments. AIS scoring also differentiates between closed and open fractures, where open fractures involve penetrating injury mechanisms and are generally assigned a higher severity score. This creates a paradox in injury classification, as this study observed open fractures of the spine, yet BABT injuries do not result from penetrating trauma. This discrepancy complicates the development of an appropriate injury risk assessment framework.

4.1 Study limitations

The development of spinal cord injury due to BABT can be divided into primary and secondary phase. Primary injury is caused by BFD acting on the spinal column and the secondary injury is the resulting haemorrhage, ischemia and inflammation [17]. Critically, this PMHS experimental model is unable to resolve these secondary elements of spinal cord injury. Research has shown that impact velocity is a key factor in determining spinal cord injury severity [18]. Blunt trauma-induced spinal cord injuries can occur through direct deformation or contact with vertebral elements [19]. Animal studies suggest BABT in the thoracic spine can generate sufficient deformation inducing injurious pressures in the central nervous system and lung [8]. Although this study did not observe direct disruption of spinal dura, several cases (SPINE 03, 04, 06, 07) involved substantial vertebral fracture, leading to comminuted bone fragments invading the spinal canal and impinging the spinal cord (Figure 5). A critical component of future analysis of spine BABT injury risks will need to involve identifying mechanical pressure or shear loading in the spinal cord. This will enable development of a dose response model to compare to current models based on animal and computational studies.

This study attempted to utilise currently available injury scoring systems to characterise the severity of injuries observed in these experiments. But these systems have several fundamental shortcomings. AIS coding requires spinal cord injuries to be classified by neurological impairment and functional loss, which is not available in the PMHS model. These injury characteristics represent an emerging phenomenon associated with backface deformation in hard armour capable of stopping ballistic rounds. Characterizing these injuries and their clinical implications will be crucial for future injury risk analysis. However, the high-energy impact mechanisms involved in these cases complicate such analyses, as these injuries remain undefined within the medical community, leaving little clinical precedent for their evaluation. Current AIS-based injury severity assessments are insufficient, as they fail to capture the full spectrum of injury presentations and lack critical consideration of neurological injury risk.

5. CONCLUSIONS

This study introduced an experimental methodology and initial injury findings from nine live-fire, non-penetrating ballistic impacts on instrumented PMHS. Observations from each case revealed injuries ranging from minor skin lacerations to multiple fractures of adjacent vertebrae. The injured vertebrae exhibited distinct fracture patterns based on the affected elements: single posterior element fractures, multiple posterior element fractures, and multiple fractures involving the vertebral body. The extent of substantial vertebral fracture correlated with impact proximity, while vertebral body involvement was associated with a higher number of fractures per vertebra. These findings contribute to an ongoing effort to better understand the factors influencing BABT injury risk and to inform future armour design standards. Additionally, because of the extent of fractures in the vertebral bodies, it is recommended that future studies utilise a human surrogate with a living tissue response (i.e. animal). By quantifying the injury risks associated with BABT to the posterior thorax, this study highlights the need for armour designs tailored specifically for rear protection. These observations suggest that rear-impact BABT presents unique injury patterns that warrant considerations for modifying armour design or performance standards. Currently fielded plates are designed to a standard based on clay deformation rather than actual injury observations. Here we present a data set that facilitates shifting from traditional armour design philosophies toward an evidence-based approach that accounts for regional injury vulnerability. Integrating these findings into armour standards will enhance survivability and operational effectiveness for personnel in combat environments.

Acknowledgments

The authors would like to express their appreciation to the U.S. Special Operations Command for funding this work under Contract No. is H92222-15-D-0004.

References

[1] E. Hanlon and et al, "Origin of the 44-mm Behind-Armor Blunt Trauma Standard," *Military Medicine,* vol. 177, no. 3, pp. 333-339, 2012.

[2] N. Pratt and et al, " Contemporary body armor: technical data, injuries, and limits," *European Journal of Trauma and Emergency Surgery,* vol. 38, no. 2, pp. 95-105, 2012.

[3] V. Kote and et al, "Investigating the Impact of Blunt Force Trauma: A Probabilistic Study of Behind Armor Blunt Trauma Risk," *Annals of Biomedical Engineering,* 2025.

[4] U. D. o. Justice, "Ballistic resistance of police body armor," *National Institute of Justice,* pp. 1-13, 1986.

[5] C. Bass and et al, "Injury Risk in behind armor blunt thoracic trauma," *International Journal of Occupational Safety and Ergonomics,* vol. 12, no. 4, pp. 429-442, 2006.

[6] A. Iwaskiw and et al, "Initial injury observations in postmortem human subjects: Non-penetrating ballistic impacts over the sternum with lightweight hard armour plates," in *Personal Armour Systems Symposium*, Copenhagen, 2021.

[7] A. Merkle and et al, " Assessing behind armor blunt trauma (BABT) under NIJ Standard-0101.04 conditions using human torso models," *Journal of Trauma-Infection and Critical Care,* vol. 64, no. 6, pp. 1555-1561, 2008.

[8] B. Zhang and et al, " Neurological, Functional, and Biomechanical Characteristics After High-Velocity Behind Armor Blunt Trauma of the Spine," *Journal of Trauma-Injury Infection and Critical Care,* vol. 71, no. 6, pp. 1680-1688, 2011.

[9] J. Blair and et al, " Military penetrating spine injuries compared with blunt," *Spine Journal,* vol. 12, no. SI, pp. 762-768, 2012.

[10] K. Holly and et al, "Spinal Injuries and Spine Care in the US Military Health System (2001-Present)," *Spine,* vol. 50, no. 2, pp. 207-215, 2025.

[11] AAAM, "Abbreviated Injury Scale (AIS)," *Association fo the Advancement of Automotive medicine,* pp. xx-xix, 2015.

[12] K. Aghkhani and et al, "The association between type of spine fracture and the mechanism of trauma: A useful tool for identifying mechanism of trauma on legal medicine field," *Journal of Forensic and Legal Medicine,* vol. 56, no. 56, pp. 80-82, 2018.

[13] J. Holmes and et al, "Epidemiology of thoracolumbar spine injury in blunt trauma," *Academic Emergency Medicine,* vol. 8, no. 9, pp. 866-872, 2001.

[14] U. Spiegl and et al, "Traumatic Fractures of the Thoracic Spine: Narrative Literature Review," *Zeitschrift fur Orthopadie und Enfallchirurgie,* vol. 159, no. 4, pp. 373-382, 2021.

[15] N. Szuflita and et al, "Spine Injuries Sustained by US Military Personnel in Combat Are Different From Non-Combat Spine Injuries," *Military Medicine,* vol. 181, no. 10, pp. 1314-1323, 2016.

[16] A. Vaccaro and et al, "The surgical algorithm for the AOSpine thoracolumbar spine injury classification system," *European Journal,* vol. 25, no. 4, pp. 1087-1094, 2016.

[17] N. Nishida and et al, "Tensile mechanical analysis of anisotropy and velocity dependence of the spinal cord white matter: a biomechanical study," *Neural Regeneration Research,* vol. 16, no. 12, pp. 2557-2562, 2021.

[18] S. Mattucci and et al, "Basic biomechanics of spinal cord injury - How injuries happen in people and how animal models have informed our understanding," *Clinical Biomechanics,* vol. 64, no. SI, pp. 58-68, 2019.

[19] C. Jones and E. Clarke, "Engineering approaches to understanding mechanisms of spinal column injury leading to spinal cord injury," *Clinical Biomechanics,* vol. 64, no. SI, pp. 69-81, 2019.

Appendix 1: AIS Injury Coding (fx = fracture, w/o = without)

Identifier	AIS Code	AIS Description	NISS	MAIS
SPINE 01	410202.1; 410602.1	Skin; abrasion; laceration	12	2
	640484.1	Spinous ligament injury, thoracic spine		
	650410.2	Vertebra, thoracic, dislocation no cord involvement, facet, unilateral		
	650417.2	Vertebra, thoracic, fx w/o neurologic deficit, multiple fx of same vertebra		

 https://doi.org/10.52202/081621-0004

	650418.1	Vertebra, thoracic, fx w/o neurologic deficit, spinous process		
SPINE 02	410202.1; 410602.1	Skin; abrasion; laceration	12	2
	450202.2	Rib cage, fx w/o flail, any location unilateral or bilateral, two ribs		
	640484.1	Spinous ligament injury, thoracic spine		
	650417.2	Vertebra, thoracic, fx w/o neurologic deficit, multiple fx of same vertebra		
	650418.1	Vertebra, thoracic, fx w/o neurologic deficit, spinous process		
	650420.1	Vertebra, thoracic, fx w/o neurologic deficit, multiple fx of same vertebra		
SPINE 03	410202.1; 410602.1	Skin; abrasion; laceration	12	2
	450202.2	Rib cage, fx w/o flail, any location unilateral or bilateral, two ribs		
	640484.1	Spinous ligament injury, thoracic spine		
	650417.2	Vertebra, thoracic, fx w/o neurologic deficit, multiple fx of same vertebra		
	650418.1	Vertebra, thoracic, fx w/o neurologic deficit, spinous process		
SPINE 04	410202.1; 410602.1	Skin; abrasion; laceration	17	3
	441800.2	Pleura laceration		
	450203.3	Rib cage, fx w/o flail, any location unilateral or bilateral, >=3 ribs		
	650417.2	Vertebra, thoracic, fx w/o neurologic deficit, multiple fx of same vertebra		
	650418.1	Vertebra, thoracic, fx w/o neurologic deficit, multiple fx of same vertebra		
SPINE 05	410202.1; 410602.1	Skin; abrasion; laceration	2	1
SPINE 06	410202.1; 410602.1	Skin; abrasion; laceration	17	3
	450203.3	Rib cage, fx without flail, any location unilateral or bilateral, >=3 ribs		
	650417.2	Vertebra, thoracic, fx w/o neurologic deficit, multiple fx of same vertebra		
SPINE 07	410202.1; 410602.1	Skin; abrasion; laceration	17	3
	450203.3	Rib cage, fx w/o flail, any location unilateral or bilateral, >=3 ribs		
	640484.1	Spinous ligament injury; thorax, laceration, minor; superficial		
	650417.2	Vertebra, thoracic, fx w/o neurologic deficit, multiple fx of same vertebra		
	650418.1	Vertebra, thoracic, fx w/o neurologic deficit, spinous process		
SPINE 08	410202.1; 410602.1	Skin; abrasion; laceration	17	3
	450203.3	Rib cage, fx w/o flail, any location unilateral or bilateral, >=3 ribs		
	640484.1	Spinous ligament injury; thorax, laceration, minor; superficial		
	650417.2	Vertebra, thoracic, fx w/o neurologic deficit, multiple fx of same vertebra		
	650418.1	Vertebra, thoracic, fx w/o neurologic deficit, spinous process		
SPINE 09	410202.1; 410602.1	Skin; abrasion; laceration	9	2
	640484.1	Spinous ligament injury, thoracic spine		
	650417.2	Vertebra, thoracic, fx w/o neurologic deficit, multiple fx of same vertebra		
	650418.1	Vertebra, thoracic, fx w/o neurologic deficit, spinous process		

Appendix 2: Vertebral Fractures (Sup = Superior, Inf = Inferior)

Identifier	Vertebra	Description
SPINE 01	T4	Spinous process
	T5	Spinous process
	T6	Lamina (comminuted); Spinous process (comminuted)
	T7	Lamina (comminuted); Spinous process (comminuted)
SPINE 02	T4	Spinous process
	T5	Lamina; spinous process
	T6	R pedicle; R transverse process; Lamina; Spinous process
	T7	R transverse process
SPINE 03	T4	Spinous process
	T5	Laminal; spinous process
	T6	Vertebral body; R pedicle; R transverse process; Lamina; Spinous process (comminuted/avulsed)
	T7	Vertebral body; R pedicle; R/L transverse process; Lamina; Spinous process (comminuted/avulsed)
SPINE 04	T4	Spinous process tip
	T5	R lamina into spinous process; Spinous process (comminuted/displaced)
	T6	Vertebral body; L lamina; L/R transverse processes; Spinous process
	T7	Vertebral body; L/R Sup articular facet; L/R transverse process; Spinous process; L/R Inf articular facet
	T8	Vertebral body; L/R Sup articular facet; L/R transverse process; Spinous process; L/R Inf articular facet
SPINE 05	T5	Spinous process; L Sup articular facet fracture; Vertebral body; L/R lamina (comminuted)
	T6	Vertebral body; Spinous process (comminuted); R Inf articular facet; L/R lamina
	T7	R Sup articular facet; spinous process
SPINE 06	T4	Spinous process
	T5	Vertebral body; L/R lamina; Spinous process
	T6	Vertebral body; L/R lamina; Left transverse process; Spinous process
	T7	Vertebral body; L/R lamina; Left transverse process; Spinous process
SPINE 07	T3	Spinous process
	T4	Spinous process
	T5	R pedicle; R lamina; R transverse process; R Inf facet; Spinous process
	T6	L/R Sup articular facet; Vertebral body; Midline lamina; Spinous process
	T7	R Sup facet; L/R Sup articular facet; Lamina (comminuted); Vertebral body
SPINE 09	T4	Spinous process
	T5	Vertebral body; Midline lamina; Spinous process (comminuted)
	T6	Vertebral body; Midline lamina; L Inf facet; Spinous process (comminuted)

Behind armour blunt trauma lung injury risk curves from different backface deformation profiles

N. Yoganandan[1], A. Shah[1], J. Koser[1], L. Somberg[1], D. Wilson[1], B. Stemper[1], R. S. Salzar[2], C. R. Bass[3], V.C. Chancey[4], B. Johnson[4], J. McEntire[4]

[1]*Department of Neurosurgery, 8701 Watertown Plank Road, Medical College of Wisconsin, Milwaukee, WI 53226, USA*
[2]*Center for Applied Biomechanics, University of Virginia,1215 Lee Street, Charlottesville, VA 22908, USA*
[3]*Biomedical Engineering, Wayne State University, 656 West Kirby Street, Detroit, MI 48202, USA*
[4]*U.S. Army Aeromedical Research Laboratory, 6901 Farrel Road, Fort Novosel, AL 36362, USA*

Abstract. Behind armour blunt trauma (BABT) occurs from backface deformations of the body armour. Different armour designs induce different backface deformation profiles. Different profiles may induce different injuries to the thoracoabdominal complex at the same impact level. It is ideal to expose armour-covered surrogates (human cadavers) and record backface deformation profiles using high-speed radiography; however, such studies are limited. This research is about the development of injury criteria from different backface deformation profiles. The design of the first indenter was based on BFDs from an earlier human cadaver study that included a hardbody armour, with data obtained from flash x-rays. The second indenter was designed with a different expected backface deformation profile. While the first indenter was spherical, the second indenter was conical and covered approximately one-third of the indenter-surrogate thorax surface contact area covered by the first indenter. Live swine tests were conducted with both indenters. Transducers were inserted into the lungs and heart. After administering anesthesia, impacts were delivered to the left lung at 30 to 90 m/s, swine were monitored for six hours, vitals were obtained, animals were euthanized, and autopsies were conducted. One former military surgeon and another civilian trauma surgeon scored injuries into mild, moderate, and severe categories. The spherical indenter induced uniform loading to the thorax. The conical indenter induced focal loading to the thorax. Survival analysis was used to develop viscous criterion injury risk curves for both indenter designs and for all grades of injuries. Mean and 95% confidence interval bounds are given in the body of the paper. Their quality indices are presented. The main conclusion of the study is that generalized injury criteria can be developed for improved body armour design, and they can be used to enhance medical readiness of our Warfighters.

1. INTRODUCTION

1.1 Literature goat data and multiparameter injury risk curves

The current BABT criterion of 44 mm deformation depth in Roma Plastilina No. 1 clay for body armour evaluations is based on the following considerations [1-5]. Experiments were conducted in the 1970s at the United States government and other research facilities Edgewood Arsenal Laboratory, Law Enforcement Assistance Agency, Land Warfare Laboratory, Lovelace Foundation, and Calspan Corporation. Seventy-four goat tests were conducted in three groups. Thirty animals in the first group were impacted at varying velocities at the thorax region using noncompliant projectiles of different mass, diameter, and weight. Surviving animals were euthanized after a 24-hour survival period. Using projectile physical properties, impact velocity and animal mass and thickness at the impact site, multiparameter models were explored to differentiate between lethal and non-lethal outcomes. Sixteen animals in the second group were tested under similar impacting conditions to validate the optimum model from the first group. Twenty-eight tests were added to this group to further validate the model.

Gelatin and different types of clay and foam were used as simulants to determine the optimum surrogate for standardization tests. Initial tests were done with a 20% gelatin block placed behind a soft body armor and high-tenacity nylon. The test was captured using a high-speed film from which the transient peak deflections were estimated. Clay experiments did not use the film camera because of opacity. Deflections from the simulant tests were compared with additional goat experiments that used the same soft body armour and projectiles. Mean peak deflections and areas of penetration were presented for different projectiles and velocities for simulant experiments. The deflections in the Roma Plastilina No.1 clay (85 mm) and 20% gelatin were (91 mm) were greater than the goat deflection (82 mm). Because of

https://doi.org/10.52202/081621-0005

complexities in recording deflections and elastic properties of the gelatin, and the closeness of the Roma Plastilina No. 1 clay deflection to the goat, clay was suggested as a simulant for BABT evaluations. Goat deflection-based injury risk curves obtained from film analysis were assumed to represent clay risk curves. Figure 1 shows the mean deflection-based injury risk curve along with 95% confidence interval bounds [6]. A deflection value of 44 mm is associated with ten percent lethality risk.

1.2 Rationale for regional injury criteria

The deflection of 44 mm was suggested as a preliminary criterion by the original authors of the study. It still serves as the lethality-based legacy BABT injury criterion to assess the performance of personal protective equipment, i.e., body armour. The original authors emphasized that additional experiments should be conducted to improve the criterion; however, only recently, advances are being made to improve the criterion [2, 6]. The material composition, anatomy and geometry of the structures covered by the body armour (rib cage, lungs, liver, heart, spine, etc.) are different. Their physiological functions also vary. Taken together, these factors play a role in the biomechanical responses, and injury mechanisms of the musculoskeletal structures and thoraco-abdominal organs to BABT. It should be noted that standardized tests in the automotive field use different limits and metrics for thoracoabdominal injury criteria specifications [7-9]. The heterogeneity of the human system should be considered in any injury criteria aimed at injury prevention and mitigation. In other words, it is important to develop injury risk curves on a regional basis. This is an overall objective of the research project.

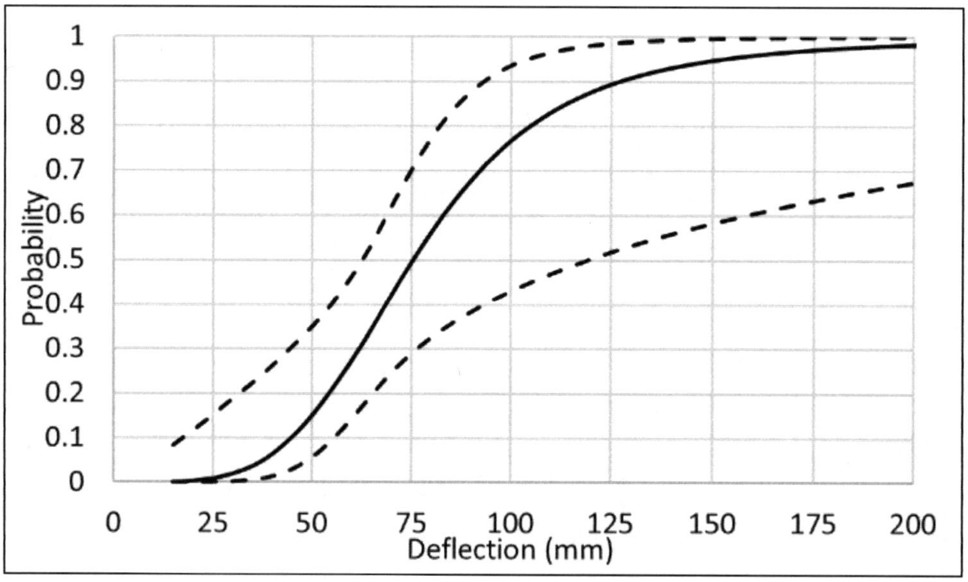

Figure 1: Deflection-based injury risk curves. Shaded areas show 95% confidence interval bounds. See text for details.

1.3 Objectives of the study

Specifically, the present study is focused on the lung region. The objectives were to develop physiologic Weibull distribution injury risk curves using the viscous criterion and survival analysis for statistical modeling for different severities of injuries from two indenter profiles using a swine model.

2. METHODS

2.1 Animal testing and instrumentation

Local institutional approval was obtained from the institutional animal care and use committee, and the same protocol was submitted to obtain approval from the United States Department of Defense, Army

https://doi.org/10.52202/081621-0005

Care and Use Review Office. Live swine were obtained from a local vendor and kept in the veterinary unit for at least 48 hours before preparation that included instrumentation and impact loading to the lung region. Trachea tubes and an intravenous line were inserted before anesthetizing the swine. Telazol and Xylazine were administered. Pressure transducers were placed in the lungs (Millar Houston, TX, USA). Another transducer was inserted into the right atrium by making an incision in the neck to isolate the jugular vein. The vessel was clamped, and a small incision was made to insert the transducer. Radiographs were taken to ensure placement of the transducers before the impact test. Isoflurane ventilation was used for continuous anesthesia delivery, and the protocol included maintaining the animal for six hours after impact loading. Temperature, pulse rate, heart rate, respiratory rate, blood pressure, peripheral capillary oxygen saturation, and electrocardiogram signals were recorded. The trauma surgeons of the study identified injuries to the skeletal and soft tissues including the lung organ. Figure 2 shows the schematic of the swine with pressure transducers and launching system with the indenter outside the launch tube.

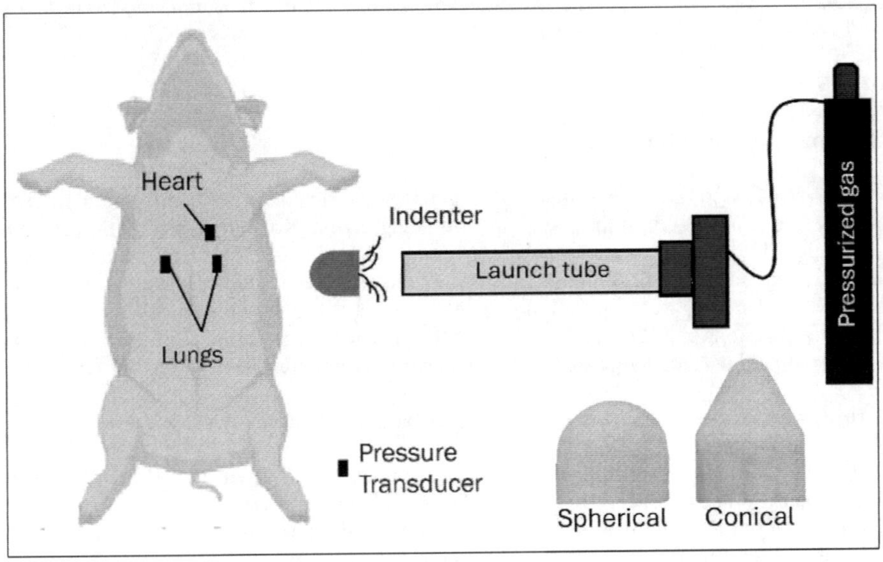

Figure 2: Schematic of the test setup for BABT impact. Launch tube and transducers are shown. Also shown are the two indenters, labelled spherical and conical.

2.2 Impact loading

The impact loading was accomplished using custom impactors (also termed as indenter) with a mass of ~230 g. Two indenters were used. The impact surface of the first indenter was based on previously documented hard body armour backface deformations from human cadaver tests [10]. The first indenter had a spherical shape. The second indenter was conical, and its impact surface area was approximately one-third of the first indenter. The angle of the cone was 57 degrees. The mass of the first and second indenters were ~230g and ~260g. Both indenters were printed with durable resin. A gas-driven launching system capable of producing a range of backface deformation velocities propelled the indenter to impact the lung region around the fourth and fifth intercostal spaces. A high-speed video camera was used to capture the impact loading in a plane parallel to the longitudinal axis of travel of the indenter, and video images were obtained at a rate of 10,000 f/s. Three parallel circumferential lines were drawn on the outside surface of the indenter to be visible for video imaging. The lines were tracked digitally to determine off-axis motion of the indenter before and after BABT impact loading. A triaxial accelerometer (Endevco, Depew, NY, USA) was mounted to the indenter. A digital data acquisition system (DTS Inc., Seal Beach, CA, USA) was used to gather data at ≥100 kHz and filtering was done at 2 kHz.

2.3 Development of injury risk curves

Injuries were graded as mild, moderate, or severe by the trauma surgeon authors. Peak magnitudes of the viscous criterion obtained using acceleration signals from the indenter were used to develop injury risk

https://doi.org/10.52202/081621-0005

curves using parametric survival analysis and Weibull distribution. Non-injury and injury data were assigned as right and uncensored variables. The confidence interval size was calculated as the width of the injury risk curve between plus and minus 95% confidence intervals. Normalised confidence interval size was calculated using the following equation at 10%, 25%, 50%, 75% and 90% levels of injury probabilities. The quality index was based on the following criteria:

$$\text{Normalised confidence interval size} = \frac{UL_p - LL_p}{VC_p} \quad (1)$$

Where p represents the probability of injury, VC_p is the mean or estimated value of the viscous criterion, and UL_p and LL_p represent the upper and lower limits of the 95% confidence intervals of the viscous criterion. A normalized confidence interval size value less than 0.5 was considered good, a value between 0.5 and 1 was considered fair, a value from 1 to 1.5 was considered marginal, and a value greater than 1.5 was considered unacceptable. This was according to established norm in biomechanical studies [11]. Viscous criterion injury risk curves are presented for both indenters and mild, moderate and severe injuries. The rationale for selecting this metric is provided in the discussion section.

3. RESULTS

3.1 Sample size and injuries

Forty-six live swine were used in the study. The sample size included four control animals, i.e., they were not subjected to any BABT impact loading to the lung region. No injuries were found at autopsy in control animals. Thus, any injuries from the impact-tested live animals were attributed to the blunt mechanical loading to the thorax. All impact-tested animals survived the six-hour period, i.e., there were no lethal outcomes, and this was true for both indenter impacts. Rib fractures and/or injuries to the lung in the form of contusions and lacerations were identified. There were six animals with no, 14 animals with mild, 10 with animals moderate, and 12 animals with severe injuries.

3.2 Injury risk curves from the spherical indenter series of tests

At 10%, 25% and 50% risk levels, the magnitudes of the viscous criterion for mild injuries were 1.65 m/s, 2.52 m/s, and 3.66 m/s, respectively. The quality indices ranged from good to fair to acceptable for all three types of injuries. Other data including 95% upper and lower confidence intervals are given in Table 1. Figure 3 shows injury risk curves from this series of tests.

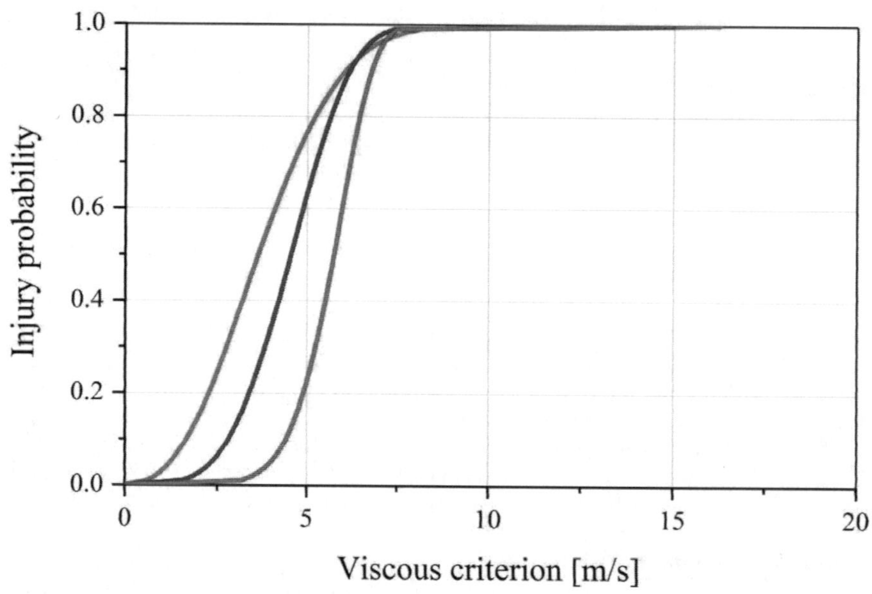

Figure 3: Lung injury risk curves from the spherical indenter. Left to right curves represent mild (green), moderate (blue) and severe (red) injuries.

Table 1: Summary of viscous criterion (m/s) values a from the spherical indenter for mild, moderate and severe injuries.

Risk	Mean value	Confidence interval upper bound	Confidence interval lower bound	Quality index
Mild injury				
5%	1.21	1.87	0.56	Acceptable
10%	1.65	2.37	0.92	Fair
25%	2.52	3.30	1.74	Fair
50%	3.66	4.45	2.85	Good
Moderate injury				
5%	2.34	3.18	1.50	Fair
10%	2.81	3.64	1.99	Fair
25%	3.64	4.40	2.88	Good
50%	4.55	5.25	3.86	Good
Severe injury				
5%	4.00	5.04	2.95	Fair
10%	4.42	5.36	3.48	Good
25%	5.08	5.86	4.30	Good
50%	5.74	6.42	5.07	Good

3.3 Injury risk curves from the conical indenter series of tests

At 10%, 25% and 50% risk levels, the magnitudes of the viscous criterion for mild injuries were 2.58 m/s, 3.97 m/s, and 5.67 m/s, respectively. The quality indices ranged from good to fair to acceptable for all three types of injuries. Other data including 95% upper and lower confidence intervals are given in Table 2. Figure 4 shows a comparison of the mean injury risk curves from this series of tests.

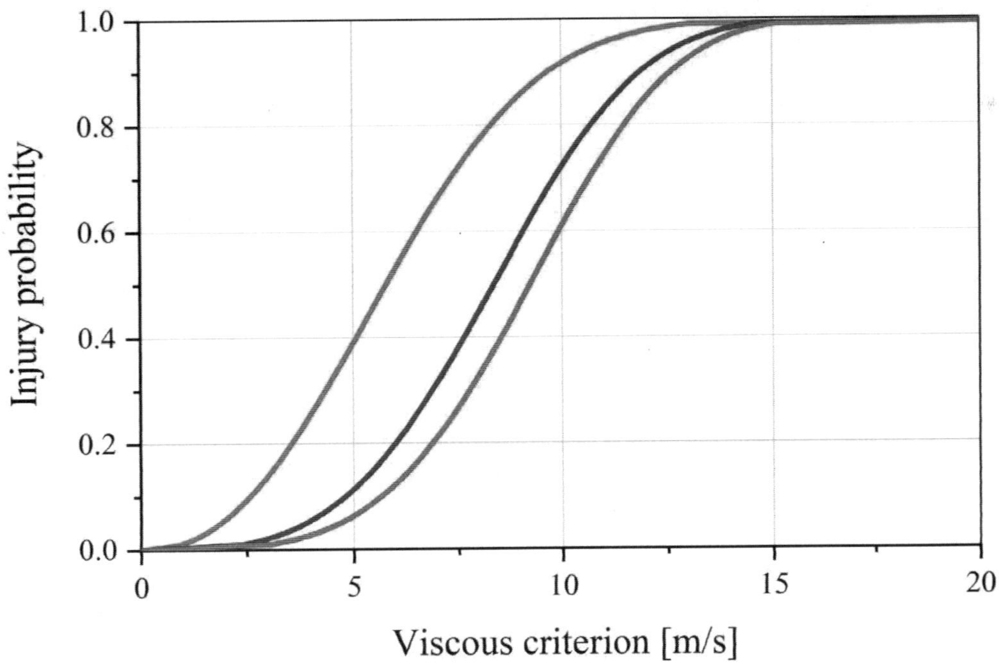

Figure 4: Lung injury risk curves from the conical indenter. Left to right curves represent mild (green), moderate (blue) and severe (red) injuries.

https://doi.org/10.52202/081621-0005

Table 2: Summary of viscous criterion (m/s) values from the conical indenter for mild, moderate, and severe lung injuries from survival analysis.

Risk	Mean value	Confidence interval upper bound	Confidence interval lower bound	Quality index
Mild injury				
5%	1.90	2.92	0.88	Acceptable
10%	2.58	3.73	1.44	Fair
25%	3.97	5.25	2.69	Fair
50%	5.78	7.16	4.40	Good
Moderate injury				
5%	3.90	5.63	2.17	Fair
10%	4.82	6.59	3.04	Fair
25%	6.46	8.24	4.69	Fair
50%	8.36	10.18	6.54	Good
Severe injury				
5%	4.71	6.75	2.66	Fair
10%	5.67	7.72	3.62	Fair
25%	7.36	9.35	5.36	Fair
50%	9.24	11.26	7.22	Good

3.4 Injury risk curves from both indenters

At 10%, 25% and 50% risk levels, the magnitudes of the viscous criterion for mild injuries were 1.90 m/s, 2.96 m/s, and 4.56 m/s, respectively. The quality indices ranged from good to fair to acceptable for all three types of injuries. Other data including 95% upper and lower confidence intervals are shown in Table 3. Figure 5 shows a comparison of the mean injury risk curves from this series of tests.

Table 3: Summary of viscous criterion (m/s) values from both indenters for mild, moderate, and severe lung injuries from survival analysis.

Risk	Mean value	Confidence interval upper bound	Confidence interval lower bound	Quality index
Mild injury				
5%	1.27	1.81	0.73	Fair
10%	1.90	2.54	1.25	Fair
25%	2.96	3.71	2.21	Fair
50%	4.56	5.40	3.71	Good
Moderate injury				
5%	2.25	3.07	1.43	Fair
10%	3.08	3.97	2.18	Fair
25%	4.35	5.31	3.39	Good
50%	6.09	7.14	5.04	Good
Severe injury				
5%	3.61	4.81	2.4	Fair
10%	4.60	5.84	3.36	Good
25%	6.02	7.29	4.75	Good
50%	7.82	9.18	6.46	Good

4. DISCUSSION

4.1 Rationale for the choice of thoracoabdominal organ

As stated in the introductory text, the objectives of the study were to develop injury risk curves on a regional basis to improve or enhance medical injury criteria (EMIC) for Warfighter protection and medical readiness and assist in improved body armour designs. The lung region was selected in the present study because of its vitality to the human system and bilateral symmetry. Lungs cover a large anatomical portion of the anterior torso in the human. Any injury criteria developed for the right lung is equally applicable to the left lung (physiological functions and injury tolerances do not differ from a biomechanical perspective). As an organ with air as primary occupier, its mechanisms of injury and tolerance are expected to be different from other organs that are also housed in the anterior thorax, i.e., heart and liver. Similar tests are necessary to develop injury risk curves for other organs, and the authors of this consortium are pursuing such efforts.

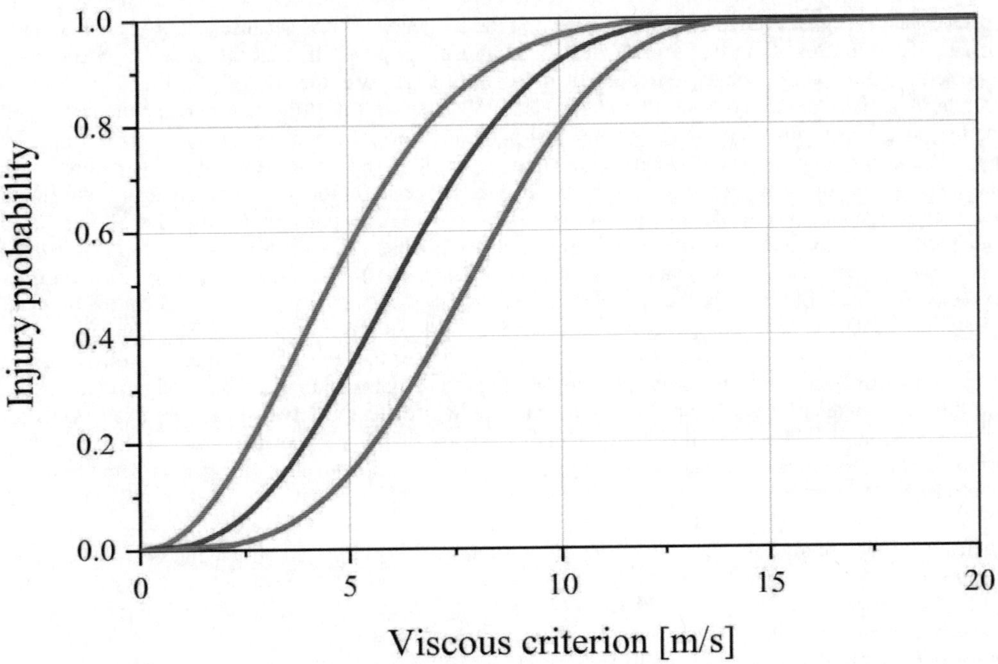

Figure 5: Comparison of lung injury risk curves from both indenter tests. Left to right curves represent mild (green), moderate (blue) and severe (red) injuries. Data are

4.2 Survival analysis and Weibull distribution

Parametric survival analysis was used to develop injury probability curves in this study. This type of analysis is becoming more popular for automotive, aviation, and military applications to define human tolerances for different body regions [12-16]. Recommendations from the International Standard Organization and earlier automotive dummy-based studies was one of the primers for the change from conventional binary regression to survival regression models [11]. The improvement in the previously recommended methodology are given elsewhere [17]. As stated in the objectives, the Weibull distribution was used as it is commonly used in impact biomechanics to develop injury risk curves; however, additional analysis can be done in the future to determine the optimal distribution. Standard textbooks provide details such as equations and the underlying theory of the Weibull function. Statistical metrics such as the Akaike information criterion, or its modifications thereof, can be used, and this is considered as a future study [18]. The use of the quality index via the normalized confidence interval size at discrete probabilities of injury is also commonly used in impact biomechanics [11]. The adjectival rating was based on the accepted. From these considerations, the present set of risk curves serve as initial injury criteria for lungs for BABT.

4.3 Justification for the viscous criterion

The legacy metric for standardized tests is based on a single peak residual deflection in the Roma Plastilina No.1 clay. This is based on the goat thorax deflection from thorax tests. Previous studies have shown that peak force best correlates for musculoskeletal system injuries such as compression fractures of spinal vertebrae [19]. The viscous criterion is shown to be a better predictor for organ injuries under impact loading [20-22]. It represents the energy dissipated by soft tissue deformation and is considered an underlying soft tissue/organ injury mechanism. The criterion includes the response of the torso unlike

https://doi.org/10.52202/081621-0005

a priori/external input metrics such as energy. Hence, the viscous criterion was used. Additional metrics such as force, impulse, energy density, and loading rate can also be used to develop risk curves and evaluate their correlatability to BABT injuries. They are future topics.

4.4 Rationale for the impacting system and injury risk curves from both indenter designs

The present study used an instrumented indenter to apply the BABT loads to the unprotected specimen, not live rounds being defeated by armour protecting the specimen. This is because, using any actual body armour will limit results to that armour design, i.e., results will not be generalizable. A recent study reported the following: "while a strength of the current study was the use of a single *body armour* (emphasis added) helmet type for all ballistic data collected with post-mortem human surrogates and load-sensing headforms to ensure strong matched-pair conditions, these injury outcomes are relevant for a single *body armour* (emphasis added) helmet and projectile type. Care must be taken to interpret these results in context of other helmet systems [23]." The same is true for torso coverage. The use of a single indenter design may be not adequate for similar reasons. While the spherical indenter design was based on backface deformation profiles from hard armor-covered human cadaver tests, to develop generalized risk curves, it was deemed necessary to include other designs [10]. The conical indenter was based an expected profile, different from the spherical indenter. It was intened to induce more focal loading and greater concentrated deflections, with a greater loading on local skeletal anatomy. This contrasted with a more distributed/uniform loading from the spherical indenter that engaged a larger area of the thorax with a wider different skeletal loading pattern. While skeletal deformations in both designs compressed the thoracic contents, the mechanism of load transfer was different between the two designs. Other indenter designs will likely lie within the boundaries of these two shapes; however, tests with actual body armor are needed. Once those data are available, it would be possible to check the generalizability of the risk curves.

Dual peaks representing the penciling effect may suggest that the body armor is less effective. Hard armors generally do not induce dual peaks. Additional tests are needed to not account for any twin peaks in the loading pattern to the thorax with other types of indenter designs. The delineation of pure pressure wave transmission to the torso resulting in lung injuries without the deformations of the skeletal and internal contents is a possibility; however, to the best knowledge of the authors, published studies developing graded injury risk curves are not available. This remains as an active research area and the authors are also pursuing this topic.

A comparison of risk curves and metrics shows that moderate injuries lie within the confines of the mild and severe injuries for both indenters and combined data from both indenters. This suggests that the bounding of the risk curves can be assumed for different backface deformations; however, the moderate injury risk curve shifts more to the right with the conical indenter. The moderate injury risk curve is approximately in the middle of the mild and severe curves with the spherical indenter. The rightward shift demonstrates the role of shape on the injury severity. The second indenter being conical induces a more focal loading to the anterior thorax, resulting in greater compressions of the thorax, while the larger area/volume of coverage of the spherical indenter applies a more distributed loading to the thoracic contents. In other words, stress concentration effects are likely more pronounced with the conical indenter. This implies that a body armour with greater local backface deformations around the lung region may impart more severe injuries than a design with more uniform backface deformations. Despite this difference, as the confidence intervals overlap, it may be reasonable as a first step to combine these data for a potential recommendation. Figure 6 shows the data at discreet levels. The developed injury risk curves are associated with mild, moderate and severe lung injuries. Greater levels of external insult to produce lethality are needed to develop lethality-based risk curves. The authors are currently pursuing those experiments.

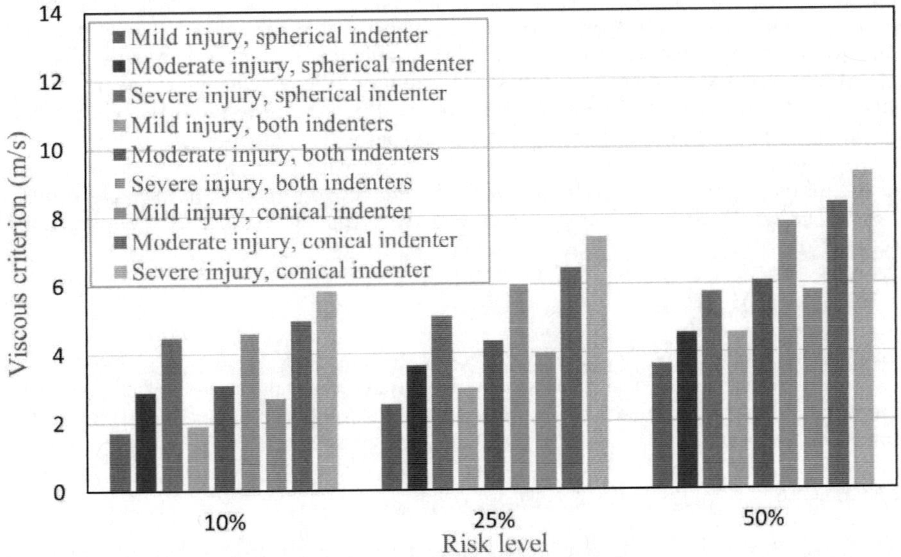

Figure 6: Mean values of the injury metric at discrete risk levels for all injuries. Data include from each indenter and combined data from both indenters.

4.5 Injury classifications

Many classifications are available to score injuries to bones and organs and soft tissues. The Abbreviated Injury Scale, AIS, is widely used for automotive applications [24, 25]. It is undergoing revisions since 1980 [24, 25]. The AAST is used for clinical treatment and classifies injuries separately based on chest wall and organ injuries. Other approaches have included the use of a four grade scale: return to duty within 72 hours, after 72 hours, after 72 hours with potential for duty-limiting conditions, and potentially fatal [26]. This classification scheme was used for inferring injuries sustained from autopsy findings from human cadaver tests. In contrast, the present study used three classifications for injuries based on treating trauma surgeons and physiological swine experiments with monitoring of vitals. Because the injuries were scored based on clinical treatment, and, as reiterated, tests were from live swine instead of cadavers, the chosen classification scheme is more applicable.

4.6 Limitations

As stated earlier, the human thoracoabdominal region houses complex structures: (a) materially they range from bony ribcage to soft tissues with widely varying mechanical properties and injury tolerances; (b) functionally also they are different: lungs being an air-related organ compared to heart, a muscle, etc.; (c) anatomically placements, i.e. rib cover, is also different: lungs fully covered by the rib cage while liver is partially exposed. Thus, injury tolerances are expected to differ. The present series of tests were focused on BABT lung impacts. It is important to conduct similar tests to the heart and liver obtain regional injury risk curves. Posterior anatomical coverage is important as threats to Warfighter's include all areas covered by the armour. The presence of neural structures in the spinal column poses challenges as monitoring physiological functions after injuries to bony elements (e.g., lamina[e]) are complex. One human cadaver (protected with experimental armour) test reported AIS=5 level injuries with spinal impact from a successfully defeated ballistic threat [10, 24]. Injuries included a laceration between thoracic vertebrae T9 and T12 and disintegration of their spinous processes into the spinal canal. The same cadaver test under the same initial conditions with sternal impact produced AIS=4 injuries. Similar studies with live swine are not available and would be necessary to define posterior impact BABT injury risk curves. This is a future study.

4. SUMMARY AND CONCLUSIONS

The present study developed physiology-based lung injury risk curves applicable to BABT loading. They were developed using an accepted animal model (live swine) with six-hour monitoring of all animals.

https://doi.org/10.52202/081621-0005

Dynamic loading was applied using an impacting system and indenter. The first/spherical indenter design was based on backface deformations obtained from hard body armour testing with human cadavers. The second/conical indenter design was based on expected profile that was distinct from the first/spherical indenter. The conical indenter induced more focal thoracic loading with potential increased stress concentrations near the impacted region than the spherical indenter. The two indenters represented a range of deformations profiles from current and future armour designs. Injury risk curves were developed using survival analysis and the viscous criterion metric, known to depict the injury mechanism for organs, and account for the temporal factor. The curves were developed for three severities, based on clinical assessments. The developed set of injury risk curves serve as a preliminary set for defining lung injury tolerance to BABT.

5. ACKNOWLEDGMENTS

This research was supported by the U.S. Army Medical Research and Development Command contract W81XWH-21-9-0015 and the Department of Veterans Affairs Medical Research. The views expressed in this material are those of the authors, and do not reflect the official policy or position of the U.S. Government, the Department of Defense, or the Department of the Army.

6. REFERENCES

[1] Clare, V. R., Lewis, J. H., Mickiewicz, A. P., and Sturdivan, L. M., 1975, "Blunt trauma data correlation," No. EB-TR-75016, Edgewood Arsenal, Aberdeen Proving Ground, MD.

[2] Prather, R., Swann, C., and Hawkins, C. E., 1977, "Backface Signatures of Soft Body Armors and the Associated Trauma Effects. ARCSL-TR-77055," U.S. Army Armament Research and Development Command Technology Center. Aberdeen Proving Ground, MD.

[3] Goldfarb, M. A., Weinstein, M. A., and Metker, L. W., 1975, "A method for soft body armor evaluation: medical assessment," U.S. Army Limited War Laboratory.

[4] Metker, L. W., Prather, R. N., and Johnson, E. M., 1975, "Methods for determining backface signatures of soft body armors. Washington, DC," U.S. Department of Justice.

[5] Montanarelli, N., Hawkins, C. E., Goldfare, M. A., and Ciurej, T. F., 1973, "Protective garments for public offiicals," U.S. Army Welfare Laboratory, Aberdeen Proving Grounds, MD.

[6] Yoganandan, N., Shah, A., Somberg, L., Baisden, J., Stemper, B. D., Bass, C., Salzar, R. S., Chancey, V. C., and McEntire, J., 2023, "A Novel Paradigm to Develop Regional Thoracoabdominal Criteria for Behind Armor Blunt Trauma Based on Original Data," Mil Med, 188(Suppl 6), pp. 598-605.

[7] Kuppa, S., Eppinger, R. H., McKoy, F., Nguyen, T., Pintar, F. A., and Yoganandan, N., 2003, "Development of Side Impact Thoracic Injury Criteria and Their Application to the Modified ES-2 Dummy with Rib Extensions (ES-2re)," Stapp Car Crash J, 47, pp. 189-210.

[8] FMVSS-214, 2008, "Federal Motor Vehicle Safety Standards," US Government, Washington, DC USA, https://www.gpo.gov/fdsys/granule/CFR-2011-title49-vol6/CFR-2011-title49-vol6-sec571-214.

[9] Prasad, P., 2015, "Injury crietria and motor vehicle regulations," Accidental Injury: Biomechanics and Prevention, N. Yoganandan, A. M. Nahum, and J. W. Melvin, eds., Springer, New York, pp. 793-809.

[10] Bass, C. R., Salzar, R. S., Lucas, S. R., Davis, M., Donnellan, L., Folk, B., Sanderson, E., and Waclawik, S., 2006, "Injury risk in behind armor blunt thoracic trauma," Int J Occup Saf Ergon, 12(4), pp. 429-442.

[11] Petitjean, A., Torsseille, X., Yoganandan, N., and Pintar, F. A., 2015, "Normalization and scaling for human response corridors and development of risk curves " Accidental Injury: Biomechanics and Prevention, N. Yoganandan, A. M. Nahum, and J. W. Melvin, eds., Springer,, NY, pp. 769-792.

[12] Rooks, T. F., Chancey, V. C., Brozoski, F. T., Salzar, R. S., Pintar, F. A., and Yoganandan, N., 2018, "Development of an injury risk curve for pelvic fracture in vertical loading environments," Traffic Inj Prev, 19(sup2), pp. S178-S181.

[13] Humm, J., and Yoganandan, N., 2021, "Lower Neck Injury Assessment Risk Curves Based on Matched-Pair Human Data for Anthropomorphic Test Devices," Mil Med, 186(Suppl 1), pp. 639-644.

[14] DeVogel, N., Banerjee, A., and Yoganandan, N., 2019, "Application of resampling techniques to improve the quality of survival analysis risk curves for human frontal bone fracture," Clin Biomech (Bristol), 64, pp. 28-34.

[15] Chirvi, S., Pintar, F., Yoganandan, N., Banerjee, A., Schlick, M., Curry, W., and Voo, L., 2017, "Human Foot-Ankle Injuries and Associated Risk Curves from Under Body Blast Loading Conditions," Stapp Car Crash J, 61, pp. 157-173.
[16] Abraczinskas, A., Olszko, A., Beltran, C., Baisden, J. L., Yoganandan, N., McGhee, J., McGovern, S., Lidl, G., Vasquez, K. B., and Chancey, V. C., 2021, "A Comparison of Nonhuman Primate Injuries in Horizontal Versus Vertical Sled +Gz (Head-to-Foot) Impact Accelerations," Mil Med, 186(Suppl 1), pp. 610-618.
[17] Yoganandan, N., Banerjee, A., Hsu, F. C., Bass, C. R., Voo, L., Pintar, F. A., and Gayzik, F. S., 2016, "Deriving injury risk curves using survival analysis from biomechanical experiments," J Biomech, 49(14), pp. 3260-3267.
[18] Akaike, A., 1974, "A new look at the statistical model identification," IEEE Transactions on Automatic Control, 19, pp. 716-723.
[19] Yoganandan, N., Moore, J., DeVogel, N., Pintar, F., Banerjee, A., Baisden, J., Zhang, J. Y., Loftis, K., and Barnes, D., 2020, "Human lumbar spinal column injury criteria from vertical loading at the base: Applications to military environments," J Mech Behav Biomed Mater, 105, p. 103690.
[20] Viano, D. C., and Lau, I. V., 1988, "A viscous tolerance criterion for soft tissue injury assessment," J Biomech, 21(5), pp. 387-399.
[21] Viano, D. C., and Lau, I., 1985, "Thoracic impact: a viscous toelrance criteria," Tenth International Conference of the Enhanced Safety of Vehicles, NHTSA, WAshinton, DC.
[22] Viano, D. C., Kroell, C. K., and Warner, C. Y., "Comparative thoracic impact response of living and sacrificed porcine siblings," Proceedings of the 21st Stapp Car Crash Conference.
[23] Iwaskiw, A., Howes, C., Hingorani, R., Bradfield, C., Mazuchowski, E., Clark, M., Luong, Q., and Drewry, D., "Loads assocaited helmet blunt trauma," Proceedings of the Personal Armour Systems Symposium. pp. 146-156.
[24] AIS, 1998, Abbreviated Injury Scale: 1998 update, American Association for Automotive Medicine, Arlington Heights, IL.
[25] AIS, 2015, Abbreviated Injury Scale: American Association for Automotive Medicine, Arlington Heights, IL.
[26] Howes, C., Mazuchowski, E., Rohrer, A., Clark, M., Luong, Q., and Drewry, D., "Injuries and Operational Implications Caused By Behind Armour Blunt Trauma Across Various Impact Locations " Proceedings of the Personal Armour Systems Symposium. pp. 168-176.

 https://doi.org/10.52202/081621-0005

Progression of Soft Tissue Contusions in Behind Armor Blunt Trauma

A. V. Stotka[1], J. A. McMahon[1], P. R. Berthelson[1], C. R. D. Bass[2], N. Yoganandan[3], B. J. McEntire[4], R. S. Salzar[1]

[1]*Center for Applied Biomechanics, University of Virginia, 4040 Lewis and Clark Drive, Charlottesville, VA 22911, USA, qhb9aw@virginia.edu*

[2]*Biomedical Engineering, Wayne State University, 656 West Kirby Street, Detroit, MI 48202, USA*

[3]*Biomedical Engineering, Medical College of Wisconsin, 8701 Watertown Plank Road, Milwaukee, WI 53226, USA*

[4]*U.S. Army Aeromedical Research Laboratory, 6901 Farrel Road, Fort Novosel, AL 36362, USA*

Abstract. Behind Armor Blunt Trauma (BABT) occurs when body armor is struck by an incoming ballistic munition, causing a high-rate loading event between the body armor's back face and the wearer's torso. Previous studies have demonstrated the risk of BABT injuries to thoracic organs in terms of impact energy. Further study of soft tissue injuries produced during BABT impacts is needed to complete the understanding of BABT. This study performed 35 BABT impacts on the interior right thigh of live porcine specimens as a surrogate model for humans to further understand the injury response of soft tissue. A blunt indenter with a similar shape and recruited mass to previously performed studies was used to impact the live porcine specimens across a range of energies. Each impact location was carefully chosen to avoid overlapping skeletal or abdominal regions, only impacting the skin and underlying tissues. The porcine specimens were then survived from 1 to 30 days after the impact to monitor the progression of injury across differing impact energy levels and post-impact survival periods. Injuries ranged from minor contusions that fully healed by 12 days after a low energy impact of 50 Joules [J], to severe injuries that produced significant swelling, skin necrosis, volumetric muscle loss, and large hematomas at the impact location. At the highest energies of 270 to 330 J, these injuries did not heal by the end of the 30-day survival period, with visible scarring on the skin, volumetric muscle loss, and unorganized muscle growth. Observations throughout the survival periods and subsequent mechanical testing of the muscle tissue indicate that most low to medium-energy impacts to muscle can heal within one month after impact; however, high-energy impacts showed potentially permanent changes to the muscles and skin.

1. INTRODUCTION

Body armor is designed to stop an incoming ballistic munition by absorbing and dispersing the impact energy. This is achieved through deformation of the body armor. While body armor is designed to prevent penetrative injuries, the resulting body armor deformation creates a high rate loading event to the underlying anatomy. This can cause physiological and structural injuries to the bones, soft tissue, and organs resulting in an injury modality referred to as behind armor blunt trauma (BABT) [1-3]. These injuries, though not as severe as penetrative injuries, can result in substantial pain, inflammation, and functional impairment in the region of impact. In some severe cases, there are potential long-term health consequences, such as localized tissue necrosis, hematoma formation, myositis ossificans, and volumetric muscle loss [4-13].

Understanding the pathophysiological progression of deep tissue contusions from behind armor blunt trauma is critical for improving protective gear design, clinical interventions, and rehabilitation strategies. Studies on blunt injuries from sports, such as American Football, may provide a path forward, as they have documented player recovery timelines, the type and severity of injury observed, and the initial treatment of these injuries [7,8,11,13-17]. While these studies help provide the injury patterns and recovery process of contused muscles, many are retrospective studies with unknown impact parameters or conditions. Studies utilizing live animal and human volunteer models that control the impact location, impact severity, and the mechanism of impact can provide further insight into these types of injuries [11,15,16].

Live porcine specimens have been shown to be a viable human surrogate for studying injuries due to their anatomical and physiological similarities to humans [18-22]. However, while porcine specimens serve as a useful model for initial injury research, an important consideration is that, unlike humans, they continuously grow throughout their lives [23]. This growth may influence the mechanical properties and healing response of tissues, including the rate and extent of tissue regeneration following an injury that

https://doi.org/10.52202/081621-0006

would not be seen in a human throughout a longitudinal study. Despite this, the pig remains a valuable model for studying the initial stages of tissue damage and recovery, providing important information that can guide the development of better protective armor and treatment strategies for human injuries.

Simulating a BABT injury in a live porcine model serves as a valuable tool for understanding the severity and recovery process of such injuries in humans. When a porcine specimen is impacted, it results in a controlled injury that mimics the physical damage that might occur in a combat setting. The extent of functional impairment following the contusion provides critical data regarding the impact on mobility and overall physical performance for humans. This information is essential when assessing a soldier's readiness to return to duty, as it helps determine how the injury affects their ability to perform tasks requiring physical strength, coordination, and endurance. Evaluating recovery through this model ensures that soldiers are not prematurely cleared for duty, promoting both their safety and operational effectiveness.

The importance of proper return to duty is imperative as the return to duty rate of military personnel evacuated due to musculoskeletal injuries was approximately the same as those evacuated due to combat related injuries (3.3% vs 2.8% return to duty, respectively) [24]. Additionally, muscle contusions have been shown to be the fourth most common injury treated among infantry brigade teams [25]. With musculoskeletal injuries being highly prevalent in the military, there is a need for new methods to understand, prevent, and reduce their impact. Recent research has identified the regions of the body that are most susceptible to injury during deployment but still lack detailed analysis that clarifies the mechanisms behind them [26,27]. The next step is to study how deep tissue contusions present themselves in musculature to elucidate new diagnostic procedures. The objective of this study was to quantify how BABT soft tissue contusions evolve as a function of impact severity and time post-impact.

2. METHODS

2.1 Live porcine preparation

Experimental testing on thirty-five (35) live Yorkshire porcine specimens was performed at the University of Virginia (UVA) with live animal testing protocols approved by the UVA Animal Care and Use Committee (protocol number 4379-12-21) and U.S. Army Animal Care and Use Review Office (protocol number MT21006.031.e002). Porcine were allowed to acclimate to the testing facility for at least three days before experimental testing was performed. Live porcine were induced with propofol through an intravenous catheter prior to intubation. Once intubated, the porcine was kept under anesthesia with Isoflurane and medical air. A local analgesic, Bupivacaine, was administered at the target impact location prior to impact. Additionally, a long-acting 96-hour analgesic, Buprenorphine, was administered subcutaneously in the thin flank skin. Further administration of analgesics was not used unless discomfort was noticed by the veterinarian care staff. Analgesics that might aid in the recovery process or reduce the inflammatory response to the injury were not employed in this study to study how a soft tissue contusion would change over time in the event of an untreated injury.

2.2 BABT impact testing

A high-rate pneumatic impactor was designed to propel a 3D-printed indenter at live porcine to study the change of musculoskeletal properties due to BABT injuries. The profile of the indenter was modeled after the backface deformation profile of hard body armor [2] and had a diameter of 10.2 centimeters [cm]. By using an indenter with this profile, injuries were characterized based on a generic threat as opposed to a specific munition or body armor. The indenter was equipped with an onboard accelerometer (Endevco 7270A) to measure the acceleration-time history of the impact with a combined mass of 247 $\pm$ 2.8 grams [g].

Thirty-five (35) total porcine specimens (mass = 43.02 $\pm$ 3.80 kilograms [kg]) were impacted for this study with survival periods ranging from 1 to 34 days post impact and target impact energies ranging from 47.98 to 338.20 Joules [J], see Table 1. There were five target survival periods and seven target impact energies, allowing one impact per combination of duration and impact energy. The reference test matrix used can be seen in Table 2. The actual test matrix (Table 1) has impact energies that are not listed in the reference test matrix (Table 2)

With the porcine specimen laying supine, a target impact location was chosen on the interior of the hind right leg. Positioning of the impact location was made by palpation of the hind leg to find where the center of the indenter would make maximal contact with the underlying soft tissues. Once a target location was found, a comparable location was found on the left thigh to serve as a control. The porcine

https://doi.org/10.52202/081621-0006

specimen was then moved beneath the barrel of the impactor, ensuring that the impact would avoid any of the abdominal or skeletal regions. Further, the specimen's right hind thigh was placed in contact with the end of the barrel to ensure an orthogonal impact condition. The porcine was then recovered from anesthesia, returned to its cage, and monitored for any signs of discomfort over the post-impact survival period. On the final day of the survival period, photos were taken of the hind legs to visually compare the surface-level damage.

Table 1. Test matrix. Specimen mass is the mass of the porcine on the day of impact. Survival time is the time in days the specimen was survived after the impact.

Specimen Number	Target Impact Energy (J)	Actual Impact Energy (J)	Specimen Mass (kg)	Survival Time (days)
CP001	100	76.38	44.4	30
CP002	40	47.98	44	30
CP003	100	122.40	42	20
CP004	50	50.13	43.2	20
CP005	150	144.88	44.4	30
CP006	300	276.31	40.6	6
CP007	225	231.58	39.6	6
CP008	150	165.67	47.0	11
CP009	150	142.67	47.0	5
CP010	190	185.84	40.6	30
CP011	50	47.65	43.6	1
CP012	80	90.56	41.2	1
CP013	150	159.88	39.8	20
CP014	50	48.60	45	5
CP015	80	77.53	45.4	5
CP016	190	180.06	42.2	20
CP017	230	236.22	40.3	30
CP018	150	145.40	50	1
CP019	120	111.14	40	1
CP020	80	83.85	42.2	12
CP021	190	179.30	53	11
CP022	230	230.87	50	13
CP023	275	310.11	43.6	30
CP024	230	237.82	43.2	20
CP025	275	250.98	40	20
CP026	120	140.29	38.4	5
CP027	190	181.65	40.4	1
CP028	120	109.91	40.7	12
CP029	230	222.99	38.0	1
CP030	270	267.02	38.2	1
CP031	190	173.46	40.8	6
CP032	310	296.70	43.8	20
CP033	270	251.45	39.8	13
CP034	340	332.34	52	34
CP035	310	338.20	41.2	8

Table 2. The reference test matrix to appropriately test a range of impact energies at discrete times post-impact.

		Days Post-Impact				
		1 Day	5 Day	12 Day	20 Day	30 Day
Impact Energy (J)	50 J					
	80 J					
	120 J					
	150 J					
	190 J					
	230 J					
	270 J					

2.3 Mechanical testing

On the final day of the survival period, both hind legs were removed for mechanical testing. A custom test device (Figure 1) was used to perform confined quasi-static compression tests (compression rate of 1.3 mm/s) on excised tissue samples (diameter = 10.2 cm, depth = 4.5 cm) to assess through-the-thickness changes in the muscle tissue. The test device was equipped with a flat, cylindrical plate (diameter = 2.54 cm) attached to a linear actuator to quasi-statically compress the tissue while recording the compressive force via a load cell (Honeywell model 31 series, 50 lb.) at a sampling rate of 500 Hz. Excised tissue samples were taken from the impact location on the right thigh and an equivalent location on the left thigh to serve as direct control for the impacted tissue.

The skin and subcutaneous tissues were removed from the samples to measure the force response of the two underlying muscles, the Gracilis and Semimembranosus. Once placed in the testing cup, the travelling head was brought into contact with the surface of the Gracilis at the sample's center point. The sample was then compressed to 25% strain to assess the loading response of the tissue. Tests were conducted at the center of each impact site, as well as at a similar location on the unimpacted contralateral sample to assess the differences between the impacted tissue and healthy tissue from the same specimen. To ensure there were no edge effects from the confined quasi-static compression tests, additional tests were performed 2.54 cm radially outward from the center point and at higher strain levels. At the completion of mechanical testing, the impacted tissue samples were bisected to visually inspect the contusion capsule within the muscles.

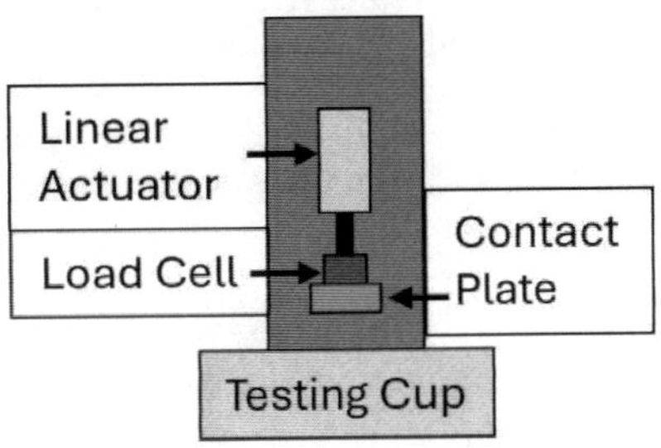

Figure 1. Diagram of the test device used for quasi-static mechanical testing.

3. RESULTS

The peak force of each quasi-static compression test was taken as an approximation for through-the-thickness stiffness of the tissue. Figure 2 shows an interpolated map of the peak compressive force responses at 25% strain from the impacted tissue as a function of the impact energy and survival time post-impact. The plot shows the peak forces in a color bar relative to the maximum force observed from this test series. Blue indicates low peak forces and low muscle stiffness while red indicates high peak forces and high muscle stiffness. The bottom right portion of the plot is uncolored because there were no tests performed within that specific impact energy and survival period. Additionally, control data is not plotted as it has an impact energy of 0 J. Control data had an average peak force 2.7 ± 1.6 Newtons

https://doi.org/10.52202/081621-0006

[N] (n = 26). Fewer data points were used as these 26 were tested within seven hours post euthanasia. Control data showed greater variance in samples tested later than 7 hours post euthanasia and were excluded due to the potential effect of rigor mortis in control tissue.

Table 3 visualizes the presence of observable contusion at each survival period end point through shaded cells in the test matrix. Cells shaded red indicate that contusion was observed in the skin and muscles. Yellow shaded cells indicate that no contusion was observed in the skin, but there was contusion in the muscles. Some high energy impacts with long survival periods were observed to have scarring on the skin at the impact location but no contusion. These tests are shaded as yellow in the test matrix as a scar is treated as a different injury than a contusion within this study. Cells shaded green did not have any observable contusion in the skin or muscles.

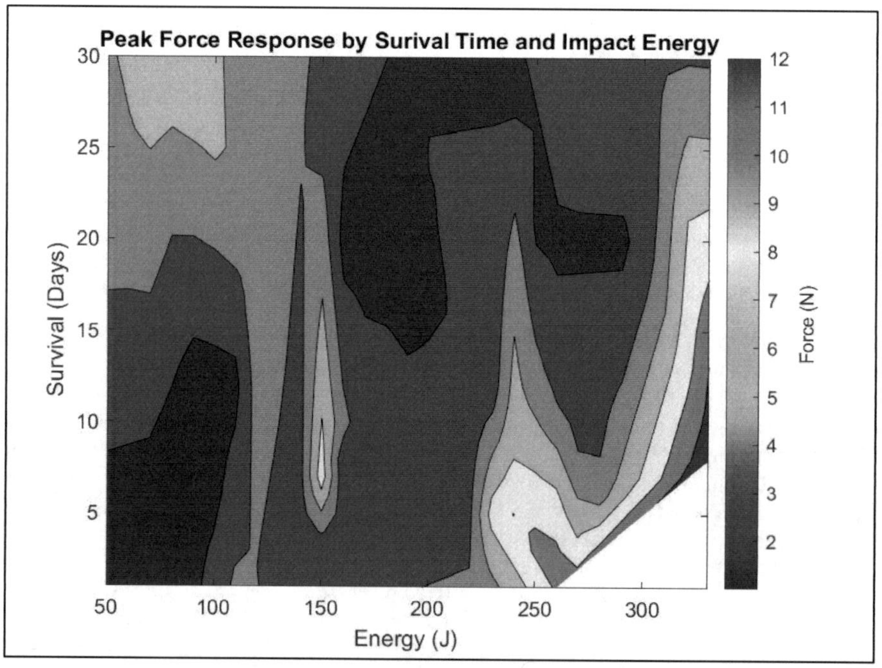

Figure 2. Interpolated peak force response of impacted muscle at its center point is by varying impact energy and survival time post-impact.

Table 3. Injury observations of surface level and deep tissue contusions. Red cells indicate contusion observed in the skin and muscle. Yellow cells indicate contusion only observed in the muscles. Green cells indicate no contusion observed.

Color Key	Impact Energy (J)	Days Post-Impact: 1 Day	5 Day	12 Day	20 Day	30 Day
RED: Contusion in skin and muscle	**50 J**					
YELLOW: Contusion in muscle	**80 J**					
GREEN: No contusion	**120 J**					
	150 J					
	190 J					
	230 J					
	270 J					

Exemplary images of the contusions within the deeper muscle, the Semimembranosus, can be seen in Figure 3. The three photos were taken from three different tests of equivalent impact energy to show the progression of a contusion from 1 day post impact to 30 days post impact. Figure 3A shows a contusion in the Semimembranosus at 1 day post impact. The dark red area of the muscle is the contused portion of the muscle. The discoloration indicates the formation of a contusion capsule as the spread of damage is mitigated. In this specimen, there was observable contusion in the skin and muscles. Figure 3B shows a contusion in the Semimembranosus at 11 days post impact. The contusion capsule has decreased in size, and the discoloration around the contusion capsule is different to that of the contusion itself. The change of color surrounding the contusion capsule shows the progression of healing and removal of the contusion. There was no observable contusion in the skin of this specimen. Figure 3C shows a fully healed Semimembranosus with no signs of contusion at the impact site 30 days post impact. No observable contusions were noted in the skin or muscles at the impact location of this porcine specimen.

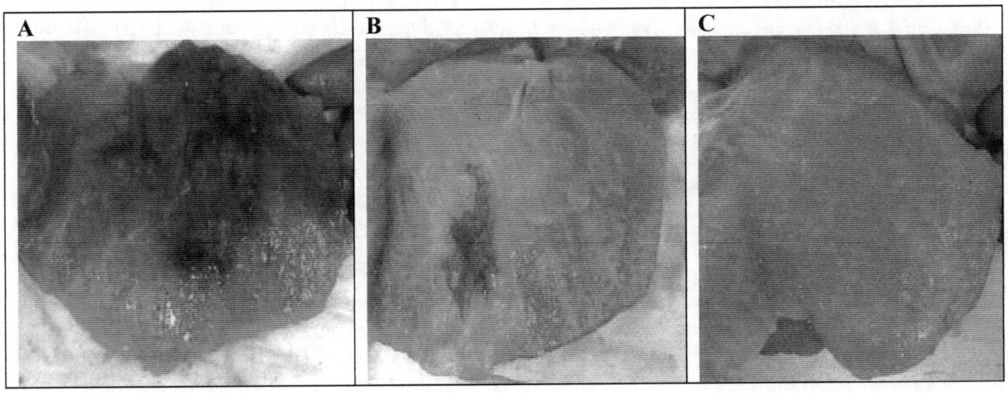

Figure 3. A) Fresh contusion in the Semimembranosus at 1 day post impact. B) Contusion capsule that is healing in the Semimembranosus at 11 days post impact. C) No contusion in the Semimembranosus at 30 days post impact.

4. DISCUSSION

The results from the quasi-static mechanical testing provide insights into the behavior of impacted versus control muscle tissue. The interpolation of peak forces from impacted tissue and the associated analysis reveals the muscle's mechanical response varied with impact energy and survival time post-impact. The variation in peak forces, as seen by the color-coded plot in Figure 2, suggests that the muscle's response to impact is sensitive to both the magnitude of the energy imparted and the time elapsed following injury. Lower peak forces are indicated by dark blue, while the highest peak forces are represented by dark red. This gradient suggests that muscle tissue undergoes varying levels of stiffness following impact, which could be indicative of differing degrees of damage or repair over time.

This change in stiffness is relevant to the study of contusions, as the mechanical properties of muscle during healing reflect underlying changes such as edema formation, necrosis, fibrosis, and scar tissue development. The confined quasi-static compression method used here allows for a qualitative assessment of these changes based on stiffness profiles. However, it does not directly quantify the specific contributions of fluid content, fibrotic tissue, or inflammation without additional imaging or histological correlation. When combined with dissection findings, this method offers strong qualitative insight into the material state of the muscle at different recovery stages.

The exclusion of certain control samples tested more than 7 hours post-euthanasia revealed an increase in the mean peak force (5.1 ± 5.7 N, $n = 35$) as compared to the samples tested within 7 hours (2.7 ± 1.6 N, $n = 26$). A two-sample t-test further highlighted the statistical significance of the difference between the two control groups ($p = 0.005$), reinforcing the idea that control muscle tissue behaves differently when considering only fresh samples. This observation suggests that post-mortem changes in muscle tissue can influence the results of mechanical tests, even when some samples were frozen. Thus, excluding these samples offers a more accurate representation of muscle tissue behavior.

For low-energy impacts (<150 J), full recovery occurred by 30 days. Peak forces increased over time, and a temporary spike was observed around day 8, attributed to the formation of a solidified contusion capsule. This capsule represented a stage in the healing process where tissue damage stabilizes before being remodeled. Following this peak, forces declined, suggesting muscle regrowth and structural restoration. These impact levels caused only mild to moderate damage and did not result in significant long-term deficits. The muscle tissue was able to repair effectively by the end of the 30-day period.

For impacts around 200 J, the results demonstrated that the peak forces remained low throughout the entire 30-day survival period. Serious contusions in the skin and muscle were noted from all

https://doi.org/10.52202/081621-0006

specimens tested at this energy level, with the injury followed by the normal recovery process. This indicates that the solidified contusion capsule and edema were not present during the end point intervals when the quasi-static mechanical tests were performed. The absence of these injuries was confirmed during muscle bisection following the mechanical testing, as shown by the lack of solidified capsule in any test in Figure 3B. Across all survival periods, the tissue appeared to follow the recovery trend shown in Figure 3, but the lack of granularity of end points between 12- and 30-days hides some of the contusion mechanics. By the end of the 30-day survival period, only a small contusion remained within the muscle with nearly complete recovery from initial injury.

In contrast, impacts greater than 230 J caused more extensive and persistent muscle injury. Stiffness remained elevated through 30 days, and healing was incomplete. Scarring at the impact site and deeper muscle contusions indicated more severe tissue disruption. These injuries likely damaged muscle fibers beyond the threshold for rapid repair, resulting in structural changes that could impair function. The observed skin scarring further suggests a traumatic impact, and the extent of damage does not bode well for recovery of the underlying muscle tissue.

While quasi-static compression of excised muscle is not a standard clinical tool, previous studies have used unconfined compression tests on muscle to assess its properties by using small, excised samples [28-30]. The approach used in this study, while qualitative, complements these methods by providing mechanical variation of the tissue's state, allowing correlation between tactile stiffness and structural injury, particularly when imaging is not feasible or when evaluating internal mechanical properties directly. Used as palpation analogs, quasi-static tests can help assess whether muscle is healing or if deeper damage, like edema or necrosis, is present. By combining tactile stiffness data with visual cues, researchers can estimate injury severity and recovery trajectory. However, this method cannot directly measure internal pathology. As shown in Table 3, while injuries are often obvious on day 1, they become more difficult to evaluate by day 5 and beyond without invasive or imaging-based confirmation.

Muscle samples from short survival periods after high-energy impacts exhibited large contusions, swelling, and necrosis. Dissections at 1 and 5 days post-injury revealed disorganized, blackened necrotic tissue, and edema pockets. These injuries were not found at lower energy levels or beyond day 5, indicating early onset and resolution of acute tissue death. The edema was often encased in a hardened capsule distinct from surrounding tissue and sometimes contained pus, suggesting localized infection. Without histological analysis, the composition of this capsule remains speculative, but its tactile and visual differences imply a pathological response.

Specimens with longer survival times showed unorganized regrowth around the contusion zone. Muscle fiber misalignment was especially apparent near the injury site and contrasted with the more uniform structure of distal muscle fibers. This disorganization, often indicated by tan discoloration (Figure 3B), likely corresponds to active tissue remodeling and capsule removal. Bisecting these muscles showed poor integration between new and existing fibers, suggesting a lack of structural continuity. Such misalignment could limit muscle contractility or strength and may contribute to longer recovery timelines. Imaging tools like ultrasound could help verify these findings in vivo.

The highest energy impact of 338.20 J resulted in the specimen heavily favoring the contralateral leg, resisting use of the impacted leg for support, and requiring additional Buprenorphine days after the impact. This condition persisted for nearly the entire survival period of 8 days with symptoms appearing after the first day. The inability for this porcine specimen to articulate the impacted hind leg and the accompanying symptoms met the humane endpoint criteria within the protocol. This was not the desired outcome from this specimen as it was intended to survive for 12 days post-impact. This result indicates that this energy level should be treated as an upper bound for allowable energy transfer.

By day 30, visible contusions on the skin had resolved for the highest energy impacts, but significant internal scarring remained. This scarring suggests deeper tissue damage not visible externally. Although muscle did not show overt fibrosis, persistent contusions imply that healing was incomplete. Even without visible scarring, disrupted fibers and prior necrosis could result in lasting structural or functional impairment. These effects may manifest as reduced strength, flexibility, or coordination in affected regions.

The absence of visible muscle damage does not guarantee full recovery. High-energy impacts may trigger maladaptive healing responses, such as fibrosis, which can alter normal muscle architecture. Given the observed skin scarring, it is plausible that similar structural changes also occurred within the muscle, even if they were less obvious. For individuals exposed to repeated high-energy impacts, these internal alterations could lead to chronic muscle damage, limited mobility, or higher injury susceptibility. Although live porcine models offer significant insight into muscular injury and recovery, they have limitations when extrapolating to human function.

Since live porcine specimens lack the ability to replicate such complex, task-specific behaviors, they limit our understanding of how muscular injuries truly impact functional performance in human daily life or work environments. Although there were occasional signs of favoring the contralateral side, the animals generally exhibited minimal outward signs of distress, even in the presence of significant leg swelling. This lack of observable discomfort and task performance capabilities in the porcine model underscores the challenge of correlating mechanical damage to pain perception in animals with that which a human might experience. This highlights the need for further research to better understand how these injuries affect the animal's functional capacity and how that might be extrapolated to human models.

5. CONCLUSION

This study created a methodology for studying the progression of soft tissue contusion healing in live porcine specimens as a surrogate model for humans. Visual inspection of the impact location indicated that skin contusions across all energy levels tested would heal by 30 days post impact. In all tests, contusions in the muscle were observed even after there was no contusion in the skin. However, some of the high energy impacts resulted in scarring in the skin after the contusion healed within the skin. The highest energy impacts still had muscle contusion after the 30-day survival period, indicating a longer recovery period for complete healing. Scar tissue formation in the skin is an injury that was not an intended consequence of this study but illuminates the potential complications of severe BABT impacts. Since no porcine survived longer than one month post impact, it is unclear whether there would also be permanent changes in the underlying musculature. It should be noted that these injuries should have healed at some point after one month, but this study did not elect to study these injuries beyond one month.

In the event of a BABT injury, contusions in the skin that healed within one week of impact should expect the muscle contusions to be healed within two to three weeks of impact. Skin contusions that last at least two weeks after the impact substantially reduce the likelihood of the muscle contusion healing within one month. Since skin contusions healed before the muscle contusions in this study, visual inspection cannot be used independently of another diagnostic measure to assess injury. Quasi-static mechanical testing of excised muscle tissue offers a promising approach to quantify BABT induced muscle contusions and their healing processes. The combination of palpation and visual inspection of the impact site is a good preliminary indicator of the severity of injury within the soft tissues.

However, the lack of complex, task-specific behavior in the porcine model restricts the direct comparison to human recovery, emphasizing the need for further studies that bridge this gap to better predict functional outcomes following muscle injuries in humans. Further studies, particularly those involving imaging and long-term monitoring, will help refine the quantification of recovery from muscle injury. Overall, this research provides essential information on the mechanical properties and recovery dynamics of muscle tissue following impact, offering a foundation for future studies aimed at improving injury management and rehabilitation strategies, particularly for high-energy impacts.

Acknowledgements

The authors would like to recognize the support of our consortium partners at the Medical College of Wisconsin (MCW), Wayne State University, and the United States Army Aeromedical Research Laboratory (USAARL). The authors would also like to recognize the support of the University of Virginia Center for Comparative Medicine team, in particular Jeremy Gatesman and Deanna Zielinski. Additionally, the authors would like to thank Trevor Leite, Ignacio Gomez-Cambronero, Riley Tufts, Kevin Kopp, Joey White, and Brian Overby from the University of Virginia Center for Applied Biomechanics.

Funding

The research was funded by the United States Army Medical Research and Development Command contract #W81XWH-15-9-001.

https://doi.org/10.52202/081621-0006

References

[1] Carr, D. J., Horsfall, I., & Malbon, C. (2016). Is behind armour blunt trauma a real threat to users of body armour? A systematic review. *BMJ Military Health*, *162*(1), 8-11.

[2] Bass, C. R., Salzar, R. S., Lucas, S. R., Davis, M., Donnellan, L., Folk, B., ... & Waclawik, S. (2006). Injury risk in behind armor blunt thoracic trauma. *International journal of occupational safety and ergonomics*, *12*(4), 429-442.

[3] Cooper, G. J., Pearce, B. P., Stainer, M. C., & Maynard, R. L. (1982). The biomechanical response of the thorax to nonpenetrating impact with particular reference to cardiac injuries. *Journal of Trauma and Acute Care Surgery*, *22*(12), 994-1008.

[4] Corona, B. T., Rivera, J. C., Owens, J. G., Wenke, J. C., & Rathbone, C. R. (2015). Volumetric muscle loss leads to permanent disability following extremity trauma. *Journal of Rehabilitation Research & Development*, *52*(7).

[5] Garg, K., Ward, C. L., Hurtgen, B. J., Wilken, J. M., Stinner, D. J., Wenke, J. C., ... & Corona, B. T. (2015). Volumetric muscle loss: persistent functional deficits beyond frank loss of tissue. *Journal of Orthopaedic Research*, *33*(1), 40-46.

[6] Rivera, J. C., & Corona, B. T. (2016). Muscle-related Disability Following Combat Injury Increases With Time. *US Army Medical Department journal*.

[7] Beiner, J. M., & Jokl, P. (2001). Muscle contusion injuries: current treatment options. *JAAOS-Journal of the American Academy of Orthopaedic Surgeons*, *9*(4), 227-237.

[8] Järvinen, T. A., Järvinen, T. L., Kääriäinen, M., Kalimo, H., & Järvinen, M. (2005). Muscle injuries: biology and treatment. *The American journal of sports medicine*, *33*(5), 745-764.

[9] Beiner, J. M., & Jokl, P. (2002). Muscle contusion injury and myositis ossificans traumatica. *Clinical Orthopaedics and Related Research®*, *403*, S110-S119.

[10] Devilbiss, Z., Hess, M., & Ho, G. W. (2018). Myositis ossificans in sport: a review. *Current Sports Medicine Reports*, *17*(9), 290-295.

[11] Rothwell, A. G. (1982). Quadriceps hematoma: a prospective clinical study. *Clinical Orthopaedics and Related Research®*, *171*, 97-103.

[12] Ryan, J. B., Wheeler, J. H., Hopkinson, W. J., Arciero, R. A., & Kolakowski, K. R. (1991). Quadriceps contusions: West point update. *The American Journal of Sports Medicine*, *19*(3), 299-304.

[13] Walton, M., & Rothwell, A. G. (1983). Reactions of thigh tissues of sheep to blunt trauma. *Clinical Orthopaedics and Related Research®*, *176*, 273-281.

[14] Ueblacker, P., Mueller-Wohlfahrt, H. W., & Ekstrand, J. (2015). Epidemiological and clinical outcome comparison of indirect ('strain') versus direct ('contusion') anterior and posterior thigh muscle injuries in male elite football players: UEFA Elite League study of 2287 thigh injuries (2001–2013). *British journal of sports medicine*, *49*(22), 1461-1465.

[15] Crisco, J. J., Jokl, P., Heinen, G. T., Connell, M. D., & Panjabi, M. M. (1994). A muscle contusion injury model: biomechanics, physiology, and histology. *The American journal of sports medicine*, *22*(5), 702-710.

[16] Desmoulin, G. T., & Anderson, G. S. (2011). Method to investigate contusion mechanics in living humans. *Journal of forensic biomechanics*, *2*.

[17] Jackson, D. W., & Feagin, J. A. (1973). Quadriceps contusions in young athletes: relation of severity of injury to treatment and prognosis. *JBJS*, *55*(1), 95-105.

[18] Sullivan, T. P., Eaglstein, W. H., Davis, S. C., & Mertz, P. (2001). The pig as a model for human wound healing. *Wound repair and regeneration*, *9*(2), 66-76.

[19] Summerfield, A., Meurens, F., & Ricklin, M. E. (2015). The immunology of the porcine skin and its value as a model for human skin. *Molecular immunology*, *66*(1), 14-21.

[20] Meyer, W., Schwarz, R., & Neurand, K. (1978). The skin of domestic mammals as a model for the human skin, with special reference to the domestic pig1. In *Skin-drug application and evaluation of environmental hazards* (Vol. 7, pp. 39-52). Karger Publishers.

[21] McMahon, J., Berthelson, P., Eaton, M., Lorente, A., Leite, T., McEntire, B. J., & Salzar, R. (2025). Use of a Porcine Cadaver Model as a Human Surrogate for Behind Armor Blunt Trauma. *Journal of Engineering and Science in Medical Diagnostics and Therapy*, *8*(4).

[22] Yoganandan, N., Shah, A., Baisden, J., Stemper, B., Otterson, M., Somberg, L., ... & McEntire, J. (2024). Matched-pair hybrid test paradigm for behind armor blunt trauma using an experimental animal model. *Trauma Surgery & Acute Care Open*, *9*(1), e001194.

[23] Bell, J. M. (1964). A study of rates of growth of Yorkshire, Lacombe, Landrace, and Crossbred pigs from birth to 200 lb. *Canadian Journal of Animal Science*, *44*(3), 315-319.

[24] Cohen, S. P., Brown, C., Kurihara, C., Plunkett, A., Nguyen, C., & Strassels, S. A. (2011). Diagnoses and factors associated with medical evacuation and return to duty among nonmilitary personnel participating in military operations in Iraq and Afghanistan. *Cmaj*, *183*(5), E289-E295.
[25] Roy, T. C. (2011). Diagnoses and mechanisms of musculoskeletal injuries in an infantry brigade combat team deployed to Afghanistan evaluated by the brigade physical therapist. *Military medicine*, *176*(8), 903-908.
[26] Sanders, J. W., Putnam, S. D., Frankart, C., Frenck, R. W., Monteville, M. R., Riddle, M. S., ... & Tribble, D. R. (2005). Impact of illness and non-combat injury during Operations Iraqi Freedom and Enduring Freedom (Afghanistan). *American Journal of Tropical Medicine and Hygiene*, *73*(4), 713.
[27] Rhon, D. I. (2010). A physical therapist experience, observation, and practice with an infantry brigade combat team in support of Operation Iraqi Freedom. *Military medicine*, *175*(6), 442-447.
[28] Van Loocke, M., Lyons, C. G., & Simms, C. K. (2006). A validated model of passive muscle in compression. *Journal of biomechanics*, *39*(16), 2999-3009.
[29] Van Loocke, M., Lyons, C. G., & Simms, C. K. (2008). Viscoelastic properties of passive skeletal muscle in compression: stress-relaxation behaviour and constitutive modelling. *Journal of biomechanics*, *41*(7), 1555-1566.
[30] Böl, M., Ehret, A. E., Leichsenring, K., Weichert, C., & Kruse, R. (2014). On the anisotropy of skeletal muscle tissue under compression. *Acta biomaterialia*, *10*(7), 3225-3234.

https://doi.org/10.52202/081621-0006

Preliminary Lethality Risk as a Result of BABT to the Heart

J. A. McMahon[1], G. F. Glass[1], A. V. Stotka[1], P. R. Berthelson[1], C. R. D. Bass[2], N. Yoganandan[3], B. J. McEntire[4], **R. S. Salzar[1]**

[1]*University of Virginia, 1215 Lee St, Charlottesville, VA 22908, USA*
[2]*Wayne State University, 656 West Kirby Street, Detroit, MI 48202, USA*
[3]*Medical College of Wisconsin, 8701 Watertown Plank Road, Milwaukee, WI 53226, USA*
[4]*U.S. Army Aeromedical Research Laboratory, 6901 Farrel Road, Fort Novosel, AL 36362, USA*

Abstract. Behind armour blunt trauma (BABT) results from a ballistic projectile interacting with protective body armour, leading to local deformation of the armour into the torso. The risk of a serious BABT injury increases when the anterior thorax is impacted due to the proximity to critical organs. In the case of an upper thorax impact where interaction with the heart occurs, a range of injuries are possible. In this study, 27 BABT impacts to the heart were performed using a live porcine model as a surrogate for the human. The porcine specimens were impacted with a range of energies using a generalized blunt indenter with the shape and recruited mass of the backface of deformed body armour. Injuries ranged from short-term changes in the electrocardiogram (ECG) and temporary apnea to long-term disturbances in the ECG, caused by a loss of coordinated blood movement, and unrecoverable respiration. Generally, increasing energy resulted in an increased risk of serious cardiac contusion and loss of organized electrophysiological function. An alternate injury mechanism was also observed at low- to mid-range energies that resulted in immediate ventricular fibrillation (as seen in commotio cordis) if the impact coincided with the rise of the "T-wave" in the normal cardiac cycle. Injury risk functions were developed exploring the risk of lethality based on energy. Preliminary thresholds of injury coinciding with the range of energy levels were observed; impacts below 100 J generally resulted in no or short-term changes to the cardiac cycle and less than a 5% risk of lethality. An input energy of 220 J showed a 50% risk of lethality with test cases generally resulting in serious cardiac arrhythmias, apnea, and death. This and future injury risk functions are vital to yield functional protective armour designs based on required level of performance and individual combat situations.

1. INTRODUCTION

Body armour is designed to protect military personnel from injury or death resulting from ballistic munition penetration into the human body. This protection is achieved by the dissipation of energy of the munition through deformation of armour material, which often results in backface deformation (BFD). The BFD of the armour may cause localized high rate loading onto the body surface, which can result in severe non-penetrating injuries. The BFD mechanism may result in structural and physiological damage to the underlying anatomical structures through an injury modality referred to as behind armour blunt trauma (BABT) [1-3].

Current body armour standards are derived from studies performed in the 1970s. These studies dictated the use of Roma Plastilina #1 clay as the standard backing material for body armour evaluation and determined a BFD acceptance criterion of less than 44 mm into the clay [4-6]. From a medical standpoint, the clay deformation criterion does not correlate to any particular injury modality or severity. Due to this, current body armour systems do not account for differences in injury tolerances across various thoracoabdominal regions. Since the human torso is not a homogenous structure, the location of a BABT impact may result in injuries of widely varied severities for similar inputs. Therefore, the use of a uniform level of protection across the human torso likely leads to armours that are overdesigned, heavy, and cumbersome, limiting soldier mobility on the battlefield [7]. By quantifying the risk of injury across the various thoracoabdominal regions, it may be feasible to optimize the level of protection to lower risk body regions as a means for minimizing weight. However, improved evaluation tools and standards must first be developed to accommodate these competing design goals.

Post-mortem human subjects (PMHS) have historically been used as surrogates to represent live humans during assessments of skeletal fracture and tissue damage due to high-rate stimuli seen in military theatre, the automotive industry, and more. However, PMHS lack the physiology that is necessary to investigate injuries associated with the underlying anatomical systems that constitute a physiological response. To incorporate physiological damage metrics, live animal models are often chosen, as they can be monitored to observe the progression of injury response over time. Large animal models, such as the porcine

https://doi.org/10.52202/081621-0007

model, are often used as a representation of the human torso for physiology and biomedical responses [8-12].

Thoracic trauma is a significant mechanism of injury in both civilian and military casualties—thoracic injury is a contributing factor in up to 75% of all trauma-related deaths and was the primary cause of over 1/3 of battlefield deaths during World War II [13, 14]. Thoracic trauma can cause morbidity and mortality through several mechanisms, including 1) airway obstruction through direct tracheal or bronchial injury, 2) loss of pulmonary function through contusion, rib fractures or hemothorax/pneumothorax, 3) exsanguination through direct vessel injury, 4) cardiac failure due to contusion, valve rupture, or dysrhythmia, and 5) cardiac tamponade due to direct cardiac wall or vessel injury [14]. Of these injury patterns, rib fractures, pulmonary injury, and blunt cardiac injury (BCI) are the most relevant injury patterns encountered in BABT, with fatalities reported from BABT pulmonary and cardiac injuries (15).

Cardiac injury, while less common than pulmonary injury in the context of blunt force trauma, conveys significant risk—up to 25% of traumatic deaths occur due to cardiac related injuries [16]. Cardiac contusion is the most common injury pattern observed, with a wide range of possible initial clinical presentations including chest pain, shortness of breath, or arrythmia. Subclinical myocardial contusion, an elevation of cardiac biomarkers (i.e. troponin) without symptoms or electrocardiographic changes, is likely of little clinical consequence. However, early development of arrythmia or cardiac failure symptoms is associated with both significant early and late complications including pump failure, complications due to clinically significant arrythmias, and myocardial necrosis with scar formation or delayed heart wall rupture [17]. Commotio cordis, the development of sudden cardiac death due to arrythmia after blunt chest trauma, is also potentially relevant to BABT. In prior porcine modeling, this phenomenon was largely dependent on impact timing relative to the cardiac cycle, as well as force of impact within the vulnerable time window [18].

While the potential for cardiac injury in BABT is apparent, current data on BCI is almost entirely from motor vehicle impacts (occupant and pedestrian) or falls [16]. As such, the spectrum of BCI risk from BABT is unknown. The objective of this study is to use a porcine cardiac model to simulate BABT and to develop injury risk functions from these results.

2. METHODS

2.1 Porcine Testing

Simulated BABT impact tests were conducted in vivo with 27 live Yorkshire pigs (mass = 43.44 ± 3.73 kg, length = 108.92 ± 5.03 cm) using a custom-designed, high-rate linear impacter system and ballistic indenter. The ballistic indenter was three-dimensionally (3-D) printed with carbon fiber-reinforced nylon (Markforged, Watertown, MA) where the impacter mass and shape was based on flash X-ray measurements of a 7.62-mm test round (7.62 ball round – 9.72 g) impacting a thick Ultra-High Molecular Weight Polyethylene (UHMWPE) armour plate selected for high deformations and low probability of penetration for round velocities of approximately 670-800 m/s armour[3]. From these flash X-ray images, the backface velocity of this type and thickness of armour was measured in the range of 50 to 60 m/s. This indenter was designed to characterize the onset of BABT injury from a generic threat rather than any specific type of ammunition or body armour. The indenter was 101.6 mm in diameter with a mass of 250 grams to represent the approximate recruited mass of the ammunition and body armour into the torso. The indenter was equipped with an Endevco 7270A-20k accelerometer to measure the impact event. The impacter system was pneumatically driven and capable of propelling the indenter at velocities greater than 50 m/s.

https://doi.org/10.52202/081621-0007

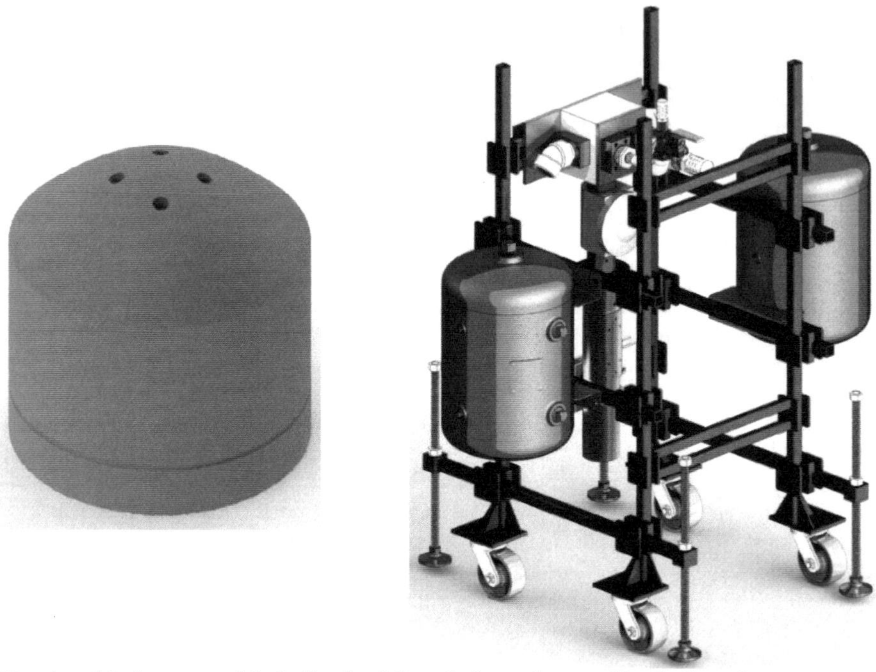

Figure 1. 3D-printed indenter modeled after backface deformation profile (left) and custom pneumatic impacting system for simulating BABT with live porcine specimens (right)

For each impact test, the impact site was chosen to cover the greatest surface area of the heart while maintaining a fully orthogonal impact between the indenter and the thorax. This impact location was lateral to the midline to avoid the sternum and typically spanned left ribs 3-6, overlapping with the cranial lobe of the left lung. The impact site was confirmed using ultrasound to ensure positioning was consistent between tests. Impact timing was controlled to avoid the phenomena known as commotio cordis, where ventricular fibrillation is induced due to blunt impact during the upstroke of the T wave [18]. This impact timing was initially performed manually with the goal of impacting during the QRS complex of the ECG. However, this proved to be too inconsistent and impact timing was modified to utilize a custom, automated impact timing control system. The control system used the ECG signal to impact consistently within the QRS complex for the remaining tests. Impact energy was varied from 20 Joules to over 300 J to get a range of injuries for the purpose of developing injury risk functions. A complete test matrix is provided in Table 1.

All specimens were procured through the UVA Center for Comparative Medicine and were acclimated upon arrival for a full three days (72 hours). On the day a specimen was tested, they were induced using Telazol and kept under anaesthesia with isoflurane and medical air for the duration of the impact event. Local analgesics (Bupivacaine) were also administered to the impact site prior to testing. This anaesthesia is not expected to influence the injury response of the specimens, especially since supplemental oxygen was avoided in the delivery of the isoflurane. Specimens were instrumented with Millar pressure transducers (PT) in the descending aorta via an introduced femoral catheter to measure arterial pressure during the impact event. Live patient monitoring data such as Electrocardiogram (ECG), heart rate, blood oxygen (Sp02), and respiration rate were also measured. Following each impact, specimens were monitored until they were determined to be recoverable (i.e., in a medically stable condition), at which point they were recovered from anaesthesia and returned to their housing area for continuous monitoring (up to 8 hours) for signs of discomfort or respiratory distress. Each porcine specimen was imaged using CT with contrast. Once radiology was complete, the specimen was euthanized, and the imaging data was transferred to a radiologist for diagnosis. A comprehensive post-euthanasia necropsy was performed immediately following euthanasia to determine and document injuries to the tissue. If a specimen reached a critical endpoint (i.e., signs of medical distress) prior to recovery from anaesthesia, the specimen was immediately euthanized followed by post-euthanasia necropsy and injury documentation. CT was not conducted for these specimens. All testing was completed under approved protocols from the University of Virginia Animal Care and Use Committee (protocol: 4379-12-21) and the U.S. Army Animal Care and Use Review Office (protocol: MT21006.031.e002).

Table 1. Test matrix with specimen mass and input conditions

Specimen Test ID	Specimen Mass (kg)	Input Velocity (m/s)	Impact Energy (J)
H01	43.6	12.79	20.04
H02	42.3	15.31	28.71
H03	45.3	19.18	45.99
H04	48.0	22.61	63.89
H05	49.0	28.04	98.28
H06	41.8	32.74	134.01
H07	42.8	36.56	167.07
H08	42.8	37.71	177.73
H09	38.8	38.90	189.17
H10	42.8	31.99	125.32
H11	47.0	31.37	123.02
H12	44.2	31.73	125.83
H13	44.0	39.38	193.87
H14	48.0	45.31	256.57
H15	45.0	47.58	282.98
H16	41.4	44.12	243.33
H17	42.8	48.53	294.37
H18	39.4	25.79	83.15
H19	43.4	38.24	182.80
H20	47.0	41.99	220.44
H21	48.0	38.72	187.45
H22	51.0	37.87	179.30
H23	37.0	11.66	17.34
H24	42.4	21.83	60.75
H25	41.0	26.92	92.41
H26	36.4	34.45	151.28
H27	37.6	38.03	184.38

Once euthanasia was complete and confirmed by the veterinary staff, the necropsy was performed by first creating a Y incision starting from the sternal notch, bifurcating at the xiphoid process, and following the costal cartilage dorsally towards the spine. The tissue was reflected bilaterally away from the ribcage to reveal skeletal injuries to the sternum, ribs, and costal cartilage. The diaphragm was then reflected to reveal the pleural cavity. Any injuries to the heart were documented *in situ*, before removing the organs for further examination and documentation.

2.2 Data Acquisition and Processing

Accelerometer and pressure transducer data were sampled at 500 kilohertz (kHz) with a 100-kHz antialiasing filter using a High-Speed SLICE data acquisition system (DTS, Seal Beach, CA). Following data acquisition, sensor offset was removed from the indenter acceleration trace, and the data were filtered using a phaseless 4-pole Butterworth low-pass filter with a cutoff frequency of 2 kHz [19]. The filtered acceleration data was

https://doi.org/10.52202/081621-0007

then used to calculate impact energy (Joules; J) which was then normalized based on specimen mass using the equal-stress equal-velocity scaling method [12, 20-22]. Here the porcine data were normalized to a mass of 39.55 kg, representing the mass of the torso of a 50th percentile male human [12, 23].

2.3 Data Censoring, Predictor Selection, and Injury Risk Function (IRF) Generation

Using survival analysis, injury risk functions (IRFs) were evaluated for lethality, and input energy was used as the predictor metric. All injurious tests were included as right censored data, and all non-injurious test data was left censored. IRFs were generated using the Survival package [24] within RStudio (RStudio PBC, Boston, MA, USA). The statistical distributions considered for each IRF included the Weibull, log-logistic, and log-normal distributions, with their cumulative distribution functions defined by Equations 1 through 4:

$$Weibull: F(x;\lambda;k) = \begin{cases} 1 - e^{-\left(\frac{x}{\lambda}\right)^k} & x \geq 0 \\ 0 & x < 0 \end{cases} \tag{1}$$

$$log-logistic: F(x;\alpha;\beta) = \frac{1}{1+\left(\frac{x}{\alpha}\right)^{-\beta}} \tag{2}$$

$$log-normal: F(x;\mu;\sigma) = \frac{1}{2} + \frac{1}{2} erf\left(\frac{\ln(x)-\mu}{\sqrt{2}\sigma}\right) \tag{3}$$

where x is the response variable (e.g., impact energy, etc.), λ, α, and σ are the scale parameters, k and β are the shape parameters, and μ is the location parameter. The erf term in the log-normal distribution is the standard Gaussian error function which is defined as:

$$erf(z) = \frac{2}{\sqrt{\pi}} \int_0^z e^{-t^2} dt \tag{4}$$

3. RESULTS

Twenty-seven (27) simulated BABT impact tests were performed for this study with input energies ranging from 15 J – 295 J. Injuries were observed to the heart, lungs, ribs, and superficial soft tissues. More specific to the heart, conduction abnormalities, arrythmias, apnea, and contusions were observed. However, special consideration was given to the instances where the impact resulted in a critical endpoint as it was desired to determine the limits of lethality. Of the 27 impacts, 4 were excluded due to impacting on the T wave of the ECG, inducing commotio cordis. 5 resulted in the specimen meeting the humane endpoint criteria and were euthanized immediately following impact. In these cases, the impact was recorded as resulting in lethality, and injury risk functions were developed for this specific outcome. The Weibull based distribution was selected as the optimal distribution since it had the lowest Brier Score Metric (BSM) and Kolmogorov–Smirnov (KS) test statistic (Dmax) values as well as a comparable AIC value (Table 2). The Weibull distribution is also ideal to use in this situation since extrapolation to smaller risks is likely to be important and the distribution passes through (0, 0).

Table 2. Lethality IRF model fit statistics

Distribution	Dmax	AIC	BSM
Weibull	0.413	21.1	1.7
loglogistic	0.429	21.2	1.94
lognormal	0.427	21	1.93

The Weibull based lethality IRF with energy as the predictor metric is shown in Figure 2, where the shape parameter k is 3.66182 and the scale parameter λ is 243.24161. Mean risk levels are shown in Table 3 with a 50% risk of lethality at 220 J.

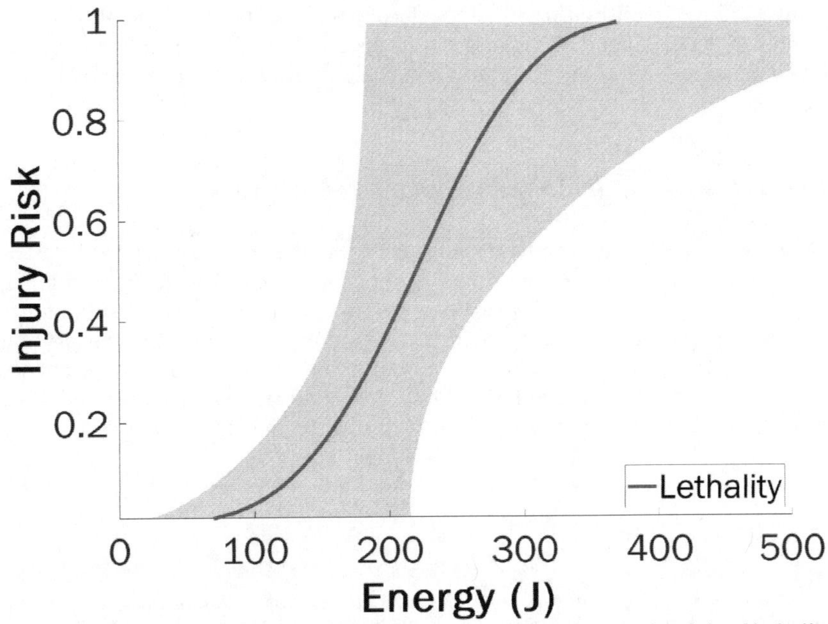

Figure 2. Injury risk function with 95% confidence intervals describing the risk of lethality in a live porcine model using input energy as the predictor metric and a Weibull distribution.

Table 3. Mean risk of lethality based on input energy

Risk Level	5%	10%	25%	50%	75%	90%	95%
Energy (J)	108.1	131.6	173.1	220.1	265.9	305.5	328.2

4. DISCUSSION

The lethality risk curve presented in Figure 2 considers factors that may affect the subsequent care of the injured warfighter. The injuries that are represented by the risk curve include trauma to the heart that leads to a fatality despite the presence of medical assistance or proximity to a trauma center. Cases of impact that lead to apnea are very much dependent on the ability of the wounded to start breathing again on their own or the resuscitative efforts of another. For the tests performed in this study, apnea was allowed to continue for several minutes before external ventilation was delivered. Still, one cannot say definitively if the specimen would have started breathing again on its own.

Cardiac arrythmia of varying severity was often induced by the impacts. In most cases, dysrhythmias were transient, but longer-term outcomes from cardiac contusion and risk for persistent or recurrent arrythmia, heart failure, and other sub-acute complications are not well-characterized by these data. Though these impacts resulting in arrythmias are not to be trivialized, they are being interpreted in this context as survivable. There were instances of sinus arrest followed by ventricular escape with a few instances of subsequent R on T phenomenon and ventricular tachycardia or fibrillation. These rhythms are significantly hemodynamically compromising and represent higher risk impacts. Separate injury risk functions could be developed for less-than-lethal impacts which would be useful in more conservative armour designs.

As the ultimate purpose of this research is to develop a set of physiologically based injury criteria for the development of new body armour designs, the analysis of the collected data will address the factors that may be important to designers. As a preliminary analysis, the focus is developing a risk function based on lethality of a backface armour blunt thoracic impact. Other risk functions based on serious, but less lethal injuries, including cardiac contusion and varying arrhythmia can be assessed and yield functional protective armour designs based on required level of performance and individual situations.

Protection against commotio cordis is not being considered in the analysis because of the apparent low energy required to induce it and reliance on a non-modifiable risk factor (time of impact relative to cardiac

https://doi.org/10.52202/081621-0007

cycle). Effective protection may require extremely heavy/stiff armour over the heart region which may be impractical for many current ballistic threats. In addition, because the risk window for a high-risk-timed impact event is limited, the likelihood of a commotio cordis event from a ballistic impact is small compared to the likelihood of an impact event resulting in a non-commotio-related cardiac arrhythmia, contusion, or laceration.

5. CONCLUSION, ACKNOWLEDGEMENTS, AND REFERENCES

Injury risk functions were developed for lethality using energy as the response variable for BABT. A Weibull-based IRF was generated from 27 impact tests performed on live porcine specimen. Preliminary injury risks coinciding with the range of energy levels were observed; impacts below 100 J resulting in a less than 5% risk of lethality and energies above 220 J resulted in a 50% risk of lethality. Other heart injury modalities were observed such as cardiac arrythmias, apnea, and heart contusions, but they were not analyzed further for this study. Future risk functions based on these less lethal injuries may be generated to yield functional protective armour designs based on required level of performance and individual situations.

Acknowledgements

The authors would like to recognize the support of our consortium partners at the Medical College of Wisconsin (MCW), Wayne State University, and the United States Army Aeromedical Research Laboratory (USAARL). The authors would also like to recognize the support of the University of Virginia Center for Comparative Medicine team, in particular Jeremy Gatesman and Deanna Zielinski. Additionally, the authors would like to recognize Ana Lorente, Trevor Leite, Kevin Kopp, Joey White, and Brian Overby from the University of Virginia Center for Applied Biomechanics.

Funding

The research was funded by the United States Army Medical Research and Development Command contract #W81XWH-15-9-001.

Disclaimer

The views, opinions, and/or findings contained in this presentation are those of the author(s) and should not be construed as an official Department of the Army position, policy, or decision, unless so designated by other official documentation. Citation of trade names in this presentation does not constitute an official Department of the Army endorsement or approval of the use of such commercial items.

References

[1] Carr, D. J., Horsfall, I., and Malbon, C., 2016, "Is Behind Armour Blunt Trauma a Real Threat to Users of Body Armour? A Systematic Review," J. R. Army Med. Corps, 162(1), pp. 8–11.
[2] Cooper, G. J., Pearce, B. P., Stainer, M. C., and Maynard, R. L., 1982, "The Biomechanical Response of the Thorax to Nonpenetrating Impact with Particular Reference to Cardiac Injuries," J. Trauma Acute Care Surg., 22(12), pp. 994–1008.
[3] Bass, C. R., Salzar, R. S., Lucas, S. R., Davis, M., Donnellan, L., Folk, B., Sanderson, E., and Waclawik, S., 2006, "Injury Risk in Behind Armour Blunt Thoracic Trauma," Int. J. Occup. Saf. Ergon., 12(4), pp. 429–442.
[4] Clare, V. R., 1976, "Body Armour: Blunt Trauma Data," U.S. Department of Justice, Law Enforcement Assistance Administration, National Institute of Law Enforcement and Criminal Justice.
[5] Goldfarb, M. A., Ciurej, T. F., Weinstein, M. A., and Metker, L. W., 1975, "A Method for Soft Body Armour Evaluation: Medical Assessment: Defense Technical Information Center," Edgewood Arsenal, Fort Belvoir, VA, Report No. EB-TR-74073.

[6] Prather, R., Swann, C., and Hawkins, C., 1977, “Backface Signatures of Soft Body Armours and the Associated Trauma Effects,” U.S. Army Armament Research and Development Command, Aberdeen Proving Ground, MD, Technical Report No. ARCSL-TR-77-55.
[7] National Research Council, 2012, Testing of Body Armour Materials: Phase III, National Academies Press, Washington, DC.
[8] Arborelius, U. P., Rocksen, D., Gustavsson, J., and G€unther, M., 2021, “Pulmonary Hypoxia and Venous Admixture Correlate Linearly to the Kinetic Energy from Porcine High Velocity Projectile Behind Armour Blunt Trauma,” Exp. Lung Res., 47(7), pp. 323–333.
[9] G€unther, M., Sonde n, A., Gustavsson, J., Arborelius, U. P., and Rockse n, D., 2020, “Feasibility of Pleural and Perilesional Subcutaneous Microdialysis to Assess Porcine Experimental Pulmonary Contusion,” Exp. Lung Res., 46(5), pp. 117–127.
[10] Prat, N., Rongieras, F., de Freminville, H., Magnan, P., Debord, E., Fusai, T., Destombe, C., Sarron, J.-C., and Voiglio, E. J., 2012, “Comparison of Thoracic Wall Behavior in Large Animals and Human Cadavers Submitted to an Identical Ballistic Blunt Thoracic Trauma,” Forensic Sci. Int., 222(1–3), pp. 179–185.
[11] Gryth, D., Rockse n, D., Persson, J. K. E., Arborelius, U. P., Drobin, D., Bursell, J., Olsson, L.-G., and Kjellstr€om, T. B., 2007, “Severe Lung Contusion and Death After High-Velocity Behind-Armour Blunt Trauma: Relation to Protection Level,” Mil. Med., 172(10), pp. 1110–1116.
[12] McMahon, Justin, Parker Berthelson, Madelyn Eaton, Ana Lorente, Trevor Leite, B. Joseph McEntire, and Robert Salzar. "Use of a Porcine Cadaver Model as a Human Surrogate for Behind Armour Blunt Trauma." Journal of Engineering and Science in Medical Diagnostics and Therapy 8, no. 4 (2025).
[13] Meredith, J. Wayne, and J. Jason Hoth. "Thoracic trauma: when and how to intervene." Surgical Clinics 87, no. 1 (2007): 95-118.
[14] Khandhar, Sandeep J., Scott B. Johnson, and John H. Calhoon. "Overview of thoracic trauma in the United States." Thoracic surgery clinics 17, no. 1 (2007): 1-9.
[15] Carr, Debra J., I. Horsfall, and C. Malbon. "Is behind armour blunt trauma a real threat to users of body armour? A systematic review." BMJ Military Health 162, no. 1 (2016): 8-11.
[16] El-Andari, Ryaan, Devin O'Brien, Sabin J. Bozso, and Jeevan Nagendran. "Blunt cardiac trauma: a narrative review." Mediastinum (Hong Kong, China) 5 (2021): 28-28.
[17] Bock, Jeremy S., and R. Michael Benitez. "Blunt cardiac injury." Cardiology clinics 30, no. 4 (2012): 545-555.
[18] Link, Mark S., Barry J. Maron, Paul J. Wang, Brian A. VanderBrink, Wei Zhu, and NA Mark Estes. "Upper and lower limits of vulnerability to sudden arrhythmic death with chest-wall impact (commotio cordis)." Journal of the American College of Cardiology 41, no. 1 (2003): 99-104.
[19] Shah, A., Yoganandan, N., Stemper, B., Martin, J., Salzar, R., Op 't Eynde, J., Bass, D., and McEntire, B. J., 2023, “Data Filtering for the Analysis of Biological Tests for Behind Armour Blunt Trauma Studies,” International Personal Armour Systems Symposium (PASS), Dresden, Germany, Sept. 11–15.
[20] Eppinger, Rolf H. "Prediction of thoracic injury using measurable experimental parameters." In Proc. 6th International Technical Conference on the Enhanced Safety of Vehicles, vol. 6, pp. 770-779. 1976.
[21] Yoganandan, Narayan, Mike WJ Arun, and Frank A. Pintar. "Normalizing and scaling of data to derive human response corridors from impact tests." Journal of biomechanics 47, no. 8 (2014): 1749-1756.
[22] Nie, Bingbing, Jason L. Forman, Hamed Joodaki, Taotao Wu, and Richard W. Kent. "Scaling approach in predicting the seatbelt loading and kinematics of vulnerable occupants: How far can we go?" Traffic injury prevention 17, no. sup1 (2016): 93-100.
[23] Clauser, Charles E., John T. Mc Conville, and John W. Young. Weight, volume, and center of mass of segments of the human body. No. AMRL-TR-69-70. 1969.
[24] Therneau, T. M., and Lumley, T., 2015, “Package ‘Survival’,” R. Top. Doc., 128(10), pp. 28–33

https://doi.org/10.52202/081621-0007

The Effects of Armour Fit and Body Habitus on Behind Armour Blunt Trauma Loading

A. Iwaskiw[1], K. Ott[1], N. Hahne[1], M. Vignos[1], N. Steiner[1], R. Hingorani[1], M. Clark[2]

[1]*The Johns Hopkins University Applied Physics Laboratory (JHU/APL), 11100 Johns Hopkins Road, Laurel, MD 20723, alexander.iwaskiw@jhuapl.edu*
[2] *U.S. Special Operations Command, 10 General Greene Avenue, Natick, MA 01760*

Abstract. Injury due to behind armour blunt trauma (BABT) is a concern to the warfighter. An existing hypothesis related to BABT is that the distance between the backface of the armour and the wearer, as well as the physical anatomical composition of the impact site, could influence BABT risk. This study utilised two separate experiments, one measuring impact force and the other measuring clay deformation to investigate this hypothesis. Live ammunition impacted hard armour plates backed by air, flesh, and either clay or a load cell system. Flesh was excised from the torso of untested postmortem human surrogates (PMHS). Sixteen load cell backed non-penetrating ballistic tests, measuring peak force and impulse, were conducted across various flesh thicknesses (0-20 mm), air gaps (0-25 mm) and velocities. In addition, 15 clay backed tests, with similar air gap and flesh thickness conditions, were tested evaluating depth left in clay. After the tests, the results were fit with models that developed a relationship between striking velocity, air gap, flesh thickness and the loading metrics of interest. Air gap and flesh thickness were found to be protective, where loading decreased with increased air gap and flesh thickness. Measured force had a decaying exponential relationship with these parameters, whereas clay deformation had a decaying linear relationship. To verify the relationship developed in these experiments, loading conditions were predicted using fit and body habitus variables measured in a separate study where BABT injuries were induced in full PMHS models. The estimated loading conditions, when using specific air gaps and flesh thicknesses, were better predictors of injury than using striking velocity alone in these experiments. This supports the hypothesis that flesh and air gap contribute to reducing the loading on the anatomy and should be considered when assessing risk of BABT.

1. INTRODUCTION

Behind armour blunt trauma (BABT) occurs when the backface of ballistic armour strikes the wearer without full penetration of the projectile, and such injuries are a concern for warfighters. [1]. The mechanics of BABT events are characterised by high rate and focal loading. When this type of event is centred over the sternum, it can result in injuries ranging from skin lacerations and abrasions to rib fractures and organ damage [2]. While historically rare, recent conflicts involving advanced personal armour systems have led to documented BABT cases [3]. These cases have confirmed the types of injuries seen in biomechanical models used to study BABT and emphasised the need to further understand BABT effects.

Military body armour optimization would benefit through development of injury-based standards. Current standards utilise a clay backing to both mimic the presence of tissue as well as provide a metric, i.e. permanent deformation, for understanding injury risk [4]. However, there have been many limitations observed in literature of this standard driving the need for a more accurate metric [5], [6]. In response, the past two decades have seen numerous efforts to refine injury risk assessment beyond clay testing. Recent studies involving postmortem human surrogates (PMHS) and animal models use various BABT loads to induce and study BABT injuries experimentally [2], [7], [8], [9], [10]. Simulated BABT injuries enable the development of continuous injury risk models using approaches such as logistic regression and survival analysis. However, significant uncertainty exists in these models, likely due to variability in biological models, which can manifest in multiple ways [11]. In studies using PMHS, biological factors such as age, height, and weight can be controlled through exclusion criteria to reduce variation. However, other factors that cannot always be screened or controlled may influence BABT loading. One such factor is body habitus—the overall physical build and shape of an individual—which can affect both armour plate fit (including the air gap between the armour backface and the wearer) and the amount of soft tissue between the armour backface and the skeleton [12], [13].

Body habitus has been recognised as a key variable in injury risk assessment across various fields. In automotive injury biomechanics, its effects have been extensively studied, with more recent research focusing on its role in blunt injury risk [14], [15], [16], [17]. In BABT research, the velocity of the armour backface is known to decrease rapidly within the first few centimetres, suggesting that the gap between the armour and the body could significantly reduce the force transmitted to the wearer [18], [19]. Body habitus is hypothesised to influence injury risk by altering armour fit and modifying

https://doi.org/10.52202/081621-0008

anatomical composition at the point of impact. This study aims to investigate its effects on BABT loading correlates (force response and clay deformation) by varying air gap distance and flesh thickness through isolated load cell and clay testing. These laboratory findings are then compared with injury outcomes from prior PMHS BABT tests to assess the influence of air gap and flesh thickness on BABT outcomes.

2. METHODS

2.1 Armour and threat

The same projectile and armour combination was used for all tests in this study. The projectile chosen was the 7.62 x 51 mm M80 Ball, while the striking velocity, flesh thickness and standoff ranges were all derived from previously conducted PMHS tests [2]. The armour used for this study was the same as armour used during the PMHS tests, namely an ultra-high molecular weight polyethylene (UHMWPE) composite multi-curved plate with a Small Arms Protective Insert (SAPI) cut geometry. As with the previous PMHS studies, each plate was shot once at a location over the centre of the load cell. The plates utilised in this study did not have a soft armour backing behind the hard armour plate as these plates were designed to be stand alone.

2.2 General test setup

To examine the effect of body habitus and specifically flesh thickness, excised tissue from PMHS were placed in a custom fixture over the load cell or clay block and positioned between the armour plate and backing material. A diagram of the setup can be found in Figure 1. As indicated in the setup, an approximate square section of PMHS flesh, 300 mm x 300 mm, was excised from PMHS across the anterior thorax and abdominal region. The PMHS flesh components used for this harvesting were all male, fresh frozen specimen procured under institutional review board exemption approval through Johns Hopkins Medical Institutions and were all untested prior to flesh harvesting. Custom fixtures were used to position the flesh sample and trim the tissue thickness to the targeted value. When the flesh was excised, an initial measurement was taken using callipers at the impact site, to determine which thickness range was closest to minimize the amount of flesh trimmed for each sample. After trimming, final calliper measurements were made at the impact site and were used as the main flesh thickness parameters to be analysed in the study.

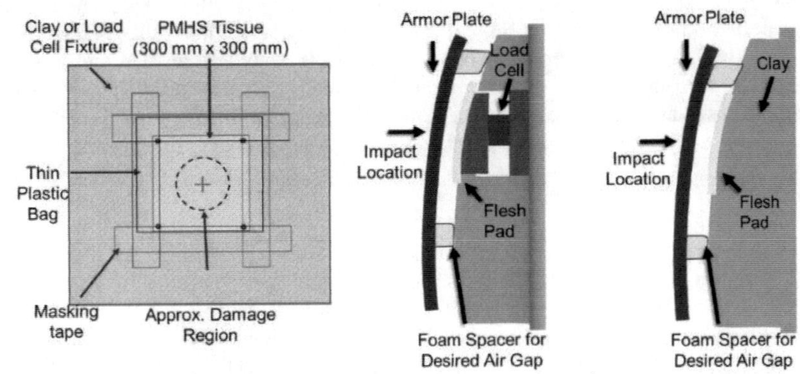

Figure 1. Diagram of tissue placement over backing fixture from a front on view (left), sagittal view of the load cell test setup (middle) and the clay setup (right).

The other variable that was controlled for was the air gap between the armour plate and the backing material. This distance was representative of the offset that an armour plate might naturally sit at due to the different body habitus of individuals. In order to both achieve the desired air gap while also approximating the material response of the structures that were driving this air gap to occur, open cell urethane foam of different thicknesses was used to achieve two different standoff positions. These spacers were placed above and below the impact location to ensure a uniform offset from the load cell and clay block. A FARO (FARO Technologies Lake Mary, Florida) coordinate measurement machine was used to determine the air gap by digitizing a point on the backing material, donning the amour, then digitizing a point on the surface of the armour. The thickness of the armour was then subtracted from

https://doi.org/10.52202/081621-0008

the measurement, resulting in an air gap measurement at the impact site along the direction of the ballistic trajectory.

2.3 Load cell test setup

Prior to testing, a custom 3D printed fixture (glass filled nylon) was created and was rigidly mounted to a 55 cm x 55 cm steel plate to provide a fixed boundary condition. The geometry of the 3D printed portion was designed to have the same curvature as the plate used for testing, with an obliquity of approximately 6 degrees at the site of impact. Within the contoured mounting fixture, a single piezo-electric load cell (PCB Model 200C50, range capacity of 222.4 kN) with a 152.4 mm diameter 4140 steel alloy impact cap was used. The impact cap was contoured to follow the curvature of the backface of the armour geometry as well. The fixture and setup can be seen in Figure 2. A total of 16 load cell backed tests were conducted.

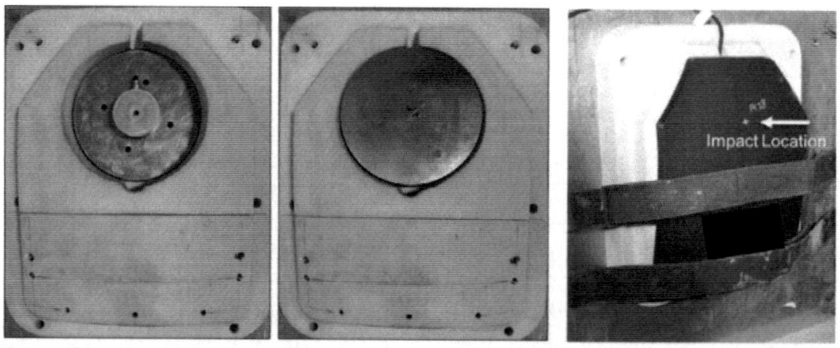

Figure 2. View of load cell fixture without cap (left), with 152.4 mm diameter impact cap (middle) and view of armour on fixture before test (right).

2.4 Clay test setup

Clay ballistic testing generally followed the clay calibration methods described in NIJ 0101.03 [5]. The armour was placed on a conditioned and tempered clay block, with a hand rolled template used to ensure no air gaps behind the curved armour plate. The clay was scanned before and after each test using a FARO laser scanner to create a point cloud representation of the clay surface. Geomagic Studio software (3D Systems, Rock Hill, South Carolina) was used to transform the point cloud into a surface model and calculate the clay deformation geometry (i.e., deformation left in clay). This resulting clay deformation value is termed adjusted backface deformation (BFD) due to the fact that air gap and flesh thickness were included in the test setup resulting in the armour not being pressed up directly to the clay. This value should not be used as a backface signature value (BFS) as it has different setup conditions, but it can be used in relative comparisons. A total of 15 clay backed tests were conducted.

2.5 Test matrix

All tests were conducted at an NIJ certified ballistic laboratory (Element U.S. Space & Defence, Belcamp, MD). Pre-test, the armour plate was placed in position, using elastic straps, over the backing material and fixture with the resulting air gap collected using a FARO arm. Data from the load cell was collected using a high-rate data acquisition system (Dewetron Inc., East Greenwich, RI, model DEWE-801) at a 1 MHz sampling rate. Three striking velocities (600, 762 and 915 m/s), three different flesh thickness values (0, 5-10 and 15-20 mm) and three different air gaps (0, 12 and 25 mm) were targeted. Table 1 shows the test matrix used for this study with the achieved striking velocities, tissue thickness, and achieved air gap.

Table 1. Experimental Test Matrix

Test #	Backing Condition	Striking Velocity (m/s)	Air Gap (mm)	Tissue Thickness (mm)
1	Load Cell	606	12.5	0
2	Load Cell	611	25.1	0
3	Load Cell	595	12.2	8.8
4	Load Cell	612	27.9	10.3
5	Load Cell	605	0	17.5
6	Load Cell	612	1.3	18.6
7	Load Cell	763	24.5	0
8	Load Cell	764	1.6	6.4
9	Load Cell	763	14.8	8.8
10	Load Cell	765	23	9.6
11	Load Cell	765	14.5	18.3
12	Load Cell	918	11.9	0
13	Load Cell	908	26.1	6.4
14	Load Cell	928	14.2	8
15	Load Cell	914	25.1	11.8
16	Load Cell	907	3.2	10
17	Clay	604	11.4	0
18	Clay	606	24.3	0
19	Clay	594	12	4.8
20	Clay	604	23.6	7.1
21	Clay	612	4.3	12.9
22	Clay	611	0	17
23	Clay	611	0	22.7
24	Clay	758	28.3	0
25	Clay	758	3.1	7.5
26	Clay	760	12.1	8.5
27	Clay	763	23.6	2.8
28	Clay	760	9.9	11.8
29	Clay	759	0	0
30	Clay	926	0	0
31	Clay	926	10	0

3. RESULTS

3.1 Load cell results and parametric model

Representative traces for two tests demonstrating the effects of each of the variables examined in this study are found in Figure 3. Each test exhibited similar bi-modal responses with a large initial peak followed by a generally smaller secondary peak. The response decays rapidly after this initial response, oscillating around 0 N within 0.5 ms. As expected, the force measured at the load cell decreased with lower striking velocities. Similarly, increasing the air gap or flesh thickness led to a reduction in recorded force. However, variations in air gap had a more pronounced effect on force response compared to changes in flesh thickness.

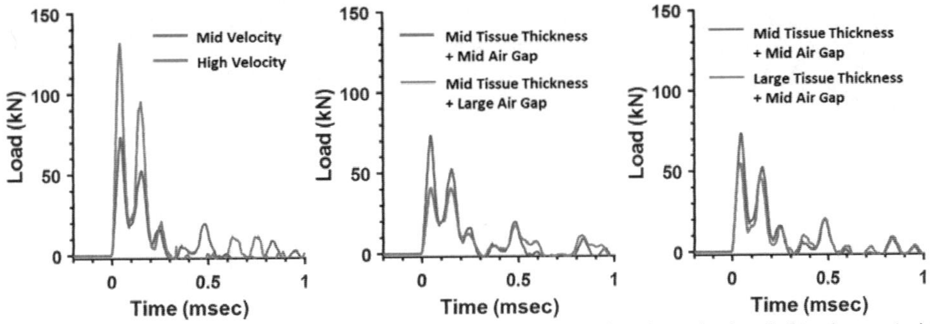

Figure 3. Force-time curves for two different ballistic tests varying the velocity (left), air gap (middle) and flesh thickness (right) while keeping the other two variables constant

Since the force response was similar across the different tests, the initial peak force was selected as a potential metric to correlate to injury. This initial peak primary force response was investigated as a function of air gap and flesh thickness as shown in Figure 4 for each of the striking velocities tested. As seen in the peak response, there was an implied correlation between air gap and the initial peak force response with less of a clear relationship between this peak force response and the flesh thickness.

https://doi.org/10.52202/081621-0008

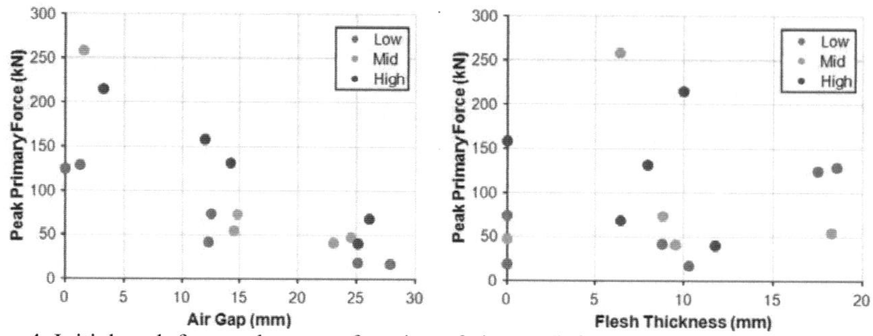

Figure 4. Initial peak force values as a function of air gap (left) and flesh thickness (right) with the three striking velocities represented by the red, green and blue dots

These results implied that the force response could be modelled using an exponential decaying model. This proposed relationship is shown in equation 1 where the peak primary force is a function of the striking velocity, air gap and tissue thickness. The model was fit by using a standard least squares fit between the natural log of peak force versus striking velocity, air gap and tissue thickness. The fit coefficient values were calculated using JMP Pro (Version 17.0.0, JMP Statistical Discovery LLC, Cary, NC). These resulting coefficient values and their associated p-values are shown in Table 2. These parameters resulted in a model that was able to predict actual peak force values with an R^2 value of 0.95.

$$\textit{Peak Primary Force (kN)} = e^{b_0+b_1*Striking\ Vel.\ (m/s)+b_2*Air\ Gap\ (mm)+b_3*Tissue\ Thick.\ (mm)} \qquad (1)$$

Table 2. Coefficients calculated for the exponential decaying model fit (equation 1)

Term	Coefficient Estimate	P value
Intercept	3.33	<0.0001
Striking Velocity (m/s)	0.003	<0.0001
Air Gap (mm)	-0.078	<0.0001
Tissue Thickness (mm)	-0.021	0.0345

3.2 Clay results and parametric model

In addition to tests conducted over load cells, the same test series was conducted over clay blocks with flesh and air gaps incorporated. Since clay BFS is used in the evaluation of the effectiveness of armour systems, it was important to see how this metric would be affected by the presence of standoff and flesh thickness. An evaluation of the adjusted clay BFD as a function of the different variables tested is shown in Figure 5. While the tissue thickness and air gap individually start to predict adjusted clay BFD, the combination of both was found to be the best predictor.

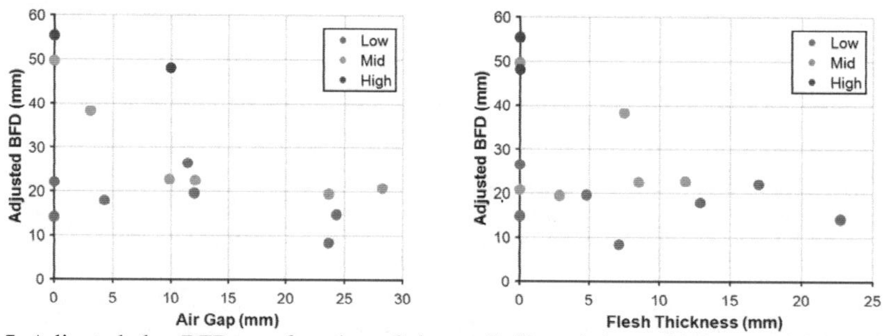

Figure 5. Adjusted clay BFD as a function of air gap (left), and tissue thickness (right) with the three striking velocities represented by the red, green and blue dots

Unlike the load cell results, the clay results were found to be best explained using a multiple linear regression to model the relationship between striking velocity, air gap, tissue thickness, and adjusted clay BFD. Utilizing the data shown in Figure 5, parameters were calculated for the linear regression model shown in equation 2. Overall, it was found that for each 1 mm of air gap or tissue thickness, there was a reduction of adjusted clay BFD values of approximately 1 mm.

$$ClayBFD(mm) = 6.295 + 0.0547 * Striking\ Vel.\left(\frac{m}{s}\right) - 1.03 * AirGap(mm) - 1.16 * TissueThick.(mm) \quad (2)$$

Table 3. Coefficients calculated for the multiple linear model fit (equation 2)

Term	Coefficient Estimate	P value
Intercept	6.295	0.2269
Striking Velocity (m/s)	0.0547	<0.0001
Air Gap (mm)	-1.03	<0.0001
Tissue Thickness (mm)	-1.16	<0.0001

3.3 PMHS outcome comparison and injury reference curve

While the models developed in equations 1 and 2 were able to accurately fit the force and clay back face deformation response respectively, investigating how this model could explain the effects of body habitus on BABT was the overall goal of this study. Using a previously collected PMHS data set, the effects of flesh thickness and air gap as it related to injury prediction was then investigated utilizing this model [2], [20]. This data set, shown in Table 4, provided 13 PMHS test data points which could be compared to the above results. From each PMHS test, the air gap, flesh thickness and striking velocity was calculated based on information collected during the test series. Similar techniques were used from the bench-top testing to collect the air gap, and the flesh thickness was measured with a pre-test CT scan at the desired anatomical impact site. Based on these values, an estimated peak primary force and adjusted clay BFD was also calculated as seen in Table 4.

Table 4. PMHS Test Matrix (Tests from Iwaskiw et al)

APL ID	Air Gap (mm)	Flesh Thickness (mm)	Striking Velocity (m/s)	Estimated Peak Primary Force (kN)	Estimated Adjusted clay BFD (mm)	Projected Operational Injury Outcome
APL01	26.0	9.1	854.35	39.53	15.69	II
APL02	12.1	10.4	756.51	84.81	23.15	II
APL03	13.2	21.5	865.33	85.19	15.09	II
APL04	12.5	15.6	657.76	54.64	11.30	I
APL05	2.9	4.5	712.62	172.34	37.07	IV
APL06	11.8	12.2	716.89	74.15	19.20	II
APL07	9.9	10.5	659.59	74.94	20.00	II
APL08	10.3	14.1	588.57	54.46	11.52	II
APL09	9.8	7.1	634.59	75.27	22.68	III
APL10	9.9	5.3	900.99	173.01	39.23	IV
APL11	10.6	18.3	917.14	130.36	24.32	IV
APL12	9.6	8.0	909.22	171.19	36.86	IV
APL13	9.2	21.1	912.57	135.55	22.26	III

As was discussed in Howes et al. 2023, a Projected Operational Injury Outcome (POIO) was calculated for each of the PMHS tested based on input from an expert panel consisting of military pathologists, trauma and general surgeons to categorise injuries based on a return to duty metric. This classification scale ranges from level I which indicates the warfighter could return to duty within 72 hours, to level IV which is thought to be a potentially fatal injury. For this study, we chose to evaluate how body habitus metrics could affect the probability of sustaining a level IV POIO as a function of striking velocity, estimated peak primary force, and estimated adjusted clay BFD. A survival analysis (SA) was applied, classifying fatal injuries as 'injured' and all other cases as 'not injured'. This approach enabled a continuous mapping of dose to injury probability, and the resulting injury curves are presented in Figure 6. The SA was unable to converge for estimated adjusted clay BFD, as some zone of mixed results (ZMR) is necessary (i.e. a dose range where, due to inherent variability, some outcomes at a given dose may result in injury while others might not). This might suggest this metric helps to differentiate between cases of non-fatal versus fatal. To evaluate the accuracy of the injury risk curves, we calculated three goodness-of-fit metrics: the Akaike Information Criterion (AIC), the Area Under the Receiver

https://doi.org/10.52202/081621-0008

Operating Characteristic Curve (AROC), and Balanced Accuracy. All three metrics improved (AIC 16.7→7.5, AROC 0.528→0.667, Balanced Accuracy 0.514→0.819) when incorporating estimated peak primary load (as derived from our fit models), indicating that injury prediction accuracy increases when body habitus is considered.

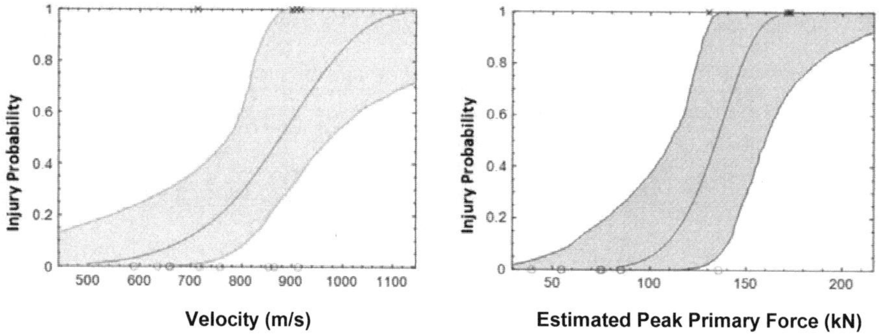

Figure 6. Probability of a fatal injury as a function of ballistic striking velocity (left) and estimated peak primary force (right)

4. DISCUSSION

To explore the effects of body habitus on BABT, this study collected data from BABT events in order to correlate loading transmitted through body armour at varying levels of air gap and flesh thickness with injury outcomes from previously conducted PMHS BABT tests. Practically, it is a challenge to collect injury data from a representative warfighter population, due to the nature and limitations of PMHS acquisition through use of specimens with great differences in age, which can be associated with body composition, to the desired population. In addition, stature, gender, and body habitus can even vary greatly across a warfighter population. There is a great utility in individually accounting for differences in body habitus and fit so that results and insights can be translated to different populations. Understanding these effects could lead to more effective standards and enable the development of more effective personal armour solutions.

4.1 Load cell response

Based on the recorded force response, we hypothesised that initial (primary) peak force correlates with injury severity. To investigate this, we examined how peak primary force varies as a function of air gap and flesh thickness. The peak forces measured in this study (17–258 kN) were greater than those observed in other blunt ballistic impact scenarios with some overlap in forces from our study's lower magnitude ballistic conditions and higher air gap and flesh thickness conditions. This may be expected due to our study's focus on ballistic velocities that induced high levels of BFD loading through a purpose-built UHMWPE plate. Bir [21] conducted tests on PMHS using law enforcement PVC batons, estimating force from impactor acceleration and mass. The values seen in that study spanned 2.6 to 4.4 kN. We observed much greater loading conditions which were likely due to the increased energy from the higher backface velocities seen in BABT [19]. Other live fire BABT testing was conducted by Bass et al on PMHS (15 kN to 30 kN) and on the AUSMAN dummy (30 kN to 70 kN) [22]. This group utilised thin profile stress gauges to obtain contact forces behind UHMWPE plates. The striking velocities in this study spanned a wider range than those in Bass et al. While force-sensing methods in BABT studies vary [23], comparing our results with previous findings highlights the role of armour fit and body habitus in force reduction. Additionally, the load magnitudes observed in this study align with simulated BABT studies where sternum loading ranges from 2.9 to 62.2 kN [24] although due to differences in the ballistic threat and soft armour modelled, direct comparisons may be inappropriate. Across all test conditions, the load cell data exhibited a bimodal peak force response, characterised by a larger initial peak followed by a lower secondary peak. This response pattern is consistent with prior ballistic force response studies [8] where the bimodal force response was observed in different ballistic testing setups and with different instrumentation packages. This bimodal response has been noted in many previous studies [22], [25]with the first peak attributed to the impedance mismatch between the armour and flesh while the second peak is associated with the local bulk motion. In these studies, the first peak was noted to have an order of magnitude more energy than the second peak. Additionally, sustained oscillations in the load cell response suggest the possible excitation of a vibrational mode within the setup. A frequency analysis of the force response identified a distinct frequency peak at approximately 8.5 kHz, well below the 30 kHz frequency limit of the PCB 200C50 load cell used in these experiments which indicates that these measurements are still valid. Overall, this study clearly showed that the air gap had a stronger correlation to peak force, both in the individual traces as well as the comparison of the overall peak response, as observed in the change of the initial peak as a function of air gap. Both air gap and flesh thickness contributed to a reduction in initial peak force, highlighting their importance in overall protection.

4.2 Clay response

Current armour systems are evaluated using clay, where maximum residual deformation is measured after ballistic impact. While challenges exist with this standard [5], no alternative has been formally established. Therefore, understanding how body habitus, specifically flesh thickness and air gap, affects clay displacement is critical. Numerous studies have used clay to assess physics-based loading parameters [6], [26] with some correlating clay BFS to kinetic energy transfer [27]. Findings from this test series indicate a strong correlation between body habitus variables and clay response, showing that each additional millimetre of offset (due to air gap or flesh thickness) reduces BFS by approximately 1 mm. If armour is typically worn with some degree of standoff from the body, performance testing standards should account for this factor. The current clay standard does not incorporate standoff, potentially representing a worst-case scenario for BABT loading. This is particularly important because armour systems optimised solely for the clay standard (without air gaps) may differ from those designed for operational wear conditions. Some designs that consider realistic air gaps and flesh thickness may offer better protection for the warfighter while minimizing weight.

4.3 Validation of body habitus effects using whole body PMHS tests

When applied to PMHS test results and injury outcomes, loading parameters adjusted for air gap and tissue thickness provided more accurate injury predictions than velocity alone. Examining the survival analysis curves in Figure 6, injury prediction models that incorporate air gap and flesh thickness show a smaller ZMR (the range where both injury and non-injury may occur) compared to those based solely on striking velocity. Additionally, when using estimated adjusted clay BFD, the potentially fatal injury group was completely separated from the non-fatal cases, suggesting that this metric effectively distinguishes injury severity. Given this complete separation, alternative binomial regression techniques could be applied for continuous injury prediction and should be considered for future analysis. These observations align with expectations, as directly measuring loading to the body and including the variations due to body habitus, should provide a stronger correlation to injury outcomes. Additionally, as was shown in equation 1, the primary peak force follows a decaying exponential function indicating that the influence of this gap diminishes as the gap increases. These results suggest that the initial few centimetres have the most critical impact on peak force which is critical for injury prediction based on the survival analysis curve and for protection. While this initial decay was best represented by an exponential fit for the load cell-based model, the clay model in equation 2 was best represented by a linear fit. Unlike the load cell-based model, the clay model appears to weight all standoff equally. Understanding the appropriate injury mechanism and injury metrics for BABT is crucial, as these two testing methods yield different conclusions regarding loading dynamics.

4.4 Limitations

While this study demonstrated strong correlations between the variables studied and injury predictions, several limitations should be considered. First, the load cell stiffness itself may have influenced the load response. The peak loads observed were higher than other studies utilizing more biofidelic surrogates. These higher forces could be attributed to armour performance (i.e. these ballistic conditions created a more severe loading condition) or due to the stiffness of the load cell system. Further testing should be done to understand the influences of the design of the backface material and armour performance on loading metrics collected. Second, although a range of PMHS specimen was used to evaluate the injury prediction, the overall sample size is small compared to the warfighter population. Since flesh thickness was found to have a protective effect, future studies should incorporate a representative range of body habitus based on the warfighter, either through additional testing or models capable of varying body composition and anthropometry. Third, while specialised fixtures were used to prepare the flesh specimens, some variance in overall thickness was expected, as is typical in PMHS testing. Further, these

https://doi.org/10.52202/081621-0008

specimens also did not include any musculature which would be expected in a living warfighter population. Lastly, this study presents air gap and flesh thickness as a single value for each impact test, where likely these values vary over the impact site. Converting these simple one-dimensional metrics to three-dimensional values may better explain these parameter effects on loading, especially when considering body regions with more spatial variance of body shape and body tissue composition. However, the contact area for this loading event is small, and the spatially varying air gap and flesh thickness may vary less than in the contact region in more broadly distributed loading events. These limitations warrant further investigation in the effect of body habitus on BABT injury response.

5. CONCLUSION

This study demonstrated that body habitus, as described by varying levels of air gap and flesh thickness, was highly correlated to the BABT loading, as measured using load cell and clay backed methods. Additionally, incorporating these body habitus parameters improved injury outcome predictions in previously conducted PMHS BABT tests. Specifically, increased air gap and flesh thickness were associated with lower measured and predicted force responses, suggesting that these factors play a critical role in reducing injury risk. These results are essential for developing more effective standards and more effective personal armour solutions.

Acknowledgments

The authors would like to express their appreciation to the U.S. Special Operations Command for funding this work under Contract No. H92236-10-D-001, Task Order 0521.

References

[1] A. W. Carroll and C. A. Sonderstrom, ""A new nonpenetrating ballistic injury," *Annals of Surgery,* vol. 188, no. 6, p. 753, 1978.

[2] A. Iwaskiw, et al., "Initial injury observations in postmortem human subjects: Non-penetrating ballistic impacts over the sternum with lightweight hard armour plates," in *Personal Armour System Symposium*, Copenhagen, Denmark, 2021.

[3] T. Talmy, et al., ""Close-range fire inflicting behind armor blunt trauma: case-series and implications for battlefield care," *Military Medicine,* vol. 189, no. 1-2, pp. e448-e453, 2024.

[4] US Department of Justice, "Ballistic resistance of police body armor NIJ Standard-0101.07.2024," *National Institute of Justice - Office of Law Enforcement Standards,* pp. 1-13, 1986.

[5] National Research Council, "Testing of body armor materials: Phase III," *National Academies Press,* 2012.

[6] M. Eaton, et al., "Testing the validity of the NIJ clay standard for approving body armor with a preliminary insight into injury correlation," *ASME International Mechanical Engineering Congress and Exposition,* 2020.

[7] C. R. Bass, "Helmet behind armor blunt trauma," in *Specialist Conference on Human Factors in Medicine*, Koblenz, Germany, 2003.

[8] A. Iwaskiw and et al, "Loads Associated with Behind Helmet Blunt Trauma: Matched-Pair Load Sensing Headform Tests Correlated with Skull Fracture Severity," in *Personal Armour Systems Symposium*, Dresden, Germany, 2023.

[9] J. McMahon, et al., "Development of Impulse-Based Rib Fracture Injury Criterion for Behind Armor Blunt Trauma," in *IRCOBI Conference*, 2023.

[10] J. McMahon, et al., "Use of a Porcine Cadaver Model as a Human Surrogate for Behind Armor Blunt Trauma," *Journal of Engineering and Science in Medical Diagnostics and Therapy,* vol. 8.4, 2025.

[11] R. Kraft, et al., "Modeling Skeletal Injuries in Military Scenarios," *The Mechanobiology and Mechanophysiology of Military- Related Injuries,* pp. 3-35, 2016.

[12] Kim and Han, “Virtual fit assessment of US army body armor using NASA spacesuit techniques,” *Applied Ergonomics 120,* vol. 104339, 2024.

[13] M. Reed and D. William, “Creating human figure models for ergonomic analysis from whole-body scan data,” *Proceedings of the Human Factors and Ergonomics Society Annual Meeting,* vol. 45, 2001.

[14] S. Moran, et al., “Injury rates among restrained drivers in motor vehicle collisions: the role of body habitus,” *Journal of Trauma and Acute Care Surgery,* pp. 1116-1120, 2002.

[15] S. Wang, et al., “Increased depth of subcutaneous fat is protective against abdominal injuries in motor vehicle collisions,” *Annual Proceedings/Association for the Advancement of Automotive Medicine,* vol. 47, 2003.

[16] J. Choi, et al., “Body mass index and mortality in blunt trauma: the right BMI can be protective,” *The American Journal of Surgery,* vol. 220, no. 6, pp. 1475-1479, 2020.

[17] S. Dogruyol and T. Dogruyol, “Adipose tissue provides a cushioning effect in low-energy isolated blunt thoracic trauma: a prospective observational,” *Acta Chir Belg,* vol. 123, no. 4, pp. 396-404, 2023.

[18] D. Hisley, J. Gurganus and A. Drysdale, “Experimental Methodology Using Digital Image Correlation to Assess Ballistic Helmet Blunt Trauma,” *J Appl Mech,* vol. 78, no. 5, p. 051022, 2011.

[19] Y. K. Wen, “Analysis of dynamic back face deformation of a body armor impact by a rifle bullet using 3D-DIC,” *Journal of Physics: Conference Series,* vol. 1507, no. 5, 2020.

[20] C. Howes, et al., “Injuries and Operational Implications Caused By Behind Armour Blunt Trauma Across Various Impact Locations,” in *Personal Armour Systems Symposium*, Dresden, Germany, 2023.

[21] C. Bir, D. Viano and A. King, “Development of biomechanical response corridors of the thorax to blunt ballistic impacts,” *Journal of Biomechanics,* vol. 37, no. 1, pp. 73-79, 2004.

[22] C. Bass and et al, “Injury Risk in behind armor blunt thoracic trauma,” *International Journal of Occupational Safety and Ergonomics,* vol. 12, no. 4, pp. 429-442, 2006.

[23] B. Anctil, et al., “Evaluation of impact force measurement systems for assessing behind armour blunt trauma for undefeated ballistic helmets,” in *Personal Armour Systems Symposium,*, The Hague, The Netherlands, 2004.

[24] M. C. Bustamante and D. S. Cronin, “Impact location dependence of behind armor blunt trauma injury assessed using a human body finite element model,” *Journal of biomechanical engineering,* vol. 146, no. 3, 2024.

[25] J. van Bree and P. Gotts, “The "twin peaks" of BABT,” in *Proceeding from the 5th Personal Armour Systems Symposium*, Clochester, UK, 2000.

[26] D. Bhattacharjee, A. Kumar and B. Ipsita, “Energy absorption and dynamic deformation of backing material for ballistic evaluation of body armour,” *Defence Science Journal,* vol. 63, no. 5, pp. 462-466, 2013.

[27] M. Bevan, et al., “Correlation between Round Kinetic Energy and Displaced Clay Volume for Three Classes of Armour,” in *Personal Armour Systems Symposium*, Washington DC, USA, 2018.

 https://doi.org/10.52202/081621-0008

Experimental modelling of blunt ballistic impact on a male thorax surrogate: study of rib fractures and lung injuries predictions

E. Dancerel-Bourlon[1], R. Delille[1], B. Bourel[1], O. Mauzac[2], N. Prat[3], C. Bir[4], D. Sherman[4], S. Roth[5] and F. Lauro[1]

[1]*Laboratory LAMIH UMR 8201 CNRS, Univ. Polytechnique Hauts-de-France 59313 Valenciennes, France*
elodie.dancerelbourlon@uphf.fr

[2]*French Ministry of the Interior, CREL/SAILMI, Place Beauvau, Paris, France*

[3]*Institut de Recherche Biomédicale des Armées (IRBA), Brétigny sur Orge, France*

[4]*Department of Biomedical Engineering, Wayne State University, 818 W. Hancock, Detroit, MI 48201, USA*

[5] *Laboratoire Interdisciplinaire Carnot of Bourgogne, site UTBM, UMR 6303, CNRS /Université de Technologie de Belfort-Montbéliard, 90010 Belfort, France*

Abstract. Modelling of body injuries caused by non-penetrating ballistic impacts is a major challenge. Experiments on biological models are difficult to conduct due to ethical and logistical constraints and on the data that can be collected. These limitations can be overcome by using physical or numerical surrogates, but their results must be correlated with real injuries recorded by ballistic injury databases. The objective of this study is to establish injury prediction curves for rib fractures and pulmonary injuries based on the applied energy levels. Twenty ballistic impact scenarios from French and American reports were analysed. These cases were replicated using a physical and biofidelic thoracic surrogate, SurHUByx. The surrogate used represents the torso of a 50th percentile male and it is designed to replicate the biomechanical responses of the human body to ballistic impacts. SurHUByx is equipped with different sensors on the ribs and in the lungs measuring deformations and pressures during impacts. The data collected were then analysed, retaining only the maximum value for each internal organ. A normalisation process adjusted the data according to body mass index to account for individual variability. Then, a regression model was used to analyse the relationship between injury probability and sensor values. The optimal injury diagnosis threshold was determined to use the Youden method, prioritising the reduction of false negatives. The results enabled the calculation of injury probability based on sensor data. This study highlights the interest of sharing ballistic injury cases to constitute a more reliable database. It also highlights the use of surrogates to replicate cases and develop injury prediction models.

1. INTRODUCTION

Law enforcement officers and military personnel are exposed to the threat of ballistic injuries in occupational situations [1]. To reduce these risks, protective equipment, such as bulletproof vests, are used. However, these do not guarantee absolute protection against injuries. Even in the absence of projectile penetration through the armour, ballistic impacts can still cause severe injuries, called Behind Armour Blunt Trauma (BABT). These injuries result from the interaction between impact energy and dynamic deformation at the back of the bulletproof vest [2].

Currently, Roma Plastilina N°1, an oil-based modelling clay, is the standard material for the evaluation of body armour. The armour is placed on the clay and impacted with test projectiles specified by the standard. Several impacts are made per panel. The acceptability of the results is determined based on one of two conditions: either each individual Back Face Deformation (BFD) measurement in this material not exceed 44 mm, or individual BFD values fall between 44 and 50 mm, and the average of all recorded impact measurements for a sample of the same model remains below 44 mm. Any measurement exceeding 50 mm is considered a failure [3].

The human torso is not a homogeneous structure, each region of the body responds differently to impacts, requiring specific studies for each area of interest. Modelling of injuries caused by non-penetrating ballistic impacts is a major challenge [4,5]. To address this issue, significant research has been carried out to better understand the effects of blunt impacts on the human body. Animals and Post Mortem Human Subject (PMHS) experiments are a classic method for assessing ballistic injuries [2,6]. However, ethical and logistical constraints severely limit these experiments, as do the difficulties associated with collecting and using the data obtained. These limitations can be overcome by using physical or numerical surrogates. Among these surrogates, Styrene-Ethylene-Butylene-Styren (SEBS)

https://doi.org/10.52202/081621-0009

blocks are commonly used to assess the ballistic protection effectiveness [7,8]. Additionally, digital anthropomorphic surrogates such as the HUByx, SurHUByx FEM, ARL, and ATBM models are used [9-11], and physical surrogates such as "Gelman" [12] and "SurHUByx" [5]. The SurHUByx surrogate is specifically designed to represent the anthropometry of a 50th percentile male torso, facilitating the recreation and analysis of ballistic impacts. However, the data obtained from these surrogates must be correlated with real injury cases.

Thus, an experiment with the use of a male physical surrogate of average build is justified, in ordor to develop new injury criteria based on medical records from officers shot in the line of duty reports, and post test records from tests on PMHS. These criteria should account for the variability in injury risks across different regions of the torso.

In ballistics research, collecting BABT cases remains a major challenge. For experimental reproduction to be relevant, several parameters must be precisely known. Several studies have already reported recreations of cases in the literature [5,13,14]. Given the difficulty of grouping these cases, the physical build of individuals described was not a criterion for exclusion. Therefore, standardisation based on Body Mass Index (BMI) was conducted to minimise these differences.

The objective of this study is to establish injury risk curves for rib fractures and lung injuries as a function of the applied energy levels. The development of these risk curves will allow for modelling of the relationship between the probability of injury and the impact applied.

This study aims to use a physical surrogate to recreate ballistic impact scenarios from French and American reports of law enforcement officers who were shot in the line of duty. Case reports were analysed and grouped. Only cases involving a thoracic ballistic impact occurring in the sagittal plane of a male individual were retained. Missing data necessary were identified. Then, the different cases obtained were recreated on the physical surrogate under conditions identical to those described in the case report. Exemplar bulletproof vests were used to ensure the vest model was the same as what was reported in the cases, as well as the projectiles specified in the reports. Finally, the standardisation of the data with respect to the BMI of average male was carried out on the data in order to perform a statistical processing of the data to generate injury probability curves corresponding to a male of the 50th percentile.

2. MATERIAL AND METHODS

The following section outlines the ballistic recreation cases forming the basis of this study, the experimental methods employed, the data processing procedures, and the overall experimental evaluation approach. In addition, the normalisation of data relative to the size of a 50th percentile male for result assessment is discussed and the principles used to construct risk curves are presented.

2.1 Case reports

In this study, ballistic impact scenarios described in French and American BABT injury reports were analysed. The objective was to gather relevant data to reproduce these situations under experimental conditions. Key information included physical subject characteristics, the type and model of bullet-resistant vest worn, precise impact locations, injury descriptions, weapons, and projectiles, including their velocities. In order to minimise variability among the cases studied, the analysis focused on impacts involving the thoracic region of male subjects. The collected data were organised into Table 1, which summarises the selected cases, including subject characteristics and identified injury types. In each case, a lesion value was assigned as 0 when no lesion was observed on the studied organ and 1 when a lesion was present. Additionally, Figure 1 illustrates the location of bullet impacts relative to the ribs, as mentioned in different medical records. Each number corresponds to the case detailed in Table 1.

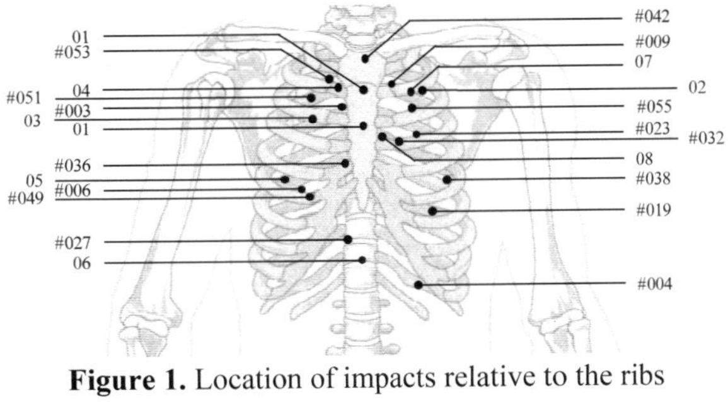

Figure 1. Location of impacts relative to the ribs

https://doi.org/10.52202/081621-0009

Table 1. Characteristics of individuals and their lesions

Case reference	Case reports type	Subject characteristics		Projectile	Impact velocity [m/s]	Body armour		Rib Fractures	Lung Lesion	Additional information
		Height [m]	Weight [kg]							
BABTID #003 [14]	US case	1.73	72.5	Lawman 9 mm 124 gr FMJ (Full Metal Jacket)	315	Second Chance	Monarch 329-MON-II	0	0	-
BABTID #004 [13]	US case	-	-	Remington .40 calibre S&W 180 gr	301	Second Chance	Ultima SMU-II + 001221	1	0	9th rib fracture
BABTID #006 [14]	US case	1.75	74.5	Lawman 9 mm 124 gr FMJ	356	Second Chance	MON-II + LSC	0	0	-
BABTID #009 [13]	US case	1.80	88.45	RWS .38 special 158 gr FMJ	263	PACA	KSG IIIA	0	0	-
BABTID #019 [14]	US case	1,75	-	Speer 9 mm 115 gr HP (Hollow Point)	366	ABA	XT2-1	1	0	8th rib fracture
BABTID #023 [14]	US case	-	-	American Eagle 9 mm 124 gr FMJ	355	ABA	XT2-2	0	0	-
BABTID #027 [13]	US case	-	-	Winchester .357 mag JHP (Jacketed Hollow Point)	379	ABA	SII-6.0	0	0	-
BABTID #032 [15]	US case	1.85	113.4	.38 calibre	296	ABA	XT3A-2	0	0	-
BABTID #036 [14]	US case	1.83	100	Winchester .45 calibre, 230 gr, FMJ	250	Safariland	GX-II.0	0	0	-
BABTID #038 [13]	US case	1.70	63.5	.38 calibre, FMJ	259	Safariland	SII-6.0	0	1	Underlying pulmonary contusion in anterior left lower lobe
BABTID #042 [14]	US case	1.65	64	Federal Premium .40 calibre S&W 180 gr HP	305	Point Blank	CIIA-1	0	1	Bruised lungs
BABTID #049 [14]	US case	1.90	100	Winchester .45 calibre 230 gr JHP	243	ABA	XT3A-2	0	0	-
BABTID #051 [14]	US case	1.87	131.5	12 gauge	406	Monarch	SUM IIA R02 6010	0	1	Pulmonary contusion
BABTID #053 [13]	US case	1.88	70.31	Lawman 9 mm 124 gr TMJ (Total Metal Jacket)	322	Safariland	BA-2000S-SM01	0	0	-
BABTID #055 [13]	US case	1.85	68.04	Winchester .357 mag JHP	374	Safariland	BR01	1	0	2nd and 4th rib fracture
01	PMHS case	1.74	72	calibre 12 Brenneke	385	CDX 2x10 layers		1	0	6 rib fractures
02	PMHS case	1.77	50	calibre 12 Brenneke	435	CDX 2x10 layers		1	0	1 rib fracture
03	PMHS case	1.65	52	Gévelot 9 mm Para. 8.18g	378	CDX 10 layers		0	0	-
04	PMHS case	1.60	47	Gévelot 9 mm Para. 8.18g	370	CDX 20 layers		0	0	-
05	PMHS case	-	51	Gévelot 9 mm Para. 8.18g	373	CDX 20 layers		0	0	-
06	French case	-	-	COP (Cartouche Opérationnelle de Police) 9 mm Parabellum HP	328	PN2002		0	0	-
07	French case	1.76	80	.38 calibre special JHP	250	COMODITEX GN 2001		0	0	-
08 [16]	Bibliographic reference	-	-	FlashBall	120	-		0	1	-

https://doi.org/10.52202/081621-0009

2.2 Anthropomorphic surrogate

This study aims to recreate BABT cases by an experimental approach. A surrogate, previously designed and validated, which allows for the development of injury risk curves. In this context, SurHUByx was used.

2.2.1 SurHUByx features

The SurHUByx physical surrogate was used to represent the torso of the 50th percentile male (Figure 2a). It was designed to replicate the biomechanical responses of the human body to ballistic impacts. The surrogate was composed of various materials specific to anatomical regions, allowing for a realistic reproduction of human physical characteristics. The external layer simulating skin was made of vinyl. The muscle tissue, the mediastinum, and vital internal organs, such as heart, lungs, liver, and spleen were replicated using SEBS gel with different concentrations as previously described. The skeletal structure is constructed from a two-component resin, and the costal cartilage was made from an elastomeric resin. SurHUByx was equipped with sensors placed on the ribs and within the lungs to measure deformations and pressures during ballistic impacts [5].

2.2.2 SurHUByx instrumentation

Strain gauges were installed on the ribs (Figure 2d). Gauges had a sensitivity of 3% with a resistance of 350 Ω. All data were recorded at a sampling rate of 1 MHz using a Yokogawa DL750 recorder oscilloscope. Fifteen gauges, positioned on the ribs, close to the impact zone, were connected to the oscilloscope. Piezoelectric pressure sensors, placed at the centre of the internal organs, measured the pressure generated during the impacts (Figure 2b). These uniaxial sensors required frontal shooting tests to ensure reliable data. The activation sensitivity of these sensors was approximately 0.2 N. Figure 2c illustrates the acquisition schema used to collect specific information about the ribs and internal organs.

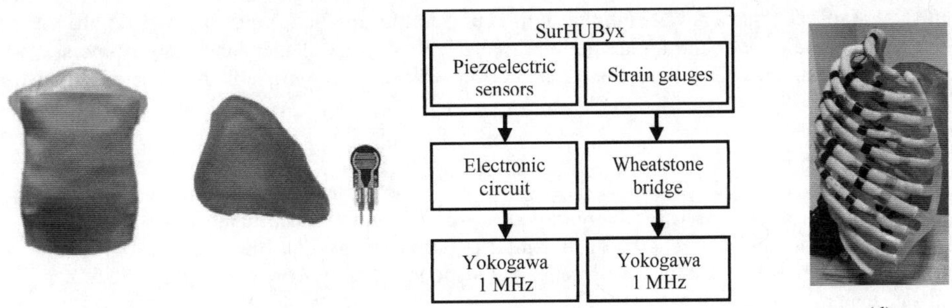

(a) (b) (c) (d)

Figure 2. SurHUByx (a) Piezoelectric sensors used in internal organs (b) SurHUByx data acquisition chain (c) and strain gauges on the ribs (d)

2.2.3 Validation of SurHUByx in the ballistic field

The surrogate SurHUByx was used in this study to simulate ballistic impacts on the thoracic region. Specifically designed to replicate the response of the 50th percentile male thoracic response to ballistic impacts, its effectiveness in the ballistic domain has been validated [5]. The results presented in the study by Chaufer [5] are consistent with the experimental values reported by Bir [17]. These data validated the global behaviour of the SurHUByx thoracic surrogate during ballistic impacts.

2.3 Test protocol

Figure 3 shows the experimental setup used to recreate ballistic impacts on the SurHUByx surrogate. SurHUByx was positioned 7.5 m from the gun. Before each impact, a calibration shot was performed to ensure that the projectile was launched at the desired velocity to replicate the specific case. After aligning the intended point of impact using a laser sight, the projectile was fired at the surrogate. The projectile velocity was recorded using an IR velocity gate. The impact on the ballistic vest was captured by a high-speed camera (Phantom v2012) at a rate of 22,626 frames per second. The body armour and projectiles were consistent with the descriptions in the case reports. Due to the limited surface area available on the

https://doi.org/10.52202/081621-0009

ballistic panels, each test case was replicated only once on the SurHUByx. To recreate a case, the same type of projectile specified in the case reports was used, along with bulletproof vest models models manufactured within five years of the one worn by the officer. Tests were conducted from the least severe to the more severe. A complete assessment of the surrogate was conducted between each impact to ensure its physical integrity. If any damage was detected on the SurHUByx surrogate, the necessary repairs were carried out before continuing the experimental tests.

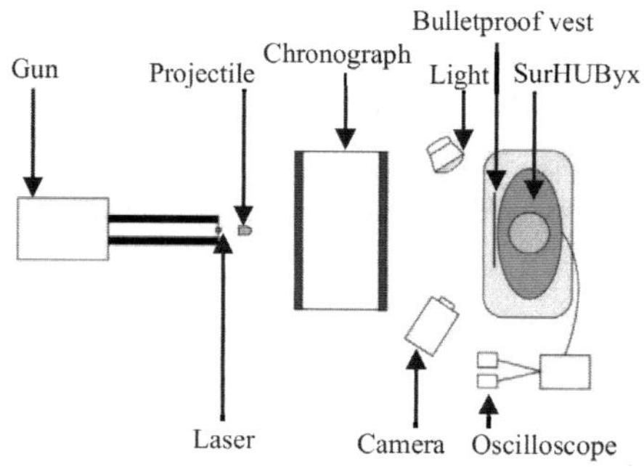

Figure 3. The setup of the experimental device

2.4 Data analysis

2.4.1 Extract data

For each case studied, only the maximum value recorded by each strain gauge was retained. Regarding the piezoelectric pressure sensors, the method was similar to that applied to the strain gauges. Only the maximum value of each sensor was retained, thus representing one datum per internal organ. For the lungs, the retained value was the highest of the two measurements. Thus, each datum obtained was associated with a lesion value: 0 if no lesion was observed on the organ studied or 1, if a lesion was present.

2.4.2 Data normalisation

In the case reports presented in Table 1, individuals had different physical characteristics in terms of mass and stature. These differences made the interpretation of the data more difficult. Given that, individual variations in these parameters influence the response of body structures to ballistic forces [18], a normalisation approach was carried out. Thus, the data from each individual were adjusted according to the BMI of a 50th percentile male, with reference values of 77 kg for mass and 1.78 m for stature, as explained by Serre et al. [19]. BMI is used to assess the body composition of an individual, as defined by Equation 1. This approach made it possible to calculate a specific normalisation factor for each case (Equation 2) [20]. Once the factor was obtained, it was applied to the collected data by multiplying the data by the corresponding normalization factor. This method allowed to for normalise normalisation of the results, facilitating their interpretation. The trends are independent of the morphological characteristics of the individuals.

$$BMI = Mass\ (kg) / Stature^2\ (m) \quad (1)$$

$$\lambda = BMI_{50th} / BMI_{subject} \quad (2)$$

2.4.3 Creation of injury probability curves

The data were analysed using survival analysis to assess the relationship between the probability of injury occurrence and the value of an independent variable representing a technical parameter. In this study, the analysis was performed using MATLAB R2024b. The proposed approach was applied to establish risk curves for ribs and lung BABT resulting from ballistic impacts on the thorax. The International Organization for Standardization (ISO) has developed a step-by-step process for developing injury risk functions [21], providing a foundation for the use of a unified approach to constructing injury risk curves. This standard has been used in several previous studies [2,22].

The steps of the ISO approach are as follows: (1) data collection, (2) determination of censoring status, (3) verification of multiple injury mechanisms, (4) sample separation based on injury mechanisms, (5) estimation of distribution parameters, (6) identification of highly influential observations, (7) verification of the distribution hypothesis, (8) selection of the appropriate distribution, (9) validation of predictions against existing results, (10) calculation of 95% confidence interval, (11) evaluation of the quality index, and (12) recommendation of a curve per body region [21].

The 12 steps of the ISO procedure provide a structured approach, but they do not address all the nuances of the problem. Subsequent research has refined this process to bridge existing gaps [18]. The ISO approach did not provide a method for determining which biomechanical metric from the experiments was most correlated with injury outcomes. To address this, the selection process for the metric with the highest area under the curve (AUC) was proposed [22]. The Receiver Operating Characteristic (ROC) curve evaluates the discriminative ability of biomechanical variables the concerning injury outcomes. It is obtained by plotting sensitivity against 1-specificity for different classification thresholds. The AUC represents the predictability measure of metrics in terms of specificity and sensitivity. A higher AUC indicates a more predictive measure. The AUC was calculated for all potential criteria, and the metric with the highest AUC was selected for further analysis.

Once the best criterion was identified, survival analysis was conducted by plotting a Kaplan-Meier curve, a nonparametric method used to estimate the survival function. The inverse survival curve was analysed to illustrate the cumulative probability of injury. The parametric survival models studied were those recommended by the ISO standard. The fitted models included Weibull, log-normal, and log-logistic distributions. For each distribution, the Akaike Information Criterion (AIC) was computed to select the most suitable model.

To complement the evaluation of the model, an influence analysis of the observations was conducted using the DFbeta statistic. DFbeta measures the impact of each observation on the survival model coefficients. A high DFbeta value indicates a potentially influential observation.

Confidence intervals were then computed to quantify estimation uncertainty. The confidence interval 95% estimates a range in which the true parameter value is expected to lie with a given probability.

Finally, in order to determine the optimal threshold to maximise the effectiveness of injury diagnostic testing, the Youden index was applied. It is denoted as J (Equation 3). In this study, misclassification of a positive individual was considered more detrimental than that of a negative individual.

$$J = \text{Sensitivity} + \text{Specificity} - 1 \tag{3}$$

3. RESULTS

A total of 23 BABT impact tests were conducted on the SurHUByx surrogate, with all impacts located in the thoracic region. The output measurements from the sensors were normalised based on the BMI of the individuals. However, detailed individual characteristics were not consistently available in the case reports. Moreover, several reports did not specify whether lung injury was present, and cases where the impact occurred too far from the lung - and thus from the relevant sensor - were excluded.

To establish rib injury risk curves, 17 cases were considered. For pulmonary injury risk assessment, 12 cases were used. However, as there is only one case of injury in this study, generating risk curves for the lungs was not possible (Figure 4). The sample size across groups is uneven. Only one case represents the group “lesion”.

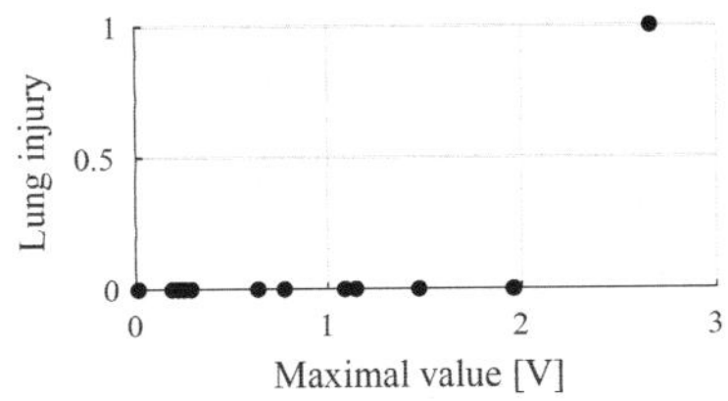

Figure 4. Lung data

https://doi.org/10.52202/081621-0009

3.1 Normalised rib injuries predictions

Figure 5 illustrates the distribution of data based on whether a rib injury was reported in the case. The "No injury" sample has on average lower maximum strain values. It is 0.27% compared to 0.54% for the "Injury" sample. However, the sample size across groups is uneven.

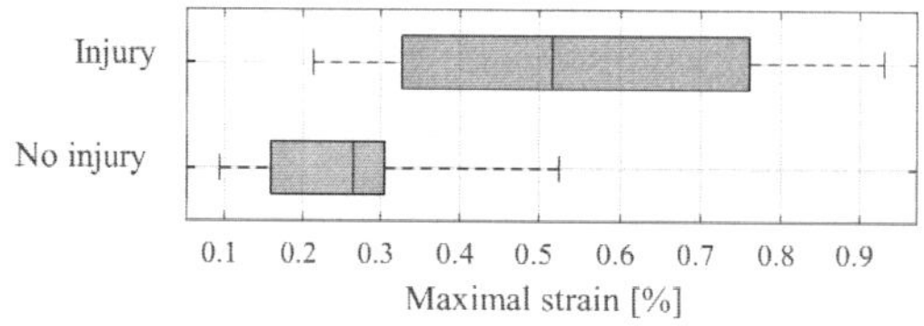

Figure 5. Boxplot for the normalised rib fracture data

The ROC curve was then plotted (Figure 6). The AUC for the maximum strain variable, after the data were normalising the data to BMI, was 0.7884. The decision threshold was determined to use the Youden index and the ROC curve. For the ribs, the threshold was 0.3462, with a specificity of 0.8462 and a sensitivity of 0.5.

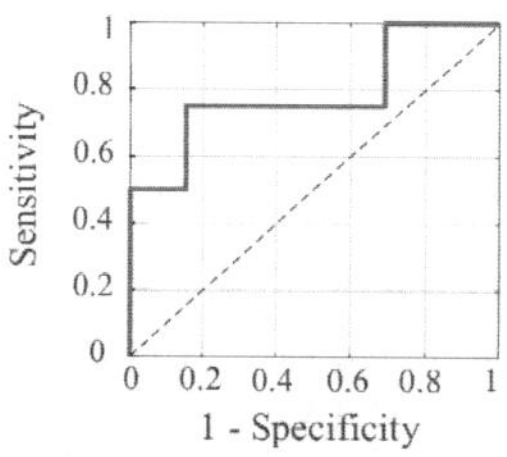

Figure 6. ROC curve of the ribs data

The Matlab routine output the inverse Kaplan-Meier survival curve (Figure 7a). Next, the three fitted models were plotted and analysed. Compared with the AIC value obtained in the analysis, the Weibull distribution was most appropriate in our study (Table 2). The risk of rib injury following a frontal ballistic impact is highlighted in Figure 7b which illustrates the results of the rib injury risk curve with the Weibull distribution. For a 50% risk of injury, the maximum rib strain was 0.65%.

Table 2. Injury risk curve final predictor and distribution selections with test statistics

Injury	Predictor	distribution	AIC
Ribs injury	maximum strain normalised with BMI	Log-normal	36.4
		Log-logistic	33.1
		Weibull	30.3

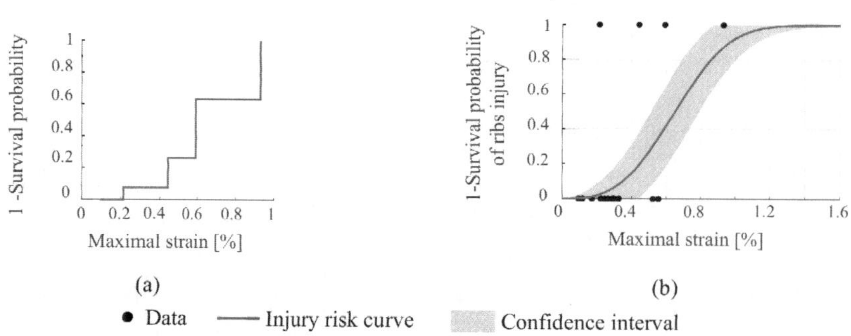

Figure 7. Survival curve (a) and best fit model injury risk curve for the ribs data (b)

4. DISCUSSION

One of the strengths of this study is the use of a physical surrogate for a male thorax to estimate the risk of certain BABT injuries. Using pressure transducers and strain gauges installed in the surrogate, experimental data were collected. These measured the stresses experienced during a ballistic impact on the SurHUByx. These data were then used for BABT injury probability analyses. However, generating distribution curves for the lung data was not possible. The inverse Kaplan-Meier survival curve had only one step (Figure 8), meaning that it cannot represent a Weibull, log-normal or log-logistic distribution. This limitation is due to the insufficient number of injury cases observed for this type of injury, making statistical estimates imprecise and the distribution curves unusable.

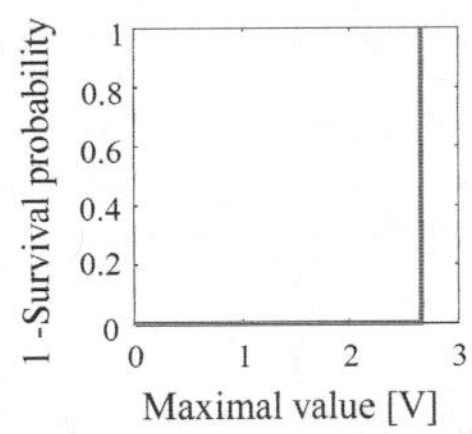

Figure 8. Survival curve

This experimental approach provided an ethical and reproducible alternative to testing on human or animal subjects [5]. However, the sensors integrated into the SurHUByx have certain limitations. For example, the sensors placed near the lungs are unidirectional and local, preventing accurate recording of impacts that do not occur in the sagittal plane. Furthermore, when an impact occurs at the tip of the lung, the distance between the sensor and the impact zone can lead to biased measurements.

The construction of injury risk curves from survival analysis is an approach widely used in the injury biomechanics [22] fields. These curves are beneficial, as they enable risk to be represented continuously as a function of an input variable [23]. This study draws on ISO recommendations and previous work [22] to establish injury risk curves. The approach adopted here aims to assess injury risks based on deformation predictions compared with observed injuries, in order to anticipate future injury probabilities.

Risk curves had already been established in the literature. McMahon et al. [2] assessed the risk of injury to the sternum, lungs and liver as a function of impact energy and penetration depth. Bass et al. [1] focused on the sternum (peak sternal impact), while Bir et al. [17] studied impact depth. However, comparison of risk levels with these studies is impossible due to the difference in the variables used. The SurHUByx does not allow quantification of the maximum depth reached on impact.

Next, the study took into account individual variability in case reports. The physical characteristics of the individuals mentioned in the reports were integrated with a scaling factor based on BMI. This BMI-based method seems to reduce morphology-related variations and reduce case exclusion. Chest depth was not used for normalisation, as suggested by the work of [17,24], as the data was not available from case reports. Normalised data from American and French ballistic reports was used to estimate the risk of rib injuries in men of average build. However, normalisation with BMI has certain limitations. For example, the mechanical properties of ribs vary according to an individual's size and morphology [25].

Finally, the collection of actual injury cases in the ballistics field remains difficult. The cases collected and analysed in this study constitute a significant database, with each reported case helping to refine BABT injury probability estimates. However, the total number of cases studied remains limited, particularly for lung injuries, where only one case of injury could be analysed in this study. This methodological weakness prevents the accurate modelling of a risk curve for this type of injury. A resampling approach, such as that proposed by DeVogel et al. [26] could be considered to balance the data classes and improve the robustness of the results.

https://doi.org/10.52202/081621-0009

5. CONCLUSION

This study established injury risk curves for rib fractures following blunt ballistic impact, using the biofidelic surrogate SurHUByx. The study highlighted the correlation between measured strain levels and rib fracture probabilities from known BABT injury cases, validating the approach based on integrated sensors. However, the results of lung injuries remain limited due to the small number of injury cases. Nonetheless, the case presenting a lung injury corresponds to the one where the piezoelectric sensor signal was the most important, which indicates a potential interest in this instrumentation to detect this type of injury.

The integration of databases of real BABT events has helped to improve the reliability of predictive models, although constraints remain, notably in terms of the morphological variability of individuals. Standardisation of data by BMI has reduced these disparities.

The perspectives of this research include the expansion of the BABT database to establish injury risk curves for all vital organs. Furthermore, since this study was based exclusively on male subjects corresponding to the 50th percentile, future work will need to take into account morphological diversity, particularly that of women, in order to improve the representativeness of predictive models.

Acknowledgments

This study was conducted as part of a PhD thesis research project supported by the French "Direction Générale de l'Armement" (DGA). The authors acknowledge the Wayne State University (WSU) for sharing BABT cases and the French Ministry of Interior for its financial and material support.

References

[1] Bass, C. R., Salzar, R. S., Lucas, S. R., Davis, M., Donnellan, L., Folk, B., Sanderson, E., & Waclawik, S. (2006). Injury Risk in Behind Armor Blunt Thoracic Trauma. *International Journal Of Occupational Safety And Ergonomics*, *12*(4), 429-442. https://doi.org/10.1080/10803548.2006.11076702

[2] McMahon, J., Berthelson, P., Eaton, M., Lorente, A., Leite, T., McEntire, B. J., & Salzar, R. (2024). Use of a Porcine Cadaver Model as a Human Surrogate for Behind Armor Blunt Trauma. *Journal Of Engineering And Science In Medical Diagnostics And Therapy*, *8*(4). https://doi.org/10.1115/1.4066934

[3] National Institute of Standards and Testing, "Ballistic resistance of body armor NIJ standard-0101.07," NIJ Standard-0101.07, 2024.

[4] Dos Santos, L. M. P., Da Cunha, M. R., Reis, C. H. B., Buchaim D. V., Da Rosa A. P. B., Tempest, L. M., Da Cruz, J. A. P., Buchaim, R. L., & Issa, J. P. M. (2025). The use of human tissue surrogates in anatomical modeling for gunshot wounds simulations: an overview about "how to do" experimental terminal ballistics. *Frontiers In Bioengineering And Biotechnology*. https://doi.org/10.3389/fbioe.2025.1536423

[5] Chaufer, M., Delille, R., Bourel, B., Marechal, C., Lauro, F., Mauzac, O., & Roth, S. (2023). A new biomechanical FE model for blunt thoracic impact. *Frontiers In Bioengineering And Biotechnology*, *11*. https://doi.org/10.3389/fbioe.2023.1152508

[6] Prat, N., Rongieras, F., De Freminville, H., Magnan, P., Debord, E., Fusai, T., Destombe, C., Sarron, J., & Voiglio, E. J. (2012). Comparison of thoracic wall behavior in large animals and human cadavers submitted to an identical ballistic blunt thoracic trauma. *Forensic Science International*, *222*(1-3), 179-185. https://doi.org/10.1016/j.forsciint.2012.05.022

[7] Gilson, L., Rabet, L., Imad, A., & Coghe, F. (2019). Experimental and numerical assessment of non-penetrating impacts on a composite protection and ballistic gelatine. *International Journal Of Impact Engineering*, *136*, 103417. https://doi.org/10.1016/j.ijimpeng.2019.103417

[8] Bracq, A., Delille, R., Maréchal, C., Bourel, B., Lauro, F., Roth, S., & Mauzac, O. (2021). On the use of a SEBS polymer gel block as a new ballistic target to assess blunt ballistic impacts: Application to a wide range of LLKE projectiles. *International Journal Of Impact Engineering*, *153*, 103874. https://doi.org/10.1016/j.ijimpeng.2021.103874

[9] Chaufer, M., Delille, R., Bourel, B., Marechal, C., Lauro, F., Mauzac, O., & Roth, S. (2023b). Advancements in numerical modelling of synthetic gel for ballistic impact assessment. *Materials Letters, 351*, 135077. https://doi.org/10.1016/j.matlet.2023.135077

[10] Chaufer, M., Delille, R., Bourel, B., Maréchal, C., Lauro, F., Mauzac, O., & Roth, S. (2024). Review of non-penetrating ballistic testing techniques for protection assessment: From biological data to numerical and physical surrogates. *Proceedings Of The Institution Of Mechanical Engineers Part H Journal Of Engineering In Medicine, 238*(4), 383-402. https://doi.org/10.1177/09544119241232122

[11] Sarhan, A. A. R., Franklyn, M., & Lee, P. V. S. (2023). The use of finite element models for backface deformation and body armour design: a systematic review. *Comput Methods Biomech Biomed Engin*. Nov 14:1 23. https://doi.org/10.1080/10255842.2023.2281275.

[12] Roberts, J. C., Biermann, P. J., O'connor, J. V., Ward, E. E., Cain, R. P., Carkhuff, B. G., & Merkle, A. C. (2005). Modeling Nonpenetrating Ballistic Impact on a Human Torso. *Johns Hopkins APL Technical Digest, 26*(1), 84-92. http://techdigest.jhuapl.edu/TD/td2601/Roberts.pdf

[13] Sherman, D., Rajaram, P., Davis, A., MacDonald, R., & Bir, C. (2023). Assessment of the f-BTTR for BABT injury risk prediction.

[14] Bir, C., Lance, R., Stojsih-Sherman, S., Cavanaugh, J. (2017). Behind Armor Blunt Trauma: Recreation of Field Cases for the Assessment of Backface Signature Testing. *DEStech Publications, Inc. ISBN 978-1-60595-419-6; 20*. doi :10.12783/ballistics2017/16912.

[15] Cronin, D. S., Bustamante, M. C., Barker, J., Singh, D., Rafaels, K. A., & Bir, C. (2020). Assessment of Thorax Finite Element Model Response for Behind Armor Blunt Trauma Impact Loading Using an Epidemiological Database. *Journal Of Biomechanical Engineering, 143*(3). https://doi.org/10.1115/1.4048644

[16] Wahl, P., Schreyer, N., Yersin, B. (2006). Injury pattern of the Flash-Ball®, a less-lethal weapon used for law enforcement: Report of two cases and review of the literature. *The Journal of Emergency Medicine*, 31(3), 325–330. doi: 10.1016/j.jemermed.2005.09.022.

[17] Bir, C., Viano, D., & King, A. (2003). Development of biomechanical response corridors of the thorax to blunt ballistic impacts. *Journal Of Biomechanics*, 37(1), 73-79. https://doi.org/10.1016/s0021-9290(03)00238-0

[18] Robbe, C. (2013). *Evaluation expérimentale de l'impact thoracique des projectiles non-létaux*. [PhD thesis]. University of Liège.

[19] Serre, T., Brunet, C., Bruyère-Garnier, K., Verriest, J. P., Mitton, D., Bertrand, S., Skalli, W. (2006). HUMOS (Human Model for Safety) geometry: From one specimen to the 5th and 95th Percentile. SAE Technical Paper Series. https ://doi.org/10.4271/2006-01-2324

[20] Eppinger, R. H., Marcus, J. H., & Morgan, R. M. (1984). Development of Dummy and Injury index for NHTSA's Thoracic Side Impact Protection Research Program. SAE Technical Papers On CD-ROM/SAE Technical Paper Series. https://doi.org/10.4271/840885

[21] International Organization for Standardization. Road vehicles - procedure to construct injury Risk Curves for the Evaluation of Occupant Protection in Crash Tests. Geneva, Switzerland: Tech. Rep. ISO/TS 18506. 2014.

[22] Yoganandan, N., Banerjee, A., Hsu, F., Bass, C. R., Voo, L., Pintar, F. A., & Gayzik, F. S. (2016). Deriving injury risk curves using survival analysis from biomechanical experiments. Journal Of Biomechanics, 49(14), 3260-3267. https://doi.org/10.1016/j.jbiomech.2016.08.002

[23] Hostetler, Z. S., Hsu, F., Yoganandan, N., Pintar, F. A., Banerjee, A., Voo, L., & Gayzik, F. S. (2020). An Improved Method for Developing Injury Risk Curves Using the Brier Metric Score. Annals Of Biomedical Engineering, 49(11), 3091-3098. https://doi.org/10.1007/s10439-020-02686-8

[24] M. Wilhelm. (2003). *A Biomechanical Assessment of Female Body Armor*. [Dissertation]. Wayne State University, Detroit, Michigan, USA, 2003.

[25] Holcombe, S. A., Wang, S. C., & Grotberg, J. B. (2017). The effect of age and demographics on rib shape. Journal Of Anatomy, 231(2), 229-247. https://doi.org/10.1111/joa.12632

[26] DeVogel, N., Banerjee, A., & Yoganandan, N. (2018). Application of resampling techniques to improve the quality of survival analysis risk curves for human frontal bone fracture. Clinical Biomechanics, 64, 28-34. https://doi.org/10.1016/j.clinbiomech.2018.04.013

https://doi.org/10.52202/081621-0009

Shot-to-shot consistency of trauma measurements on a BABT rig and Roma Plastilina #1 clay

D. B. Rahbek[1], G. Roberson[2], T. Thorvaldsen[1], A. Wien[2]

[1]Norwegian Defence Research Establishment (FFI), Instituttveien 20, NO-2007 Kjeller, Norway (dennis-bo.rahbek@ffi.no)
[2]NFM Technology, Glynitveien 15, 1400 Ski, Norway

Abstract. NFM Technology and the Norwegian Defence Research Establishment (FFI) are developing a Norwegian test rig for evaluating behind armour blunt trauma, the N-BABT rig. Part of the development process has been selecting a suitable material to use a surrogate chest wall. This membrane needs to exhibit both representative and consistent performance over consecutive impacts while showing minimal part and batch variation. Stereo high-speed cameras were used to capture the dynamic impact on the membranes. These images were analysed using digital image correlation (DIC). The shot-to-shot consistency and printing durability were evaluated using a 144 gram, 62 mm diameter smooth plastic sphere at 48 m/s, with nominally 50 impacts at the centre of each candidate membrane. Within this paper, we present the results of one candidate membrane and a baseline test series using Roma Plastilina clay #1 (RP1). The candidate membrane proved consistent, and the speckle print proved robust for impacts at the same impact point, giving low variations in measurements of maximum membrane displacement (D_{max}), velocity (V_{max}) and viscous criterion (VC_{max})
The membrane displacement was found to be affected by whether the membrane was straight or had a slight bend at the impact point prior to impact, however, neither V_{max} nor VC_{max} was affected by this. All three metrics were found to be sensitive to impact velocity. Normalising for membrane baseline bending and taking impact velocity into account, the shot-to-shot consistency (standard deviation) was as low as 0.9 % for D_{max} and V_{max}, and 2.3 % for VC_{max}. The back-face signature (BFS) values for the RP1 clay were found to correlate with clay temperature in the impact crater, while the sensitivity towards impact velocity was limited. The shot-to-shot consistency was found to be 3.1 % when accounting for temperature and impact velocity. Overall, the N-BABT rig was found to be robust for many consecutive impacts, and this with a very low inherent shot-to-shot variation.

1. INTRODUCTION

Most body armour ballistic test procedures use a witness material (often a modelling clay [1]) to support the body armour and provide a quantification of behind armour trauma, the back-face signature (BFS). BFS is a simple residual static measurement of a dynamic impact and does not accurately address the injury mechanisms of a blunt impact on the body [2]. The widely used Roma Plastilina #1 (RP1) clay is for instance temperature and work sensitive, and its rheological properties might change with time when not being regularly used [3]. The development of the temperature insensitive clay ARTIC [4] solves some of the issues, but it does not allow for capturing the dynamics of the impact event. Moreover, this will not resolve the lab-to-lab and technician-to-technician variability of this style of test procedure, that can affect the outcome of an impact to RP1 and other similar witness materials.

Several behind armour blunt trauma (BABT) rigs have been developed to allow dynamic measurements of impacts on body armour [5-8]. Common for many of these rigs is that they consist of a flexible membrane, made of a silicone or polyurethane, that mimics the response of the torso, i.e. provides a biofidelic response. Some membranes are validated against test on porcine [6], while others against data from tests on post-mortem human subjects (PMHS) [5,7,8]. Depending on the overall rig and set-up, some membranes are cylindrical, whereas others are flat.

There are also variations between the various BABT rigs with respect to the test parameter to be used as injury criteria. Both the membrane velocity and the viscous criterion of the membrane has been utilized. The viscous criterion is used to predict the risk of thorax skeletal injuries and expresses the product of the compression depth and the compression velocity of the torso, adjusted for the chest depth.

As part of the European Defence Agency Category B project CERAMBALL II, NFM Technology and the Norwegian Defence Research Establishment (FFI) have been developing a Norwegian version of a BABT rig, i.e. the N-BABT rig. The aim is to provide more informed advice on injury risk and improved capability to efficiently discern differences between protective materials and armour solutions based on injury criteria metrics.

In this paper, part of the membrane evaluation for the N-BABT rig is presented. The aim with the tests presented here is to evaluate if the properties of the membrane candidates would change as a result of repeated impacts, to assess the inherent variability of membranes and the data capturing methodology, and to compare this with similar tests performed on RP1 clay.

https://doi.org/10.52202/081621-0010

2. EVALUATION OF MEMBRANES

Several polyurethane chemistries and thicknesses have been evaluated in the development of the rig membrane in order to evaluate the stability, performance and endurance of the material and the applied printing method for the speckle pattern needed for the digital image correlation (DIC). The objective is to find a membrane with biofidelic properties, with similar properties as reported by Bir et al. [9] and in STANREC 4744 AEP-99 [10]. Comparisons to the biofidelity criteria in AEP-99 indicates that the membranes are slightly too soft. This will be addressed in next iterations of the set-up.

Three white polyurethane membranes, M30-1, M30-2 and M30-3, were investigated in this work. A speckle pattern was printed on the membrane surface, see Figure 1(b). Several iterations with printing different techniques were needed to obtain a robust print. The speckle print is required to enable the utilisation of DIC, which provides a global noncontact measurement methodology to capture the dynamic impact event. The nominal size of the membranes is 650 mm x 650 mm x 30 mm, with a printed area of 500 mm x 500 mm.

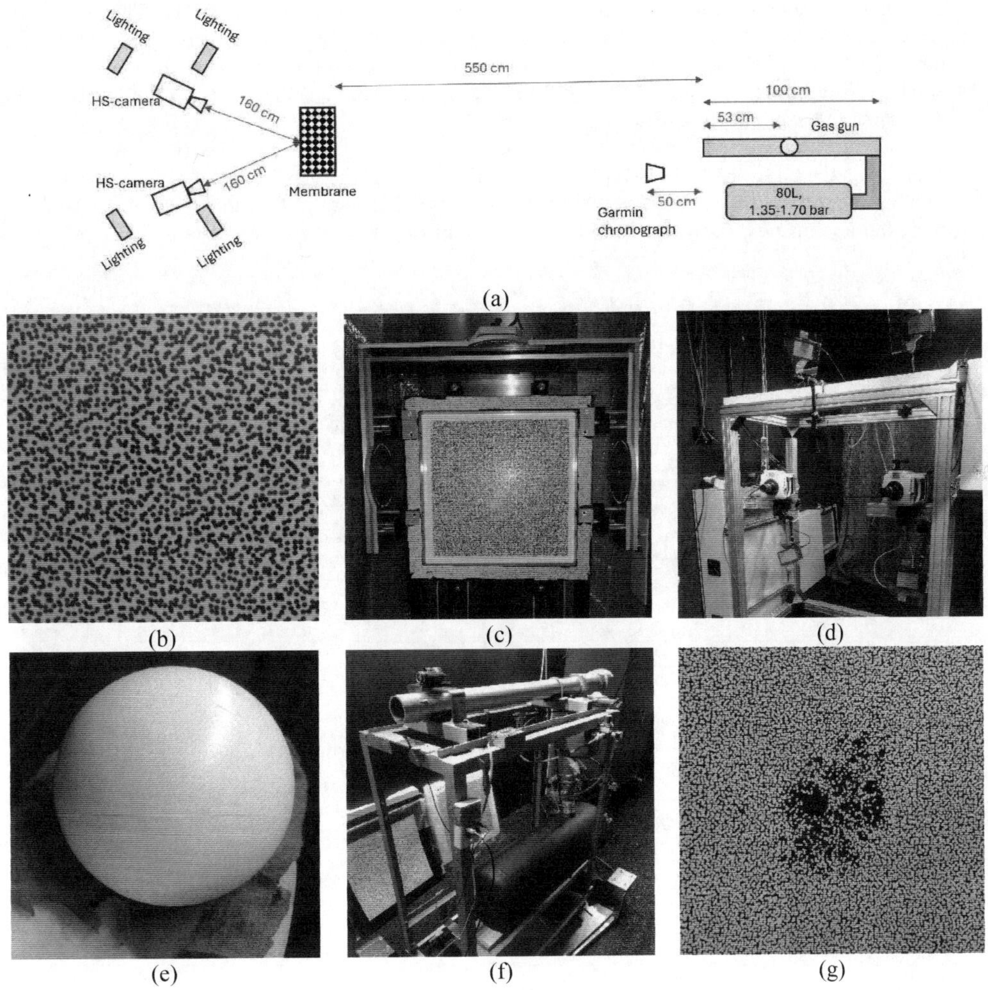

Figure 1. **(a)** The layout of the experimental set-up for the tests on the membranes. The N-BABT rig: (b) speckle print on a membrane, (c) a membrane in the membrane fixture as seen from the camera position, and (d) the frame with two high-speed cameras and four LEDs, (e) nylon ball with cotton gun patch to create a gas seal (f), the gas gun used to propel the nylon ball and (g) an example of damage to a poorly performing print.

3. CURRENT N-BABT SET-UP AND EXPERIMENTAL METHOD

https://doi.org/10.52202/081621-0010

The impact testing on the N-BABT rig was performed at NFM Technology. An overview of the set-up is shown in Figure 1(a). The membrane was clamped in a rigid metal frame, Figure 1(c), which shows the membrane from the camera side. In order to capture the impact event, two high speed cameras, Phantom V2512, were utilized together with four high power LEDs for illumination, Figure 1(d).

To evaluate the inherent properties of the N-BABT rig and the RP1 clay, nylon balls with a diameter of 62 mm and a mass of 144.17 g (with a 0.14 g standard deviation), Figure 1(e), were employed. The desired impact velocity was 48 m/s since the momentum of 6.9 kg·m/s at this velocity is very close to the momentum of 6.8 kg·m/s for the NIJ HG2 threat .44 MAG JHP, 240 grain at 436 m/s, as given in the NIJ Standard 0123.00 [11]. This was expected to give a rather large deformation, which is useful for evaluating if the membrane and print could withstand many impacts. The nylon ball was propelled using a gas gun, shown in Figure 1(f). The velocity was measured using a Garmin Xero C1 Pro radar chronograph. The distance from the gas gun muzzle to the target front, clay or membrane, was 550 cm. Light-screen chronographs were found to be inconsistent and unreliable at low velocities.

3.1 Tests on the N-BABT rig

For the three membranes, the impact position was at the centre position of the membrane. The high-speed cameras were placed 160 cm from the back side of the membrane, Figure 1(a). The cameras resolution was 768 pixels x 768 pixels with a field of view 30 cm x 30 cm around the centre on the "body side" of the membrane. The frame rate was 39100 fps. To find the membrane displacement, all the sets of image pairs were analysed using the VIC-3D DIC software.

The membrane displacement was filtered using a CFC 1000 standard time filter [12], as recommended in the STANREC 4744 AEP-99 [10], and from this the displacement at the impact point, D(t), was found. The membrane velocity, V(t), was found by deriving D(t) with respect to time. Using an average thorax thickness (D0) of 0.236 m, the maximum viscous criterion was computed as explained in [10]

$$VC_{max} = \max((D(t)/D0)*V(t)) \qquad (1)$$

An example of the time evolution of the membrane deflection, velocity and viscous criterion of an impact event is presented in Figure 2. The maximum displacement (D_{max}), the maximum velocity (V_{max}) and maximum viscous criterion value (VC_{max}) was then found for each impact event.

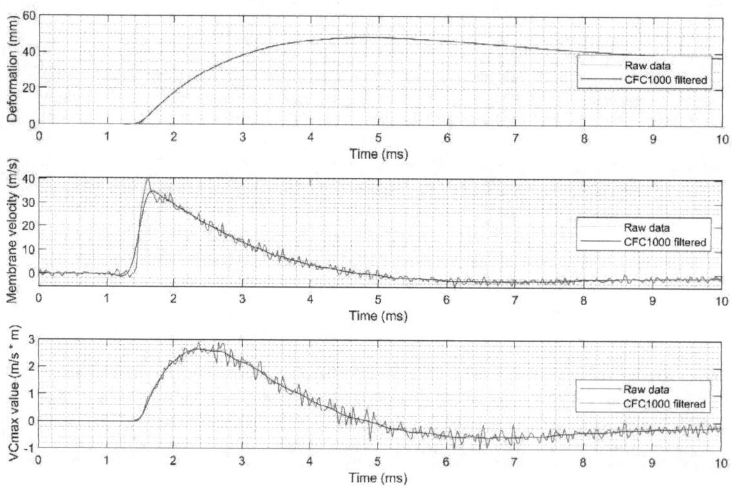

Figure 2. Example of the time evolution of the raw data (red) and the filtered (blue) outcome of an impact on a membrane (not a M30 membrane, but one discarded membrane). Membrane displacement is on top, velocity in the middle and viscous criterion on the bottom. The difference between the raw and filtered data is small for the deformation, but becomes clear when derived with respect to time, as is the case for the velocity.

3.2 Tests on Roma Plastilina No. 1

For the tests on RP1 clay, a test procedure based on the NIJ Standard 0101.06 [1] was applied. Two clay boxes, A and B, were evaluated, both used on almost daily basis. The boxes were kept in an oven and heated to 33 °C. Before use, a drop tests with a 1043 g steel sphere with diameter 63.5 mm was performed. The average drop depth was 18.4 mm for box A, and 20.4 mm for box B.

For the impact testing, the set-up was the same as for the N-BABT, exchanging a clay box for the membrane and without cameras and LED lights. The impact position was in the centre of the clay box. A single 4 ply sheet of soft armour grade ultra-high molecular weight polyethylene (UHMWPE) was placed on the clay before impact, Figure 3(a). This allowed the removal of the ball from the clay without damaging the BFS. Post impact, the UHMWPE sheet and ball were embedded in the clay, Figure 3(b), but was easily removed without disturbing the impact crater, Figure 3(c). Excess clay was scraped off, and the BFS depth was measured using a vernier depth gauge.

Following the measurement of BFS, an infra-red (IR) image was captured of the clay box utilising a FLIR E6 Pro camera. The clay temperature was also measured inside the impact crater at a depth of 10 mm below the surface using a thermocouple thermometer. The clay box was then repaired with clay from the oven, and a new IR image was taken prior to the next test shot. The before and after impact IR images of the clay surface temperature were used to measure the temperature in the crater, which combined with the thermocouple measurements provided an indication of the effects of cooling and locally imparting work into the clay on the BFS. All temperature measurements were performed before adding the UHMWPE layer or after removing the layer. The UHMWPE layer is not expected to affect the temperature of the clay significantly.

(a) (b) (c)

Figure 3. (a) The clay box with the UHMWPE interlayer in position, (b) the ball has impacted the interlayer material and is stuck, and (c) the indentation in the clay after removal of the interlayer.

3.3 Test summary

Table 1 provides an overview of the total number of tests performed for each type of target. There were no velocity readings for the membrane M30-2. Moreover, a number of tests was conducted with the other two membranes, but without velocity readings. The average impact velocities were in the range from 47.2 m/s to 49.1 m/s. The impact velocity was slightly higher for the membrane test, while the standard deviation was slightly lower for the RP1 tests.

Table 1. Overview of the number of tests performed for each type of target.

Type	Target	Accepted tests	Tests with velocity reading	Mean velocity (m/s)	Std. dev. of velocity (m/s)
RP1	Box A	17	17	47.8	0.3
RP1	Box B	38	38	47.2	0.5
Membrane	M30-1	46	39	48.6	1.1
Membrane	M30-2	46	0	-	-
Membrane	M30-3	50	28	49.1	0.6

https://doi.org/10.52202/081621-0010

4. MEMBRANE RESULTS

4.1 Effect of impact number

For the N-BABT rig to be useful for many tests, the membrane performance should not vary significantly during the course of the membrane life. Figure 4 shows the performance for the first 50 impacts. The results for V_{max} and VC_{max} are consistent for all impacts. There are a few impacts with lower values for M30-1, shots 5-10, but this is explained by slightly lower impact velocities, as will be described in section 4.3.

For D_{max}, the results are stable for each individual membrane, especially after around 10-15 shots. The D_{max}, however, appears slightly smaller for the first 10-15 shots. This could possibly be related to the Mullins effect, which describes a softening of elastomers during the first load cycles. Additionally, the displacements on M30-2 are consistently smaller than for the other two, in principle identical, membranes. This was found to be caused by a slight bending of the membrane prior to the impact.

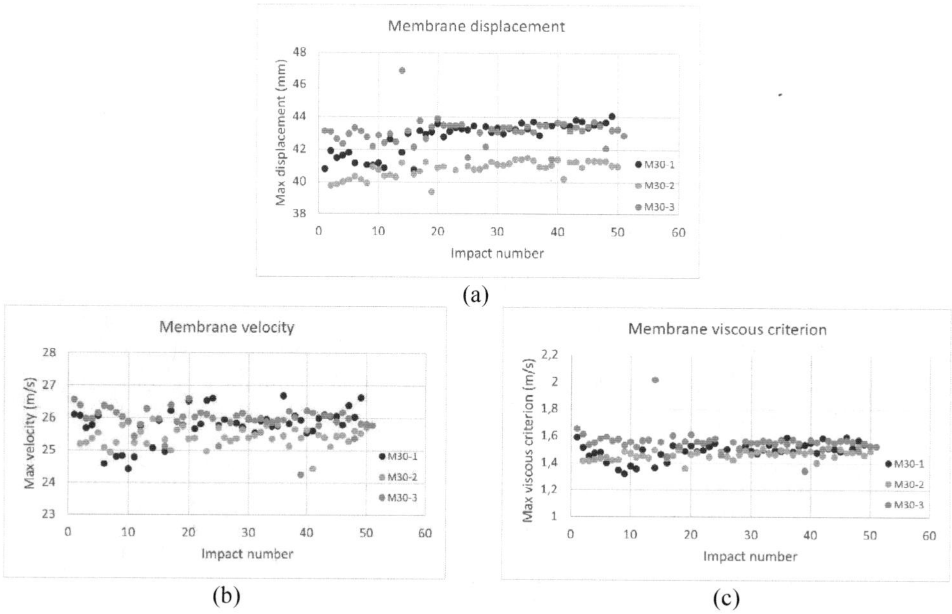

Figure 4. Results for the three membranes. (a) D_{max}, (b) V_{max}, and (c) VC_{max}.

4.2 Effect of initial membrane position

A metal frame is used to mount and secure the membrane in the test position. When the membrane is fixed in the metal frame, the frame is placed horizontally on a table, and the membrane is placed in the fixture while being supported by the table. This is done in order to keep the membrane as flat as possible with respect to the impact direction, as illustrated in Figure 5(a), where the arrow depicts the Z-plane (out-of-plane direction) of the DIC analysis, as well as the direction of the ball. That is, the membrane deformation is in the positive Z-direction.

It was, however, observed during the data analysis, that there was a slight variation in the starting position of the membrane for the different impacts. The membrane Z-position ranged from roughly -1 mm to -3 mm at impact, Figures 5(b)-(d). This variation in starting position is not due to overall movements of the membrane in the frame, but due to the membrane is beding slightly towards or away from the incoming ball. The negative Z-position means that the membrane is bending slightly towards the incoming nylon ball and away from the cameras, Figure 5(a).

This bending can affect the outcome in two ways. First, the displacement is calculated with the membrane starting position as a zero-point reference. Second, because the initial bending is towards the ball, the membrane can be displaced more with respect to the starting point before it reaches the same level of tension as if the membrane was not bending.

The max displacement and the initial Z-position of the membrane impact point, Figure 5(b), is clearly correlated. This correlation partly explains the slightly smaller displacements observed for the first 10-15 impacts.

For the V_{max} and the VC_{max}, the influence of the initial Z-position appears much smaller. This is likely because both the velocity and the viscous criterion of these measures reach their maximum value long before the max displacement is reached, Figure 2.

In order to assess the influence of the Z-position on the shot-to-shot variation, the data was adjusted for Z-position by fitting a straight line to the data shown in Figures 5(b)-(d) and then adjusting for the slope of this line. The data are presented in Table 2.

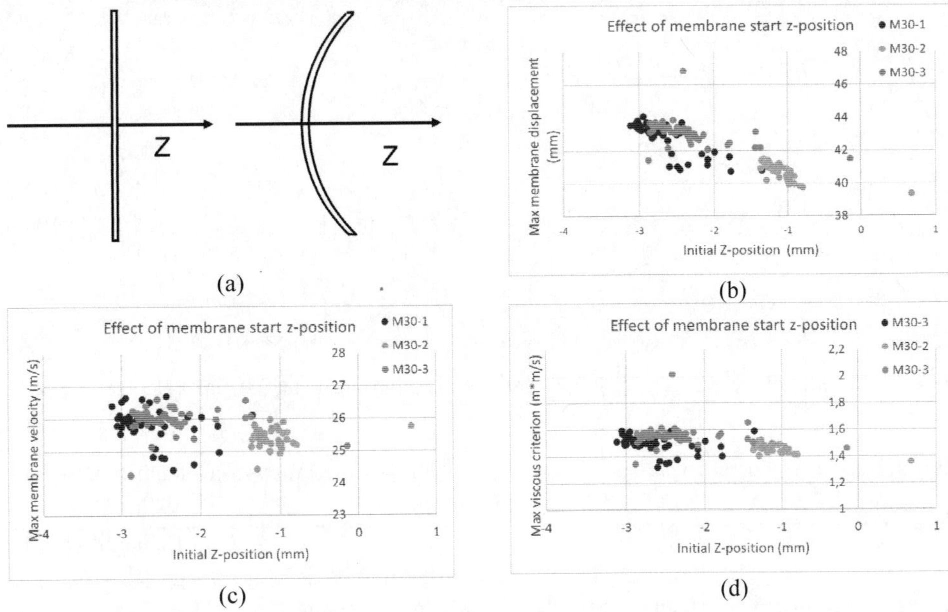

Figure 5. (a) Initial Z-position of the membrane – to the left the membrane is straight and starts in the Z = 0 position, while to the right the membrane is curved against the impact side, and it starts with a negative Z-position. Z-position effect on: (b) D_{max}, (c) V_{max}, (d) VC_{max}.

4.3 Effect of impact velocity

Despite not capturing the impact velocity for all membrane impacts, it is of interest to evaluate how large portion of the shot-to-shot variation arises from the impact velocity variations. In order to evaluate this, the membrane data (adjusted for the initial Z-position, as described in the previous section, and only for the impacts with velocity measurements), have been plotted against the impact velocity in Figures 6(a)-(c).

It is evident from the figures that all the outcome parameters, D_{max}, V_{max} and VC_{max}, correlate with the impact velocity. Given that we want the rig to be sensitive to impact conditions, such as velocity, this is positive. The effect of the velocity on the shot-to-shot variation has been estimated in the same way as for the Z-position, by fitting a straight line to the data and adjusting for the data for the slope. The resulting standard deviations are given in Table 2.

In Table 2, the mean value for D_{max}, V_{max} and VC_{max} are presented for membranes M30-1 and M30-3 for the shots with velocity measurements, and for the two membranes combined as one. The values for the two membranes are very similar for all three measures. From a testing perspective, this is positive since this indicates that it should be possible to compare data from impacts performed on different membranes (that are meant to be identical).

Table 2 also includes the relative standard deviation of the mean, as well as the relative standard deviations when the data are adjusted for the initial Z-position and velocity. Overall, the relative standard deviations are small, from 1.6 % for unadjusted V_{max} to 3.9 % for unadjusted VC_{max}. When adjusted for Z-position, the relative standard deviation drops from 1.9 % to 1.4 % for D_{max}, while the V_{max} and VC_{max} are essentially unaffected. Adjusted for impact velocity the relative standard deviation for both D_{max} and V_{max} is 0.9 %.

 https://doi.org/10.52202/081621-0010

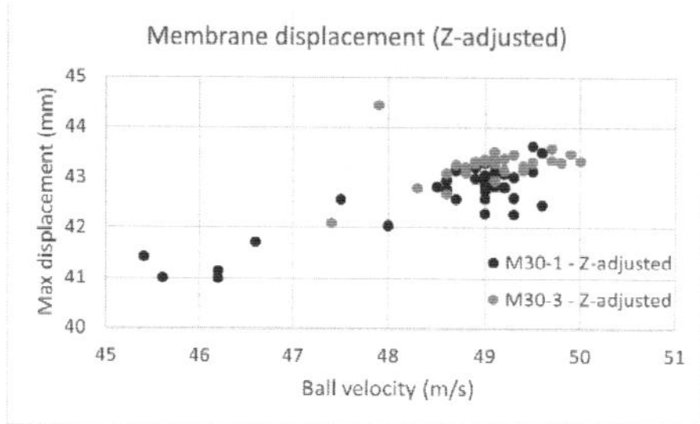

(a)

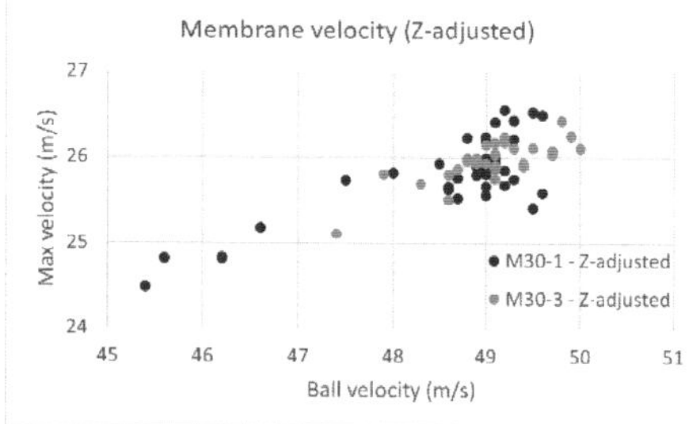

(b)

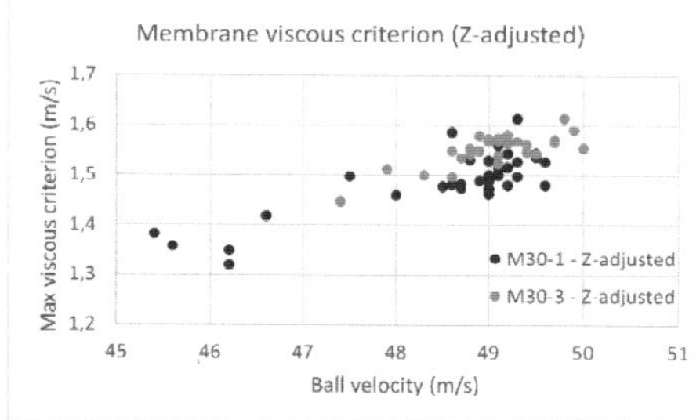

(c)

Figure 6. Effect of ball velocity on the membrane: (a) max displacement, (b) max velocity, (c) VC_{max}.

Table 2. Relative standard deviations for D_{max}, V_{max} and VC_{max} criterion for the membranes, M30-1 and M30-3, with and without adjusting for initial Z-position and impact velocity.

				Z-position adjusted		**Z-position and velocity adjusted**	
Output	**Membrane**	**Mean**	**Std. dev.**	**Mean**	**Std. dev.**	**Mean**	**Std. dev.**
D_{max}	M30-1	42.8 mm	2.3 %	42.7 mm	1.5 %	42.8 mm	0.8 %
	M30-3	43.1 mm	1.2 %	43.3 mm	0.9 %	43.1 mm	0.9 %
	Combined	**42.9 mm**	**1.9 %**	**42.9 mm**	**1.4 %**	**42.9 mm**	**0.9 %**
V_{max}	M30-1	25.8 m/s	1.9 %	25.8 m/s	1.9 %	25.9 m/s	1.1 %
	M30-3	25.9 m/s	1.1 %	26.0 m/s	1.0 %	25.9 m/s	0.6 %
	Combined	**25.9 m/s**	**1.6 %**	**25.9 m/s**	**1.6 %**	**25.9 m/s**	**0.9 %**
VC_{max}	M30-1	1.49 m/s	4.2 %	1.49 m/s	4.1 %	1.50 m/s	2.4 %
	M30-3	1.55 m/s	2.3 %	1.55 m/s	2.1 %	1.54 m/s	1.4 %
	Combined	**1.52 m/s**	**3.9 %**	**1.52 m/s**	**3.9 %**	**1.52 m/s**	**2.3 %**

5. ROMA PLASTILINA RESULTS

5.1 Effect of impact number

The BFS values for the impacts are presented in Figure 7. The BFS values are scattered around 48 mm, from roughly 44 mm to 52 mm. The BFS for the two clay boxes appear to be similar. This means that tests performed with body armour on the two clay boxes can be compared. There does not appear to be a clear trend for the BFS with respect to impact number. One concern is that the clay temperature would change during the testing, as the laboratory room temperature is below the oven temperature. However, given that the impact position was the same for all shots, and that new clay from the oven was used to fill the crater, the temperature was fairly stable in that region during testing.

https://doi.org/10.52202/081621-0010

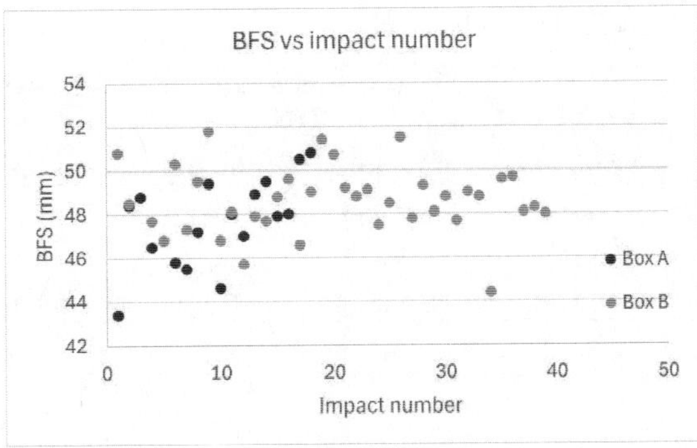

Figure 7. BFS values as function of impact number for the two clay boxes, A and B.

5.2 Effect of clay temperature

Of the measured temperature data, the BFS and the IR-based surface temperature after impact showed the highest degree of correlation (the trends were the same for the other measures).

Figure 8(a) shows the BFS values as a function of the clay surface temperature after impact for both clay boxes. The temperature span was larger for box A than for box B. This is most likely explained by the fact that box A was used in the first tests, and that the measurement routines were more established for box B. The test frequency was around 7 min for box A and 5 min for box B. The data show that the BFS is higher when the clay temperature after testing is higher. It is also clear that the BFS varies for the same temperature.

The RP1 BFS data have been adjusted for temperature by fitting a straight line to the data and adjusting for the slope. The relative standard deviations are given in Table 3.

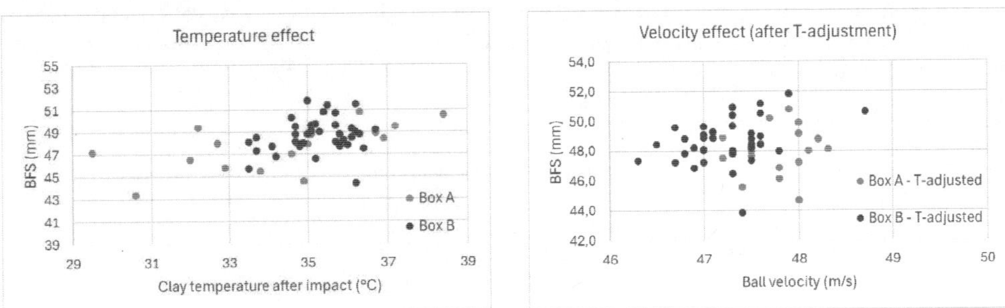

Figure 8. (a) BFS values as a function of temperature inside the impact crater after impact and (b) BFS values (adjusted for clay temperature) as a function of impact velocity.

5.3 Effect of impact velocity

The effect of impact velocity on the BFS values (temperature-adjusted) are shown in Figure 8(b). There is a slight tendency towards higher BFS values for higher velocities, but the data scatter for identical impact velocities is large compared to the variation in impact velocity, 0.7 % and 1.0 % for box A and B (based on Table 1), respectively.

The results in Table 3 show that the mean BFS values for the two boxes are not very different, as Figure 7 also showed. The relative standard deviation for the unadjusted data was quite low, 4.2 % and 3.2 % for box A and B (on the order of 1.5-2.0 mm for the measured BFS values). When adjusted for the temperature, the relative standard deviation became lower for box A, where the temperature ranges were the largest, while there was no substantial effect for box B. The adjustment for impact velocity only lowers the relative standard deviation slightly, which means that for these tests, the temperature and impact velocity was well controlled, and that the inherent shot-to-shot variation was on the order of 3 %. It is worth noting here, that the use of the UHMWPE layer placed on the clay box for making ball extraction easier could have contributed to the variation from shot-to-shot.

https://doi.org/10.52202/081621-0010

Table 3. BFS results both with and without adjusting for clay temperature and impact velocity.

	BFS		Temperature adjusted BFS		Temperature and velocity adjusted BFS	
Clay box	**Mean**	**Std. dev.**	**Mean**	**Std. dev.**	**Mean**	**Std. dev.**
Box A	47,7 mm	4,2 %	48.0 mm	3.4 %	47.7 mm	3.4 %
Box B	48.6 mm	3.2 %	48.5 mm	3.1 %	48.6 mm	2.9 %
Combined	**48.3 mm**	**3.6 %**	**48.4 mm**	**3.2 %**	**48.3 mm**	**3.1 %**

6. DISCUSSION

The data from the membrane and the RP1 clay testing is summarised in Figure 9. The relative standard deviations for the combined data for the RP1 and for the membranes are compared with and without adjusting for temperature/Z-position and impact velocity. The data shows that the shot-to-shot variation of the membrane are lowest for D_{max} and V_{max}. It is roughly half of the clay variation and the VC_{max} variation. The higher VC_{max} variation is not surprising since it is the product of the other two measures. The shot-to-shot variation of VC_{max} and the RP1 clay is similar. The VC_{max} value was essentially insensitive to variation in the initial Z-position (bending) of the membrane, something that could prove problematic from an experimental point of view if it had. We do not plan on correcting for initial Z-position in the future.

Moreover, the VC_{max} was found to be velocity sensitive, which is not surprising since it is the product of the velocity and the displacement, which were both found to be velocity sensitive. Given that the VC_{max} value is the value that is meant to inform about the risk of injury, it is positive that the shot-to-shot variation of this parameter is comparable to the existing test method, RP1 clay, if not even slightly better.

For further analysis, it would be interesting to compare the correlation between D_{max}, V_{max} and VC_{max} for identical impacts. If the correlation is strong, then for instance the V_{max} value, which has low Z-position sensitivity and displays a low shot-to-shot variation, might be a useful measure in for instance product development, where a small variation would allow for more accurate measurement with fewer shots.

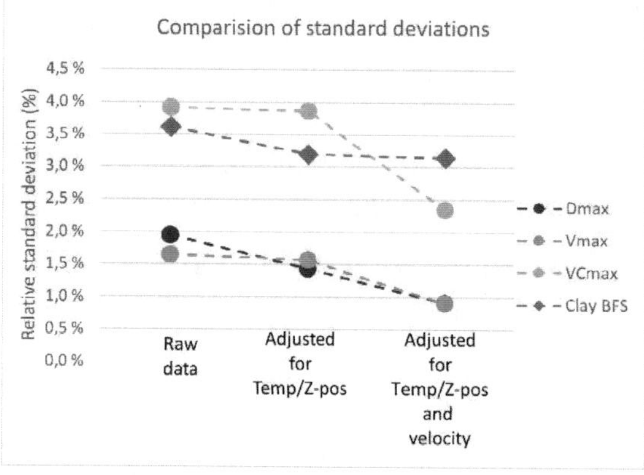

Figure 9. Relative standard deviations for the membrane data: D_{max}, V_{max}, VC_{max} compared with clay BFS. The raw data are compared to the data adjusted for temperature/Z-position and velocity.

7. CONCLUSION

The current state of the N-BABT rig has been described together with experimental tests performed to evaluate the endurance and inherent variation of the rig. The main objective with the testing was to evaluate 1) whether the selected membrane material would give consistent results for a substantial number of impacts, 2) if the three tested and "identical" membranes would yield identical results, and 3) that the variation from shot-to-shot was not inferior to the commonly used clay witness material, i.e. Roma Plastilina #1.

The results showed that the membrane-to-membrane variation was small, that the membranes showed an overall low shot-to-shot variability with respect to all measures, D_{max}, V_{max} and VC_{max}, with relative standard deviations of 1.9 %, 1.6 % and 3.9 %, respectively. For the RP1 clay the value was 3.6 %. It was also observed that the initial position of the membrane (caused by bending of the membrane) had a substantial effect on the measured D_{max}, but not on V_{max} and VC_{max}. When the data was adjusted for initial membrane bending and impact velocity, to assess the inherent variability, the relative standard deviations for the measures from the membrane was 0.9 %, 0.9 % and 2.3 % for D_{max}, V_{max} and VC_{max}, respectively. For the RP1 clay, the adjustment with respect to clay temperature and impact velocity resulted in an estimate for the inherent relative standard deviation of 3.1 %.

Acknowledgments
The authors would like to acknowledge Rune Mellem, Sondre Gabrielsen and August B. Strengen, NFM Technology, for their assistance during the experiments. Susanne Thomesen and Toivo Horvei, NFM Technology, are thanked for valuable comments and discussions, and Bernt B. Johnsen, FFI, is acknowledged for proofreading the manuscript.

References

[1] NIJ Standard 0101.06, Ballistic Resistance of Body Armor, National Institute of Justice, July 2008.

[2] Cannon L., Behind Armour Blunt Trauma - an emerging problem, J R Army Med Corps, 147 (2001), 87-96.

[3] Gardner J., Cora Cruz J., Mrozek R., Hopping J., Beaudoin E., Lombardo L., Ballistic Assessments of the ARL Reusable Temperature Insensitive Clay (ARTIC) as a Ballistic Backing Material, Proceedings of the Personal Armour Systems Symposium, Dresden, Germany, 2023, pp. 208-217.

[4] Roberson C., Perciballi W., Rogers D., Fleming D., Consistently Depressed?, Personal Armour Systems Symposium, Quebec, Canada, 2010

[5] Anctil, B., Bayne, T., Bourget, D., Pageau, G., Binette, J.-S., Rice, K., and Toman, A., An Alternative to Plastilina for Evaluating the Performance of Body Armours, Personal Armour Systems Symposium, Brussels, Belgium, 2008.

[6] Tam, W., Rozant, O., Thoral-Pierre, K., Pope, D., Softley, I., Baker, L., Cook, W., Scaramuzzino, P., and Ferry, E. The UK Behind Armour Blunt Trauma Rig – 20 years on, Proceedings of the Personal Armour Systems Symposium, Amsterdam, Netherlands, 2016.

[7] Andrei A., Robbe C., Penders, T., Papy, A., Development of a new thoracic surrogate for KENLW and BABT impacts, Proceedings of the Personal Armour Systems Symposium, Dresden, Germany, 2023, pp. 321-340.

[8] N. Shewchenko, E. Fournier, T. Bayne, S. Magnan, and D. Bourget, The Development of the f-BTTR and its use for Hard Armour Testing, Proceedings of Personal Armour Systems Symposium, Copenhagen, Denmark, 2020, pp. 481-490.

[9] Bir C., Viano D., King A., Development of biomechanical response corridors of the thorax to blunt ballistic impacts, J. Biomech, 37 (2204), 73-79.

[10] NATO, STANREC 4744 edition 5 - AEP 99 ed: b ver 1, Thorax injury risk assessment of non-lethal projectiles, 2021.

[11] NIJ Standard 0123.00, Specification for NIJ Ballistic Protection Levels and Associated Test Threats, National Institute of Justice, October 2023.

[12] SAE, Part 1 - Electronic Instrumentation, in Surface Vehicle Recommended Practice, SAE-J211-1 Instrumentation for Impact Test, Warrendale, PA, USA, The Engineering Society For Advancing Mobility Land Sea Air and Space International, 1995.

 https://doi.org/10.52202/081621-0010

Study of the influence of the padding in the helmet's impact response

A. Miranda-Vicario[1], J. Martín-de León[2,3], M. del Val-Armesto[4] , I. Sánchez-Calderón[4], F. Lizalde-Arroyo[2,3], F. Coghe[1] and M.A. Rodríguez-Pérez[2,3]

[1] *Royal Military Academy, Department of Weapon Systems and Ballistics, Renaissance Avenue, 30, B-1000 Brussels, Belgium, angel.mirandavicario@mil.be*

[2] *CellMat Laboratory, Universidad de Valladolid, Paseo de Belén 7, 47011 Valladolid, Spain*

[3] *BioEcoUVA Research Institute on Bioeconomy University of Valladolid, Calle Dr. Mergelina, 47011 Valladolid, Spain*

[4] *CellMat Technologies S.L, C. del Argón 1, Valladolid 47012, Spain*

Abstract. Upon a ballistic impact on a helmet, the shell deforms to absorb the kinetic energy from the projectile. This deformation plays a crucial role in reducing the direct penetration of the projectile, but it can also transfer some energy to the wearer's head, which may lead to injuries. These injuries, commonly referred to as Behind Helmet Blunt Trauma (BHBT), occur when the force of the impact is transmitted to the skull without penetration, causing damage to the brain or surrounding tissues. To mitigate this risk, helmets are designed with a gap between the skull and the helmet. This gap not only serves as a protective buffer against blunt trauma but also enhances comfort and ergonomics for the user. The gap between the skull and the helmet can be filled with various materials, typically liners or padding. These inner components are critical in distributing the impact energy more evenly, reducing the risk of localized injuries. Different materials and configurations can have significant effects on the helmet's performance in a ballistic impact scenario.

This study explores different types of cellular polymers, designed in the CellMat Laboratory at the University of Valladolid, used as the inner padding of the helmet and evaluates their effectiveness in reducing blunt trauma compared to commercial solutions. The focus is on assessing how these materials respond when subjected to a ballistic impact from a 9 x 19 mm Full Metal Jacket (FMJ) projectile. In terms of forces transmitted to the user, a cellular material made of thermoplastic polyurethane (TPU) was able to produce similar results than a commercial solution, showing the path to enhance current solutions.

1. INTRODUCTION

Generally, the helmet has to protect from high-mass, low-velocity impacts, such as in sports or automotive fields. The literature has already shown solutions that are able to reduce the load for the user [1], [2], [3]. In order to avoid injury from ballistic impacts, to the head, it is required to stop the threat typically using the shell of the helmet, and reduce the associated load for the user, known as Behind Helmet Blunt Trauma. Some studies have addressed this issue. The main threat studied is the 9 x 19 mm Full Metal Jacket (FMJ) [4], [5], or steel projectiles at low velocities [6], [7]. Nowadays higher levels of ballistic protection are considered for ballistic helmets. More powerful threats, such as rifle rounds have been studied, despite the increasing difficulty of defeating them within the constraints of a personal protection system [8]. The helmet has to dissipate the energy to defeat the ballistic threat and keep the user safe. The liner plays an important role as it has to accommodate the load of the impact transmitted to the skull.

Cellular polymers are biphasic materials presenting a solid phase and a porous phase, when the pore size is reduced up to the micrometric or nanometric scale, these materials are known as micro and nanocellular polymers. The reduction of the pore size to these scales allows these materials to present an innovative combination of properties such as a reduction in weight together with excellent mechanical performance [9].

The goal is to study for the first time the response of these micro and nanocellular polymers as liners for protecting the head from the deforming cone of the composite armour in comparison to commercial liners. As there are no clear biomechanical limits, a standard military helmet was mounted on a headform, and was fired at with pistol ammunition in order to estimate the values of the current loads. After this, different liner configurations were tested and compared with the previously determined loads for the helmet.

https://doi.org/10.52202/081621-0011

2. EXPERIMENTAL SET-UP

2.1 Ballistic test set-up

In order to perform the ballistics tests, a universal receiver with interchangeable barrel was used to fire the projectiles. The projectile velocity was measured with a double optical base, DRELLO LS19, mounted on a frame. The target was positioned 5 m ahead. The optical bases were positioned 2.5 m in front of the target (Figure 1). The measurement of the velocity was corrected to take into account the deceleration of the projectile from the centre of the bases to the impact point [10].

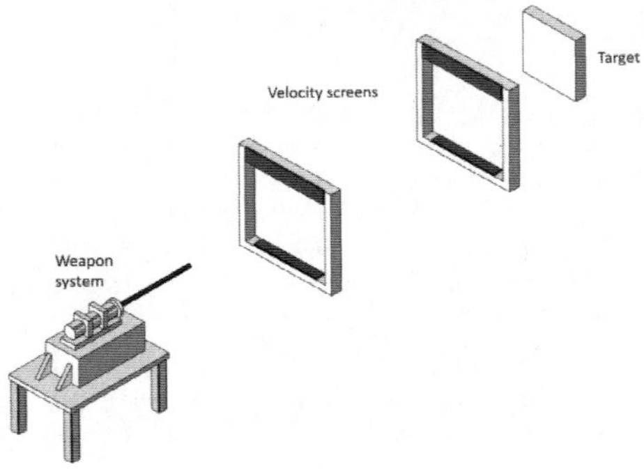

Figure 1.- Ballistic test setup.

2.2 Ammunition

The 9 x 19 mm FMJ projectile (Figure 2) weighs 7.9 g. It has a brass jacket and a lead alloy core. This pistol round is very popular and is specified in several standards as a testing round from organisations such as the Association of Testing Bodies for Attack Resistant Materials and Constructions - best known by its acronym VPAM (Vereinigung der Prüfstellen für angriffshemmende Materialien) and the National Institute of Justice (NIJ), and the Home Office Scientific Development Branch (HOSDB) ([11], [12], [13]).

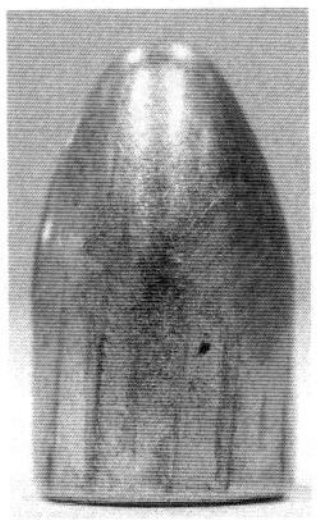

Figure 2.- 9 x 19 mm FMJ projectile.

2.3 Ballistic Load Sensing Headform

The surrogate used for testing was the Ballistic Load - Sensing Headform (BLSH) [14]. This system consists of a metallic headform equipped with different force and acceleration sensors to assess the ballistic impact protection performance of helmets (Figure 3). The headform enables a direct measurement of the dynamic loads imparted to the head by the deformation of a ballistic helmet caused by non-penetrating projectiles, and the global acceleration of the head.

https://doi.org/10.52202/081621-0011

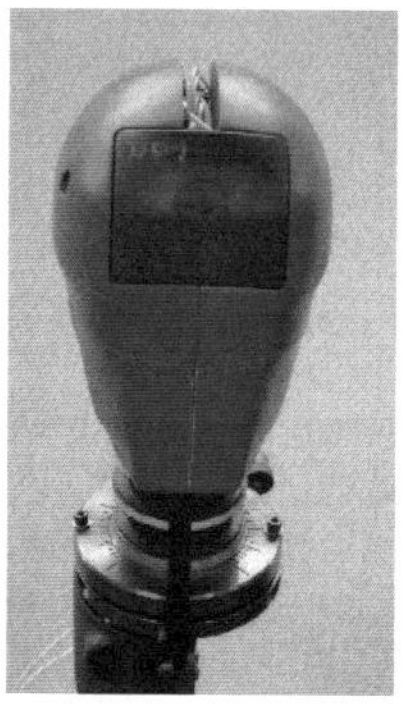

Figure 3.- Ballistic load sensing headform.

The BLSH is used to evaluate the instantaneous force produced during the impact of a projectile on a ballistic helmet. For this purpose, it is equipped with an array of seven load sensors in the impact zone. There are two different headforms, the first one has one array in the front and another in the back, and the second one has one array on each side. A skin-simulating pad covers the sensors to better simulate the actual head response [14], [15], [16]. The headform is mounted on a flexible neck, the Hybrid III anthropomorphic test device (ATD). This ATD is the most widely used in the world. It accurately simulates the human dynamic response during a crash event. As explained in [17], the head was designed to simulate a real human head. The mechanical properties of the human skull are known and described in the literature [18], and could be related to the force measurement obtained from the BLSH. However, there is no suggested correlation between the BLSH force data and brain injury data [19]. Therefore, the direct force measurement is more closely related to skull damage. To assess brain injury, the Head Injury Criterion (HIC) was used.

3 MATERIALS AND METHODS

3.1 Test procedure

First, four helmets from the Belgian Army were tested in order to determine the actual limits and response of the system. The first type of liner tested is a net system. Secondly, different liner configurations were tested to see the load transmitted to the headform from the projectile impact. Ten different configurations were tested: two commercial solutions (Skydex and Koroyd) and eight cellular materials produced for this work.

The 9 x 19 mm projectile was fired at the helmet with a velocity ranging from 328 to 361 m/s in the four positions of the helmet (front (F), right (R), left (L) and back (B)). The impact velocity does not relate to any ballistic resistance standard but to the resistance of the helmet, as when it was designed this threat was not included. The large variation of the impact velocity during the tests was attributed to the manual loading of the quantity of powder in the cartridge. The target was aligned and placed on the surrogate as explained in [28], and they were tested at least three times in different positions of the helmet (Table 1). Names given to the samples correspond to the used polymer and its density. No complete perforations of the shell were observed for this range of velocities.

Table 1.- Number of tests for each configuration.

Tested sample	Number of fair impacts
Actual liner	16
Skydex	4
Koroyd	4
PMMA 390	4
PMMA 191	4
PEI	3
TPU 265	4
TPU 329	4
TPU 367	3
TPU 414	3

https://doi.org/10.52202/081621-0011

TPU 492	4

3.2 Liner

Their main function is to hold the helmet in the correct position in an ergonomically acceptable way. Ballistic helmets typically employ one out of two types of liners, foam padded and strap-netting liners. The two liner systems (net systems and pads) have different performances with regards to thermal comfort and reducing the pressure field around the head when a blast or impact event occurs. Pads allow for the distribution the impact forces over larger areas, reducing the loads on the head of the user [24], [25]. The key to the foam's energy-absorbing properties is the cellular structure and its crushable structure. During compression, the cells compress and fracture which prevents the impulse from being transferred to the body directly. Low density foam with a small cell size is more effective at reducing the transfer rate of this impulse [26], [27].

Micro- and nanocellular polymers are biphasic materials with a solid phase and a gaseous phase in the micro and nanometric range respectively. These materials have already been proven to present excellent mechanical performance under conventional laboratory loads and speeds [9]. However, as far as the authors know, these materials have not been investigated as liners.

For the testing campaign, different micro and nanometric polymeric foams have been designed by the CellMat Laboratory at the University of Valladolid, to be tested as liners. The produced materials from different polymeric matrices, present a wide range of relative densities and cell sizes. The materials have been produced through the gas dissolution foaming technique which comprises three steps, saturation, desorption, and foaming. During saturation, the polymer is saturated under a CO_2 atmosphere at a certain saturation pressure and temperature (P_{sat}, T_{sat}). Once the sample is fully saturated the gas is rapidly released to atmospheric pressure creating a thermodynamical instability that produces phase separation and the creation of nucleation points. Finally, samples are submitted to a foaming temperature (T_f) above the glass transition temperature of the gas-polymer system, during the foaming time (t_f) to expand the nucleation points into the final cells.

To produce the foams, three polymeric matrices with different properties have been selected: a common thermoplastic polymethylmethacrylate (PMMA), a high-performance polymer polyetherimide (PEI) and an elastomeric thermoplastic polyurethane (TPU). An example of the obtained cellular structures is shown in Figure 5.

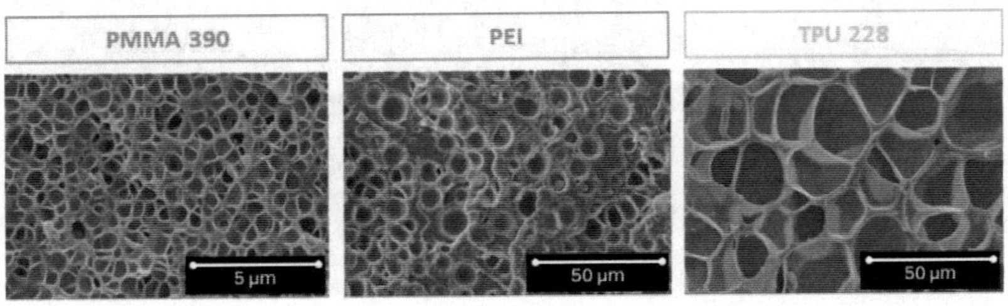

Figure 5. SEM micrographs of some of the produced cellular materials: PMMA, PEI, and TPU.

The combination of the different polymeric matrices with a wide range of production parameters allows the production of foams with very different characteristics (Table 2).

Table 2.- Production parameters and characteristics of the produced foams.

	Saturation Parameters		Foaming Parameters		Cellular Materials Characteristics	
Material	Psat (MPa)	T sat (°C)	T f (°C)	t f (s)	Areal Density (AD) (g/m²)	Cell size (nm)
PMMA	31	25	60	120	7800	300-600
			100		3820	500-800
PEI	30	225	170	180	19000	1000
TPU	30	150	150		4560	2000
			140		6160	
			130		7040	
			120		7760	
			110		8880	

https://doi.org/10.52202/081621-0011

PMMA with relative areal densities (AD) from 3820 to 7800 g/m^2 and cell sizes from 40 to 800 nm have also been obtained. PEI presents the highest AD, 19000 g/m^2, combined with a 1-micron cell size. Finally, foams based on TPU have been produced with a wide range of AD from 4560 to 8880 g/m^2 and a fixed cell size of around 2 microns.

The produced liners have also been compared with commercially available liners: the originally mounted strap-netting system, Skydex and Koroyd. The commercial solution Skydex consists of four layers: in contact with the helmet, a hook-and-loop fabric layer; followed by the Skydex impact layer, a patented polymer-based geometry designed to absorb energy. Next is a comfort layer made from reticulated urethane, and finally a moisture-wicking fabric layer in contact with the head. The Koroyd solution consists of a collection of welded polymer tubes that deform plastically under impact. The first two commercial solutions have already been implemented in ballistic helmets available in the market. The third is present in several types of helmets, mostly motorcycle and industrial helmets, but it can be designed to withstand specific mechanical loads. The tested configuration was designed to limit the transmitted force to the skull to values below 5000 N.

3.3 Setup for sample testing

For the commercial solutions (Skydex and Koroyd), it was possible to place the material directly in its correct place (front, back, right, and left) for each type of sample in the helmet, a former helmet from the Belgian Army manufactured by Schuberth.

Due to manufacturing limitations in thickness, several layers of the produced cellular materials were stacked to produce the testing samples with a thickness of around 20 mm. (Figure 4.a). This stack was located in the desired position of the helmet (Figure 4.b.). Finally, the helmet was fixed to the load-sensing headform aligning the helmet position with the desired impact point (Figure 4.c.) and ensuring the correct gap between the headform and the helmet.

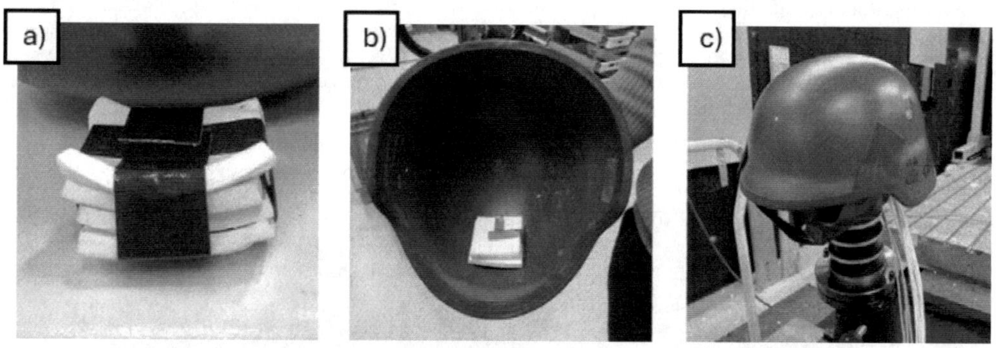

Figure 4.- a) Stacked cellular samples to obtain the desired thickness b) Cellular samples positioned in the helmet c) Helmet positioned on the load-sensing headform.

3.4 Head Injury Criterion (HIC)

Some experimental measures have been developed by the automotive and sports industries to assess injury risks for low-speed, high-mass impacts. Criteria such as HIC, viscous criterion, angular rotation thresholds, translational acceleration limits, and head impact power have been proposed. It is commonly accepted that skull fracture can be related to the maximum dynamic force; however, there is no consensus on the criterion to use to cover the full spectrum of possible blunt impact head injuries.

The Head Injury Criterion is a measure of the likelihood of head injury arising from an impact, taking into account the duration of the event and the mean value of the acceleration. The HIC can be used to assess safety related to vehicles, and sports equipment [20]. It is deemed appropriate for low-speed, non-penetrating impacts [21]. The impacts assessed here are outside of the scope of this criterion, but the goal is to check the feasibility of an armour and liner system and to compare their response with an actual helmet.

The HIC is defined as:

$$HIC = MAX\left(\frac{1}{T_2 - T_1}\int_{T_1}^{T_2} a(T)dT\right)^{2.5}(T_2 - T_1) \qquad \text{Eq 1}$$

where T is time. $T_2 - T_1$ is limited to a specific value, usually 15 ms or 36 ms (HIC-15 or HIC-36) [22] and a is the acceleration of the head measured in G's.
The HIC value can be related to an injury threshold [23]. These limits are used as guidelines in the automotive field. In that field, impacts happen at relatively low velocities and a timescale of 15 ms is appropriate. For a ballistic impact, it can last a couple of ms, deviating from the initial assumptions of HIC. So, no direct link to injury should be derived from this data. But it is still relevant to be able to compare the different systems, as the only objective is to compare the outcome after the impact and not to assess the risk of injury.

4 RESULTS

From the BLSH, it is possible to extract the information on the force and acceleration applied to the head during the impact. One result of the force measurement for each tested material is represented in Figure 6.

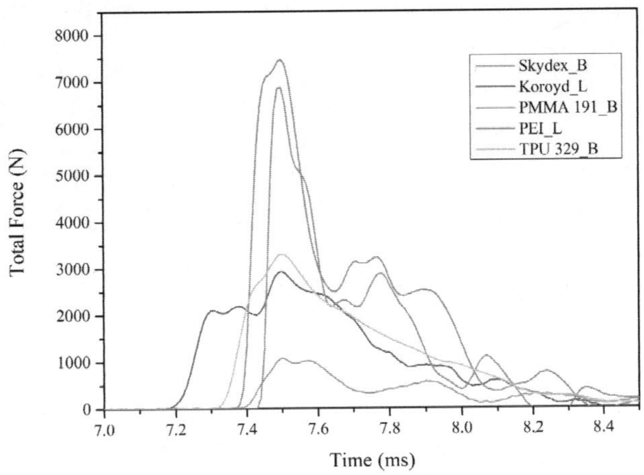

Figure 6. Force versus time.

The peak value is presented in the Figure 7:

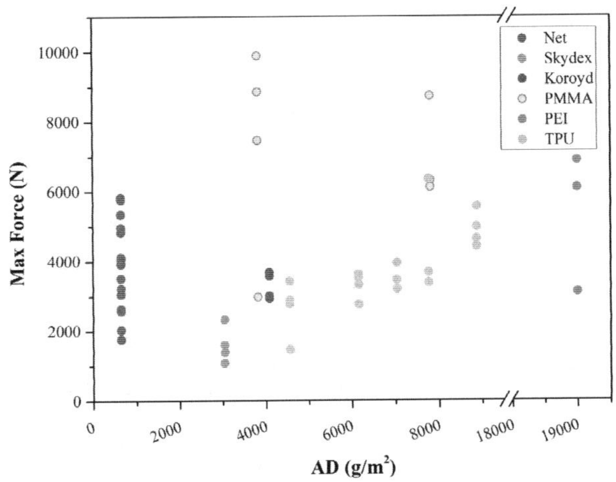

Figure 7. Maximal force versus AD.

https://doi.org/10.52202/081621-0011

Comparing the obtained values of the maximum force for the different materials in Figure 7, it is possible to observe that the net solution offers a lighter solution, with a measured force between 1763 and 5756 N. The commercial Skydex is the next lightweight solution that offers reduced force levels, with a maximum of 2320 N.
From the cellular materials manufactured for these tests, a correlation is observed between the areal density with the transmitted force. On the one hand PMMA, while offering lightweight solutions reaches force values up to 9880 N. PEI material presents a medium force around 5000 N but with a very high AD. Finally, the TPU in the lighter version offers force values of 3422 N. The measured force in TPU increases with AD.
Regarding the acceleration values, Figure 8 shows the values for the HIC.

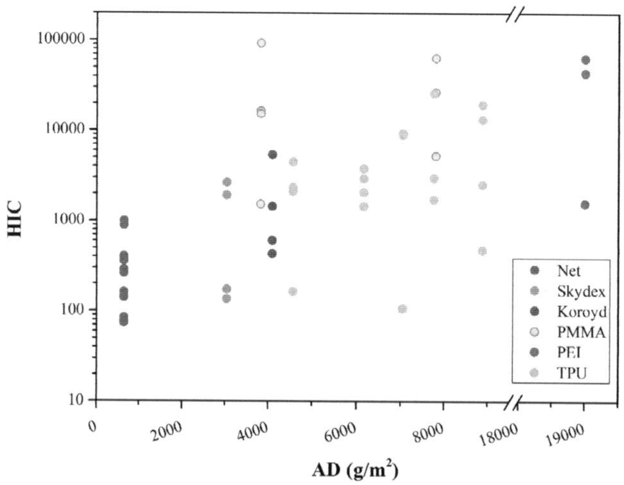

Figure 8. HIC versus AD

5 DISCUSSION

For all the tested solutions, the net liner seems to be the lighter one, but it offers a lower degree of protection, as the results for this particular case show a potential for injury, as the skull can support around 5000 N [16], [29]. Skydex and Koroyd are able to reduce to force to the user to a satisfactory level, with Skydex being lighter than Koroyd.
Regarding the produced cellular materials, the ones based on PMMA and PEI offer a very high transmitted force to the headform. As shown in Figure 6, PMMA collapsed and crumbled during the impact (Figure 9), not being able to absorb the incoming energy and transferring a significant load to the headform. Most probably the steep line is produced by the direct contact of the back face of the helmet with the headform.

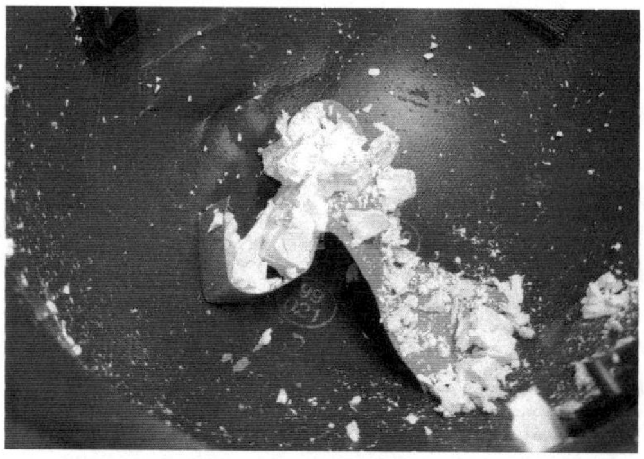

Figure 9. PMMA sample after the impact.

https://doi.org/10.52202/081621-0011

On the other hand, cellular materials based on TPU show better results being able to reduce the load to the head with a reduced mass. Due to the fact that TPU is an elastomeric polymer, it can handle this load several times without affecting its mechanical properties. This characteristic might be useful for other types of severe loads that do not affect the integrity of the shell, as the overpressure of an explosion. As can be seen in Figure 10, TPU 329 shows a similar behaviour in the second impact (B_2) as in the first impacts at different positions.

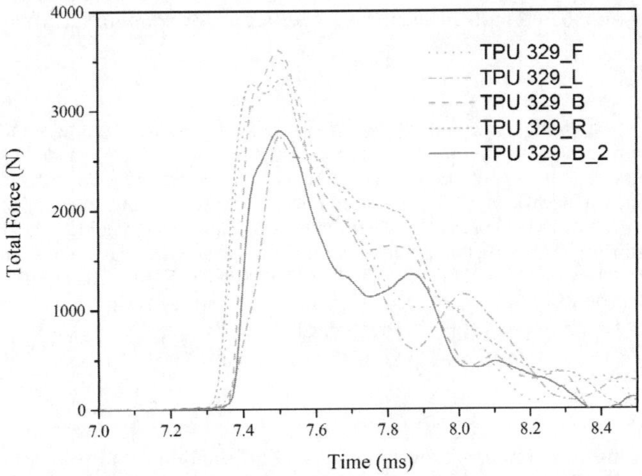

Figure 10. Force for the firsts and second impact for the TPU.

These tests show the potential of the produced cellular materials for pads for helmets, or other protective systems with high mechanical constraints as they could offer high levels of protection with a lightweight structure.

The time between the signal starting to increase and its maximum is important to understand the transfer of the force. A long time delay, such as the one observed in the commercial Skydex and the produced TPU foam means that the liner is able to transfer force to the headform for a longer period, and therefore it will allow to effectively reduce the maximum of the curve. Surprisingly, the TPU material worked in the elastic region, being able to withstand more impacts with similar characteristics. On the other hand, a short time delay means a short time to transfer the force, increasing the value of the maximum as is the case with PMMA or PEI.

For the values of the HIC, all of them are quite high as the net, with the lowest values, exceeds 900. The Skydex performs slightly better than the Koroyd. And from the manufactured samples it is possible to observe a tendency for the TPU to reduce the HIC with lower AD. More samples would determine the real shape of the curve as it seems to achieve a flat region below the 6000 g/m^2. The results of the PMMA and the PEI are very high and inconsistent, and the PMMA shows a response independent from the AD.

It can be concluded that elastomeric polymers are the ones able to reduce both the total force as well as HIC. The presence of cells of around 2 microns in this elastomeric matrix not only reduces the AD of the materials but also leads to the observed mechanical performance. When the impact is received, instead of the material deforming as a whole, each of the cells compresses within the elastic range, resulting in excellent performance. Additionally, if the cell size is small enough in comparison to the volume occupied by the polymeric chain, the mechanical properties of the cellular polymers are even better, due to 3D confinement [31].

On the other hand, the manufacturing production method is cheap and very versatile, because the control of each production parameter allows for a wide range of cell sizes and relative densities. The obtained preliminary results show that elastomeric matrices with smaller AD and cell sizes could lead to the reduction of the total force and HIC. Materials with these characteristics will be produced for further testing.

As can be seen, cellular materials with the elastomeric matrix, TPU, are again the ones performing better, with HIC values in the same range as Skydex samples. On the other hand, cellular polymers from rigid matrices show high HIC values.

https://doi.org/10.52202/081621-0011

6 CONCLUSIONS

In this study, the loads transmitted to the headform from the impact of a 9 x 19 mm FMJ projectile into a helmet were analyzed. It compared the influence of several liners. The tests compared three commercial solutions (net, Koroyd and Skydex) and eight cellular samples manufactured for these tests. The output results were the force and acceleration measured by the headform.

Among the manufactured materials, the cellular polymers based on TPU seem to be the best as they offer a performance comparable to the commercial ones, showing the potential of this kind of material to handle severe loads. It offers very good results for the force as well as for the acceleration. Moreover, it is an elastic material, and therefore its mechanical properties are not diminished after an impact.

Acknoledgments

This research was done within the framework of the DYMASEC – DYnamic MAterial Behaviour and SECurity applications – research pole of the Royal Military Academy, Brussels, Belgium, and funded by the Department for Scientific and Technological Research of Defence of the Belgian Armed Forces (study MSP 21 - 03). Funds from the Recovery and Resilience Mechanism - Next Generation EU Funds and the Community of Castilla y León Funds. Complementary Research and Development Plans with the Autonomous Communities in R&D&I actions of the Component 17. Investment 1 (C17.I01.P02). Financial support the Science and Innovation Ministry of Spain Ministerio de Ciencia, Innovación y Universidades (RTI2018-098749-B-I00 and PID2021-127108OB-I00, TED2021-130965B-I00 and PDC2022-133391-I00) is grateful acknowledged. Financial assistance from the Junta of Castille and Leon (VA202P20) is gratefully acknowledged.

References

[1] S. F. Khosroshahi, S. A. Tsampas, and U. Galvanetto, 'Feasibility study on the use of a hierarchical lattice architecture for helmet liners', *Mater. Today Commun.*, vol. 14, pp. 312–323, Mar. 2018, doi: 10.1016/j.mtcomm.2018.02.002.

[2] B. J. Ramirez and V. Gupta, 'Evaluation of novel temperature-stable viscoelastic polyurea foams as helmet liner materials', *Mater. Des.*, vol. 137, pp. 298–304, Jan. 2018, doi: 10.1016/j.matdes.2017.10.037.

[3] S. Farajzadeh Khosroshahi, R. Olsson, M. Wysocki, M. Zaccariotto, and U. Galvanetto, 'Response of a helmet liner under biaxial loading', *Polym. Test.*, vol. 72, pp. 110–114, Dec. 2018, doi: 10.1016/j.polymertesting.2018.10.012.

[4] X. G. Li, X.-L. Gao, and S. Kleiven, 'Behind helmet blunt trauma induced by ballistic impact: A computational model', *Int. J. Impact Eng.*, vol. 91, pp. 56–67, May 2016, doi: 10.1016/j.ijimpeng.2015.12.010.

[5] M. Rodríguez-Millán, T. Ito, J. A. Loya, A. Olmedo, and M. H. Miguélez, 'Development of numerical model for ballistic resistance evaluation of combat helmet and experimental validation', *Mater. Des.*, vol. 110, pp. 391–403, 2016, doi: 10.1016/j.matdes.2016.08.015.

[6] L. B. Tan, K. M. Tse, H. P. Lee, V. B. C. Tan, and S. P. Lim, 'Performance of an advanced combat helmet with different interior cushioning systems in ballistic impact: Experiments and finite element simulations', *Int. J. Impact Eng.*, vol. 50, pp. 99–112, Dec. 2012, doi: 10.1016/j.ijimpeng.2012.06.003.

[7] G. X. Gu, M. Takaffoli, A. J. Hsieh, and M. J. Buehler, 'Biomimetic additive manufactured polymer composites for improved impact resistance', *Extreme Mech. Lett.*, vol. 9, pp. 317–323, Dec. 2016, doi: 10.1016/j.eml.2016.09.006.

[8] E. Palta, H. Fang, and D. C. Weggel, 'Finite element analysis of the Advanced Combat Helmet under various ballistic impacts', *Int. J. Impact Eng.*, vol. 112, pp. 125–143, Feb. 2018, doi: 10.1016/j.ijimpeng.2017.10.010.

[9] J. Martín de León, V. Bernardo García, and M. A. Rodríguez Pérez, *Nanocellular Polymers*. in From Microscale to Nanoscale. De Gruyter, 2024. doi: 10.1515/9783110756135.

[10] NATO, 'STANAG 2920 - Procedures for the evaluation and classification of personal armour-Bullets and fragmentation threats', Nov. 2012.

[11] National Institute of Justice, 'NIJ Standard 0108.01 Ballistic Resistant Protective Materials', *Law Enforc. Stand. Lab. Natl. Bur. Stand.*, Sep. 1985.

[12] VPAM, 'VPAM APR 2006 General basis for ballistic material, construction and product testing', 2009.

[13] J. Croft and D. Longhurst, 'Body Armour Standards for UK Police', *HOSDB*, 2007, Accessed: Jul. 18, 2017. [Online]. Available: http://www.worldcat.org/title/hosdb-body-armour-standards-for-uk-police-2007/oclc/212430754

[14] M. Philippens, B. Anctil, and K. Markwardt, 'Results of a round robin ballistic load sensing headform test series', presented at the Personal Armour Systems Symposium, Cambridge, United Kingdom, 2014.

[15] D. E. Raymond, G. S. Crawford, C. A. Van Ee, and C. A. Bir, 'Biomechanics of Temporo-Parietal Skull Fracture From Blunt Ballistic Impact', in *SBC2008*, ASME 2008 Summer Bioengineering Conference, Parts A and B, Jun. 2008, pp. 1065–1066. doi: 10.1115/SBC2008-192966.

[16] D. Raymond, C. Van Ee, G. Crawford, and C. Bir, 'Tolerance of the skull to blunt ballistic temporo-parietal impact', *J. Biomech.*, vol. 42, no. 15, pp. 2479–2485, Nov. 2009, doi: 10.1016/j.jbiomech.2009.07.018.

[17] M. Franklyn and P. V. S. Lee, *Military Injury Biomechanics: The Cause and Prevention of Impact Injuries*. CRC Press, 2017.

[18] F. A. Pintar, M. M. G. M. Philippens, J. Zhang, and N. Yoganandan, 'Methodology to determine skull bone and brain responses from ballistic helmet-to-head contact loading using experiments and finite element analysis', *Med. Eng. Phys.*, vol. 35, no. 11, pp. 1682–1687, Nov. 2013, doi: 10.1016/j.medengphy.2013.04.015.

[19] National Academies Press, *Testing of Body Armor Materials: Phase III*. Washington, D.C., 2012. Accessed: Jul. 15, 2016. [Online]. Available: http://www.nap.edu/catalog/13390

[20] D. Marjoux, D. Baumgartner, C. Deck, and R. Willinger, 'Head injury prediction capability of the HIC, HIP, SIMon and ULP criteria', *Accid. Anal. Prev.*, vol. 40, no. 3, pp. 1135–1148, May 2008, doi: 10.1016/j.aap.2007.12.006.

[21] M. Frank *et al.*, 'Blunt Criterion trauma model for head and chest injury risk assessment of cal. 380 R and cal. 22 long blank cartridge actuated gundog retrieval devices', *Forensic Sci. Int.*, vol. 208, no. 1–3, pp. 37–41, May 2011, doi: 10.1016/j.forsciint.2010.10.023.

[22] R. Eppinger, E. Sun, S. Kuppa, and R. Saul, 'SUPPLEMENT: DEVELOPMENT OF IMPROVED INJURY CRITERIA FOR THE ASSESSMENT OF ADVANCED AUTOMOTIVE RESTRAINT SYSTEMS - II', Mar. 2000, Accessed: Aug. 12, 2016. [Online]. Available: https://trid.trb.org/view.aspx?id=691729

[23] R. M. Greenwald, J. T. Gwin, J. J. Chu, and J. J. Crisco, 'Head Impact Severity Measures for Evaluating Mild Traumatic Brain Injury Risk Exposure', *Neurosurgery*, vol. 62, no. 4, pp. 789–798, Apr. 2008, doi: 10.1227/01.neu.0000318162.67472.ad.

[24] J. Li, H.-N.-S. Tan, and K. Y. Seng, 'A Biomechanical Computational Study of the Role of Helmet Pads in Mitigating Blast-Induced Traumatic Brain Injury', in *6th World Congress of Biomechanics (WCB 2010). August 1-6, 2010 Singapore*, C. T. Lim and J. C. H. Goh, Eds., in IFMBE Proceedings, no. 31. , Springer Berlin Heidelberg, 2010, pp. 1518–1521. doi: 10.1007/978-3-642-14515-5_387.

[25] M. Salimi Jazi, A. Rezaei, G. Karami, F. Azarmi, and M. Ziejewski, 'A Study on the Impact of Helmet Padding Materials on the Brain Pressure Under Shock Loads', ASME, Jun. 2012, p. 641. doi: 10.1115/SBC2012-80136.

[26] G. H. Liaghat, A. A. Nia, H. R. Daghyani, and M. Sadighi, 'Ballistic limit evaluation for impact of cylindrical projectiles on honeycomb panels', *Thin-Walled Struct.*, vol. 48, no. 1, pp. 55–61, Jan. 2010, doi: 10.1016/j.tws.2009.07.008.

[27] R. Nasirzadeh and A. R. Sabet, 'Study of foam density variations in composite sandwich panels under high velocity impact loading', *Int. J. Impact Eng.*, vol. 63, pp. 129–139, Jan. 2014, doi: 10.1016/j.ijimpeng.2013.08.009.

[28] National Research Center, 'Review of Department of Defense Test Protocols for Combat Helmets', Mar. 2014.

[29] J. A. Motherway, P. Verschueren, G. Van der Perre, J. Vander Sloten, and M. D. Gilchrist, 'The mechanical properties of cranial bone: The effect of loading rate and cranial sampling position', *J. Biomech.*, vol. 42, no. 13, pp. 2129–2135, Sep. 2009, doi: 10.1016/j.jbiomech.2009.05.030.

[30] T. A. Gennarelli and E. Wodzin, 'AIS 2005: A contemporary injury scale', *Injury*, vol. 37, no. 12, pp. 1083–1091, Dec. 2006, doi: 10.1016/j.injury.2006.07.009.

[31] J. Martín-de León, J. L. Pura, M. L. Rodríguez-Méndez, and M. A. Rodríguez-Pérez, 'Viscoelastic property enhancement of polymethylmethacrylate molecularly confined within 3D nanostructures', *Eur. Polym. J.*, vol. 214, p. 113181, Jun. 2024, doi: 10.1016/j.eurpolymj.2024.113181.

 https://doi.org/10.52202/081621-0011

Behind Helmet Blunt Trauma: Pathophysiological understanding and relevance of specifying helmets

R. Manet[1,2], S. Lecre[3], A. Lambourg[3], A. Lefranc[3], A. Dagain[2,4], **C. Crepin[3]**

[1] *Department of Cranial Neurosurgery, Neurological Hospital Pierre Wertheimer, University Hospital of Lyon (Hospices Civils de Lyon), France*
[2] *Department of Neurosurgery, Military Teaching Hospital Sainte-Anne, Toulon, France*
[3] *General Directorate for the Armament – Land Systems (DGA Techniques terrestres), Bourges, France*
clement-m.crepin@intradef.gouv.fr (presenting author email address)
[4] *French Military Health Service Academy, Ecole du Val-de-Grâce, Paris, France.*

Abstract. A helmet is one of the most important elements of a soldier's personal protective equipment (PPE). Its primary purpose is to protect the head from fragments and bullets, and secondary, to protect against blunt impacts. In France, as well as other countries, helmets are still specified to prevent ballistic penetrating traumatic brain injuries (TBI) and attenuate blunt trauma. The pathophysiology of this type of blunt TBI, described as behind helmet blunt trauma (BHBT), remains scarcely studied.
In the first part of this study, we provide a detailed overview of the basic pathophysiology of TBI and review current literature on the specific pathophysiological consequences of BHBT, which range from superficial skin abrasions to extensive skull and brain damage. The severity of these injuries depends on the mechanical properties of the projectiles, as well as those of the helmet, which theoretically absorbs part of the energy at impact, thereby attenuating TBI.
In the second part of this study, we discuss the mechanical phenomena (acceleration, rotational motion, etc.) happening on impact, which lead to BHBT-related injuries. Then, we provide a review of literature on the various methods used to quantify these mechanisms through multiple criteria, the use of physical surrogates – instrumented or not – and numerical models. Some of these methods are currently employed by certain countries to assess the protecting capabilities of helmets against BHBT.
Finally, based on this review, we critically assess the relevance of these specifications, considering the architectural constraints they impose on the system and their actual protective effect.

1. INTRODUCTION

Nowadays, combat helmets are designed to stop 9 mm bullets. But by doing so, the energy of the bullet is transmitted to the helmet, causing a deformation of its shell. Even though this bullet is not a realistic threat on military ground, it is still used to specify helmets, as it allows us to compare them. Up until now, military helmets were not able to stop bullets from shoulder weapons. However, new helmets have been developed with the aim of stopping this kind of threat. In addition, to improve helmets, their mass must be reduced without compromising the protection. In order to do so, the tendency is to switch the shell's material, from aramid to ultra-high molecular weight polyethylene (UHMWPE), which has a lower areal density (AD) than the former. This implies an increase of the backface deformation (BFD), which in turn raises the risk of behind helmet blunt trauma (BHBT), therefore explaining why evaluating the BFD seems necessary. This is even more important to take into consideration now that shoulder weapons are becoming a tested threat.

Loftis et al. reported 92 989 US civilians injured at the head from 2013 to 2015 [1]. They highlight the fact that even if penetrating wounds are not so common (6%), they lead to complex injuries (involving skull, meninges, brain and vessels), which tend to be particularly lethal (54% of them). On the other hand, blunt injuries are more common (90%), but less lethal (3%) of them. Plaisted reports that among the 400 000 US soldiers who suffered traumatic brain injuries (TBI) due to blunt impacts (car accident, fall, parachuting, etc.) between 2000 and 2019, 83% presented mild TBI, 10% presented moderate TBI, 1% presented severe TBI and only 1.3% presented a penetrating wound [2]. This reflects the fact that head injury is still a major threat for the soldier's health, therefore emphasizing the importance of protecting the head.

Regarding French feedback on injured soldiers, over an 11-year period (2012-2023, covering conflicts such as operations Pamir in Afghanistan (2002-2014), Barkhane in Sahel (2014-2022) and Chammal in Iraq and Syria (since 2024)), 408 soldiers were listed as injured, and 52 (13%) of them succumbed to their wounds [unpublished personal data]. Regarding those 52 soldiers, 7 (13%) of them only had an injury on the head and neck area, and 25 (48%) others suffered at least an injury on the head and neck area. Even though blast injuries are the most common threat on modern battlefields, accounting for 87 recorded skull injuries, ballistic impacts still resulted in 18 skull injuries. This data highlights that bullets remain a relevant threat and that head injuries must be carefully

https://doi.org/10.52202/081621-0012

considered, especially since the number of soldiers who avoided a head injury thanks to their helmet is not given in this study. In addition, this feedback comes from a context of asymmetric conflicts, which means that the epidemiology could differ significantly in high intensity conflicts. Furthermore, since French helmets from this feedback were made of aramid, those numbers regarding injured and dead soldiers could have been higher if helmets were made of UHMWPE, assuming that the BFD generates major head injuries. However, it remains unknown whether ballistic head injuries which happened while wearing a helmet were due to bullet penetration or BFD, therefore questioning if BFD presents a significant injury risk. This brings up the question of whether or not current helmet specifications remain relevant. Indeed, this requirement seems to hinder mass reduction, which is an important aspect nowadays, especially as helmets are increasingly equipped with accessories such as night vision goggles with their batteries and communication headsets for example, all of which significantly add to the total weight. For instance, a special operations helmet's shell, which weighs approximately 700-800 g, may reach around 3 kg when fully equipped. In fact, some countries might remove this specification after observing no injury - or only marginal cases - due to BFD.

Therefore, the aim of this study is to provide a detailed overview of the basic pathophysiology of TBI as well as the specific pathophysiological consequences of BHBT, and to discuss the mechanical phenomena (acceleration, rotational motion, etc.) happening on impact, which lead to BHBT-related injuries. The aim is to critically assess the relevance of the current helmet's specifications.

2. ANATOMY AND PATHOPHYSIOLOGY OF TRAUMATIC LESIONS OF THE HEAD

Head trauma can lead to various types of injuries. Broadly, they are divided into traumatic lesions of the skull and soft tissues from TBI, which include primary injuries (direct consequence of the trauma on brain tissues) and secondary injuries (resulting from secondary pathophysiological mechanisms such as impaired cerebral blood flow and inflammation reactions). Although this work excludes penetrating ballistic injuries, BHBT also includes open injuries. In addition, head injuries are often associated with maxillo-facial trauma. This review will focus on skull injuries, TBI, and/or a combination of both. These different lesions are summarized in Table 1.

Table 1. Definitions of the main type of head injuries

Category	Type of Injury	Description
Superficial lesions 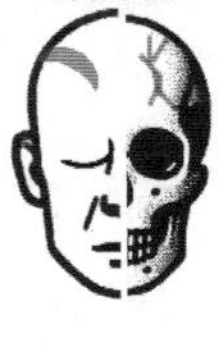	Scalp contusions / cephalhematoma	Contusion or hematoma of the scalp, without skin opening
	Lacerations of the scalp	Breaching of the skin, generally resulting un massive bleeding
	Skull fracture	Simple (one fracture line) or comminuted (multiple fracture lines) without vault deformation
	Depressed fracture	Comminuted fracture with sinking vault
	Open fracture	Fracture associated with scalp laceration
	Skull base fracture	Can result in dural opening with nasal or ear cerebrospinal fluid leak
	Penetrating TBI	Breaching of the scalp, skull, and dura mater, exposing the brain directly to the outside
Focal brain injuries 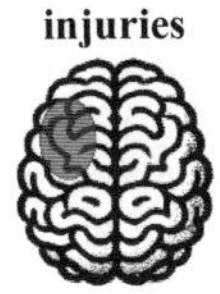	Cerebral Contusion	Localized haemorrhagic injury of the brain parenchyma, often caused by direct impact
	Epidural Hematoma	Accumulation of blood between the dura mater and the skull, usually due to arterial rupture
	Subdural Hematoma	Accumulation of blood between the dura mater and the arachnoid, often caused by venous rupture
	Subarachnoid Haemorrhage	Presence of blood in the subarachnoid space, which may result from traumatic brain injury
Diffuse brain injuries 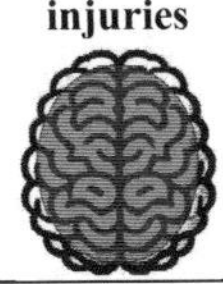	Cerebral Oedema	Focal or diffuse excessive accumulation of fluid in the brain, increasing the intracranial pressure
	Concussion	Temporary alteration of brain function without visible structural injury
	Diffuse Axonal Injury	Damage to brain axons caused by acceleration-deceleration forces, and secondary injuries, leading to prolonged coma

2.1 Skull and soft tissues injuries

https://doi.org/10.52202/081621-0012

Blunt head traumas are frequently associated with skull and soft tissues injuries. Soft tissues are highly vascularized, meaning that lacerations can often result in profuse bleeding requiring prompt haemostasis. Skull fractures increase significantly the risk of an intracranial hematoma [3]. It is important to distinguish blunt trauma from penetrating TBI which associates lacerations at the same site of soft tissues, skull fracture and the dura matter, exposing the brain parenchyma to the outside.

2.2 Traumatic brain injuries

TBI correspond to a structural injury and/or functional impairment of the brain resulting from a trauma [4].

2.2.1 Primary Brain Injuries

Primary brain injuries are classified into focal and diffuse injuries, which often occur together. They are most commonly observed in moderate and severe TBI.

Focal brain damage results from lacerations, compression, concussion and shearing forces, at the primary impact site (coup). Damage may develop in tissues opposite to the impact (contrecoup) or near the coup site due to the secondary impacts when the brain rebounds against the skull [5]. In addition to the transmitted energy of the impact, cavitation, which corresponds to the vaporization of water at a constant temperature due to reduced pressure, has also been proposed as another mechanism of focal lesion in blunt TBI [6]. This hypothesis suggests that a rapid increase in acceleration may lead to an increase of the intracranial pressure (ICP) at the impact location and a decreased pressure at the opposite side of the cerebrum (contrecoup), which can cause the development of cavitation bubbles. When they burst or collapse, they can create local tissue damage [7–9]. However, this hypothesis has not been confirmed by any clinical data to date.

In contrast, diffuse brain injuries primarily result from non-contact forces of rapid deceleration and acceleration, causing shearing and stretching injuries to neuronal axons, glial cells, and blood vessels. If diffuse insults, such as blast TBI, are a major source of diffuse brain injuries, focal impacts can also lead to these types of consequences. The diffuse axonal injury (DAI) represents the main diffuse structural brain injury. It is characterized by widespread lesions in subcortical and deep white matter, often involving the frontal and temporal white matter, corpus callosum, and brainstem. These structural and functional impairments in neurons lead to various clinical consequences, including bilateral neurological deficits and/or delayed awakening from coma, categorized according to the Adams classification. The severity of axonal injury and neuronal degeneration significantly impact long-term TBI prognosis [10]. The majority of DAI in TBI does not result from a direct lesion of the axons at the time of primary injury, but from a secondary injury mechanism [11]. Diffuse brain injuries also lead to a delayed traumatic chronic encephalopathy, a neurodegenerative disease [12].

2.2.2 Secondary Brain Injuries

Secondary brain injuries arise from mechanical, biochemical, cellular, and physiological changes that begin during the primary injury and can extend from hours to years afterward. Those injuries include excitotoxicity, mitochondrial dysfunction, release of reactive oxygen species and lipid peroxidation, neuroinflammation, axonal degeneration, glial scar and apoptotic cell death [13,14].

Post-traumatic intracranial hypertension (ICH) is the primary driver of secondary brain injuries. It is defined as an increase in ICP above 20 mmHg, resulting from an imbalance between the volumes of the intracranial steady space (the cranium is a rigid box) and its components (brain parenchyma, cerebrospinal fluid (CSF), and cerebral blood volume) [15]. Due to the exponential relationship between intracranial volume and pressure [16], additional volumes from primary injuries and secondary injuries result in an exponential increase in ICP. It may potentially lead to a vicious circle: ICH worsen secondary injuries (in particular cerebral oedema) which, in turn increase ICH. ICH results in two main processes. Firstly, pressure gradient within the cranio-spinal cavity can lead to displacement of brain parenchyma (cerebral herniations) which can lead to severe cerebral impairment. In particular, the compression of the brain stem can result in cardiac and or respiratory arrest. Secondly, the increase in ICP progressively reduces the cerebral perfusion pressure (CPP = Mean Arterial Pressure - ICP), impairing oxygen and nutrient delivery to neurons, promoting secondary brain injuries, and leading to brain death when cerebral blood flow (CBF) is severely compromised.

While the severity of TBI is related to the severity of the impact, there is no direct correlation between primary and secondary injuries [5]. Thus, head trauma can result in moderate primary brain injuries but severe secondary brain injuries.

Clinically, TBI can manifest as a reduced level or loss of consciousness, altered mental state, memory loss, neurologic deficits (persistent or transient), and/or intracranial lesions. TBI severity is classified based on clinical symptoms, in particular the consciousness level, as assessed using the Glasgow Coma Scale (GCS) which ranges from 15 (normal consciousness) to 3 (deep coma). A GCS ≤ 8 corresponds to a coma [17]. TBI is considered mild when GCS ≥13 [18] (sometimes considered ≥14), moderate when 9 ≤ GCS ≤ 12, and severe when GCS ≤ 8 [19,20].

The severity of head injury can be classified according to the Abbreviated Injury Scale (AIS), a tool used in severe trauma settings to quickly assess and communicate the overall severity of trauma and expected clinical outcomes (see Table 2).

Table 2. AIS classification in head injuries [21,22]

AIS	Head injury	Lesions
1	Headache or dizziness	Superficial lesions
2	Unconscious less than 1 h	Closed fracture
3	Unconscious 1 - 6 h	Comminuted and/or depressed fracture
4	Unconscious 6 - 24 h	Open skin (skin opening), massively depressed fracture, small focal lesions
5	Unconscious more than 24 h	Penetrating TBI (dural opening), large focal lesion, diffuse lesions
6	Non-survivable	Massive destruction of both cranium (skull), brain, and intracranial contents (crush)

3. CLINICAL CONSEQUENCES OF BHBT

Literature regarding mechanisms and clinical consequences of military BHBT is extremely scarce. BFD can lead to several types of injuries, including skull fractures, focal brain injuries and diffuse brain injuries.

Key factors can significantly influence skull responses. The predominant fracture types in BHBT are linear and depressed, influenced by factors including impact velocities, contact areas, and loading regions [22,23]. For instance, broader contact areas typically result in linear fractures, while more localized impacts often cause depressed fractures [24]. In terms of impact severity, broad contact areas larger than 13 cm² generally lead to linear fractures, whereas smaller, localized impacts usually result in depressed fractures [24]. A study on helmeted post mortem human subjects (PMHS) impacted by 9 mm bullets showed that impacts at 460 m/s primarily produced linear fractures, while impacts with velocities around 440 m/s can lead to both linear and depressed fractures [25].

Focal brain injuries likely result mainly from mechanical loading conditions. Both skull deformations and relative skull-brain motions can induce focal brain injuries, although their precise contribution in the primary mechanism of the blunt ballistic impacts remains unclear. These injuries typically occur directly beneath the impact site. Skull deformation generates compressive stresses in the brain, leading to injuries such as cerebral contusions and haemorrhages [26]. Skull fractures may also rupture meningeal vessels, potentially causing epidural hematomas. Post-impact skull movement can induce additional injuries (contrecoup) due to skull-brain collisions because of inertia and inflict additional injuries [27]. The precise contribution of skull deformation versus relative motion to injuries is not yet clear.

The specific mechanisms for diffuse injuries in BHBTs are not well-defined, necessitating more comprehensive research. Changes in ICP during impacts are probable critical indicators of TBI severity but has not been precisely studied in military BHBT, and establishing a direct correlation between initial ICP changes and injury severity in humans remains extremely challenging. Mechanical stress enhances the vulnerability of neurons to secondary injuries, even in cases of mild TBI [28].

4. HELMET EVALUATION REGARDING BHBT

4.1 Biomechanics of the head

Biomechanics is the study of a biological system using mechanics. Regarding TBI, biomechanics consists in studying the head's movements and mechanical loads applied to the brain, in order to predict head injuries based on the mechanical phenomena [29].

Upon impact, dynamic mechanical forces act on the skull and the brain, generating both linear and rotational movements of the head (see Figure 1). The overall acceleration of the head can be described using three-dimensional linear and rotational accelerations, measured at its centre of gravity [29]. Linear acceleration compresses the brain against the skull at the opposite side of the impact location. This creates a volume variation of the brain tissue, and consequently, intracranial pressure fluctuations, and eventually cavitation, which can cause brain injuries. Rotational acceleration, on the other hand, puts the brain tissue in rotation, generating internal shear, as different regions undergo varied inertial movements due to local density variations, which can also cause brain injuries [30,31, 32,33].

The mechanism behind diffuse injuries in BHBT, especially DAI and cerebral haemorrhages, probably result from a combination of angular and linear accelerations. The relative contribution of linear and rotational

https://doi.org/10.52202/081621-0012

accelerations resulting from an impact depends on multiple factors, including the direction and location of the force, and the material properties of the skull and brain [29]. If the force vector of an impact passes through the head's centre of gravity, it will mainly induce a linear movement of the head. However, if the force vector does not pass through the head's centre of gravity, it will generate an impulsive moment, leading to both a linear and rotational acceleration of the head [29]. Angular accelerations resulting from high-energy projectiles with incidental impact may be the predominant mechanism, rather than linear accelerations, in causing certain focal lesions, particularly subdural hematomas, as a result of the tearing of bridging veins [34,35].

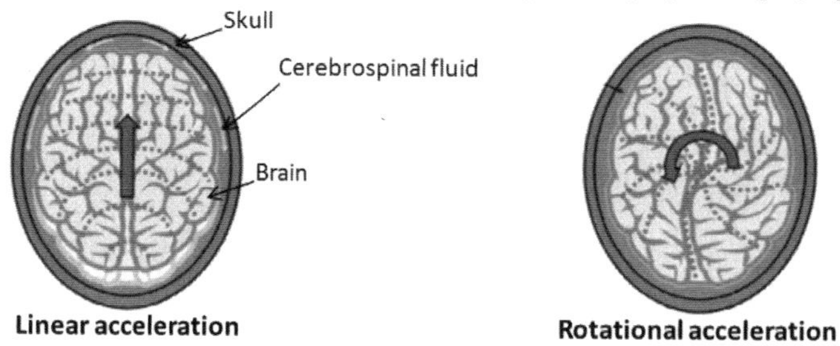

Figure 1. Brain's movement when exposed to a linear acceleration (left) or rotational acceleration (right) [31]

The head's mechanical response to loading can vary depending on the phenomenon's frequency content. The brain has a viscoelastic behaviour [36]. The skull can be considered as a rigid shell for an impact frequency content below 150 Hz. However, over 200 Hz, the skull exhibits a vibratory response that can increase or modify the skull flexure patterns produced by the force [37]. Gurdjian et al. stated that head accelerations recorded opposite to the blow do no not correlate with rigid body acceleration in short duration impacts, sometimes differing by a factor greater than 2 [37]. This can explain the differences observed between long duration impacts (low frequency) and short duration impacts (high frequency) [38]. Therefore, it is complicated to compare the long duration impacts studied in the literature, like road accidents which have a duration of around 10 ms, with ballistic impacts, which generally last less than 5 ms [39–42]. Furthermore, impact velocities in ballistic cases are way higher than those in road accidents.

Since brain injuries can result from different types of solicitation, defining a single injury criterion for the head, representative of all the different types of brain injuries, remains challenging [36].

4.2 Head surrogates

Currently, various standards attempt to assess the injury risk of BHBT in ballistic testing on helmets. Among those, the AEP 2920(A)(V2) proposes to mount helmets on a rigid headform filled partly with BackFace Signature (BFS) materials such as Roma Plastilina #1, but the tolerated indent depth (residual deformation) shall be set by the National Authority (NA). The French ministry of defence (MoD) requires the use of HERBIN SUEUR 40 plastilina with a specification on the indent depth [43]. This remains under the scope of the AEP 2920(A)(V2) which allows the NA to use other BFS materials. According to the VPAM HVN 2009, a headform made of ballistic soap shall be used and the maximum energy transferred to the head, which is calculated based on the ratio between the indent depth made by the ballistic impact and the indent depth made by the calibration test, shall not exceed 25 joules.

These methods can be used to compare the protection level of different helmets regarding BHBT, but they present some limitations. Indeed, not only the ability of these types of surrogates to represent the injuries that can occur on a human are greatly questioned, but they also don't take into consideration the head's dynamic response. OP't Eynde et al. studied the fidelity of the Roma Plastilina #1 and concluded that the rebound effect of the plastilina causes a diminution of the final deformation of 5 to 25% [44].

To address these limitations, another type of surrogate can be used, which can be instrumented with sensors at the centre of gravity of the surrogate to record the three-dimensional linear and rotational accelerations endured by the head. Indeed, the NIJ 0106.01 proposes to use an instrumented headform made of magnesium to monitor the linear acceleration which must not exceed 200-g. The Hybrid III (HIII) head, developed initially by General Motors (USA) for crash tests, is one of the surrogates that are mainly used [45]. It was designed from biometric data collected on 13 cadavers whose head circumferences were between 56 et 58 cm [46]. The Ballistic Load Sensing Headform (BLSH), designed by Biokinetics (CAN) specifically for ballistic helmet testing, is another type of surrogate that can be used to assess skull fracture risks due to backing effects. Other surrogates exist, but they are mostly used for non-ballistic impacts. The differences between the main surrogates found in the literature for ballistic testing are presented in Table 3.

Headforms' size, mass and moments of inertia are important to take into consideration and having those similar to the real values is important for the experimental values to be exploitable. Some of them exist in different sizes, but the most commonly used size is the one corresponding to the 50th percentile male, which is available for the HIII and BLSH headforms. The ATLAS headform is modular and the Pk17dynA headform was originally an XXL size (645 mm), but is now also available for helmets in L-XL sizes (425 mm). Regarding the mass, the HIII headform weighs 4.54 kg, the BLSH headform weighs 4.91 kg, the ATLAS headform is rigid, as it is fixed on a solid neck, and the Pk17dynA weighs 7.84 kg (including instrumentation devices).

Table 3. Example of different surrogates used to measure the accelerations and / or force

	Developer	Initial purpose	Material	Instrumentation	Ref.
HIII	General Motors (USA)	Crash tests	Aluminium and vinyl skin	Acceleration / velocity sensors	[45]
BLSH	Biokinetics (CAN)	Ballistic testing on helmets	Magnesium alloy (and skin pads on force sensors)	Acceleration / velocity sensors, + Force sensors	[47]
ATLAS	John Hopkins Applied Physics Laboratory	Ballistic testing on helmets	Polymer headform, steel impact cap, neoprene impact pad	Force sensors	[48, 49]
Pk17dynA	University of German Federal Armed Forces, Bundeswehr Research Institute for Materials, Institute of Legal Medicine of Munich	Ballistic testing on helmets	Polyamide headform. Sensors covered by curved steel plates	Force sensors	[50, 51]

4.3 Injury criteria

4.3.1 Main injury criteria

All of the surrogates discussed previously have the same objective, which is to measure a mechanical phenomenon upon impact on a helmeted surrogate, whether it is the accelerations and / or velocities and / or force. The signals recorded are then used to determine different injury criteria, presented in Table 4, to offer insight on different aspects of head injury risks.

The most basic criteria that can be calculated are peak translational acceleration (PTA), peak translational velocity (PTV), peak rotational acceleration (PRA), and peak rotational velocity (PRV).

The most commonly used criteria are the Head Injury Criterion (HIC) and the Brain Injury Criterion (BrIC), which come from the automotive industry. The HIC, created in 1971, is based on the linear accelerations measured at the head's centre of gravity, and comes from studies on PMHS. It succeeded the Wayne State Tolerance Curve (WSTC), which was created in 1960, as a result of the first studies done on PMHS in car accidents [29]. The WSTC is a graph that shows the tolerance threshold of a linear acceleration depending on the time of exposure. Indeed, an impact with a high acceleration can be endured for a shorter duration compared to an impact with a lower acceleration, before it becomes critical. The HIC was used by the National Highway Traffic Safety Administration (NHTSA) at first in 1972, in order to evaluate the injury risk of automotive crash tests [52]. The BrIC was invented in 2013, in order to study the rotational phenomenon of the impact, by taking into consideration the rotational velocities. Therefore, the HIC represents the linear phenomenon of the impact, while the BrIC represents the rotational phenomenon.

The Head Impact Power (HIP), created in 2000, is a mix between the HIC and the BrIC, as it takes both aspects (linear and rotational) into consideration. It is representative of the power transmitted to the head, given in kW [53].

The Blunt Criterion, developed in the early 2000s as well, is used by the U.S. Department of Defence in order to quantify the vulnerability of ballistic impacts [54]. It takes into consideration the kinetic energy of the projectile and its diameter, as well as the target's capability to tolerate the impact energy, by considering its weight and thickness. This criterion is not relevant here as it only takes into consideration the initial parameters, and not the phenomena observed upon impact, and is therefore independent of the helmet tested.

The impact force is another relevant physical parameter, which could be used to predict the risk of skull fractures, but can only be obtained with a headform that has force sensors to record the force time curves, like the BLSH.

https://doi.org/10.52202/081621-0012

Other criteria exist, based on brain tissue deformation instead of head kinematics, like the maximum principal strain (MPS), the cumulative strain damage measure (CSDM), the dilatation damage measure (DDM), and the relative motion damage measure (RMDM). The MPS and CSDM are used for the evaluation of diffuse axonal injuries, while the MPS and DDM are used to evaluate contusion, and the RMDM is used to assess subdural hematoma [55]. However, those criteria are computed, which means they come from simulation and cannot be determined with experimental tests.

Table 4. Injury criteria for assessing the risk of head injuries

Criterion	Formula	Unit	Main injury studied	Ref.
Peak Translational Acceleration	$PTA = \max\left(\sqrt{a_x{}^2 + a_y{}^2 + a_z{}^2}\right)$	g *or* m/s²	Skull fracture	[54]
Peak Translational Velocity	$PTV = \max\left(\int \left(\sqrt{a_x{}^2 + a_y{}^2 + a_z{}^2}\right) dt\right)$	m/s	/	/
Peak Rotational Velocity	$PRV = \max\left(\sqrt{\omega_x{}^2 + \omega_y{}^2 + \omega_z{}^2}\right)$	rad/s	Diffuse axonal injury	[56]
Peak Rotational Acceleration	$PRA = \max\left(\frac{d}{dt}\left(\sqrt{\omega_x{}^2 + \omega_y{}^2 + \omega_z{}^2}\right)\right)$	rad/s²	Subdural hematoma	[56]
Force	$F = \sum F_i$ *where i is the number of sensors*	N	Skull fracture	[57]
Blunt Criterion	$BC = \ln\left(\frac{\frac{1}{2} \times m \times v^2}{W^{1/3} \times T \times D}\right)$	/	Skull fracture	[58]
Head Injury Criterion	$HIC = \left[\frac{1}{(t2 - t1)} \int_{t1}^{t2} a_r \, dt\right]^{2,5} (t2 - t1)$ where $a_r = \sqrt{a_x{}^2 + a_y{}^2 + a_z{}^2}$	/	Skull fracture, Brain injury, Concussion	[59]
Brain Injury Criterion	$BrIC = \sqrt{\left(\frac{\omega_x}{\omega_{xc}}\right)^2 + \left(\frac{\omega_y}{\omega_{yc}}\right)^2 + \left(\frac{\omega_z}{\omega_{zc}}\right)^2}$	/	Diffuse axonal injury	[60]
Head Impact Power	$HIP = \sum m.a.v + \sum I.\alpha.\omega$	kW	Concussion	[53]

To obtain the different criteria, the recorded signals need to be filtered before they are used in the equations. The filtering process is a fundamental aspect, which is rarely mentioned according to the bibliographic researches made for this publication. Regarding the HIC for example, the cut-off frequencies have been established from sport and automotive impacts. The duration of these impacts is much longer and the deformation velocities are less important than for ballistic impacts, which implies a much higher frequency content for the latter. The use of filtering values that are based on long duration impacts may potentially underestimate the real effect in ballistic impacts. This would mean that the values of the filter cut-off frequency need to be increased. In addition, the resonance frequency of the system composed of the surrogate head and the helmet also needs to be taken into consideration.

The linear accelerations are filtered with a CFC (Channel Frequency Class) 1000 [36, 62]. The angular velocities are sometimes filtered with a CFC 180 for motorcycle helmets [62], and sometimes with a CFC 600 according to the HFM-271 [36]. Human Factors and Medicine (HFM) are NATO groups working on defence science and technology activities. Regarding the HIP, Newman et al. filtered the signals with a CFC 1000, and later re-filtered digitally with a CFC 180 [53]. According to the Biokinetics software, the force signals obtained with the BLSH seem to be filtered with an equivalent to a CFC 2166. No other publication was found explaining which filters were used and why.

4.3.2 Injury criteria thresholds

In order to be able to use these criteria, a tolerance threshold has to be determined. The helmet can be considered having a sufficient protection level when the value obtained for a criterion is under the threshold.

However, the thresholds used usually come from long duration impacts, and can therefore be questioned as those types of impacts are very different from ballistic ones. This question can also be raised regarding the criteria's formulas. For instance, the HIC formula corresponds to the integral of a linear approximation for a time duration impact between 5 and 50 ms of the WSTC log-log plot [61]. The slope of this approximation might be different under 5 ms.

For the acceleration of the head, apart from the NIJ 0106.01 which specifies a threshold of 200-g, as previously discussed, the European standard for motorcycle safety ECE 2206 specifies a threshold of 275-g tolerance for the head, which corresponds to an AIS of 5, which is quite high [62].

For the HIC and the BrIC, the thresholds are determined as 10% of AIS 2+ by the HFM-271 [36]. Thus, the HIC's threshold used for protection of mounted soldiers is 250 (Figure 2), which is the value recommended as a concussion threshold for football players, and is lower than the one used in crash tests (700) [36,63]. HFM-271 proposes a threshold of 0.27 for the BrIC when the signals are filtered with a CFC 600 [36].

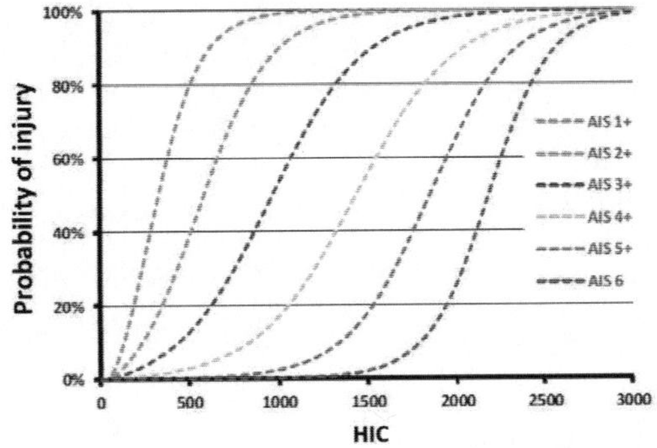

Figure 2. Probability of injury depending on the HIC and the AIS [36]

There is no official threshold regarding the HIP. Newman et al. found that there is a 50% probability of concussion for a HIP of 12.8 kW, based on 12 cases involving 24 football players, which have been re-enacted [53]. Marjoux et al. re-enacted 61 accidents, including pedestrian, motorcyclist and footballer accidents. They found that a 50% risk of skull fracture was associated to a HIC of 667 and a HIP of 38 kW, a 50% risk of subdural hematoma was associated to a HIC of 1 429 and a HIP of 55 kW, and that a 50% risk of moderate neurological injury (AIS 3-4) was associated to a HIC of 533 and a HIP of 24 kW [64]. Once again, these values do not come from a ballistic context, and it is therefore questionable whether or not they can be used for the evaluation of combat helmets.

Several thresholds have been proposed for the force applied to the skull, depending mainly on the age of the subject, but on the impact's location on the skull as well. Raymond et al. proposed a force threshold, on the parietal region, where fractures have a 50% probability of appearing, at 5 970 N, based on impact tests performed on PMHS at low velocities (20-35 m/s) [22]. Anctil et al. set this 50% probability of fracture threshold at 4 256 N for the temporal region [65]. Bolduc and Anctil proposed, in a study on the BLSH headform, that 6.0 kN corresponds to a 25% risk of skull fracture, identified as a threshold for head blunt trauma associated with ballistic impact, and a force of 7.2 kN for a 50% risk of skull fracture [66]. Finally, regarding the standards and as summarized by Philippens [67], FprCEN/TR16148:2010 [68] specifies a force of 5 to 8.5 kN for the fracture threshold. EN397 and EN12792 specify respectively 5 kN and 10 kN against the same threat: 5 kg spherical impactor at 1 m/s impact velocity on a helmeted rigid mounted metal headform. In a literature review from Li et al., they reported that the use of a non-lethal projectile generates a loss of consciousness for a force between 2.5 kN and 5 kN, and meningeal injuries for a force between 5 and 7.5 kN [69]. The variability in those thresholds highlights the need for further research to establish robust injury risk models.

4.4 Numerical simulation

Simulation can be useful, for different reasons. First, numerical simulation is more affordable than real ballistic tests. Also, a numerical model can represent the several parts of the head more precisely than the physical surrogates. Simulation avoids the use of animals or PMHS, that are usually representative of an older population than soldiers and causes ethical issues. Furthermore, one can have access to metrics that cannot be available experimentally, such as internal skull energy (see Figure 3). Finally, this solution enables parametric studies to better understand the consequences of ballistics impacts on (un)helmeted heads.

https://doi.org/10.52202/081621-0012

The first ballistic helmet simulation studies, such as the one conducted by Tham et al., aimed to demonstrate the feasibility of reliably modelling the dynamic deformation of the shell impacted by a 9 mm FMJ bullet [70].

Some, like Palta et al., used simulation as an intermediate verification step. If BFD values are consistent with and without a clay headform, the modelling is deemed reliable, allowing for parametric research [71].

Simulation has also enabled the evaluation of configurations incorporating headform simulators (like HIII and BLSH headforms) and digitally biofidelic human head surrogates. For instance, Miranda-Vicarion et al. establish relationships between BHBT and impact velocity [72], while Tan et al. analysed the effects of selecting between impact-absorbing foams and suspension nets on peak skull acceleration [41]. Caçoilo et al. examined the influence of impact angle and projectile energy on the Head Injury Criterion (HIC) and found that a 0° impact is the most severe [73].

The use of biofidelic headforms allows a better understanding of injury mechanisms. Aare and Kleiven analysed the forces transmitted to the skull and brain deformations depending on the shell material and impact angle [42]. Li et al. quantified the influence of standoff, thickness and pad hardness on ICP and HIC, showing that increasing the standoff by 2 mm can reduce intracranial pressure by up to 50% in their case [74]. Deck et al. also demonstrated that the probability of skull fracture significantly decreases with a slight increase in standoff. Their parametric study with a 9 mm bullet found that skull fracture probability drops from 100% at 22.5 mm to 22% at 23 mm, while the risk of DAI decreases from 86% at 19 mm to 6% at 20 mm [40]. Li et al. conducted a review on finite element method (FEM) models of biofidelic heads used for non-ballistic impacts (blunt trauma), ballistic impacts, and blast exposures, highlighting their diversity [75].

However, anatomical models remain complex to develop, limiting their widespread use. In addition, to specify helmets, ballistic testing is still necessary, and simulation cannot be sufficient by itself.

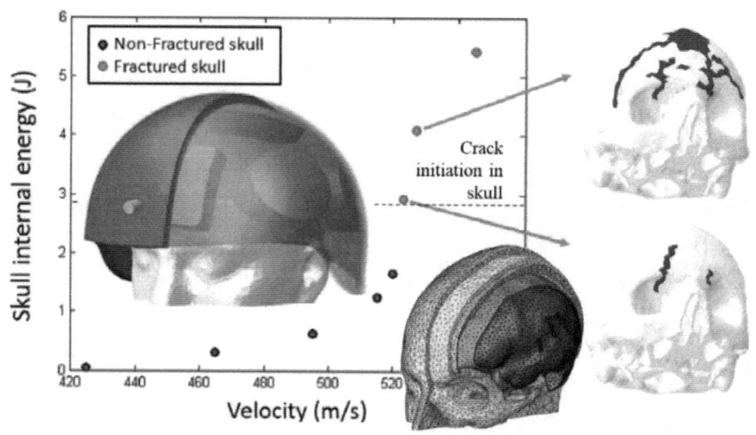

Figure 3. Example of skull fractures simulation using a model derived from MRI scans, triggered by a ballistic impact on a helmet [71]

5. DISCUSSION

Theoretically, protection from BHBT is important to take into account in helmet specifications, especially with the worldwide integration of UHMWPE as a protective material instead of para-aramid. However, there is not a lot of clinical feedback regarding the current protection, in particular on whether head injuries on the battlefield are mainly due to penetrating bullets or to BHBT.

Studies currently conducted by the French MoD with different types of helmets and headforms highlight the multiplicity of parameters that consistently influence the ability of helmets to protect the head against BHBT. These factors must be thoroughly understood and controlled to effectively minimize the BHBT against different types of projectiles, up to the 7.62 mm bullet. Some of these parameters are the standoff distance [76], the impact location (frontal/occipital/lateral), the presence of a pad at the impact location, the surrogate used, and other parameters directly linked to the helmet tested, such as the material used, the thickness of the helmet, its mass and the friction coefficient.

Clay headforms are convenient for assessing a helmet's ballistic protection due to their affordability and ease of use. However, evaluating the ability of a helmet to minimize BHBT solely by measuring the BFS in clay headforms is insufficient, as it does not account for the dynamic response of the head, which can lead to various types of injuries.

Instrumented headforms offer a more relevant method for determining real injury risks. Dynamic phenomena measured during a ballistic impact test on helmeted instrumented headforms can be used to determine different injury criteria which can be associated to various types of injuries, to a certain extent.

Despite their advantages, instrumented headforms are relatively expensive and require multidisciplinary knowledge for effective use. Moreover, the correlation between the residual measurements in a clay headform, which is yet the only standardized material to assess BHBT, and the dynamic response of instrumented headforms is not straightforward [43]. Additionally, the method for filtering the data before calculating the different criteria significantly impacts the results, and it is crucial to dissociate high-velocity (ballistics) from low-velocity (road accidents) impacts, as they have distinct characteristics. Some filtering values, criteria formulas and associated thresholds currently used are derived from the automotive field and may need reconsideration for ballistic applications. Furthermore, the variety of injury assessment metrics and thresholds poses challenges in establishing definitive criteria, indicating a need for more standardized research approaches to define injury thresholds more accurately. This inconsistency arises from differences in skull material behaviour, mechanical properties across various regions of the skull, biological variations between specimens, and test conditions (the projectile, helmet if there is one, and surrogate used). The lack of identical impact conditions in biomechanical tests further complicates the processing of the results.

Therefore, it seems important to consider the different studies conducted regarding this subject, to have a more complete approach in determining injury criteria and their thresholds, like in the literature review done by Nsiampa and Coghe [77]. In another review on the mechanisms and research methods for blunt ballistic head injury, Li et al. stated that to their knowledge no existing headform is suitable for the study of both skull and brain injuries [69].

From a clinical perspective, the transposition of laboratory tests to current protection from brain injuries appears hazardous. Firstly, the links between different criteria and specific types of superficial injuries (such as soft tissue and skull injuries) and primary brain injuries are difficult to confirm. Under the same loading conditions, consequences can vary significantly, and for the same mechanism, clinical outcomes can be very different. For example, a significant impact on the frontal bone, associated with a depressed fracture, can result in a mild TBI, whereas a lighter impact resulting in a simple closed fracture on the temporal bone can lead to a severe TBI with catastrophic epidural hematoma. Secondly, the role of helmets in preventing secondary brain injuries appears even more challenging to assess, as these injuries result from indirect pathophysiological mechanisms and are not necessarily linked to the primary injuries. A moderate primary injury can lead to severe secondary injuries, depending on associated injuries, particularly in haemorrhagic polytrauma, which may result in ischemia with major secondary brain injuries.

Military personnel are also particularly exposed to other types of threats like the blast overpressure (primary blast) coming from high-explosive ammunitions (IED, Artillery shell, etc.) but also from cumulative effect of low intensity blast overpressure coming from allied weapons systems (artillery canons, mortars, etc.). As the former can cause direct injuries, the latter can cause secondary injuries from repetitive mild TBI, resulting in deficits in cognition, memory, sleep, vision, and hearing, which cannot be studied with the same methods and materials than for BHBT [78]. The need to protect the soldiers against this threat can bring incompatibilities with the specifications regarding the protection against ballistic impacts [79].

6. CONCLUSION

Epidemiological data indicates that BHBT is a serious threat to soldiers. Understanding the blunt injury mechanisms caused by the BFD remains crucial to improve their prevention. However, assessing the clinical consequences of BHBT is challenging due to the interactions between various complex pathophysiological mechanisms. While it is possible to describe wounds post-injury, predicting the type of wounds that may occur from a non-penetrating impact remains very difficult.

To predict this risk, standardized plastilina headforms offer affordability and ease of use but do not account for dynamic phenomena. In contrast, instrumented headforms can measure accelerations, velocities, and impact forces to study the dynamic response of the head and determine various injury criteria. However, the lack of correlation between these injury criteria and actual clinical consequences, the numerous factors involved in their calculation, and the reliance on low-velocity impact thresholds make it difficult to use these injury criteria to specify helmet effectively in a ballistic case.

https://doi.org/10.52202/081621-0012

Acknowledgments

The authors acknowledge the Center for Epidemiology and Public Health of the French Armed Forces (CESPA) for their contribution to the epidemiological study of head injuries among French soldiers.

References

[1] Loftis K. Assessment of Head Injuries: Blunt Versus Penetrating. Personal Armour Systems Symposium (PASS 2020), Copenhagen, Denmark: Royal Military Academy (Belgium); 2025, p. 253–60. https://doi.org/10.52202/078352-0027.

[2] Plaisted, T. A. J., and R. Neice. "Test and assessment methods to evaluate combat helmets for rotation-induced injury." Proceedings of the Personal Armour Systems Symposium, Copenhagen, Denmark. 2020. n.d.

[3] Edna TH. Risk factors in traumatic head injury. Acta Neurochir (Wien) 1983;69:15–21. https://doi.org/10.1007/BF02055848.

[4] National Academies of Sciences, Engineering, and Medicine; Health and Medicine Division; Board on Health Care Services; Committee on the Review of the Department of Veterans Affairs Examinations for Traumatic Brain Injury. Evaluation of the Disability Determination Process for Traumatic Brain Injury in Veterans. Washington (DC): National Academies Press (US); 2019.

[5] Maas AIR, Stocchetti N, Bullock R. Moderate and severe traumatic brain injury in adults. Lancet Neurol 2008;7:728–41. https://doi.org/10.1016/S1474-4422(08)70164-9.

[6] Ward JW, Montgomery LH, Clark SL. A Mechanism of Concussion: A Theory. Science 1948;107:349–53. https://doi.org/10.1126/science.107.2779.349.

[7] Amauger J. Cavitation induced by acceleration and possible link to brain concussion. Fluid mechanics [physics.class-ph]. Institut Polytechnique de Paris, 2022. English. NNT : 2022IPPAX068. tel-04105395 n.d.

[8] Recent Research in Behind-Armor Blunt Trauma and Traumatic Brain Injury, ARL-MR-864, Army Research Lab Aberdeen Proving Ground MD Weapons and Materials Research Directorate. https://www.govinfo.gov/content/pkg/GOVPUB-D101-PURL-gpo125140/pdf/GOVPUB-D101-PURL-gpo125140.pdf. Retrieved February 17th, 2025. n.d.

[9] Gross AG. A New Theory on the Dynamics of Brain Concussion and Brain Injury. Journal of Neurosurgery 1958;15:548–61. https://doi.org/10.3171/jns.1958.15.5.0548.

[10] van Eijck MM, Schoonman GG, van der Naalt J, de Vries J, Roks G. Diffuse axonal injury after traumatic brain injury is a prognostic factor for functional outcome: a systematic review and meta-analysis. Brain Inj 2018;32:395–402. https://doi.org/10.1080/02699052.2018.1429018.

[11] Krieg JL, Leonard AV, Turner RJ, Corrigan F. Identifying the Phenotypes of Diffuse Axonal Injury Following Traumatic Brain Injury. Brain Sci 2023;13:1607. https://doi.org/10.3390/brainsci13111607.

[12] Hugon J, Hourregue C, Cognat E, Lilamand M, Porte B, Mouton-Liger F, et al. Chronic traumatic encephalopathy. Neurochirurgie 2021;67:290–4. https://doi.org/10.1016/j.neuchi.2021.02.003.

[13] Jamjoom AAB, Rhodes J, Andrews PJD, Grant SGN. The synapse in traumatic brain injury. Brain 2021;144:18–31. https://doi.org/10.1093/brain/awaa321.

[14] Ray SK, Dixon CE, Banik NL. Molecular mechanisms in the pathogenesis of traumatic brain injury. Histol Histopathol 2002;17:1137–52. https://doi.org/10.14670/HH-17.1137.

[15] Cushing H. Studies in intracranial physiology & surgery : the third circulation, the hypophysis, the gliomas / by Harvey Cushing. Wellcome Collection n.d. https://wellcomecollection.org/works/j9mbyy5z (accessed February 9, 2025).

[16] Langfitt TW, Weinstein JD, Kassell NF. CEREBRAL VASOMOTOR PARALYSIS PRODUCED BY INTRACRANIAL HYPERTENSION. Neurology 1965;15:622–41. https://doi.org/10.1212/wnl.15.7.622.

[17] Teasdale G, Jennett B. Assessment of coma and impaired consciousness. A practical scale. Lancet 1974;2:81–4. https://doi.org/10.1016/s0140-6736(74)91639-0.

[18] Carroll LJ, Cassidy JD, Holm L, Kraus J, Coronado VG, WHO Collaborating Centre Task Force on Mild Traumatic Brain Injury. Methodological issues and research recommendations for mild traumatic brain injury: the WHO Collaborating Centre Task Force on Mild Traumatic Brain Injury. J Rehabil Med 2004:113–25. https://doi.org/10.1080/16501960410023877.

[19] Carney N, Totten AM, O'Reilly C, Ullman JS, Hawryluk GWJ, Bell MJ, et al. Guidelines for the Management of Severe Traumatic Brain Injury, Fourth Edition. Neurosurgery 2017;80:6–15. https://doi.org/10.1227/NEU.0000000000001432.

[20] Geeraerts T, Velly L, Abdennour L, Asehnoune K, Audibert G, Bouzat P, et al. Management of severe traumatic brain injury (first 24hours). Anaesthesia Critical Care & Pain Medicine 2018;37:171–86. https://doi.org/10.1016/j.accpm.2017.12.001.

[21] Wilde EA, Wanner I-B, Kenney K, Gill J, Stone JR, Disner S, et al. A Framework to Advance Biomarker Development in the Diagnosis, Outcome Prediction, and Treatment of Traumatic Brain Injury. J Neurotrauma 2022;39:436–57. https://doi.org/10.1089/neu.2021.0099.

[22] Raymond D, Van Ee C, Crawford G, Bir C. Tolerance of the skull to blunt ballistic temporo-parietal impact. J Biomech 2009;42:2479–85. https://doi.org/10.1016/j.jbiomech.2009.07.018.

[23] Raymond D, Crawford G, Van Ee C, Bir C. Development of biomechanical response corridors of the head to blunt ballistic temporo-parietal impact. J Biomech Eng 2009;131:094506. https://doi.org/10.1115/1.3194751.

[24] Vaughan PE, Vogelsberg CCM, Vollner JM, Fenton TW, Haut RC. The Role of Interface Shape on the Impact Characteristics and Cranial Fracture Patterns Using the Immature Porcine Head Model, ,. Journal of Forensic Sciences 2016;61:1190–7. https://doi.org/10.1111/1556-4029.13111.

[25] Rafaels KA, Cutcliffe HC, Salzar RS, Davis M, Boggess B, Bush B, et al. Injuries of the head from backface deformation of ballistic protective helmets under ballistic impact. J Forensic Sci 2015;60:219–25. https://doi.org/10.1111/1556-4029.12570.

[26] Jirlow U, Hossain I, Korhonen O, Depreitere B, Rostami E. Cerebral contusions - Pathomechanism, predictive factors for progression and historical and current management. Brain Spine 2024;4:103329. https://doi.org/10.1016/j.bas.2024.103329.

[27] Gurdjian ES. Cerebral contusions: re-evaluation of the mechanism of their development. J Trauma 1976;16:35–51.

[28] Arundine M, Aarts M, Lau A, Tymianski M. Vulnerability of central neurons to secondary insults after in vitro mechanical stretch. J Neurosci 2004;24:8106–23. https://doi.org/10.1523/JNEUROSCI.1362-04.2004.

[29] Namjoshi DR, Good C, Cheng WH, Panenka W, Richards D, Cripton PA, et al. Towards clinical management of traumatic brain injury: a review of models and mechanisms from a biomechanical perspective. Disease Models & Mechanisms 2013:dmm.011320. https://doi.org/10.1242/dmm.011320.

[30] Roberts VL, McElhaney JH. Review of head injury tolerance to direct impact. Proc IRCOBI 1973:241–58.

[31] Laporte S. « Biomécanique lésionnelle du cerveau », La médecine du sport. https://www.lamedecinedusport.com/biomecanique-lesionnelle-du-cerveau/. Retrieved march 12th, 2024. n.d.

[32] Meaney DF, Smith DH. Biomechanics of concussion. Clin Sports Med 2011;30:19–31, vii. https://doi.org/10.1016/j.csm.2010.08.009.

[33] Sabet AA, Christoforou E, Zatlin B, Genin GM, Bayly PV. Deformation of the human brain induced by mild angular head acceleration. J Biomech 2008;41:307–15. https://doi.org/10.1016/j.jbiomech.2007.09.016.

[34] Gennarelli TA, Thibault LE. Biomechanics of acute subdural hematoma. J Trauma 1982;22:680–6. https://doi.org/10.1097/00005373-198208000-00005.

[35] Kleiven S. Why Most Traumatic Brain Injuries are Not Caused by Linear Acceleration but Skull Fractures are. Front Bioeng Biotechnol 2013;1:15. https://doi.org/10.3389/fbioe.2013.00015.

[36] NATO Science and Technology Organization (NATO). (2023) Injury Assessment Methods for Vehicle Occupants in Blast-Related Events (HFM-271). n.d.

[37] Gurdjian ES, Hodgson VR, Thomas LM. Studies on mechanical impedance of the human skull: Preliminary report. Journal of Biomechanics 1970;3:239–47. https://doi.org/10.1016/0021-9290(70)90025-4.

[38] CHINN B, CANAPLE B, DERLER, EMPA, DOYLE D, OTTE D, SCHULLER E, et al. COST 327 - Motorcycle Safety Helmets - Final Report of the Action 2001.

[39] Dingelstedt KJ, Rowson S. Characterizing Natural Frequencies of the Hybrid III and NOCSAE Headforms. Ann Biomed Eng 2024;52:2831–43. https://doi.org/10.1007/s10439-024-03498-w.

[40] Deck C, Daelen EV, Bourdet N, Philippens MMGM, Willinger R. Assessment of Behind Armour Blunt Trauma risk for a combat helmet by a combined experimental and numerical method. PASS2018 - 14th Personal Armour Systems Symposium 2018, Washington, DC, USA, 1-5 October 2018, 2018, p. 516–26.

[41] Tan LB, Tse KM, Lee HP, Tan VBC, Lim SP. Performance of an advanced combat helmet with different interior cushioning systems in ballistic impact: Experiments and finite element simulations. International Journal of Impact Engineering 2012;50:99–112. https://doi.org/10.1016/j.ijimpeng.2012.06.003.

[42] Aare M, Kleiven S. Evaluation of head response to ballistic helmet impacts using the finite element method. International Journal of Impact Engineering 2007;34:596–608. https://doi.org/10.1016/j.ijimpeng.2005.08.001.

[43] Cordeau B, Barbillon F, Richard B. Assessment of Behind Head Blunt Trauma with an equivalent projectile. 2019.

[44] Op't Eynde J, Pang DY, Morino CF, Abrams MZ, Kait JR. The Fundamental Limitations of Clay for Assessing Human Response for Behind Armour Blunt Trauma, 2023.

[45] Humanetics Innovative Solutions. USER MANUAL Harmonized Hybrid III 50th Male — 78051-9900-H [Rev. A] 2017.

https://doi.org/10.52202/081621-0012

[46] Yoganandan N, Pintar FA, Zhang J, Baisden JL. Physical properties of the human head: Mass, center of gravity and moment of inertia. Journal of Biomechanics 2009;42:1177–92. https://doi.org/10.1016/j.jbiomech.2009.03.029.
[47] Biokinetics Test Equipment Datasheets - V07 (2023). https://biokinetics.com/wp-content/uploads/2023/09/BLSH-Force-Verification-Datasheet-May-2023.pdf. Retrieved March 14th, 2025. n.d.
[48] Iwaskiw A, Howes C, Hingorani R, Bradfield C, Mazuchowski E, Luong Q, et al. Loads Associated with Behind Helmet Blunt Trauma: Matched-Pair Load Sensing Headform Tests Correlated with Skull Fracture Severity, 2023.
[49] Trexler MM, Luong QT, Alphonse V, Perrino K, Athey M, Storck S, et al. Development of a Modular Load-Sensing Headform, 2018.
[50] Grobert S, Gedon H, Peldschus S, Peschel O. Comparative Tests with "Pk17dynA" for an Integrated Assessment of Ballistic Backface Deflections on Combat Helmets, 2018.
[51] Grobert S, Gedon H, Peschel O. Development of a Test Setup for an Integrated Assessment of Ballistic Backface Deflections on Combat Helmets, 2016.
[52] Eppinger R, Sun E, Bandak F, Haffner M, Khaewpong N, Maltese M, et al. Development of Improved Injury Criteria for the Assessment of Advanced Automotive Restraint Systems - II. NHTSA 1999. https://doi.org/PB2001-103435.
[53] Newman JA, Shewchenko N. A Proposed New Biomechanical Head Injury Assessment Function - the Maximum Power Index, 2000, p. 2000-01-SC16. https://doi.org/10.4271/2000-01-SC16.
[54] Frank M, Bockholdt B, Peters D, Lange J, Grossjohann R, Ekkernkamp A, et al. Blunt Criterion trauma model for head and chest injury risk assessment of cal. 380 R and cal. 22 long blank cartridge actuated gundog retrieval devices. Forensic Science International 2011;208:37–41. https://doi.org/10.1016/j.forsciint.2010.10.023.
[55] Wang F, Wang Z, Hu L, Xu H, Yu C, Li F. Evaluation of Head Injury Criteria for Injury Prediction Effectiveness: Computational Reconstruction of Real-World Vulnerable Road User Impact Accidents. Front Bioeng Biotechnol 2021;9:677982. https://doi.org/10.3389/fbioe.2021.677982.
[56] Yu X, Halldin P, Ghajari M. Oblique impact responses of Hybrid III and a new headform with more biofidelic coefficient of friction and moments of inertia. Front Bioeng Biotechnol 2022;10:860435. https://doi.org/10.3389/fbioe.2022.860435.
[57] Bolduc, M., and B. Anctil. "Improved test methods for better protection: a BABT protocol proposal for STANAG 2920." Proc personal armour system symp. Vol. 7. 2010. n.d.
[58] Sturdivan LM, Viano DC, Champion HR. Analysis of injury criteria to assess chest and abdominal injury risks in blunt and ballistic impacts. J Trauma 2004;56:651–63. https://doi.org/10.1097/01.ta.0000074108.36517.d4.
[59] Versace J (1971) A review of the severity index. In Proceed- ings of the 15th Stapp Car Crash Conference, Technical Paper 710881, pp 771–796. Warrendale, PA: Society of Automotive Engineers n.d.
[60] Takhounts EG, Craig MJ, Moorhouse K, McFadden J, Hasija V. Development of Brain Injury Criteria (BrIC). Warrendale, PA: SAE International; 2013. https://doi.org/10.4271/2013-22-0010.
[61] Gadd CW (1966). Use of a Weighted-Impulse Criterion for Estimating Injury Hazard n.d. https://www.sae.org/publications/technical-papers/content/660793/ (accessed March 13, 2025).
[62] United Nations Economic Commission for Europe (UNECE). 2021. UN Regulation No. 22 – Rev 5 – 06 Series." https://unece.org/transport/documents/2021/08/standards/un-regulation-no-22-rev5-06-series". Retrived March 14th, 2025. n.d.
[63] Eppinger R, Sun E, Bandak F, Haffner M, Khaewpong N, Maltese M et al. 1999. Development of Improved Injury Criteria for the Assessment of Advanced Automotive Restraint Systems – II. National Highway Traffic Safety Administration's. https://www.nhtsa.gov/sites/nhtsa.gov/files/criteria.pdf. Retrieved March 14th, 2025. n.d.
[64] Marjoux D, Baumgartner D, Deck C, Willinger R. Head injury prediction capability of the HIC, HIP, SIMon and ULP criteria. Accident Analysis & Prevention 2008;40:1135–48. https://doi.org/10.1016/j.aap.2007.12.006.
[65] Anctil B, Keown M, Bourget D, Pageau G. A Novel Test Methodology to Assess the Performance Ballistic Helmets 2005.
[66] Bolduc M, Anctil B. Improved Test Methods for Better Protection, a BABT Protocol Proposal for STANAG 2920, 2010.
[67] Philippens M. Combat helmet blunt impact performance and test method and injury criteria., 2018.
[68] FprCEN/TR 16148 2010 Head and neck impact, burn and noise injury criteria - A Guide for CEN helmet standards committees, EUROPEAN COMMITTEE FOR STANDARDIZATION Management Centre: Avenue Marnix 17, B-1000 Brussels n.d.

[69] Li Y, Adanty K, Vakiel P, Ouellet S, Vette AH, Raboud D, et al. Review of Mechanisms and Research Methods for Blunt Ballistic Head Injury. J Biomech Eng 2023;145:010801. https://doi.org/10.1115/1.4055289.
[70] Tham CY, Tan VBC, Lee HP. Ballistic impact of a KEVLAR® helmet: Experiment and simulations. International Journal of Impact Engineering 2008;35:304–18. https://doi.org/10.1016/j.ijimpeng.2007.03.008.
[71] Palta E, Fang H, Weggel DC. Finite element analysis of the Advanced Combat Helmet under various ballistic impacts. International Journal of Impact Engineering 2018;112:125–43. https://doi.org/10.1016/j.ijimpeng.2017.10.010.
[72] Miranda-Vicario A, Bravo PM, Coghe F. Experimental study of the deformation of a ballistic helmet impacted with pistol ammunition. Composite Structures 2018;203:233–41. https://doi.org/10.1016/j.compstruct.2018.07.012.
[73] Caçoilo A, Mourão R, Teixeira-Dias F, Azevedo A, Coghe F, Valente RAF. Modelling ballistic impact on military helmets: The relevance of projectile plasticity. Defence Technology 2021;17:1699–711. https://doi.org/10.1016/j.dt.2020.09.011.
[74] Li XG, Gao X-L, Kleiven S. Behind helmet blunt trauma induced by ballistic impact: A computational model. International Journal of Impact Engineering 2016;91:56–67. https://doi.org/10.1016/j.ijimpeng.2015.12.010.
[75] Li Y, Fan H, Gao X-L. Ballistic helmets: Recent advances in materials, protection mechanisms, performance, and head injury mitigation. Composites Part B: Engineering 2022;238:109890. https://doi.org/10.1016/j.compositesb.2022.109890.
[76] Sarron J-C, Dannawi M, Faure A, Caillou J-P, Da Cunha J, Robert R. Dynamic effects of a 9 mm missile on cadaveric skull protected by aramid, polyethylene or aluminum plate: an experimental study. Journal of Trauma and Acute Care Surgery 2004;57:236–43. https://doi.org/10.1097/01.TA.0000133575.48065.3F.
[77] Nsiampa N, Coghe F. Review of Literature: Behind Helmet Blunt Trauma Mechanisms. Hum Factors Mech Eng Def Saf 2023;7:6. https://doi.org/10.1007/s41314-023-00063-6.
[78] Kim SY, Yeh P-H, Ollinger JM, Morris HD, Hood MN, Ho VB, et al. Military-related mild traumatic brain injury: clinical characteristics, advanced neuroimaging, and molecular mechanisms. Transl Psychiatry 2023;13:1–15. https://doi.org/10.1038/s41398-023-02569-1.
[79] Huang X, Chang L, Zhao H, Cai Z. Study on craniocerebral dynamics response and helmet protective performance under the blast waves. Materials & Design 2022;224:111408. https://doi.org/10.1016/j.matdes.2022.111408.

https://doi.org/10.52202/081621-0012

Preliminary study of Behind Helmet Blunt Trauma Assessment using a Numerical Head Model

N. Nsiampa[1], F. Coghe[1]

[1] *Royal Military Academy, Renaissancelaan 30, Brussels 1000, Belgium*
nestor.nsiampa@dymasec.be

Abstract. Modern combat helmets, made of flexible composite materials with high ballistic performance and low weight, have significantly improved both ballistic protection and ergonomic comfort for soldiers. However, these lightweight helmets have a drawback: upon impact, they exhibit increased backface deformation, which raises the risk of blunt trauma injury, also known as Behind Helmet Blunt Trauma. Such injuries can impair a soldier's performance and, in severe cases, pose a threat to his/her life.
The study of the injury mechanisms associated with the Behind Helmet Blunt Trauma is challenging due to the highly dynamic and complex nature of ballistic impact events. To investigate this type of trauma, we used a finite element head model, originally developed for non-lethal impact applications. For this purpose, we replicated the experiments conducted by Freitas using a numerical approach. Due to the limitations in the modeling approach, only a qualitative analysis was performed. The preliminary results indicate that the model offers valuable insight into the impact dynamics. The role of helmet pads in mitigating injury is highlighted with off-pad impacts showing a notably higher risk of cranial injury compared to on-pad impacts. This trend is consistent with the experimental findings reported by Freitas.

1. INTRODUCTION

Most modern helmets are made of aramid material or of Ultra-High Molecular Weight Polyethylene (UHMWPE) material. These new composite flexible materials have improved soldier protection against fragments and small caliber rounds, and as well as comfort. However, they have also increased the risk of Behind Helmet Blunt Trauma (BHBT) due to an increase of helmet back face deformation (BFD), which may impair the soldier's performance and even endanger his/her life.

BHBT is a type of blunt trauma characterized by an induced non-penetrating injury to the head caused by helmet impact-related events. A ballistic impact on a helmeted soldier can be divided into two impact phases: a primary impact caused by the interaction between the threat and the helmet shell and a secondary impact between the helmet dynamic BFD and the head. The former impact can be considered as a high speed and low mass impact and the latter as a lower speed and slightly higher mass [1,2]. Therefore, the strain rates generated during the initial bullet-helmet impacts are higher than those induced by the helmet's BFD on the head.

Numerous studies have been conducted, particularly in the civilian world, to better understand the injury mechanisms associated with blunt impacts characterized by lower rate loadings and develop safety equipment such as helmets in order to ensure the protection and safety of people and reduce severe or fatal injuries in road, sports, automotive and workplace accidents [3-7]. For ballistic impact events characterized by high rate loadings, only a few studies have investigated this type of impacts and the resulting injuries. Tests on human beings are ethically unacceptable, so mechanical or biological surrogates are used to provide baseline biomechanical response, to investigate BHBT injury and injury tolerance levels and to assess the risk of injury associated to such events [8-12]. Unfortunately, there is still a lack of standardized procedures for testing methods, clear injury criteria, and injury tolerance levels for injury assessment.

Today, with the development of computing techniques, finite element (FE) models offer an additional alternative method [12-16]. However, there are still multiple challenges to overcome. On the one hand, anthropometric human body FE models are built with great levels of anatomical details, but the validation is still incomplete because biomechanical test data are generally complicated and expensive to obtain. On the other hand, material models that describe the model response under high rate strain loadings require dynamic characterization experiments in order to obtain the model parameters. However, this is not straightforward due to the anisotropic behavior of human tissues and the short time scale involved.

In the context of performing BHBT risk assessment using numerical simulations, an in-house 50th percentile finite element head model, originally developed for kinetic energy non-lethal applications, is used to investigate to which extent this model can also be implemented for BHBT applications. This preliminary study aims to compare the results of the head model's results with experimental data

https://doi.org/10.52202/081621-0013

reported in Freitas'paper [11]. The present study is limited to the impact of a 9x19mm Full Metal Jacket (FMJ) projectile on a helmeted head under frontal 0° NATO impact conditions.

The next sections give an overview of Freitas' experiments and a brief summary of head injury criteria.

1.1 Freitas experiments

Freitas et al. conducted non-perforating ballistic tests using various projectile types on a helmeted Human Head Surrogate (HHS) to gain insights into injury mechanisms within the scope of BHBT studies [11]. The HHS was composed of refreshed human craniums, a cerebral spinal fluid (CSF) surrogate, and soft tissue surrogates representing the skin, dura, and brain. Perma-gel was used to simulate the skin and was molded to the cranium, while the brain surrogate consisted of a mixture of perma-gel and iron powder. The CSF surrogate was water pressurized to 2 MPa, and the dura surrogate was modeled as a thick layer of silicone. The head surrogate was supported by a Hybrid III 50th Male Neck assembly.

The helmet used for the tests was the US Marine Corps' large-sized Lightweight Combat Helmet (LWH). This helmet is made of a para-aramid shell, and a pad system and a suspension system [17-18]. The protection system for lower energy projectiles consisted of the helmet only. For higher energy projectiles, the protection system consisted of the helmet with a ceramic applique layup. The pad system consisted of 7 paired sets of hard and soft pads that were distributed as follows: one crown pad set, four lateral pad sets, and one front and one back pad sets. Two different pad systems configurations were used. The 7-pad configuration consisted of paired pads located at the front center, front right, front left, back center, back right, back left, and on the crown and the 5-pad configuration was derived from the 7-pad configuration where the front and the back center paired pads were removed. Two types of frontal impacts were considered: the front on-pad impacts for the 7-pad configuration and the front off-pad impact (no pad support on the impact location) for the 5-pad configuration.

On-pad impacts with 9x19 mm FMJ bullets resulted in lower intracranial pressures and skull strains and consequently, reduced likelihood of injuries when compared with the off-pad impacts. The authors concluded that in general, the pads provided an effective energy dissipation mechanism that helped to mitigate injury. Moderate skull fractures were observed for front center off-pad impacts while there was no injury or cranial fracture for on-pad impacts. The mean peak intracranial pressure and the mean peak strain value they reported for the occurrence of moderate cranial fracture was 255 kPa and of -0.21% respectively.

1.2 Injury criterion

A variety of head injury criteria are referenced in the literature for assessing head trauma. Generally, the type and severity of blunt impact injuries depend on the rate and magnitude of the impact load, as well as the specific region of the body affected. The main head injuries are skull fracture, skin lacerations or contusions, concussion, hemorrhaging, hematoma (blood clots), anoxic injury (lack of oxygen), and diffuse axonal injury (DAI) which is a damage to the brain neurons [19].

The injury can be mild, severe, or even fatal. A variety of physical parameters are used to predict head lesions: the maximum head force, the intracranial pressure (ICP), the head accelerations (linear or rotational), the strain, the stress and the energy. A review of injury criteria can be found in [20,21].
It is worth mentioning that there is still a lack of standardized procedures for testing methods, clear injury criteria, and injury tolerance levels for injury assessment from ballistic impacts. For example, Table I gives tolerance values found in literature for the occurrence of skull injury for blunt impacts relative to intracranial and peak maximum principal strain.

Table 1. Reported skull fracture tolerances

Skull fracture			
Injury criterion	***Tolerance***	***Risk of injury***	***Refs.***
Intracranial pressure			
	51.2 kPa	50% skull injury	[22]
	560 kPa	50% risk of severe fracture injury	[23]
	150 kPa	Significant risk of fracture	[24]
	255 kPa	Moderate cranial injury	[11]
Peak maximum principal Strain			
	-0.5 %	50% skull injury	[25,26]
	-0.4 %	Fracture	[11]

https://doi.org/10.52202/081621-0013

2. METHODOLOGY

To gain deeper insights into BHBT and assess the associated injury risks, a finite element model of a human head protected with a combat helmet was used to perform ballistic impact simulations using 9x19 mm FMJ bullets. The 50th percentile head model was developed within the framework of injury risk assessment for kinetic energy non-lethal impacts and the validation is presented in Section 3.2. The aramid helmet shell model was developed and validated against experimental ballistic tests on BFD in [27]. The material properties of the 9 x19 mm FMJ bullet were retrieved from [28]. Section 3.3 presents the validation of the foam pads of the helmet.

The setup of the impact of 9x19 mm FMJ on the protected head was similar to the experiment one with two configurations, the 7-pad configuration and the 5-pad configuration (see section 1.1). No head support was modeled. Two types of impacts were considered as in the experiments: the on-pad impact and the off-pad impact. Only frontal normal (0° NATO) impacts were considered in this study. All the simulations were performed with LS-Dyna software [29].

To assess the risk of skull fracture, intracranial pressure and cranial strain were used as injury criteria and to compare the numerical results with those from Freitas' experiments, intracranial pressure was measured within the CSF and the cranial strain was measured in the skull at the same location as in the experiments.

3. FE modelling

3.1 Setup

Figure 1 shows the setup for the impact of 9x19 mm FMJ against a protected head. Experimental tests were carried out using large-sized aramid helmets while in the simulations, medium sized aramid helmets were considered to be consistent with the 50th percentile head model. Unlike the experiments, the numerical simulations did not include a suspension system. Additionally, unlike in the experiments where the head was supported by the Hybrid III neck assembly, the head was not supported and could move freely. Finally, material properties of the head soft tissues were also different from the experiments as the actual soft tissue material models defined in the 50th percentile model for non-lethal impacts were used (see section 3.2).

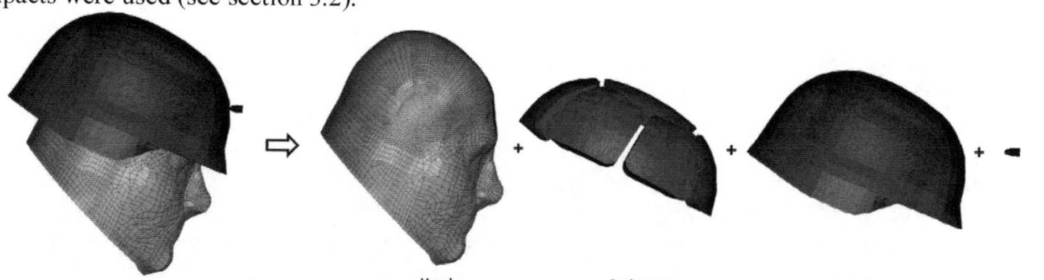

Figure 1: Setup of the impact of 9x19 mm against a protected head

Due to symmetry considerations, a half-model was used in the simulations. The analysis focused on two distinct helmet-only impact scenarios: (i) a frontal on-pad impact with the 7-pad configuration, and (ii) a frontal off-pad impact with the 5-pad configuration. Both cases involved normal (0° NATO) impacts using a 9 x19 mm FMJ projectile at a velocity of 435 m/s. Numerical sensors were placed within the CSF and the skull at the frontal impact site, matching the experimental setup, to record intracranial pressure and strain time histories. The next two sections briefly describe the validation of the head model and the foam models.

3.2 The head model validation

The 50th percentile head model developed in the framework of kinetic energy non-lethal applications was used in this study. Table 1 summarizes the different organs/tissues of the head that were modeled and the corresponding material properties. All the head organs except the CSF were modeled with linear elastic material models. A fluid-like material model was used for the CSF. It is important to note that no failure model was implemented in the current head model.

Table 2. Material properties of the head model

Organ	Material model	ρ (g/mm^3)	E (MPa)	ν (-)	K (Mpa)	G^0 (Mpa)	G^∞ (MPa)	β (ms^-1)	Refs.
Skull	Elastic	2.00E-3	6000	0.21	-	-	-	-	[30]
Falx Tentorium Meninges	Elastic	1.14E-3	31.5	0.45	-	-	-	-	[31]
Mandible Ethmoïd Face	Elastic	2.00E-3	5000	0.23	-	-	-	-	[31]
Scalp	Elastic	1.24E-3	16.7	0.42	-	-	-	-	[31]
Cerebrum Cerebellum Brainstem Callosum	Viscoelastic	1.04E-3	-	-	1125	0.049	0.162	0.0162	[31]
						VC			
CSF	Fluid	1.04E-3			2190	0.1			[32]

Where ρ is the material density, E, the Young modulus, ν, the Poisson's ratio, K, the bulk modulus, G^0 and G^1 are short-time shear modulus and infinite shear modulus, and VC, the viscous coefficient.

The head model was validated against the experimental data from low dynamic impact tests performed on cadavers by Yoganandan et al. [33] and Nahum et al. [34]. Figure 2 presents a comparison between the numerical and experimental results. In both cases, satisfactory agreement was achieved when predicting the impact force and intracranial pressure (ICP).

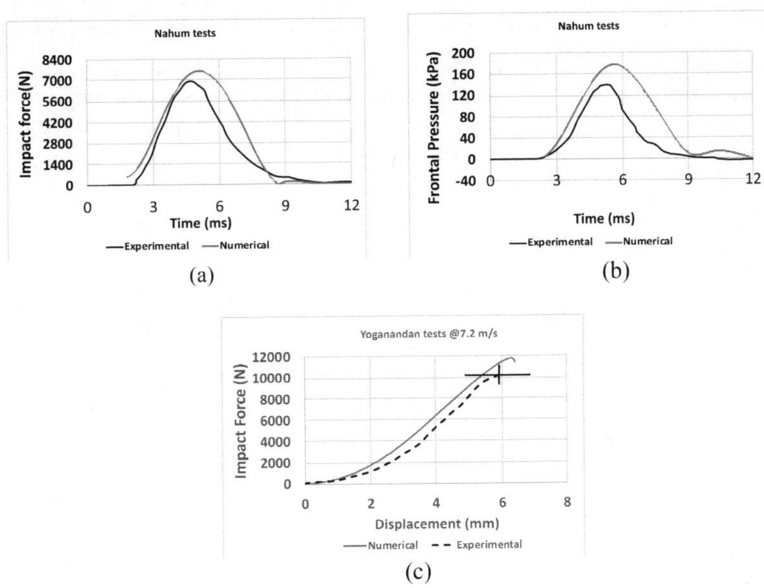

Figure 2: Head model validation: (a)-(b) Nahum tests[34]: Impact force and intracranial pressure ICP (c) Yoganandan tests [33]: impact force-displacement

3.3 The foam model validation

The helmet incorporated a foam pad system, consisting of a combination of hard foam and soft foam layers to optimize both energy absorption and wearer comfort. To validate the pad system, quasi-static tests performed in [35] are numerically replicated.

The quasi-static tests (Figure 3) involved a 5 kg-cylinder, representing a human head, impacting a bilayer foam pad composed of paired hard and soft foam pads wrapped in a water-resistant, airtight covering (not modelled in this study). Each foam pad had a diameter of 127 mm and a thickness of 9.5 mm. The foam pad materials were modeled using hyperelastic models, with compressive stress-strain curves at two different strain rates implemented as input curves for each foam type. (Figure 4). The impact velocities were 3.1 m/s and 4.3 m/s.

https://doi.org/10.52202/081621-0013

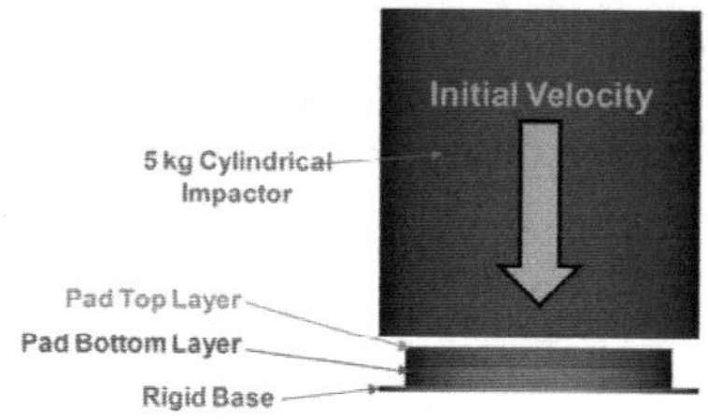

Figure 3. Experimental cylinder impact test setup [35]

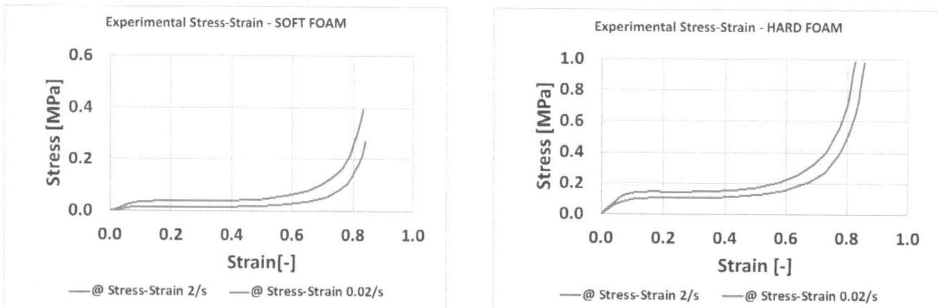

(a) (b)

Figure 4. Experimental stress-strain loading curves: (a) Soft foam - (b) Hard foam [35]

Comparison between numerical results and experimental results are given in Figure 5. The simulation results show good correspondence with the experimental data especially for the impact velocity of 3.1 m/s. The model overestimates the cylinder acceleration for the impact velocity of 4.3 m/s i.e. the material tends to exhibit an increased stiffness.

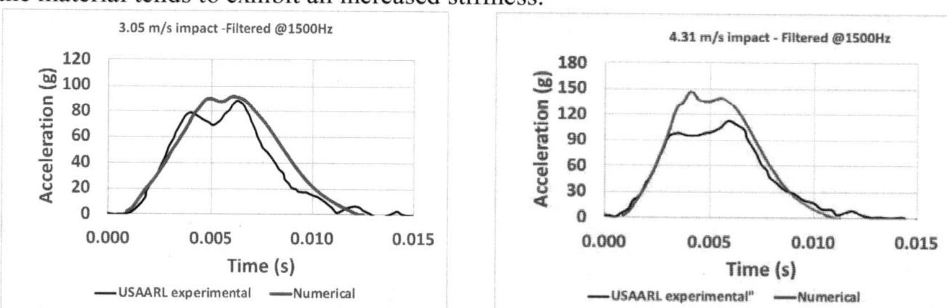

(a) (b)

Figure 5. Comparison between numerical and experimental results: (a) 3.1 m/s - (b) 4.3 m/s

4. SIMULATIONS

This section describes the results of the impact of 9mm FMJ bullet on the protected head.

4.1 Results

4.1.1 Seven-pad and five-pad configuration results

Figures 6(a) and 6(b) illustrate the evolution of intracranial (CSF) pressure and cranial strain at the impact site for both the 7-pad (on-pad) and 5-pad (off-pad) configurations. The intracranial pressure curves in Figure 6(a) exhibit high-frequency oscillations in both cases, with a noticeably greater damping effect observed in the on-pad configuration. This damping effect is further emphasized in the auto power spectrum (energy distribution) analysis of the two pressure signals shown in Figure 6(c). Additionally, Figure 7(d) presents the impact force curves for both configurations, revealing that the maximum impact force in the on-pad configuration is lower than that in the off-pad configuration.

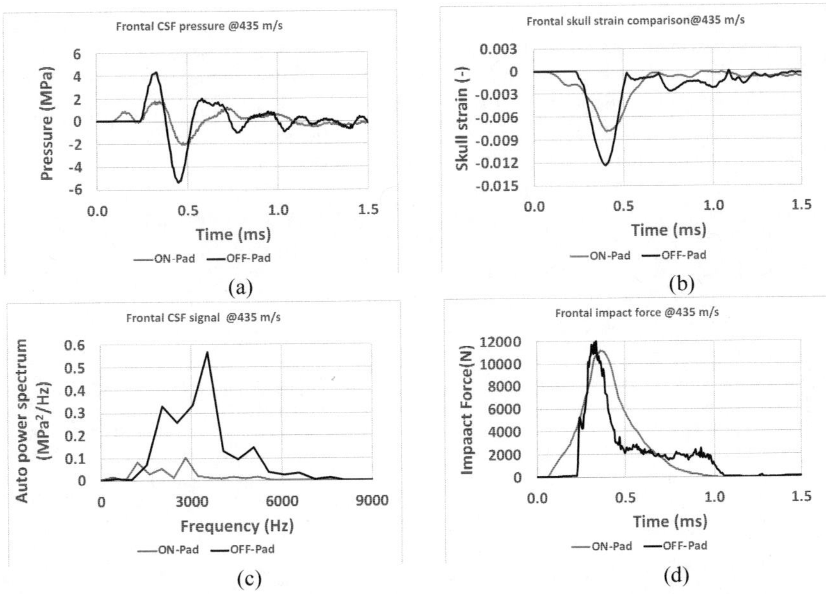

Figure 6. Numerical results corresponding to the 7-pad (on-pad impact) and 5-pad (off-pad impact) configurations at the impact zone.

A closer examination of the on-pad impact dynamics, especially during the loading phase shows how the energy is transferred and/or absorbed, contributing to a deeper understanding of head injury mechanisms. Figure 7 provides insight into the complex impact dynamics by illustrating the chronological evolution of the impact event within the impact zone. This evolution is effectively highlighted through the skull strain time history, which clearly delineates the distinct phases of the impact loading dynamics.

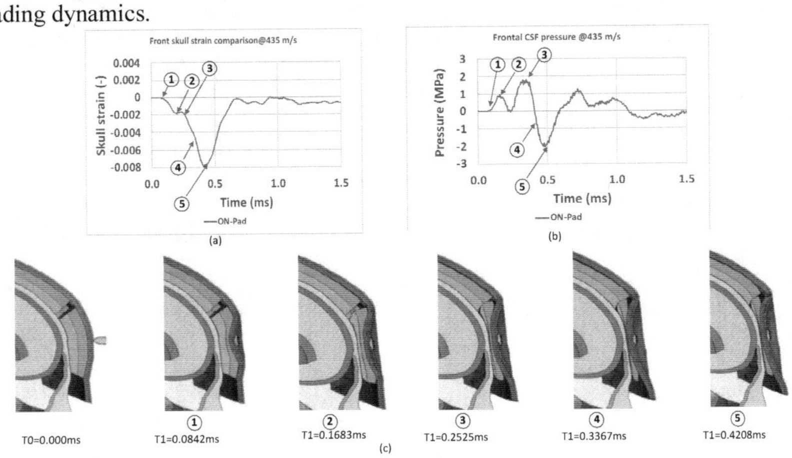

Figure 7. Different phases of the frontal impact dynamics

These distinct phases are globally characterized by the change in the slope of the strain time history, indicating a transition to another impact loading phase as observed in Figure 7(a) and Figure 7(c).

Up to point 1: The initial impact leads to the compression of the hard foam pad (pink color), the soft foam pad (blue color) is only slightly compressed;

Up to point 2: The soft foam pad is almost fully compressed under the loading of the hard foam and the helmet shell, but the skin (light blue color) is only slightly compressed;

Up to 3: Both foam pads continue to be compressed by the action of the helmet and act as a rigid body leading to a significant loading of the skin layer and the head;

Up to point 4: The skin layer becomes fully compressed under the loading of the protective system which continues to act as a rigid body;

https://doi.org/10.52202/081621-0013

Up to point 5: at this stage, the head has reached a full loading, and the unloading process begins. Beyond point 5, the unloading process continues.

After identifying these key points in the strain time history, corresponding points can also be identified in the pressure time history as shown in Figure 7(b).

4.1.2 Numerical vs experimental results for the 7-pad configuration

Figure 8 presents the numerical and experimental curves of the front CSF pressure and the front skull minimum principal strain within the impact location. There is a difference between the two curves of the two metrics in term of profile and value.

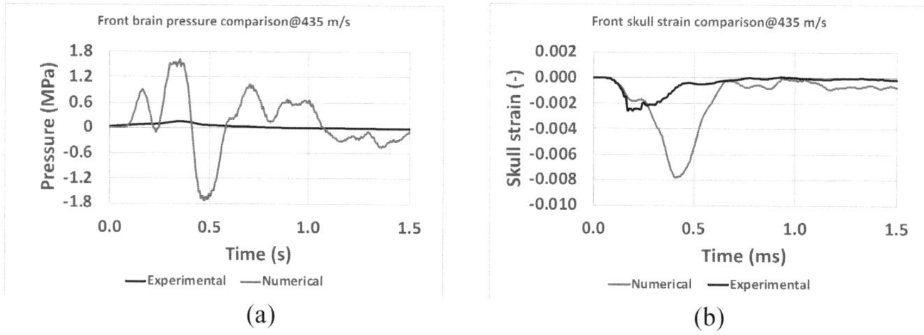

Figure 8. Numerical and experimental results: (a) front pressure and (b) front skull minimal principal strain at the impact zone in the 7-pad (on-pad) configuration

4.2 Discussions

4.2.1 Comparison between the 7-pad and 5-pad configurations

Figures 6(a)-(c) show the differences in the head dynamic response associated with the 7-pad and 5-pad configurations. These differences are particularly highlighted in Figure 6(d) when we compare the impact force. Indeed, the on-pad impact results in a smoother mechanical loading of the head due to the cushioning effect of the pads (mainly the center pad), which gradually absorb and distribute the impact force. In contrast, the off-pad impact configuration, characterized by a lack of a front center padding leads to a more sudden and direct impact of the helmet shell backface to the head. This difference in loading mechanisms significantly affects the mechanical response of the head, as reflected in the variations in intracranial pressure and cranial strain observed between the two configurations. Moreover, the intracranial pressure of 4.2 MPa and the minimum principal strain of (-1.2%) for the off-pad impact are greater than the intracranial pressure of 1.9 MPa and the minimum principal strain of (-0.75%) for the on-pad impact, and similarly, the maximum impact force for the off-pad impact (12 kN) is greater than for the on-pad impact (11kN). These results indicate that there is a higher risk of skull fracture associated with the off-pad configuration than the on-pad configuration. A similar trend was observed in the experimental data.

The unloading process in Figure 6(d) differs between the two configurations. In the on-pad configuration, unloading is smooth and continuous. In contrast, the off-pad configuration shows an initial phase of quick unloading of the helmet shell followed by a plateau, during which the residual force exerted on the head remains nearly constant. This plateau may correspond to the ongoing relaxation of the pads. Finally, a terminal phase occurs where the force drops to zero, indicating that there is no more contact between the helmet backface and the head.

A frequency analysis using the auto power spectrum performed on the pressure time histories (Figure 7(c)) reveals a significant attenuation in amplitude (or energy) for the on-pad configuration, along with a leftward shift (towards lower frequencies) in the spectrum indicating a reduced frequency spectrum compared to the off-pad configuration. The on-pad configuration compared to the off-pad impact demonstrates a clear damping effect on high-frequency oscillations. This suggests that the pads effectively absorb and dissipate impact energy, thereby reducing the intensity of the interaction between the BFD and the head.

On the other hand, the head is a complex structure and the material properties of the head model play also a key role in the head dynamics as well as the wave reflections at different interfaces. Moreover, the intracranial content (brain, cerebellum …) constituted of soft tissues surrounded by the CSF (modelled as a fluid) may oscillate relative to the skull.

4.2.2 Comparison between the numerical and experimental results

Regarding skull injury severity in BHBT assessment, the maximum intracranial pressure and minimum principal strain values predicted by the numerical head model for both on-pad and off-pad impacts are significantly higher than the injury tolerance thresholds reported by Freitas et al. [11] and other values summarized in Table 1. This indicates that the numerical model may overestimate injury severity. However, this observation should be interpreted with caution regarding the actual occurrence of skull fracture injuries based solely on these results. The discrepancy likely stems from the fact that the numerical model does not fully replicate the dynamics of the experimental tests particularly because the helmet model was validated using data from helmets different than those in Freitas' experiments. This highlights the need for further investigation and refinement of the helmet material models to enhance the fidelity and predictive accuracy of the simulations assuming that material data for the specific type of LWH helmets used by Freitas can be obtained.

Aside from the helmet considerations, two additional factors may be investigated regarding the head model:

(i) The soft tissue models used in the head model may behave differently from the experimental soft tissue surrogates, as they were not validated against these surrogates.

(ii) The current numerical head model was initially developed for non-lethal impact applications, which may not be suitable for BHBT applications. Therefore, it may be necessary to further investigate and adapt the head model to improve its predictive capability for BHBT. This will require more biomechanical data specifically derived from ballistic tests.

In Section 4.1.1, key points were identified corresponding to the onset or transition between different phases of the impact loading process, as observed in both the strain and pressure time histories in Figure 7 and Figure 9(a). When these key points are compared to the experimental pressure time history, similar features can also be identified that delineate the distinct loading phases previously described (Figure 9(b). This similarity across the various impact phases suggests that the numerical model can qualitatively capture the characteristic impact dynamics observed in the experimental tests.

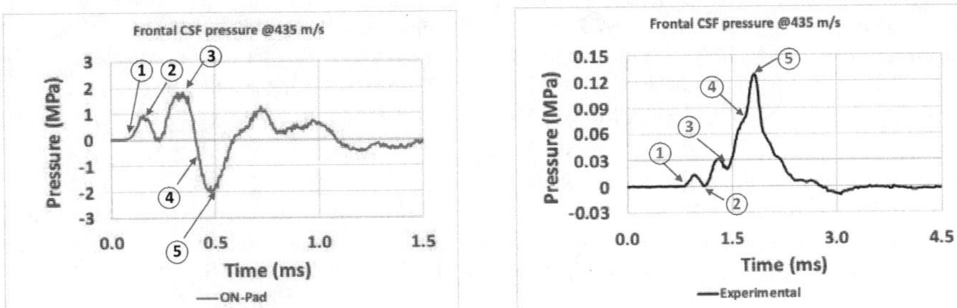

(a) (b)

Figure 9. Pressure histories -Idenfication of transition corresponding points of the impact phases: (a) Simulation – (b) Freitas' experiment [11]

4.2.3 Limitations

There are few limitations to our numerical approach. There was a difference in helmet material properties between the helmets consistent with the 50th percentile head model from which the material model used in our study was derived and the large sized helmets used in Freitas' experiments. Likewise, the head soft tissues material properties of the head model were different from those of the experimental surrogates which might not accurately reflect the mechanical response. Therefore, the impact dynamics may not have been correctly captured.

The pad model did not incorporate high strain rate dependency.

The numerical head was free to move while the HHS was fixed to the Hybrid III neck.

The finite element head model used was originally developed for non-lethal impact applications and might not fully capture the complex biomechanical response of head tissues under high-rate ballistic loading.

https://doi.org/10.52202/081621-0013

5. CONCLUSIONS

An in-house finite element model of the head, originally developed for non-lethal impact applications, was used within the BHBT framework to assess its applicability to BHBT scenarios. Freitas' experiments were replicated to see at which extent the head model could be used to investigate the BHBT and to gain a better understanding of the impact dynamics. The intracranial pressure and the cranial strain were used as metrics for cranial injury risk assessment.

The results showed that the model could qualitatively capture the main phases of the impact dynamics and provide valuable insight into the mechanical behavior observed during ballistic impacts. Key features which delineated the different phases of the loading process were identified that helped to better understand the impact dynamics. Similar features were also identified in the experiments.

A clear distinction was observed between the 7-pad (on-pad) and 5-pad (off-pad) impact configurations, with the off-pad impacts generating higher pressure, strain, and impact forces showing the role of the pads in dissipating the energy. Therefore, the model predicted a higher likelihood of cranial injury with the 5-pad configuration compared to the 7-pad configuration. A similar trend was observed in experiments.

When compared to the experimental tolerance thresholds, the numerical results consistently showed higher values for all the two metrics, indicating a risk of high predicted severity of cranial injury. However, this observation should be taken with cautious.

A further investigation to enhance the predictive capability of the head model by using additional ballistic biomechanical experiment or by incorporating strain rates dependent material models to see at which extent the model developed for non-lethal applications is suited for BHBT. Nevertheless, through this qualitative analysis, the model showed promising and potential capability for BHBT study.

References

[1] Hisley, D. M., Gurganus, J. C., and Drysdale, A. W. (August 8, 2011). "Experimental Methodology Using Digital Image Correlation to Assess Ballistic Helmet Blunt Trauma." ASME. J. Appl. Mech. September 2011; 78(5): 051022. https://doi.org/10.1115/1.4004332

[2] Young L, Rule GT, Bocchieri RT, Walilko TJ, Burns JM, Ling G. When physics meets biology: low and high-velocity penetration, blunt impact, and blast injuries to the brain. Front Neurol. 2015 May 7;6:89. doi: 10.3389/fneur.2015.00089. PMID: 25999910; PMCID: PMC4423508.

[3] Determining Estimates of Lives and Costs Saved by Motorcycle Helmets, (NHTSA, March 2011. DOT HS 811 433, National Highway Traffic Safety Administration, Washington, DC), http://www-nrd.nhtsa.dot.gov/Pubs/811433.pdf

[4] Yang K.H., Hu J., White N.A., King A.I., Chou C.C., Prasad P., Development of numerical models for injury biomechanics research: a review of 50 years of publications in the Stapp Car Crash Conference. Stapp Car Crash J. (2006); 50:429-90. doi: 10.4271/2006-22-0017. PMID: 17311173.

[5] Global Status Report on Road Safety: Time for Action, WHO, (2009). The World Health Organization, Geneva, ISBN 9789241563840 https://iris.who.int/bitstream/handle/10665/44122/9789241563840_eng.pdf

[6] Abbas A.K., Hefny A.F., Abu-Zidan F.M., Does wearing helmets reduce motorcycle- related death? A global evaluation. Acct Anal Prev (2012) 49:249–252, doi:10.1016/j.aap.2011.09.033

[7] Snyder R.G.; Foust D.R.; Bowman B.M., Study of impact tolerance through free-fall investigations. Technical report (1977). Corporate Author: Highway Safety Research Institute, Ann Arbor, Mich, https://hdl.handle.net/2027.42/725

[8] Bass C., Bolduc M., and Waclawik S., Development of a Non-Penetrating, 9mm Ballistic Helmet Trauma Test Method. Proceedings of the Personal Armour Systems Symposium, (2002) pp. 18-22

[9] Raymond D.E. Biomechanics of blunt ballistic temporo-pariental head impact. PhD thesis (2008), Wayne State University, Detroit, Michigan.

[10] Rafaels K.A., Cutcliffe H.C., Salzar R.S., Davis M., Boggess B., Bush B., Harris R., Rountree M.S., Sanderson E., Campman S., Koch S., Dale Bass C.R., Injuries of the head from backface deformation of ballistic protective helmets under ballistic impact. J Forensic Sci. 60(1):219-25. doi: 10.1111/1556-4029.12570. Epub 2014 Jul 17. PMID: 25039407

[11] Freitas C.J., Mathis J.T., Scott N., Bigger R.P., and MacKiewicz J. "Dynamic response due to behind helmet blunt trauma measured with a human head surrogate," International Journal of Medical Sciences, 08 Mar 2014, 11(5):409-42, https://doi.org/10.7150/ijms.8079

[12] Merkle A.C., Wing I.D., Roberts J.C., et al. Human Surrogate Head Response to Dynamic Overpressure Loading in Protected and Unprotected Conditions. 26th Southern Biomedical Engineering Conference SBEC 2010, IFMBE Proceedings. 2010; 32: 22-25.

[13] Rodriguez-Millan M., Rubio I., Burpo F.J., Olmedo A.,. Loya J.A, Parker K.K., Miguélez M.H, Impact response of advance combat helmet pad systems, International Journal of Impact Engineering, Volume 181,023,104757, ISSN 0734-743X, https://doi.org/10.1016/j.ijimpeng.2023.104757.
[14] Rodriguez-Millan M., Rubio I., Burpo F.J., Tse K.M., Olmedo A., Loya J.A., Parker K.K., Miguélez M.H., Experimental and numerical analyses of ballistic resistance evaluation of combat helmet using Hybrid III headform, International Journal of Impact Engineering, Volume 179, 2023, 104653, ISSN 0734-743X, https://doi.org/10.1016/j.ijimpeng.2023.104653
[15] Pinnoji, P.K., Mahajan, P. Finite element modelling of helmeted head impact under frontal loading. Sadhana 32, 445–458 (2007). https://doi.org/10.1007/s12046-007-0034-6
[16] Palomar-Toledano, M.; Lozano-Mínguez, E.; Rodriguez-Millán, M.; Miguélez, MH.; Giner Maravilla, E. (01-1). Relevant factors in the design of composite ballistic helmets. Composite Structures. 201:49-61. https://doi.org/10.1016/j.compstruct.2018.05.076
[17] https://ciehub.info/spec/PD/FQ-PD-06-35A.pdf
[18] Moss W.C., King M.J., Blackman E.G.. Towards reducing impact-induced brain injury: lessons from a computational study of army and football helmet pads. Comput Methods Biomech Biomed Engin. 2014;17(11):1173-84. doi: 10.1080/10255842.2012.739162. Epub 2012 Dec 18. PMID: 23244512.
[19] Review of Department of Defense Test Protocols for Combat Helmets (2014); Committee on Review of Test Protocols Used by the Department of Defense to Test Combat Helmets. National Academies Press (US): Washington, DC, USA
[20] Li Y., Adanty K., Vakiel P., Ouellet S., Vette A.H., Raboud D., Dennison C.R. (2023) Review of mechanisms and research methods for blunt ballistic head injury. J Biomech Eng 145(1): 010801 (13 pages) Paper No: BIO-22–1088. https://doi.org/10.1115/1.40552 89
[21] Nsiampa, N., Coghe, F. Review of Literature: Behind Helmet Blunt Trauma Mechanisms. Hum Factors Mech Eng Def Saf 7, 6 (2023). https://doi.org/10.1007/s41314-023-00063-6
[22] Bass C., Bolduc M., and Waclawik S. (2002) Development of a Non-Penetrating, 9mm Ballistic Helmet Trauma Test Method. Proceedings of the Personal Armour Systems Symposium, pp. 18-22
[23] Sarron JC, Dannawi M, Faure A, Caillou JP, Da Cunha J, Robert R (2004) Dynamic effects of a 9 mm missile on cadaveric skull protected by aramid, polyethylene or aluminum plate: an experimental study. J Trauma. 2004 Aug;57(2):236-42; discussion 243. doi: 10.1097/01.ta.0000133575.48065.3f. PMID: 15345967
[24] NATO – AEP-103 Head injuries assessment of non-lethal projectiles. July 2021
[25] Raymond DE (2008) Biomechanics of blunt ballistic temporo-pariental head impact. PhD thesis, Wayne State University, Detroit, Michigan.
[26] Raymond D., Van Ee C., Crawford G., Bir C. (2009) Tolerance of the skull to blunt ballistic temporo-parietal impact. J Biomech. Nov 13;42(15):2479-85. doi: 10.1016/j.jbiomech.2009.07.018. Epub 2009 Aug 11. PMID: 19674749
[27] Nsiampa N., Coghe F., Modeling of Schuberth helmet under ballistic impact for head injury assessment, 34th International Symposium on Ballistics, May 2025
[28] Krishnan K., Rajan S.D., Belegundu A.D., (2008) A general optimization methodology for ballistic panel design. Paper presented at the Eng Opt 2008 conference, Rio de Janeiro, Brasil
[29] https://www.ansys.com/products/structures/ansys-ls-dyna
[30] Tan L.B., Chew F.S., Tse K.MM, Chye Tan V.B., Lee H.P.. Impact of complex blast waves on the human head: a computational study. Int J Numer Method Biomed Eng. 2014 Dec;30(12):1476-505. doi: 10.1002/cnm.2668. Epub 2014 Sep 17. PMID: 25132676.
[31] Deck C., Willinger R. Head injury prediction tool for protective systems optimisation. In 7th European LS-DYNA Conference, DYNAmore GmbH, Salzburg, Austria, May 14-15, 2009.
[32] Willinger R., Taleb L., Kopp C. Modal and temporal analysis of head mathematical models. J Neuro- trauma. 1995; 12: 743–754. https://doi.org/10.1089/neu.1995.12.743 PMID: 8683626
[33] Yoganandan N., Pintar F.A., Sances A., Walsh P.R., Ewing C.L., Thomas D.J., Snyder R.G., Biomechanics of skull fracture. J Neurotrauma 1995;12(4):659-668.
[34] Nahum A.M., Smith R., Ward C.C. Intracranial pressure dynamics during head impact, SAE Tech. Pap. 770922 (1977).
[35] Moss W. & King M. (2011). Impact Response of US Army and National Football League Helmet Pad Systems. 60. 10.2172/1021058.

 https://doi.org/10.52202/081621-0013

Behind Helmet Blunt Trauma Due to Ballistic Mandible Protectors

A. Iwaskiw[1], N. Hahne[1], C. Howes[1], S. Gravelyn[1], A. Lurski[1], A. Rohrer[2], N. Tsantinis[3], M. Clark[3]

[1]*The Johns Hopkins University Applied Physics Laboratory (JHU/APL), 11100 Johns Hopkins Road, Laurel, MD 20723, alexander.iwaskiw@jhuapl.edu*
[2] *Armed Forces Medical Examiner System, Defense Health Agency, Dover Air Force Base, DE*
[3] *U.S. Special Operations Command, 10 General Greene Avenue, Natick, MA 01760*

Abstract. Ballistic mandible protectors (BMPs) have recently been integrated with combat helmets to protect the face from primarily blunt and ballistic threats. Advancements in armour materials, design, and fabrication have enabled some BMP to stop small arms rounds from penetrating. In preventing penetration, armour backface can deform and strike the wearer causing injury, termed behind helmet blunt trauma (BHBT). It is of great interest to assess the risk of BHBT involving BMP. In this study, live ammunition striking an ultra-high molecular weight polyethylene BMP was used to create BHBT loading. Four postmortem human surrogate (PMHS) tests were conducted at ballistic conditions near muzzle velocity, using one BMP per PMHS. Two anatomical impact locations were tested; directly over the maxilla intersecting the mid-sagittal plane and over the mandible oblique from the mid-sagittal plane. X-rays and computed tomography scans were acquired before and after the test and were used in aiding characterization of the BHBT injury. An anatomical dissection, facilitated by a board-certified forensic pathologist, was conducted to physically evaluate the characteristics and severity of the injuries identified in the medical imaging. Backface armour deformation of the BMP was found to be substantial. The distance from the inner surface of the BMP and the face was greater (≈45 mm) for the maxilla shot versus the mandible shot (≈25 mm). Injuries observed in the maxilla impact location included lacerations of the upper and lower lips, teeth avulsions and fractures, and maxilla, hard palate, and additional facial bone fractures. Injuries observed in the off-centre mandible impact location included upper lip, lower lip, and tongue lacerations, teeth avulsions and fractures, and comminuted maxilla and mandible fractures. BHBT injuries from BMP provided better outcomes than penetrating injuries at the same locations. These BHBT injuries are thought to be potentially survivable as determined by a forensic pathologist, justifying the use of BMP.

1. INTRODUCTION

Combat helmets represent a multifaceted tool available to the warfighter. The helmets' primary function is to protect the wearer from various threats including protection from blunt loading from falls and large objects, protection from threats that could penetrate the helmet, such as ballistic rounds and fragmentation, as well as protection from the helmet system deforming and striking the wearer from non-penetrating events (Behind Helmet Blunt Trauma or BHBT). BHBT is a more recently developing threat that is not well understood and is the focus of recent research [1], [2]. While this research and knowledge in characterizing the risk associated with these helmet systems is helpful, there is no generalizable injury risk function yet to be used for new BHBT scenarios. As new form factors and subcomponents are designed and incorporated in helmet systems, there is a need to characterise the risk to the wearer of those new systems. Ballistic mandible protectors (BMPs) have recently been integrated with combat helmets to protect the face. Advancements in armour materials, design, and fabrication have enabled some BMP to stop small arms rounds from penetrating. There is an unknown risk to the wearer from BHBT from the backface of these BMPs.

BMPs primarily protect the face, which consists of soft tissue supported by skeletal structures. The upper facial bones support the eyes, nose and sinuses. The maxilla, mandible and associated teeth support structures of the mouth including the cheeks, tongue, gums and lips. Existing research in injury biomechanics examines facial injuries resulting from blunt trauma [3], [4]. These include test methodologies specifically developed to understand and correlate the loads to different structures and injuries to the face [5]. However, the high rate and focal nature of injuries induced by BHBT loading to the skull and face are distinct from typical blunt head trauma observed in the general population [6], [7]. For this reason, there is a need to understand the BHBT loading from BMP and resulting induced injuries. Postmortem human surrogate (PMHS) testing has been used in literature to correlate BHBT loading to injury [7], [8], [9]. In order to start to characterise the unique nature of BHBT loading from BMP involved events, this study presents the results from PMHS testing using live ammunition striking

https://doi.org/10.52202/081621-0014

BMP and inducing BHBT. The results of this study can help inform future performance standards, necessary medical readiness if these systems are used, and future armour design.

2. METHODS

2.1 Projectile and armour combination

The same projectile and armour combination was used for all tests in this study. The projectile chosen is relevant to current operational threats, while velocities were near muzzle velocity to represent a worst-case scenario. The armour was an ultra-high molecular weight polyethylene (UHMWPE) composite BMP which was affixed to a size medium helmet by being clipped on to the rail mounting system (Figure 1). Each BMP was shot once, either centred on the mid-sagittal plane 70 mm from the bottom edge or off-centre 50 mm from the mid-sagittal plane 48 mm from the bottom edge. In addition to the helmet and BMP, communication equipment, including a boom microphone placed directly beneath the impact location, were donned to create a realistic situation in which a boom microphone might influence the loading from the backface of the armour to the face. The boom microphone was 15 mm thick in the direction of the shot, taking up some of the air gap between the backface of the BMP and the wearer. All ballistic events had 0 degrees of obliquity from the strike face.

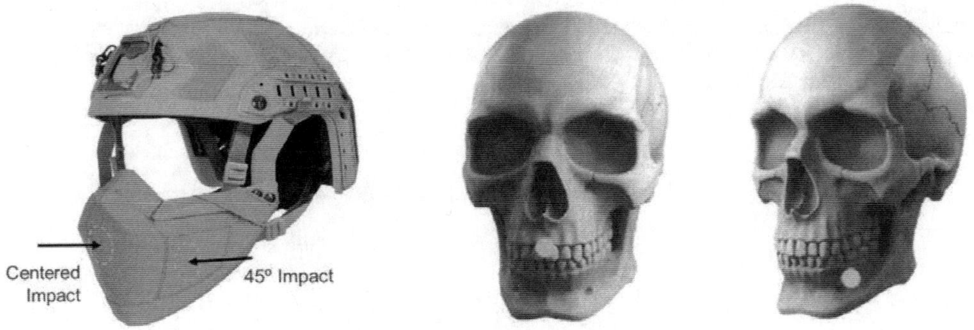

Figure 1. (Left) Helmet with BMP, showing striking locations; (middle) mid-sagittal plane maxilla impact location; (right) off-centre mandible impact location

2.2 PMHS methods

This study investigated the biomechanical response of PMHS and associated characteristic injuries incurred by BHBT centred over the face. Four head/neck specimens were selected based on a stringent criterion to reduce variation and avoid conditions that could influence injury outcomes (Table 1), including head sizes to ensure fitment in a size medium helmet. All PMHS were acquired under institutional review board exemption approval through Johns Hopkins Medical Institutions. Once received, high-resolution computed tomography (CT) scans (slice thickness of 0.625 mm) were acquired for specimens at the Johns Hopkins University Applied Physics Laboratory to ensure specimens had no apparent abnormalities or signs of trauma that could influence interpretation of injury, including detailed review of dental quality. Specimens were stored at −20 °C until being thawed at 4 °C 6 days prior to testing for instrumentation and test preparation. PMHS specimens were prepared through dissection of the neck, removing tissue to expose the cervical spine, and potted using polymethyl methacrylate, fixing the cervical spine at the level of C7. This potting fixture was used to position the neck and head/neck system during the live fire test.

Two anatomical locations were identified as sites of interest: (1) the maxilla on the mid-sagittal plane (centred between the two central incisors) and (2) the mandible approximately 45 degrees rotated from the mid-sagittal plane directly on the mandible below the teeth (Figure 1).

https://doi.org/10.52202/081621-0014

Table 1. Test specimen information

Test ID	Age	Head Width (cm)	Head Length (cm)	Impact Location
BMP_01	55	15.75	20.00	Mid-sagittal maxilla
BMP_02	46	16.00	20.30	Mid-sagittal maxilla
BMP_03	64	16.25	18.50	Off-centre mandible
BMP_04	67	15.50	21.60	Off-centre mandible

2.3 PMHS BHBT experimental setup

Ballistic testing was conducted at a National Institute of Justice (NIJ)-certified ballistic range [10]. Inverted specimens were positioned to correspond to the tested impact location on the anatomy. Helmets and BMP were then donned, positioned to collocate the intended BMP and anatomical locations. Communication sets (including headphones and a boom microphone) were included in as-worn conditions, where the boom microphone was placed directly over the anatomical impact location, almost in contact with the wearer, representing a potential worst-case impact scenario. Pre-test photos and x-rays were taken, the latter of which was used to measure the air gap between the armour backface and the anatomical impact site. Specimens were then impacted, using live ammunition striking the BMP, with the armour backface from a nonpenetrating ballistic event (Figure 2). Projectile velocity measurements were obtained using two sets of Oehler Research model No. 57 infrared screens with Hewlett-Packard (HP) counter chronographs (universal counters, HP model No. 53131A). Velocity was measured 1.5 m from the helmet and BMP, and then velocity loss equations were applied specific to the projectile to calculate striking velocity.

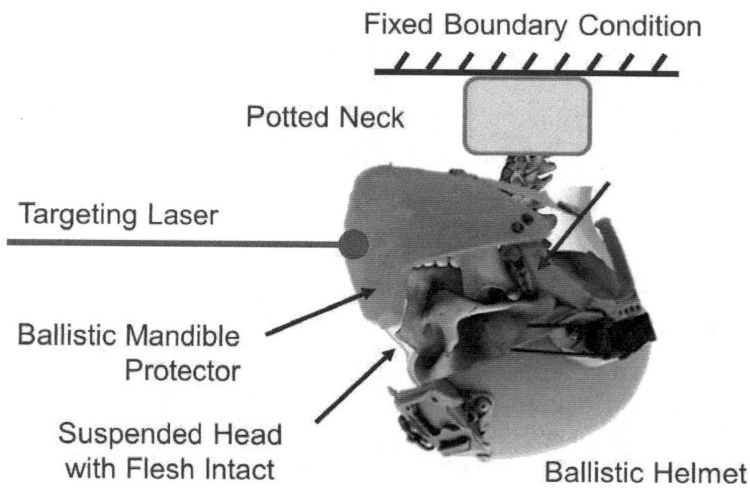

Figure 2. Graphic describing test setup, including fixed/potted neck, suspended PMHS, ballistic helmet and BMP and impact location (showing midsagittal maxilla impact)

Various instrumentation was used to better understand the loading event. A frangible break screen (Whitner, BS07-BCD, Baltimore, MD) was used at two locations, one at the last velocity gate (1.5 m from the helmet and BMP) to trigger data collection, and one adhered to the outer surface of the BMP to mark the timing of penetration of the outer shell of the BMP. Thin-profile piezoresistive force sensors (FlexiForce A301, Tekscan, Norwood, MA) were adhered to the front face of the boom microphone and to the surface of the skin at the targeted anatomical location. These sensors were used to gather the timing of loading events, including backface striking the boom, and backface/boom striking the PMHS. High-speed cameras (ACS-1-M60, NAC Image Technology, Tokyo, Japan) were used to capture the bullet entry and the system kinematics. The high-speed imaging was laterally and oblique to the shot direction and recorded at 100,000 and 50,000 frames per second, respectively. Sensor data were acquired at 1-MHz sample frequency using a Dewesoft data acquisition system (Dewe R4, HS-STG, Trbovlje, Slovenia), and cameras and data acquisition were synchronised together on the first break screen signal.

Injury outcomes were assessed using radiological evaluation of micro-computed tomography (μCT) images, using pre- and post-event scans, and anatomical dissections performed by a board-

https://doi.org/10.52202/081621-0014

certified forensic pathologist. This body of work uses dissection findings as ground truth to identify and characterise injury outcomes, as not all injuries were evident through μCT.

3. RESULTS

Typical post-test x-rays showed the round embedded in the ballistic armour, the deformation of the backface of the armour, and the injured anatomy (Figure 3). Injuries observed in the x-ray, while not definitive, showed facial and dental fractures.

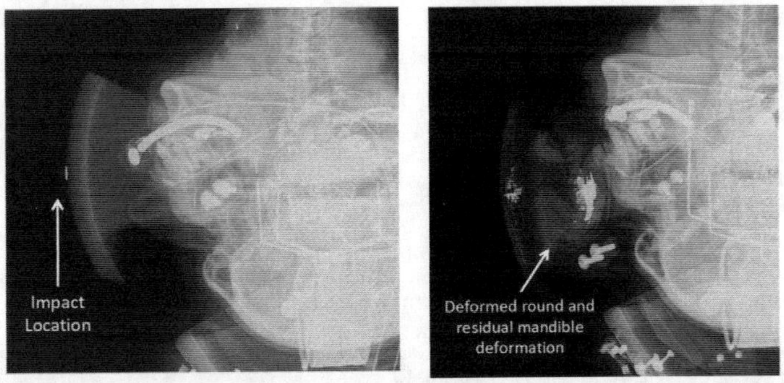

Figure 3. Lateral X-ray image of BMP and PMHS before impact (left) and image of BMP and PMHS directly after impact (right) for mid-sagittal maxilla impact (BMP_02)

Backface deformation of the BMP was substantial for this series of tests, appearing to fill the air gap present before impact. The air gap measured before impact was larger in the mid-sagittal centred striking location (approximately 45 mm) versus the oblique striking location (approximately 25 mm) because of the unique geometries of the BMP and the face (Figure 4).

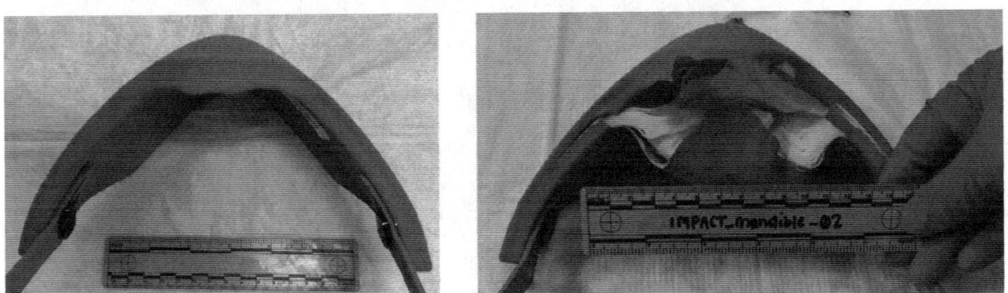

Figure 4. Top-down view of intact BMP (left) and residual backface deformation on the inner surface of the BMP (right) for mid-sagittal maxilla impact (BMP_02)

The impact location on the PMHS had foam and plastic fragments from the boom microphone embedded or adhered to the skin surface/wound. Injuries observed in the maxilla impact location included lacerations of the upper/lower lips, dental avulsions/fractures, maxilla fractures, and additional facial bone fractures (Figure 5).

https://doi.org/10.52202/081621-0014

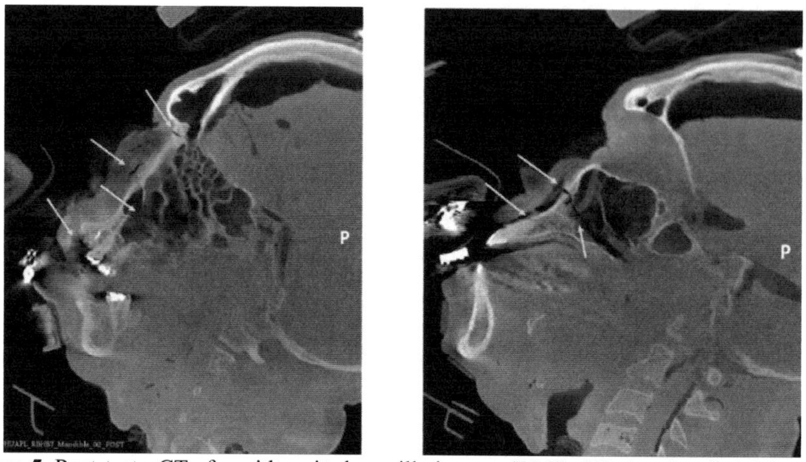

Figure 5. Post-test μCT of a mid-sagittal maxilla impact (BMP_02) showing facial fractures

Injuries observed in the off-centre mandible impact location included upper/lower lip and tongue lacerations, dental avulsions/fractures, and comminuted maxilla/mandible fractures (Figure 6).

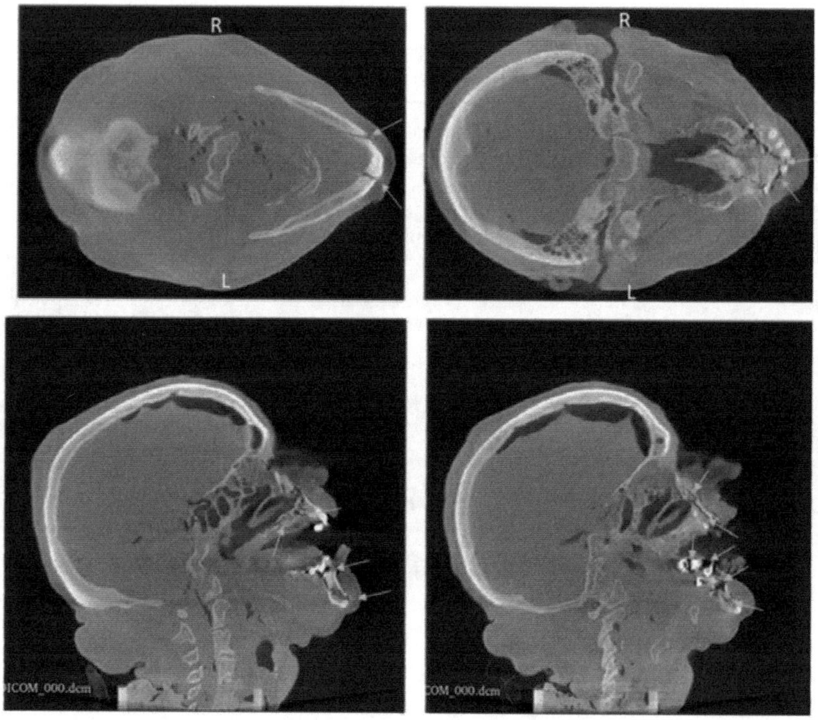

Figure 6. Post-test μCT of PMHS in axial (top) and sagittal (bottom) planes showing damage to the mandible and maxilla for an off-centre mandible shot (BMP_04)

The maximum abbreviated injury scale (MAIS) was used to quantify the most consequential injury. An alternative way to assess the severity of a complex injury is to use the New Injury Severity Score (NISS) injury model [11]. The NISS is calculated from MAIS scores and is the sum of the squares of most severe MAIS-coded injuries, regardless of body region. These injury outcomes can be found in Table 2. Individual injury codes can be found in Appendix A.

Table 2. Test conditions and injury results

Test ID	Impact Location	Air Gap (mm)	Injuries	Max. AIS	NISS
BMP_01	Mid-sagittal maxilla	45.48	a. Full-thickness laceration of the lower lip and lacerations of the oral mucosa b. Multiple tooth fractures/avulsions c. Multiple comminuted fractures of the maxilla with involvement of the lateral pterygoid plates, nasal sidewalls and nasal orbital ethmoid complexes; associated instability of the maxillary complex (Le Fort I-II) d. Right medial and inferior orbital wall fractures	2	9
BMP_02	Mid-sagittal maxilla	45.33	a. Laceration of the lower lip and lacerations of the oral mucosa b. Left zygomatic process fracture c. Left nasal sidewall and left medial orbital wall fractures d. Right nasal sidewall fracture e. Right inferior orbital wall fracture f. Multiple right-sided maxilla and hard palate fractures with involvement of right lateral pterygoid plate	2	12
BMP_03	Off-centre mandible	29.59	a. Laceration of the chin and oral mucosa b. Avulsion fractures of multiple upper and lower teeth c. Tongue laceration d. Comminuted maxilla fractures, involving left posterior aspect (partial Le Fort I) e. Multiple mandible fractures (left and right parasymphyseal, left ascending ramus)	3	17
BMP_04	Off-centre mandible	23.36	a. Laceration of the skin and soft tissue of the chin and oral mucosa b. Avulsion fractures of multiple upper and lower teeth c. Tongue laceration d. Comminuted maxilla fractures, with involvement of left posterior aspect (partial Le Fort I) e. Comminuted mandible fractures (symphyseal, left parasymphyseal, and dentoalveolar)	3	17

4. DISCUSSION

Injury results included soft tissue lacerations, facial bone fractures, and dental fractures/avulsions. All cases had complex midface fractures best described as a partial Le Fort fracture [12], ranging from Level I to II, indicating near separation of the hard palate or maxilla from the skull base. In addition, the comminuted mandible fractures of the off-centre impacts have the potential to adversely affect one's ability to maintain a patent airway, which would require immediate medical intervention.

Results show that injuries varied according to impact location, which may be attributed to regional anatomic differences, regional differences in contact conditions such as air gap, and/or armour performance. While this study was intended to gain a preliminary understanding of the types of injuries due to BHBT that were possible using relevant rounds and BMPs, it is of interest to explore the likelihood of these sources of regional sensitivity, as well as next steps, to support the potential of regional BHBT performance metrics.

https://doi.org/10.52202/081621-0014

4.1 Regional tolerances

In the absence of the relationship between the as-worn conditions of the BMP and BHBT loading, it is important to evaluate previous work into characterizing load tolerances in facial trauma. Extensive research in the automotive safety industry has focused on understanding the risk of facial fractures from occupants impacting large structures like the dashboard or steering column. In simulations of dashboard impacts, Nyquist [13], observed that the mandible had a lower minimum fracture load (685 N) than the maxilla (1468 N). However, this study applied lower impact rates and forces over larger contact areas than those characterised in BHBT. In contrast, Nahum [14] using a 6.5 cm diameter cylindrical impactor on PMHS, observed that the maxilla (625–1980 N) exhibited lower regional tolerance to loading compared to the mandible (1900–4100 N). This difference in loading tolerances between studies highlights the importance of contact area because the maxilla exhibited lower regional loading tolerance than the mandible during highly focal impacts. The impact energies and velocities observed in Nahum's study ranged from 4.9–46.7 J and 3–5.4 m/s, respectively. More recently, Daniel [15] conducted anterior–posterior impacts to the mandible, centred on the symphysis of the mandible. These studies used a constrained mandible, in order to simulate the helmeted conditions and use of a chinstrap. The impactor was smaller in diameter (2.87 cm) and similar impactor velocities and energies (2.5-4.4 m/s and 9.8-31.8 J) than earlier automotive studies. Daniel found using this impactor that the mean force to cause fracture of the mandible was 2683 N which compares favourably to Nahum's study. The ballistic energy associated with the BMP tests conducted for this study were 2113–2164 J. Ballistic energies cannot be directly compared with previous impactor energies, due to the fact that energy is lost due to the partial penetration of the round in the ballistic shell. However, backface deformation velocities have been measured in similar BHBT events to estimate energy delivered to the head. The instantaneous kinetic energies at impact in other BHBT events [16] have ranged from 4–24 percent of the corresponding input ballistic energies at standoffs ranging from 12.7–19.1 mm. This approximation of BHBT kinetic energy contends that the impact conditions in Nahum's and Daniel's studies may be substantially less than the BFD energies estimated in this study. More controlled PMHS studies with faster loading conditions with greater energy, as well as realistic BHBT contact, than those observed in previous studies could ensure that true fracture tolerances are known to be used for regional BHBT performance metrics to the face due to BMP.

4.2 Air gap and contact conditions

It is a known phenomenon that the BFD of armour varies in shape over time [16], [17]. Typically, the backface of the armour begins to accelerate reaching a maximum velocity before experiencing exponential velocity decay. In Hisley's study [16], where a 9-mm round was used, the maximum velocity is observed in the first 12 mm of deformation. In Wen's study [17], which used a rifle round, it is unclear at what displacement velocity decreases, but the authors of this paper estimate it at less than 20 mm. For the BMP used in this study, there were substantial air gaps ranging from 23.36–29.59 mm in the mandible shot, and 45.33–45.48 mm in the maxilla shot conditions. Given the exponential decay of BFD velocity reported in the literature, the difference in BFD velocity between approximately 25 mm and 45 mm could be substantial. As a result, the energy delivered through BFD (which corresponds to BFD loading) may be significantly lower for the maxilla than for the mandible. This discrepancy could obscure the observation that the maxilla sustained fewer injuries under identical ballistic conditions, as the actual BFD loading may not be equivalent. It remains crucial to understand the relationships between ballistic conditions, standoff distance, and BFD loading for BMP at different impact locations. In addition to the air gap-dependent backface velocity, the presence of the boom microphone could have influenced the loading from the backface of the BMP to the wearer. The hard plastic in the boom microphone was observed to be shattered after the test, and could have influenced the soft tissue injuries in the tests. The stiffness and strength of the hard plastic is likely much less than that of the skeleton and the ballistic material of the BMP, so it is hypothesised that the overall effects of the boom mic in the load path are minimal as it relates to skeletal injuries.

4.3 BHBT performance

Armor and helmets are known to exhibit regional differences in performance [6]. In BHBT, this performance is often measured by metrics from different backing conditions. Some can measure loading dynamically, such as unbacked DIC [16], high-speed x-ray [18], and load cells. Others can measure backface deformation statically, such as deformation left in clay [10] and residual deformation of the armour [7]. In our study, the only measurable way to assess armour performance is through measuring residual backface deformation, as no prior BHBT armour performance testing was conducted nor was the backface of the armour dynamically measured during our testing. In our testing, residual backface deformation was larger in the mid-sagittal maxilla than the off-centre mandible. This is likely to be driven more from the differences in air gap between the backface of the armour and the face, then the ballistic performance differences at those two locations.

To allow for future BMP BHBT performance to be standardised, steps should be taken to (1) evaluate the loading of these ballistic conditions to compare to injury to establish a dose response relationship for BMP BHBT and facial injury, and (2) provide a repeatable, flexible platform on which to evaluate future BMP systems and ballistic threats. Unique facial anatomy-based clay template in tandem with existing clay-based headforms such as the Peepsite headform [19] could be useful in quantitively measuring BFD loading. Clay headforms, however, lack biofidelity and dynamic measurements, and may only provide comparisons between BMP. Load cell-based headforms, such as BLSH and ATLAS, may offer a link to facial fracture observed in literature [20] and provide dynamic measurements. These headforms were primarily designed to assess BHBT by using high-magnitude loadcells that are more resilient to high-rate, high-magnitude loading seen in BHBT. However, these headforms were designed to assess ballistic helmet testing locations and do not have the instrumentation to assess facial loading. The Facial and Ocular CountermeasUre Safety (FOCUS) headform [5] has a form factor that can measure the relevant loading scenarios in BMP, however it is unclear if the instrumentation can support measurements in the high-frequency, high-magnitude loading environment produced by BMP BHBT. Steps should be taken to further adapt and mature BHBT testing platforms to enable the link from BHBT dose into injury response and support future BMP performance standards.

4.4 Limitations

This study is the first known to recreate BMP BHBT loading on PMHS through a limited number of tests and impact locations. There are most likely specific influences from this BMP design and round/armour type interactions that relegate these specific conclusions to this armour and threat pairing only. These results indicate that BMP BHBT injuries are possible and include lacerations and fractures to the anatomy of the face. While PMHS models have limitations in providing a physiological response (such as haemorrhaging, contusions, and any nervous system injury), the skeletal injuries from this study should be considered appropriate given the bone and dental quality of the specimens used. In addition, other studies have shown that postmortem bone, when hydrated, can approximate the mechanical parameters of bone from a living individual, even after a freeze cycle [21],[22]. The injuries observed in this study were characterised by using (1) the Le Fort Classification and (2) AIS. The Le Fort system was developed for military use in the early 1900s [12] and was specifically designed to assess and characterise injuries from blunt trauma to a large area of the face [20], and it may be inadequate for the focal loading believed to be present in this study. Additionally, AIS was originally designed for the automotive industry and may lack the specificity to adequately capture the types of injuries seen in BHBT events. There are other maxillofacial injury classifications, such as the fracture severity ranking scale [23] and the Maxillofacial Injury Severity Score [24] that could be evaluated for their utility to use in describing BHBT injury from BMP.

5. CONCLUSION

BHBT injuries from BMP are likely to result in less severe outcomes than penetrating injuries at the same anatomical locations, given the high likelihood of lethality of penetrating wounds and the significant challenges associated with reconstructive surgery for high-velocity small arms injuries [25]. BHBT injuries are generally considered survivable, reinforcing the justification for BMP use.

Regional sensitivity to injury severity was observed, particularly in relation to facial fractures. Mandibular injuries are considered more severe than maxillary injuries likely due to the potential complications in maintaining an airway caused by unstable mandibular fractures. This information can inform the optimization of BMP to reduce injury risk through further research, the development of BHBT testing standards, and the establishment of performance thresholds.

https://doi.org/10.52202/081621-0014

Acknowledgments

The authors would like to express their appreciation to the U.S. Special Operations Command for funding this work under Contract No. H92236-10-D-001, Task Order 0521. Additionally, acknowledgement is due for the pathologists who conducted the anatomical dissections.

References

[1] N. Nsiampa, F. Coghe, "Review of Literature: Behind Helmet Blunt Trauma Mechanisms," Human Factors and Mechanical Engineering for Defense and Safety, 7:1 (2023), 6.

[2] Y. Li, et al., "Review of Mechanisms and Research Methods for Blunt Ballistic Head Injury," Journal of Biomechanical Engineering, 145:1 (2023), 010801.

[3] D. Hampson, "Facial injury: a review of biomechanical studies and test procedures for facial injury assessment," Journal of Biomechanics, 28:1 (1995), 1-7.

[4] R. Crespo, et al., "Biomechanics of midface trauma: A review of concepts," Journal of Oral and Maxillofacial Surgery, Medicine, and Pathology, 33:4 (2021), 389-393.

[5] J. Cormier, and S. Manoogian, "Biomechanical Response of the Human Face and Corresponding Biofidelity of the FOCUS Headform," SAE International, 3 (2010), 842-859.

[6] A. Iwaskiw, et al., "Loads Associated with Behind Helmet Blunt Trauma: Matched-Pair Load Sensing Headform Tests Correlated with Skull Fracture Severity," Personal Armour System Symposium (Dresden, 2023).

[7] C. Howes, et al., "Behind Helmet Blunt Trauma: A Potentially Significant Mechanism of Injury for Helmeted Warfighters," Military Health System Research Symposium (Kissimmee FL, 2022).

[8] K. Rafaels, et al., "Injuries of the Head from Backface Deformation of Ballistic Protective Helmets Under Ballistic Impact," Journal of Forensic Sciences, 60:1 (2015), 219-225.

[9] C. Freitas, et al., "Dynamic Response Due to Behind Helmet Blunt Trauma Measured with a Human Head Surrogate," International Journal of Medical Sciences, 11:5 (2014), 409.

[10] NIJ Standard 0107.01, Ballistic Resistance of Body Armor, National Institute of Justice (2024).

[11] M. Stevenson, et al., "An overview of the injury severity score and the new injury severity score," Injury Prevention, 7:1 (2001), 10-13.

[12] B. Phillips, L. Turco, "Le Fort fractures: a collective review," Bulletin of Emergency & Trauma, 5:4 (2017), 221.

[13] G. Nyquist, et al., "Facial impact tolerance and response," SAE Technical Paper 861896, 30th Stapp Car Crash Conference, (1986).

[14] A. Nahum, "The Biomechanics of Maxillofacial Trauma," Clinics in Plastic Surgery, 2:1 (1975), 59-64.

[15] Daniel, Ray W. et al., "Fracture Injury Risk of the Restrained Mandible to Anterior–Posterior Blunt Impacts," Journal of Biomechanical Engineering, 143.4 (2021): 041012.

[16] D. Hisley, et al., "Experimental Methodology Using Digital Image Correlation to Assess Ballistic Helmet Blunt Trauma," Journal of Applied Mechanics, 78 (2011), 051022.

[17] Y. Wen, et al., "Analysis of dynamic back face deformation of a body armor impact by a rifle bullet using 3D-DIC," Journal of Physics: Conference Series, 1507:3, (IOP Publishing, 2020).

[18] N. Faderl, M. Becker, "High-speed 300kev flash x-ray cinematography for investigations of the ballistic impact behavior of target materials and structures," Anomaly Detection and Imaging with X-Rays (ADIX) IV, 10999, SPIE, 2019.

[19] National Research Council (U.S.), "Committee on Testing Armor Materials," Testing of Body Armor Materials: Phase III, U.S. Army, Ed., (National Academies Press, Washington, DC, 2012).

[20] C.R. Bass, N. Yoganandan, "Skull and Facial Bone Injury Biomechanics," Yoganandan, N., A. M. Nahum, and J. W. Melvin, Eds., Accidental Injury, (Springer, New York, NY, 2015).

[21] W. Kemp,"Postmortem change and its effect on evaluation of fractures." Academic forensic pathology 6.1 (2016): 28-44.

[22] T Topp, et al., "Embalmed and fresh frozen human bones in orthopedic cadaveric studies: which bone is authentic and feasible? A mechanical study." Acta orthopaedica 83.5 (2012): 543-547.

[23] J. Rhee, et al., "Experimental trauma to the malar eminence: fracture biomechanics and injury patterns," Otolaryngology—Head and Neck Surgery, 125.4 (2001), 351-355.
[24] J. Zhang, et al., "Maxillofacial Injury Severity Score: proposal of a new scoring system," International Journal of Oral and Maxillofacial Surgery, 35:2 (2006), 109-114.
[25] T. Lew, et al., "Characterization of craniomaxillofacial battle injuries sustained by United States service members in the current conflicts of Iraq and Afghanistan," Journal of Oral Maxillofacial Surgery, 68:1 (2010), 3.

APPENDIX A

Table A-1. Complete injury coding data

APL Identifier	AIS Coding of injury observed	
BMP_01	2 [251205]	Multiple fractures of same orbit
	2 [250806]	LeFort II
	1 [251408]	Teeth, any combination of dislocation/fracture/avulsion
	1 [210602]	Skin/subcutaneous/muscle, face, [includes eyelid, lip, external ear, forehead] laceration, minor; superficial
BMP_02	2 [250800]	Maxilla fracture
	2 [251231]	Medial wall fracture
	2 [251205]	Multiple fractures of same orbit
	1 [251802]	Zygoma fracture [includes zygomaticomaxillary complex (ZMC) and malar fractures], non-displaced [KN I]
	1 [250400]	Facial bone(s) fracture
	1 [210602]	Skin/subcutaneous/muscle, face, [includes eyelid, lip, external ear, forehead] laceration, minor; superficial
BMP_03	3 [251900]	Panfacial fracture
	2 [250614]	Open/displaced/comminuted [any or combination]
	2 [243404]	Tongue, laceration, deep; extensive
	1 [251408]	Teeth, any combination of dislocation/fracture/avulsion
	1 [210602]	Skin/subcutaneous/muscle, face, [includes eyelid, lip, external ear, forehead] laceration, minor; superficial
BMP_04	3 [251900]	Panfacial fracture
	2 [250614]	Open/displaced/comminuted [any or combination]
	2 [243404]	Tongue, laceration, deep; extensive
	1 [251408]	Teeth, any combination of dislocation/fracture/avulsion
	1 [210602]	Skin/subcutaneous/muscle, face, [includes eyelid, lip, external ear, forehead] laceration, minor; superficial

https://doi.org/10.52202/081621-0014

Blast Load Profiles Over a Surrogate Human Skull With and Without a Helmet: Computational Analysis of Shock Tube Experiments

J. Magallanes[1], N. Meisner[1], S. Fu[1], J. Hamilton[1], A. Nelson[2], and P. VandeVord[2]

[1]*Karagozian & Case, 500 N. Brand Blvd., Ste. 1500, Glendale, CA, U.S.A.*
magallanes@kcse.com
[2]*Virginia Tech Center for Injury Biomechanics, Blacksburg, VA*

Abstract. This paper describes an experimental and computational study focused on quantifying the blast pressure profile over a surrogate human skull with and without an Advanced Combat Helmet (ACH). Experiments are conducted using a large shock tube to measure incident and reflected pressure-time histories at discrete points directly on a surrogate human skull surface. The skull was oriented face-on towards the inbound incident wave inside the shock tube. Pressure histories predicted using a CFD model are shown first, alongside the measured pressure histories from the shock tube experiments, to evaluate the manipulation of the blast wave by the helmet, while investigating the efficacy of the CFD model to predict those loads. The CFD model predicts pressure-time histories consistent with shock tube experiments, including a pressure increase observed at the back of the skull with the ACH mounted to the skull. There is, however, very little impulse increase observed in both the shock tube and CFD results when accounting for the test replicate scatter. Then, predictions from both the CFD model and an empirical model, based on the Kingery-Bulmash model with simple modifications to account for clearing effects, are used to predict the pressure profiles around the skull and ACH for an HE charge in the free-field, as would commonly be of interest in blast protection applications. These comparisons indicate that the empirical blast methodology has the potential to provide conservative pressures and impulses around the surrogate skull with future investigation and refinements.

1. Introduction

Methods to quantify blast loads over a human body is of interest in many protection applications. The applications of primary relevance to this paper involve understanding blast wave interactions with humans, the resulting primary injury mechanisms, and analysis methods that can be used as tools to develop or evaluate the effectiveness of Personnel Protection Equipment (PPE). An incident blast wave generated by an energetic source, like a high explosive (HE) charge, will reflect off a human body and their PPE, generating a spatially and time- varying distribution of reflected, partially reflected, incident, and dynamic pressure around the surface of the body or PPE. The loading transmitted into the body, whether directly from these impending pressure waves, or as secondary load transfer paths through the PPE contacting the body, is complex and governs the actual pressure transmission into the organs of the body. After all, it is this pressure transmission that causes primary blast injuries, whether they manifest as tympanic membrane rupture (TMR, a.k.a. ear drum rupture), traumatic brain injuries (TBI or mild TBI), lung injuries, or any other injury of concern in airblast environments. PPE needs to be designed or qualified to provide the desired level of protection against these loading mechanisms to be effective.

This paper is specifically focused on quantifying the blast pressure profile over a surrogate human skull and use of analysis tools to assess the effectiveness of a helmet to shield the head from the blast loads. This is done through the experimental and computational study described in this paper. Though there are many previous studies that evaluate the effectiveness of helmets against incident blast waves or seek to standardize their testing and evaluation (e.g., [1] [2]), this study focuses instead on quantifying the spatial and temporal variations of the loads over the human head through both experiments and analysis models. Having validated analysis models, like those explored in the study, offer means to predict loads under many more conditions than can be feasibly tested, which offers great value in the design and evaluation of effectiveness of PPE like helmets, in addition to interpreting trends observable from the extant datasets. This is done experimentally using a large shock tube to measure incident and reflected pressure-time histories at discrete points directly on a surrogate human skull surface. Measurements of the incident blast wave generated in the shock tube were obtained using incident pressure gages mounted on the shock tube walls, which when combined with the pressure sensors mounted on the skull, provide data that quantifies the actual reflected pressures generated over the surface of the skull. Measurements were taken with and without an Advanced Combat Helmet (ACH), which provide metrics to assess the effectiveness of the ACH to shield the protected side of the skull from those pressures. The forces transmitted through the helmet back into the skull, however, was outside the scope of the present effort and may be the subject of a future effort.

The analysis methods evaluated for the study focus on an empirical methodology widely used in the blast-resistant design of structures, on one end of complexity, to a much more resolved and physics-

https://doi.org/10.52202/081621-0015

based Computational Fluid Dynamics (CFD) method on the other end. The empirical methodology is based on the Kingery-Bulmash model [3] for HE charges, where the angle of incidence and clearing effects are accounted for in a method we call KB++ in this paper. This method can be reproduced from blast effects manuals available in the open literature [4]. For the CFD model, we utilize the ArunaCFD solver [5], which solves the fully compressible form of the Euler equations. CFD methods, like those described in the paper, are now more widely available and validated for predicting blast loads from HE.

The study is described and reported in several concise sections, starting with a description of the methods used for both the experiments and analysis in Section 2. Section 3 presents both the experimental and analytic results. Pressure histories predicted using the CFD model are shown first, alongside the measured pressure histories from the shock tube experiments. They are presented for the bare skull (B-Sk) and the bare skull with the helmet (Sk+H) to evaluate the manipulation of the blast wave by the helmet, while investigating the efficacy of the CFD model to predict those loads. This establishes the validity of the CFD methodology against experiments. Then, predictions from both the empirical model (KB++) and the CFD model (ArunaCFD) are used to predict the pressure profiles around the skull and PPE for a HE charge in the free-field, as would commonly occur in many blast protection problems of interest. A HE charge of one pound (0.454 kg) of TNT (trinitrotoluene) is selected at a range of 5 feet (1.524 meters), for a scaled range of 5 ft/lbs$^{1/3}$ (1.983 m/kg$^{1/3}$) for the study. This scaled range is well within the range of applicability of the KB++ methodology, as well as for the CFD model. We focus on quantifying the efficacy of the empirical model with respect to the CFD model to evaluate if a fast-running empirical methodology can provide design loads from which to design or evaluate PPE. A brief discussion of the results is provided in Section 4, which is then followed by conclusions in Section 5.

2. Methods

2.1 Experiments

A large Advanced Blast Simulator (ABS) at Virginia Tech was used for the experiments, which is reported in detail in other references [6] and journal paper in preparation. The ABS, shown in Figure 1, consists of three sections: a pressure driver, the test section, and an end-wave eliminator. The test section has a 4 x 4 ft (1.22 x 1.22 m) test section, is 45 feet (13.7 m) long, and supports mounting cadavers or other physical surrogates like the skull used in this effort (Figure 1). The driver is a gas detonation system, which can generate peak pressures up to 70 psi (480 kPa) with up to 4 msec durations. For this effort, the incident pressure generated is 24 psi (170 kPa) with a 3 msec duration for this effort (40 psi-msec or 276 kPa-msec impulse), which was measured by incident gages mounted inside the shock tube. High-speed cameras recorded the test specimen through an observation window.

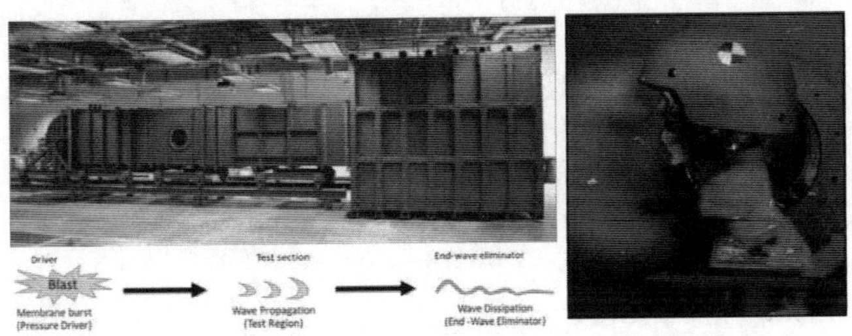

Figure 1. Experiment fixtures showing (left) the shock tube (ABS4 at Virginia Tech) and (right) instrumented surrogate human skull inside the test section stand with an ACH fixed to it.

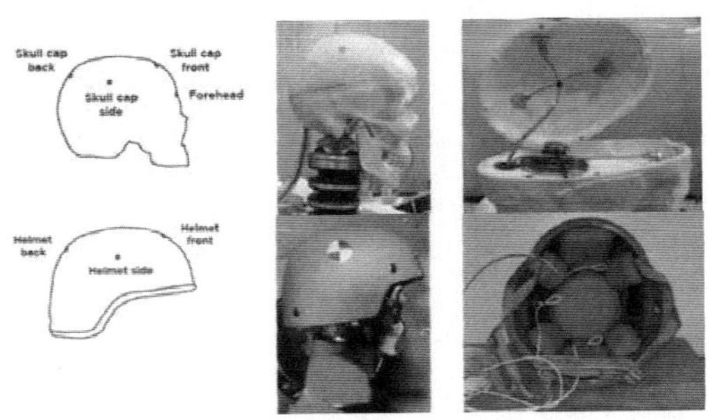

Figure 2. Pressure sensors mounted (top) on the skull and (bottom) on the helmet.

The test specimen is a synthetic human skull model (3B Scientific, Model 1020159) that was mechanically attached to a Hybrid-III neck-spine (Humanetics, Farmington Hills, MI) and secured to a test stand located mid-height of the shock tube section. The version secured with an ACH is shown in Figure 1 and can be configured in various orientations. The 0° face-on condition shown in Figure 1 is the focus in this paper. To capture pressure wave interactions within the helmet and skull, pressure sensors (Endevco, Model 8540-200) were placed on the test specimen as shown in Figure 2. For the skull, intracranial sensors were installed flush at the various locations shown in Figure 2. These

https://doi.org/10.52202/081621-0015

sensors were carefully selected and placed to measure the pressure in the air between the skull and the helmet while not interfering with the pads of the helmet. Additional sensors were placed on the helmet per Figure 2. Casting resin was poured into the skull to bond the neck bracket and seal off ocular and nasal passages.

2.2 CFD Methods

2.2.1 CFD Model and Solver

ArunaCFD is a compressible gas dynamics finite-volume solver for structured grids. The compressible gas dynamics are treated by solving the multi-species form of Euler's equations in three dimensions. The inviscid form of Euler's equations is solved using the piecewise parabolic method (PPM) for gases [7]. The PPM scheme employed in ArunaCFD is a high-order, finite-volume, monotonic upwind scheme for conservation laws (MUSCL), which is an example of a reconstruct-solve-average (RSA) method. ArunaCFD supports several Equations of State (EOS) for gases, including the ideal gas law leveraged to model air and the compressed gas inside of the shock tube.

In conservation form, these equations can be written as

$$\frac{\partial \rho}{\partial t} + \nabla \cdot (\rho v) \tag{1}$$

$$\frac{\partial \rho v}{\partial t} + \nabla \cdot (\rho v v + pI) = f_b \tag{2}$$

$$\frac{\partial \rho E}{\partial t} + \nabla \cdot (v[\rho E + p]) = \rho v \cdot f_b + \rho \dot{q}_c \tag{3}$$

$$\frac{\partial \rho X_i}{\partial t} + \nabla \cdot (\rho X_i v) = \dot{X}_i^c \tag{4}$$

where ρ is the density, v is the velocity vector, p is the pressure, X_i is the mass fraction of species i, $\dot{X}_i^c$ is the creation (or destruction) rate of species i due to any chemical or nuclear reactions present in the flow field, E is the sum of the internal ϵ and kinetic energy per unit mass,

$$E = \epsilon + \frac{1}{2}|v|^2 \tag{5}$$

$\dot{q}_c$ is the net (specific) energy deposition rate due to all reaction processes, f_b is the body force vector, and t is the time coordinate. Constraints on the mass fractions are imposed to enforce conservation of mass ($1 = \sum X_i$).

Finally, the EOS relating pressure to the internal energy, density, and composition closes this equation set,

$$p = p(\rho, \epsilon, X_i) \tag{6}$$

To treat high explosive (HE) detonation products, the Jones-Wilkins-Lee (JWL) EOS is used [8]. The form of the JWL implemented in ArunaCFD is written as follows:

$$p = A\left(1 - \frac{\omega}{R_1 V}\right)e^{-R_1 V} + B\left(1 - \frac{\omega}{R_2 V}\right)e^{-R_2 V} + \frac{\omega E}{V} \tag{7}$$

where V is the volume relative to the undetonated state (υ/υ_0). The parameters A, B, R_1, and R_2 are constant fitting parameters associated with the model, while ω is assumed-constant material parameter. E is the current internal energy per unit initial volume (e/υ_0).

Embedding complex surfaces in the CFD grid, like the skull and helmet, is accomplished using Immersed Boundary Methods (IBM) by reading faceted CAD geometry to modify the CFD grid values. The general idea with IBM is to place a complex geometry into a cartesian grid by modifying the cell values to conform to the surfaces describing the faceted geometry. The IBM is accomplished in ArunaCFD using the staircase methodology outlined by Ning et al.[9], which marks cell centers of the CFD mesh as stationary solids if they are within the boundaries of the input STL file. All other cells are treated as fluid under the gas dynamic laws described above. Any solid/gas interface in ArunaCFD is treated as reflective boundaries (slip) in ArunaCFD. The staircase methodology is binary, and results in either filled solid cells, or filled fluid cells. This may result in decreased accuracy near the boundary if grid cells are not refined enough to represent a complex surface.

https://doi.org/10.52202/081621-0015

2.2.2 Initial Conditions for Shock Tube Experiments

The CFD domain was set up to represent conditions inside the test chamber. The fixed grid, shown in Figure 3, measures 265 x 50 x 50 cm and is discretized with a resolution of 2 mm in each direction (approximately 8.5 M CFD grid cells). The dimension of the CFD were determined as a reduced model of the larger ABS shock tube. This "scaled" model is used for the computational domain, rather than modeling the exact dimensions of the ABS, to reduce the computational cost associated with the simulations. The boundary conditions included reflective conditions on the sides of the domain, reflective conditions on the driver side, and outflow conditions on the end-wave side (refer to Figure 1 for nomenclature). The air in the chamber was treated using the ideal gas law, while the driving side was initiated with a high energy gas using the same ideal gas law. The ideal gas properties of the high energy gas are computed by modifying the initial energy and temperature of the ideas gas to obtain a comparable pressure wave produced in the ABS. Computations were defined to run to 5 ms, which ran on 64 CPUs (Intel Xeon 5218, 2.3GHz, 32 CPUs, 256 GB DDR4 3200 RAM) using ArunaCFD's domain decomposition scheme leveraging the Message Passing Interface (MPI).

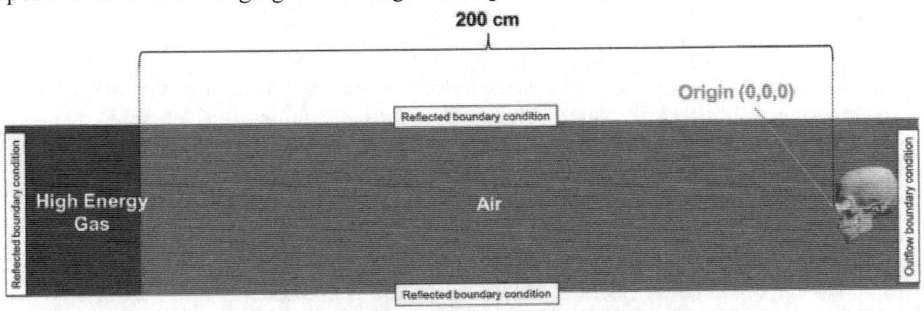

Figure 3. CFD model for the shock tube experiments.

(a)

(b)

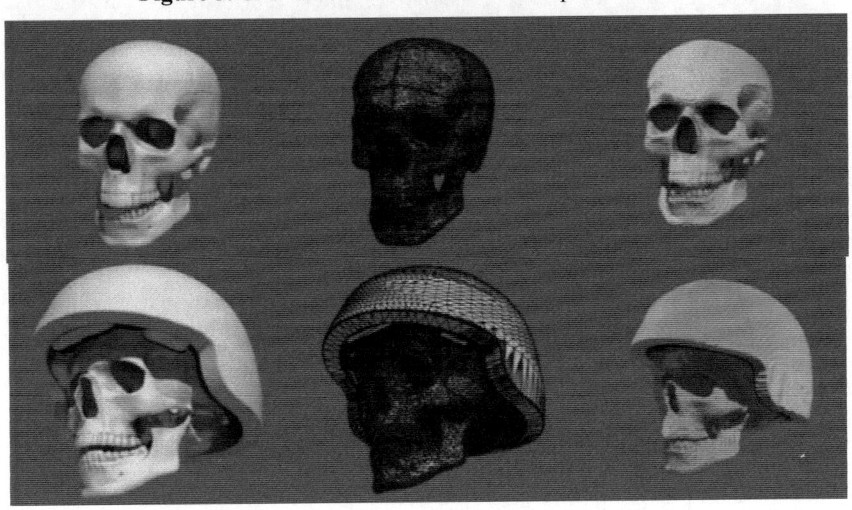

Figure 4. IBM representations in the CFD model for the (a) B-Sk and (b) Sk+H cases.

The skull and ACH were inserted into the CFD domain using the IBM described earlier using faceted data obtained using actual laser scans of the test articles. Both versions of the skull with and without the ACH are shown in Figure 4. The leftmost part of each version shows the laser-scanned point cloud generated for the actual bare skull (a) or skull + helmet (b), the centermost shows the facetized data generated from the laser-scan data to create the STL file, and the rightmost shows the realization of the skull/ACH in the CFD grid.

2.2.3 Initial Conditions for Bare Explosives

A similar grid is used for the bare explosive simulation, which is shown in Figure 5. This fixed grid also measured 265 x 50 x 50 cm and is discretized with the same resolution 2 mm in each direction (approximately 8.5M CFD grid cells). The boundary conditions are defined as all outflow to model a free

https://doi.org/10.52202/081621-0015

field. As in the previous model, the air is treated using the same ideal gas law EOS, while the HE is treated using the JWL EOS with parameters calibrated for TNT. The one pound of TNT was modeled as a sphere having a diameter of 4 cm, which was determined using the density of TNT (1.63 g/cc).

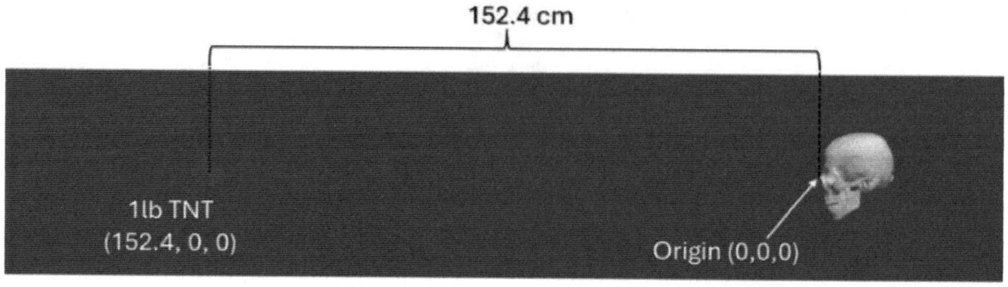

Figure 5. CFD model for the free air HE burst.

2.3 Empirical Methods

2.3.1 Kingery-Bulmash (K-B) Empirical Model

The K-B curves [3] are widely used to estimate blast loads on structures for spherical idealized free-air and hemispherical surface HE bursts [4], including those for finite-sized reflecting surfaces. They involve estimating blast wave parameters on those reflecting surfaces, including the peak reflected pressure and impulse (p_r, i_r), incident (or side-on) pressure and impulse (p_{so}, i_{so}), time of arrivals (t_{TOA}), as well as others that characterize the negative phase, all as a function of the scaled distance $Z=R/W^{1/3}$ (where R is the standoff distance and W is the equivalent weight of TNT). The method accounts for the effects of the angle of incidence (α) on the peak reflected pressure (p_{ra}) and the reflected impulse (i_{ra}), which then allows constructing pressure-time histories at any point of the reflecting surface [4].

Clearing effects are recognized in the methods to account for the reduction in total impulse that will occur at a point on a finite sized reflecting surface as compared with one having infinite size. In such cases, rarefaction waves, or clearing waves, will propagate from the edges of the reflecting surface towards the point of interest, relieving the inbound reflected wave. For finite-sized planar surfaces, like those on a building structure or component, the clearing time, t_c, is computed as:

$$t_c = \frac{4S}{(1+C_R)C_r} \tag{8}$$

where S is the smallest clearing distance, C_R is an aspect ratio parameter, and C_r is the sound velocity.

2.3.2 Application to obtain estimates of blast loads over the surrogate human skull and helmet

The KB++ procedures implement a very simple version of these methods to estimate loads on the skull and helmet. See Figure 6. First, the K-B blast wave parameters are computed in the normal way by calculating α for each point of interest (POI). For reflecting points (e.g., Pt. A in Figure 6(a)), the clearing time t_c is computed, as a first order approximation, as the distance to the closest point around the skull where α is 90 degrees (i.e., side-on). The methods described in the previous section are then used along with the assumption that clearing begins instantaneously[4]. Figure 6(b) shows the constructed pressure-time history. For non-reflecting points ($\alpha > 90$ degrees, like Pt. B in Figure 6(a)), side-on pressure is assumed with the distance computed according to the path illustrated that figure.

(a) (b)

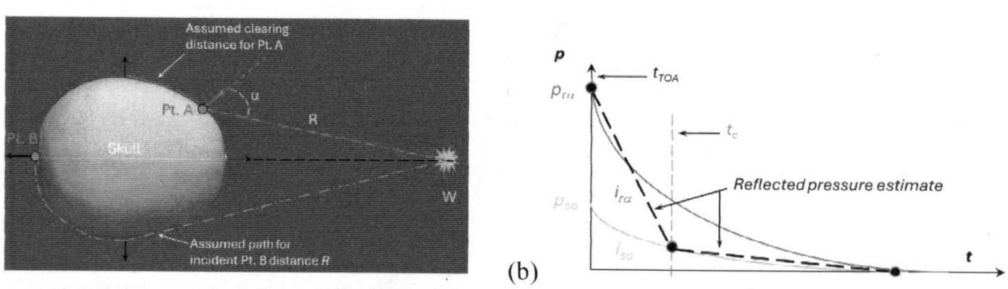

Figure 6. Schematic for KB++ pressure loads accounting for angle of incidence and clearing time.

https://doi.org/10.52202/081621-0015

3. Results

3.1 Shock Tube Experiments and CFD Comparisons

The CFD results match well against the experiments and exhibit both the reflected and diffraction waves expected over the skull and helmet. Static images showing pressure along a centerline plane of the CFD domain are shown in Figure 7, with the B-Sk case shown in Figure 7 and the Sk+H case shown in Figure 7. The case for Sk+H exhibits more complex reflections off the helmet, the most obvious ones in these figures being on the front side facing the blast.

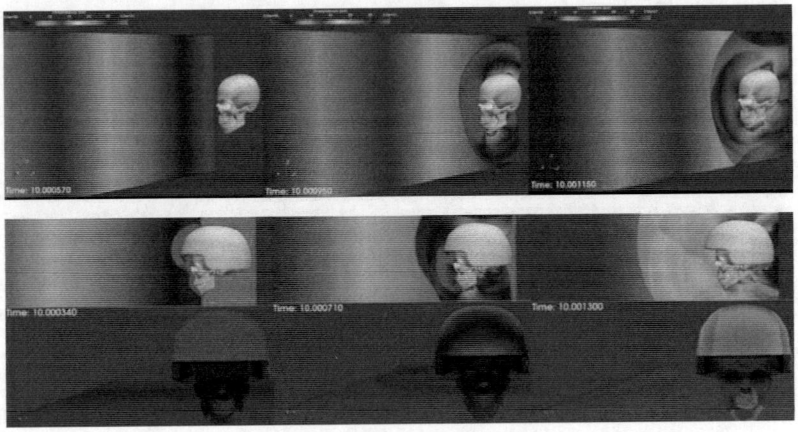

Figure 7. Pressures along the centerline of the shock tube CFD domain for the bare skull (Top: B-Sk) and the skull + helmet (Bottom: Sk+H). Note the snapshots are taken at different times in each simulation (msec).

The pressure histories measured at pressure gages shown on Figure 2(a) show very close alignment of the CFD results with the shock tube experiments. Figure 8 and Figure 9 plot pressure and impulse histories at the four gage locations on the skull for the B-Sk (Figure 8) and Sk+H (Figure 9): (a) Forehead, (b) Skull Cap Front, (c) Skull Cap Side, and (d) Skull Cap Back. Refer to Figure 2 for the illustration indicating their locations on the test articles. Shock tube data are plotted in black, which come from three replicate experiments, while the CFD results are plotted in blue. For the experimental data, one example pressure history is shown, while the envelope of impulses is shown from all three replicate experiments. However, this may result in some small discrepancies and differences between the CFD simulations and the experimental results. The One can observe that the pressure and impulse for the B-Sk case is highest at the Forehead, followed by the Skull Cap Front, Side, and Back, respectively. This behavior is observable from both the experimental and CFD results. One can also observe from the Sk+H case, that the helmet manipulates the pressure waves at the skull cap points under the helmet. The forehead traces, however, are virtually identical in both the B-Sk and Sk-H cases. It is worth noting that the CFD simulation is not an exact replica of the test facility shock tube, as modeling a 4ft x4ft x 45ft shock tube in a high-fidelity simulation would result in unacceptably long run times for the kind of analysis that is required. The CFD simulations here were configured in such a way as to reduce the overall length of the shock tube region while still maintaining consistent parameters that agreed with the initial conditions of the experiments.

Figure 10 presents this data in the form of bar plots for peak pressure (Figure 10(a)) and peak impulse (Figure 10(b)) measured at each of the skull pressure gages, with each plot (pressure (a) or impulse (b)) showing the B-Sk case on the left and the Sk+H on the right (including the three experimental replicates in each). It's worth noting here that the CFD initial conditions were configured against one specific test in the test series (Test 24a). The following plots show the comparison of the pressure profiles between both the CFD (orange), and the test series (blue) with the results from test 24a in dark blue, and the scatter of the test shown as transparent light blue. The helmet has an observable effect on the peak pressures, with the forehead and skull cap front experiencing the most significant drops, the skull cap side experiencing the least, while the skull cap rear conversely experiencing an increase, rather than a drop, in pressure. This trend is observed in both the CFD, and the experimental results. Total impulse, however, was relatively unchanged with or without the helmet, though the CFD results would imply some small drop in impulse across all the pressure gages with the helmet mounted on the skull.

https://doi.org/10.52202/081621-0015

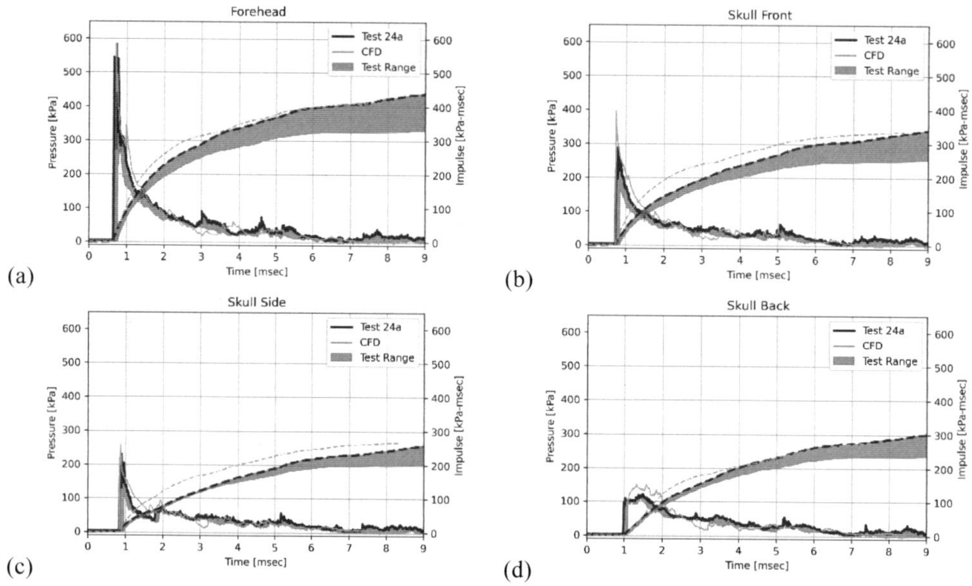

Figure 8. Shock tube CFD predictions versus experiments: Bare Skull (B-Sk).

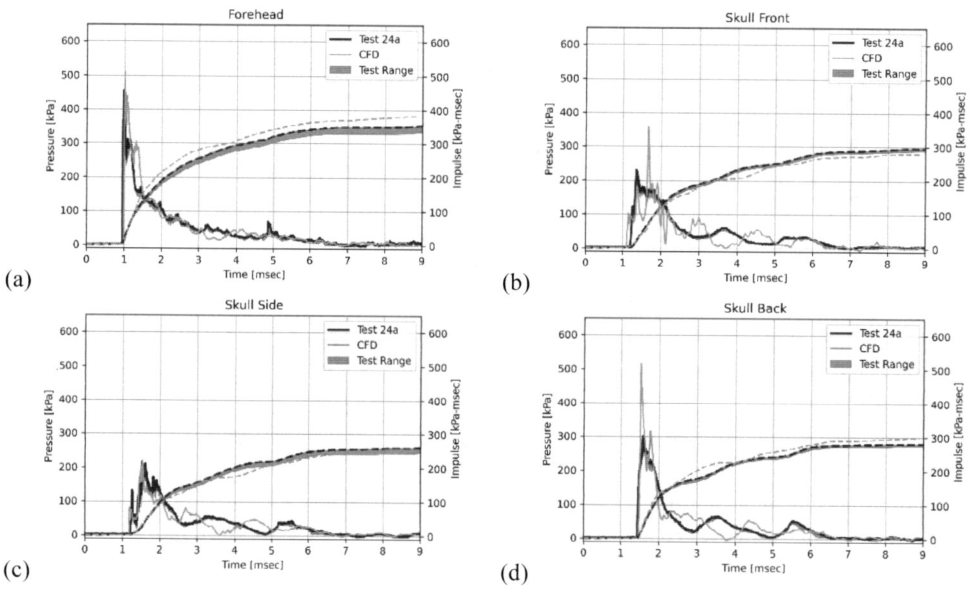

Figure 9. Shock tube CFD predictions versus experiments: Skull + Helmet (Sk+H).

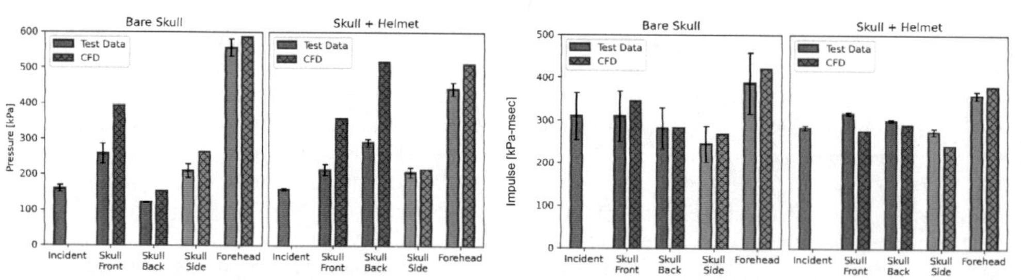

(a) Pressure

(b) Impulse

Figure 10. Comparison of peak pressure (a) and impulse (b) for the shock tube experiments and model.

https://doi.org/10.52202/081621-0015

Table 1. Summary of computed peak pressure and impulses for both the shock tube test series, and the CFD model. Pressures (p) are reported in kPa, while impulse is reported in kPa-msec.

St.	Bare Skull (B-Sk)						Skull + Helmet (Sk+H)					
	CFD		Test		%Difference		CFD		Test		%Difference	
	p	i	p	i	p_{diff}	i_{diff}	p	i	p	i	p_{diff}	i_{diff}
FH	578	415	579	431	0.2	3.7	510.0	386.0	553.0	351.0	7.8	10.0
SF	397	326	297	335	33.7	2.7	458.0	282.0	231.0	297.0	98.3	5.1
SS	268	261	233	252	15.0	3.6	229.0	240.0	229.0	256.0	0.0	6.3
SB	147	276	118	300	24.6	8.0	514.0	301.0	303.0	291.0	69.6	3.4
Average					18.4	4.5	**Average**				43.9	6.2

3.2 High-Explosive Pressure Distributions

Turning to the empirical KB++ model, which we intend to evaluate as a potential fast-running methodology for estimating pressures and impulses from HE charges, the studies for the same B-Sk and Sk+H cases show that under the one pound TNT charge (0.454 kg) at 5 feet of standoff (1.524 meters, Z=5 ft/lbs$^{1/3}$ (1.983 m/kg$^{1/3}$)), the KB++ methodology provides conservative pressure and impulse predictions, as compared with the CFD predictions

This is most effectively presented through the data shown in Figure 11, which shows computed pressure and impulse histories at four stations on the skull identified in Figure 12, and tabulated peak pressures and impulses presented in Table 1. These stations were selected slightly differently than in the shock tube experiments so that one can compare more reflected rather than side-on stations along the crown of the skull. Note from Figure 12, that Stations 1 and 3 have direct line-of-sight to the charge and would therefore experience reflected pressures proportional to, according to the K-B model, their angle of incidence (α), while Stations 4 and 6 are on the side and rear-face of the skull and would not reflect the inbound wave. Station 7, the rearmost station, was not selected because it is below and shielded by one of the ACH pads.

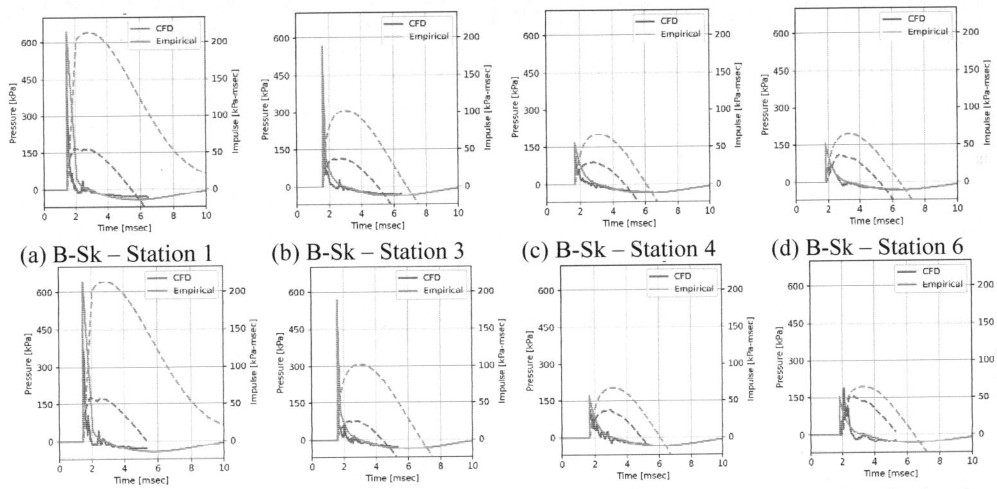

(a) B-Sk – Station 1 (b) B-Sk – Station 3 (c) B-Sk – Station 4 (d) B-Sk – Station 6

(e) Sk+H – Station 1 (f) Sk+H – Station 3 (g) Sk+H – Station 4 (h) Sk+H – Station 6

Figure 11. Calculated pressure histories at several stations designated in Figure 12 for the free-air HE burst of Z=5 ft/lbs$^{1/3}$ (1.983 m/kg$^{1/3}$) using the CFD and empirical (KB++) models.

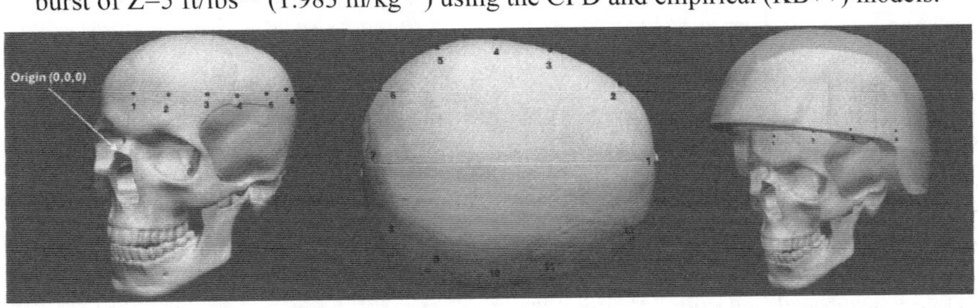

Figure 2. Stations selected along the crown of the skull for the empirical vs. CFD evaluation.

https://doi.org/10.52202/081621-0015

As can be seen in Figure 11 and Table 1, peak pressures and impulses are consistently higher than the CFD predictions for all Stations for both the B-Sk and Sk+H cases, except for Station 6 at the back of the skull for the SK+H case. For the B-Sk case, KB++ provides peak pressures and impulses that are 37.8% and 56.0% higher than the CFD predictions. For the Sk+H case, KB++ provides peak pressure and impulses that are 29.9% and 40.7% higher than the CFD predictions. On average, then, the KB++ methodology provide pressures and impulses that are higher and consistently more conservative than the CFD predictions. Station 6 is the only exception, on the other hand, where KB++ predicts a pressure less than the CFD prediction by 22.4%. This might indicate, when combined with the observations presented in the previous section concerning the shock tube data, that the KB++ methodology is missing any additional reflections in the helmet that might be contributing to the higher pressures observed consistently in the shock tube data at the back of the skull with the helmet in place (Sk+H). Refer to Figure 10 to evaluate these similarities; recall, that the helmet had appeared to amplify the peak pressure at gauge at the back of the skull, but that the total impulses were relatively unaffected by the presence of the helmet.

Table 2. Summary of computed peak pressure and impulses at Stations 1-7 for the free-air HE burst of Z=5 ft/lbs$^{1/3}$ (1.983 m/kg$^{1/3}$) using the CFD and empirical (KB++) models. Pressures (p) are reported in kPa, while impulse is reported in kPa-msec.

St.	Bare Skull (B-Sk)						Skull + Helmet (Sk+H)					
	CFD		Empirical		%Difference		CFD		Empirical		%Difference	
	p	i	p	i	p_{diff}	i_{diff}	p	i	p	i	p_{diff}	i_{diff}
1	559	55.6	642	213	12.9	74.0	461	58.4	642	213	28.2	72.6
2	284	33.7	610	145	53.4	76.8	280	35.8	610	145	54.0	75.3
3	273	38.3	569	102	52.0	62.5	168	25.6	569	102	70.5	74.9
4	130	29.7	169	66.9	23.4	55.7	111	37.9	169	66.9	34.2	43.3
5	92.2	30.7	161	65.6	42.8	53.2	N/A[1]	-	-	-	N/A[1]	N/A[1]
6	70.0	37.2	154	64.4	54.6	42.3	189	52.2	154	64.4	22.4	18.9
7	110	45.9	147	63.2	25.4	27.4	N/A[1]	-	-	-	N/A[1]	N/A[1]
Average					37.8	56.0	**Average**				29.9[2]	40.7

[1] - Station underneath helmet pad.

[2] - Average is calculated using absolute values.

4. Discussion

Investigation of the pressure fields (Figure 7(b)) present in the CFD model indicate, at least qualitatively, that these pressure rises occur from reflections generated by the incident wave interacting with the rear of the helmet. This might intuitively explain why there may be increases in reflected pressures at those rear-facing points, with very little increase, if any, on the total impulse imparted to the head. However, the complexity of the interactions of the incident, reflected, and diffractions waves around the skull and helmet makes it difficult to interpret. We have also only currently looked at the face-on orientation and would expect different orientations to display different distributions, at least on the rear-side of the skull.

Comparisons with the empirical equations and the CFD simulation show much larger average differences as compared to the average differences between the CFD and the shock tube in the previous section. Though there are many potentially explanations for this, we believe the primary one is likely due to the difficultly in predicting blast profiles over non-orthogonal surfaces using the empirical equations. Both the shock tube experiments and the CFD calculations appear to imply that geometric effects, and the complexity of the shockwaves generated around the skull are additional complexities that are not current accounted for the in the empirical equations. This is likely the cause of the better agreement of peak pressure at Station 1, which has the least geometric complexity. Additionally, the KB equations are empirically derived equations based on test data of bare high explosives. ArunaCFD uses an empirically derived equation of state (the JWL EOS), in which parameters for TNT are either gathered from test data, or thermochemical analysis tools. There are often errors association with these empirical fits for the equation of state for the CFD simulation which may be another reason there is larger discrepancies between the CFD and empirical equations, as compared to the CFD and the test data.

5. Summary and Conclusions

The experimental and computational results presented in this paper show that a CFD model predicts pressure-time histories consistent with shock tube experiments, and that an empirical blast

https://doi.org/10.52202/081621-0015

methodology, based on the K-B model called KB++, has the potential to provide conservative pressures and impulses around a surrogate skull for HE charges in the free-field with future investigation and refinements. The shock tube analysis was limited to an incident pressure wave of 24 psi and 40 psi-msec (t_d = 3 msec) but includes three replicate experiments, where the CFD results were within 34% and 8% of the pressure and impulse, respectively, and reproduced the distribution of the pressures and impulses measured in the experiments. The empirical methodology, which uses the K-B model by accounting for the angle-of-incidence and clearing time effects, in a very simple way, was shown to provide conservative pressure and impulses against the CFD from an HE charge in the free-field (Z=5 ft/lbs$^{1/3}$ (1.983 m/kg$^{1/3}$)), with exception the higher peak pressures observed at the back of the helmet, with empirical results within 34% and 48% of the pressure and impulse of CFD results, respectively. This study is, of course, limited to the shock tube conditions and HE scaled range evaluated in this study, and only in the face-on condition for the skull.

These efforts would benefit from extensions we are interested in pursuing in the future. First, the CFD analysis should be extended to experiments for different skull orientations. The face-on case evaluated here involves only a subset of conditions of interest to actual PPE applications (e.g.,[10]), and validation of the CFD model for other orientations would allow for applications in others that cannot be feasibly tested. Second, the KB++ methodology could likely be further extended to address the stated gaps and assumptions to provide fast-running conservative loads for purposes of designing and evaluating PPE. Though the KB++ method will not likely be a perfect substitute for either experiments or CFD analysis, it could certainly provide a quick design and analysis tool for evaluating effectiveness of PPE once refined and validated within known conditions. Finally, this experimental and analysis study was limited to the blast wave impinging on the skull and skull+helmet, which ignores the loads transferred through the helmet pads back into the skull. As we pointed eluded to in the introduction, this may be an important load transfer mechanism to consider for injury analysis and PPE design. Experiments and analysis tools that can provide insight into these load transfer mechanisms could allow for improved design of PPE.

References

[1] C. Carneal, A. Merkle, D. Zinn, J. Andrist, J. Clark, V. Alphonse, D. Ritzel, M. Carboni, B. DeCritofano and J. M. M. Cyganik, "Development of a Laboratory Shock Tube System for Helmet Blast Overpressure Performance Assessment," in *Proc. Personal Armour Systems Symposium (PASS) 2016*, Amsterdam, 2016.

[2] M. Maach, "Comparison of Hybrid III head response to shock tube and explosive blast loading," in *Int. Research Council on the Biomechanics of Injury (IRCOBI) Conf.*, Antwerp (Belgium), 2017.

[3] C. Kingery and G. Bulmash, "Airblast Parameters from TNT Spherical Air Burst and Hemispherical Surface Burst," U.S. Army Armament Research and Development Center, Ballistic Research Laboratory, Aberdeen Providing Ground, MD, 1984.

[4] Department of Defense, "Structures to Resist the Effects of Accidental Explosions, UFC-3-340-02," Whole Building Design Guide, http://dod.wbdg.org/, 2008.

[5] Karagaozian & Case, Inc., "ArunaCFD," [Online]. Available: https://www.kcse.com/arunacfd/. [Accessed 15 March 2025].

[6] A. Nelson, D. Ritzel, N. Showalter, D. Boppe, A. Riegel and P. VandeVord, "Characterization of an advanced blast simulator for investigation of lage scale blast traumatic brain injury studies," *Ann. Biomed. Eng.*, vol. 53, no. 1, pp. 133-142, 2024.

[7] P. Colella and P. Woodward, "The piecewise parabolic method (PPM) for gas-dynamical simulations.," *Journal of computational physics,* vol. 54, no. 1, pp. 174-201, 1984.

[8] E. L. Lee, H. C. Hornig and J. W. Kury, "Adiabatic expansion of high explosive detonation products," Lawrence Radiation Laboratory, University of California, Livermore, 1968.

[9] J. Ning, T. Ma and G. Lin, "A grid generator for 3-D explosion simulations using the staircase boundary approach in Cartesian coordinates based on STL models," *Adv. in Eng. Soft.,* vol. 67, pp. 148-155, 2014.

[10] D. Wallace and S. Rayner, "Combat Helmets and Blast Traumatic Brain Injury," *J. Mil. and Veterans Health,* vol. 20, no. 1, 2012.

https://doi.org/10.52202/081621-0015

Human repeated blast lung injury assessments – A new framework for use behind Personal Protective Equipment (PPE)

M. Panzer[1], **K. Sirhan**[2] and C. Bass[2]

[1]*University of Virginia, 4040 Lewis and Clark Dr, Charlottesville, VA 22911, USA*
[2]*Wayne State University, 818 W Hancock St., Detroit, MI 48201, USA dale.bass@wayne.edu*

Abstract. Human injury criteria from blast are fundamentally multivariate depending on at least two physical input parameters. Exposure parameters for human injury risk used by Bowen [4] and numerous others are peak overpressure and positive-phase duration. With noisy field data or complex blast, determining an unambiguous positive-phase duration is difficult. Assessments of structural blast effects generally use positive-phase specific impulse and peak overpressure as more convenient damage metrics. Recently, Panzer et al [8] developed repeated blast lung injury assessments for lung injury and fatality using 2,349 live animal tests for peak overpressure and duration. These assessments are robust for repeated ideal blast exposures but are uncertain for complex blast. Existing complex blasts risk evaluations are sparse and are often inconsistent with well-established single blast risk. To provide a more robust framework for assessments behind armor, where substantial modifications of ideal or complex blast waveforms are expected, the risk functions of Panzer were recast in terms of positive-phase specific impulse and peak positive-phase overpressure. This ideal blast risk assessment is further extended to a repeated and complex blast framework suitable for use behind PPE. Identification is based on the observation that blast lung injuries are produced by both shock wave transmission through the torso and chest wall motion. Injurious portions of complex waveforms are isolated using chest wall acceleration thresholding of input pressure waveforms. This procedure removes the non injurious portions of the time nistory and restricts input to those impulsive waveforms that produce injurious chest wall motions. This approach assumes measurements are taken in a reflected conditions (e.g. on a simulated chest behind armor) but is extensible to incident or omnidirectional sensing. The resulting assessments are consistent with single ideal blast injury risk models of Bowen [4]and Bass [7] and repeated blast evaluations of Panzer et al [8] while extending use behind PPE.

1. INTRODUCTION

Blast-related injuries have increased as the consequence of increased explosive usage in modern warfare and terrorist actions (e.g. [1, 2]). The threat posed by improvised explosive devices (IEDs) is also increasing [3], highlighting the need for further research into the injury mechanisms, particularly those related to chest injuries behind armor. Blast waves are generated when explosives are detonated causing an outward increase in pressure. Ideal blast waves or Friedlander waves have been studied extensively in the past for their simplicity and straightforward experimental models (eg. [4]). Friedlander waves are produced in open areas allowing the blast wave to continuously propagate outward unimpeded. During an ideal wave a spike of increased pressure (peak overpressure - OVP) occurs at detonation and decays exponentially (positive phase) the decay passes ambient pressure creating a vacuum (negative phase) before returning to ambient pressure. Blast wave injury response is intrinsically multivariate. Exposure parameters for human injury risk used by Bowen et al. [4] and numerous others are peak OVP and positive-phase duration. With noisy field data or complex blast, determining an unambiguous positive-phase duration is difficult. Assessments of structural blast effects generally use positive-phase specific impulse and peak overpressure as more convenient damage metrics. However, it is often difficult in practice to unambiguously identify the positive phase duration, and blast risk functions for structures often use peak OVP and positive phase impulse to categorize risk (e.g. [5]). Generally, however, for ideal waves, any two of peak OVP, positive phase impulse, or positive phase duration can be used to characterize blast.

Complex blast waves, often generated within range of buildings, debris, and other structures that cause the blast wave to reflect and bounce while impinging on the torso, may be both highly directional and highly variable. This complicates analyses using most existing injury criteria. Behind personal protective equipment (PPE), even simple blast can become complex, and complex blast can become more complex. Estimates of injury risk become uncertain behind body armor and assessments of the local blast environment on the chest is not established. However, an approach based on surface pressure transducers, modified using paired accelerometers and reflected blast time histories, has some promise for assessing injury risk while wearing PPE (c.f. [6]).

To assess short duration ideal blast, Bass *et al.* conducted a study with 2,550 large animals for short duration blasts with less than 30 milliseconds positive phase duration [7]. The curves were compared to Bowen

https://doi.org/10.52202/081621-0016

et al. [4] and matched the previously developed risk assessment for reflected ideal blast. However, the two studies did not show similar curves developed with the direction of the blast shock parallel and perpendicular to the blast. Recently, Panzer *et al* [8] extended the Bass risk assessment to repeated blast lung injury assessments for lung injury and fatality using 2,349 live animal tests. This study combines and extends short duration assessments of Bass *et al* and long duration assessments of Rafaels et al [9].

The risk model in Panzer *et al* was intended for use with ideal blast waves. Existing complex blast risk evaluations are sparse and are often inconsistent with well-established single blast risk. For example, the Axelsson/Yelverton thoracic response model assessment [10] is inconsistent with both the Bowen curves and the Panzer assessments for a single ideal blast, recommending higher injury tolerance values that are inconsistent with the existing data. Other models are often taken directly from automobile rate response models, well outside their range of applicability, and require instrumentation that is not compatible with many circumstances, including in vehicles and behind body armor. (c.f [11]).

The typical injury paradigm for the lung in automobile biomechanics (e.g. [12]) and for blast (e.g. [13]) is based on the hypothesis that displacement into the lung causes lung contusions. However, the compressive speed of sound in the lung is ~20 m/s. Since chest wall velocity and imputed shock wave propagation often exceeds 20 m/s in blast, there is a strong potential for compressive shock wave development in the lung [14] as shown in **Figure 1**. Existing thoracic response for blast lung injury models do not directly account for such compressive shocks (e.g. [10, 11]). There are several potential paths into the lung, in simple or complex scenarios either with or without body armor. In air, supersonic shocks generally propagate at velocities well over the 340 m/s quiescent speed of sound in air. When the air shock impinges on the torso, there are two alternatives, one is that the local air speed of sound is greater than 1500 m/s, the speed of sound in the flesh components of the thoracic wall. If so, a shock wave may be transmitted through the chest wall. If the speed of the incoming waves is subsonic in the chest wall, a piston-type excitation is transmitted across the torso. In both cases, the shock or high velocity piston loading may be propagated into the lung as a shock since the lung speed of sound is ~20 m/s for most normal physiological volume fractions of air and tissue. Generally, subsonic displacement requires a relatively large impulse over a long time and is not likely to produce lung injuries typical of blast. There are potential reflections and transmissions at each interface.

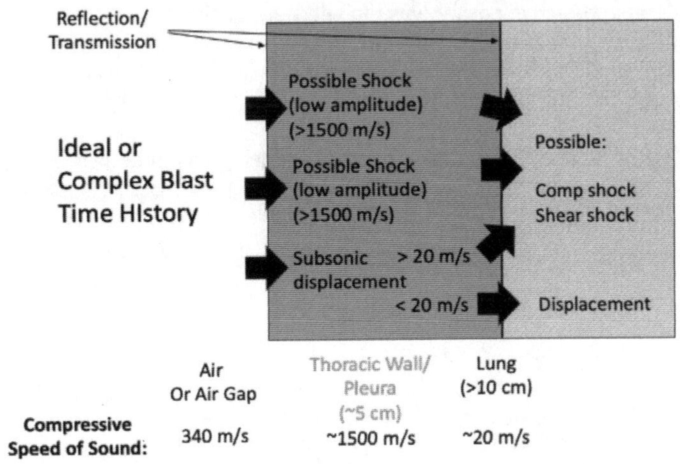

Figure 1. Routes of transmission of blast shock into the torso. Note: not drawn to scale.

Based on this paradigm of compressive shock damage in the lung, the purpose of this study is to establish a robust framework for assessments of primary blast injuries behind armor, where substantial modifications of ideal or complex blast waveforms are expected. We recast the risk functions of Panzer in terms of positive-phase specific impulse and peak positive-phase overpressure and use this analysis to framework for use in assessing injury risk based on analyses of compressive shock in the lung.

https://doi.org/10.52202/081621-0016

2. METHODS

2.1 Multiple Blast Exposure Injury Criteria

The single and multiple blast injury assessment methodology of Panzer [8] was adapted to produce both single and multiple blast injury risk assessments in terms of positive phase reflected OVP and positive phase reflected impulse. All experiments in the dataset had peak OVP rise times of less than 0.3 ms, the typical rise time expected for simple blast injury [15, 16]. The study data input included 2,349 large animal experiments with a range of positive phase OVP and impulse. Substantial differences between small and large animals led to the restriction to species of > 15 kg body mass or high phylogenetic order, such as primates. For experiments where only one of incident pr reflected pressure measurements were reported, the Rankine-Hugoniot relations provided a conversion between incident and reflected pressure. CONWEP was used to assess peak overpressure and duration when only charge characteristics were reported. To convert duration to impulse, the Friedlander approximation was used between duration, OVP, and impulse. Since Bass *et al* [7] and Rafaels *et al* [9] found that animal orientation and the presence of a reflecting surface did not influence the model results, these were excluded.

Experiments with both single and repeated blasts were used. Repeated blasts included 2-100 exposures, with elapsed interval timing between 1-30 minutes. For this assessment, a dichotomous outcome variable was used. Fatality data were classified as dead, and injury data were classified as 'surviving'. Since data from 14 species of disparate sizes were used in this analysis, the scaling procedure of Bowen [4] was used. This is equivalent to the assumption that a single length scale characterizes all interspecies differences, and that mass density and effective modulus of elasticity was constant between species. Altitude correction was not used for short duration blasts given experimental evidence that this correction has negligible influence on injuries. Long duration scaling was used based on rodent experiments suggesting the scaling was required [4].

Fatality risk functions were used a logistic regression of a piecewise linear model with peak overpressure, scaled impulse, and number of exposures as parameters. This model incorporated multiple breakpoints in impulse that provide fidelity to a continuous nonlinear fit while maintaining a simple form. The location of these breakpoints were optimized for model fit to the data.

The basis for the model is the single exposure fatality risk function. The repeated injury risk model was solved at the same time as the repeated fatality model since multiple blast fatality data were sparse. In the original analysis [8], the effect of the repeated exposure was modeled as a declination function that scaled the risk. In the current study, the results are presented as a multiple injury risk model for integrated impulse over multiple exposures. Note that this model is strictly only valid for repeated ideal blasts without further analysis. The regression models had goodness of fit characteristics that replicated the original analysis for peak OVP and positive phase duration.

The optimizations were performed a using custom Matlab gradient descent solver (The Mathworks Inc., Natick, MA). The code had been previously verified by comparing the logistic regression results of a simplified dataset and model with those calculated in commercially available statistical software (JMP 8; SAS Institute Inc., Cary, NC)[8].

2.2 Mechanism of Wounding

The paradigm developed in this study assumes that compressive shocks are the principal source of blast wounding, particularly for short durations, but generally for all sharp rising peaks of any duration. For ideal shocks to form from local material acceleration (here, chest wall acceleration, a_{cw}) a distance/peak amplitude and time are required (Friedrichs, 1948). In n half-plane domain bounded an accelerating wall, the shock path (x) and shock pressure (p) are calculated as

$$x = x_v + c_0\left\{(t-t_\gamma) + \frac{3(\gamma+1)}{32\gamma}\frac{a_{CW}(t-t_\gamma)^2}{c_0} - \frac{45(\gamma+1)^2}{2048\gamma^2}\frac{a_{CW}^2(t-t_\gamma)^3}{c_0{}^2} + \cdots\right\} \quad (1)$$

$$p = p_0\left\{1 + \frac{3(\gamma+1)}{4}\frac{a_{CW}(t-t_\gamma)}{c_0} - \frac{63(\gamma+1)^3}{512\,\gamma}\frac{a_{CW}^2(t-t_\gamma)^2}{c_0{}^2} + \cdots\right\} \quad (2)$$

https://doi.org/10.52202/081621-0016

where c_0 is the speed of sound in the material, g is the ratio of specific heats (1.4 for an adiabatic process), a_{cw} is the chest wall acceleration, p_0 is the ambient pressure, p is the shock pressure, x is the distance from the material wall, and x_g and t_g are the shock initiation distance and time from the material wall. These analyses can be combined with chest wall physical response data or models to assess the potential for lung shock injuries with accelerations that produce shocks in the lung parenchyma at distances shorter than the local depth of the lung.

Shock formation distance calculated using Equation (1) with typical parameters ($c_0 = 20$ m/s, $p_0 = 101$ kPa, $g = 1.4$) implies that shocks occur away from the pleural interface but well within typical porcine (or human) lungs (< 1 mm or less for higher severity blasts) based on typical human lung depths of ~180 mm or more. Previous experience in a pig lung model demonstrates this shock formation distance through alveolar damage [14]. Since pig lung size is approximately human lung size, this implies shock formation is occurring analytically at a distance that is plausible to explain the preliminary results. An important note is that this phenomenon does not scale with size/species, except with effects of size through the local applied acceleration and chest wall effective mass. Similarly, shock formation time calculated from Equation (2) using typical parameters ($c_0 = 20$ m/s, $p_0 = 101$ kPa, $g = 1.4$) shows that shock formation time is often 0.5 ms or less. This implies shock formation is occurring analytically at a time that shorter than that required to transit the model in large animals. Timing and wave reflections around the torso and back that are not considered in this analysis. The presence of strong reflections may make shock effects worse, but likely require strong enough shocks that the damage is fatal with or without strong reflections (i.e. the shock must still be a shock when reflected from opposite surfaces of the lung).

2.3 Complex Blast

To assess complex blasts using the hypotheses developed in this study, two aspects are important. The first is the single exposure blast assessment in terms of reflected impulse and reflected peak OVP as discussed above. The second is a relationship between applied thoracic impulse/pressure and chest wall motion, particularly local accelerations present in the shock model in Equations (1) and (2). For this aspect, we used the correlations of Boutillier et al [18] in the porcine thoracic model. These provide correlations between blast exposure (both reflected and incident impulse) at various levels and maximum chest wall acceleration, velocity for a range of severities. These severities are generally limited to lower impulse levels, but there are peak acceleration levels greater than 40,000 m/s^2. For example, for low impulse values, the correlation of max acceleration vs reflected impulse is

$$a_{max}=0.0644\, I_{ref}^2+78.225\, I_{ref}. \tag{3}$$

For complex blast, the ideal blast assessments must be adapted for the acceleration levels necessary to cause blast shock over the lung distance available without being separating discrete loading events during a complex blast time history. Based on the hypothesis in this study, this implies that the chest wall acceleration must be lower than the shock formation distance that would cause lung damage over the distance the wave must travel through the lung. Assessments are performed for both peak and mean acceleration levels from the Boutillier et al data. Assuming simple piston accelerative loading of the chest, mean acceleration is defined over a symmetric triangular rise and fall so that $a_{mean} = 0.41\, a_{max}$. Projected frontal area of the lungs is estimated using body surface area estimates assuming the torso 0.09% of total body area of 2.0 m^2 for adult males [cf. 19].

A limiting factor for shock development is the dimensions of the human lung. This is approximately 180 mm for human males [20]. Equation (1) is used to determine chest wall acceleration thresholds for alveolar injuries for 0% and 50% volume fractions damage. To assess threshold acceleration criteria and fatality criteria respectively. Effective chest wall mass, m_{eff}, is calculated from the Boutillier data for various combinations of peak OVP and positive phase impulse. So, for a given complex time history, a filter for chest wall acceleration values may be used to eliminate all non injurious thoracic wall response. Using the analyses the anthropometric parameters of adult humans, and the effective mass of the chest at various peak OVPs, the risk function for for multiple exposures in terms of peak OVP and positive phase impulse, may be used to assess the filtered complex blast including behind PPE.

3. RESULTS

The injury risk function for 5%, 50% and 95% risk of fatality in terms of peak reflected OVP and positive phase reflected impulse for single ideal blast exposure is shown in **Figure 2**. This reoptimization of the original Panzer model uses two breakpoints to obtain fidelity with the blast data at intermediate reflected impulse values between

https://doi.org/10.52202/081621-0016

1,000 kPa ms and 10,000 kPa ms. The high impulse limit of the risk function is essentially flat with little dependence on peak reflected OVP suggesting that these blast injuries are not sensitive to time dependence represented by positive phase impulse. At shorter reflected impulse values, the risk functions for peak OVP have a step slope to high OVP implying the smaller the reflected impulse, the larger the peak reflected OVP risk tolerance at each risk level displayed. The model displays good model fit with error assessments commensurate with the original model analyses.

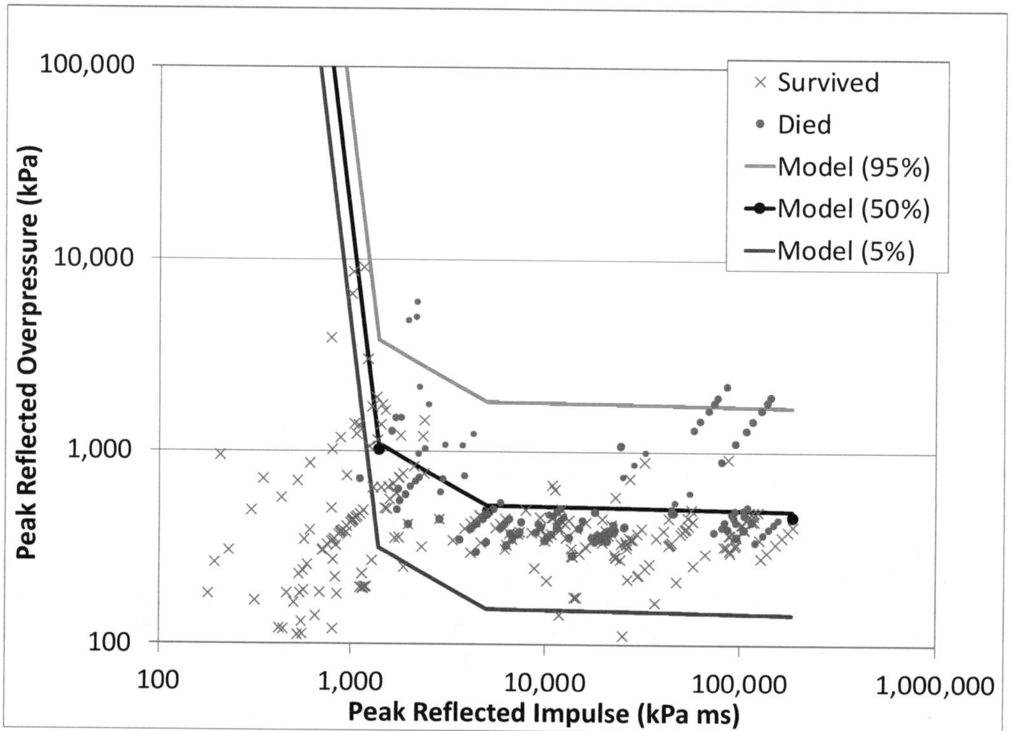

Figure 2. Single ideal blast exposure injury risk for 5%, 50% and 95% risk of fatality in terms of positive peak OVP and peak reflected impulse using the survival and fatality data in the current database. The survival and fatality experimental data for this model are shown in the plot.

The injury risk function for 5%, 50% and 95% risk of fatality in terms of peak reflected OVP and positive phase reflected impulse for multiple ideal blast exposure is shown in **Figure 3**. This reoptimization of the original Panzer model uses two breakpoints to obtain fidelity with the blast data at intermediate reflected impulse values between 1,000 kPa ms and 10,000 kPa ms. The high impulse limit of the risk function is essentially flat with little dependence on peak reflected OVP suggesting that these blast injuries are not sensitive to time dependence represented by positive phase impulse. At shorter reflected impulse values, the risk functions for peak OVP have a step slope to high OVP implying the smaller the reflected impulse, the larger the peak reflected OVP risk tolerance at each risk level displayed. This model includes the effect of the declination function from the original Panzer analyses [8] to provide analyses of complex blast through a filtering function that eliminates non injurious dynamic content. The model displays good model fit with error assessments commensurate with the original model analyses.

The shock formation distance using peak acceleration based on chest response porcine data of Boutillier [18], adult male anthropometric values, and the inversion of Equation 1 is shown in **Figure 4**. Using the 50% male lung depth as the threshold for fatality, the low risk (5%) value approach shock formation at the lung midpoint at high impulse. This suggests that the long impulse region may be conservative in this analysis.

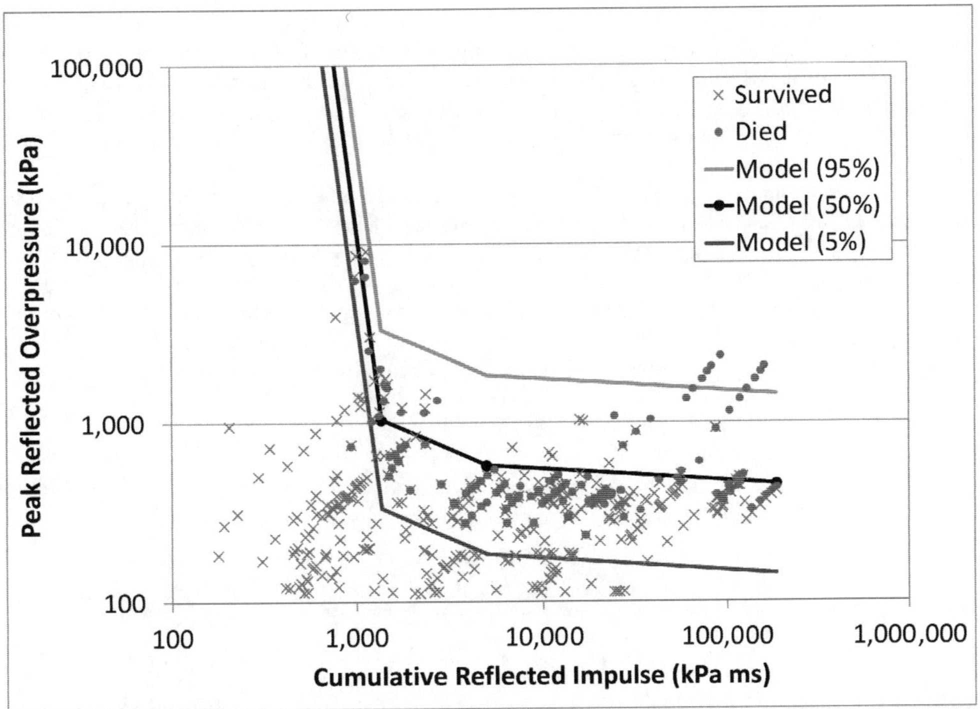

Figure 3. Multiple ideal blast exposure injury risk for 5%, 50% and 95% risk of fatality in terms of peak reflected OVP and cumulative reflected impulse using the survival and fatality data in the current database. The survival and fatality experimental data for this model are shown in the plot.

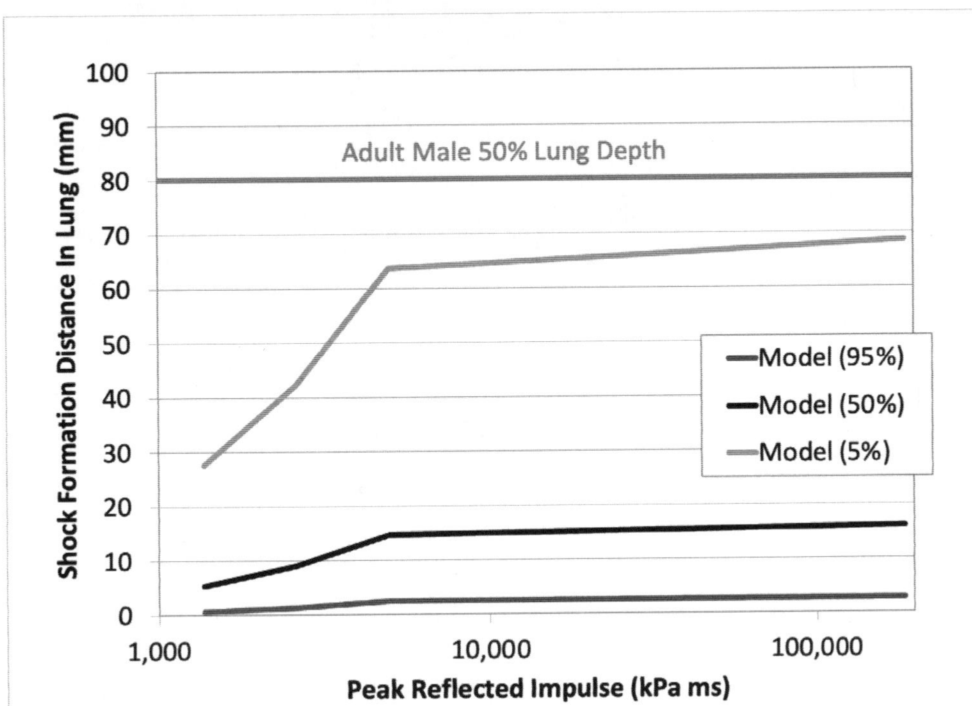

Figure 4. Internal lung shock wave development distance against peak reflected impulse. Accelerations are based on chest response porcine data of Boutillier (2017) and mid-size male anthropometric characteristics. Reference adult male lung depth is derived from Kramer et al [20].

https://doi.org/10.52202/081621-0016

Inverting Equation (1) for chest wall acceleration, this implies that chest wall acceleration values below 1850 m/s^2 will not cause lung shocks over the anterior/posterior distance of the lung. So, pressure time histories that do not produce chest wall accelerations of this magnitude will not result in shock injuries to the alveoli. This value provides a threshold for lung damage. Assuming 50% lung damage is not immediately fatal, acceleration values below 3,700 m/s^2 will not cause fatalities. There is strong evidence for this association of large lung damage with survival in experience in Israeli bus bombings [21] where up to 70% lung damage was survivable under some circumstances. By extension, for complex blast, only the accelerative loading that is > 3700 m/s^2 chest wall acceleration can cause fatalities, with wounding possible down to ~1850 m/s^2. Below the 3700 m/s^2 threshold, the limiting probability for fatalities is approximately 2.5% for long duration/low pressure, suggesting that values below 0.05% are likely conservative.

The thoracic effective accelerating mass can be estimated using the response data of Boutillier as shown in **Figure 5** in terms of peak reflected OVP. Note that the effective mass takes three ranges depending on injury risk while align with approximately logarithmic dependence on peak reflected OVP so that

$$M_{eff} = -0.448\ ln(P_{refl}) + 0.433 \tag{4}$$

For lower applied impulse, the thoracic model response has higher effective mass, and for higher impulse, the thoracic model response has lower effective mass.

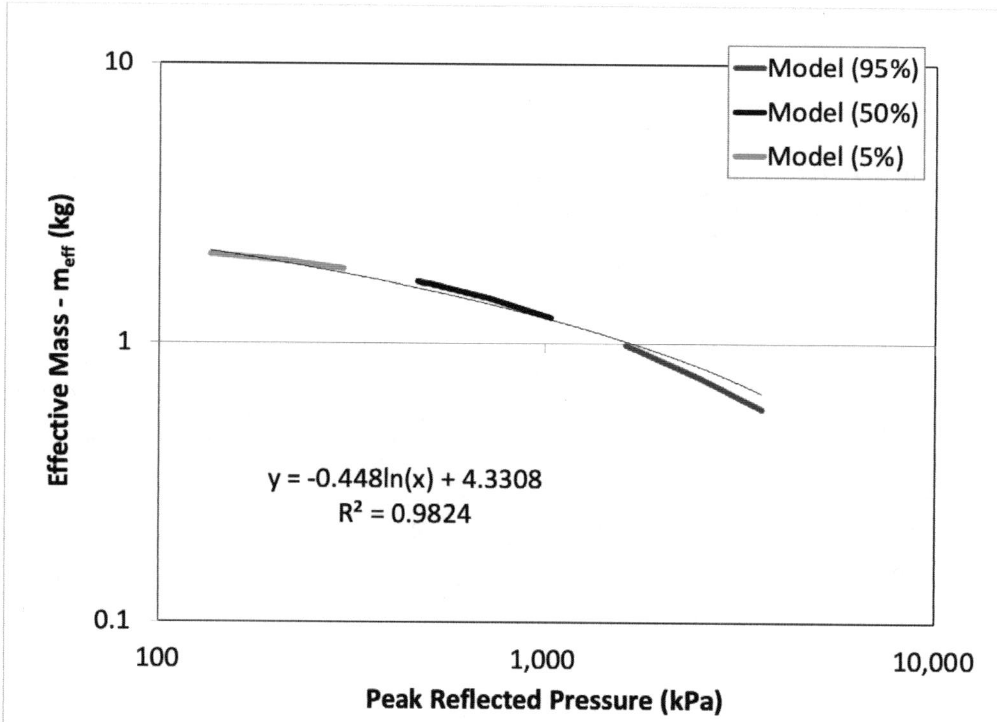

Figure 5. Effective mass of the chest wall vs peak reflected OVP based on the response data of Boutillier [18]

So, to calculate risk for complex blast fatalities, assuming a 50% shock depth to fatality and the effective mass shown in **Figure 5** with anthropometric parameters for a mid-size male we take a filter and calculate approach

1. Filter nonfatal exposures for the complex pressure/impulse time history based on threshold acceleration values derived from the thoracic response data of Boutillier [18], assuming arthrometric parameters for the appropriate human demographic such that

$$a_{threshold}(t) = P_{ref}(t)\ A_{lung}/m_{eff} \tag{5}$$

This results in a modified complex time history with non injurious components removed.

2. Use the risk assessments in **Figure 3** for cumulative impulse, $I_{cumulative}$, to assess injury risk against peak reflected OVP.

This combined assessment produces a complex blast injury assessment consistent with the single blast injury risk functions in **Figure 2** and **Figure 3** consistent with shock formation distance shown in **Figure 4**.

4. CONCLUSIONS

This study provides a generalizable paradigm for complex blast with PPE based on chest wall acceleration and velocity that predicts alveolar disruption characteristic of blast, not rib fractures and associated lung injuries associated with automobile crashes and other blunt trauma. Based on a filtering technique to remove non injurious components, it is extensible to different human anthropometries and human nonfatal injuries, as opposed to fatalities. This assessment is consistent with the available single and multiple exposure ideal blast data, in contrast with previous chest wall motion models intended for use with complex blast [10, 11]. In general, the injury risk function is insensitive to the high impulse, low overpressure regime. This is typical of nuclear blast exposures where tertiary effects become dominant.

Beyond this paradigm, we have provided injury risk assessments for multiple ideal blasts in terms of positive phase reflected pressure and reflected impulse. These are an extension of the multiple blast risk assessments of Panzer [8]. There are distinct advantages to using overpressure/impulse relations to assess blast lung risk, especially for use under PPE. Generally, reflected pressure measurements are available under PPE, but care must be taken to calibrate and account for gauge acceleration on nonrigid surfaces, and thoracic variation of input in a complex blast field. This will generally require multiple sensing locations for effective use of this technique.

In using this work in other contexts, care must be taken in interpreting response data, especially thoracic response data obtained by simple lumped mass approaches (e.g Axelsson [10]) or finite element models that have not been validated for larger accelerations. Chest wall velocity injury regimes from the Axelsson model predict blast lung injuries starting at 3-4 m/s chest wall velocity. These predictions stand in contrast with typical displacement-based automobile injury criteria for thoracic models that show no evidence of blast lung injuries [12].

The principal limitation of this study is that blast exposure is assumed to be directed towards the torso, and pressure/impulse exposure orthogonal to the local chest direction are not an important source of chest wall motion and injury. From the data, however, the differences between parallel and reflected are large, and tangential blast exposure is likely a small contributor to chest wall motion, especially behind PPE. Other limitations include limited thoracic response data in the low pressure/high impulse regime. In addition, the Boutillier response estimates are based on with local length of 9.5 mm and a density of ~9000 kg/m^3 [18]. Since the typical rib mass density is ~700 kg/m^3, this disparity may have influence on the local measurement of acceleration, reducing computed accelerations and velocities.

Future efforts will include expansion to additional anthropometries, and reduction in the approximations that lead from pressure to thoracic time histories. Other improvements include the implementation of a finite volume blast shock model for lung that can use arbitrary reflected OVP time histories to validate and extend the experimental thoracic model.

Acknowledgments

The authors gratefully acknowledge the support of the University of Virginia School of Engineering and the Wayne State University College of Engineering.

References

[1] DePalma RG, Burris DG, Champion HR, Hodgson MJ. NEJM, 352 (2005), 1335-42.
[2] Bass CR, Panzer MB, Rafaels KA, Wood G, Shridharani J, Capehart B. Ann Biomed Engr. 40 (2012), 185-202.
[3] Wolf SJ, Bebarta VS, Bonnett CJ, Pons PT, Cantrill SV. Lancet. 374 (2009), 405-15.

 https://doi.org/10.52202/081621-0016

[4] Bowen I. Estimate of man's tolerance to the direct effects of air blast. (1968) Technical Progress Report.
[5] Ngo T, Mendis P, Gupta A, Ramsay J. Blast loading and blast effects on structures–an overview. Electronic journal of structural engineering. 2007 Jan 1(1):76-91.
[6] Bass CR, Salzar RS, Lucas SR, Davis M, Donnellan L, Folk B, Sanderson E, Waclawik S. Int J Occ Safety Ergo. 12 (2006), 429-42.
[7] Bass CR, Rafaels KA, Salzar RS. Pulmonary injury risk assessment for short-duration blasts.. 2008 Sep 1;65(3):604-15.
[8] Panzer MB, Cameron R, Rafaels KA, Shridharani J, Capehart BP. J Trauma, 72 (2012), 454-66.
[9] Rafaels KA, Cameron R, Panzer MB, Salzar RS. J Trauma. 69 (2010), 368-74.
[10] Axelsson H, Yelverton JT. J Trauma. 40, (1996) 31S-7S.
[11] Stuhmiller JH, Ho KH, Vander Vorst MJ, Dodd KT, Fitzpatrick T, Mayorga M. J Biomech. 29 (1996), 227-34.
[12] Forman JL, Lopez-Valdes FJ, Duprey S, Bose D, De Dios ED, Subit D, Gillispie T, Crandall JR, Segui-Gomez M. Accident Anal & Prev. 80 (2015), 7-17.
[13] Phillips YY. Ann Emergency Med. 15 (1986), 1446-50.
[14] Op 't Eynde J. PhD Dissertation, Duke University, 2023.
[15] Richmond DR, Taborelli RV, Bowen IG, et al. Blast Biology—A Study of the Primary and Tertiary Effects of Blast in Open Underground Protective Shelters (WT-1467). (1959) Albuquerque, NM: Lovelace Foundation.
[16] De Candole CA. Blast injury. Can Med Assoc J. 1967;96:207–214.
[17] Friedrichs KO. Comm Pure Appl Math. 3 (1948), 211-45.
[18] Boutillier J, De Mezzo S, Deck C, Magnan P, Naz P, Willinger R. J Biomech.8 (2017), 169-75.
[19] Tikuisis P, Meunier P, Jubenville C. Eur J Appl Phys. 85 (2001), 264-71.
[20] Kramer GH, Capello K, Bearrs B, Lauzon A, Normandeau. Health Phys. 102 (2012), 378-83.
[21] Bala M, Shussman N, Rivkind AI, Izhar U, Almogy G. J Trauma. 69 (2010), 1022-9.

The effectiveness of PPE on blast overpressure propagation in a nonhuman primate

Y. Chen[1], K. Ondar[2], D. Fernandez[3], M. Maffeo[3], D. Wilder[1], J. Long[1], VS Sajja[1]

[1]*Walter Reed Army Institute of Research,503 Robert Grant Ave, Silver Spring, MD, USA, yungchia.chen.civ@health.mil*
[2]*Katmai Health Services, LLC, 11001 O'Malley Center, Suite 204, Anchorage, AK, USA*
[3] *U.S. ARMY Combat Capabilities Development Command Soldier Center, 10 General Green Ave, Natick, MA, USA*

Abstract. Injuries from explosive devices continue to be a major concern in military and civilian populations and personal protective equipment (PPE) could help reduce these injuries. However, the standards set by the National Institute of Justice as part of the acquisition process for explosive ordinance disposal (EOD) PPE only test the integrity and not the protectiveness of the suit. At present, there are no protection requirements against blast overpressure (BOP) for bomb suits; there are also no standardized guidelines or test methodologies to study the effectiveness or protectiveness of the EOD bomb suits and other PPE against primary BOP. The current testing strategies include surrogate testing (e.g. hybrid III), which are not validated for BOP. The objectives of this study are to determine the effectiveness of PPE based on BOP biomechanics using a custom-built EOD-PPE fabricated to fit the nonhuman primate. Two intracranial pressure (ICP) sensors (ventricle and epidural) and a thoracic pressure (ITP) sensor were placed in anesthetized animals. Animals with and without PPE were exposed to the same range of BOP levels. The presence of the vest did not affect the peak pressure in the thoracic region, but the helmet seemed to decrease peak pressures in the intracranial region. The vest did not reduce the impulse in the thoracic region, but the helmet did reduce impulse in the intracranial region. The presence of the helmet did increase the rise time of the pressure wave in the cranial region; this effect was not similarly reflected in the thoracic region. Future work conducting a similar comprehensive assessment with other species including postmortem human surrogates (PMHS) that may help define scaling laws across species. This will produce better correlations between blast parameters and injury risk, enabling the creation of guidelines for testing the effectiveness of PPE.

1. INTRODUCTION

Protecting military personnel from the devastating effects of explosions, particularly those encountered by Explosive Ordnance Disposal (EOD) specialists, remains a critical challenge. While substantial progress has been made in developing personal protective equipment (PPE) to mitigate injuries from shrapnel, fragments, and blunt force trauma, a significant gap persists in safeguarding against blast overpressure (BOP). This invisible threat poses a unique challenge, as it can induce severe, often lethal, internal injuries even when external damage is minimal. At present, there are no appropriate guidelines and test methodologies to study the effectiveness of the EOD bomb suits and other PPE against BOP [1].

Current testing standards for EOD suits and other PPE predominantly focus on maintaining structural integrity against blast events, neglecting the complex biomechanical responses of the human body to BOP. The requirements from the Maneuverability Center of Excellence to develop lighter PPE to meet the demands of the Warfighter in future multi-domain operations 2028 (MDO) includes the EOD suit, helmet and vest. This suit has been designed to protect against injuries from projectiles, shrapnel, debris and blunt object impact, but not BOP. This oversight stems from a lack of standardized testing protocols and validated injury criteria specific to BOP exposure. Consequently, the true efficacy of existing PPE to mitigate BOP-induced injuries remains largely unknown, hindering the development of next-generation PPE.

Several factors contribute to this knowledge gap. Major factors are the lack of understanding of BOP injury mechanism (e.g. contribution of peak, impulse or rise time to injury), how BOP parameters correlate to injury, and how well current PPE protects against them. This knowledge gap makes it difficult to quantify the protective capacity of PPE and establish safety standards for Service Members. Current surrogate models used to assess BOP effects, such as the Hybrid III crash test dummy, are ill-suited for replicating the intricate biomechanical interactions between blast waves and the human body. Another factor is conducting research using improper simulated blast waves which have pressure profiles that differ significantly from the "free-field" explosions experienced by personnel in the field. This discrepancy in blast wave characteristics further limits the translatability of current research findings to real-world scenarios.

https://doi.org/10.52202/081621-0017

There have been attempts to develop more sophisticated testing methods such as using human surrogate dummies (e.g. Human Surrogate Torso Model (HSTM)), which can measure internal pressures in response to blast exposure [2]. However, these models lack validation against human biomechanics data, making it even more difficult to translate response from the HSTM to human injury. Furthermore, a lack of standardized test methodology and injury risk criteria for BOP make it challenging to compare the effectiveness of different PPE designs and develop material-based countermeasures. Addressing these limitations requires a multi-pronged approach that includes developing more biofidelic test models, validating non-surrogate models, and establishing well-defined injury criteria based on the biomechanical forces associated with BOP.

To create proper methodologies to evaluate PPE protectiveness against BOP exposure will require three steps: 1) establish biomechanical responses to blast in multiple experimental animal models; 2) develop the correlation of these biomechanical responses to injury; and 3) develop transfer functions based on biomechanics to scale the findings from animals to humans and establish injury risk thresholds. The research described this paper is part of the first step, to compare the effect of PPE on BOP parameters in the non-human primate (NHP) (pre-clinical) animal model. A custom PPE was designed and made to fit the NHP in similar fashion to the EOD suit. All animals were instrumented with pressure catheters in the head and thoracic region of the body. Ultimately, this research will pave the way for designing and fielding next-generation PPE capable of providing comprehensive protection against the full spectrum of blast-related threats faced by military personnel.

2. METHODS

A total of nine Rhesus macaques were used for this study. There were two females and seven males that averaged 9.80±0.97 years old, weighing 9.29±0.70 kg. Paired housed animals were separated prior to the start of the study in side-by-side cages with their paired cage mate.

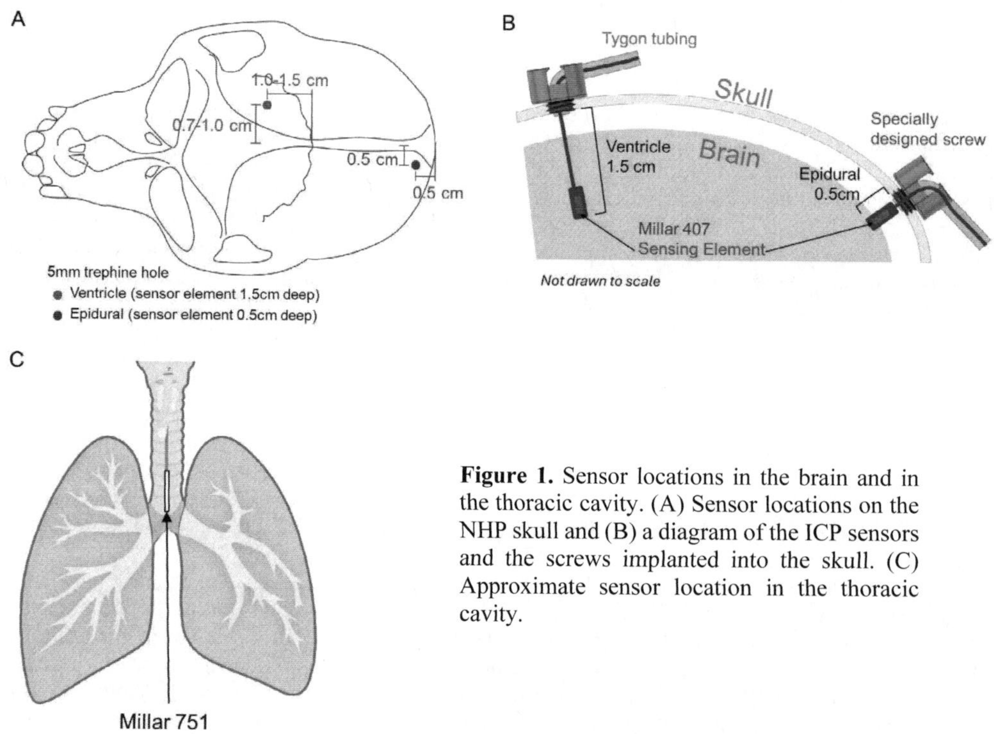

Figure 1. Sensor locations in the brain and in the thoracic cavity. (A) Sensor locations on the NHP skull and (B) a diagram of the ICP sensors and the screws implanted into the skull. (C) Approximate sensor location in the thoracic cavity.

https://doi.org/10.52202/081621-0017

2.1 Surgical procedure.

Non-survival surgery was performed to implant pressure sensors in the brain and thoracic cavity. Animals were fasted for ≥12 hours prior to surgery and fully sedated with anesthesia throughout the surgery and data collection. Routine monitoring included indirect blood pressure, heart rate, respiration rate, end-tidal CO2, pulse oximetry, and body core temperature to ensure all animals were appropriately sedated.

2.2 Pressure sensor implantation

After shaving the fur, a local anesthetic (bupivacaine 0.5%) was administered to the incision site. The animals were placed in ventral recumbency, and an incision was made in the scalp. Trephines (5mm diameter) were used to drill two holes at the following locations, ventricle and epidural (Figure 1A). The coordinates for the ventricle sensors were 1-1.5 cm rostral, 0.7-1 cm lateral to bregma; epidural coordinates were 0.5cm rostral and lateral to lambda, opposite to the ventricle sensor. A specially designed screw secured the ventricle and epidural sensors (Figure 1B) to the skull. The ventricle sensor (Millar SPR-407 catheter, 1.5F body size) was placed in the lateral ventricle at 1.5cm deep; the epidural sensor (Millar SPR-407 catheter) was placed in the occipital lobe at 0.5cm deep. A third pressure sensor (Millar SPR-751S, 5F body size) was inserted into the esophagus and advanced approximately 22 cm into thoracic cavity and taped to the endotracheal tube (Figure 1C). The 751 Millar had two sensing elements (thoracic and distal), measuring 2.5 cm apart, with the distal caudal to the thoracic sensor. Anesthesia was maintained for the duration of the blast procedure until euthanasia.

2.3 Blast exposure

Animals were secured in an upright seated position within the Advanced Blast Simulator (ABS; ORA Inc., Marion, NC), which comprises of a 5.8 m long system with a driver chamber, divergent transition section, and a 61-cm² diameter test section (Figure 2). Compressed air was used to rupture acetate membranes (Graphix Inc, Ohio) or Valmex® membranes (Low and Bonar, Martinsville, Inc., Martinsville, VA) of varying thicknesses to generate blast waves with peak incident pressure levels of 32, 68, 104, or 145 kPa and associated impulses of 46, 117, 219, or 282 kPa•sec, respectively. The incident pressures chosen spanned the range produced by the ABS. This range of pressures has also been used in other animal species to study the BOP propagation through the brain and body; using the same ranges will enable direct comparison across species.

Figure 2. Advanced blast simulator (ABS)

A pitot probe (Stumptown Research and Development, Marion, NC) measured peak reflective, stagnation and incident pressures near the animal. Nine Rhesus macaques underwent multiple blast exposures, yielding ≥9 data points per pressure group. Data was acquired using a TMX-18 system (Astronova Inc., West Warwick, RI) at a sample rate of 800,000 Hz. Following data collection, animals

https://doi.org/10.52202/081621-0017

were euthanized according to the 2020 AVMA Guidelines for the Euthanasia of Animals using an intravenous injectable barbiturate.

2.4 Data analysis

Pressure profiles were analyzed using MATLAB (R2020a, Natick, MA). Pressure-time histories were analyzed for peak pressure, rise time, positive duration and impulse. The peak pressure was defined as the peak that is present after the initial rise of the shock, not accounting for the reflective peaks. Rise time is defined as the first time point where the pressure exceeds 2 kPa. Positive duration was defined from the beginning of the rise time to the next time the pressure trace crossed zero (zero crossing). Finally, impulse was defined as the integral of pressure (area under the curve of the positive phase) across the positive duration. Negative phase data was collected but not reported for this paper. All Millar sensors were filtered with a 4th order Butterworth filter with a 10,000Hz cutoff frequency. This filter was selected to remove the presence of cable whip and other high frequency noise events while keeping intact the majority of the signal. Analysis was performed as described above in all sensors: pitot, ventricle, epidural, thoracic, and distal.

The data were then imported into JASP (Version 0.19.3) where all required assumptions were assessed before completing the analysis. The Shapiro-Wilks test was implemented to test for multivariate normality. Multivariate outliers were examined with the Cook's Distance. Linearity and homoscedasticity of each of the predictor variables was assumed and analyzed through the residuals plot of the fitted data. Independence of residuals was analyzed through the Durbin-Watson test between observations and finally, the assumption of multicollinearity was also analyzed. Q-Q plots tested the data for normality and analysis of covariance (ANCOVA) tested for significant interaction between variables. When p value is less than 0.05, comparison in the data is considered statistically significant.

2.5 Personal Protective Equipment (PPE)

The custom-tailored body armour was constructed using the same or similar materials that are used in the MedEng EOD-8 bomb suit, which is widely fielded across military and law enforcement organizations. The PPE was tailored to fit the average sized NHP for this study. The suit only included the ballistic materials and a cover fabric. The other external accessories used on a typical bomb suit were not replicated. Every attempt was made to use identical ballistic materials if available. If substitutions were necessary, similar materials at the same areal density were used. The helmet was constructed using Tensylon HA 120 for the outer shell, Spectra Shield 4232 for the mandible, and an 18 mm thick Plexiglass visor. The outer shell material used for the suit was a TenCate Defender M fabric. The soft armor for the torso, arms, and legs were constructed using Twaron Style 1028 with the same layering as is used in the EOD-8. The rigid plates in the center neck, upper torso, groin, and thighs were constructed using Spectra Shield 4232 at the same areal density as the equivalent items in the bomb suit. The armour plates were shaped and sized to fit the NHP body in a manner to provide equivalent coverage as provided by the full-sized bomb suit. A Skydex attenuation panel was removed from an existing bomb suit torso plate and trimmed and mounted to the NHP torso plate.

3. RESULTS

The pressure-time histories for the incident, intracranial (ventricle and epidural), and intrathoracic (thoracic and distal) pressures were analyzed for the effect of PPE on the following blast parameters: peak pressure, impulse, rise time, and positive duration.

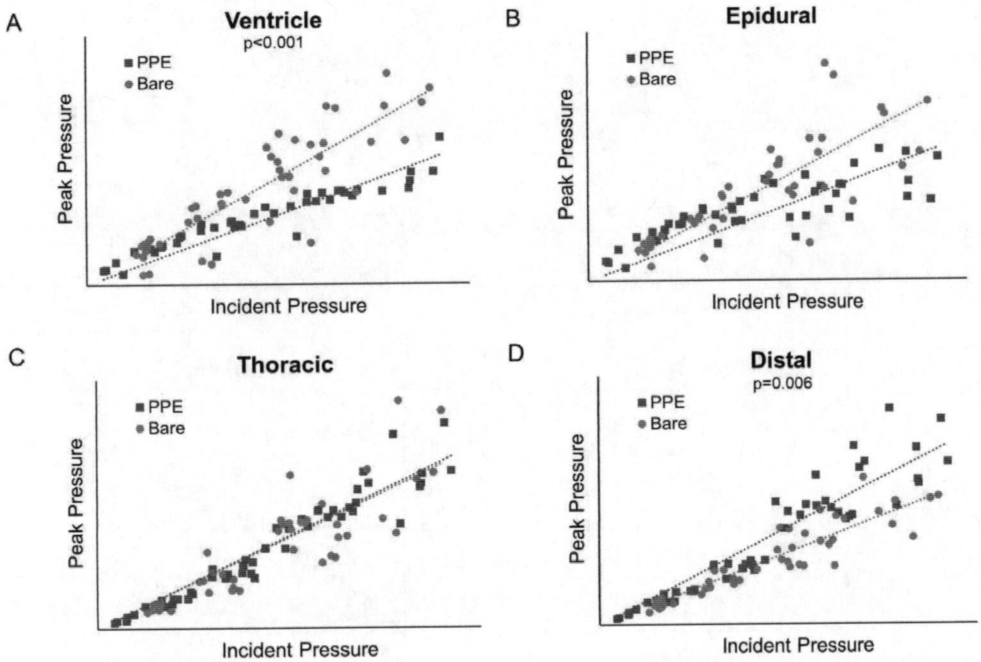

Figure 3. Comparison of the peak pressure data between PPE and bare configuration in Millar sensors showed significant changes in some of the sensors. (A) Ventricle ($p<0.001$) but not (B) epidural saw significant differences. (C) Thoracic was not significantly different but (D) distal ($p = 0.006$) did show significant differences.

3.1 PPE effect on peak pressures

A comparison of pressure data between the Millar and incident sensors showed that PPE produced differing effects between the brain and the thoracic region (Figure 3). The helmet significantly changed the peak pressures in the ventricle ($p<0.001$) (Figure 3A) but not the epidural. In contrast, the PPE had no effect on the thoracic sensor (Figure 1C) but significantly ($p < 0.01$) changed the peak pressures at the distal sensor (Figure 3D) suggesting that the PPE possibly amplified the transmission of pressure into the thoracic region.

3.2 PPE effect on impulse

The effect of PPE on impulse is illustrated in Figure 4. The PPE configuration saw lower impulse in the head, though this decrease is only significant ($p<0.001$) for the ventricle sensor (Figure 4A, B). Similar to the epidural, the presence of protection did not affect impulse in the intrathoracic region (Figure 4C, D).

https://doi.org/10.52202/081621-0017

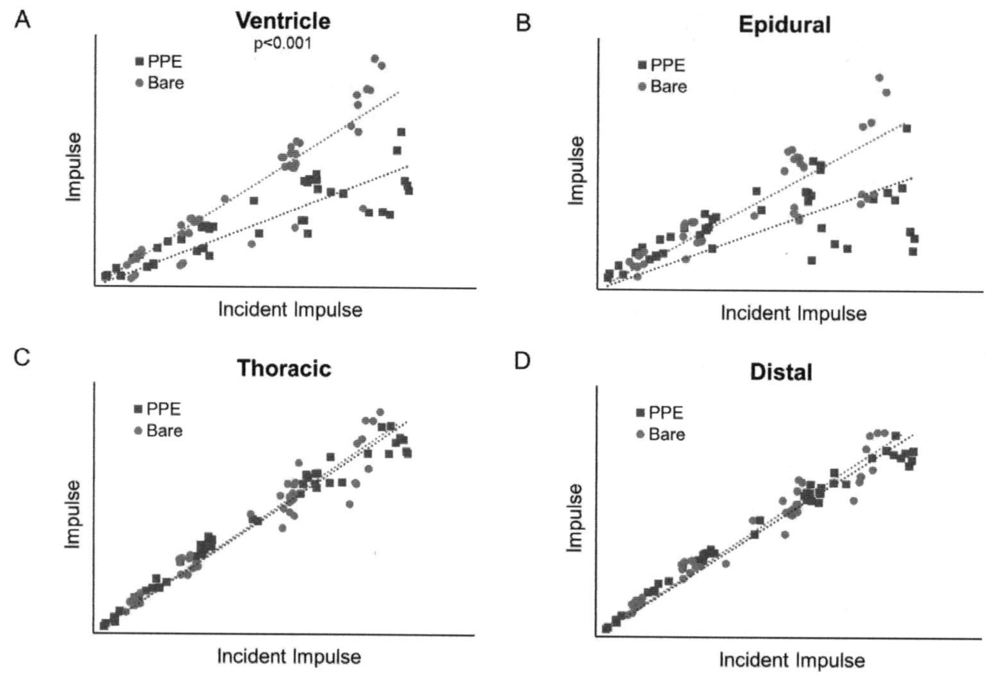

Figure 4. Comparison of the positive pressure impulse data between PPE and bare configurations showed only the (A) ventricle sensor ($p<0.001$) had significant differences. The pressures measured with the (B) epidural, (C) thoracic, (D) and distal sensors were not significantly altered by the presence of PPE.

3.3 PPE effect on rise time

The presence of the PPE was found to significantly impact the rise time at all four sensor locations (Figure 5). In the thoracic region, PPE seem to have opposing effects on the rise time. The thoracic rise time ($p<0.001$) increased with PPE and whereas it decreased for the distal sensor ($p=0.005$). Note, the ANCOVA only tested for statistical changes but not for statistical increases or decreases. For the sensors in the head, both ventricle ($p<0.001$) and epidural ($p<0.001$) locations yielded similar behavior; the rise times were significantly different when the PPE was worn versus when it was not worn. As there were no significant differences in the rise time of the incident pressures across all kPa's tested, the intracavity rise times were graphed in box and whisker plots to better display the group differences between PPE and bare configurations.

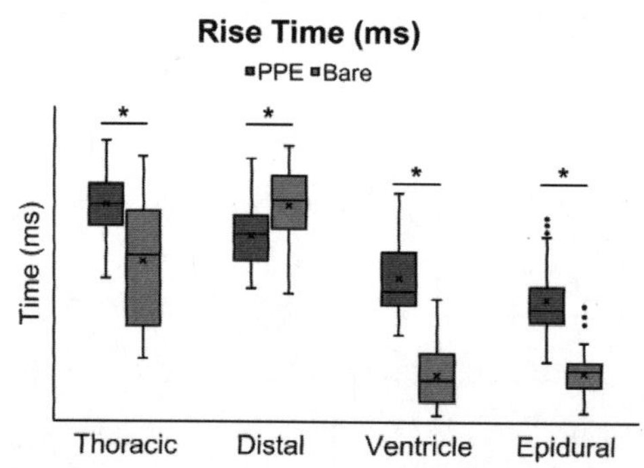

Figure 5. PPE had a significant effect on rise time on each of the four sensor locations: thoracic ($p<0.001$), distal ($p=0.005$), ventricle ($p<0.001$), and epidural ($p<0.001$).

3.4 PPE effect on positive duration

Figure 6 compares the effect of PPE on positive duration for each of the four sensor locations. The only sensor location that was significantly altered by the PPE was the distal sensor ($p<0.001$) (Figure 6D). The PPE did not affect the positive duration in the other three sensor locations. These findings suggest that the presence of PPE somewhat affects the duration of the positive pressure phase, although the exact nature of this relationship requires further investigation.

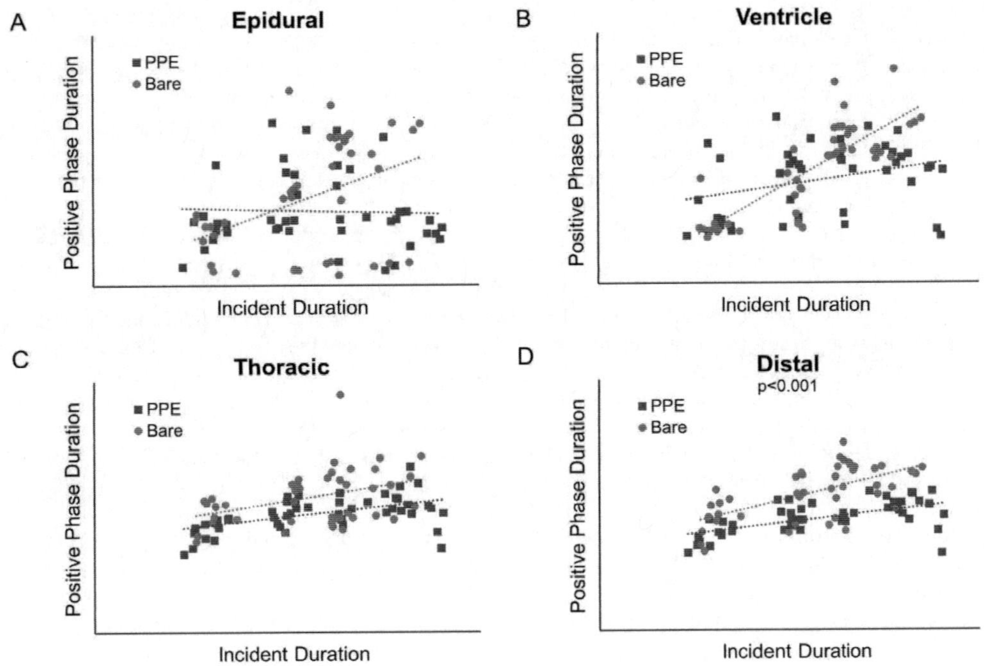

Figure 6. Comparison of the positive durations between PPE and bare configurations at the four sensor location: (A) ventricle, (B) epidural, (C) thoracic, (D) distal locations ($p<0.001$). The PPE seemed to have an impact on pressures only at the distal sensor location.

4. DISCUSSIONS

The current study investigated the biomechanical responses from BOP loading with and without the presence of PPE on intracranial (ventricle and epidural) and thoracic (thoracic and distal) pressures in a non-human primate model. Overall, the results show that the helmet significantly reduced the peak pressure and impulse in the ventricle region of the brain. This reduction in impulse was primarily due to a reduction of peak pressure but not due to a reduction in positive duration. In the thoracic region, the distal sensor saw an increase in peak pressures but a decrease in positive duration when the PPE was worn. This may explain why the calculated positive impulse saw no change with PPE. The rise time of the intracranial pressure in the brain for the bare configuration was found to be shorter than in the thoracic region, implying that thoracic surge is not the mechanism of intracranial pressure transmission in the cranium. The PPE significantly altered the rise time of for all four pressure sensors suggesting that the shockwave properties of the BOP were altered.

A number of research articles that have published studies on the biomechanical response to BOP in multiple species such as rodent [3, 3, 5], swine [6, 7], and other animals [8, 9]. The findings in this study are generally consistent with previous studies, showing increasing intracranial and intrathoracic pressures with increasing BOP levels. The intracavity rise times are also increased compared to the incident pressure rise times (data not shown). Some studies have reported higher intracranial pressure than the exposed incident pressure, a finding that was not observed in this study and could be due to a difference in species.

https://doi.org/10.52202/081621-0017

When comparing the biomechanical performance of PPE, only a handful of studies exist. Some studies used animal models to study the effectiveness of body worn PPE [10, 11], others used physical anthropomorphic test dummies (ATD) [12, 13]. ATDs [14, 15, 16, 17] or computational modeling [18, 19] can be used to study the biomechanical responses of helmet and eye protection but does not allow for an injury response comparison. One of the significant limitations of these data from ATDs and computational models are lack of validation from experimental data. This research is unique in that a custom-made, scaled-down PPE made from the similar materials used for human PPE to protect both head and body was created to fit the NHPs. Data from this study could be used to compare to PMHS tests with PPE, validate and verify computational modeling results, and develop transfer function to define injury risk criteria.

Previous studies have shown that body worn PPE can have varying effects on intrathoracic pressures from BOP exposure. For example, Philips et al. 1988 [9] found that a cloth ballistic vest increased intrathoracic pressures and mortality in sheep. Young et al. 1985 [20] found that protective vests (e.g. ballistic vest, ceramic vests, field jacket) generally did not affect intrathoracic pressures compared to wearing fatigues for living humans standing in front of a shock tube, exposed to 18.6 kPa overpressure, with one exception. The ballistic vests they tested increased the intrathoracic pressures compared to wearing fatigues alone.

Research with more modern ballistic vests (hard and soft) were overall found to reduce peak pressure when used on instrumented testing rigs [21]. Sekine et al 2021 [11] showed that swine wearing ballistic vests observed no mortality at the 3-hour survival time point whereas the control group (that did not wear ballistic armor) saw a mortality rate of 55%. In contrast to the more recent studies, the findings in this research are more similar to the Young et al 1985 [20] study where they found most PPE, except the ballistic vest, had little effect on ITP compared to fatigues. The custom NHP PPE did not significantly alter most of BOP parameters analyzed (peak pressure, impulse, and positive duration) for the thoracic sensor. It is interesting to note that the PPE affected the distal sensor readings significantly compared to the thoracic, despite only being 2.5cm apart. This suggest that placement of the sensors in the thoracic region is important, due to potential complex interactions in the chest cavity. Studies with survival NHPs will need to be conducted to compare the injury response between PPE and bare conditions.

Similar to past research, the NHP helmet in this study was effective in reducing the peak pressures in the brain. Azar et al. 2020 [14] found that a tactical ballistic helmet with a visor combination reduced peak pressures in a surrogate headform. Dionne et al, 2018 [15] used a Hybrid III model to test the efficacy of an EOD suit exposed to blast. However, their analysis focused solely on using acceleration as a metric, as opposed to looking at reduction in BOP propagation. While some studies have shown increased pressure under the standard issue helmet, the data from this is more relevant to the helmets that have visors, similar to EOD helmets and could be a contributing factor in the reduction of pressure inside the cranium.

There are several limitations to this study. The first is that only pressure was measured. The difference in outcome between the thoracic and distal sensors demonstrate how PPE may affect BOP parameters. This suggests it would be valuable to include accelerometers and or strain gages to determine if body worn PPE could affect chest wall velocity. Another limitation is the size of the animal model with respect to the size of the ABS used. It is estimated that approximately 45% of the cross-sectional area of the ABS was occupied by the specimen, and more if the specimen was in the suit. As the cross-sectional area of the ABS is increasingly occluded, the dynamic pressure can increase, resulting in reflections [22]. The exact level of acceptable occlusion to minimize reflections is experiment dependent. Furthermore, free flight studies in the ABS showed the drag forces did not create sustained accelerations with larger objects compared to smaller diameters objects [23]. This suggests the maximum allowed blockage in the ABS permitted to accurately assess the impact on the flow conditions is not clear.

5. CONCLUSIONS

The results of this study provide valuable insights into the effects of PPE against BOP exposure and will shed light on the primary mechanisms underlying blast injury. Furthermore, these findings can be leveraged to validate various non-experimental models such as surrogates and computational models of the brain and body. Comparing the results from NHP with post-mortem human subjects (PMHS), will enable the development of scaling factors to facilitate interspecies comparisons. Direct comparisons between rodent and swine ICP data to humans are confounded by differences in head size, shape, skull thickness, and other physiological differences. Correlating the outcome from survival animals with and without PPE to the blast parameters in this paper, can inform the development of an injury risk curve as well as provide greater insight on the primary mechanism of blast injury. Altogether, these efforts will potentially guide designs of more effective protective measures and enhance our understanding of blast physics and its effects on people.

Acknowledgements

The authors would like to thank the hard work of former and current colleagues Drs. Elizabeth McNeil and Venkata R. Kakulavarapu for their technical expertise and assistance. We also would like to thank the WRAIR Veterinary Services Program for their exceptional care and management of the animal subjects. This research was funded by USAMRDC Military Operational Medicine Research Program, authors appreciate the financial support and resources provided by this program, which has enabled the pursuit of this important research question.

Disclaimer

Material has been reviewed by the Walter Reed Army Institute of Research. There is no objection to its presentation and/or publication. The opinions or assertions contained herein are the private views of the author, and are not to be construed as official, or as reflecting true views of the Department of the Army or the Department of Defense. Research was conducted under an IACUC-approved animal use protocol in an AAALAC International - accredited facility with a Public Health Services Animal Welfare Assurance and in compliance with the Animal Welfare Act and other federal statutes and regulations relating to laboratory animals.

References

[1] Review of Department of Defense Test Protocols for Combat Helmets, Washington (DC): National Academies Press (US), ISBN-13: 978-0-309-29866-7, (2014).
[2] Wickwire A., Carneal C., Merkle A., Jackson T., Carboni M., DeCristofano B., Maffeo M., Effect of Torso Armor on Surface and Internal Pressure Response of a Human Surrogate. Personal Armour Safety Symposium 2014, Cambridge, UK.
[3] Leonardi A.D.C., Bir C.A., Ritzel D.V., VandeVord P.J., Intracranial Pressure Increases during Exposure to a Shock Wave. Journal of Neurotrauma, 28 (2011), 85–94.
[4] Rubio J.E., Unnikrishnan G., Sajja V.S.S.S., Van Albert S., Rossetti F., Skotak M., Alay E., Sundaramurthy A., Subramaniam D.R., Long J. B., Chandra N., Reifman J., Investigation of the direct and indirect mechanisms of primary blast insult to the brain. Sci Rep, 11 (2021), 16040.
[5] Norris C., Murphy S.F., Talty C.E., VandeVord P.J., Spatial Intracranial Pressure Fields Driven by Blast Overpressure in Rats. Annals of Biomedical Engineering, 52 (2024), 2641–2654.
[6] Bauman R.A., Ling G., Tong L., Januszkiewicz A., Agoston D., Delanerolle N., Kim Y., Ritzel D., Bell R., Ecklund J., Armonda R., Bandak F., Parks S., An Introductory Characterization of a Combat-Casualty-Care Relevant Swine Model of Closed Head Injury Resulting from Exposure to Explosive Blast. Journal of Neurotrauma, 26 (2009), 841–860.
[7] Feng K., Zhang L., Jin X., Chen C., Kallakuri S., Saif T., Cavanaugh J., King, A., Biomechanical Responses of the Brain in Swine Subject to Free-Field Blasts. Frontiers in Neurology, 7 (2016).
[8] Romba J.J. and Martin P., The Propagation of Air Shock Waves on a Biophysical Model, Armed Services Technical Information Agency No. AD264932 (1961), 1–25.
[9] Phillips Y.Y., Mundie T.G., Yelverton J.T., Richmond D.R. Cloth ballistic vest alters response to blast. The Journal of Trauma, 28 (1988), S149-152.
[10] McNeil E.M., Reilly M.J., Wilder D.M., Benton M.A., Long J.B., Sajja V.S.S.S. Soft-armor Vest Effectiveness and Intrathoracic Biomechanics in Rodents Exposed to Primary Blast. Annals of Biomedical Engineering, 51 (2023), 1616–1626.
[11] Sekine Y., Saitoh D., Yoshimura Y., Fujita M., Araki Y., Kobayashi Y., Kusumi H., Yamagishi S., Suto Y., Tamaki H., Ono Y., Mizukaki T., Nemoto M. Efficacy of Body Armor in Protection Against Blast Injuries Using a Swine Model in a Confined Space with a Blast Tube. Annals of Biomedical Engineering, 49 (2021), 2944–2956.
[12] Boutillier J., Cardona V., Magnan P., Ogier M., De Mezzo S., Montespan F., Menini W., Mosnier J., Naz P., Prat N.J. A New Anthropomorphic Mannequin for Efficacy Evaluation of Thoracic Protective Equipment Against Blast Threats. Frontiers in Bioengineering and Biotechnology, 9 (2022)

 https://doi.org/10.52202/081621-0017

[13] Ouellet S. and Philippens M. The multi-modal responses of a physical head model subjected to various blast exposure conditions. Shock Waves, 28 (2018), 19–36.
[14] Azar A., Bhagavathula K.B., Hogan J.D., Ouellet S., Satapathy S., Dennison C.R. Protective Headgear Attenuates Forces on the Inner Table and Pressure in the Brain Parenchyma During Blast and Impact: An Experimental Study Using a Simulant-Based Surrogate Model of the Human Head. Journal of Biomechanical Engineering. 142 (2020).
[15] Dionne J.P., Levine J., Makris A. Acceleration-based methodology to assess the blast mitigation performance of explosive ordnance disposal helmets. Shock Waves, 28 (2018), 5–18.
[16] Skotak M., Salib J., Misistia A., Cardenas A., Alay E., Chandra N., Kamimori G.H. Factors Contributing to Increased Blast Overpressure Inside Modern Ballistic Helmets. Applied Sciences, 10 (2020).
[17] Alphonse V., Carneal C., Luong Q., Clark J., Andrist J., Townsend K., Maffeo M., Carboni M., Cyganik J. Effect of Helmet and Eyewear on Headform Kinematic Response to Primary Blast Overpressure Exposure. Personal Armour Safety Symposium 2018, Washington DC, US.
[18] Sundar S. and Ponnalagu A. Biomechanical Analysis of Head Subjected to Blast Waves and the Role of Combat Protective Headgear Under Blast Loading: A Review. Journal of Biomechanical Engineering, 143 (2021), 100801.
[19] Tan L.B., Tse K.M., Tan Y.H., Sapingi M.A.B., Tan V.B.C., Lee H.P. Face shield design against blast-induced head injuries. In International Journal for Numerical Methods in Biomedical Engineering, 33 (2017).
[20] Young A.J., Jaeger J.J., Phillips Y.Y., Yelverton J.T., Richmond D.R. The influence of clothing on human intrathoracic pressure during airblast. Aviation, Space, and Environmental Medicine, 56 (1985), 49–53.
[21] Wood G.W., Panzer M.B., Shridharani J.K., Matthews K.A., Capehart B.P., Myers B.S., Bass C.R. Attenuation of blast pressure behind ballistic protective vests. Injury Prevention, 19 (2013), 19–25.
[22] Needham C.E., Ritzel D., Rule G.T., Wiri S., Young L. Blast Testing Issues and TBI: Experimental Models That Lead to Wrong Conclusions. Frontiers in Neurology, 6 (2015).
[23] Ritzel D.V., Van Albert S., Sajja V., Long J., Acceleration from short-duration blast. Shock Waves, 28 (2018), 101–114.

https://doi.org/10.52202/081621-0017

Analysis of Injuries Sustained by Law Enforcement Officers Wearing Body Armour

D. Watts[1,2], MS, L. Alejandro de Leon, MD[3], K. Inaba, MD[4], FRCSC, FACS, and C. Bir, PhD[1]

[1]*Department of Biomedical Engineering, Wayne State University, 818 W. Hancock, Detroit, MI 48201, USA*
doriwatts@wayne.edu
[2]*Explico Engineering, 40028 Grand River Avenue, Suite 300, Novi, MI 48375*
[3]*Grady Memorial Hospital, 80 Jesse Hill Jr Drive, Atlanta, GA, 30303*
[4]*Department of Surgery, Keck School of Medicine, USC, Los Angeles, CA, USA*

Abstract. To further understand Behind Armor Blunt Trauma (BABT), a database of real-world cases has been established. Data were collected related to the threat, body armour worn at the time of the incident, armour penetration injury, and overall resulting injuries. A questionnaire was completed by participants describing their injuries and post-incident limitations. Medical records were obtained, when possible, including radiographic imaging results. A total of 101 cases were reviewed and characterized with the assistance of a trauma surgeon. Although the majority were minor injuries such as bruises, over 40 cases were considered clinically significant. These cases include rib and sternal fractures, pulmonary and hepatic contusions, hemothorax, pleural effusion, spleen laceration, and hemoperitoneum. A statistical analysis was performed to assess the relationship between variables related to the incident (i.e. vest level, weapon type, standoff distance) to variables related to injury (i.e. penetration injury status, internal injury status, maximum AIS (MAIS) score). Most cases resulted in injuries with a MAIS score of 1 (n=62). Penetration injuries were found in 13 cases, and internal injuries were not found in 8 of those 13 cases. Internal injuries were noted in 17.8% (n=18) of cases, with the shot location in 5 of the 18 cases being at the outer limit of the vest or an extremity. Vest level, standoff distance, and weapon type did not have significant relationships with whether or not there was a penetration injury, internal injury, or MAIS score. There was a notable (but not significant) relationship between vest level and clinical significance (P=0.011). Many cases resulting in MAIS scores of 2 or higher involved shots to areas not protected by armour. However, pulmonary contusions, small bowel wounds, and hepatic contusions occurred due to BABT. This data analysis improves understanding of vest performance in real-world scenarios, therefore providing information for the improvement of armour.

1. INTRODUCTION

Law enforcement officers (LEOs) are exposed to significant occupational hazards that result in both fatal and nonfatal injuries. Their job often requires physical engagement with suspects, vehicular operations, and exposure to environmental risks and they experience some of the highest rates of work-related injuries compared to other professions [1]. Between 2012 and 2017, over 300,000 LEOs were treated in emergency departments across the United States [2]. According to the most recent Law Enforcement Officers Killed and Assaulted (LEOKA) data, a total of 13.2% of all LEOs were assaulted in 2023 with 33.7% of these officers sustaining injuries [3]. In addition, there has been an increase injuries related to assaults with firearms over the past ten years with the total going from 9.4% in 2014 to 13.8% in 2023 [3].

Body armour plays a crucial role in modern personal protection systems, safeguarding not only LEOs, but military personnel from ballistic threats. The design and effectiveness of body armor have evolved significantly over the years, incorporating advanced materials and structural innovations to enhance protection while maintaining wearability. However, despite these advancements, the phenomenon of behind armor blunt trauma (BABT) remains a critical concern given the energy transfer that can cause an injury even if the vest physically stops the bullet from penetrating. In addition, there are issues with body armour fit and coverage ([4]) that may also lead to officer injury based on a round not impacting the armour.

In an effort to determine the type of injuries sustained by LEOs in the field, a BABT database has been established [5]. The characterization and understanding of real-world BABT cases is helpful in establishing a link between armour use, threat level, and injury outcome. In addition, it can help to determine appropriate thresholds for certification standards. As such, Bir et al. (2017) published their findings regarding back face signature (BFS) assessment in a laboratory setting in an attempt to correlate the deformation into clay to known injuries that resulted from real-world cases. Results indicated a 50%

https://doi.org/10.52202/081621-0018

risk of severe injury with a deformation of 43.6 mm [5] which aligns with the current National Institute of Justice 0101.07 standard [6]. However, additional work to determine how real-world injuries correlate to the current certification standards is warranted.

In an effort to understand the types of real-world injuries being sustained by LEOs, the current research explores a dataset of medical outcomes related to BABT and penetrating trauma. Characterization of the injuries reported in medical records and self-reports are presented. In order to report trends in the incidence of injury in real-world scenarios where body armour was involved, 101 cases where officers were shot while wearing body armour were compiled and analyzed.

2. METHODOLOGY

Cases were identified from the IACP/DuPont Kevlar Survivor's Club database along with databases held by various body armor manufacturers. After the identification of potential cases, the officers were contacted, and consent was obtained. A survey was provided to each officers along with a request for their medical records. Whenever possible, police reports were also obtained. Key variables were extracted from the medical records by a level I trauma surgeon and from the police reports by trained researchers.

Data extracted from medical records for the analysis includes age, gender, vital signs, anatomical location of the gunshot, wound description, imaging studies, surgeries performed, and time spent in the hospital. Data extracted from police reports, surveys, and interviews for the analysis includes type of weapon used, type of vest worn, level of protection of the vest, and standoff distance. Injuries were coded according to the Abbreviated Injury Scale (AIS). A total of one hundred and one (101) cases had adequate data and were included in the analysis.

Data were analyzed using the Statistical Package for Social Science (SPSS), version 23 (SPSS INC., Chicago, IL). Two-sided Fisher's exact test was used to assess the relationships between variables to improve understanding in scenarios where officers are shot while wearing body armour.

3. RESULTS

3.1 Demographics

Regarding the officers and their vests, of the 101 cases analyzed, 98 (97.0%) involved officers that were male and 3 (3.0%) that were female. The average age was 34.45 +/- 7.36 years. The most common armour noted in the dataset were level II armour (32 or 31.7% of cases) and level IIIA armour (20 or 19.8% of cases) (Table 1). The level of enhanced protection was unknown for a large majority of cases (n=87 or 86.1% of cases).

Regarding incident information, the most common standoff distance was 0 to 5 feet (1.5 m) (n=35, or 34.7% of cases). The second-most common standoff distance was 6-10 feet (1.8-3 m)(n=22, or 21.8% of cases), and the distance was unknown for 23 or 22.8% of cases (Figure 1). The most common weapon type noted was handgun (n=65, or 64.4% of cases), followed by shotgun (n=11, or 10.9% of cases) (Figure 2). The most common type of firearm used (of the cases in which the specific firearm was known) was a 9mm (n=18). Cases involving a single shot made up a majority of the cases (n=55, or 54.5% of cases) and there were 46 cases that involved multiple shots. The most common case within the multiple shots category was 2 shots (n=38, or 37.6% of cases). The most common type of bullet noted was lead (n=15), but a for a majority of cases, the bullet type was unknown (n=62, or 61.4% of cases).

https://doi.org/10.52202/081621-0018

Table 1. Gender, age, and vest level demographics.

Gender		
Male	98	97.0%
Female	3	3.0%
Age		
Mean	34.45	
Standard Deviation	7.355	
Min	22	
Max	53	
Vest Level		
Unknown	27	26.7%
IIA	16	15.8%
II	32	31.7%
IIIA	20	19.8%
III	5	5.0%
IV	1	1.0%

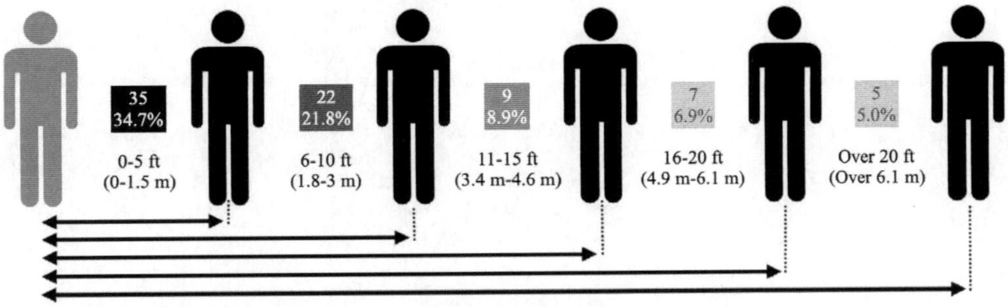

Figure 1. Standoff distances noted in the dataset.

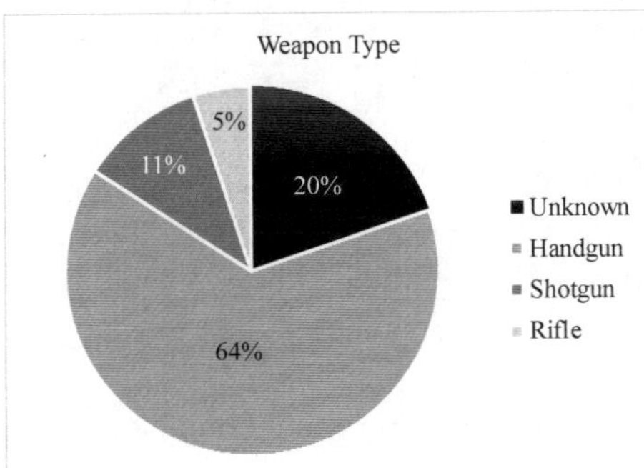

Figure 2. Weapon types observed in the dataset.

3.2 Injury Outcomes

Regarding injury and clinical outcome, 43 cases (42.6%) were found to be clinically significant. Internal injuries, which were defined as any injury within a cavity that did not include skin or soft tissue, were found in 18 cases (17.8%). The types and number of internal injuries thereof are reported in **Error! Reference source not found.**. Penetration injuries were noted in 13 cases (12.9%). Note – these injuries include those that are considered backface signature injuries where the bullet is stopped by the vest but there is still an open wound behind the armour. Of the 82 cases in which the patient disposition was known, 45 patients (44.6%) were sent home, and 37 patients (36.6%) were admitted. The mean hospital length of stay was 2.76 days. In 59 cases (58.5%), the hospital length of stay was noted as 0 days or 1 day.

All 101 officers had at least 1 external injury, 56 (55.4%) had at least 2 external injuries, 24 (23.8%) had at least 3 external injuries, and 9 (8.9%) had at least 4 external injuries. Of the 78 officers who underwent some form of imaging, 36.6% (n=37) received positive findings which included fractures of the rib and clavicle, bullets and/or bullet fragments, hemothorax, and pulmonary contusion. A total of 27 officers (26.7%) underwent some form of surgical procedure. Regarding AIS scores, a majority of cases

https://doi.org/10.52202/081621-0018

resulted in injuries with an MAIS score of 1 (n=62, or 61.4%) (**Error! Reference source not found.**). Shot locations are provided in Figure 4 with injuries coded with a maximum AIS of 2, 3, or 4.

To determine if there was a statistically significant association between the variables of Internal Injury, Single vs Multiple Shots, Vest Level, Max AIS Score and Clinical Significance, a two-sided Fisher's exact test was conducted. The variables of Max AIS had a significant association with all variables examined except Vest Level. There was also a statistically significant association between Single Shot vs Multiple Shot and Clinical Significance (Table 3).

Table 2. Types of internal injuries reported.

Internal Injury Reported	# Cases
Spleen laceration	1
Hemoperitoneum	1
Grade II liver injury - Hepatic contusion	1
Pleural effusion	1
Retroperitoneal hematoma	1
Sternal fracture/hematoma	1
Hemothorax	1
Pneumothorax	1
Pulmonary vein rupture	1
Rib fractures	3
Lower extremity artery/mucle tear/severe fracture/ injury	2
Severe facial fractures/injuries	2
Upper extremity artery/mucle tear/severe fracture/ injury	3
Small bowel/colon/intestine laceration	4
Pulmonary contusion	5

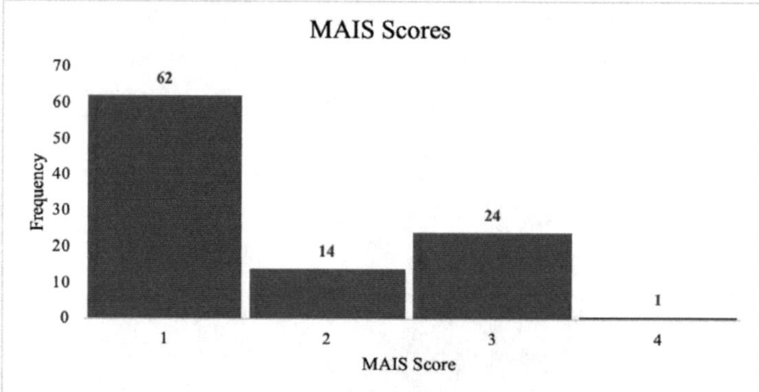

Figure 3. AIS score histogram.

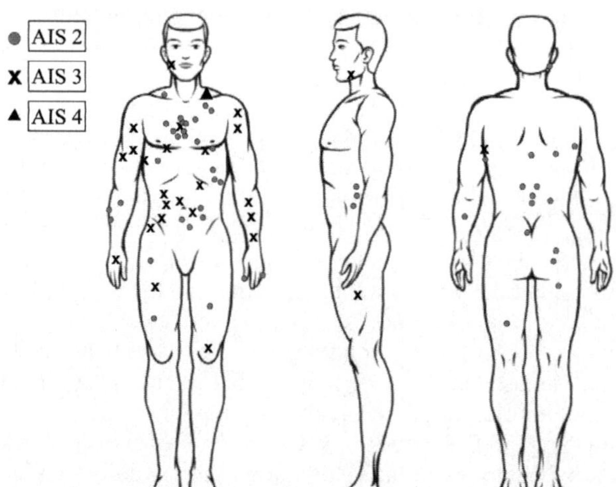

Figure 4. Shot locations of cases that had injuries that were categorized as AIS 2, 3, or 4.

Table 3. P values resulting from the two-sided Fischer's exact test.

	Internal Injury	Single Shot v Multiple Shots	Vest Level	Max AIS Score	Clinical Significance
Internal Injury		0.049	0.447	<0.001	<0.001
Single Shot v Multiple Shots	0.049		0.628	<0.001	<0.001
Vest Level	0.447	0.628		0.446	0.011
Max AIS Score	<0.001	<0.001	0.446		<0.001
Clinical Significance	<0.001	<0.001	0.011	<0.001	

P Values
P < 0.001
0.05 < P < 0.001
0.1 < P < 0.05
0.2 < P < 0.1
P > 0.2

3.3 AIS 2 Injuries

Fourteen of the 101 total cases documented injuries that had a maximum AIS score of 2. Half of the cases were determined to be clinically significant, and the other half not clinically significant (n=7 for each category). Cases 2, 11, 38, 44, 58, 84, and 95 were clinically significant.

Case 2 involved an officer wearing a Level II vest who was shot once with a .40 caliber handgun at a distance of approximately 10 feet (3 m). He was diagnosed with rib fractures to the left 8th and 9th ribs, a spleen laceration, and hemoperitoneum. Case 11 involved an officer wearing a Level II vest who was shot once with a 9 mm handgun at a distance of approximately 5 feet (1.5 m). He had a deep abrasion and laceration to his left chest inferior to the nipple, and was diagnosed with a fracture of the left 8th rib with a small effusion. Case 38 involved an officer wearing a Level II vest who was shot twice with an unknown weapon at a distance of approximately 8 feet (2.4 m). He reportedly had small contusions to the upper right chest, a large contusion to lower right back, and was diagnosed with a pulmonary contusion. The officer indicated he believed he was shot 2 to 3 times in the same spot in his back. Case 44 involved an officer who was shot twice with a .45 caliber handgun at an unknown distance wearing a vest of an unknown level. The injuries involved the left shoulder and left flank, and injuries included a left clavicle fracture and a bruise to the epigastrium area.

Case 58 involved an officer wearing a Level II vest who was shot once in the shoulder with a .40 caliber handgun at an unknown distance. Multiple foreign bodies were found in his shoulder, which required debridement and removal of the foreign bodies. Case 84 involved an officer wearing a Level IIIA vest who was shot once with a .22 caliber weapon at a distance of approximately 6 to 10 feet (1.8 to 3 m). He had a contusion on his back and soft tissue damage to the right hip. A pelvis X-ray indicated an L3 fracture. Case 95 involved an officer wearing a Level III vest who was shot four (4) times with a .45 caliber pistol at a distance of approximately ten (10) feet (3 m). Two of the shots impacted the armour in the area of his chest, which resulted in severe bruises, but records indicated the right arm was "blown away" and needed to be completely reconstructed, and his left arm required an elbow arthroscopy with bullet removal.

Case 2 (described above) was the only MAIS2 case that was coded as a penetration injury. Four cases were coded as having an internal injury (Cases 2, 18, 38, and 86), which included diagnoses of a splenic laceration, a pulmonary contusion, and a sternal fracture, respectively. Case 18 involved an officer wearing a level II vest who was shot once in the chest from a distance of two (2) feet (0.6 m). He had a large abrasion to the left chest and a pulmonary contusion. Case 86 involved an officer wearing a level IIIA vest who was shot twice in the chest with a .25 caliber handgun at a distance of 6 to 10 feet (1.8 to 3 m). He presented with a bruise to the left side of his chest from one shot and was diagnosed with a sternal fracture as a result of the 2nd shot, which reportedly missed the vest and struck the sternal notch.

Cases 19, 58, and 95 required surgical procedures, all to extremities. Case 19 involved an open reduction and internal fixation (ORIF) on the 5th metacarpal, Case 58 (described above) required debridement and removal of foreign bodies in the shoulder, and Case 95 (described above) required a left elbow arthroscopy with bullet removal.

https://doi.org/10.52202/081621-0018

Table 4. Circumstances surrounding AIS 2 injuries sustained by officers.

STUDY	Armor Type	Threat	Standoff Distance	Anatomical Structures Injured
2	Level II	.40	6 - 10 feet (1.8-3 m)	Skin defect, hemoperitoneum, 8th and 9th left rib fractures, grade I spleen laceration
4	Level II	9 mm	0 - 5 feet 0-1.5 m	Abrasion, entrance wound at the right auxilla area, exit wound inferiolateral to the right scapula
11	Level II	9 mm	0 - 5 feet 0-1.5 m	Left chest deep abrasion and laceration, small effusion, left 8th rib fracture
18	Level II	Unknown	0 - 5 feet 0-1.5 m	Large abrasion to left chest, pulmonary contusion in anterior lower left lobe
19	Unknown	Unknown	Unknown	Left 5th metacarpal fracture
27	Unknown	.45	Unknown	Severe bruise and laceration to right abdomen that turned necrotic
33	Level IIIA	Handgun	0 - 5 feet 0-1.5 m	Communted fracture to right mastoid air cells without intracranial injury, and chest abrasion
38	Level II	Unknown	6 - 10 feet (1.8-3 m)	Contusions to right upper chest and right lower back, right lower lobe pulmonary contusion
44	Unknown	.45	Unknown	Left clavicle fracture, bruise in epigastrium, left flank abrasion
58	Level II	.40	Unknown	Multiple foreign bodies in left shoulder with skin defect above the clavicle without bleeding
84	Level IIIA	.22	6 - 10 feet (1.8-3 m)	Back contusion, L3 fracture, and soft tissue damage in the hip
86	Level IIIA	.25	6 - 10 feet (1.8-3 m)	Chest bruise and sternal fracture
95	Level III	.45	6 - 10 feet (1.8-3 m)	Right arm "blown away" and needed to be completely reconstructed, bruises to center of chest
101	Level IIA	.40	0 - 5 feet 0-1.5 m	Left pectoral muscle tear

In summary, of the 14 cases that had injuries with an MAIS score of 2, five of the cases involved shots to areas not covered by the vest; one case was to the axilla (Case 4), and four were to an extremity (Cases 19, 33, 58, and 95). Four of the cases involved an internal injury (Cases 2, 18, 38, and 86).

3.4 AIS 3 Injuries

Twenty-four cases of the analyzed 101 cases involved injuries with a MAIS score of 3 (Table *5*). Twenty-one of the 24 were clinically significant. Cases 1, 21, and 34 were found to not be clinically significant.

Case 10 involved an officer wearing who was shot twice with a shotgun at a distance of approximately 15 feet (4.6 m). The left abdomen and left thigh were affected areas, with the left flank sustaining multiple entrance wounds (specifically the left costal area) and the left thigh sustaining a wound to the lateral portion. Internal injuries to the small bowel, superficial stomach, and mesocolon were sustained and required suture repair. Case 15 involved an officer wearing a Level II vest who was shot twice (once in the chest and once in the left forearm) at a distance of approximately 17 feet (5.2 m). The chest sustained an abrasion, and the proximal left forearm sustained extensive soft tissue disruption and required surgical repair of ulnar artery and forearm muscles. This officer also sustained a hepatic contusion.

Case 22 involved an officer wearing a Level IIA vest who was shot twice with a handgun at a distance of approximately 5 feet (1.5 m). This officer was shot in the chest and left upper extremity; he sustained a 4 cm contusion on his right chest above and medial to the nipple, and a through-and-through injury to his wrist, which shattered the distal end of the radius, requiring an ORIF surgery. Case 24 involved an officer wearing a Level IIIA vest who was shot four times with a handgun at a distance of approximately 13 feet (4 m). This officer was shot in the anterior chest, knee, and sacrum. The chest impact resulted in a bruise, and the knee sustained a through-and-through injury. The right thigh below the buttock was another reported area of injury. This officer sustained an iliac wing fracture with bullet fragment lodged just to the right of his sacrum and also required knee surgery.

Case 35 involved an officer wearing a Level IIA vest who was shot twice with a shotgun at a distance of 5 feet (1.5 m) in his right chest. The chest wound involved a severe laceration and a large right anterior superior pulmonary contusion. Case 36 involved an officer wearing a Level II vest who was shot six times with a pistol at a reported distance of 15 feet (4.6 m). The injured regions included the left shoulder, back, right hand and right thigh. The officer diagrammed two shots to the back of the vest, and reported injuries included erythema and penetration to the skin at the L2-L3 level. A foreign body was noted lateral to the pelvis. The left humerus sustained a comminuted fracture, and his right hand sustained a through-and-through wound, which resulted in a comminuted fracture of metacarpal heads 4 and 5. The open fractures required surgery. This officer did not sustain an internal injury of the thorax.

Case 40 involved an officer who was shot twice with a shotgun at a distance of 2 feet (0.6 m). A contusion was noted in the sternal area, a right humerus fracture was noted as well as a left shoulder injury with a deformity, and an internal injury to the popliteal artery and vein was noted, as there was reportedly no pulse in the right popliteal vasculature. Surgical procedures were required for fractures and a vascular graft. Case 41 involved an officer who was shot twice with a shotgun at a distance of 10 feet (3 m). The back and the left forearm were the reported areas affected. The officer was reportedly diagnosed with left forearm compartment syndrome, which included a radial fracture and required a fasciotomy. Case 42 involved an officer who was shot twice with a 9 mm handgun at a distance of 8 feet (2.4 m). This officer sustained lacerations to their left ear and scalp, and injuries to the right buttock, right forearm, and left lower abdomen. The officer was diagnosed with L5 and sacral fractures, as well as colon and small bowel injuries. Case 43 involved an officer who was shot twice with a rifle at a distance of 20 feet (6.1 m). The right shoulder, right posterior chest, and left flank were affected. Chest and pelvis x-rays revealed multiple bullet fragments and a right rib fracture. The humerus was fractured which required an ORIF surgery.

Case 45 involved an officer who was shot twice with a .45 caliber weapon at a distance of 5 feet (1.5 m). The affected areas were listed as mid-abdomen, upper left chest, right back, left superior limb, and right buttock. A chest x-ray revealed right hemothorax, and an ORIF surgery was required on the forearm. Case 46 involved an officer who was shot twice with a handgun at a range of 5 feet (1.5 m). The officer presented with abdominal pain and a pulseless left arm, with injury regions being noted as left neck, upper chest, left arm, and inferior limb. Imaging revealed bullet fragments in the chest region, as well as a left kidney laceration and left hand fracture. Case 50 involved an officer who was shot once with a 9mm handgun at a distance of approximately 3 feet (0.9 m). The affected area was noted as under the right armpit, where there was a hole. A bullet was found in this officer's intercostal space of the 11th rib above the diaphragm. The officer was also diagnosed with a liver contusion.

Case 54 involved an officer wearing a level IIIA vest who was shot seven times with a rifle at a distance of approximately 21 to 50 feet (6.4 to 15.2 m). Three of the shots were to the torso and abdomen, which resulted in severe bruising and a laceration. A six inch section of intestine needed to be removed as a result of the abdominal impact. There was also a through-and-through injury to the right bicep-triceps area and a grazing wound to the rear scalp. Case 59 involved an officer who was shot two times at point blank range with a 12-guage shotgun. The chest, left eye, and right forearm were affected. The left eye injury required a skin graft and vitrectomy. The chest was noted as having a bruise. Case 60 involved an officer wearing a level IIA vest who was shot once with a 12-guage shotgun at a distance of approximately 6 feet (1.8 m). In addition to burns, this officer sustained serious thoracic injuries due to a shot location under the left armpit that stuck the edge of the armor. This officer was reportedly diagnosed with a shattered lateral rib, a collapsed left lung, broken left shoulder girdle, and two ruptured deep pulmonary veins. Depending on the degree of the pneumothorax, this injury could potentially be an AIS 4 injury.

Case 69 involved an officer wearing a Level IIA vest who was shot twice with a .38 caliber weapon at a distance of 1 to 5 feet (0.3 to 1.5 m). The neck, chest, abdomen, and wrist were noted as affected areas. The operative findings reportedly pertained to the small bowel and colon, which indicated this officer had an internal injury. Case 71 involved an officer wearing a Level II vest who was shot twice with a .40 caliber weapon. This officer sustained a face abrasion, chest bruise, and an abdominal penetrating injury. A pelvis x-ray revealed a bullet in the abdominal cavity, and the officer was diagnosed with a retroperitoneal hematoma. Case 76 involved an officer wearing a level IV vest who was shot once with a 9mm weapon. The impact area was noted as the right lower quadrant, and operative findings indicated a comminuted left iliac wing fracture and a large left retroperitoneal hematoma.

https://doi.org/10.52202/081621-0018

Case 77 involved an officer wearing a Level II vest who was shot twice with a .38 caliber weapon at a distance of one foot. This officer reportedly had a bruised chest as well as lost soft tissue in the face, which had bone fractures and required multiple surgeries. This officer was also noticed to have an internal injury. Case 93 involved an officer wearing a Level II vest who was shot once with a .44 caliber weapon in her left chest below the breast at a distance of approximately 2 feet (0.6 m). This officer sustained a pulmonary contusion and broken ribs, as well as a ½" hole in the skin and extensive bruising. The officer was also diagnosed with pleural effusion.

Table 5. Circumstances surrounding AIS 3 injuries sustained by officers.

STUDY	Armour Type	Threat	Standoff Distance	Anatomical Structures Injured
1	Level II	Unknown	11 - 20 feet 3.4 m-6.1 m	Severe bruise to the right side of chest, below and inside of nipple, and 2 holes above left knee
10	Unknown	Shotgun	11 - 15 feet 3.4 m-4.6 m	Multiple left flank entrance wounds, small bowel, superficial stomach, and mesocolon wounds, and left lateral thigh wound
15	Level II	Unknown	16 - 20 feet 4.9 m-6.1 m	Chest abrasion, hepatic contusion, injury to ulnar artery and forearm muscles that required surgery
21	Unknown	Unknown	Unknown	Highly comminuted left mandible fracture, superficial bruises to the 3 shots that impacted the vest, and shots to the right buttock and left thigh
22	Level IIA	Handgun	0 - 5 feet 0-1.5 m	Through-and-through injury to wrist, resulting in comminuted fracture to distal end of radius, which required ORIF, and chest contusion
24	Level IIIA	Handgun	11 - 15 feet 3.4 m-4.6 m	Through-and-through injury to left knee with a laceration below the patella, resulting in knee surgery, iliac wing fracture with bullet fragment, and chest contusion
34	Level II	Unknown	11 - 15 feet 3.4 m-4.6 m	Through-and-through injury to the right humerus, extensively comminuted fracture to proximal humerus resulting in ORIF, abrasion to chest
35	Level IIA	Shotgun	0 - 5 feet 0-1.5 m	Large anterior superior pulmonary contusion with open wound from chest injury
36	Level II	Pistol	11 - 15 feet 3.4 m-4.6 m	Through and through injury to right hand, comminuted fracture to humerus and metacarpal heads 4 and 5, two gunshot wounds lateral to the spinal column between L2 & L3 with erythema and penetration to the skin and minimal bleeding, right thigh injury
40	Unknown	Shotgun	0 - 5 feet 0-1.5 m	Contusion in sternum, right humerus fracture, right thigh injury with injury to popliteal artery and vein - no pulses
41	Unknown	Shotgun	6 - 10 feet (1.8-3 m)	Left forearm compartment sydrome, radial fracture, fasciotomy left forearm
42	Unknown	9 mm	6 - 10 feet (1.8-3 m)	L5 fracture, transverse colon and small bowel injuries, ear and scalp lacerations
43	Unknown	Rifle	20 feet (6.1 m) and over	Right rib fracture, humerus injury resulting in ORIF
45	Unknown	.45	0 - 5 feet 0-1.5 m	Right hemothorax, right forearm injury resulting in ORIF
46	Unknown	.380	0 - 5 feet 0-1.5 m	Left arm injury resulting in pulselessness, left kidney laceration, left hand fracture
50	Unknown	9 mm	0 - 5 feet 0-1.5 m	Liver contusion, hole in right auxilla
54	Level IIIA	Rifle	20 feet (6.1 m) and over	Abdominal injury requiring 6 inches of intestine removal, through-and-through injury to right arm, bruises and lacerations in area covered by armour
59	Unknown	12-Gauge Shotgun	Unknown	Left eye laceration requiring a skin graft, chest bruise
60	Level IIA	12-Gauge Shotgun	6 - 10 feet (1.8-3 m)	Lateral rib shattered, left lung collapse, 2 deep pulmonary veins ruptured, left shoulder gurdle broken
69	Level II	.38	0 - 5 feet 0-1.5 m	Small bowel and colon injuries
71	Level IIA	.40	Unknown	Face abrasion, chest bruise, abdominal penetrating injury, retroperitoneal hematoma
76	Level IV	9 mm	Unknown	Small bowel and colon injuries
77	Level IIIA	.38	0 - 5 feet 0-1.5 m	Bruised chest, facial bone fractures and lost tissue resulting in multiple surgeries
93	Level II	0.44	0 - 5 feet 0-1.5 m	Extensive left chest bruising with large ulcerations, pleural effusion, rib fractures, and lung contusion

In 13 of the 24 cases that had injuries with an MAIS score of 3, the injury that resulted in the MAIS 3 score involved shots to areas not covered by the vest; one case was to the axilla (Case 50), and twelve were to an extremity (Cases 1, 15, 21, 22, 24, 34, 36, 40, 41, 46, 59, and 77). Twelve of the cases (50%) involved an internal injury (Cases 10, 15, 35, 40, 54, 59, 60, 69, 71, 76, 77, and 93).

3.5 AIS 4 Injury

There was one case with a MAIS 4 injury. The officer was wearing a Level IIIA vest and was shot twice with a .38 caliber weapon from 20 feet (6.1 m); one shot was to zone 1 of the neck (**Figure 5**), and one was to the pectoral region of the chest. The injury to the chest was listed as a bruise. The operative findings stated left subclavian artery and venous injuries (Figure 5). The hospital length of stay was 5 days, and the case was considered clinically significant.

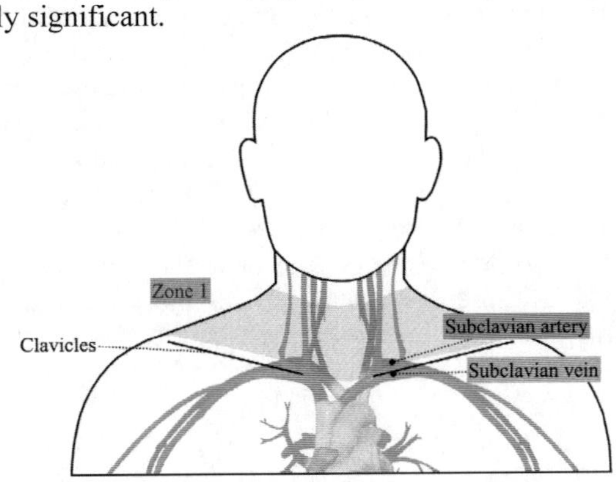

Figure 5. Description of the anatomy involved in the injury from Case 73, MAIS 4.

4. DISCUSSION AND CONCLUSION

The relationships between scenario variables and injuries have been described, and a thorough explanation of the cases resulting in MAIS 2, 3, and 4 injuries has been provided, as well as a breakdown of the internal injuries observed and shot locations with relation to the armour. Vest level, standoff distance, and weapon type did not have significant relationships with whether or not there was a penetration injury, internal injury, or MAIS score. There was a notable (but not significant) relationship between vest level and clinical significance (P=0.011) and internal injury and single shot versus multiple shot scenarios (P=0.049) (Table 3). Not surprisingly, there was a significant relationship between MAIS score and internal injury and clinical significance and internal injury (P<0.001 for both). There was also a significant relationship between MAIS score and single shots versus multiple shots and also clinical significance and single shot versus multiple shots (P<0.001 for both), suggesting that officers who are impacted more than once have a higher rate of sustaining more serious injuries.

A majority of the shot locations (61%) fell within the bounds of the generally covered armour area. However, as is shown in Figure 4, a considerable number of shots resulting in MAIS 2+ injuries (approximately 39, or 46%) fell outside or at the edge of the region that would be protected by body armour. Many AIS 3 injuries were the result of shots to the extremities that caused extensive vascular damage. The AIS 4 injury was a shot to region 1 of the neck (i.e. on the edge of the area covered and protected by armour). With regards to the shot locations of AIS 2+ injuries (Figure 4), one injury was to the eye and required a skin graft, but the shot locations appeared to be on the vest; the eye injury was a potential secondary injury.

There are several considerations when examining whether a vest offers the best protection, both coverage and fit are essential. One of the complaints related specific to body armor is the lack of mobility and comfort of the vest. Making the vest as small as possible will often add to the comfort level, however this ultimately would decrease the protective ability of the armor. Modifications by the officer to how the vest is worn in the field can also lead to areas of poor coverage. The current data support the need to ensure proper coverage to avoid life threatening injuries. This analysis shows that when shots impact armour, the resulting injury is minor a majority of the time (i.e. AIS level 1).

https://doi.org/10.52202/081621-0018

Even with shots on the armour, serious injuries still occurred as a result of behind armour blunt trauma without penetration. Internal injuries noted that were not coded as penetration injuries (i.e. BABT injuries) included pulmonary contusion (Cases 18, 35, and 38), small bowel injury, superficial stomach wound, and mesocolon wound (Cases 10 and 69), and hepatic contusion (Case 15). AIS 2 injuries that were not coded as internal or penetrating injuries included a lumbar vertebrae fracture (Case 84), muscle tear (Case 101), rib fracture (Case 11), clavicle fracture (Case 44), and substantial hematoma (Case 27). All AIS 3 injuries that were not coded as internal or penetrating injuries were injuries to the extremities or were shots to areas not covered by armor (i.e. axilla). The only AIS 4 injury in the dataset was the result of an impact to an area not covered by the body armour.

It is notable that many of the cases (n=62) that resulted in minor injuries (MAIS score of 1) such as bruises and abrasions despite the serious circumstances of the threat. Twenty-one (33.9%) of the MAIS 1 cases had standoff distances in the range of 0 to 5 feet (1.5 m), and 14 (22.6%) had standoff distances in the 6 to 10 feet (1.8 to 3 m) range.

The discussed analysis of the dataset of 101 documented incidents provides insight into the injuries sustained by officers in the field. However, a limitation of analyzing any event in hindsight is the available data. In the subject analysis, police reports, interviews, medical records, photographs, and other documents were replied upon. Unfortunately, very few cases contained all the data required to fully reconstruct and understand the incident and injuries. The analysis was performed with the data obtained to the most reasonable extent possible. Another aspect of this analysis is that it only included LEOs that survived their injuries, the analysis of those who sustained fatal injuries is ongoing. Even given these limitations, the current effort provides insight into real-world injuries sustained by officers wearing armour which can be used to compliment both experimental and laboratory testing.

Acknowledgements

Kathryn Loftis for assisting with AIS coding.

References

Ballistic Resistance of Body Armor NIJ Standard 0101.07.

[1] H. M. Tiesman, D. I. Swedler, S. Konda, and K. M. Pollack, "Fatal occupational injuries among U.S. law enforcement officers: A comparison of national surveillance systems," *Am. J. Ind. Med.*, vol. 56, no. 6, pp. 693–700, Jun. 2013, doi: 10.1002/ajim.22182.

[2] H. M. Tiesman, S. Konda, J. Grieco, M. Gwilliam, J. Rojek, and B. Montgomery, "Resistance-Related Injuries Among Law Enforcement Officers: Addressing the Empirical Gap," *Am. J. Prev. Med.*, vol. 59, no. 6, pp. e231–e238, Dec. 2020, doi: 10.1016/j.amepre.2020.05.015.

[3] United States Federal Bureau of Investigation, "Law Enforcement Officers Killed and Assaulted (LEOKA)." 2023. Accessed: Mar. 13, 2025. [Https://cde.ucr.cjis.gov/LATEST/webapp/#]. Available: https://cde.ucr.cjis.gov/LATEST/webapp/#

[4] C. Bir, "US Body Armor Fit and Coverage," Final Report TSWG Task PP-IP-4206. Contract Number: N4175619C3033, 2022.

[5] C. Bir, R. Lance, S. Stojsih-Sherman, and J. Cavanaugh, "Behind Armor Blunt Trauma: Recreation of Field Cases for the Assessment of Backface Signature Testing," in *30th International Symposium on Ballistics*, DEStech Publications, Inc., Dec. 2017. doi: 10.12783/ballistics2017/16912.

[6] "Ballistic Resistance of Body Armor NIJ Standard 0101.07," Oct. 2023.

Relative effect of ballistic plate coverage on the protection system performance

G. St-Onge[1], A. Bouamoul[2], and S. Ouellet[2]

[1]*DRDC Valcartier, 2459 Pie-XI Blvd North, Quebec, Qc, Canada,*
Correspondence: *Gabriel.St-Onge@drdc-rddc.gc.ca*

Abstract. Ballistic plates are designed and fitted to protect vital organs from injuries while maintaining the ability of soldiers to perform their mission effectively. Thus, the design of a ballistic protection system should represent the best compromise between the protection it offers (i.e., the ability to stop a threat and protect organs from life-threatening injuries) and the physical burden it adds to the soldier. The optimization of plate coverage to minimize impact on soldier mobility and comfort is often difficult due to the lack of robust tools to study the effect of changes in coverage on soldier vulnerability. The aim of the study reported herein was to inform this design optimization process by evaluating the impact of changes in ballistic plate geometry on the ability to minimize the risk of life-threatening injuries. The Canadian Operational Research Environment for Vulnerability and Lethality (CORE V/L) software, a numerical Vulnerability and Lethality tool, was used in conjunction with various protection performance metrics (i.e., Average Anatomical Profile (AP) Score, Average New Injury Severity Score (NISS), Maximum Abbreviated Injury Scale (MAIS) mapping, etc.) to evaluate parametrically the effect of changes in the length and width of a plate and its general shape. A range of ballistic plate configurations were therefore simulated and compared in a way that informs designers of the effect of small geometric changes on protective performances. Small changes in the geometry of ballistic plates are shown to affect its ability to protect soldiers from life-threatening injuries. The study also identifies which changes have the largest and smallest influence on soldier survivability. All simulations were performed on an anatomically accurate representation of a 50th percentile human male. Future research will aim to extend recent findings to male and female of different statures.

1 INTRODUCTION

In military settings, ballistic protection protects soldiers against injuries caused by projectile or fragments. Whether it is bullets from a firearm, fragment from shell or grenade, or rock and soil from IED (Improvised Explosive Device) those projectiles can cause significant injury to military personnel. In military conflicts, it was reported that 90% of all injury sustained during military conflict since World War I are caused by penetrating projectiles [1]. This is still relevant in more recent conflict as it was reported that 83% of healthcare specialists in Ukraine have observed penetrating injury pattern from bullets and fragments [2]. Conflicts can also provide some level of information on the protection performances of fielded protective equipment. For example, a study from the Armed Forces Institute of Pathology revealed that during the Operation Iraqi Freedom, multiple casualties were caused by lethal penetrating injury in the thoracic region [3]. Of these lethal penetrating injuries 32.9% occurred just outside of the area protected by a ballistic plate. Wound entries were observed above (12.3%), on the sides (15.1%), and below (5.5%) the plate, with 19% of the injuries occurring near the edge of the plate (i.e., 4% near the top edge, 12.3% near side edge, and 2.7% near the bottom edge). Only 6.8% of the injuries observed were in the plated area. Similar observations were made in Ukraine, where the incidence of penetrating thoracic trauma outside of the protected region is significantly higher than in other conflicts [4]. Thus, there might be opportunities to further optimize the shape and dimensions of individual ballistic protection.

Designing an optimal ballistic plate is not a simple task since compromises must be made between the protection it offers (i.e., the ability to stop a threat and protect organs from life-threatening injuries) and the physical burden it adds to the soldier. In most research, fit and comfort are typically assessed separately from coverage protection performance. Studies on fit and comfort are usually done on soldiers wearing actual or prototype protective equipment. Such studies can help understand the effect of plate dimensions on the range of motion and operability offered by a protection system [5-7], but give little to no information on the actual protection provided by the equipment. Fit studies can also be performed with soft tissue simulation tools where pressure points and air gap can be correlated to the level of fit and comfort [8]. Studies on coverage protection are more commonly performed through Vulnerability and Lethality (V/L) analysis where an anatomical model that corresponds to the anatomy of a single individual is used to assess potential injuries sustained by soldiers struck by a projectile. Such study helps to understand the vulnerabilities of the human body when impacted by ballistic projectiles and to identify critical regions that require protection [9-12]. Analysis of protection coverage using Computerized Tomography (CT) scans and organ positions can also provide information on the effect of stature and morphology, but this information can be difficult to relate to generic V/L model anatomy [13-15]. Because mobility and comfort metrics are naturally different from vulnerability metrics, it is difficult to

 https://doi.org/10.52202/081621-0019

define a mobility-vulnerability trade space to optimize ballistic protection. To optimize ballistic protection, a robust methodology to study coverage performance, fit and comfort must be developed. The aim of the current study was to inform such design optimization process by evaluating the impact of changes in ballistic plate geometry on its ability to minimize the risk of life-threatening injuries. A methodology to assess coverage performance is presented where the Canadian Operational Research Environment for Vulnerability and Lethality (CORE V/L) software was used in conjunction with various protection performance metrics (i.e., Average Anatomical Profile (AP) Score, Average New Injury Severity Score (NISS), Average Injury Severity Score (ISS), etc.) to evaluate parametrically the effect of changes in the length and width of a plate. A range of ballistic plate configurations were therefore simulated against relevant threat and compared in a way that can inform designers of the effect of small geometric changes on protective performances.

2 METHODOLOGY

All analysis was performed using the CORE V/L tool. This software allows the prediction of injuries from projectile impact on a dismounted combatant. The anatomical model can be equipped with Personal Protection Equipment (PPE), and the effect of PPE on projectile performance can also be evaluated. Different types of simulation can be performed with the software depending on the objective of the study. For the current study, the relative performance of various PPE design was investigated. Different models and methods are needed to perform this type of analysis.

2.1 Anatomical Model and Injury Prediction

To model the effect of projectiles on the human body, an anatomical model of the human male is used. The model is composed of approximately 2500 structures divided into seven subgroups: circulatory system, connective tissue, lymphatic system, muscle, nervous system, organ, and skeleton. Figure 1 shows the CORE V/L male avatar in a standing pose with its different subgroups (i.e., skin and protection, organs and circulatory system, and skeleton). For simplicity, all simulations were conducted with the anatomical model in a standing posture. To help with the comparison of the coverage offered by different ballistic protection to the thoracic region of the human body, anthropometric measurements were collected on the model according to measurement locations presented in ANSUR II and CFAS anthropometric survey [16, 17]. Measurements were specific to the thoracic region. Anatomical land mark (i.e., Suprasternale notch (A), chest point (B), tenth rib (C), and iliac crest (D)) and measurements are illustrated on Figure 2, and the corresponding values for the anatomical model are presented in Table 1.

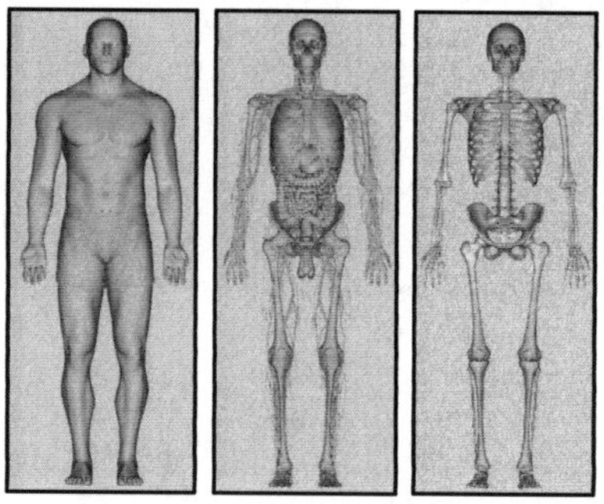

Figure 1. CORE V/L anatomical model of the human male.

From these measurements, a representative reference area of the thoracic region can be computed. This region represents the ideal medical coverage proposed by Breeze [12], spanning from the suprasternal notch to the iliac crest in length, with the chest breadth at the top and the bicristal breadth at the bottom. The reference area can be computed with Equation (1), and the representative surface is presented in Figure 2. For the anatomical model, the reference area is 1104 cm^2.

$$A_{ref} = \frac{(M_3 + M_4)}{.2} M_2 \quad (1)$$

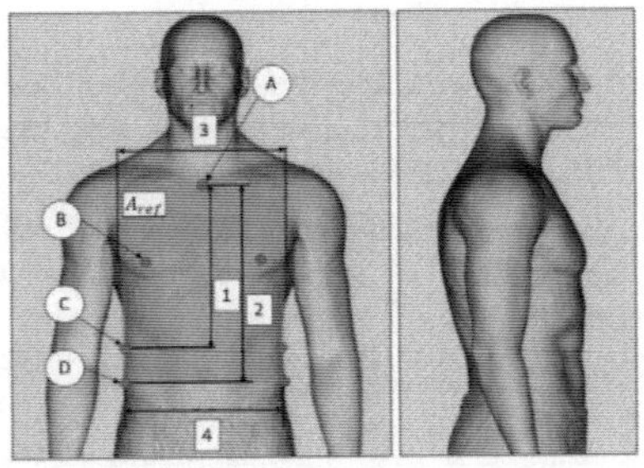

Figure 2. Identification of landmarks and measurements on the anatomical model.

Table 1: Anthropometric measurements for CORE V/L anatomical model.

ID	Measurement	Values (m)
1	Suprasternale-Tenth Rib Length	0.311
2	Suprasternale-Iliocristale Length	0.384
3	Chest Breadth	0.299
4	Bicristal Breadth	0.276

2.2 Ballistic plate

To evaluate the effect of length and width on coverage performance, multiple ballistic plate geometries were generated using a custom model generator in the CORE V/L toolkit. For this study, the general shape of the plate was inspired by the medium SAPI size, and only the width and length of the ballistic plates were modified between geometries. Dimensions of the different plate geometries were defined relative to a reference length measurement (i.e., Suprasternale-Tenth Rib Length), and a reference width measurement (i.e., Chest Breadth). The length of the plate varied from 120% to 60% of the reference, and the width varied from 100% to 40%. For this study, 49 permutations of the plate geometry were generated and analyzed. Figure 3 shows all the plates and their fit on the anatomical model. All plates were centred on the torso, fitted to the body, and the top edge was aligned with the suprasternal notch.

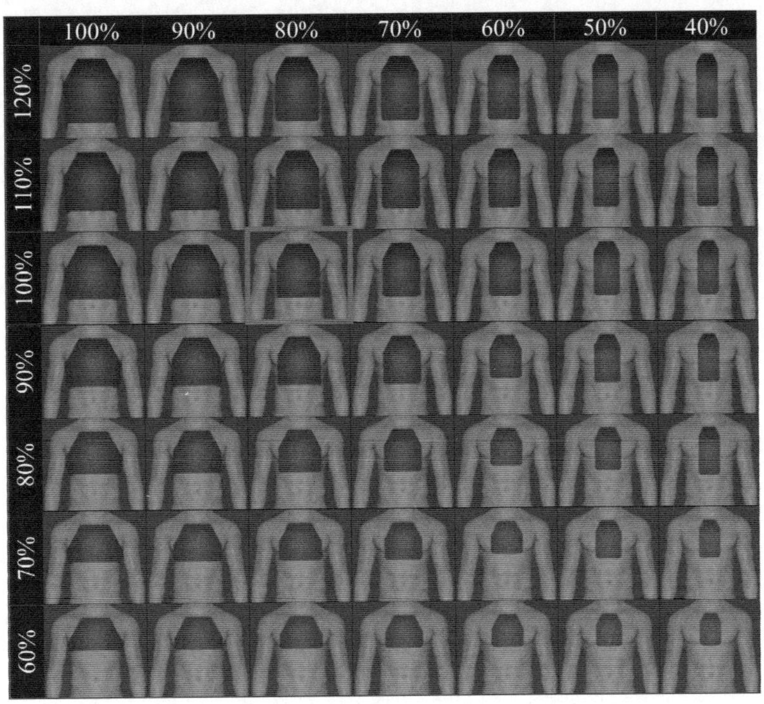

Figure 3. Placement and geometry of all 49 ballistic protections on the anatomical model. Relative dimension of the medium SAPI size is identified with a red rectangle.

https://doi.org/10.52202/081621-0019

A medium SAPI size was also simulated since it is the standard plate size that fit the anatomical model. The relative size of the medium plate corresponds to approximately 100% of the reference length and around 80% of the reference width. Since a coverage performance analysis of the various plate geometries is performed for this study, interaction between projectiles and the ballistic protection are not considered. Once an impact with the plate geometry is detected, the projectile is considered blocked. Also, to normalize result based on body shape and size, the coverage of ballistic protections normalized by the anatomical model reference area (Eq. (2)).

$$Relative\ Coverage\ Area = \frac{A_{plate}}{A_{ref}} \tag{2}$$

2.3 Projectile model

To assess the performance of ballistic protections, two relevant projectiles were chosen (i.e., 5.56 mm and 7.62 mm). The two projectiles were modelled based on generic properties and do not necessarily represent a specific threat. Mass (m), density (ρ_p), shape factor (γ), and impact velocity (v_s) are presented in Table 2. To simplify the modelling process, tumbling, degradation, and deviation of the projectile after impact are not modelled. Retardation of the projectile and damage tissue volume along the shot line are modelled using a penetration model adapted and upgraded from the Computer MAN model [18].

Table 2. Projectiles properties.

Projectile	Mass (gr)	Density (kg/m^3)	Shape Factor (-)	Impact Velocity (m/s)	Kinetic Energy (kJ)
5.56 mm	62	8960	0.360	950	1.813
7.62 mm	154	8960	0.406	850	3.605

To assess coverage performance, multiple independent shot lines were fired at the CORE V/L anatomical model. For each engagement orientation, 3600 shot lines were modelled, and distributed on a uniform equidistant grid with 10 mm spacing. Three azimuth angles were chosen to assess the coverage performance (i.e., Az. -45°, 0°, and 45°). The location of all shot lines can be observed on Figure 4. Each dot corresponds to the starting point of a shot line, and their directions are perpendicular to the corresponding grid.

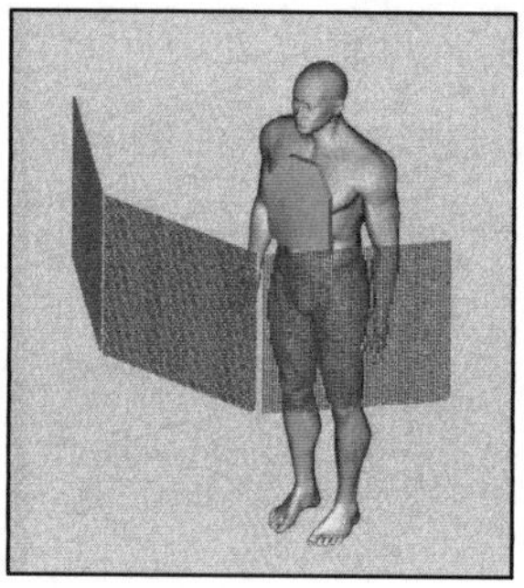

Figure 4. Shot line location and orientation fired at the anatomical model.

2.4 Injury Prediction and Coverage Performance Metric

To assess injuries caused by projectiles, CORE V/L correlate tissue damage evaluated by the penetration model to anatomical based injury description from the Abbreviated Injury Scale (AIS) 2015 revision [19]. For each shot line, multiple injury descriptions are recorded and a severity score is assigned to each one. Severity score for injuries varies from minor (AIS 1) to maximal (AIS 6). From the injury assessment different scoring methodologies can be used to assess the likelihood that a soldier will survive the injury cause by a projectile. The three scoring methodologies tested for this study are the Injury Severity Scale (ISS) [20], the New Injury Severity Scale (NISS) [21], and the Anatomical Profile (AP) [22]. Those survivability metrics compute a probability to survive (P_s) based on the injury descriptions related to a shot line.

To attribute a unique score for each ballistic protection tested, an average P_s ($\overline{P_s}$) can be computed over all shot lines. However, since ballistic protections are meant to protect the thoracic region, only shot line that causes injuries in that region will be used to compute the average score. The exclusion of shot lines in the averaging process will be performed based on the AIS region related to the maximum AIS (MAIS) injury description over that shot line. The AIS body regions that are included in the computation are presented in Table 3. The resulting $\overline{P_s}$ score will be used in this study to rank the coverage performance offered by each configuration tested.

Table 3. Specific AIS regions included in the $\overline{P_s}$ computation.

AIS Code	Region
4XXXXX.X	Thorax
5XXXXX.X	Abdomen
6X04XX.X	Spine - Thoracic
6X06XX.X	Spine - Lumbar

3 RESULTS

From the terminal ballistic simulations, injury assessment of 10,800 shot lines in three different orientations was performed. Figure 5 presents an example of the data generated for an unprotected and protected soldier. Results are presented according to the MAIS severity measured for a particular shot line. It can be observed that for this ballistic protection (i.e., Medium SAPI), all maximal injuries are being protected from frontal engagement, and most critical injuries in the upper thoracic region are also being protected. However, for engagements with azimuth angles varying from -45° to 45° more critical and some maximal injury severities are being exposed. This supports the importance of assessing different engagement angles to evaluate coverage performance of ballistic protection. With results from each independent shot line, severity metrics can also be computed. Figure 6 present result for all severity metrics (i.e., AP, ISS, NISS) in the Az. 0° orientation. It can be observed that all severity metrics behave differently. The distribution of ISS metric seems to match with the distribution of maximal and some critical severity regions, which is not the case for NISS and AP. The NISS, and AP do not necessarily match MAIS metric, but seem to represent a cumulative effect of all injuries caused by the projectile. It can also be observed that for the NISS and AP the lowest probability of surviving an injury is located at the midline of the body.

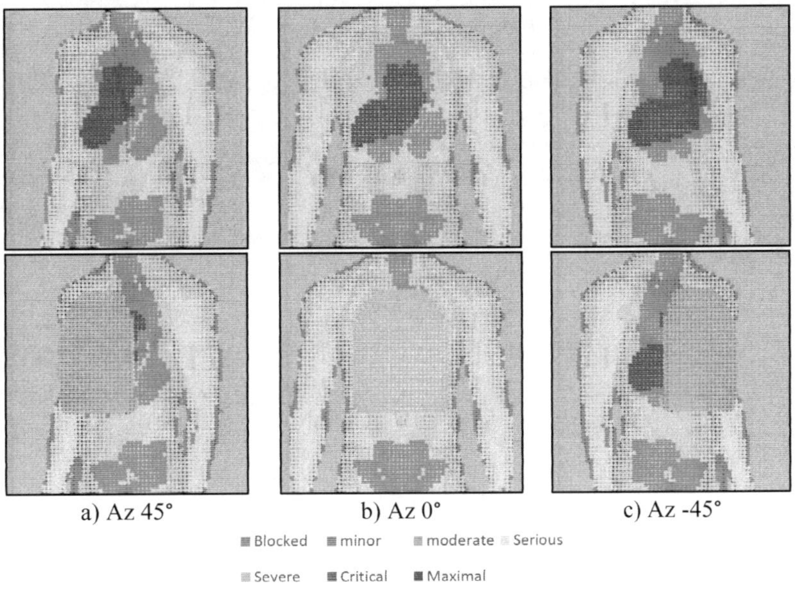

Figure 5. Protected and unprotected simulation result for Az. 45°, 0°, and -45°.

From the results presented in Figure 6, the computation of $\overline{P_s}$ was performed for each configuration tested and each survivability metrics. Those results are presented in Figure 7. It can be observed that all injury scoring systems give different results for a same configuration. For example, the average probability of surviving for an unprotected soldier impacted by a 5.56 mm projectile is 0.39 with the AP

https://doi.org/10.52202/081621-0019

score, 0.50 with the NISS score, and 0.73 with the ISS score. The offset seems to stay relatively constant between NISS and AP results, but it is not the case between the ISS and AP. In that scenario, the offset between the two scoring methods reduces as the relative coverage area increases. Furthermore, the behaviour between the scoring system seems to stay consistent when two projectiles of different masses and impact velocity are used. However, the use of the heavier projectile systematically resulted in lower average protection performance metrics. Result from Figure 7 also shows a good correlation between the probability to survive an injury and the relative coverage area. This correlation is most significant for the AP score as there is less dispersion than with the other metrics. This also indicates that changes in length or width alone may not have a significant influence on survivability scores. Given that the AP score demonstrated the highest sensitivity among the other metrics, all subsequent results will rely exclusively on this metric to assess coverage performance of ballistic protection.

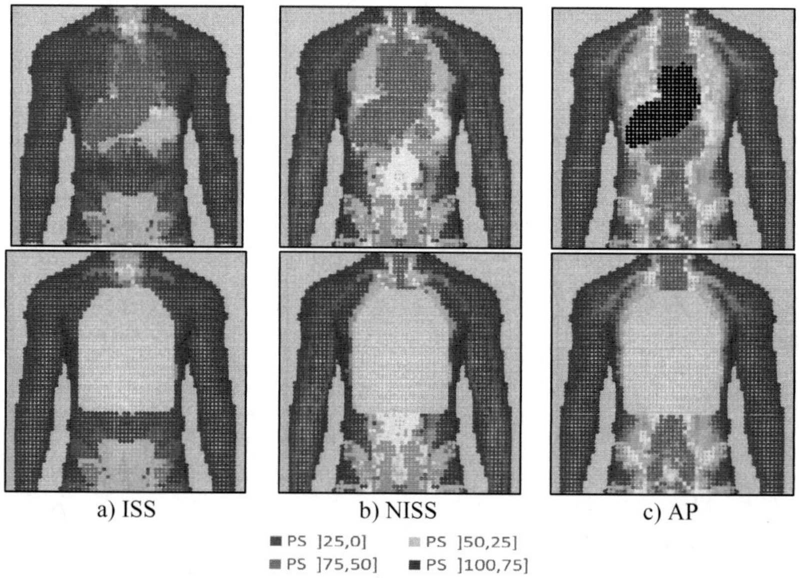

Figure 6. Protected and unprotected simulation results for three survivability metrics.

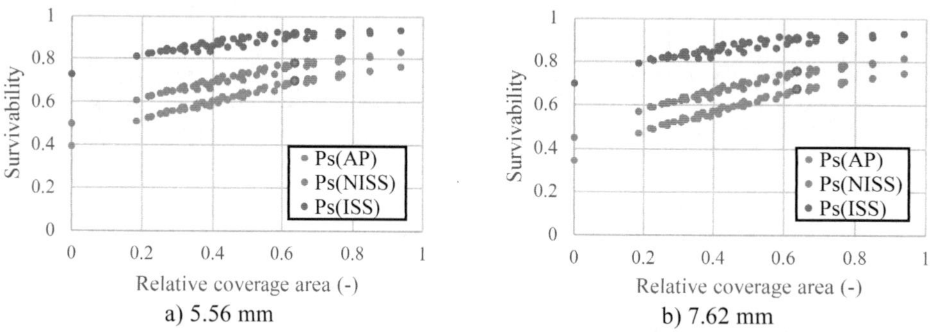

Figure 7. Average probability to survive injuries over relative coverage area based on all configurations tested. The red circle indicates the medium SAPI size result.

To better understand the relationship between length, width, and coverage area an isosurface plot was computed. From Figure 8, isolines correspond to a change in relative coverage, whereas colour gradient corresponds to change in $\overline{P_s}$. From the results, it can be observed that for a specific relative coverage area, there exist a combination of length and width that maximize the survivability metric. Linear interpolation was performed between all the results to find those optimized combination for the isolines presented in Figure 8. Optimized configurations are identified with red dots on Figure 8 and the specific values are presented in Table 4. From those results, it can be observed that for both projectiles larger plates seem to maximize the $\overline{P_s}$, when the full width is used. However, for smaller plates an equal proportion of length and width seem to maximize the $\overline{P}_s$. It can also be observed that changes in plate length and width have a relatively minor effect on the $\overline{P}_s$, as the range between minimum and maximum values for a specified relative coverage is considerably smaller than the variation in the coverage performance metrics across different coverage areas.

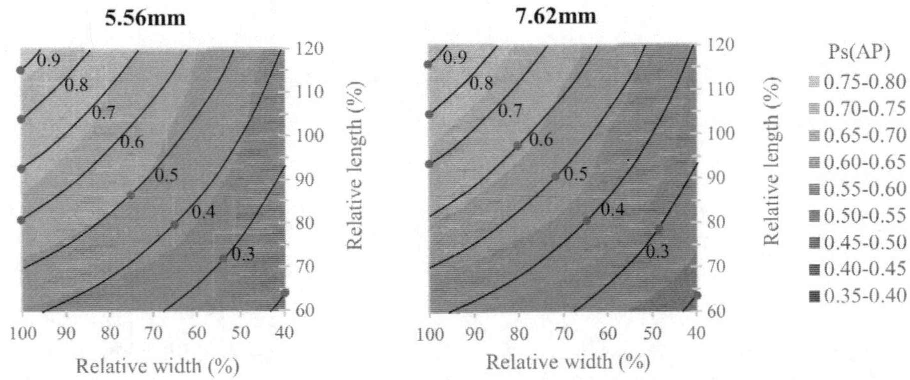

Figure 8. Average probability (i.e., AP) to survive injuries over the relative dimensions of the plate (i.e., length, and width) for all configurations tested. Relative coverage area is presented with iso lines, and the colour gradient represents the probability of surviving injuries. Optimized interpolated configurations are presented with red dots on the iso lines.

Table 4. Min and max $\overline{\boldsymbol{P}_s}$ for a given relative coverage area.

Rel. Coverage	0.9	0.8	0.7	0.6	0.5	0.4	0.3	0.2
5.56mm								
Max $\overline{P}_s$	0.7560	0.7289	0.6995	0.6604	0.6257	0.5841	0.5582	0.5171
Min $\overline{P}_s$	0.7565	0.7362	0.7144	0.6856	0.6515	0.6140	0.5685	0.5175
7.62mm								
Max $\overline{P}_s$	0.7373	0.7098	0.6775	0.6385	0.5990	0.5525	0.5192	0.4768
Min $\overline{P}_s$	0.7380	0.7147	0.6910	0.6575	0.6233	0.5813	0.5359	0.4802

4 DISCUSSION

4.1 Effectiveness of survivability metrics and relative coverage area to quantify ballistic protection's coverage

This study introduced novel metrics that quantitatively characterize the performance of different protection systems, demonstrating their potential as effective tools for comparative evaluation. The average probability to survive ($\overline{P}_s$) was computed for three injury scores (i.e., ISS, NISS, and AP). Significant differences were observed as presented in Figure 7, with the most significant differences being observed between the ISS and the two other metrics. This difference can be explained by the use of three MAIS severities identified for a given shot line to assess the survivability of soldiers. Another limitation is that the three severities used must be from different body regions. Since penetration of projectiles usually cause injuries in a single body region, it was observed that outcomes were underestimated. Those limitations of the ISS metric were also reported in other studies [23, 24]. The other two metrics (i.e., NISS, and AP) are not limited by body regions thus allow for a better representation of every injury identified for a given shot line. The AP metric being a more complex method, that includes all injuries from a given shot line seem to significantly predict lower survivability than the simpler NISS metric, which indicates that the AP is more sensitive. For this reason, the AP metric was chosen for most results presented in this study.

To quantify the coverage of ballistic protection, a relative coverage area was computed. This metric is a measure of the area ratio between the ideal medical coverage and the surface of ballistic protection. However, this metric does not necessarily consider the additional protection provided by a curved plate for angled shots. To demonstrate the correlation between the ability of a protection to stop projectiles in the thoracic region and the relative coverage area metric, the percentage of projectiles successfully stopped was recorded and presented in Figure 9. Results indicated that the relative coverage area of a plate is generally a reliable predictor of the number of blocked shot lines. However, this relationship weakens for a protection system that has larger coverage. This indicates that for larger plates, a more pronounced curvature has a larger effect on the ability of a protection system to block projectiles.

https://doi.org/10.52202/081621-0019

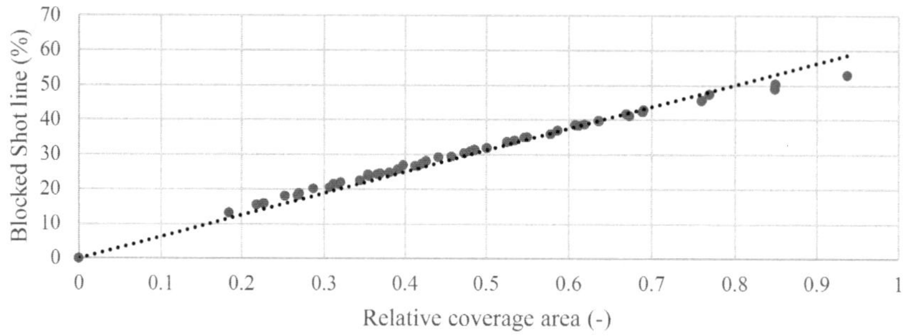

Figure 9. Correlation between relative coverage area and the ability to block shot lines in the thoracic region.

4.2 Effect of Ammunition Properties on the Coverage Performance of Ballistic Protection

In the current study, the coverage performance of 49 ballistic protections was assessed with two relevant projectiles (i.e., 5.56 mm, and 7.62 mm). The data generated by CORE V/L and presented in Figure 7 and Figure 8 shows differences between the projectiles tested. By comparing results obtained with the AP metric in Figure 10, it can be observed that the larger projectile present significantly lower $\overline{P_s}$. This is due to the larger projectile having more kinetic energy at the impact than the other one. Since damage caused to live tissue is most often proportional to the energy deposited by the projectile, it is to be expected that the larger projectile will cause more damage resulting in lower average probability to survive.

It can also be observed from Figure 10 that the difference in $\overline{P_s}$ between the two projectiles reduces as the relative coverage area increase. Similar differences also seem to be observed in Figure 8, where the behaviour of the optimal configuration changes at a relative coverage area of 0.6 for the 7.62 mm as opposed to 0.5 for the 5.56 mm projectile. Further investigation is required to fully understand the differences observed between the two projectiles as the coverage increase. However, preliminary analysis suggests that projectile size, shape and kinetic energy all contribute to the observed variations in the data.

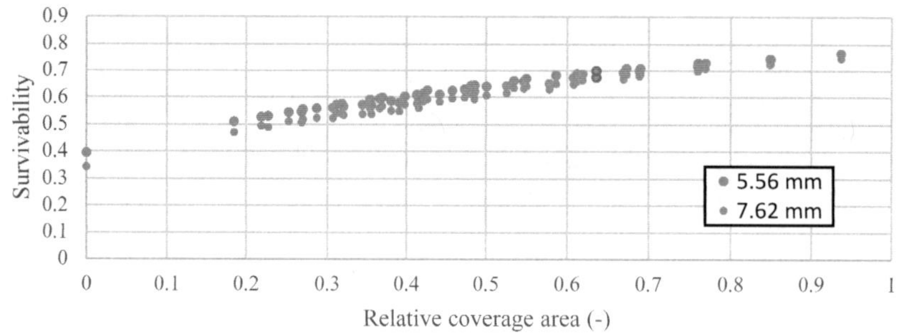

Figure 10. Effect of projectile properties on the $\overline{P_s}$ metric.

4.3 Extending results to military personnel

In an attempt to extrapolate the generated data to a larger military population, plate dimension, and coverage area were normalized based on CORE V/L anatomical model anthropometry. Length has been normalized by the suprasternal to 10th rib length; width has been normalized by the chest breadth; coverage area has been normalized by the reference area of the thoracic region. Thus, by comparing anthropometric measurements taken on a soldier and the size of a plate, one would be able to assess the coverage performance of a specific plate on an individual. It would also be possible to estimate the gain or decrement in coverage performances if a larger or smaller plate size is being used. By interpolating between results and extending those results to a larger population, a linear scaling assumption would need to be made. This would indicate that organ size and placement are assumed to scale linearly based on external anthropometric landmark. However, there is evidence in the literature that organ location may not correlate to external body markers [15, 25, 26], and there also exist significant difference between male and female anatomy that would need to be taken into consideration. More investigation would be needed to better understand the effect of stature, gender, and organ location on the coverage performance of ballistic protection.

The current study also proposed that an optimization of the length and width of a ballistic protection can be performed for a given relative coverage area in order to maximize the $\overline{P_s}$ metric. However, other factors would also need to be considered to optimize the geometry of a ballistic protection for the threat and environment that soldiers are exposed to. For example, fit and comfort, weight, and thermal load are other factors that play an important role in this optimization of protection system. To evaluate the interaction between those parameters, genetic or other AI algorithm could help in that process. There exist studies, like the one presented by Khoe, where a soft tissue protection was optimized based on V/L analysis, thermal loading, and weight [27]. However, no study presented in the literature was found to optimize protection system based on all factors.

5 CONCLUSION

This study used the CORE V/L tool set to assess the effect of length, width, and coverage area on the coverage performances of ballistic protections. The analysis was performed on a series of 49 ballistic plate models, with varying length and width. Plate dimensions were all referenced to anthropometric measurements of the thoracic region. Simulation of the impact of 10,800 independent projectiles oriented in three different directions was performed with two distinct projectiles (i.e., 7.62 mm and 5.56 mm). Injury prediction and survivability metrics (i.e., ISS, NISS, AP) were assessed on each shot line. From that assessment, an average probability to survive ($\overline{P_s}$) was computed, which allowed ballistic protection to be ranked based on their shape and size. From those results, survivability metrics were found to perform differently, with the ISS presenting large differences compared to the others, and the AP presenting the most sensitivity to injuries identified for each shot line. Differences were also observed between the two projectiles simulated as size, shape and kinetic energy seem to influence damage caused to the tissues model. By comparing the effect of length, width and coverage area, it was found that the coverage area has a larger influence on the $\overline{P_s}$ compared to length and width alone. However, it was also found that for a given coverage area a combination of length and width can be found to further optimize the coverage performance. The methodology presented in this paper allowed for a ranking of different ballistic plate shapes and sizes to be made and inform how relative changes to the plate can influence the risk of life-threatening injuries. Since plate dimensions are referenced to anthropometric measurements on the anatomical model, it would be possible to extend those results to a military population and estimate the gain or loss in coverage performance offered by ballistic protection to an individual. However, this presents some limitations as differences in organ placement, stature, and gender are not taken into account. Subsequently, to optimize the geometry of a ballistic protection, other factors also need to be considered, like fit and comfort, weight, and thermal load. A potential way forward could be the development of a genetic or another AI algorithm that would optimize plate geometry based on V/L model to assess coverage performances, soft tissue models to assess fit and comfort, and other loading models that would affect the operational capability and survivability of a soldier.

References

1. Roberts, P., *Patterns of injury in military operations.* Current Anaesthesia & Critical Care, 2002. **13**(5): p. 243-248.
2. Lawry, L.L., et al., *Qualitative assessment of combat-related injury patterns and injury prevention in Ukraine since the Russian invasion.* BMJ Mil Health, 2025.
3. *Marine Lethal Torso Injuries: Preliminary Findings.* 2005, Armed Forces Institute of Pathology: Rockville,MD.
4. Epstein, A., et al., *Putting Medical Boots on the Ground: Lessons from the War in Ukraine and Applications for Future Conflict with Near-Peer Adversaries.* J Am Coll Surg, 2023. **237**(2): p. 364-373.
5. Coltman, C.E., B.R. Brisbine, R.H. Molloy, and J.R. Steele, *Can smaller body armour improve thoracolumbar range of motion and reduce interference when female soldiers perform dynamic tasks?* Appl Ergon, 2022. **98**: p. 103602.
6. Molloy, R., S. Laing, M. Jaffrey, and A. Furnell, *Determining the Maximum Acceptable Length of a Hard Ballistic Plate*, in *Personal Armour Systems Symposium (PASS 2020).* 2020. p. 58-67.
7. Choi-Rokas, H., T. Garlie, and K. Blake Mitchell, *Theoretical Framework for the Sizing of Body Armor Plates to Optimize Fit*, in *Digital Human Modeling and Applied Optimization.* 2022.

https://doi.org/10.52202/081621-0019

8. Kim, K.H., et al., *Virtual fit assessment of U.S. army body armor using NASA spacesuit techniques.* Appl Ergon, 2024. **120**: p. 104339.
9. Bret, V., J. Caneau, E. Teil, and F. Barbillon. *Protected Surface Rate And Injury Level Reduction With 3d Cad Protection Model.* in *Protective Armor System Symposium, PASS 2016.* 2016. Amsterdam, NL.
10. Fryer, R., J. Breeze, and E. Lewis, *Area of Coverage and sizing definitions of personal armour for UK Armed Forces personnel*, in *Protective Armor System Symposium, PASS 2021.* 2021: Copenhagen, DK.
11. Eberius, N. and P. Gillich, *Survivability Analysis for the Evaluation of Personel in Body Armor.* 2010, Army Research Laboratory.
12. Breeze, J., et al., *Defining the essential anatomical coverage provided by military body armour against high energy projectiles.* J R Army Med Corps, 2016. **162**(4): p. 284-90.
13. Breeze, J., et al., *Using CT scans to determine the optimal sizes of hard armour plates to protect the torso for UK female Armed Forces personnel.* BMJ Mil Health, 2023. **169**(1): p. 69-74.
14. Breeze, J., E.A. Lewis, and R. Fryer, *Determining the dimensions of essential medical Coverage required by military body armour plates utilising Computed Tomography.* Injury, 2016. **47**(9): p. 1932-8.
15. Coltman, C.E., et al., *Can thoraco-abdominal organ boundaries be accurately determined from X-ray and anthropometric surface scans? Implications for body armour system coverage and design.* Appl Ergon, 2024. **119**: p. 104311.
16. Gordon, C.C., et al., *2012 ANTHROPOMETRIC SURVEY OF U.S. ARMY PERSONNEL: METHODS AND SUMMARY STATISTICS.* 2014, U.S. Army Natick Soldier Research: Massachusetts, USA.
17. Keefe, A.A., H. Angel, and B. Mangan, *2012 Canadian Forces Anthropometric Survey (CFAS).* 2015, Defence Research and Development Canada: Toronto, Canada.
18. Saucier, R. and H.M. Kash III, *ComputerMan Model Description.* 1994, Army Research Laboratory: Aberdeen Proving Ground, MD.
19. *Abbreviated Injury Scale 2015 Revision*, in *AAAM.* 2015: Des Plaines, IL.
20. Baker, S.P., B. O'Neill, W. Haddon, and W.B. Long, *The injury severity score: A method for describing patient with multiple injuries and evaluating emergency care.* The Journal of Trauma, 1974. **14**(8).
21. Osler, T., S.P. Baker, and W. Long, *A modification of the injury severity score that both improves accuracy and simplifies scoring.* J Trauma, 1997. **43**(6): p. 922-5; discussion 925-6.
22. Copes, W.S., et al., *Progress in Characterizing Anatomic Injury.* Journal of Trauma and Acute Care Surgery, 1990. **30**(10).
23. Lavoie, A., et al., *The New Injury Severity Score: a more accurate predictor of in-hospital mortality than the Injury Severity Score.* J Trauma, 2004. **56**(6): p. 1312-20.
24. Kilgo, P.D., T.M. Osler, and W. Meredith, *The worst injury predicts mortality outcome the best: rethinking the role of multiple injuries in trauma outcome scoring.* J Trauma, 2003. **55**(4): p. 599-606; discussion 606-7.
25. Summers, S.J., et al., *Do thoracoabdominal organ boundaries differ between males and females? Implications for body armour coverage and design.* Appl Ergon, 2023. **106**: p. 103891.
26. Laing, S. and M. Jaffrey, *Thoraco-abdominal Organ Locations: Variations due to Breathing and Posture and Implications for Body Armour Coverage Assessments.* 2019, Defence Science and Technology Group: Australia.
27. Khoe, Y.S. *Generative Design Of Soft-Armour.* in *Protective Armor System Symposium, PASS 2023.* 2023. Dresden, Germany.

Optimising coverage, size and inclusivity of future body armour: The UK's new, innovative Tri-Service anthropometric survey

E A Lewis[1], **R H Pringle**[2], L Evans[3], S C Bloodworth-Race[3], R N Fryer[4]

[1]QinetiQ, Building 240, Bristol Business Park, Bristol, BS16 1EJ
[2]QinetiQ, Cody Technology Park, Ively Rd, Farnborough GU14 0LS
rhpringle@qinetiq.com
[3]Defence Equipment and Support (DE&S), Ministry of Defence (MOD), Bristol BS34 8JH
[4]Defence Science and Technology Laboratory (Dstl), Portsdown West, Portsdown Hill Rd, Fareham PO17 6AD

QinetiQ has completed an up-to-date tri-service anthropometric survey of UK Armed Forces personnel. This survey, conducted on behalf of the UK Ministry of Defence is unique, as it contains new measurements specifically for optimising the coverage, size and inclusivity of future body armour.

A Stratified Sample Design was used to gather sufficient data in the following primary demographic groups: Sex, Ethnicity and Age; and in the following secondary groups: Service (e.g. Army, Royal Air Force [RAF], Royal Navy [RN]), Ranks (e.g. Officers, Others) and Service Groups (e.g. Infantry, Aircrew, Submariners). This strategy was developed to ensure statistical validity.

The main survey was preceded by a pilot survey to refine the methodology and to de-risk the main survey, conducted between October 2023 and May 2024. Two new state-of-the-art VITRONIC VITUS Bodyscan 3D Scanners were used to capture the individual measurements and used in tandem with Anthroscan software to deduce measurements.

1,886 Armed Forces personnel (1,510 males and 376 females) including 32 nationalities were measured from across 15 Army, RN and RAF bases and captured 194 measurements per person. The survey includes 20% female representation across all services, a 21% increase in females measured over the last UK anthropometric survey conducted in 2006/07. Fifty-four new measurements were captured specifically for optimising coverage, size and inclusivity of body armour. Examples include suprasternal notch to tenth rib, suprasternal notch to iliac crest and suprasternal notch to breast point (standing and sitting).

This new anthropometric survey will benefit all Users of defence equipment for the next ten to twenty years. It provides important data to immediately benefit UK Armed Forces body armour acquisition projects, particularly for women, by including 54 new measurements specifically for body armour. The 3D scanners will enable an enduring capability to measure anthropometry within the Armed Forces through longitudinal analyses.

1. INTRODUCTION

In 2024, QinetiQ completed an up-to-date tri-service anthropometric survey of UK Armed Forces personnel. This survey, conducted on behalf of the UK Ministry of Defence is unique, as it contains not only 'traditional' anthropometric measurements but also includes new measurements specifically for optimising the coverage, size and inclusivity of future body armour.

1.1 Body armour: medical coverage and anthropometry

The purpose of body armour is to prevent mortality (death) and significant long-term morbidity. It must not restrict movement as far as practicable and must not incur an unacceptable thermal or physical burden. Body armour must also provide coverage and protection to key anatomical structures. For the torso, coverage of the heart, great vessels, liver and spleen, by hard body armour plates, has been determined as essential to prevent mortality [1]. Subsequent work [2][3][4] demonstrated that the positions of internal organs and structures could be related to external anthropometric landmarks and

https://doi.org/10.52202/081621-0020

used to describe and optimise body armour coverage for Armed Forces personnel. For example, the arch of the aorta corresponds to the suprasternal notch, the lower border of the liver corresponds to the lower border of the ribcage (tenth rib) and the bifurcation of the aorta corresponds to the iliac crest (Figure 1).

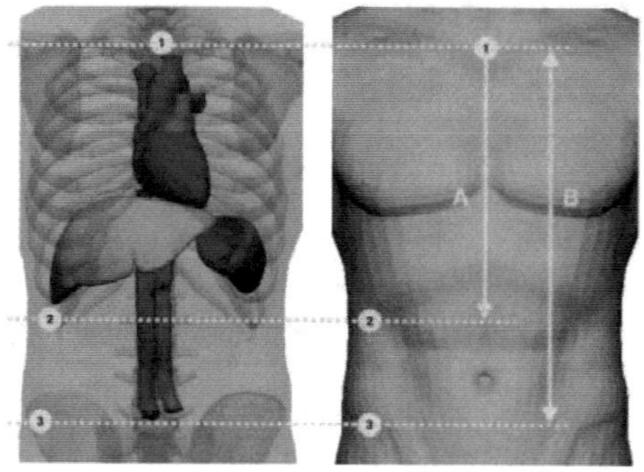

Figure 1 Anthropometric landmarks to describe coverage of the torso: (1) suprasternal notch (representing the top of aortic arch); (2) lower border of 10th rib (representing the lower border of the liver) and (3) iliac crest (representing the bifurcation of the aorta). A = Threshold Coverage (measurement of suprasternal notch to tenth rib), B = Objective Coverage (measurement of suprasternal notch to iliac crest) [3]

The same studies demonstrated that stature (total height) is not an accurate measure to enable sizing, fitting and scaling of body armour [2] (for example; fitting body armour using solely the traditional 'chest circumference' and 'height' measurements is not deemed adequate) therefore 'torso height' has been recommended as a measure instead. Torso height (defined as suprasternal notch to iliac crest) provides a more accurate measure than suprasternal notch to umbilicus (belly button), for example, as the position of the umbilicus can vary with weight loss or gain [2] or post-pregnancy [5]. Furthermore, to articulate medical coverage for new body armour acquisitions, the terms 'Threshold Coverage' and 'Objective Coverage' have been recommended [3]. Threshold coverage is defined as the 'Absolute minimum area that must be afforded coverage by personal armour. Less than the threshold would not be considered to provide an improvement over current capabilities and is not medically recommended'. Threshold coverage relates to coverage of 'anatomical structures (in the torso) that if injured would result in life-threatening haemorrhage and lead to death within 60min without surgical intervention' [3]. The surface (anthropometric) landmarks used to determine 'Threshold coverage' are Suprasternal notch to lower border of ribcage (10th rib). Similarly, the landmarks used to determine 'Objective Coverage' are Suprasternal notch to superior border of the iliac crest [3] (shown in Figure 1).

1.2 Previous UK Anthropometry Survey and the need for updated data

The last comprehensive anthropometry survey of UK Armed Forces personnel was conducted in 2006–07 [6] with the survey data freely available in the MOD's Human Factors Integration (HFI) Technical Guide for Anthropometry: People Size [7]. The data gathered were predominantly based on white Caucasian, male Armed Forces personnel; 2470 personnel were sampled, 88% were white Caucasian and 87% were male. Whilst this may have been appropriate in 2007, this is no longer the case, as the composition of the UK Armed Forces has changed significantly in the past 18 years [5]. Since late 2018, for example, all roles in the UK Armed Forces have been opened to women, including Ground Close Combat roles [4] and hence the number of women, as a percentage of the total number of Armed Forces personnel, has increased. Opening up all roles to women has meant that all equipment, including body armour, must be usable by both men and women [5]. Women in the Armed Forces have been wearing hard body armour plates that were optimised for large men, partly because specific anthropometric measures, that are now known to be required, were not included in previous anthropometry surveys. Suprasternal notch to iliac crest, therefore, is only one of the new measures incorporated into this new, updated anthropometry survey[1] as a result of recommendations from previous work[2][3][4][5]. Other specific measures are also required to enable the optimisation of the torso body armour system, as well

[1] NB. Suprasternal notch to umbilicus is also included as a measure in this study

https://doi.org/10.52202/081621-0020

as specific measures for the optimisation of armour for the extremities, such as arm protection and pelvic protection. Furthermore, new measurements are required to optimise body armour sizing and fitting regimes, particularly acknowledging the differences between male and female Armed Forces personnel as well as for other minorities. Without new underpinning data, it is not possible to quantify the differences between sexes, but also not possible to implement any meaningful changes to body armour coverage, sizing and fitting. Up-to-date anthropometric data has the potential to revolutionise all military equipment, but for body armour, it also has the potential to increase the number of sizes to optimise the medical coverage and fit, particularly for women. One size does not fit all when it comes to armour and women are not small men, hence increasing the range of sizes and shapes will impact on fit, form, function, interoperability, integration and reduce mass of body armour. Body armour not only saves lives, its optimisation will reduce the incidence of musculoskeletal injuries which are the most common cause of medical downgrading and medical discharge in both Service men and women [9].

2. METHODS AND EQUIPMENT

2.1 Internal and external anthropometry requirements

MOD's intent was to procure new, innovative 3D scanners to conduct a new, tri-service, anthropometric survey of UK Armed Forces personnel (to capture 'external' anthropometric measures), coupled with using Magnetic Resonance Imaging (MRI) scanners to capture 'internal' organ boundary and anthropometric data [9]. This paper's focus is the 'external' anthropometry survey only.

2.2 Stratified sample design for the main survey

For the main external anthropometry survey, a Stratified Sample Design (SSD) was devised by MOD statisticians, provided to QinetiQ, and used to gather sufficient data in the following primary demographic groups: Sex, Ethnicity and Age; and in the following secondary groups: Service, Ranks, and Service Groups (see 2.5). This strategy was developed to ensure statistical validity [5]. An SSD was selected for the Tri-Service anthropometric survey, rather than a simple random sample, for the following reasons: (a) A simple random sample may not include sufficient members from minority groups for accurate parameter estimation; (b) The data was required to be of a known precision for the strata; and (c) Stratified sampling produces lower variance estimates for the whole population.

The Stratified Sample was based on Serving Military Population data from the Defence Statistics, Joint Personnel Administration system (JPA), (from Jan. 2021 to Jan. 2022) and the 2006/2007 Tri Service Anthropometry Survey data [6]. Following the calculation of strata sample size the JPA data was subsequently used to identify UK military locations that provided the greatest coverage across all strata. This approach aimed to provide a current Anthropometry dataset that is proportional to the demographic of the entire Armed Forces. The number of participants required for each strata sample was calculated using a method derived from BS EN ISO 15535-2012 [10], which also suggests a similar approach, stating that the source for the variance should be that of a previous anthropometric measurement with the highest Coefficient of Variation (CoV). The variable selected from analysis of the 2006/2007 Survey data [6] was "Weight". From these calculations, the generated strata sample sizes were produced and predicted to be sufficient to estimate the average weight for each strata to +/- 4 kg with 90% confidence.

2.3 Selection and descriptors of individual measurements

Stakeholder workshops were conducted with a range of MOD disciplines and functions represented, to capture requirements for anthropometry data [5]. After careful consideration and investigating the art of the possible within the programme bounds, 194 measurements were selected, and a new, more descriptive style of naming the anthropometry measurements was adopted e.g. left/right, sitting position, midpoint, maximum prominence, vertical height or contour etc. to add clarity to the measurement definitions.

2.3 Survey parameters

The survey design and execution included: (a) Demographic information from the participants; (b) Equipment required to take the measurements; (c) Postures in which the participants were scanned (in

https://doi.org/10.52202/081621-0020

the 3D scanner); (d) Manual measurements taken from the participants; (e) Measurements to be extracted from the scan data; and (f) Landmarks needed in the measurement process (scan or manual). The parameters described in (a) – (f) were subsequently used to define the measurement process both physically (manual measurements taken directly from the participant) and virtually (by automatically extracting located landmarks or by manually picking landmarks from the 3D scans) [11].

2.5 Demographic information

Demographic data was collected from participants as part of the consenting process for participating in the anthropometric survey and in accordance with the requirements of the MOD Research Ethics Committee (MODREC) stipulations. The demographic data collected comprised of: Participant Number, First Name, Middle Name, Surname, Service Number, Service, Rank, Officer (Yes/No), Date of Birth, Sex (Now), Sex (At Birth), Ethnicity, Nationality, Role, Pilot (Yes/No), Aircrew (Yes/No), Submariner (Yes/No), Association, Length of Service, Handedness and any Physical impairments were recorded [11]. All data produced by the Survey after the initial collection were anonymised with the Participant Number being the only reference used to identify an individual. Anthropometry databases created by MOD will not contain the names and dates of birth of individuals.

2.6 Equipment Used

2.6.1 3D Whole body scanners

Two new state-of-the-art VITRONIC VITUS Bodyscan 3D Scanners [11] were procured for this survey, used to capture the individual measurements and used in tandem with Anthroscan software to deduce measurements [11]. The VITRONIC VITUS Bodyscan 3D Scanners comprise of an aluminium box section structure onto which are mounted four sensor sleds. The sensor sleds are made up of two cameras and use optical triangulation with an eye-safe infrared laser light. These sleds are on a motorised carriage that move them in a vertical motion from top to bottom to capture imagery of the participant. The scanner specifications are shown in Table 1. Data is transferred from the scanning head to a control Personal Computer (PC) via a Wi-Fi Router [11].

Measurement principle	Optical double triangulation
Number of sensor sleds	4
Interface	USB port and LAN
Measurement accuracy	Average max. circumference errors significantly < 1 mm
Scan time	approx. 6 to 10s
Point density	~300 points / cm^2

Table 1 Specifications for the VITRONIC VITUS Bodyscan 3D (Whole Body Scanner)

2.6.2 Traditional measurement tools

'Traditional' measurement tools were also used, such as tape (used to take circumferences or lengths along the body surface), small callipers (used to take measurements from point to point, e.g. the head and hand length), a stadiometer (used to measure heights from the ground), a measuring rod (used for measuring fingertip to fingertip span), a pupilometer (used for measuring interpupillary distance), scales (Withings Bodyscale) (used for measuring the weight of the participants, can also be used as a BMI/Body fat device) and a foot box (used for measuring foot length and breadth). Suitably qualified personnel took all measurements and the data entered directly into a PC [11].

2.6.3 Anthroscan software

The Anthroscan software provided with the VITRONIC VITUS Bodyscan system has the functionality to automatically extract ('auto extraction') a number of measurements. These measurements are calculated from scans in a standardised pose by mathematically identifying a series of landmarks, which are then combined to extract the measurements. Anthroscan can also be programmed to assist the operator in identifying landmarks on screen by automatically selecting the correct scan posture, zooming into the approximate area of the body where the landmark is to be placed and providing tools for modifying or editing landmark position. Following landmarking, measurements can be defined as point to point (3D line distance), height/length, breadth and depths (single axis measurements), contours across the body surface or circumferences (virtual tape measures) [11].

https://doi.org/10.52202/081621-0020

2.7 Pilot Studies

Two Pilot Studies were conducted to ensure that the scanners and the associated software could capture 3D data from participants and extract the required measurements accurately, consistently and with sufficient fidelity. The Pilot Studies also sought to assess and optimise all aspects of the measurement protocol, including the accuracy and reliability of manual measurements – both within and between those taking the measurements [5][11].

2.8 Main Survey

The main survey was conducted between October 2023 and May 2024 across 15 different Army, Navy and Royal Air Force bases [12] [13]. The 2006–07 survey measured 92 dimensions from 2470 personnel [6], and all of these dimensions were included in the new survey. Fifty-four new measurements were specifically added to support the development of new body armour. Twelve extra measurements were included to support 3D modelling systems e.g. Knee-ankle length and a further seven 'new' useful measurements were added that had not been captured previously e.g. Buttock-heel length, Sitting.

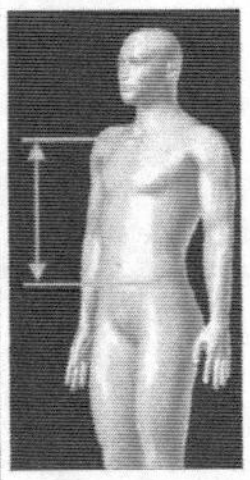

Ref	M155
Name	Suprasternal notch to Iliac Crests midpoint, Vertical Height
Method	Scanned
Posture / Pose	Stand 1
Landmarks	Suprasternal notch; Plane level with Iliac Crests
Type	Height (mm)
Topography	Single Axis
Standard	None

Measurement Description:

Vertical distance between the suprasternal notch and a plane level with the Iliac Crests.

Notes for measurers.

Participant stands in Stand 1 position.

Figure 2 Measurement definition for 'Suprasternal notch to Iliac Crests midpoint, Vertical Height' [11]

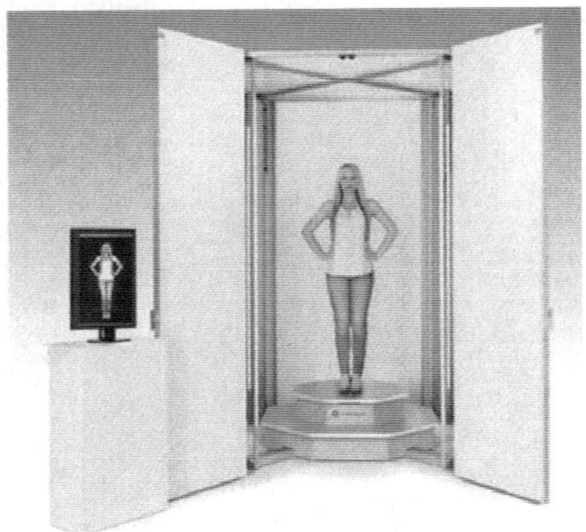

Figure 3 VITRONIC VITUS Bodyscan system [5]

3. RESULTS

3.1 Survey headlines

3.1.1 Numbers and demographics

One thousand, eight hundred and eighty-six (1,886) Armed Forces personnel, including 32 different nationalities, were measured from across 15 Army, Royal Navy (RN) and Royal Air Force (RAF) bases over a period of 20 weeks. This included 1,510 male (M) and 376 female (F) participants: 793 Army ((141 F, 652 M), 587 Army Male [excluding Nepalese], 65 Army Nepalese), 627 RAF (147 F, 480 M), 329 RN (77 F, 252 M), 137 Royal Marines (RM) (11 F [band members], 126M) [12]. One hundred and ninety-four measurements were captured per person [12] [13]. The survey includes 20% female representation across all services, a 21% increase in females measured over the last UK anthropometric survey conducted in 2006/07 [6].

3.1.2 Measurement methods

For the total 194 measurements per person; two 'standing' scans captured 115 measurements, three 'sitting' scans captured 53 measurements, 23 manual measurements were taken and three measurements

https://doi.org/10.52202/081621-0020

were derived by calculations (e.g. Sitting/Standing ratio). The Anthroscan software was used to create each of the scanned measurements. Twenty-two of the Anthroscan automated measurements [13] were used and another 146 measurements were coded within the software (e.g. applying functions for Virtual Height calculations and Virtual tape measures used for contours).

3.2 New measurements captured for optimising body armour

Fifty-four (54) new measurements were captured specifically for optimising coverage, size and inclusivity of body armour [12]. The full list is provided in Table 2.

#	ID	Measurement Type
1	M022	Axilla height, anterior, Sitting
2	M023	Axilla height, posterior, Left
3	M024	Axilla height, posterior, Sitting, Left
4	M042	Cervicale (C7) height, Sitting
5	M045	Cervicale (C7) to Iliac crests midpoint, Contour
6	M046	Cervicale (C7) to Iliac crests midpoint, Vertical Height
7	M047	Cervicale (C7) to Iliac crests midpoint, Sitting, Contour
8	M048	Cervicale (C7) to Iliac crests midpoint, Sitting, Vertical Height
9	M049	Cervicale (C7) to posterior Axillae midpoint, Contour
10	M050	Cervicale (C7) to posterior Axillae midpoint, Vertical Height
11	M051	Cervicale (C7) to posterior Axillae midpoint, Sitting, Contour
12	M052	Cervicale (C7) to posterior Axillae midpoint, Sitting, Vertical Height
13	M053	Cervicale (C7) to level with top of Thigh, Sitting, Left, Contour
14	M054	Cervicale (C7) to level with top of Thigh, Sitting, Left, Vertical Height
15	M055	Cervicale (C7) to level with Umbilicus, Top, Contour
16	M056	Cervicale (C7) to level with Umbilicus, Top, Vertical Height
17	M057	Cervicale (C7) to level with Umbilicus, Top, Sitting, Contour
18	M058	Cervicale (C7) to level with Umbilicus, Top, Sitting, Vertical Height
19	M059	Cervicale (C7) to Tenth ribs midpoint, Anterior, Contour
20	M060	Cervicale (C7) to Tenth ribs midpoint, Anterior, Vertical Height
21	M061	Cervicale (C7) to Tenth ribs midpoint, Anterior, Sitting, Contour
22	M062	Cervicale (C7) to Tenth ribs midpoint, Anterior, Sitting, Vertical Height
23	M194	Chest breadth
24	M105	Iliac crest height, Left
25	M106	Iliac crest height, Sitting, Left
26	M107	Iliac crests midpoint to Umbilicus, Top, Contour
27	M108	Iliac crests midpoint to Umbilicus, Top, Vertical Height
28	M109	Iliac crests midpoint to Umbilicus, Top, Sitting, Contour
29	M110	Iliac crests midpoint to Umbilicus, Top, Sitting, Vertical Height
30	M150	Suprasternal notch height
31	M151	Suprasternal notch height, Sitting
32	M152	Suprasternal notch to Breast point, Left, Contour
33	M153	Suprasternal notch to Breast point, Left, Vertical Height
34	M154	Suprasternal notch to Breast point, Sitting, Left, Vertical Height
35	M155	Suprasternal notch to Iliac crests midpoint, Vertical Height
36	M156	Suprasternal notch to Iliac crests midpoint, Sitting, Contour
37	M157	Suprasternal notch to Iliac crests midpoint, Sitting, Vertical Height
38	M158	Suprasternal notch to level with anterior Axillae midpoint, Contour
39	M159	Suprasternal notch to level with anterior Axillae midpoint, Vertical Height
40	M160	Suprasternal notch to level with anterior Axillae midpoint, Sitting, Contour
41	M161	Suprasternal notch to level with anterior Axillae midpoint, Sitting, Vertical Height
42	M162	Suprasternal notch to level with Tenth ribs midpoint, Contour
43	M163	Suprasternal notch to level with Tenth ribs midpoint, Vertical Height
44	M164	Suprasternal notch to level with Tenth ribs midpoint, Sitting, Contour
45	M165	Suprasternal notch to level with Tenth ribs midpoint, Sitting, Vertical Height
46	M166	Suprasternal notch to level with top of Thigh, Sitting, Left, Contour
47	M167	Suprasternal notch to level with top of Thigh, Sitting, Left, Vertical Height
48	M168	Suprasternal notch to Top of Umbilicus, Contour

49	M169	Suprasternal notch to Top of Umbilicus, Sitting, Contour
50	M170	Suprasternal notch to Top of Umbilicus, Sitting, Vertical Height
51	M171	Suprasternal notch to Top of Umbilicus, Vertical Height
52	M172	Tenth rib height, Left
53	M173	Tenth rib height, Sitting, Left
54	M179	Thigh height, Sitting, Left

Table 2 The Fifty-four (54) new measurements captured specifically for optimising coverage, size and inclusivity of body armour, from [12]

3.3 Initial comparison of some new measurements

Three of the fifty-four (54) new measurements captured specifically for optimising coverage, size and inclusivity of body armour are included in Table 3. This compares the measurements for 'Chest Breadth', 'Suprasternal notch to tenth rib' and 'Suprasternal notch to iliac crest' for the 1st, 5th, 25th, 50th, 95th and 99th percentiles for All (all Female and all Male combined), All Female, All Male, Army Female, Army Male, RAF Female, RAF Male, Royal Navy Female and Royal Navy Male.

Measurement & ID (see Table 2)	%ile	All mm	All Female (F) mm	All Male (M) mm	Army (F) mm	Army (M) mm	RAF (F) mm	RAF (M) mm	RN (F) mm	RN (M) mm
Sample size (n)		*n=1880*	*n=375*	*n=1505*	*n=141*	*n=649*	*n=147*	*n=478*	*n=76*	*n=252*
Chest Breadth (M194)	1st	270	261	280	258	273	267	285	267	290
	5th	288	272	296	272	293	278	297	274	298
	50th	338	309	344	306	341	313	347	306	339
	95th	394	362	396	350	394	366	397	362	407
	99th	413	387	405	381	410	372	421	398	427
Sample size (n)		*n=1876*	*n=375*	*n=1501*	*n=141*	*n=646*	*n=147*	*n=479*	*n=76*	*n=250*
SSN to 10th rib (M163)	1st	251	247	255	246	253	255	255	249	261
	5th	267	258	271	254	270	262	273	260	278
	50th	305	289	310	289	310	288	308	292	310
	95th	346	323	350	325	352	322	345	316	348
	99th	364	336	356	335	366	335	365	329	364
Sample size (n)		*n=1878*	*n=375*	*n=1503*	*n=141*	*n=649*	*n=147*	*n=478*	*n=76*	*n=250*
SSN to IC (M155)	1st	319	306	336	309	340	321	333	292	340
	5th	339	324	353	321	354	329	354	314	352
	50th	390	362	396	362	399	363	397	362	395
	95th	444	409	448	404	450	410	447	406	449
	99th	473	424	474	423	475	423	462	426	490

Table 3 Key measurements for torso body armour, 'Chest Breadth', 'Suprasternal notch (SSN) to 10th rib', 'SSN to Iliac crest (IC)' comparing between (some of the) datasets. 'n' = numbers measured [12]

4. DISCUSSION

4.1 Opportunities for data analysis

There is a vast amount of analysis possible from this updated dataset; from the comparison with the previous UK survey dataset to gain an understanding of the secular change and improvements in the accuracy of data collection methods, to the specific data comparisons between male and female data, between ethnicities, between services and between internationally available datasets. Furthermore, there are 54 new measures specifically for the optimisation of body armour that require in-depth analysis to determine the optimum coverage, fit and number of sizes of hard ballistic plates, soft body armour vests, arm protection, neck protection and pelvic protection for both men and women in the UK Armed Forces population. It has been impossible therefore, to present all these data permutations within this one paper. It is also important to highlight that as well as the opportunity to analyse the data tables, it is also possible to further interrogate the 3D scans themselves to obtain even more specific data, should there be a particular measurement required now or in the future. For example, at an MOD/QinetiQ Working Group,

 https://doi.org/10.52202/081621-0020

(held in August 2024, and subsequent to the completion of the main survey), two new measurements were added to the dataset: M195 – Axilla to Iliac Crest and M196 – Chest Depth (sternum to vertebrae). The measurement dataset should, therefore, be seen as dynamic, and updated as and when an anthropometric need arises.

This new data can be exploited into new equipment programmes; accurate internal data can be used to underpin realistic virtual battlefield simulation and surgical training; and could inform personalised, individual survivability modelling and simulation. Other applications include aircrew cockpit sizing, vehicle crew compartments and blast survivability modelling. The full range of exploitation pathways for anthropometric data has yet to be discovered [9].

4.1 Overall differences between this survey and last survey

Some statistically significant differences were identified between data gathered during the 2023/2024 study [12] [13] and the previous 2006/2007 tri-service anthropometric survey [6]. These differences can be attributed to a number of factors including secular change, changes in data collection methods, the removal of deduced measures and improvements in the accuracy of three-dimensional scanners [14]. A summary of the key differences identified between the surveys is provided in [14] but include Hip Breadth, Waist Circumference, Chest Depth, Thigh Clearance Height and Buttock-Popliteal length. It is not possible to compare data of new measurements (the 54 measures specifically for body armour) as the data was simply not gathered in the previous 2006/2007 anthropometric survey [6].

4.2 (Some) new measurements captured specifically for optimising coverage, size and inclusivity of body armour

One size does not fit all when it comes to body armour plates and this new anthropometric data further reinforces that a range of plate sizes would better fit the UK Armed Forces population compared to a single sized front and rear plate [2][3][4]. Whilst it is not possible to compare data of the 54 measures specifically for body armour with the previous 2006/2007 anthropometric survey [6], it is possible to make a simple comparison with previously published male [2] and female Computed Tomography (CT) scan data [4] for two key measurements: Suprasternal notch to 10th rib ('Threshold coverage') and suprasternal notch to iliac crest ('Objective coverage'). Table 4, for example, shows the 50th percentile data for these studies. For reference, the length of the existing front UK hard ballistic plate is 334mm; which is longer than the Suprasternal notch to 10th rib measurements ('Threshold coverage'), for over half of today's Armed Forces population.

Measurement	Percentile	All F [12] *n=375* (mm)	All M [12] *n=1501-1503* (mm)	CT (F) [4] *n=45* (mm)	CT (M) [2] *n=120* (mm)
SSN to 10th rib	50th	289	310	290	300
SSN to IC	50th	362	396	355	371

Table 4 50th percentile data for 'SSN to 10th rib' and 'SSN to IC' for 'All females' and 'All males' [12] compared with CT scan data of UK Armed Forces females [4] and males [2]

Whilst the CT data in Table 4 were captured with the subjects supine (lying down), this is not thought to affect the measurements of the (bone) anthropometric landmarks, but may affect internal organ positions and boundaries [1] [15]. Further work is underway as part of the initial MOD requirement that is collecting internal organ anthropometric data from (standing and supine) MRI scans of UK Armed Forces personnel.

Table 3 in the 'Results' section shows the comparison between 'Chest Breadth' percentile data. This measure was initially not included in the list of measurements (although 'Chest Breadth (maximum prominence), Contour' was included) but was later added as a key measure for optimising body armour and provides a useful comparator to other published work [15].

4.3 New Standardised Measurement Set

Significant effort went into creating the master list of individual measurements [11]; taking great care to standardise the descriptions of each measurement into a template (Figure 2). It was important to create this 'master list' to enable consistency of measurement and nomenclature and to clearly differentiate between other (similar-sounding) measures. This is a key milestone in the development of national and international standards of measurement descriptions for future anthropometric surveys.

4.4 Transition Plan

MOD's intention is to transition the capability to measure anthropometry to the three services – Army, RN and RAF. QinetiQ delivered a Transition Plan [13] with nineteen key associated documents addressing not only the hardware and software required by the three Services but also identifying the establishments to locate the equipment to best support further anthropometric studies. The training required by those responsible for the equipment is outlined as well as the protocols (e.g. to measure personnel and manage their data) to follow to ensure that data are recorded in a consistent manner by each Service; ensuring reliability and validity of data measurements and management. An enduring capability is being implemented based on this Transition Plan, with the 3D scanners procured for this survey already transferred to SoldierWorks for the Army and The Institute of Naval Medicine (INM) for the Royal Navy; complementing the 3D scanner in the Royal Air Force, based at the Centre of Aviation Medicine (RAFCAM). It is anticipated that this will result in the measuring of Armed Forces personnel becoming 'business as usual', dynamic, responsive and populating a 'living anthropometric database'.

5. CONCLUSIONS

This is the first Tri-Service anthropometric survey of the UK Armed Forces to be conducted in the UK in 18 years. One hundred and ninety-four measurements were captured per person from 1,510 males and 376 females and from 32 nationalities. The data are applicable across Defence and include 54 new measurements specifically for body armour optimisation, 12 extra measurements required for 3D modelling packages and a further 7 measurements deemed useful that were not captured in previous anthropometric surveys.

The 3D body scanners procured for this survey have already transitioned to their permanent locations, to meet the MOD's aspiration for measuring Armed Forces personnel as part of 'business as usual' and populating a 'living database'.

The HFI Technical Guide for Anthropometry, TG1.1 V6 [14] has been updated by MOD to include some of the new data and is freely available to all via the MOD's Defence Gateway portal.

At the time of writing, a database solution containing all the new data is still in development by MOD. In the future, it is anticipated that this new anthropometric database tool will be made available through the MOD's Defence Gateway portal.

6. RECOMMENDATIONS

All future UK military equipment should be developed using this up-to-date, representative and relevant anthropometric data, thereby ensuring that Defence is diverse, inclusive and consistent in its approach to equipment acquisition, whilst at the same time maximising the likelihood that defence equipment is safe, comfortable and usable under extreme operational conditions and in the most severe environments.

Logistics and resources did not allow the scope of this project to include Reservists. It is recommended that with the transition of the 3D scanners to military establishments, this will provide the opportunity to measure this population.

It is recommended that other nations and User groups (such as Police and Security) adopt the additional 54 measurements to optimise their body armour systems.

It is recommended that the measurement definitions, postures and measurement protocols created for this anthropometry survey are adopted internationally and in new standards e.g. ISOs and NATO standards.

It is recommended that this new anthropometry data is shared freely and widely by MOD to maximise exploitation opportunities.

https://doi.org/10.52202/081621-0020

There is an opportunity to further exploit the current dataset; this should be seen as dynamic and updated as and when an anthropometric need arises. For other specific defence research, applications and equipment, several measurements could be further analysed/deduced from the captured 3D scans to optimise equipment design and workplace integration. Furthermore, the 3D scanning capability provides the ability to measure all types of User-borne equipment ensembles in situ for future research and trials.

References

[1] Breeze J, Lewis EA, Fryer R, Hepper A E, Mahoney P F, Clasper JC. Defining the essential anatomical coverage provided by military body armour against high energy projectiles. J R Army Med Corps. 2016 Aug;162(4):284-90

[2] Breeze J, Lewis EA, Fryer R. Determining the dimensions of anatomical coverage provided by military body armour plates utilising Computed Tomography. Injury. Int. J. Care Injured 47 (2016) 1932–1938

[3] Breeze J, Bowley DM, Naumann DN, Marsden M, Fryer R, Keene D, Ramasamy A, Lewis EA. Torso body armour coverage defined according to feasibility of haemorrhage control within the prehospital environment: a new paradigm for combat trauma protection. BMJ Mil Health 2022 Oct;168(5):399-403.

[4] Breeze J, Fryer RN, Bowley D, Gibb I, Lewis E. Using CT scans to determine the optimal sizes of hard armour plates to protect the torso for UK female Armed Forces personnel. BMJ Mil Health 2023;169:69–74.

[5] Evans, L., Pringle, R., & Lewis, E. (2024). Women are not small men: the UK's new, comprehensive Tri-Service Anthropometry survey. Ergonomics. 2024 Jul 31:1-10.

[6] Pringle, R. H., A. J. Puxley, K. P. Puxley, G. M. Turner, and A. K. Tyrrell. 2011. Anthropometry Survey of UK Military Personnel 2006-7 (Issue 3). QINETIQ/07/01821/3.0.

[7] Cummings, R. 2022. Human Factors Integration Technical Guide for Anthropometry: People Size. Version 5.3, 18th January 2024

[8] Ban on women in ground close combat roles lifted. https://www.gov.uk/government/news/ban-on-women-in-ground-close-combat-roles-lifted

[9] Lewis EA. Digital mannequins for our diverse Armed Forces. ID: 12887. Funding proposal to MOD Defence Innovation Fund. TLB Ideas Scheme (Round 9). 27 October 2020.

[10] ISO 15535:2023, General requirements for establishing anthropometric databases. Edition 4 (2023).

[11] Pringle R, Newell K, Huggins A. Measurement Definitions for the Tri-service Anthropometry Survey 2022/3. QINETIQ/EMEA/EO/TN220691. 22 July 2022

[12] Anthropometry Survey 2024 Percentiles Tables Master V3.0. October 2024. QINETIQ/24/03482

[13] Huggins A, Tri-Service Anthropometry Survey 2023/4 - Transition Report. 07 August 2024. QINETIQ/EMEA/AS/TR240288

[14] Cummings R. Human Factors Integration Technical Guide for Anthropometry: People Size. Version 6.0, 22nd January 2025

[15] Summers S J, Laing S N, Davidson R A, Jaffrey M A, Zhou A, Coltman C E. Do thoracoabdominal organ boundaries differ between males and females? Implications for body armour coverage and design. Applied Ergonomics 106 (2023) 103891

Recent developments of an active ventilated vest for military use

S.B. Ballak[1], B. Vos[2], E. Bakkers[2], E. Hesen[1], F. Vuijk[1], L.L. Ahsmann[1], S. Golembiewski[1], M. Catoire[1], L. Klous[1]

[1]*Netherlands Organization for Applied Scientific Research (TNO), Department of Human Performance, Unit Defense, Safety and Security, Kampweg 33, 3769 DE, Soesterberg, The Netherlands (sam.ballak@tno.nl)*
[2] *Inuteq B.V., Deventer Osnabrückstraat 30, 7418, BN, The Netherlands*

Abstract. Previous research has shown that an active ventilated vest can provide physiological and perceptual benefits regarding the thermal strain of a soldier. In the present article further design integration and technical developments of a ventilated vest worn in combination with Personal Protective Equipment (PPE) are described: step 1; technical improvements, step 2; improved functionality, step 3; comparison of subjective experience with commercially available passive ventilation vest during field study. To improve integration with the current ballistic system, the controller of the ventilated vest and its cables were simplified. Additionally, the new version of the ventilated vest was connected with the battery of the existing energy supply system. To improve and provide extra functionality intermittent ventilation was added. When using this intermittent ventilation, battery life was extended from ~10 to ~25 hours, while maintaining similar skin temperature compared to continuous ventilated modes, and lower skin temperature compared to a control condition ($P<0.05$). An additional benefit of intermittent cooling was an improved cooling sensation ($P<0.05$) compared to continuous cooling at similar fan speeds. To evaluate operational usability and perceptions for the newest version of the ventilated vest, a 5-day field study was performed during a military training in the heat (daily max. temperature: 45.2 ± 1.2°C, daily mean relative humidity: 25.5 ± 3.6%). The ventilated vest was compared to a commercially available passive ventilated vest. It was found that the active ventilated vest was perceived as more operationally usable ($P<0.04$), more comfortable ($P<0.001$), less warm ($P=0.002$), and its use resulted in a higher perception of cooling ($P<0.001$), compared to the passive vest. About 80% of the time soldiers chose full fan speed continuous ventilation during the field study. To conclude, the updated ventilated vest was subjectively evaluated as more operational usable, compared to the passive ventilated vest, especially during light intensity work, transport and breaks.

1. BACKGROUND

Military personnel are often deployed in warm and/or humid conditions. The nature of military deployment in such environmental conditions often poses high heat strain on its personnel. The combination of physical performance, personal protective clothing (PPE) [1], load carriage (together up to 50-70kg), in combination with long and unpredictable exercises [2] leads to the accumulation of this heat strain. Together, this can lead to decreased physical and cognitive performance, heath illness, and in extreme circumstances even death [3].

Therefore, heat mitigation (cooling) strategies that can effectively dissipate heat produced by the soldier in these circumstances, could improve performance and decrease discomfort, while simultaneously reducing the risk of heat illness casualties. There has been research in various occupational workers and elite athletes, where mitigation strategies have proven effective in decreasing heat strain [4,5,6]. However, because of the specific demands of military deployment (e.g. electricity availability and duration of exercise) the translation of these mitigation strategies proves to be challenging.

The body's most effective method to lose heat is by evaporation of sweat [7]. However, in a military context this method may be hampered by wearing PPE [8,9]. Therefore, for soldiers wearing PPE a mitigation method specifically targeting improvement of sweat evaporation, may prove effective in decreasing heat strain. One method that could prove effective in increasing evaporation of sweat from the skin is to allow airflow between the skin and PPE, for example by forced ventilation. The principle behind this concept is explained in Figure 1. In essence the clothing and ballistic layers are by-passed, allowing evaporation of sweat from the skin and transportation to the environment. Since this forced ventilation method requires an energy source, a key advantage is its ability to function for extended durations (i.e., several hours). The ability to function for extended durations is essential in military operations, where alternative mitigation strategies, such as phase-change material vests, have a relatively limited cooling duration [6].

https://doi.org/10.52202/081621-0021

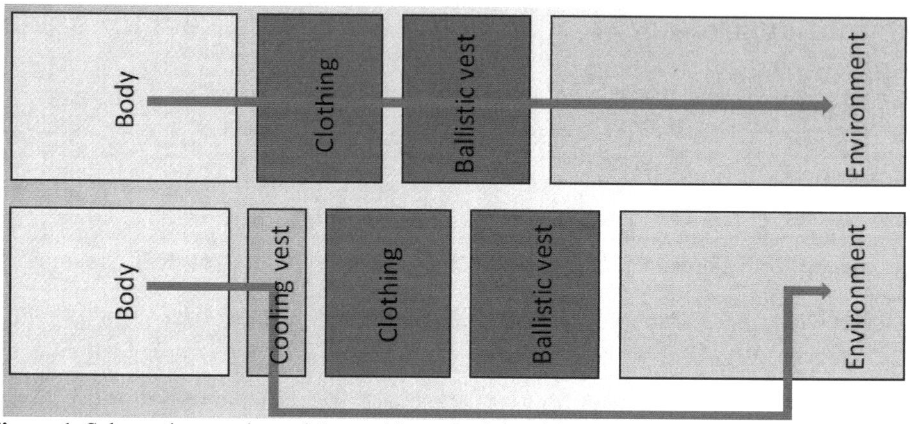

Figure 1. Schematic overview of the working principle of forcing air via the ventilated vest worn underneath the PPE system

Previous studies have published on the effectiveness of forced ventilation in decreasing heat strain in personnel wearing protective clothing [9, 10]. Glitz et al. [9] showed that wearing air-diffusing garments (in 25°C, 50% RH, performing a walking activity at 3km/h, 5% incline on a treadmill) improved performance, strain loss and thermal relief (~0.6°C lower body temperature). In addition, another cooling method 'Dry Air Comfort' (DAC), in which dry air is insufflated into the garment's microclimate, was tested previously and found effective for stationary use, because it required a flexible tube connecting personnel with an air compressor [10]. This kind of forced ventilation was also applied to helicopter pilots (unpublished data). Though this work shows that forced ventilation is effective in cooling personnel, lowering core and skin temperature, during walking exercise, in demanding thermophysiological situations, it is not a solution for a dismounted soldier due to the requirement of connection to an air compressor.

For this reason, attempts have been made to design an effective ventilated vest for the dismounted soldiers. In 2023 a ventilated vest, especially designed for the dismounted soldier of the Dutch MoD, was developed at Technology Readiness Level (TRL) 7. This was done through careful and standardized testing with a thermal manikin and climate room testing [11], however also field tests were carried out with military personnel.

The TRL 7 vest was described and illustrated in previous work from the same lab [11]. In short, the ventilation vest consists of foam parts. These parts are inserted in the front and back of an adjusted battle shirt from the Dutch MoD (Figure 2). The fabric used to adjust the battle shirt has characteristic of absorbing sweat from the skin and allowing for easy dissipation it into the air channels [11], more so than the current battle shirt. In addition, to force air through the channels four ventilators (EBMPapst, RLF35 type, St. Georgen, Hungary) were inserted in the lower area of the foam parts (Figure 2), two in the front part and two in the back part. The front and back parts are connected via electrical cables. Furthermore, in the TRL 7 iteration the foam parts were connected with two additional modules: a separate controller and an external battery. More specifications like thermal insulation and evaporation resistance were reported previously as well ([11].

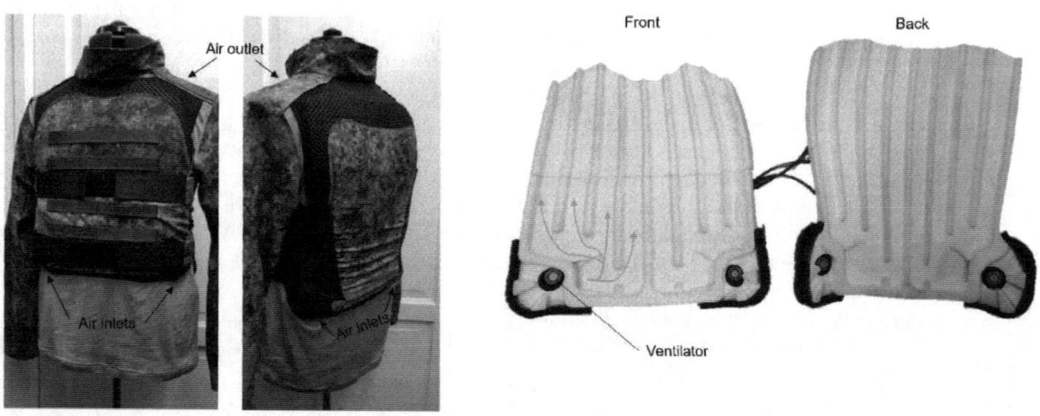

Figure 2. Illustration, description of (the inside of) the ventilation vest (previous iteration). In the left penal air inlets and outlets are shown. The right penal shows the inside of the panel with the locations of the ventilators and channels through which the air is forced, for both the front and back part. Taken with permission from [11]. Patent pending by industry partner

The present paper describes developments for further integration and optimization towards a TRL 8 of this active ventilated vest. Part of these developments are: 1) technical improvements, 2) functional improvements and 3) comparison to a commercially available ventilation vest during a field study.

2. STEPS OF DEVELOPMENT

To further develop the ventilation vest, aiming for TRL 8, prioritization of improvements were listed. Some technical issues were raised in previous field studies, mostly focusing on lessening burden for the soldier (e.g. fewer cables and improved ventilators) and the ventilated vest should be easier in use (dummy proof). This report describes both the technical and functional developments as well as the field testing conducted to reach TRL 8. Confidentiality of the (ballistic) testing facilities and about technical information prohibits communication of some of the protocol and results in full detail. If more details are warranted, contact with the corresponding author can be made.

2.1 Technical developments

2.1.1. Ballistic testing

The ventilated vest is meant to be worn underneath PPE, but it is not acceptable that the addition of the ventilated vest would decrease the functionality of PPE. Therefore, it was imperative to test the ballistic properties of the PPE in combination with ventilated vest. Ballistic testing was performed at the ballistic testing facility of TNO. The primary research question was whether wearing the ventilation vest would influence the protection and function of the PPE. With that goal in mind a series of testing was conducting using different ammunition, different PPE configuration with and without the ventilation vest, and hitting the PPE at sites were both the foam parts and the ventilator parts of ventilation vests were located. The exact protocol and results are classified, however the main results can be provided here.

The addition of the ventilated vest underneath the PPE configuration worn by the Dutch MoD resulted in a comparable or slightly increased ballistic resistance and Behind Armour Blunt Trauma (BABT) as measured with the 'V_{50} method', when hit by different categories of ammunition. Although bullets directly aimed at the region of the ventilators are stopped, direct impact on the location of the ventilators may increase the risk of (non-lethal) secondary injury. However, the probability of this happening is considered very small. Therefore, ballistic protection remains similar (or is even slightly increased) while wearing the ventilated vest underneath the PPE configuration.

2.1.2. Hardware changes

During previous field testing military personnel provided feedback about the ventilators. Its sound was supposedly too loud (~60 dB at 3cm distance at full fan speed) and its efficiency (battery life of 6-8 hours while during full use) needed to be improved for extended use in the military context. Therefore, the ventilators type RLF35 used in the previous iteration were replaced by a new generation of the same type, which lead to a power consumption reduction of 30-40% when set at similar rotational velocity, as well as a reduction of the high-frequency sound (~10 dB at 3cm both at full and lower percentage fan speed). In addition, an extra coating was added for protection to water and salts (sweat).

In previous iterations, cables were not fully hidden, and therefore soldiers provided feedback that the cables of the ventilated vest were an annoyance or could even be a safety hazard (field study Curaçao 2023). Therefore, in the current iteration, cables were minimized as the stand-alone controller was now integrated in the cable, decreasing extra cables (see Figure 3). Also, loose cables were managed by adding appropriate holes in the fabrics.

https://doi.org/10.52202/081621-0021

Figure 3. Picture of the ventilated vest worn underneath the PPE system. In this picture it can be seen that loose hanging cables are minimized. The arrow points towards the new controller, incorporated in the cable connecting the battery

As mentioned above, another technical development was the integration of the controller in the cable. In previous iterations of the ventilated vest a stand-alone controller was part of the system. However, after feedback on the complexity of the system and the wish for dummy proof use, this new model was introduced. In addition, the controller was directly linked to the battery (Figure 3).

A development which was very important for the Dutch MoD was the integrating of the ventilated vest with the battery of the PPE system used by the MoD (Bren-Tronics, Li-ION battery). The realization of this development improved the functionality of the ventilated vest in multiple ways. As no external battery was needed anymore, it simplified the system. Additionally, by using the MoD's own power source it is anticipated that logistical problems are less likely to occur. A possible disadvantage is that the PPE-battery is used for more purposes and may therefore run out of energy earlier. This may increase the need for battery life management. This issue was addressed by performing tests on power usage of the ventilated vest. Depending on the condition used the battery life of the ventilated vest was between 11 hours (full fan speed) to 64 hours (40% of fan speed). The need to use the ventilated vest on full fan speed is only during specific moments, e.g. highly physical activities, therefore it is anticipated that the current battery life using the ventilated vest is sufficient for operational use. This however, remains to be confirmed.

2.2 Functional developments

One of the most important goals was to improve the operational useability of the ventilated vest. Therefore, it was important to improve battery life (as described above), but also to reduce (subjective) heat strain, to increase acceptance of the system by military personnel. Physiological effectiveness of the ventilated vest was shown previously by reducing core and skin temperature by -0.3±0.2 ˚C and -1.1±1.3 ˚C in 32 ˚C and 33 % relative humidity, 5 km/h walking speed [11]. Additionally, a climate room study (Soesterberg, The Netherlands) was performed to test the effect of the cooling method (thermal alliesthesia, i.e., change of perception depending on the body's thermal state) on cooling sensation and skin temperature of the ventilation vest [12]. The different cooling conditions used in that study were 70% and 100% continuous fan speed and 70% intermittent fan speed (alternating fan speed during cycles of 30 sec).

From that study it was concluded that the ventilated vest caused a reduction in subjective thermal strain. More specifically, the intermittent condition seemed to have an additional benefit on cooling sensation (the sense of cooling, which is scored from none to extremely high), compared to the other two continuous ventilated conditions. For thermal sensation (which is scored from cold to hot), the additional benefit of intermittent was not observed. Additionally, while the intermittent version used less power (18 hours vs. 11 hours at full fan speed, data not shown) and blew less air through the vests, it still resulted in similar scapula skin temperature under the vest as the other two ventilated vest conditions, which was lower than the control conditions. Therefore, for the further development of the ventilated vest towards TRL 8, it was concluded that the intermittent functionality was beneficial concerning reducing subjective thermal strain.

2.2 Comparison to a commercially available product

A field study was conducted in Senegal, Africa. The main objective of this field study was to test and assess the operational usability of the ventilation vest during semi-controlled military conditions. In addition, a second aim was to compare its functionality with a commercially available passive ventilation system. Claims of that passive vest regarding cooling were similar or even better than the results of the recently developed active ventilation vest [11]. Therefore, it was of interest to compare the two ventilation vests on subjective cooling, thermal comfort and sensation (Figure 4).

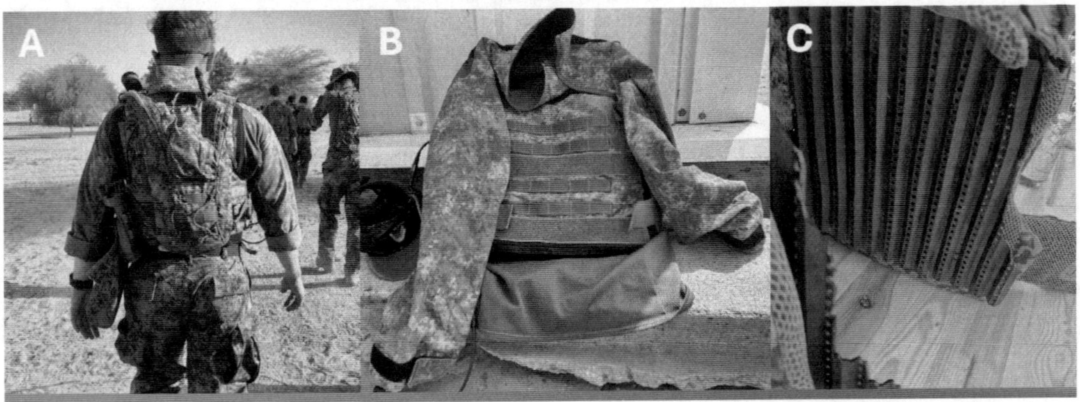

Figure 4. Example of active ventilated vest worn (A) and stand-alone (B) and the passive commercially available product on the inside of the ballistic configuration (C)

The protocol was approved by the local ethical committee of TNO (2024-040). To this end 15 male subjects of the regiment were included (43rd Mechanized Brigade). Due to practical issues outside of the researchers control a total of 24 observations were completed (13 with the active ventilated vest and 11 with the commercially available product) over 4 days. An additional group of 15 subjects of the regiment was included as a control group, This group did not wear the ventilation vests, only PPE, during identical exercises.

At the end of each day the passive and active ventilation vests were distributed among the subjects. Additionally, the configuration and set-up of both the ventilation vests and PPE were visually checked. At the end of the testing day subjects filled in a questionnaire which consisted of 27 questions, evaluating specific parts of functionalities, thermal comfort scored from 0 (comfortable) to 4 (extremely uncomfortable) (ISO10551, 2001), thermal sensation scored from -3 (cold) to +3 (hot) (ISO10551, 2001), cooling sensation scored from 0 (none) to 6 (extremely high) (TNO), and overall rating of functionality/usability scored from 1 (extremely bad) to 7 (extremely good) [13].

Training sessions performed during the measurement period were a speed march, static shooting, and active shooting, ranging from 1.5 to 4.5 hours. The maximum air temperature during the measuring period was between 43.6-47.0 °C, relative humidity ranged between 22-31.8%, and average UV index was 9.7.

Thermal sensation was significantly lower ($P<0.05$) in the (active) ventilation vest (1.9 ± 0.6), compared to the commercially available vest (2.8 ± 0.4, Figure 5A). In addition, the control group, only wearing PPE, was asked the same questions, during the same training. The thermal sensation of the active ventilation group was also significantly lowered compared to the thermal sensation scored by the control group (2.6 ± 0.4, $P<0.05$). In addition, also the thermal comfort (Figure 5B) and cooling sensation (Figure 5C) were significantly improved by the ventilation vest (1.0 ± 0.4 and 3.5 ± 0.7) compared to the commercially available product (2.4 ± 1.1 and 0.8 ± 0.9 ,$P<0.05$).

https://doi.org/10.52202/081621-0021

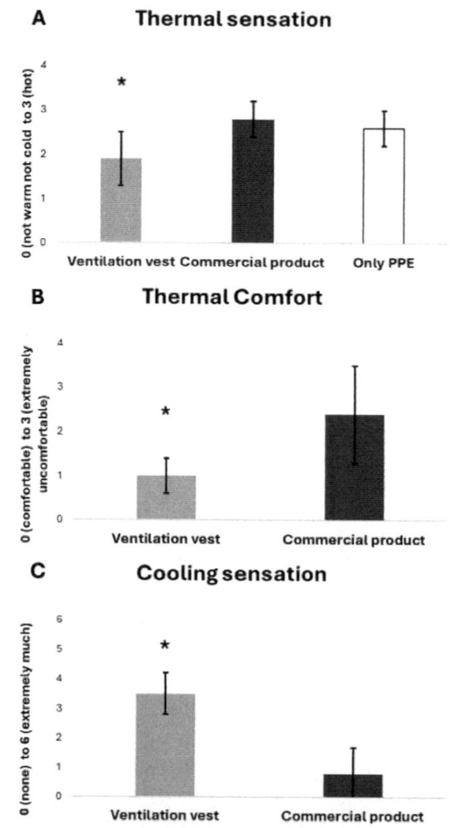

Figure 5. Thermal sensation (A) for the ventilation vest compared to the commercially available product and a control group wearing only PPE (* = P<0.05). Also (B) thermal comfort and (C) cooling sensation were significantly improved by the ventilation vest compared to the commercially available product (* = P<0.05)

Additionally, while wearing the ventilation vest, the evaluation on 'usability of vests for military tasks', was also significantly improved, compared to the commercially available product (Figure 6, * = P<0.05).

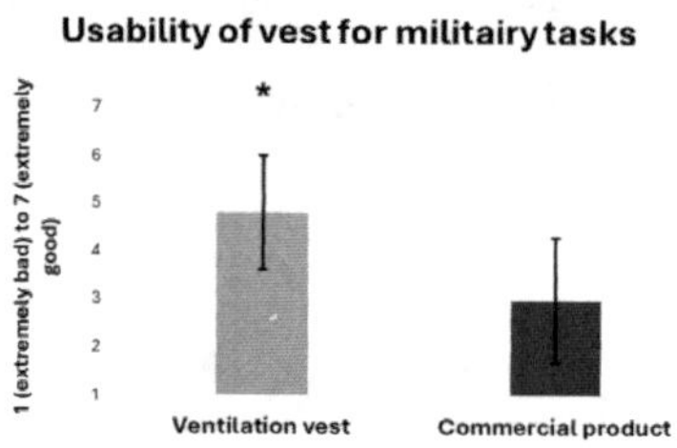

Figure 6. Usability of the vests was evaluated. The (active) ventilation vest was scored significantly higher than the commercially available product (* = P<0.05)

During the training the subjects were free to use the active ventilated vests in the modality they preferred (continuously or intermittent, full fan speed - 100% - or a lower percentage). Interestingly, during about 80% of the observations the subjects used the continuous, full fan speed setting of the active ventilation vest. This possibly means that there is a difference between cooling sensation (see intermittent developments) and the overall preference of the modality. Therefore, more evaluations specifically on the perception of the different ventilation vest modalities in the field is recommended for the near future.

Since the training was performed in low relative humidity levels (22.0-31.8%), the subjects were informed about the possibility of 'pre-wetting' the vests, to support the natural wetting of the fabrics by sweat. This was done a lot (though not systematically noted) with both vests by the subjects. Especially in the active ventilated vests this was perceived as an immediate cooling or relief.

Therefore, it can be concluded that the ventilated vest improved thermal sensation over wearing PPE only in the field and was scored useful by the subjects. In addition, on thermal sensation, comfort and cooling sensation the (active) ventilated vest scored significantly better than the commercially available product, which works on passive ventilation. Hence, the developed active ventilated vest provide a (subjective) advantage during operational use in the heat.

3. FUTURE DEVELOPMENTS AND CONCLUSION

This project was designed to produce a functional and usable cooling solution for the dismounted soldier wearing PPE. In addition to previous work [14, 15], the specific goal for this project was to better integrate the system with the PPE system used by the Dutch MoD and to test it in a realistic setting.

Hence, with both technical and functional updates of the active ventilated vest abovementioned goal was met. This cooling solution is not meant to cool aggressive and quickly for a short time, but moderate for a long period needed for operational use (>8 hours). Results from climate room testing and field tests, support this.

The product is effective in cooling military personnel during operational use in the heat (measured both physiologically as well as subjectively) and its ballistic protection is at least comparable to wearing only PPE. In addition, technical improvements have made the ventilation vest less complex and more integrated in the PPE system currently used by the Dutch MoD. The product was also evaluated during field testing (both in Curaçao and Senegal), and assessed as usable by the subjects. In addition, the technical and functional improvements of the current iteration have led to a product at a TRL 8. Obviously small improvements can still be made (e.g. positioning of cables and interaction with shooting position of gun). Other developments foreseen for the near future include creating suitable and usable cooling solutions, using the same technology, for other parts of the MoD.

In conclusion, the developments and positive evaluation of the active ventilated vest have demonstrated its effectiveness in reducing heat strain for dismounted soldiers wearing PPE in warm conditions.

REFERENCES

[1] Hunt A.P. Potter A.W. Linnane D.M. Xu X. Patterson M.J. Stewart I.B. Heat stress management in the military: Wet-bulb globe temperature offsets for modern body armor systems. Human Factors, Vol. 64(8) (2021) 1306-1316.

[2] Ashworth E.T. Cotter J.D. Kilding A.E. Methods for improving thermal tolerance in military personnel prior to deployment. Military Medical Research (2020), 7:58.

[3] Periard J.D Degroot D. Jay O. Exertional heat stroke in sport and the military: epidemiology and mitigation. Experimental Physiology, 107 (2022): 1111-1121.

[4] Bongers C.C.W.G. Hopman M.T.E. Eijsvogels T.M.H. Cooling interventions for athletes: an overview of effectiveness, physiological mechanisms, and practical considerations. Temperature Vol 4 (2017) No. 1, 60-78.

[5] Ciuha U. Valencic T. Mekjavic I.B. Cooling efficiency of vests with different cooling concepts over 8-hours trials. Ergonomics 64 (5) (2021): 625-639.

[6] Filingeri D. Fournet D. Hodder S. Havenith G. Mild evaporative cooling applied to the torso provides thermoregulatory benefits during running in the heat. Scand J Med Sci Sports. Jun 25 (2015) suppl 1: 200-10.

[7] Gagnon, D. Crandall C. Handbook of Clinical Neurology; Chapter 13 Sweating as a heat loss thermoeffector. (Elsevier 2018) volume 156, pages 211-232.

[8] Larsen B. Netto K. Aisbett B. The effect of body armor on performance, thermal stress, and exertion: A critical review. Military Medicine (2011) 176(11), 1265-1273.

[9] Glitz K.J. Seibel U. Rohde U. Gorges W. Witzi A. Piekarksi C. Leyk D. Reducing heat stress under thermal insulation in protective clothing: microclimate cooling by a 'physiological' method. Ergonomics (2015) 58(8), 1461-1469.

[10] Glitz K.J. Seibel U. Freitag S Rohde U. Gorges W. Witzki A. Piekarski C. Leyk D. Interval microclimate cooling with the Dry Air Comfort (DAC) method: an effective strategy to sustain evaporative heat loss. JSAMS (2017) Vol 20 S2 S45.

 https://doi.org/10.52202/081621-0021

[11] Linssen L., Klous L., Reinten J., Psikuta A., Catoire M., Kingma B. Impact of a ventilated vest on cardiac recovery and thermal perception of soldiers during an intermittent activity in a warm environment. 2025. *Manuscript accepted for publication in Journal of Biometeorology*.

[12] Hesen E., Catoire M., Ballak S., van der Hout L., Bastiaansen B, Bakkers E., Intermittent cooling of a ventilated vest decreases subjective thermal load. ICEE conference Jeju 2024.

[13] Likert R. A technique for the measurement of attitudes. (1932) Archives of Psychology, 22 140, 55.

[14] Kingma B., Impact assessment of load bearing vests, combat armour and ventilated vest configurations on thermal strain. PASS conference Dresden 2023.

[15] Linssen L., Klous L., Ahsmann l., Reinten, J., Ballak S., Vos B., Bakkers E., Kingma B., Potential impact of a ventilated vest on perception, physiology and cognition of soldiers in controlled and uncontrolled warm environments. ICSPP conference London 2023.

A Machine Learning Model to Optimize Bomb Suit Size Selection

J.P. Dionne, J. Levine, M. Keown, and A. Makris

Med-Eng Holdings ULC, 2400 St. Laurent Blvd, Ottawa, Ontario, Canada, K1G 6C4
jean-philippe.dionne@safariland.com

Abstract. Given their high complexity and cost, bomb suits are often shared among bomb squads or military units. For any given suit size, personalized fit is possible through various adjustment systems, to accommodate a wide range of body sizes and types. While sizing tables based on height and weight are provided by manufacturers to facilitate size selection, variances in body shapes, or morphotype, are such that sizing charts do not always provide the best sizing recommendation. A total of 177 volunteers were sized (height and weight) and then fit in the most appropriate size of bomb suit by a sizing expert, taking wearer preferences into account. The predictions from the standard sizing chart were first compared with the users' preferences, with a relatively low success rate (68%). A decision tree algorithm was then applied to optimize the predictions based on solely height and weight, yielding an improved success rate of 85%. Considering that bomb suit size selection is governed by more than just height and weight, a smartphone application was then used to generate an avatar and estimate 83 body measurements, based on two photos from each volunteer (front and profile views). The output from this app was then fed into a multinomial regression machine learning model to provide suit size recommendations that were more precise than those obtained from the basic sizing chart (prediction success rate of 92%). The multinomial regression model was thus found to outperform both the standard legacy sizing chart and the decision tree that had been based on solely the height and weight of an individual. In addition, the multinomial regression model provided independent predictions for the trousers and jacket, whereas the legacy chart was developed to predict the overall suit size.

1. INTRODUCTION

Personal Protective Equipment (PPE) for military applications is typically designed to accommodate the widest possible range of male and female personnel, while minimizing the number of sizes [1]. This approach balances substantial development costs and logistical challenges, which are particularly significant for military use. These cost challenges are further amplified in the case of bomb suits due to the high expense of moulds for hard armour and the lower production volumes for such specialized equipment compared to standard soldier PPE.

Bomb suits are rarely issued as personal items and are instead often shared among multiple operators. This shared usage necessitates a broader fit for each suit size to accommodate a wider range of users. The US National Institute of Justice (NIJ), through its NIJ 0117.01 standard for public safety bomb suits [2], mandates that manufacturers provide at least three distinct sizes to accommodate users, while specifying maximum weights for each suit size.

Bomb suits are more complex than traditional military fragmentation vests, often consisting of three main components: a helmet, jacket, and trousers. These elements must not only fit properly but also function seamlessly together. To accommodate diverse body types within a limited number of sizes, bomb suits include adjustable features like hook-and-loop closures, straps, and other attachments. Despite these features, some bomb technicians may not fit into any standard size. While customized sizing for such outliers is theoretically possible, as discussed by Flynn et al. [3], the economic feasibility of this solution is often limited, leaving some technicians without adequately fitted PPE.

Manufacturers typically provide sizing charts based on height and weight to assist users in selecting an appropriate size. However, these two parameters fail to capture significant variations in body shapes. Individuals with the same height and weight may have vastly different body dimensions and require different suit sizes. For example, Kozey et al. [4] found that limb lengths and circumferences play a critical role in proper sizing for immersion suits used by law enforcement and the military. Consequently, sizing charts based solely on height and weight might lead to suboptimal recommendations.

As Choi et al. [1] emphasize, PPE development should address sizing issues early in the design process to ensure proper functionality and fit [5]. Their proposed approach includes identifying design problems related to the item's concept, function, and fit based on anthropometric data, followed by iterative prototype development and testing. The final step involves producing products in all necessary sizes, along with a comprehensive sizing chart and tariff. This process aims to achieve high accommodation rates for both males and females while minimizing the number of sizes.

https://doi.org/10.52202/081621-0022

Figure 1. An NIJ-certified bomb suit ensemble from Med-Eng

However, not all products are developed using this systematic approach. In such cases, sizing charts must be refined retrospectively. This study focuses on a Med-Eng (Ottawa, Canada), bomb suit (Figure 1), an NIJ 0117.01 certified system which is currently offered in four predefined sizes. The objective is to develop optimized sizing tools for this bomb suit using anthropometric measurements and sizing trials from a large sample of volunteers and applying classification algorithms. The analysis will focus on the suit's jacket and trousers, excluding the helmet.

2. METHODS

A fitting exercise was conducted for the bomb suit, involving 177 volunteers whose height and weight were measured (Figure 2). Under the guidance of a bomb suit sizing expert, the best-fitting sizes for the bomb suit jacket and trousers were determined for each participant through dressing trials. The trials also assessed whether individuals would benefit from the trousers' "leg expanders" feature (Figure 3) to account for larger waist and lower body dimensions (larger leg diameter).

In addition to manual height and weight measurements, further sizing data were collected using an Android-based sizing application, XpertFit (Safariland, Jacksonville, FL). This app captures two photos of each participant (frontal and profile views) while they wear tight-fitting clothing. After height and weight are entered manually, the app's algorithm generates an avatar (Figure 4) and estimates 83 body measurements, including arm length, chest circumference, thigh length, and more.

Initially, the preferred user sizes obtained from dressing trials were compared with the manufacturer's existing sizing chart (Figure 5). Using the preferred sizes, an improved sizing chart was created through a decision tree model. The CART decision tree algorithm, as described by Duda et al. [6], was implemented in MATLAB and used height and weight as the sole input variables. A separate decision tree was constructed to determine the need for leg expanders, also based exclusively on height and weight.

Recognizing that bomb suit sizing might depend on more than just height and weight, a multinomial regression machine learning model inspired by Dionne [7] was developed. This model utilized the 83 body measurements collected via the XpertFit app. The model generates a linear equation with weighted coefficients for each measurement, optimizing the fit selection based on the preferred suit sizes identified during the dressing trials.

To evaluate the accuracy of the original sizing chart, decision trees, and the multinomial regression model, confusion matrices were generated. These matrices compared the preferred sizes identified during the trials with the sizes recommended by each model, providing a quantitative measure of prediction performance. While the analysis could be extended to provide separate recommendations for the jacket and trousers, only an overall bomb suit size will be discussed here.

Figure 2. Fitting exercise of the bomb suit

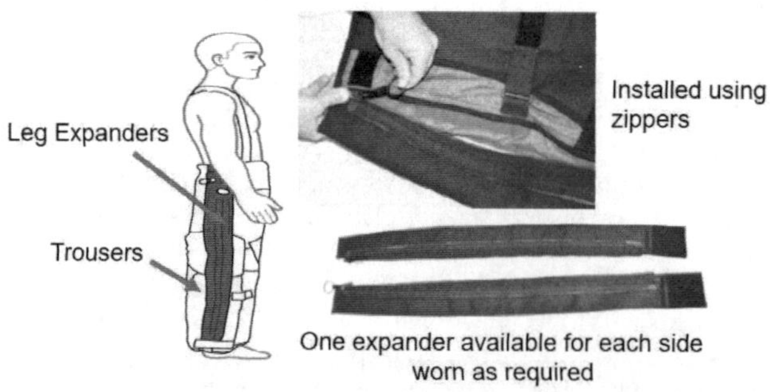

Figure 3. The leg expander sizing feature from the bomb suit, worn on both legs, adds 10 cm to the circumference of each leg

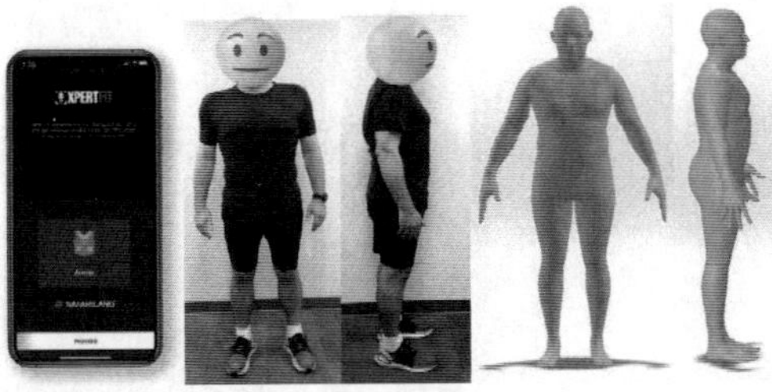

Figure 4. Photos and corresponding avatars from the XpertFit Android application

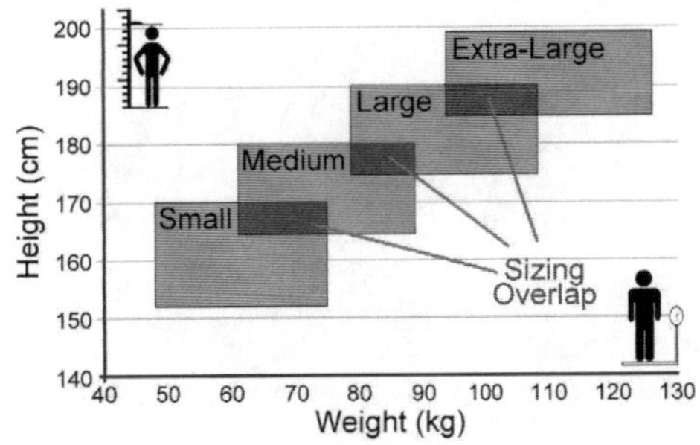

Figure 5. Original sizing chart for the bomb suit as provided by the manufacturer

3. RESULTS

The best-fitting options (preferences) identified by the 177 volunteers are presented in Figure 6, categorized by suit size selection and the use of leg expanders. The "X" marks on the graph denote cases where volunteers could not properly don any suit, based on evaluations by a bomb suit sizing expert. The suit size preferences shown in Figure 6 were compared to the original sizing chart provided by the manufacturer (Figure 5) to evaluate the chart's effectiveness in predicting the best fit. These comparisons are summarized in Figure 7 using a confusion matrix. The matrix displays the actual best-fit results across each row and the manufacturer's sizing chart predictions across each column. Green cells represent correct predictions by the sizing chart, while pink cells indicate incorrect predictions.

Performance metrics are provided for additional clarity. The percentages listed in the last column of the confusion matrix represent "recall," which reflects the proportion of volunteers correctly classified for each size. For example, only 63% of the Small volunteers were correctly classified (24 correct classifications out of 38 total Small volunteers: 24+4+10). Similarly, the percentages in the last row indicate "precision," representing the accuracy of the sizing chart's predictions. For instance, 69% of the predictions for Extra-Large were correct (9 accurate predictions out of 13 total predictions: 4+9).

The overall performance of the sizing chart was 68%, as indicated in the bottom-right corner of the matrix. This value was calculated by dividing the total number of correct predictions (green cells) by the overall number of volunteers (177).

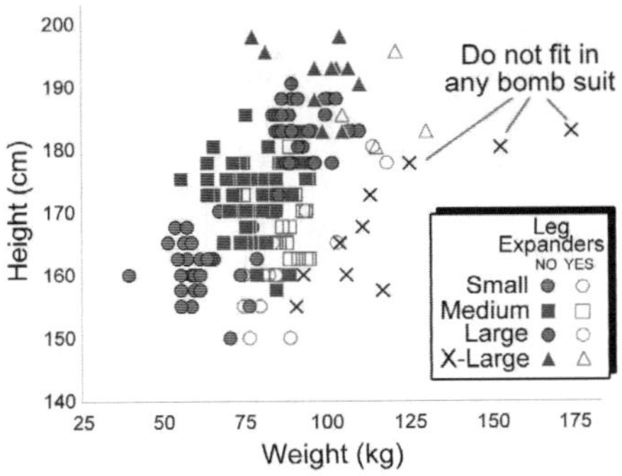

Figure 6. User preferences for suit sizes and the use of leg expanders for the 177 volunteers

ACTUAL BEST FIT	PREDICTED - SIZING CHART: Small	Medium	Large	X-Large	No-Fit	Recall ↓
Small	24	4			10	63%
Medium		44	14		13	62%
Large		1	34	4	4	79%
X-Large			2	9	4	60%
No-Fit					10	100
Precision →	100%	90%	68%	69%	24%	68%

Figure 7. Confusion matrix: 177 users vs. the original sizing chart from the manufacturer

In the next step, the results from Figure 6 were input into the MATLAB CART algorithm to construct the most accurate decision tree for bomb suit sizing. Some pruning of the tree (removal of lower-level branches) was applied to balance performance and clarity, as shown in Figure 8. The data from each of the 177 volunteers were then passed through the tree to generate predictions. As with the evaluation of the manufacturer's sizing chart, a confusion matrix was created (Figure 9) to compare the predictions from the decision tree with the volunteers' preferences. The decision tree achieved an overall accuracy of 85%, significantly outperforming the 68% accuracy of the manufacturer's sizing chart. It is also worth noting that the precision of the "No-Fit" cases improved substantially, rising from 24% in Figure 7 (with the sizing chart) to 60% in Figure 9 (with the decision tree).

https://doi.org/10.52202/081621-0022

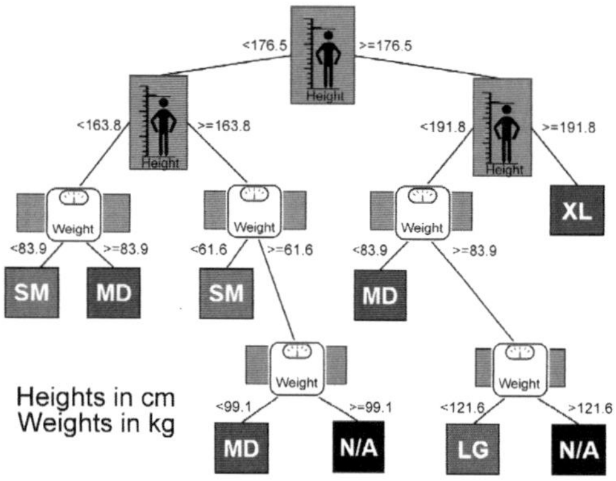

Figure 8. Decision tree for the bomb suit size generated based on height and weight only for the 177 volunteers

ACTUAL BEST FIT	PREDICTED - DECISION TREE: Small	Medium	Large	X-Large	No-Fit	Recall
Small	32	2				94%
Medium	6	63	2		4	84%
Large		6	40	4		80%
X-Large				10		100%
No-Fit			1	1	6	75%
Precision →	84%	89%	93%	67%	60%	85%

Figure 9. Confusion matrix: 177 users vs. the bomb suit size decision tree

While the manufacturer's sizing chart (Figure 5) does not include any recommendations for the use of leg expanders, a decision tree incorporating the data from the user's preferences (Figure 6) was built to provide such recommendations. Once again, the MATLAB CART algorithm was applied, and the resulting decision tree is shown on Figure 10. It must be noted that this decision tree can only be built once the decision tree for the bomb suit size has been built, since some decision steps in the leg expander decision tree depend on the suit size selected. Figure 11 then presents the confusion matrix specific to the leg expander recommendations. A relatively low recall performance was observed for the cases where users felt more comfortable with leg expanders. Indeed, in 13 cases, the decision tree failed to arrive at the same outcome. Despite this, a high overall accuracy of 90% was obtained for this decision tree.

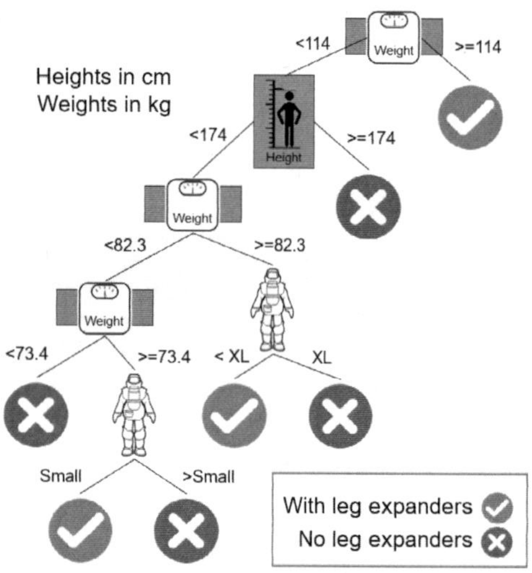

Figure 10. Decision tree for the use of leg expanders generated based on height and weight only for the 177 volunteers

Leg Expanders — ACTUAL BEST FIT	PREDICTED DECISION TREE: No	Yes	Recall
No	124	5	96%
Yes	13	35	73%
Precision →	91%	88%	90%

Figure 11. Confusion matrix: 177 users vs. the leg expanders decision tree

An improved sizing chart can be created by integrating the volunteers' preferences for bomb suit size (decision tree from Figure 8) and the use of leg expanders (decision tree from Figure 10). Figure 12 illustrates the data from all 177 individuals, plotted against the size boundaries predicted by the two decision trees. Shaded areas indicate cases where the use of leg expanders is recommended. This figure provides a clear graphical representation of the overall prediction accuracy, effectively highlighting both correct and incorrect predictions based on the placement of each data point on the chart.

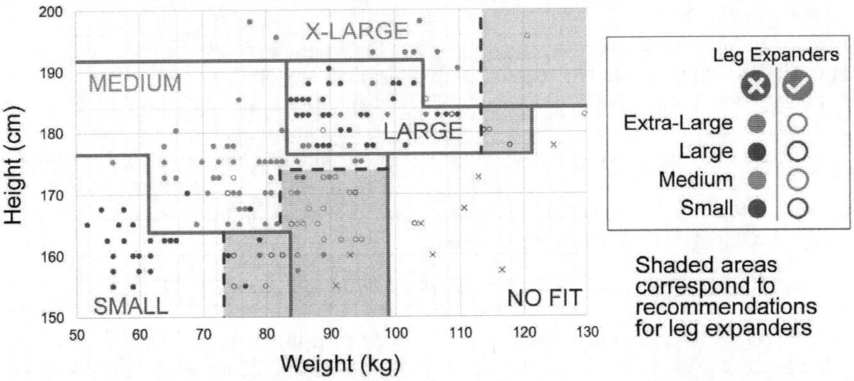

Figure 12. Sizing chart combining bomb suit size and leg expanders, derived directly from the two decision trees – individual preferred sizes displayed

The sizing chart in Figure 12 highlights a notable limitation: the area corresponding to individuals taller than 185 cm and weighing less than 75 kg (top left) contains no data points. This absence is likely due to the rarity of such individuals (very tall and slim) in the general population. Consequently, predictions for this region are unreliable and would likely fail to align with actual data, should such individuals exist. It is also doubtful that all individuals with these characteristics could properly fit into a bomb suit. Furthermore, this gap causes the sizing chart to predict a jump directly from Medium to Extra-Large for tall individuals weighing less than approximately 85 kg, bypassing the Large size entirely. Additionally, Figure 12 shows volunteers who were deemed unable to fit properly into a bomb suit by the bomb suit sizing expert (marked as "x" in Figure 12) but were incorrectly placed in the "Medium with leg expanders" area. These individuals were too large for a Medium suit yet not tall enough to fit into a Large suit. This observation indicates the need for adjustments in this area of the chart.

To overcome these shortcomings, a revised sizing chart was developed, incorporating targeted manual adjustments to address the identified weaknesses in the automatically generated version. Two key modifications were implemented: first, subtle changes were introduced to better accommodate tall, slim individuals as well as shorter, broader individuals; second, boundary lines were simplified to enhance clarity and readability. Notably, these adjustments were applied to sizing regions with minimal or no collected data points. The refined chart, shown in Figure 13, excludes the explicit display of the 177 individual data points for improved visual simplicity.

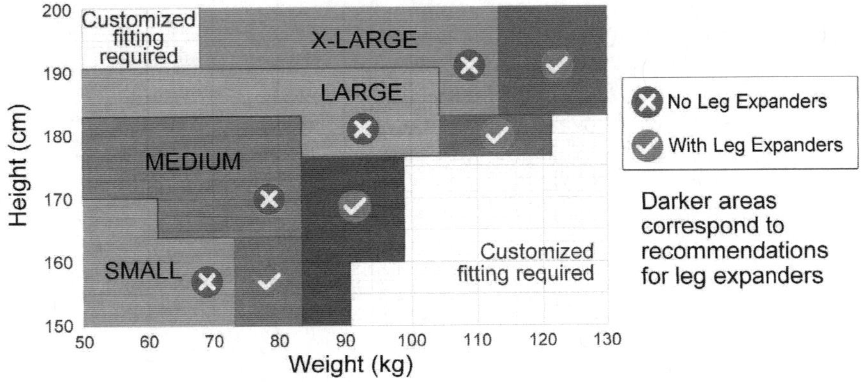

Figure 13. Sizing chart, revised and simplified to introduce regions requiring customized user fitting

https://doi.org/10.52202/081621-0022

But while the revised sizing chart (Figure 13) is much more effective at predicting bomb suit size, while at the same time offering recommendations for the usage of leg expanders, it must be kept in mind that it only takes height or weight of individuals into consideration. As other body measurements can play a major role in determining sizing, a different approach incorporating the 83 measurements provided by the XpertFit app was then followed. A machine learning algorithm, proposed by Dionne [7] was applied to the entire data set. This algorithm consisted in considering five binary linear logistic regressions, each corresponding to a specific bomb suit size (including the "No Fit" case). These five equations are respectively used to answer the following five questions:

- Is the individual more likely to be size SMALL than any other size?
- Is the individual more likely to be size MEDIUM than any other size?
- Is the individual more likely to be size LARGE than any other size?
- Is the individual more likely to be size X-LARGE than any other size?
- Is the individual more likely to not fit into any bomb suit, as opposed to be able to fit?

In each case, a linear equation (Equation 1) is generated by associating a multiplying factor to each of the 83 body measurements from the XpertFit app:

$$F = f_0 + f_1 \cdot x_1 + f_2 \cdot x_2 + (\dots) + f_{83} \cdot x_{83} \quad \dots \qquad \text{............(1)}$$

where each x corresponds to the value of a body measurement, and each f is a multiplying factor. Such a linear equation is built for each of the five cases highlighted above. The various factors are then optimized to offer the best possible accuracy in size predictions, when compared with the preferences from the 177 individuals. A sigmoid function (Equation 2) is then applied to the results of the five linear equations:

$$\sigma(x) = \frac{1}{1 + e^{-x}} \qquad (2)$$

This sigmoid function results in a value between 0 and 1. When the value is below a threshold of 0.5, it is generally considered that the statement is true, otherwise it is considered as false. But for this specific problem, the highest value from the five sigmoid functions was sought. The size corresponding to this highest value was then considered as the most likely size for the individual.

Size predictions were generated using the machine learning algorithm for all 177 individuals, and the results were compared to their preferred sizes in the form of a confusion matrix (Figure 14). This approach achieved an overall accuracy of 92%, outperforming the decision tree method, which had an accuracy of 85% (Figure 9). Notably, both the recall and precision for the "No Fit" case reached a perfect 100%, a significant improvement over the decision tree approach, which achieved recall and precision values of 75% and 60%, respectively.

The same machine learning algorithm proposed by Dionne [7], utilizing Equations 1 and 2, was also applied to predict whether volunteers required the use of leg expanders. This analysis produced another confusion matrix (Figure 15), with an overall accuracy of 92%. This is slightly higher than the 90% accuracy obtained using the CART decision tree.

ACTUAL BEST FIT	PREDICTED - XPERT FIT					Recall
	Small	Medium	Large	X-Large	No-Fit	
Small	37	1				97%
Medium		67	4			94%
Large		4	38	1		88%
X-Large			4	11		73%
No-Fit					10	100
Precision →	100%	93%	83%	92%	100%	92%

Figure 14. Confusion matrix: 177 users vs. the bomb suit size Multinomial Linear Regression based on the XpertFit data

Leg Expanders	PREDICTED MULTINOMIAL REG		Recall
ACTUAL BEST FIT	No	Yes	
No	130	6	96%
Yes	9	31	78%
Precision →	94%	84%	92%

Figure 15. Confusion matrix: 177 users vs. the Leg Expanders Multinomial Linear Regression based on the XpertFit data

4. DISCUSSION

This study was initiated in response to some issues observed with the existing manufacturer's sizing chart for the bomb suit ensemble, with the goal of improving sizing recommendations. The original sizing chart (Figure 5) was developed without extensive trials involving volunteers, relying instead on approximations and common sense. Additionally, the original chart included overlapping areas, which caused confusion among end users whose height and weight fell within these ambiguous zones.

The experimental bomb suit fitting trials involving 177 volunteers confirmed that the legacy sizing chart could be improved (accuracy of 68%). Using the height and weight of the volunteers, along with their known size preferences, a decision tree model was developed, resulting in a significantly improved sizing chart with an accuracy of 85% (Figure 13). Unlike the legacy chart, this new sizing chart also provided recommendations on the use of leg expanders. Furthermore, a machine learning approach based on multiple linear logistic regressions achieved an even higher overall accuracy of 92% for both suit size and leg expander recommendations. These results are summarized in Figure 16.

Based on these findings, the logistic regression approach appears to be the most accurate and should be prioritized. However, the decision tree approach remains valuable due to its simplicity and usability. The decision tree allows for a clear graphical representation, enabling end users to quickly determine their recommended size—a practical feature for inclusion in product specifications. In contrast, the logistic regression model, which relies on 83 body measurements, does not lend itself to the creation of a simple sizing chart. Moreover, access to the XpertFit app is required to generate these measurements, a tool that may not be available to all users. By contrast, height and weight remain the most accessible and universally known measurements, making the decision tree approach highly practical despite its slightly lower accuracy. Both approaches are therefore valuable.

This study focused on the bomb suit ensemble as a whole (jacket and trousers), but the same methodology could be applied to provide separate sizing recommendations for jackets and trousers. This could enable "mix-and-match" sizing for individuals, a direction that warrants exploration in future research. In the case of decision trees, variables other than height and weight could be selected.

The accuracies reported in this study for both the decision tree and multinomial regression approaches could be further improved with a larger sample size of volunteers. Machine learning models generally benefit from larger datasets, and increasing the number of participants beyond the 177 considered here could enhance the predictive performance. Additionally, this study used static body measurements, including height, weight, and other parameters calculated through the XpertFit app. Future research could incorporate "dynamic anthropometry," as suggested by Avadenei et al. [8], by analysing volunteers' movements and conducting representative motions during the fitting process.

Finally, it is important to emphasize that this study focused on fitting individuals to pre-existing, predefined bomb suit sizes. For the development of future bomb suits, a more advanced approach can be adopted from the outset, following the Sustainable Product Evaluation, Engineering, and Design (SPEED) process described by Robinette et al. [9]. This approach will incorporate iterative sizing loops throughout the design process. Additionally, unsupervised learning techniques will be employed to better define user groups and establish optimal sizing categories for future bomb suits.

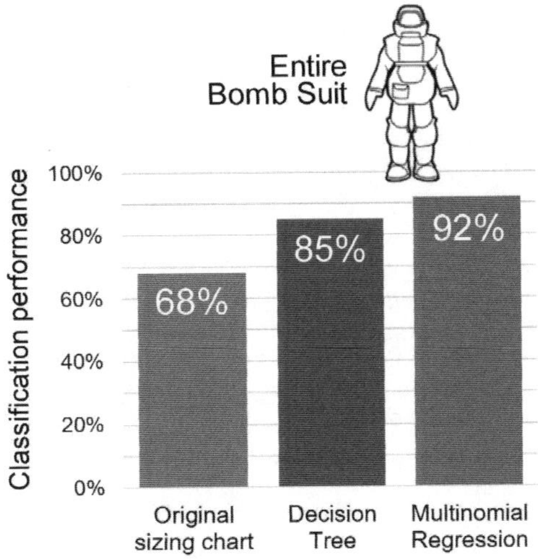

Figure 16. Comparison of the overall classification performance for the entire bomb suit for the three cases considered: original sizing chart, decision tree, and multinomial regression

https://doi.org/10.52202/081621-0022

5. CONCLUSION

This study demonstrated that bomb suit sizing accuracy can be significantly improved through data-driven approaches. The original sizing chart, while useful, achieved limited predictive performance. By applying a decision tree based on height and weight, accuracy improved substantially. A further enhancement was achieved using a machine learning model incorporating detailed anthropometric data, resulting in a 92% success rate for both suit size and leg expander recommendations. These findings suggest that integrating advanced modelling techniques into PPE size selection can enhance both fit and user satisfaction. While the decision tree offers a practical and accessible solution for most users, the machine learning approach provides the greatest accuracy and should be adopted where feasible. Future efforts should focus on larger datasets and dynamic anthropometry to further refine these models and support the development of next-generation bomb suits.

References

[1] Choi, H.J., Garlie, T., Mitchell, K.B. (2016). Effects of anthropometrics and body size changes on the development of personal protective equipment (PPE) sizing systems in the US Army. *US Army Natick Soldier Research, Development and Engineering Center Natick*

[2] *U.S. National Institute of Justice (NIJ)*, "Public Safety Bomb Suit Standard", NIJ-0117.01, March 2016.

[3] Flynn, M.A., Keller, B., & DeLaney, S.C. (2017). Promotion of alternative-sized personal protective equipment. *Journal of safety research*, 63,43-46

[4] Kozey, J.W., Brooks, C.J., Dewey, S.L., Brown, R.C., Howard, K.A., Drover, D., MacKinnon, S., & McCabe, J. (2009). Effects of human anthropometry and personal protective equipment on space requirements. *Occupational Ergonomics, 8* (2-3), 67-79

[5] Brisbine, B.R., Radcliffe, C.R., Jones, M.L., Stirling, L., & Coltman, C.E. (2022). Does the fit of personal protective equipment affect functional performance? A systematic review across occupational domains. *Plos one, 17* (11), e0278174.

[6] Duda, R.O., Hart, P.E., Stork, D.G., (2000) *Pattern Classification, Second Edition.* Hoboken: Wiley

[7] Dionne, J. P. (2024). Générateur de textes aléatoires automatiquement lemmatisés (Master's Degree dissertation, Université TÉLUQ).

[8] Avadenei, M. (2020). Dynamic anthropometry – a solution for improving the shape of individual protective garments. *International Journal of Engineering Research and Technology, 9*, 162-165

[9] Robinette, K., Veitch, D., Alemany, S., & Bredenkamp, K. (2025). *Product Fit and Sizing: Sustainable Product Evaluation, Engineering, and Design.* CRC Press

Development of an Application to Evaluate the Effect of Headborne Equipment on Warfighter Burden

M. Vignos[1], M. Tumperi[1], M. Yates[1], C. Hanley[1], B. Lindsey[1], N. Thomas[1], G. Holt[1], K. Ott[1], H. Kowpak[1], J. Moramarco[1], A. Gonzalez[1], B. Tate[1], C. Pyles[1], J. Hopping[2], and Q. Luong[1]

[1]*The Johns Hopkins University Applied Physics Laboratory, 11100 Johns Hopkins Rd, Laurel, MD, USA, 20723*
mike.vignos@jhuapl.edu
[2]*Program Executive Office Soldier, 10170 Beach Road, Fort Belvoir, VA, USA, 22060*

Abstract. The added mass of personal protective equipment can negatively affect warfighter injury risk and performance. Specifically, the mass properties of headborne equipment (HBE) influence neck mechanics, marksmanship performance, and neck pain and discomfort. This indicates the importance of considering the effect of HBE mass properties on musculoskeletal burden in the design and evaluation of HBE. Thus, the objective of this work was to develop a software application to allow users to evaluate the effect of HBE on musculoskeletal burden, while considering physiological variations across warfighters. This application uses musculoskeletal models to evaluate changes in biomechanical metrics of musculoskeletal burden. The models used in this application were developed from a previously validated whole-body musculoskeletal model. This model was then scaled to the statures, weights, and strengths of a 50th percentile female and male. The ability to vary neck muscle strength from a 5th percentile female to a 95th percentile male was then incorporated using the neck muscle strength of healthy individuals. A model of HBE with the ability to vary mass properties was rigidly attached to the skull. A model of skeletal muscle fatigue was then calibrated using neck muscle fatigue data. This allowed users to model neck muscle fatigue that develops while wearing HBE. Simulations of walking, running, and static prone were implemented using previously collected data. These models and simulations were then packaged into a software application to simulate variations in HBE mass properties, biological sex, activity, strength, and muscle fatigue. This application was then used to investigate the effects of varying HBE mass properties and fatigue. It was found that increasing HBE mass, shifting the centre of mass anterior, and increasing levels of fatigue each led to increased neck joint impulse. This indicates that each of these parameters should be considered in the design and use of HBE.

1. INTRODUCTION

Soldiers wear personal protective equipment (PPE) to protect from blunt, ballistic, and blast events. Of this PPE, helmets serve as a mounting point for performance-enabling equipment, such as heads-up displays, communication devices, night vision devices, and additional armour. However, this headborne equipment (HBE) fundamentally alters the biomechanics of the neck. While HBE is designed to improve performance, HBE may also put warfighters at risk for acute and chronic injury and, potentially, decreased performance, particularly during long-duration operations in which fatigue and discomfort can start to affect warfighters. Specifically, variations in HBE mass properties can negatively impact neck pain and discomfort [1–3], marksmanship performance [3, 4], task concentration and focus [5], and neck muscle fatigue [6]. Much of the prior research has focused on short-duration timescales and, thus, the potentially increasing effects of HBE during long-duration missions are unknown. These risks associated with HBE indicates the need for improved guidance for HBE mass properties.

Recent HBE research has transitioned from a focus on design of flight helmets [7, 8] to the effects of HBE during dismounted operations to inform improved HBE guidance. This research has found that changing HBE mass properties can affect marksmanship performance, neck fatigue, pain/discomfort, user acceptance, and muscle activity [2, 4]. The information from these studies can be used to start to develop guidance for HBE mass properties for dismounted operations [2, 3]. However, experimental limitations make it challenging to use these data alone to develop complete guidance for HBE mass properties. In particular, certain biomechanical and physiological metrics, such as joint reaction forces, are very challenging or impossible to quantify experimentally. This limits our knowledge on the effect of HBE on the musculoskeletal system. Experiments with human participants can also only be used to evaluate the effect of HBE to a limit due to ethical limitations with inducing severe fatigue and high injury risk, which limits the ability to establish a threshold for HBE guidance. Lastly, the time required for experiments poses a challenge for rapid evaluation of new or exploratory HBE.

Computational modelling provides a complementary capability to experiments and overcomes many of the above limitations. In particular, musculoskeletal (MSK) models can be used to evaluate the

Distribution Statement A: Approved for Public Release; Distribution is Unlimited

https://doi.org/10.52202/081621-0023

effect of external loads on biomechanical response and, thus, provides a method to directly assess the effect of HBE on MSK burden. MSK models have been used to determine that cervical spine motion during target acquisition and head-neck posture may affect neck pain, joint mechanics, and muscle fatigue more than HBE mass properties [9]. This indicates that the scenario is important to consider in HBE mass property guidance for warfighters. However, this prior research focused on a 50th percentile male and simulated relatively simple movements. This limits our understanding of how variations across warfighters during operational movements interact with HBE to affect MSK burden. Additionally, existing MSK models do not incorporate muscle fatigue, which prevents investigations into the effect of mission duration on MSK burden. Further, use of MSK models requires expertise in MSK modelling software and biomechanics. This makes it challenging for operationally-focused individuals to use these tools to evaluate the effect of HBE on MSK burden when assessing equipment trade-offs.

The overall goal of this project was to develop a software application to enable users to evaluate the effect of HBE on warfighter burden while overcoming the limitations of existing MSK models. To achieve this goal, the objectives were to (1) improve existing MSK models to more easily allow for an understanding of the effect of HBE, biological sex, activity, strength, and muscle fatigue on MSK burden, (2) integrate the updated MSK models and analyses into a software application, and (3) use this software application to investigate the effect of HBE mass properties and muscle fatigue on MSK burden.

2. DEVELOPMENT OF THE SOFTWARE APPLICATION

2.1 Development of 50th Percentile Female and Male MSK Models

A previously validated MSK model implemented in OpenSim (Simbios, Stanford, CA) with a detailed representation of the cervical spine was selected as the starting model [10–13]. This model was selected based on its prior validation for analyses used to evaluate MSK burden with MSK models. A rigid model of HBE was then attached to the skull of this model through a fixed joint. The hyoid muscle group was added to the model using previously established muscle definitions to improve the flexion moment-generating capacity of the baseline model [10]. The height and weight of this model was then scaled to 50th percentile male and female models using uniform scaling and preserved mass distribution to align with measurements previously reported for the United States Army [14].

The neck muscle strengths of the 50th percentile models were also scaled to that of an average female and male [15]. This was done by first evaluating the models' maximum isometric strength in flexion and extension about T1-C7 using a modified version of the approach developed by Mortensen et al. [16]. The kinematics of the model were first constrained for all joints except those in the cervical spine. The kinematics of the cervical spine were prescribed to maintain a neutral, upright posture. A 3D linear reserve actuator was applied at the T1-C7 joint to prevent translation of the base of the cervical spine. The strength of the model's neck flexor muscles was measured by applying a 300 N posterior force to the front of the skull and a residual actuator in the opposite direction. The residual force and muscle activations necessary to maintain a neutral cervical spine were then calculated using computed muscle control (CMC) [17]. The difference between the 300 N force and residual actuator force was used to compute the resultant force on the skull. This resultant force vector was then used with a vector from the centre of the T1-C7 joint to the point of force application to compute the models' flexion moment-generating capacity. This process was repeated with a 300 N anterior force and a residual actuator applied to the back of the skull to compute the extension moment-generating capacity of the models. This provided the models' maximum isometric neck flexion and extension strengths, which were then scaled to match the average strength of healthy males and females [18]. This scaling was done by first multiplying the maximum isometric force of the neck muscles by a scale factor computed as the target strength of the model divided by the initial strength. The flexion and extension strengths of the model were then recalculated with the scaled neck muscles. This process was repeated until the model strength was within 0.001% of the target strength. The final 50th percentile male model strength was 52 Nm and 20 Nm in extension and flexion, respectively. The final 50th percentile female model strength was 21 Nm and 15 Nm in extension and flexion, respectively. This provided a female and male model with 50th percentile height, weight, and neck muscle strength.

The ability to model variations in neck muscle strength was then implemented by developing strength-scaling functions to continuously scale the 50th percentile models between 5th and 95th percentile individuals. These functions were developed by first creating normal distributions of the maximum isometric strength for neck flexion and extension of healthy females and males [18]. These distributions were then used to compute target strengths in 5-percentile increments. The maximum isometric force of the neck flexor and extensor muscles of the 50th percentile models were then scaled to match the

https://doi.org/10.52202/081621-0023

maximum isometric strength of the model for each of the target strengths. This was done using the same approach as was used to scale the models' neck muscle strengths to average strengths [16]. This provided the scale factors needed to scale the 50th percentile models to each of the target strengths. An 8th-order polynomial was then fit to the relationship between muscle scale factor and target strength. This produced a set of equations to compute the scale factor needed to continuously scale the flexor and extensor muscles of the male and female models to a target strength percentile.

Simulations of walking, running, and a static prone posture were then implemented. These activities were selected to capture a range of common warfighter activities. The kinematics and ground reaction forces for walking were obtained from an open-access repository [19]. The kinematics and ground reaction forces for running were collected from participants running at a self-selected speed on a level treadmill [20]. For both activities, data were selected from participants that were closest to the height and weight of the 50th percentile models. The 50th percentile models were then were scaled to the representative participant anthropometry [13]. Synthetic data were generated to simulate static prone. The kinematics for this posture were defined by positioning the 50th percentile models into a generic prone posture. Joint definitions at the elbow, wrist, hip, knee, and ankle joints were then locked, keeping only the necessary joints in the head-neck region active. The ground reaction force was calculated as the weight of the model and applied to the whole body's centre of mass (CoM). For all three activities, the joint angles and ground reaction forces were used as input to a residual reduction algorithm and CMC simulations in OpenSim [13, 17]. The CMC simulations generated estimates of muscle activations and forces for the neck muscles and joint reaction forces and torques for the cervical spine. Walking and running simulations were conducted over a complete gait cycle. The prone posture was simulated for one second to be similar in duration to the walking and running simulations.

2.2 Muscle Fatigue Model

The ability to model muscle fatigue was implemented to simulate the effect of long-duration missions on MSK burden. The selected muscle fatigue model is based on compartment theory and divides the state of muscle fibres into three categories: "active" and able to exert force, "fatigued" and not exerting force, and "resting," but able to be recruited [21–24]. This model was implemented by defining a new muscle type in an OpenSim plugin that computes the proportion of muscle fibres in each state using the muscle fatigue model and uses the Millard muscle model to simulate muscle mechanics [25]. A "fatigued" and "resting" state were added to the model, in addition to the default "active" state. The changes in muscle state were translated to a change in muscle activation with the following equation:

$$\frac{da}{dt} = \frac{dM_A}{dt} * a_t + M_A * \frac{da_t}{dt} \quad (1)$$

where $\frac{da}{dt}$ is the activation rate of the muscle, $\frac{dM_A}{dt}$ is the rate of change in active muscle fibres, a_t is the target activation from the first-order activation dynamics in OpenSim, M_A is the proportion of muscle fibres in an "active" state, and $\frac{da_t}{dt}$ is the target activation rate. Activation is modelled in OpenSim as a first-order differential equation that "relates the rate of change of muscle activation to the muscle excitation (i.e., the firing of motor units)" [26]. The target activation is, therefore, related through the first-order differential equation to the excitation target that is computed by CMC. Importantly, this modification to account for muscle fatigue does not change the underlying mechanics of the muscle model meaning each muscle still behaves mechanically the same as a Millard muscle model.

2.2.1 Neck Muscle Fatigue Experiments

The implemented muscle fatigue model was designed to generalize to all types of skeletal muscle. This is done by calibrating the fatigue and recovery parameters to accurately capture the muscle fatigue dynamics for a given body region. The parameters of this fatigue model were previously calibrated for multiple body regions, including the shoulder, knee, and elbow, but not the neck [24]. The parameters for modelling neck muscle fatigue were determined through calibration to data from fatigue experiments.

Twenty young, healthy participants volunteered for the neck muscle fatigue experiments (10 male/10 female, 25.7 ± 3.4 years, 172.2 ± 7.5 cm, 70.6 ± 12.0 kg). Eligible participants had no neck pain in the past 3 months (as rated by scoring 4 or less on the Neck Disability Index [27]), performed at least 150 minutes total of moderate exercise per week, and performed strength training at least two days per week. Interested participants were excluded if they had conditions that may have interfered with their ability to safely perform exercise testing. The protocol was approved by the Johns Hopkins University Institutional Review Board and the US Army Human Research Protection Office (IRB00335161). All

https://doi.org/10.52202/081621-0023

participants provided informed consent.

Six isometric maximum voluntary contractions (MVCs) were performed for both neck extension and neck flexion to establish a baseline maximum torque per motion. Participants performed the isometric neck exercises on a Biodex dynamometer (System 4 Pro, Biodex Medical Systems, Inc., Shirley, NY, USA) while seated in a rigid chair that was weighted to the ground. The centre of the dynamometer was aligned with the palpated C7 spinous process of each participant. A custom rigid padded attachment for testing neck muscle strength was developed using a Biodex ankle attachment and a padded fixture. The custom attachment was fit to the Biodex dynamometer arm so that the attachment arm rested against the occipital protuberance for extension and the brow line for flexion (Figure 1). MVC trials consisted of six 10-second cycles. Participants pushed with as much force as possible for the first 7 seconds of each cycle and rested for the last 3 seconds. Biodex data were acquired at 2000 Hz.

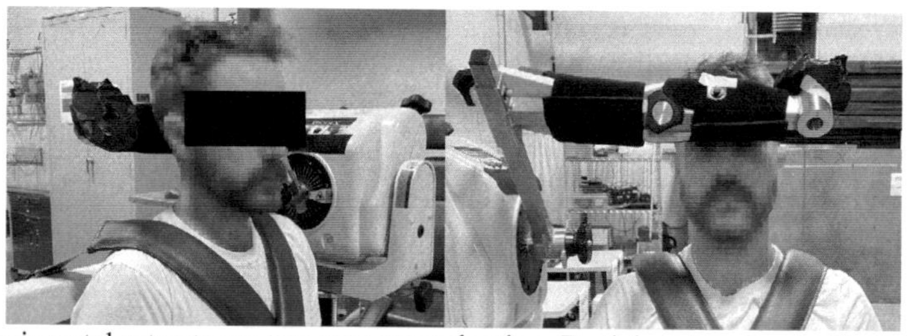

Figure 1. Experimental setup to capture torque production. Participants were positioned in the Biodex System 4 in different orientations for extension (left) and flexion (right).

Participants then performed a fatigue protocol that consisted of repeated MVC cycles in extension. Each MVC cycle consisted of a 7-second maximum contraction followed by 3 seconds of rest. Participants repeatedly performed these MVC cycles until either (a) fifteen minutes had passed or (b) participants decided to end the protocol due to fatigue. The Biodex dynamometer and fixture were set up the same as during the MVC extension trials. Before the start of the fatigue protocol, participants were shown a target line of the maximum torque achieved during their baseline extension MVC trials and were encouraged to practice matching peak torque production. Participants were asked to report their rating of perceived exertion (RPE) using the Borg scale [28] immediately before and at the end of each minute of the protocol. At the conclusion of the extension fatigue protocol, participants rested for ten minutes and then were encouraged to complete a neck flexion fatigue protocol. The Biodex dynamometer and fixture were set up the same as for the MVC flexion trials. The flexion fatigue protocol was performed the same as the extension fatigue protocol, but with participants pressing their forehead against the fixture in flexion. Fourteen of the twenty participants performed both the extension and flexion fatigue protocols.

For each trial, the torque produced by the weight of the Biodex arm was subtracted from the measured torque. The torque data were then filtered with a 10th-order low-pass Butterworth filter with a 10 Hz cut-off frequency [29]. A moving average filter with a window size of 200 samples was then applied. For each 7-second cycle, the peak torque and time of peak were extracted. The maximum torque achieved for each fatigue protocol was then used to calculate the percent MVC for each cycle. Percent MVC data were then regularized to 10 Hz by linear interpolation to enable use within the fatigue model solver and comparison with simulated fatigue model results.

2.2.2 Fatigue Model Calibration

The regularized percent MVC data from the fatigue experiments were then used to determine a set of optimal model parameters to describe neck muscle fatigue. This parameter optimization used a simulated annealing routine (MATLAB, Mathworks, Natick, MA, USA) to optimize the fatigue (F), recovery (R), and recovery accelerator (r) parameters within the muscle fatigue model to best fit the percent MVC data for each participant. The optimal model parameters were determined by minimizing the sum of squared errors between the model predicted and experimentally measured normalized torque time histories (Figure 2). This optimization was performed for both the flexion and extension fatigue data for each participant, which resulted in a set of flexor and extensor muscle fatigue parameters for each participant.

Unpaired t-tests were performed to test for differences in the fatigue parameters across loading conditions (i.e., flexion versus extension) and across biological sex (i.e., female versus male). The threshold for significance was set at $p < 0.025$ to account for multiple comparisons. This statistical analysis was performed to evaluate if different fatigue parameters should be used for the flexor and

https://doi.org/10.52202/081621-0023

extensor muscles and for the 50th percentile male and female models within the software application.

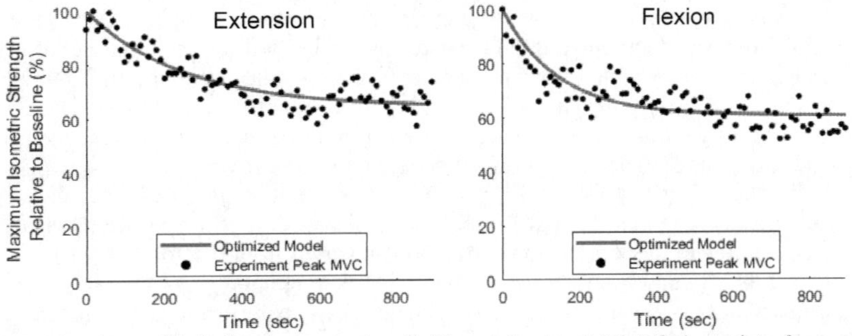

Figure 2. Example model fit for the extension (left) and flexion (right) fatigue data for one participant. The black dots are the peak torque for each MVC as a percent of the maximum torque and are plotted at the time the peak was achieved. The line is the fatigue model with optimized parameters.

A significant difference was found in the extension and flexion fatigue factors ($p < 0.025$, Figure 3). There was no significant difference in the recovery factor and the recovery accelerator across fatigue protocols. Additionally, there was no significant difference in each of the fatigue parameters across biological sex within both the extension and flexion fatigue protocols. This indicated that different fatigue parameters should be used for the extensor and flexor muscles, but that the same fatigue parameters could be used for the 50th percentile female and male models. The final fatigue factor, recovery factor, and recovery accelerator were computed by averaging each of these parameters across participants for the flexors and extensors separately (Table 1).

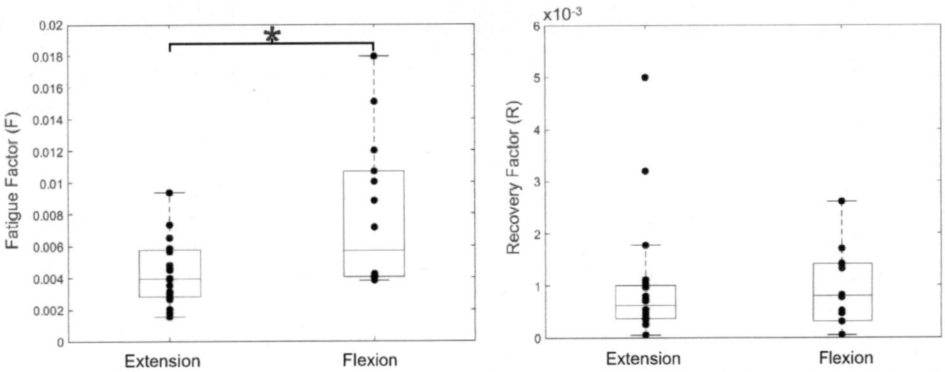

Figure 3. Fatigue factor (left) and recovery factor (right) determined by optimizing the model to the extension and flexion fatigue data. The black dots are the fatigue and recovery factor for each participant. The red lines are the median of the optimized parameters. * indicates a significant difference ($p < 0.025$) in the fatigue factor for the extensors and the flexors.

Table 1. Final fatigue model coefficients calculated for the extensor and flexor muscles

Neck Region	Fatigue Factor (F)	Recovery Factor (R)	Recovery Accelerator (r)
Extensor Muscles	0.003604	0.000347	25.02088
Flexor Muscles	0.006766	0.000867	14.971507

2.2.3 Defining Levels of Fatigue for Software Application

The resultant neck muscle fatigue model was then used to model levels of fatigue within the software application. Four levels of fatigue were selected through input from subject matter experts and potential end-users. The levels of fatigue were modelled by varying the proportion of muscle fibres in the fatigue compartment for each of the neck muscles. Each of the neck muscles are first grouped into a flexor or extensor compartment. The proportion of muscle fibres in the fatigue compartment for each of the muscles are then varied to simulate each of the different levels of fatigue. The remainder of the muscle fibres start in the active compartment at the beginning of a simulation (i.e., 100 minus the proportion of muscle fibres in the fatigue compartment). None of the muscle fibres start in the resting state.

https://doi.org/10.52202/081621-0023

The mapping of proportion of muscle fibres in the fatigue compartment to level of fatigue was developed to align with the neck muscle fatigue experiments. Severe fatigue was set to a 50% reduction from the models' baseline strength based on participants' average maximal reduction in torque generating capacity. This was determined to be a reasonable upper limit of neck muscle fatigue as the fatigue experiments were a very severe loading case for the neck muscles and participants typically reached an "extremely hard" level of exertion (i.e., 18+ on the 20-point Borg RPE scale [28]). Mild and moderate fatigue were then set as equal increments from no fatigue to severe fatigue, with mild and moderate set to reductions of 83% and 67% of baseline strength, respectively.

Mapping from the target reduction in strength to proportion of muscle fibres in the fatigue compartment was then determined through a sensitivity analysis. In this sensitivity analysis, the proportion of muscle fibres in the fatigue compartment was varied from 0% to 75% for the extensors and 0% to 70% for the flexors in 5% increments (Figure 4). The remaining muscle fibres were in the active compartment and none of the muscle fibres were in the resting compartment. The maximum torque generating capacity of the model was then assessed for each of these fatigue levels. A linear regression model was then developed with proportion of muscle fibres in the fatigue compartment as the independent variable and the change in model strength relative to baseline as the dependent variable. This regression model was then used to compute the proportion of muscle fibres in the fatigue compartment needed to induce the target change in model strength for each of the levels of fatigue. This produced the final model states used for each level of fatigue in the software application.

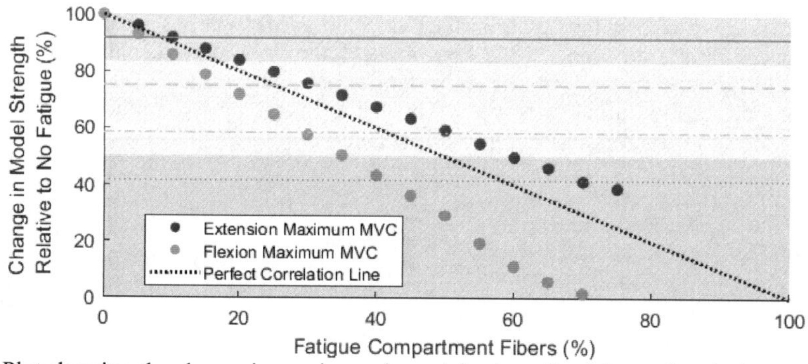

Figure 4. Plot showing the change in maximum isometric strength of the neck, relative to no fatigue, due to changing the proportion of muscle fibres in the fatigue compartment for both the extensors (blue dots) and flexors (turquoise dots). A perfect correlation (black dotted line) and the target changes in strength for each level of fatigue are shown. These results were used to map from target level of fatigue to proportion of muscle fibres in the fatigue compartment.

2.3 Integration into HTAT Software Application

The developed models and analyses were integrated into a software application to evaluate the effect of HBE mass properties, activity, biological sex, and neck muscle strength and fatigue on MSK burden. This software is called the Headborne Equipment Tradespace Assessment Tool (HTAT) and is comprised of simulation (Figure 5), analysis (Figure 6), and help pages. The Simulation Page contains three panels: Setup Properties, Simulation Queue, and Simulation Details. The Setup Properties panel is used to set up and run a model by selecting a biological sex, activity, HBE configuration, neck muscle strength, and level of muscle fatigue. The Simulation Queue is used to visualize the status of simulations that are completed, running, or queued to run. For simulations in the queue, additional information on when the simulation started running, ended (if applicable), and the setup properties are shown to track simulations. Immediately above the Simulation Queue is a progress indicator with a spinning circle to show that the indicated simulation is actively running. The Simulation Details panel contains information on the simulation that is currently running and the simulation that has most recently completed running. This panel also contains a video clip of the activity selected when a simulation starts running.

The Analysis Page allows users to visualize results from successful simulations. Users first load simulations into the Active Simulations Table to use during an analysis session. Users then move through the Add Plot workflow to generate line, bar, or scatter plots based on any subset of the active simulations. Plots may then be modified to adjust the colours, labels, or scaling. Plots can be saved individually or as a group to a user selected folder. Data in the plots may also be exported to a Microsoft Excel file for inspection or development of different plots. Together, the simulation and analysis page allow users to evaluate the effect of HBE on warfighter burden.

https://doi.org/10.52202/081621-0023

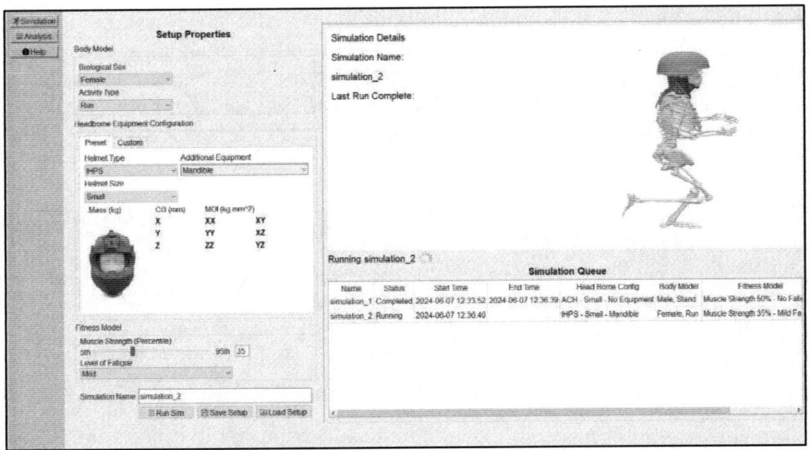

Figure 5. The simulation page within HTAT is used to setup, run, and monitor progress of simulations

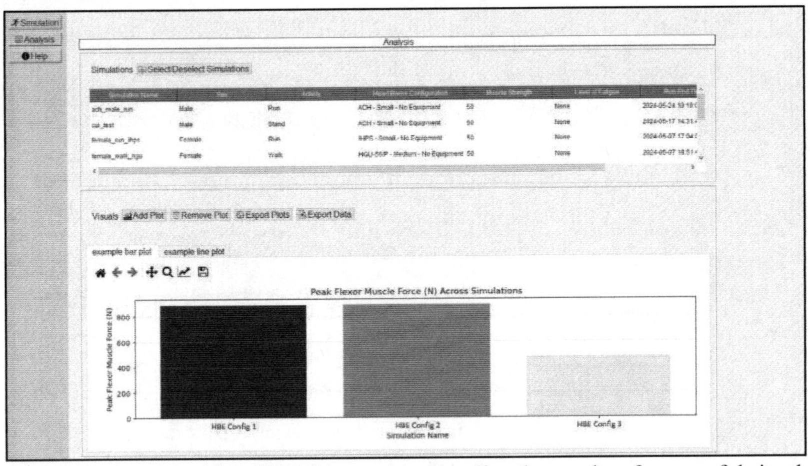

Figure 6. The analysis page within HTAT is used to visualize the results of successful simulation runs. Visualizations and the associated data can also be exported within this page.

3. INVESTIGATING THE EFFECT OF HBE DESIGN ON WARFIGHTER BURDEN

The effect of HBE mass properties on warfighter burden was then investigated using HTAT. Previous studies have found that HBE mass properties and activity being performed affect neck pain, cervical spine mechanics, and muscle fatigue [9]. However, this research focused on a well-rested 50th percentile male and analysed relatively simple movements. This limits our understanding of how variations across warfighters during functional activities interact with HBE to further impact MSK burden. Thus, the goal of this investigation was to perform a preliminary analysis of the effect of HBE mass properties and muscle fatigue on MSK burden, while considering warfighter variations as a potential use-case of HTAT.

3.2 Methods

Using HTAT, we analysed the effect of variations in HBE mass, CoM, and muscle fatigue for male and female walking. A HBE configuration fielded within the United States Army was selected as the base configuration for varying HBE mass properties. The following variations were simulated relative to this configuration: (1) varying the HBE mass from 1 kg to 4 kg with a constant CoM and (2) varying the CoM from 2 cm posterior to 10 cm anterior relative to the head CoM with a constant HBE mass of 2 kg. These ranges were selected to encompass the mass and anterior-posterior CoM location of fielded equipment [30]. Walking simulations were performed with each of the variations in HBE mass properties for the female and male models. For variations in fatigue, walking simulations were performed at each level of fatigue with the female and male 50th percentile models and the base HBE configuration.

The joint reaction forces were computed within HTAT as the result of all forces acting at each

https://doi.org/10.52202/081621-0023

joint, including muscle forces [31]. The resultant joint reaction force at T1-C7 was then integrated across time for each simulation to compute the joint impulse. The change in joint impulse relative to the lowest mass configuration was computed for simulations with varying HBE mass. The change in impulse relative to a CoM aligned with the head CoM was computed for simulations with varying CoM. The change in impulse relative to no fatigue was computed for simulations with varying levels of fatigue.

3.3 Results

Increasing the HBE mass and shifting the CoM location anterior led to an increase in joint impulse at T1-C7 for both the male and female models (Figure 7). Joint impulse increased by 105 Ns and 57.2 Ns for the male and female models, respectively, across the range of simulated HBE mass. Joint impulse increased by 18.6 Ns and 61.1 Ns for the male and female models, respectively, across the range of CoM locations. However, the change in joint impulse was relatively small for the male model (i.e., less than 5 Ns) when the CoM location was between 2 cm posterior to 4 cm anterior. Shifting the CoM posterior led to a decrease in joint impulse for the female model.

Increasing fatigue increased the joint impulse for both the female and male model (Figure 8). Joint impulse increased by 61.4 Ns for the male model and 43.0 Ns for the female model from no fatigue to severe fatigue.

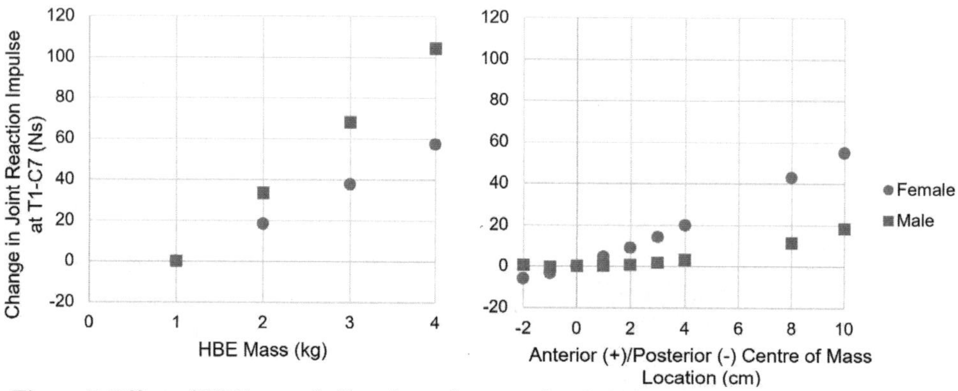

Figure 7. Effect of HBE mass (left) and anterior-posterior CoM location on the change in joint impulse at T1-C7 during simulated walking. This change was computed relative to the lowest HBE mass (i.e., 1 kg) or a CoM aligned with the head CoM. Blue dots are simulations with the 50th percentile female model and red squares are simulations with the 50th percentile male model.

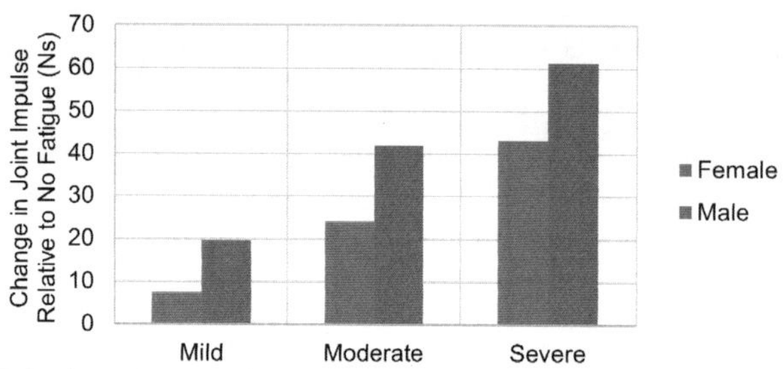

Figure 8. Change in joint impulse relative to no muscle fatigue for mild, moderate, and severe fatigue

4. DISCUSSION AND CONCLUSIONS

The goal of this project was to develop a software application to evaluate the effect of HBE on warfighter burden across a range of warfighters. This was done by first extending an existing MSK model to simulate differences across equipment and warfighters by incorporating variations in HBE, biological sex, activity, strength, and muscle fatigue. These updated MSK models and analyses were then integrated into the Headborne Equipment Tradespace Assessment Tool (HTAT). This application evaluates the effect of simulated variations in HBE mass properties, 50th percentile height and weight (through variations in biological sex), activity, neck muscle strength, and neck muscle fatigue on MSK burden. This provides a number of use cases, including (1) evaluating differences in MSK burden across existing

https://doi.org/10.52202/081621-0023

HBE configurations or new HBE designs, (2) evaluating differences in MSK burden across a population with varying strength and stature for a given HBE configuration, (3) understanding changes in MSK burden across a mission due to increased muscle fatigue, and (4) assessing how activities performed by warfighters may elevate MSK burden when wearing HBE. Each of these use cases can be used to help inform guidance for HBE mass properties during dismounted operations.

The effect of HBE mass and anterior-posterior CoM location on MSK burden was investigated using HTAT. Increased HBE mass and a more anterior CoM led to increased joint impulse in the cervical spine. This is consistent with previous studies indicating that varying HBE mass and CoM location can alter neck pain, discomfort, and muscular fatigue [1–4]. The results from HTAT expand upon these previous findings to allow the change in joint impulse with varying HBE mass properties to be quantified, which could inform objective guidance on HBE mass properties. Improved understanding of the relationship between these biomechanical metrics and risk of injury or changes in performance would help establish thresholds for HBE design. The change in joint impulse was relatively small for the male model when the CoM was between 2 cm posterior to 4 cm anterior. This may be due to greater strength of the male model resulting in a reduced sensitivity to variations in HBE CoM. Shifting the HBE CoM posterior led to reduced joint impulse for the female model. However, we cannot extrapolate this effect to a CoM beyond the range that was simulated. The range in joint impulse was greater for variations in HBE mass than for CoM. While a single unit change in HBE mass and CoM are not equivalent, this suggests that reducing HBE mass may provide greater benefit to reducing warfighter burden.

Increasing the level of fatigue increased cervical spine joint impulse for both 50th percentile models. This indicates that, as warfighters become more fatigued during a mission, biomechanical changes occur that may increase likelihood of neck injury. Additionally, the range in joint impulse was similar magnitude when varying levels of fatigue and varying HBE mass properties. While there is no way to directly compare changes in muscle fatigue and changes in mass properties, this suggests that level of fatigue and, thus, duration a warfighter is wearing a HBE configuration, could contribute to neck injury risk and is likely important to consider when developing guidance for HBE mass properties.

When interpreting results from HTAT, it is important to consider the following limitations. First, the current models do not account for changes in movement patterns that may occur due to differences in HBE or fatigue. This allows users to understand biomechanical changes that would occur if a warfighter maintained the same movement pattern across different HBE. However, the ability to evaluate changes in movement patterns could be implemented through use of predictive MSK models [32]. Next, the male and female models differ in height, weight, neck muscle strength, and self-selected walking and running speeds. This makes it challenging to assess the specific factor driving differences in outcomes between these models. Implementing the ability to control variations in anthropometrics across a range of movement dynamics would improve understanding of the effect of these parameters on MSK burden. Finally, there is no direct mapping between the metrics evaluated in HTAT and injury risk or changes in performance. Future work will investigate the relationship of biomechanical and other model-based metrics with neck injury risk and changes in performance to improve interpretability.

In conclusion, we developed a software application to evaluate the effect of HBE mass properties on MSK burden across variations in warfighter stature, weight, strength, and fatigue and across a range of activities. This application can be used within armour acquisition, design, and test and evaluation to support assessments of the tradespace of warfighter protection, performance, and injury risk offered by HBE. Using this application, we found that variations in HBE mass properties and level of fatigue led to increases in MSK burden. This indicates that HBE mass properties, variations across warfighters, and mission duration should all be considered when developing guidance for design and use of HBE.

Acknowledgments

The authors would like to thank Program Executive Office Soldier Product Manager Soldier Protective Equipment for sponsoring this effort. The authors would also like to thank Marina Carboni at US Army Combat Capabilities Development Command Soldier Center for measuring the mass properties of the HBE configurations in HTAT. Any opinions, findings and conclusions or recommendations expressed in this material are those of the author(s) and do not necessarily reflect the views of PEO Soldier, DEVCOM Soldier Center, or Naval Sea Systems Command.

References

[1] Tack D. W., Nakaza E. T., McKee K. W., MacEachern C. A., and Marrao C. C., Investigation of the Preferred Mass Properties for Infantry Headwear Systems, Humansystems Incorporated, DRDC Toronto CR-2005-230, (2006).

[2] Knight J. F. and Baber C., Neck muscle activity and perceived pain and discomfort due to variations of head load and posture, *Aviat Space Environ Med*, 75(2), pp. 123–131, (2004).

 https://doi.org/10.52202/081621-0023

[3] Madison A., Preliminary Head-Supported Mass (HSM) Performance Guidance for Dismounted Soldier Environments, USAARL Technical Memorandum 2019-11, (2019).
[4] Madison A. M. *et al.*, Preliminary Head-Supported Mass Performance Guidance for Dismounted Soldier Environments, *Military Medicine*, 188(Supplement_6), pp. 520–528, (2023).
[5] Lim J. *et al.*, Additional helmet and pack loading reduce situational awareness during the establishment of marksmanship posture, *Ergonomics*, 60(6), pp. 824–836, (2017).
[6] Gallagher H. L., Caldwell E., and Albery C. B., Neck Muscle Fatigue Resulting from Prolonged Wear of Weighted Helmets, AFRL-RH-WP-TR-2008-0096, (2008).
[7] Butler B. P. and Allen N. M., Long-Duration Exposure Criteria for Head-Supported Mass, ARMY AEROMEDICAL RESEARCH LAB FORT RUCKER AL, (1997).
[8] McEntire B. J. and Shanahan D. F., Mass Requirements for Helicopter Aircrew Helmets, ARMY AEROMEDICAL RESEARCH LAB FORT RUCKER AL, (1997).
[9] Barrett J. M., Healey L. A., Fischer S. L., and Callaghan J. P., Cervical Spine Motion Requirements From Night Vision Goggles May Play a Greater Role in Chronic Neck Pain than Helmet Mass Properties, *Hum Factors*, p. 001872082210906, (2022).
[10] Cazzola D., et al., Cervical Spine Injuries: A Whole-Body Musculoskeletal Model for the Analysis of Spinal Loading, *PLOS ONE*, 12(1), (2017), doi: 10.1371/journal.pone.0169329.
[11] Silvestros P. *et al.*, Musculoskeletal modelling of the human cervical spine for the investigation of injury mechanisms during axial impacts, *PLoS ONE*, 14(5), p. e0216663, (2019).
[12] Silvestros P., *et al.*, Electromyography-Assisted Neuromusculoskeletal Models Can Estimate Physiological Muscle Activations and Joint Moments Across the Neck Before Impacts, *Journal of Biomechanical Engineering*, 144(3), p. 031011, (2022), doi: 10.1115/1.4052555.
[13] Seth A. *et al.*, OpenSim: Simulating musculoskeletal dynamics and neuromuscular control to study human and animal movement, *PLOS Computational Biology*, 14(7), p. e1006223, (2018).
[14] Gordon C. C. *et al.*, 2012 Anthropometric Survey of U.S. Army Personnel: Methods and Summary Statistics, U.S. Army Natick Soldier Research, Development and Engineering Center, Natick, (2014).
[15] Roos P. E., Vasavada A., Zheng L., and Zhou X., Neck musculoskeletal model generation through anthropometric scaling, *PLoS ONE*, 15(1), pp. 1–21, (2020).
[16] Mortensen J. D., Vasavada A. N., and Merryweather A. S., The inclusion of hyoid muscles improve moment generating capacity and dynamic simulations in musculoskeletal models of the head and neck, *PLOS ONE*, 13(6), p. e0199912, (2018), doi: 10.1371/journal.pone.0199912.
[17] Thelen D. G., Anderson F. C., and Delp S. L., Generating dynamic simulations of movement using computed muscle control, *Journal of Biomechanics*, 36(3), pp. 321–328, (2003).
[18] Vasavada A. N., Li S., and Delp S. L., Influence of Muscle Morphometry and Moment Arms on the Moment-Generating Capacity of Human Neck Muscles, *Spine*, 23(4), p. 412, (1998).
[19] Moore J. K., Hnat S. K., and Bogert A. J. van den, An elaborate data set on human gait and the effect of mechanical perturbations, *PeerJ*, 3, p. e918, (2015), doi: 10.7717/peerj.918.
[20] Vignos M. F. *et al.*, Physical Fatigue State Classification through Fusion of Physiological and Biomechanical Metrics, presented at the 2023 International Congress on Soldiers' Physical Performance, London, UK, Sep. 12, 2023.
[21] Liu J. Z., Brown R. W., and Yue G. H., A Dynamical Model of Muscle Activation, Fatigue, and Recovery, *Biophysical Journal*, 82(5), pp. 2344–2359, (2002).
[22] Xia T. and Frey Law L. A., A theoretical approach for modeling peripheral muscle fatigue and recovery, *Journal of Biomechanics*, 41(14), pp. 3046–3052, (2008).
[23] Looft J. M. and Frey-Law L. A., Adapting a fatigue model for shoulder flexion fatigue: Enhancing recovery rate during intermittent rest intervals, *J Biomech*, 106, p. 109762, (2020).
[24] Frey-Law L. A., Looft J. M., and Heitsman J., A three-compartment muscle fatigue model accurately predicts joint-specific maximum endurance times for sustained isometric tasks, *Journal of Biomechanics*, 45(10), pp. 1803–1808, (2012), doi: 10.1016/j.jbiomech.2012.04.018.
[25] Millard M., *et al.*, Flexing Computational Muscle: Modeling and Simulation of Musculotendon Dynamics, *Journal of Biomech Eng*, 135(021005), (2013), doi: 10.1115/1.4023390.
[26] Millard M., Seth A., and Loan P., First-Order Activation Dynamics, 2012. https://simtk-confluence.stanford.edu:8443/display/OpenSim/First-Order+Activation+Dynamics.
[27] Vernon H., The Neck Disability Index: State-of-the-Art, 1991-2008, *Journal of Manipulative and Physiological Therapeutics*, 31(7), pp. 491–502, (2008), doi: 10.1016/j.jmpt.2008.08.006.
[28] Williams N., The Borg Rating of Perceived Exertion (RPE) scale, *Occupational Medicine*, 67(5), pp. 404–405, (2017), doi: 10.1093/occmed/kqx063.
[29] Nordez A., Casari P., and Cornu C., Accuracy of Biodex system 3 pro computerized dynamometer in passive mode, *Medical Engineering & Physics*, 30(7), pp. 880–887, (2008).
[30] Estep P. N. *et al.*, Mass Properties Comparison of Dismounted and Ground-Mounted Head-Supported Mass Configurations to Existing Performance and Acute Injury Risk Guidelines, *Military Medicine*, 184(Supplement_1), pp. 245–250, (2019), doi: 10.1093/milmed/usy342.
[31] Steele K. M., DeMers M. S., Schwartz M. H., and Delp S. L., Compressive tibiofemoral force during crouch gait, *Gait & Posture*, 35(4), pp. 556–560, (2012).
[32] De Groote F. and Falisse A., Perspective on musculoskeletal modelling and predictive simulations of human movement to assess the neuromechanics of gait, *Proceedings of the Royal Society B: Biological Sciences*, 288(1946), p. 20202432, (2021), doi: 10.1098/rspb.2020.2432.

Body Armour Comfort & Mobility Assessment and Implementation

L. Boogh[1], C. Djololian[1]

[1]DuPont Specialty Products Operations Sàrl, 154 Route du Nant-d'Avril, 1217 Meyrin, Geneva, Switzerland
louis.boogh@dupont.com, christophe.djololian@dupont.com

Abstract. Comfort issues associated with body armour have become a crucial focus for military and law enforcement. As the demand for reliable personal protective equipment increases, the user comfort and mobility constraints induced by body armour have emerged as essential factors influencing wearability and performance. Despite advancements in material technology allowing weight reductions, many users experience discomfort due to rigidity and the restrictive nature of traditional body armour.

Comfort and mobility testing of body armour is thus a pivotal aspect of developing effective personal protective equipment, as it directly influences user acceptance and operational performance. Body armour, while designed primarily to offer protection, must also consider the physical and psychological impact on the wearer.

This paper discusses body armour comfort testing methods addressing issues identified during wear trials and considering feed-back from end-users. Aside from fitting issues, the most prominent discomforts and mobility constraints were related to either a high rigidity of the armour construction reducing mobility and accelerating exhaustion, or pains occurring at localized pressure areas. Three test methods were developed allowing to characterize these aspects. The fundamental consideration of the test method designs is to replicate in-use movements and anatomical aspects observed during wear and were shown to correlate well with user perception.

The test methods, first presented at PASS 2023, have been further optimized and are now available to the industry. They allow to include quantified comfort and mobility criteria in the design of ballistic protective materials and solutions. End-users, for whom comfort is most critical, can now use and eventually include comfort and mobility criteria in tender processes in parallel to ballistic protective performance, weight and cost. Ultimately, the target for these new additional selection criteria is to reduce the likelihood of fatigue and distraction in-use. Illustrations of the implementation of these test methods to the selection of soft ballistic inserts are presented in this paper.

1. INTRODUCTION

1.1 The stake around comfort & mobility assessment

The necessity for comfort in body armour is increasingly acknowledged by end-users who require protective equipment that does not compromise operational performance. Although ballistic efficacy—shaped by the materials and design of the body armour—exhibits a strong correlation with the level of protection provided, the importance of ergonomics must not be underestimated. Enhanced comfort alleviates painful movements and pressure points, thereby promoting more consistent and effective utilization of the vest. Furthermore, improvements in mobility can mitigate movement restrictions and facilitate energy conservation, leading to enhanced operational efficiency, decreased exposure to hazardous environments, and reduced risk of musculoskeletal injuries.

A thorough literature review conducted by Pope et al. [1] demonstrated that body armour substantially impacts physical performance and introduces biomechanical constraints for the wearer. However, they emphasized that the effects of body armour on marksmanship and physiological responses remain insufficiently explored. Their findings underscored the importance of deliberate consideration in body armour selection, factoring in ergonomic aspects alongside their implications for perceived and measured exertion, work capability, balance, and stability.

Recent field surveys [2] reveal that police officers often feel compelled to remove their body armour owing to discomfort, highlighting the detrimental effects that body armour can impart on user well-being and effectiveness. Even if heat has been recognized as a significant element influencing this behaviour, similar to the instinct to discard a sweater in warm conditions, almost 50% of respondents identified discomfort as the primary reason for removing their protective gear, whereas concerns related to weight were expressed by only about 25% of participants, indicating that comfort is paramount in the selection of body armour.

Moreover, research investigating military body armour indicates that diminished mobility may adversely impact movement, posture [3], and range of motion, [4] even when the equipment weights are relatively moderate. Thus, ergonomics, defined as the harmonious integration of comfort and mobility,

https://doi.org/10.52202/081621-0024

is expected to not only affect immediate performance outcomes but also potentially heighten the long-term risk of musculoskeletal disorders.

Despite recent advances in materials designed to enhance comfort without compromising ballistic protection, a consensus on clear and objective standardized testing methods remains elusive. As a result, assessing comfort presents a significant challenge for end-users during the body armour selection process, in conjunction with other critical factors such as ballistic performance, cost, and weight. The lack of established criteria for comfort further complicates the efforts made by manufacturers to develop solutions that prioritize user comfort.

1.2 Past assessments of comfort & mobility have been undertaken through a range of methodologies which have not resulted in fully satisfactory outcomes to date

The assessment of comfort and mobility constraints in body armour through various methodologies—including weight measurement, plunger pressure tests, pole or table edge tests, and wear trials—reveals significant limitations that undermine their relevance in providing accurate evaluations.

1.2.1 Incumbent test methods

Weight measurement alone is insufficient for a comprehensive understanding of comfort and mobility, as it fails to encapsulate the nuanced experiences of wearers. The plunger pressure test, while quantifying material resistance to complex folding, often does not correlate well with subjective comfort perceptions during wear trials, primarily resulting from the original test design focused on single-layered fabric conformability rather than complex multi-layered constructions found in body armour constructions [5][6]. Similarly, the pole or table edge test, which measures the angle of deformation, lacks applicability in real-world movements during which body armour experiences multi-axial deformations, rendering its findings less representative of practical wear scenarios [7].

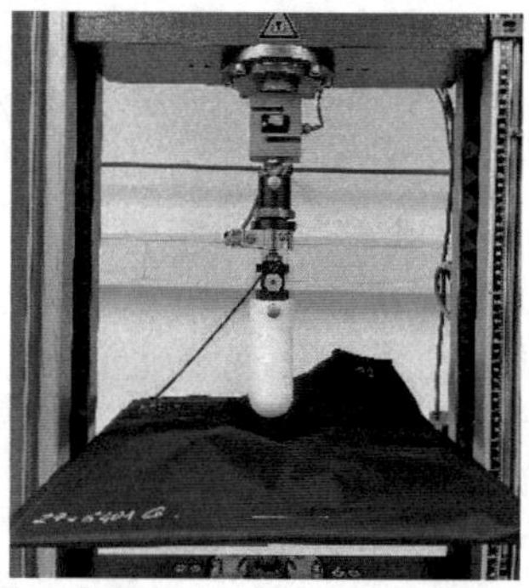

Figure 1. Example of plunger pressure test set-up used in DuPont

1.2.2 Wear trials

Although wear trials are considered the most realistic method for assessing comfort and mobility, these require significant resources and are not standardized across the different military and law enforcement agencies, complicating their implementation as benchmarks for body armour evaluation [8]. Collectively, these deficiencies indicate that current assessment methods inadequately address the multifaceted nature of comfort and mobility constraints experienced by end-users in actual operational contexts.

1.3 Three innovative Comfort and Mobility testing methods have been developed relying on real situations and body movement

In response to the pressing need for enhancing the end-user experience for military and law enforcement personnel, three innovative laboratory test methodologies have been developed using a universal tensile testing machine (UTM). These testing methods address the main comfort aspects of mobility and constraints and are inspired by real-world operational scenarios and in such are designed to closely replicate the diverse range of dynamic body movements encountered by users in various environments. By allowing for the quantification of mobility constraints and comfort levels, these methodologies provide critical insights into the interactions between body armour and the wearer's physical capabilities.

https://doi.org/10.52202/081621-0024

1.3.1 The Lower Costal Bending (LCB) test

The first of these methodologies, known as the Lower Costal Bending (LCB) test, is specifically designed to simulate the bending motions of the upper body. During active duty, such bending movements are common, yet they often lead to significant energy dissipation and accelerated physiological fatigue among wearers. The LCB test serves an essential function by quantifying the resistance to bending and thus, the amount of energy lost due to the stiffness inherent in the ballistic solution. This quantification is critical for understanding how much energy the wearer expends during routine movements, thereby informing improvements in ballistic vest design for enhanced operational endurance. The recently published "HO-DPT 67474-2" test method procedure defines the energy, En.LCB expressed in N.mm, required to flex a soft ballistic insert at the lower costal bending axis of the torso.

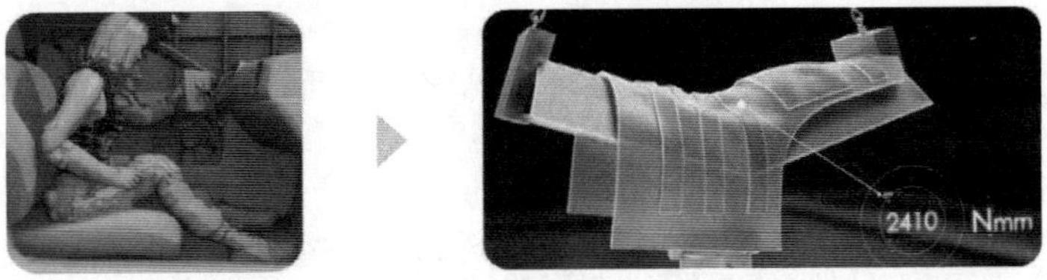

Figure 2. Mobility constraints that led to the LCB test design

1.3.2 The Double Curvature Compression (DCC) test

In terms of mobility optimization, it is imperative that ballistic solutions conform effectively to the contours and shapes of the wearer's body in authentic end-use situations. Observational studies utilizing advanced digital imaging techniques [9] reveal that body armour is frequently required to adapt dynamically to multi-curved shapes dictated by the human form. The Double Curvature Compression (DCC) test was established to replicate the anatomy of a bent torso, enabling value chain players to evaluate the conformability of various ballistic material solutions comprehensively. A high degree of conformability is intrinsically linked to enhanced mobility. The "HO-DPT 67474-3" test method procedure defines the energy required, En.DCC expressed in N.mm, to conform a soft ballistic insert to the anatomical multi-curved bent torso shape.

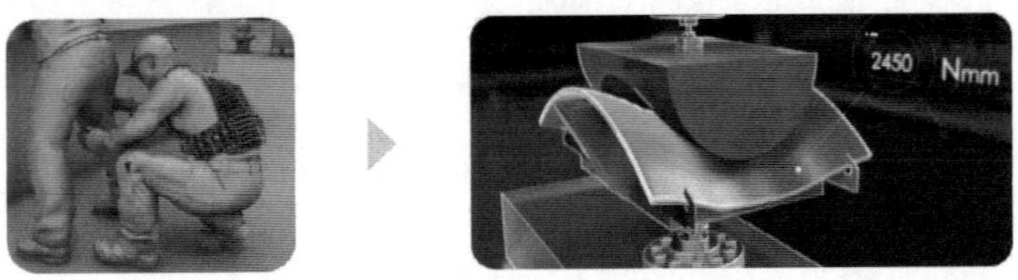

Figure 3. Mobility constraints that led to the DCC test design

1.3.3 The Edge Pressure (EP) test

Finally, the edges of the ballistic vests can contribute to localized discomfort and pains, particularly at critical pressure points such as the lower abdomen, neck, and arms. To address this issue, the Edge Pressure (EP) test has been developed. This test simulates the anatomical shape of the arm and the corresponding curvature of ballistic solutions to assess the discomfort generated at these potential pressure points. By identifying the causes of localized discomfort through the EP test, manufacturers can make informed design modifications to improve the overall comfort and functionality of body armour. The "HO-DPT 67474-1" test method procedure defines the force, F.EP expressed in N, developed at the edges of a soft ballistic insert during a given displacement.

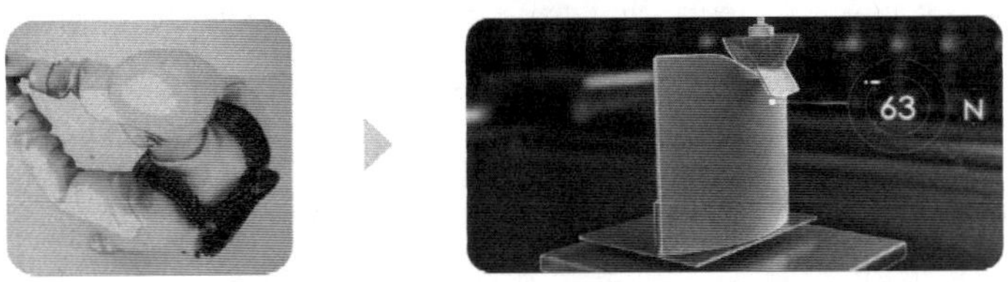

Figure 4. Edge pressure pain that led to the EP test design

https://doi.org/10.52202/081621-0024

In summary, the collective implementation of these methodologies—namely, the Lower Costal Bending test, the Double Curvature Compression test, and the Edge Pressure test aims at proposing a comprehensive understanding of the intricate relationship between body armour and the limitations imposed on wearers' mobility and comfort. Through these advancements, the ability to identify and mitigate discomfort levels significantly contributes to fostering an improved end-user experience, ultimately enhancing the performance and well-being of those who rely on such critical protective gear.

2. APPROACH TAKEN

2.1 Objectives

The aim of this study is to present a methodology to implement and use these new test methods for each stakeholder, including end-users, ballistic armour manufacturers, and fabric weavers.

2.2 Principle

In the ongoing effort to define the most relevant way to implement these testing methodologies in the industry, an extensive series of investigated exchanges have been conducted with key stakeholders within the ballistic armour value chain to define a specific use framework for each player.

2.2.1 End-users focus

The engagement with end-users allowed to articulate their specific requirements and preferences. Consequently, the identification of optimal strategies for integrating the proposed testing methodologies into tender documents has been defined, ensuring that the solutions were tailored to meet their distinct needs.

2.2.2 Ballistic Armour Manufacturer focus

Simultaneously, ballistic armour manufacturers have been approached, focusing on the assessment and comparison of under-development body armour solutions. Through this collaborative approach, manufacturers were able to gain a deeper understanding of how these testing solutions could be incorporated into the design processes. Key considerations during these discussions included the ballistic performance of the solution design, its flexibility capabilities, its overall weight, and the associated costs as main design criteria for a new body armour solution.

2.2.3 Material provider focus

Interactions with material providers such as fibres manufacturers or weavers, highlighted the application of these testing methodologies in the development of innovative and flexible structural designs. This allowed for informed decision-making regarding the selection of raw materials that would most significantly enhance the desired properties of final ballistic products.

This multifaceted approach not only underscores the collaborative nature of innovation in the field of ballistic armour but also emphasizes the importance of stakeholder engagement in defining and implementing standardized testing solutions that align with the needs of the industry.

3. EXPERIMENT

3.1 End-User experiment

To evaluate the effectiveness of end-user solutions, a dedicated testing campaign was conducted to assess four distinct solutions designed to meet the requirements of level IIIa ballistic protection, specifically through rigorous testing methodologies, including the Edge Pressure (EP) test and the Double Curvature Compression (DCC) test.

The primary objective of this assessment was to determine whether these solutions could be clearly differentiated based on their performance metrics. Furthermore, the study aimed to establish a scoring framework that would be suitable for use in a procurement tender process, ensuring that each solution could be objectively evaluated based on a standardized methodology.

3.2 Ballistic Armour Manufacturer experiment

In the context of a comprehensive evaluation conducted for ballistic armour manufacturers, a total of eight distinct solutions were rigorously assessed based on varying levels of ballistic performance. Each solution was systematically organized within a comparative table, which delineated key parameters, including ballistic performance levels, flexibility scores, weight, and pricing.

The primary objectives of this assessment were twofold: first, to ascertain whether a strong quantifiable relationship can be highlighted between the degree of flexibility in the armour solutions and their corresponding ballistic performance; and second, to identify the optimal design according to the diverse evaluation criteria. To achieve these objectives, an analysis method was employed to investigate the potential trade-offs and synergies between flexibility and ballistic resistance.

3.3 Material provider experiment

In the context of material development, four distinct solutions were formulated, all featuring identical construction parameters and a consistent range of ballistic performance levels. Each of these solutions utilized a unidirectional configuration, while varying the product manufacturing process parameters and type of resin employed in their composition.

The primary objective of this study was to evaluate the flexibility of these solutions using the Double Curvature Compression (DCC) test, with the aim of identifying any perceptible differences among them. Additionally, the analysis sought to determine whether these observed differences in performance were significant enough to inform the selection of the optimal resin based on its flexibility characteristics.

4. RESULTS & DISCUSSION

4.1 End-User results

A series of intensive exchanges conducted with end users have culminated in the identification of two examples of viable options to be incorporated within the framework of a tender.

4.1.1 Development of a Comparative Rating System

The first option involves establishing a systematic comparative rating mechanism. This approach would enable the positioning of body armour solutions based on their performance metrics, specifically aiming to minimize En.DCC or En.LCB, (measured in N.mm) and F.EP (measured in N) values.

The solution exhibiting the lowest Energy (En.DCC or En.LCB) combined with the lowest Force (F.EP) values would be awarded the highest number of points within this rating system, while all other body armour solutions would receive prorated scores accordingly. In the below example, the DCC test score of solution A is calculated by dividing its measured energy (En.DCC) by the energy measured for the sample with the highest En.DCC, sample D in this case, 4’473 / 2’027=2.21.

This metric-driven evaluation strategy aims to facilitate objective comparison among different products in the tender submissions.

Solution	Ballistic level (NIJ)	En.DCC (N.mm)	DCC test score	F.EP (N)	EP test score	Flexibility score	Flexibility rating
A	Level IIIa	2'027	2.21	49	2.33	**4.5**	**Highest flexibility**
B	Level IIIa	2'825	1.58	85	1.34	**2.9**	2nd highest flexibility
C	Level IIIa	4'313	1.04	99	1.15	**2.2**	3rd highest flexibility
D	Level IIIa	4'473	1.00	114	1.00	**2.0**	4th highest flexibility

Figure 5. Comparative Rating System table

4.1.2 Establishment of a Reference Point for Improvement

The second option proposes defining a benchmark using for example the currently deployed ballistic vest as a reference point. In this context, a scale of improvement, which may be expressed as a percentage, would be established in relation to this reference point.

The DCC test score and EP test score below would thus represent respectively the reduction in mobility constraint and the reduction of edge pressure versus a selected reference such as the system currently in-use.

https://doi.org/10.52202/081621-0024

This approach allows for a more nuanced assessment of the enhancements provided by new body armour solutions relative to existing standards, thereby enabling end users to gauge potential advancements versus a familiar system more effectively allowing end-users to select an improved solution with respect to incumbent in-use equipment.

Solution	Ballistic level (NIJ)	En.DCC (N.mm)	DCC (reduction of mobility constraint % Vs referenced solution)	F.EP (N)	EP (reduction of edge pressure % Vs referenced solution)	Reduction of edge pressure & mobility constraint (%) Vs referenced solution	Flexibility rating
A	Level IIIa	2'027	53%	49	51%	52%	
B	Level IIIa	2'825	35%	85	14%	24%	
C	Level IIIa	4'313	0%	99	0%	0%	Referenced solution
D	Level IIIa	4'473	-4%	114	-15%	-9%	

Figure 6. Comparative Rating System with a referenced solution

4.1.3 In summary

Both options outlined above can be integrated into the tender process. By implementing these systematic approaches, stakeholders can facilitate a more informed and structured evaluation of body armour solutions, ultimately aiding in the decision-making process for procurement.

4.2 Body Armour Manufacturer results

In the frame of this study, a total of ten viable solutions have been developed for body armour manufacturers and evaluated against various criteria, including the protection level, edge pressure, conformability, resistance to bending, weight, and price (Figure 7.).

As may be expected the protection level significantly influences the results of assessments related to the edge pressure, double curvature compression, and lower costal bending tests (Figure 8.). Following the test procedure recommendations, for En.DCC and En.LCB the values of the third sample solicitation are used. It is worthy to be noted that the comfort criteria and ratings are not in accordance with the respective solutions weights outlining the value of not only considering weights with respect to comfort characteristics.

Solution	Protection level	Comfort & mobility				Price
		Edge Pressure	Conformability	Bending stifness	Weight	
		F.EP_after cycle	En. DCC_3rd trial	En. LCB_3rd trial		
		N	N.mm	N.mm	kg/m2	in %
AA	Level II 9mm only	23	785	NC	3.9	122%
BB	Level II 9mm only	35	982	NC	3.9	98%
CC	Level II 9mm only	47	1'586	NC	4.5	57%
DDD	Level IIIa	49	2'027	NC	6.1	52%
EEE	Level IIIa	85	2'825	2'308	4.3	163%
FFF	Level IIIa	99	4'313	3'171	5.9	64%
GGG	Level IIIa	114	4'473	NC	5.0	100%
HHH	Level IIIa	128	5'086	4'194	5.7	66%
III	Level IIIa + Kr2	145	6'500	NC	6.8	161%
JJJ	Level IIIa + Kr2	325	16'000	NC	9.4	141%

Figure 7. Table highlighting the different design criteria for a Body Armour Solution. The En.DCC (N.mm) has been highlighted with a colour scale

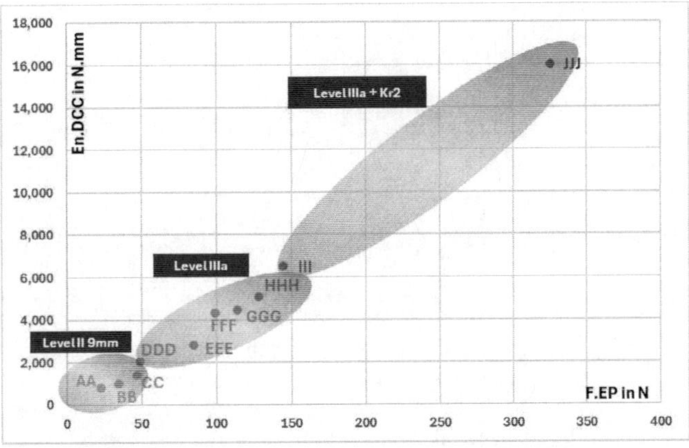

Figure 8. Graph highlighting solutions according to F.EP in N and En.DCC in N.mm (3rd trial) and their associated levels of ballistic performances

Several implications arise from this understanding:

Careful consideration is necessary when establishing thresholds for comfort below which body armour is deemed acceptable and above which it is considered uncomfortable. The primary function of body armour is to ensure protection; hence, comfort and mobility acceptance levels must be defined relative to the protection it provides. Body armour manufacturers will thus define comfort levels for different ballistic protection requirements as well as eventually end-user experience.

Furthermore, the data obtained from these three testing methods can enable assessing the trade-offs between ballistic protection area and user comfort, particularly in evaluating how the ballistic protection area can be enlarged by the body armour manufacturer thanks to the use of more flexible materials.

While data from these three different methodologies frequently align, it is important to examine their outcomes concurrently. Each method—Edge Pressure test, which evaluates pressure distribution and its impact on comfort; Lower Costal Bending test, which assesses the resistance of materials to bending forces; and Double Curvature Compression test, which investigates material behaviour under complex compressive loads—provides a distinct perspective that enriches the overall understanding of comfort and mobility. Therefore, a multifaceted approach incorporating these complementary testing methods is ideal for developing products that effectively address the diverse needs of users.

4.3 Material provider results

Within the framework of this study, four different products in the developmental phase, using similar structures but different raw materials, have been evaluated using the Double Curvature Compression (DCC) test. The primary objective of this assessment was to identify the structure exhibiting the highest degree of flexibility and to facilitate the selection of raw materials that would most effectively enhance flexibility in the final application.

The typical Force (N) versus Deformation (mm) curves, obtained through the DCC testing, are presented below. These curves demonstrate significant variances among the four evaluated solutions, which can be attributed by the analysis of the area beneath the curves. The results of this analysis are quantified in N.mm, enabling a successful guidance in the development of the structures under investigation.

Product development can at an early stage use these test methods to consider comfort aspects in addition to ballistic performance and further production and commercial constraints.

In the below example and for this specific material, two curve sections can be distinguished: 1. The energy needed (N.mm) leading to the main fold on the soft ballistic insert; 2. The energy needed (N.mm) to further conform the folded soft ballistic insert to the double curvature shape. More flexible materials may show a less pronounced first peak before propagation of folds.

https://doi.org/10.52202/081621-0024

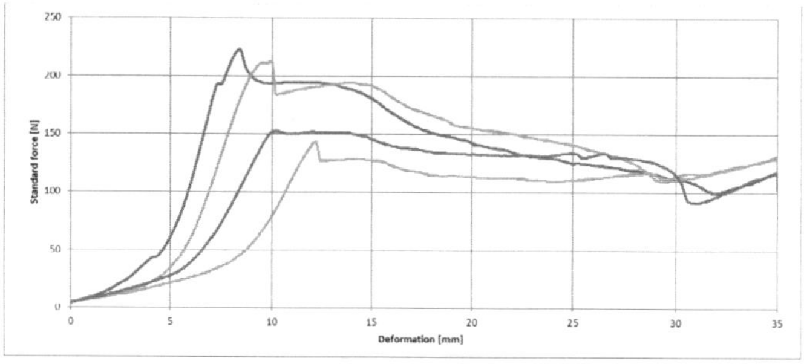

Figure 9. Force (N) Deformation (mm) curves obtained with the DCC for four different products using similar structures but different raw materials

4.4 Test methods availability & accessibility

To ensure that each participant in the value chain is provided with adequate access to these testing methods, continuous communication was fostered among all stakeholders. In this context, Hohenstein Laboratories was identified as a pivotal partner, offering independent testing solutions. Their involvement aimed to facilitate the testing process, provide comprehensive reports, and potentially equip stakeholders with the necessary methodologies for thorough evaluation and assessment of their materials and products.

4.5 Scope of use and path forward

The above test methods and approaches described are focused at evaluating the impact of the soft-armour ballistic protective material assembly on comfort and mobility. The soft ballistic panel assembly is one of the main contributor to comfort. Nevertheless, the additional impacts of rigid insert plates, body armour vest carriers as well as the shaping to enhance fitting for female end-users is to be considered.

4.5.1 Complete ballistic armour vest assessment in conjunction with plates and outer vest carrier

The Edge Pressure test indicates that while the influence of a plate is in many cases limited, the relevance of the test persists, as the localized edge pressure generated by the soft ballistic insert remains significant.

The Double Curvature Compression (DCC) test methodology is incompatible with the use of a plate and therefore does not allow to effectively assess the impact of the plate. However, to enhance mobility, the conformability of ballistic inserts continues to play a crucial role when utilized alongside rigid plates. In that context, DCC should still be capable to discriminate the relative impact of using different soft-armour ballistic solutions when used with the same insert plates.

The Lower Costal Bending (LCB) set-up on the other allows to test complete body armour systems including the insert plate and highlights the importance of the flexibility of ballistic inserts, even when used in conjunction with plates. Preliminary investigations have indeed demonstrated the feasibility of evaluating complete vest solutions, which encompass the ballistic insert, outer carrier, and insert plate. Note that outer carrier designs can also be tested independently from the soft body armour inserts for comfort optimisation of this component. Nevertheless, further validation of the testing procedures including the positioning of the samples is required to ensure that they accurately correlate with user comfort perceptions during actual use.

4.5.2 Ballistic armour female vests

The Edge Pressure test is pertinent for evaluating both male and female body armour, as it effectively assesses pain and pressure points at the vest edges while being minimally influenced by the overall shape of the vest.

Regarding mobility, the assessment becomes more complex with female vests. The Double Curvature Compression (DCC) test can however still be applied to evaluate the conformability of the assembly of constituent materials, given that the multi-axial deformation capability of the ballistic insert is likely a critical factor for the comfort of female vests. Nonetheless, this test does not provide insights into the effects of female-specific shaping techniques.

The Lower Costal Bending (LCB) test can on the other hand generally be utilized in its current form to discern the effects of female body-shaped inserts. To enhance testing methodologies tailored to female body armour, collaborative efforts with end-users will be essential to develop eventual upgrades to the test set-up that could for example leverage proprietary body shape designs.

5. CONCLUSION

The design and assessment of body armour necessitate a meticulous balance between comfort thresholds and ballistic protection levels, given that the primary function of such armour is to safeguard the user while facilitating mobility in various scenarios. This balance requires manufacturers to establish comfort parameters that are aligned with the varying levels of ballistic protection, drawing insights from the experiences of end users. A multifaceted approach is highly valuable, particularly through the application of distinct testing methodologies—namely the Edge Pressure, Lower Costal Bending, and Double Curvature Compression tests. Each of these methodologies offers a unique perspective on material performance, contributing to a comprehensive understanding of how body armour interacts with the wearer's body under different conditions.

Moreover, it is critical to acknowledge that the perspectives on comfort, flexibility, and overall performance provided by these testing methods do not necessarily correspond with traditional weight assessments which allow a more complete overall comfort evaluation from the wearer's standpoint. During the tender adoption process, the insights gleaned from these testing methodologies become instrumental for end users, equipping them with the necessary information to make informed decisions regarding the optimal balance of protection, comfort, and mobility that meets their specific operational requirements.

This analytical framework is essential for both manufacturers and users as it facilitates the evaluation of trade-offs between enhanced ballistic protection and user comfort. By optimizing material design and vest configuration in light of these findings, manufacturers can achieve superior effectiveness to align with end-user specifications. Ultimately, the nuanced understanding of these dynamics is pivotal for developing body armour solutions that effectively address the diverse and evolving needs of users, ensuring not only safety but also operational efficiency and user acceptance in the field.

Acknowledgments

We thank all the Ballistic Armour Manufacturers and End-Users for their collaboration and valuable inputs enabling this implementation methodology of Comfort & Mobility assessment test methods.
Christophe Chabert, Guillaume Coulbaux and Olivier Mauzac from the French MOI are particularly acknowledged for their feedback and support.
A special acknowledgement goes to Jeremie Maurin for the set-up of test design concepts, the assembly and manufacturing of test-rig elements, as well as for leading different testing campaigns. Nicolas Pont, Hakim Hoepffner, Philippe Caudoux and Michel Demolis are acknowledged for their support and technical contributions without which this work would not have been possible. Jan Beringer and Christine Lörcher from Hohenstein are acknowledged for their contribution in the development of the hyperbolic paraboloid test method. Ben Kennedy is warmly acknowledged for his support and interest in this project. Finally, Steven LaGanke and Joe Hovanec are acknowledged for their trust, sponsorship, and support of this project.

References

[1] Tomes C., Orr R.M., and Pope R., Annals of Occupational and Environmental Medicine, 2017, vol. 29, p. 14.
[2] Avient Corp, US law enforcement survey, 2022; www.officer.com/on-the-street/apparel/uniform-fabrics/press-release/21289606/dsm-dyneema-dyneema-releases-us-law-enforcement-survey-results.
[3] Phillips M.P., Shapiro R. and Bazrgari B., Ergonomics, 2016, vol. 59, pp. 682–91.
[4] Lenton G., Aisbett B., Neesham-Smith D., Carvajal A. and Netto K., Ergonomics, 2015, vol. 59(6), pp. 808–816.
[5] Horsfall I., Champion S.M. and Watson C.H., Applied Ergonomics, 2005, vol. 36, pp. 283–292.
[6] Kissinger C., Mitschang P. and Neitzel M., Materials for Transportation Technology Vol 1; Wiley-VCH Verlag GmbH, Germany, Winkler P.J. Ed. , 2000, pp. 176-182.
[7] ASTM D1388, Standard Test Method for Stiffness of Fabrics – Option A: Cantilever test, 2018.
[8] Karakolis T., Sinclair B.A., Kelly A., Terhaar P. and Bossi L.L.M., Human Factors, 2017, vol. 59(4), pp. 535-545.
[9] Boogh L., Djololian C., Personal Armour Systems Symposium (PASS 2023), Dresden, Germany, 11-15 September 2023, pp. 259-268.

https://doi.org/10.52202/081621-0024

Characterisation of hit point and wound tract distribution from lethal firearm violence in Stockholm 2021-2024

Jens Danielsson[1], David Andersson[1], Lisa Wallgren[2] and Lydia Kahn[2]

[1]Swedish Defence Research Agency, 164 90 Stockholm, Sweden
jens.danielsson@foi.se
[2]The Swedish Board of Forensic Medicine, Västra Järnvägsg.3, 111 64 Stockholm, Sweden

Abstract. The design of personal armour systems is based on the assumption that incoming missile threats have a roughly horizontal directivity, and that hits are concentrated to the chest area. This seems reasonable in a military context, with small arms fired at intermediate to long distances. However, in close quarters combat environments, engagements with illegal combatants, or in law enforcement situations, this may no longer be the case. This work investigates possible repercussions of these factors on modern body armour design.
Here we report the results of a multidisciplinary study of lethal shootings in the Stockholm metropolitan area in the period 2012-2023. We characterise the directionality of the observed wound tracts in terms of their corresponding entrance and exit wounds. The majority of the data is collected from civilian homicides, which to a first order approximation represents close quarters combat and intermediate range engagements. Furthermore, in case of multiple wound tracts each wound tract is assessed by an experienced forensic pathologists to characterize the wound as lethal or not.
Preliminary results indicate a presence of entrance wounds for lethal wound tracts outside the area typically covered by body armour, and furthermore that the wound tracts still cause lethal damage to vital intra-abdominal or thoracic tissues. We report observed distributions of hit points and wound tract directionality, to provide the community with an image of the threat of lethal close quarter and intermediate range combat situations.

1. INTRODUCTION

It is a well-known fact that the use of body armour reduces fatality rates in firearm engagement [1], both in a civilian and military context [2]. However, there are also studies that suggests that body armour to varying degrees impair physical performance [2, 3], due the added weight and reduced mobility. This means that the use of, and the design of, body armour systems, especially those with high-weight plates, has to balance the cost of reduced physical performance and the gain of increased survivability. In order to successfully optimize such a trade-off, ballistic threats have to be carefully characterized and evaluated.

The design strategy of body armour is to protect the most vital organs while minimizing weight to maximally maintained mobility [4]. The conventional solution is to add ceramic plates covering the front and back of the thoracic cage, to protect the heart, lungs and to some extent abdominal contents. However, this placement of said plates is made with the assumption that the angular dispersion of incoming missiles is limited, and that the target is in standard anatomical position, exposing either the chest or back areas. In line with this assumption, armed force related gunshot wounds, in abdomen and thorax, are commonly found with lateral entry wounds [5, 6]. Hence, in some cases, the assumptions on missile threat directionality will certainly be sound – such as armed forces fighting from an entrenched position – nonetheless, in close quarter combat, or in a law enforcement environment, more dynamic encounters may occur where the directionality of incoming missiles may be less predictable. In such cases, we propose, the directionality of incoming missiles may more closely resemble those of civilian shootings.

There are numerous studies describing the epidemiology and the characteristics of wounds from firearms based on civilian health care and/or forensic pathology statistics [7-10]. The overall pattern is strikingly consistent across countries and continents [11] . Overall, in civilian firearm-related fatalities head injuries are dominating in relative numbers, followed by thoracic and abdominal injuries [11-16]. This is not surprising as these body regions harbours the main vital organs and large vessels. An important deviation from this pattern is lethal gunshot wounds caused by law enforcement personnel. A retrospective study of such gunshot injuries showed that the main location of potentially lethal entry wounds was in fact the front of the chest [17], followed by the back, the head and abdomen. This highlights an important distinction between trained and untrained shooters, though it certainly also reflects the difference in context where these lethal firearm injuries occurred. However, it

https://doi.org/10.52202/081621-0025

is worth pointing out that the threat against the law enforcement typically are perpetrators, much like the ones that lie behind the statistics of civilian gunshot fatalities.

In armed forces, fatal gunshot wounds are rare, compared to injuries and fatalities from various explosive devices [18]. This may be related to the increased use of body armour, but the evolving characteristics of the battle field probably is a more plausible explanation [7, 18].

The location of the vital organs is an important component when determining where extra protection is needed, however in this paper we highlight the fact that the projectile may in principle take any number of pathways to reach these vital organs. If a projectile hits a person from an angle not normal to the chest or back plates, it can penetrate into, *e.g.*, the thoracic cavity and cause fatal injuries. As an example: A study by forensic pathologists of 40 people presumably killed by armed forces showed that only a fraction of the observed wound tracts followed a horizontal path, and conversely that a comparatively large portion of wound tracts did not originate in the central front or back of the thoracic cage [19].

All of this highlights the importance to not only account for the actual hit point (the entry wound), but also the pathway, or the directionality, of the wound tract, when characterizing firearm projectile threats.

To address this question 104 wound tracts from gun shots were analysed. They were the result from civilian shootings in the greater Stockholm metropolitan area, from 2021-2024. None of the victims carried body armour. The wound tracts were characterized in terms of entrance and exit points and the directionality of the pathway connecting the wounds. Furthermore, for victims with multiple wounds, each wound tract was evaluated by an experienced forensic pathologist to determine whether it is considered lethal or not. The observed positions of the entrance wounds in this study are in good agreement with those in other studies, with hits in the head, thorax or abdomen, in order of relative lethality. However, analysis of the directionality of the subsequent wound tracts suggests that the directions of incoming projectiles is highly stochastic. Furthermore, many lethal thoracic and/or abdominal injuries was registered in wound tracts that did not exhibit a directionality consistent with the placement of reinforcement plates in modern body armour systems.

2. METHODS

All shot lines included in this study resulted from non-suicidal civilian shootings in the greater Stockholm metropolitan area. A total of 104 shot lines were randomly selected to be included. For each case, the autopsy protocol, associated photographic documentation, and computer tomography (CT) data was analysed and the localisations of the entrance and exit wounds was determined. In the cases where the projectile was retained in the body, the position of the bullet was instead determined. Each wound tract was analysed independently and any combination of injuries are not reported here. In order to be able to compare shot lines across bodies of varying sizes and morphologies, the body was divided into 53 sub-regions (Figure 1) and the position of each entrance and exit wound was assigned a specific region. The entry wound localisation analysis was then performed at this data resolution.

To analyse the directionality of the shot line, each entry and exit wound was given a random location within the region (NX) where it was observed. The pseudo-random localisation (within a sub-region defined in Fig. 1) of the entrance and exit wounds were then projected onto a three-dimensional human model. This model was based on the geometries of the low resolution human computer model ComputerMan [20]. The directionality of every wound tract was then characterised by two angles, where the first angle refers to rotation around the longitudinal axis of a standing person, where a front-to-back direction is 0° and where positive rotation is related to increased angle value. The second angle refers to elevation where a horizontal path is defined as 0°, straight from above is +90° and from below is -90°.

3. RESULTS AND DISCUSSION

104 gunshot injuries were analysed, and these resulted in a total of 208 external wounds or retained bullet, 47 (45 %) wound tracts was considered lethal. The overall localisations and lethality are summarised in Table 1. In order to determine the mechanism of lethality in this dataset, the positioning of the entry and exit wounds were determined and analysed.

For each wound the localisation was determined with a higher resolution than in Table 1, following the anatomic division in Fig. 1. In general, the distribution of wounds (entry and exit) involves all body regions, save the distal parts of the extremities, Fig. 2.

https://doi.org/10.52202/081621-0025

3.1. Localisation of entry and exit wounds

Next, the position of entry wounds was analysed, both in order to be able to compare with other data, but also because the position of the entry wound is related to the feasibility of covering the entrance wound with body armour.

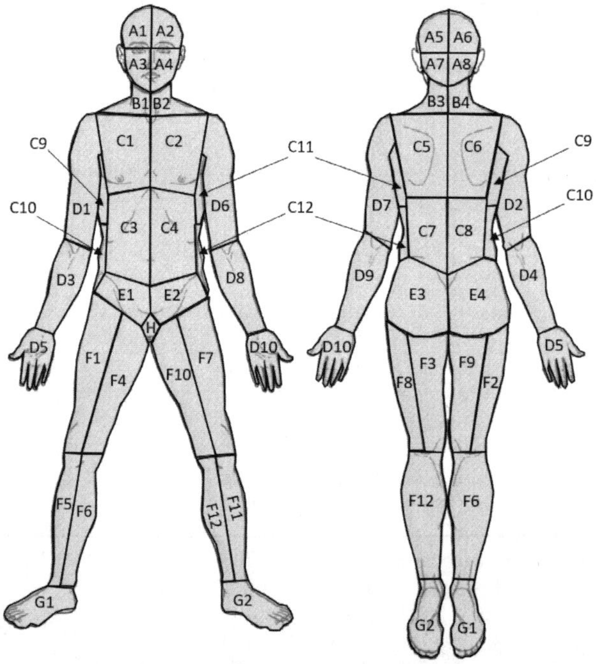

Figure 1. The division of the body into regions. There are 7 regions A: Head, B: Neck, C: Torso, D: Upper limbs, E: Pelvis and buttocks, F: Lower extremities, G: Feet and H: Outer genitalia. Each region is then further divided into sub-regions (save the outer genitalia).

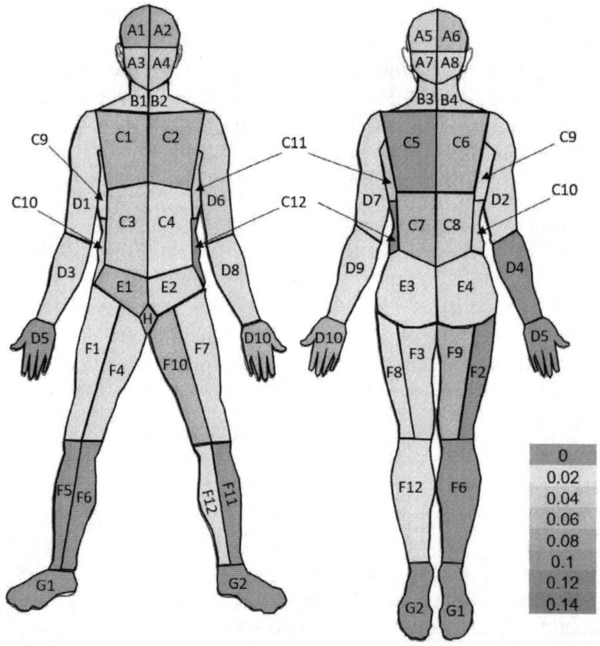

Figure 2. Relative localisation distribution of all gunshot related wounds in this study. Total number of wounds (entry and exit/retained bullet) are 208. Colour represents fraction of wounds in each area.

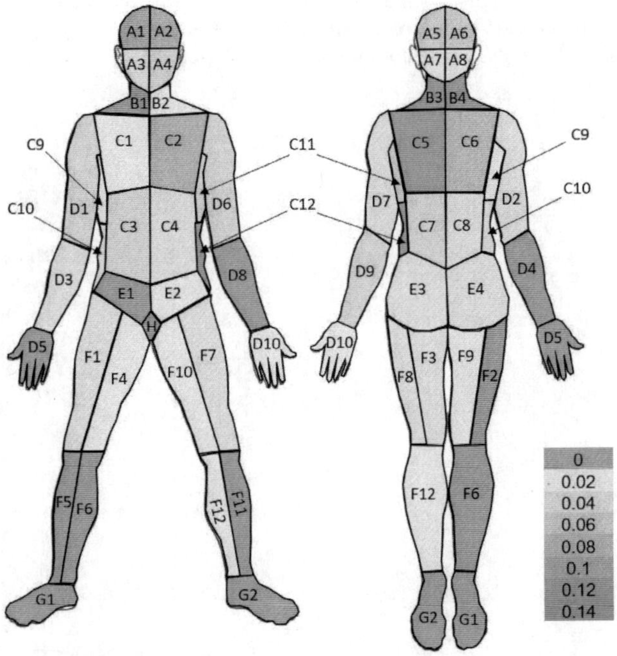

Figure 3. Relative localisation distribution of entry wounds in this study. Total number of entry wounds are 104. Note the similarity in distribution between entry wounds alone and entry and exit wounds together (Fig. 2).

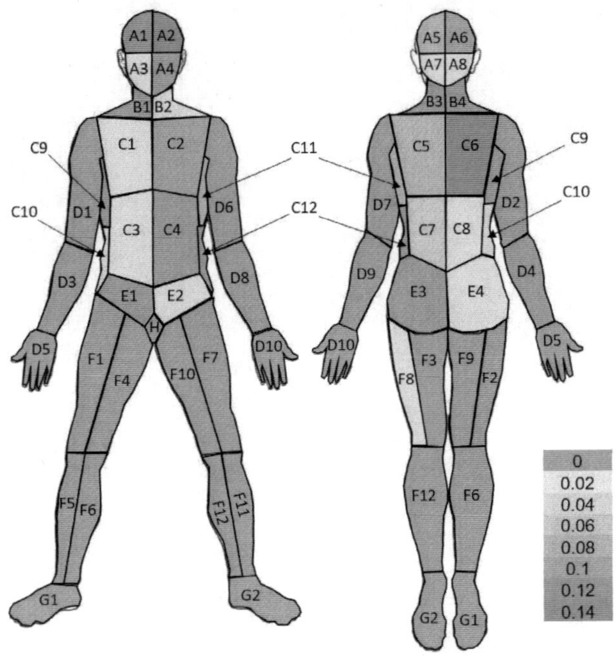

Figure 4. Relative localisation distribution of entry wounds related to lethal injuries in this study. Total number of lethal entry wounds are 47. It is clear that lethal injuries are associated to entry wounds in the proximity of the regions harbouring the vital organs.

https://doi.org/10.52202/081621-0025

Interestingly, the distribution of entry wound localisations in Figure 3 overlaps reasonably well with the general distribution of wounds in Figure 2. This indicates that there are no specific angle of incidence, but rather that the shot lines seem uniformly distributed.

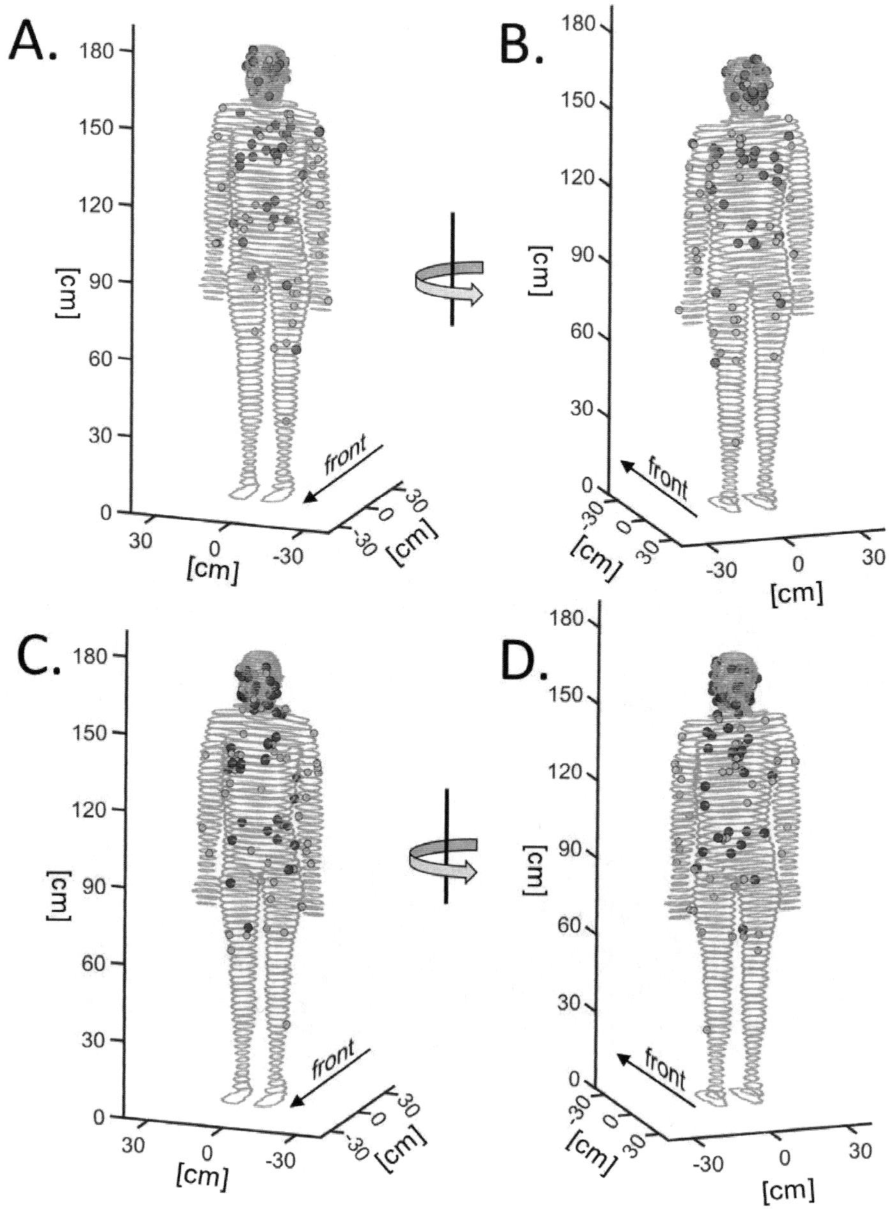

Figure 5. Entry (panel A-B) and exit (panel C-D) wounds projected onto a 3D human model. Panels A and C show the human model from a slightly elevated front side view, and panels B and D obliquely from behind. Larger, dark red and blue markers refer to lethal injuries.

Finally the position of the lethal entry wounds was identified (Figure 4 and 5), as these are important in terms of design of protection. Here, relatively large differences in distribution of locations compared to the more general wound distributions are found, Figures 2 and 3. First, the number of sub-regions that are involved in lethal shot lines are significantly lower than the number of regions involved in non-lethal shot lines. This concurs with intuition: lethal shot lines involve entry wounds in the head, neck, thorax, abdomen or pelvis. Only one entry wound (in the thigh) outside of those regions was considered lethal. This entry wound corresponded to a shot line almost parallel to the main standing axis of the person, resulting in injuries in the pelvic/abdominal areas.

Table 1. Localisation and lethality of the injuries in the dataset used for this study.

Localisation	No of wounds[1)] (N = 208)	Entry wounds (N = 104)	Lethal injury (N = 47)	Lethality[2)]
Head/neck	64	31	22	71 %
Thorax	58	29	17	59 %
Abdomen	26	12	5	42 %
Pelvis/buttocks	12	5	2	40 %
Extremities	48	27	1	4%

1) Here wounds are defined both as a physical entry or exit wound or a remaining bullet in the tissue.
2) Lethality = N(lethal injuries)/N(entry wounds)

3.2. Directionality analysis

For each pair of entry- and exit wounds an exact position in its assigned sub-region was randomly selected. This was done to introduce a slight variability of the precise localisation. The localisation of the entry and exit wounds projected onto the human model ComputerMan are shown in Figure 5. From these positions, a directionality for each shot line could be estimated, and are defined as follows: the projectile pathway starts at the entry and ends at the exit wound, and are assumed to be straight. Figure 6 shows all wound tracts, and the directionality are shown as arrows.

It is evident from Figure 6 that, in this data set, there are no obvious preferred direction to the shot lines, rather a seemingly random distribution of directions is observed. This was already indicated by the localisation distribution similarities between entrance and exit wounds (Figures 2 and 3). A question then arises: is this random distribution of directionality also a feature of the lethal wound tracts? To investigate this the directionality for the lethal wound tracts alone was plotted (Figure 7).

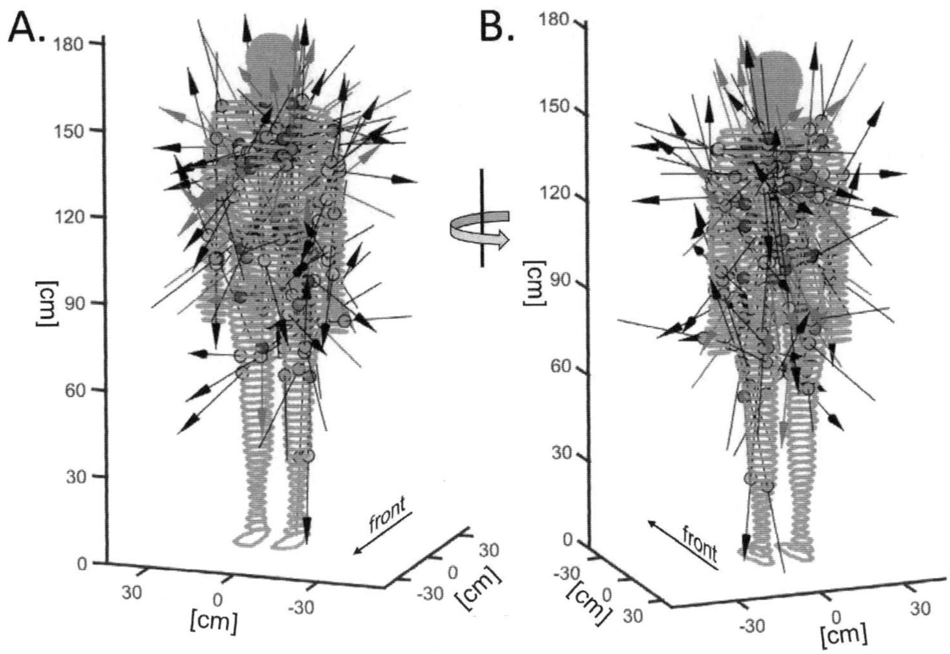

Figure 6. Directionality for all 104 shot lines. Each tract is defined by its corresponding entry and exit wound. An arrow shows the directionality, and indicates the direction of the projectile. The head injuries are removed for visibility reasons. Red arrows indicate wound tracts considered lethal; black non-lethal.

https://doi.org/10.52202/081621-0025

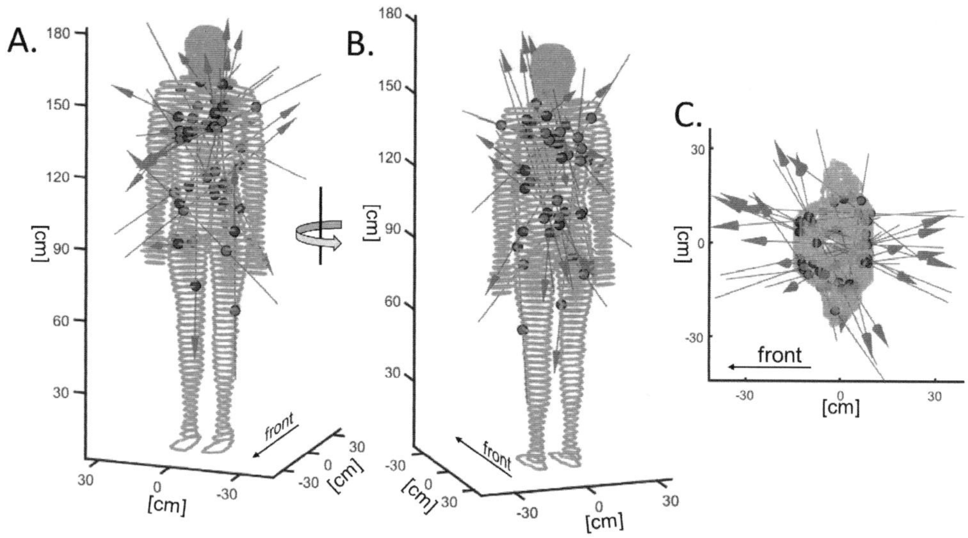

Figure 7. Directionality distribution for lethal wound tracts. The head injuries are removed for visibility reasons. The rightmost panel show the model directly from above.

Here, it is clear that the observed directionalities of the lethal wound tracts are more or less uniformly distributed along all directions. In fact, it appears that many of the wound tracts follow shot lines that does not overlap with the typical placement of hard body armour plates. Nonetheless, the full length of the lethal wound tracts outside the head are almost exclusively located within in the torso, with a preference to the thoracic area. This type of analysis is visually accessible and provides an illustration of the high degree of variability of the wound tracts from gun shots.

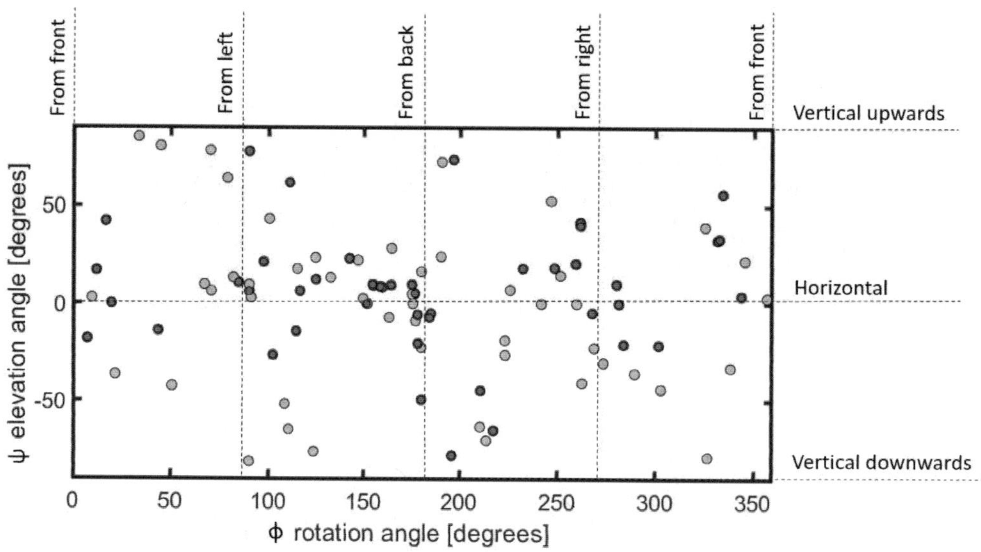

Figure 8. Angles φ, ψ (rotational and elevation angles) for each wound tract vector. Blue markers correspond to non-lethal injuries and red to injuries assessed as lethal.

In order to obtain a quantitative measure on the directionalities, each directionality vector was parametrised by its rotational angle around the human model's main axis as well as the elevation angle from the ground. The full-frontal direction was set to rotational angle $\varphi = 0°$, and a fully horizontal direction corresponds to an elevation angle $\psi = 0°$. Hence, a horizontal wound tract through the thorax directly from the front to the back, would be characterized by $\varphi, \psi = [0°, 0°]$. In Figure 8, the angles for all vectors are shown.

Although the number of data points are too few to form a definitive conclusion about the directionality, it is noted that there seem to be a slight horizontal clustering of lethal shot lines. The variability in ψ is for lethal injuries $\pm$ 32°, while for non-lethal $\pm$ 38°. More data is however needed to ensure viable statistics.

4. CONCLUDING REMARKS

In this study we present a retrospective, morphologic characterisation of wound tracts from gun shots in Stockholm. All gun shots come from civilian shootings, i.e., suicide data as well as law enforcement shooting data has been omitted. The data set includes both wound tracts individually considered lethal and non-lethal. The lethal and non-lethal injuries are separated in the analysis and significant differences in the localisation and directionality is found.

Overall, the distribution of individual entry and exit wounds agree well with the data reported in the literature on civilian shootings. It is noted that while the lethal entry wounds are mainly located in the head, thoracic and abdominal region, the directionalities of the corresponding shot lines are more or less stochastic.

This method to analyse forensic data may serve as a basis to develop and optimize body armour, both soft and hard, plate armour.

References

[1] Liu, W. and Taylor, B., *The effect of body armor on saving officers' lives: An analysis using LEOKA data.* J Occup Environ Hyg, 2017. **14**(2): p. 73-80.

[2] Schram, B., et al., *The effects of body armour on mobility and postural control of police officers.* J Bodyw Mov Ther, 2020. **24**(3): p. 190-194.

[3] Larsen, B., Netto, K. and Aisbett, B., *The effect of body armor on performance, thermal stress, and exertion: a critical review.* Mil Med, 2011. **176**(11): p. 1265-73.

[4] Danielsson, U., Bergh, U. and Grambo, S., *Body armour. Effects on performance and physical load.* FOI, Totalförsvarets Forskningsinstitut, 2003. FOI-R--0799--SE.

[5] Dilday, J., et al., *Injury Patterns and Treatment Strategies in Civilian and Military Isolated Abdominal Gunshot Wounds.* World J Surg, 2023. **47**(11): p. 2635-2643.

[6] Mabry, R.L., et al., *United States Army Rangers in Somalia: an analysis of combat casualties on an urban battlefield.* J Trauma, 2000. **49**(3): p. 515-28; discussion 528-9.

[7] Bellamy, R.F., Maningas, P.A. and Vayer, J.S., *Epidemiology of trauma: military experience.* Ann Emerg Med, 1986. **15**(12): p. 1384-8.

[8] Marri, M.Z. and Bashir, M.Z., *An epidemiology of homicidal deaths due to rifled firearms in Peshawar Pakistan.* J Coll Physicians Surg Pak, 2010. **20**(2): p. 87-9.

[9] Nyberger, K., et al., *Epidemiology of firearm injuries in Sweden.* Eur J Trauma Emerg Surg, 2022. **48**(3): p. 2349-2357.

[10] Nyberger, K., et al., *Management and outcomes of firearm-related vascular injuries.* Scand J Trauma Resusc Emerg Med, 2023. **31**(1): p. 35.

[11] Thube, H.R. and Chikhalkar, B.G., *A study of fatal gunshot injuries over two years in Mumbai, India.* Indian Journal of Forensic and Community Medicine, 2017. **4**(3): p. 154-157.

[12] De-Giorgio, F., et al., *Suicidal or homicidal sharp force injuries? A review and critical analysis of the heterogeneity in the forensic literature.* J Forensic Sci, 2015. **60 Suppl 1**: p. S97-107.

[13] Jena, S., et al., *Critical Analysis of Firearm Wound Characteristics in Victims Brought for Autopsy at LHMC, New Delhi: A 20-Year Retrospective Study.* European Journal of Medical and Health Sciences, 2023. **5**(4).

[14] Myint, S., et al., *Fatal firearm injuries in autopsy cases at central Bangkok, Thailand: a 10-year retrospective study.* J Forensic Leg Med, 2014. **28**: p. 5-10.

[15] Solarino, B., Nicoletti, E.M. and Di Vella, G., *Fatal firearm wounds: a retrospective study in Bari (Italy) between 1988 and 2003.* Forensic Sci Int, 2007. **168**(2-3): p. 95-101.

[16] Toygar, M., et al., *An analysis of firearms-related deaths between 1993-2010: a retrospective study.* Ulus Travma Acil Cerrahi Derg, 2013. **19**(6): p. 536-42.

[17] Gill, J.R. and Pasquale-Styles, M., *Firearm Deaths by Law Enforcement.* Journal of Forensic Sciences, 2009. **54**(1): p. 185-188.

[18] Belmont, P.J., Jr., et al., *Incidence and epidemiology of combat injuries sustained during "the surge" portion of operation Iraqi Freedom by a U.S. Army brigade combat team.* J Trauma, 2010. **68**(1): p. 204-10.

[19] Rainio, J., Lalu, K. and Penttila, A., *Independent forensic autopsies in an armed conflict: investigation of the victims from Racak, Kosovo.* Forensic Sci Int, 2001. **116**(2-3): p. 171-85.

https://doi.org/10.52202/081621-0025

[20] US Army lab, *A Computer Man Anatomical Model*, 1978, Army ballistic research lab Aberdeen proving ground MD.

Ballistic evaluation of flexible hard armour plates in bent and flat configurations

J. Bélanger[1], S. Ouellet[2], G. Pageau[3]

[1]*Munition Experimental Test Centre (METC) – Valcartier, Department of National Defence, 2459 de la Bravoure Rd., Québec City, Québec G3J 1X5, Canada.*
Jacob.belanger2@forces.gc.ca
[2]*Defence R&D Canada – Valcartier, Department of National Defence, 2459 de la Bravoure Rd., Québec City, Québec G3J 1X5, Canada.*
[3]*Promaxis Systems Inc., 2385 St. Laurent Blvd, Ottawa, Ontario K1G 6C3, Canada.*

Abstract. The growing recognition of the need for gender-specific protective equipment has highlighted the limitations of traditional rigid ballistic plates, particularly for female users. This emerging requirement has driven industry to develop innovative solutions to increase the fit and comfort of body armour, including high-curvature rigid plates and flexible hard armour. Flexible plates are designed to better conform to diverse body morphologies and offer increased comfort and mobility while maintaining the ballistic performance of traditional hard armour plates. Given the increasing availability of these technologies and the limited available data regarding their performance, various models of flexible plates were acquired with two objectives: 1) Adapt the ballistic test methodology to account for the new challenges posed by flexible hard armour, and 2) evaluate the ballistic performance of the acquired armour system. The plates were initially radiographed in their neutral state to establish a baseline. Subsequent bending tests revealed directional limitations in some models and a significant initial resistance that decreased with repeated flexions. This led to the development of a break-in procedure to simulate real-world usage. A second radiography series was then performed with the plates bent to match the NIJ 0101.07 applique shape. This was conducted to assess the impact of the break-in process on the plate's construction and integrity and identify potential vulnerabilities in the bent configuration. Finally, the plates were subjected to ballistic testing in the standalone condition against their specified threat in various bent configurations to evaluate their performance. While only a few complete penetrations were observed, the tests highlighted excessive backface deformation. Additional limited testing, however, suggests that using the plates in conjunction with other protective elements (e.g. as a system) may help mitigate backface deformation to a certain extent.

1. BACKGROUND

In recent years, there has been a growing requirement for body armour that can accommodate a more diverse population and a larger range of body morphologies. Flexible hard armour plates and body armour with higher curvature radii are becoming increasingly common on the market. The latest edition of the NIJ now includes test methodologies specifically tailored toward female body armour [1].

The increasing demand for inclusive ballistic plate armour has driven the industry to develop new solutions that enhance suitability and comfort for everyone. These solutions include a growing range of available sizes and shapes, such as the one in Figure 1 below.

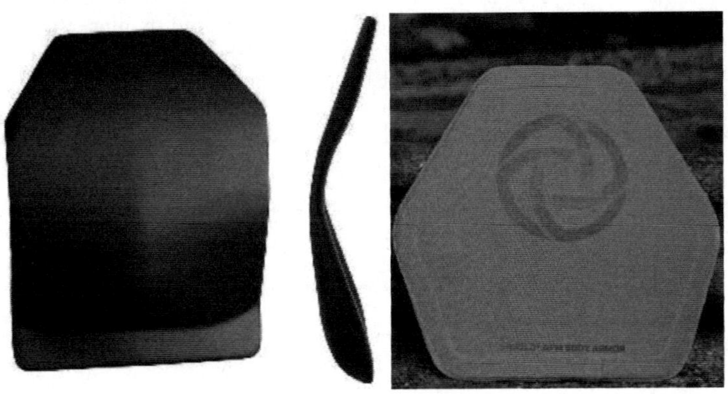

Figure 1 : Example of inclusive hard armour plates on the market [2] [3]

https://doi.org/10.52202/081621-0026

One of the first commercial flexible hard armour systems was *Dragon Skin*, proposed in 2007 by Pinnacle Armour. The system, shown in Figure 2, featured a unique design composed of overlapping ceramic discs arranged in a flexible scale-like pattern, resembling reptilian/fish skin [4].

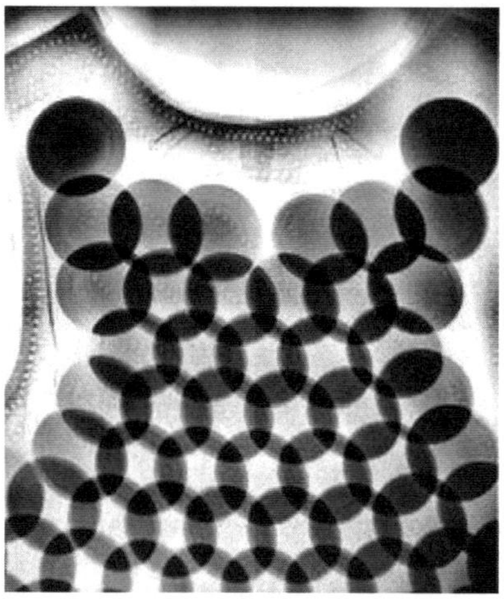

Figure 2 : X-ray of dragon skin body armour (adapted from Wikipedia)

Despite the commercial failure of the *Dragon Skin [4]*, new models of flexible hard armour systems seemingly meeting military performance requirements are now available. The added conformability of such designs accommodates various body types and hopes to enhance both comfort and mobility for users.

The increasing market availability of this technology, coupled with limited performance data, led the Department of National Defense to procure small quantities of various Commercial of the Shelf (COTS) systems with the goals to understand current technologies, investigate necessary adaptions to test methodologies and assess the performance of flexible plates compared to standard rigid plates. The levels of performance stated by the manufacturers for the product evaluated were level RF1 following the NIJ 0101.07 standard [1]. Some models were claimed to stop additional and more performant threats, such as the 7.62mm API-BZ.

2. METHOD

2.1 General observations of the technologies

The first notable difference is the thickness between standard rigid plates and flexible plates. For this level of performance, a modern plate thickness typically ranges between 19 to 25 mm. In this case, most models were around 22mm, but some were as thick as 29 and even 37mm. This difference in thickness can be explained by the level of performance of the plate and by its construction. Instead of relying on a monolithic ceramic design, all tested flexible plates used multiple tiles made of various materials and shapes (e.g. hexagonal) placed side by side in a mosaic structure (Figure 3) and bonded to semi-flexible composite backing, i.e. not consolidated. The flexible composite backing provides some shape conformability while maintaining the ability to stop projectiles. The overlapping of adjacent tiles for one particular system also contributed to making the flexible plate thicker.

https://doi.org/10.52202/081621-0026

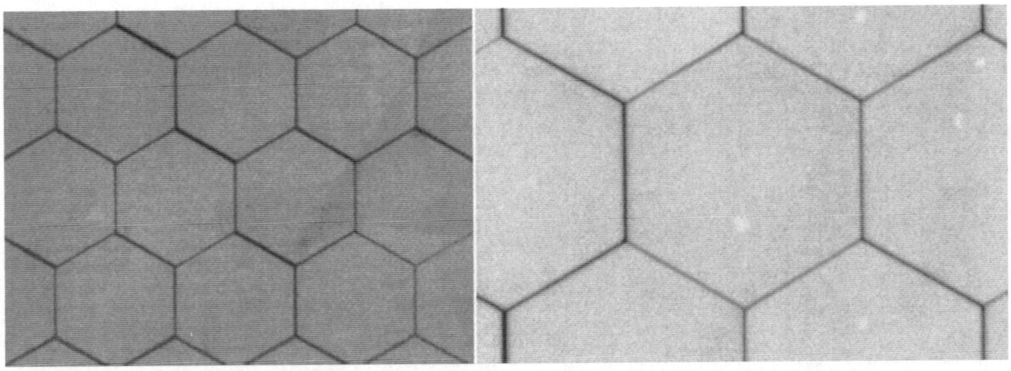

Figure 3 : Construction example of two different flexible plates seen through X-Ray. Both designs have the same scale.

Regarding areal density, the acquired plates ranged from 32 to 48 kg/m². These values were obtained based on the protected area of the plate in a neutral position (not bent) as opposed to the overall area. As a reference, a regular modern level 3 plate can have an areal density as low as 26 kg/m². The materials used in the construction of these plates have a significant influence on the areal density. For instance, the heaviest system used metallic tiles at the strike face instead of ceramic tiles. The data reveal a distinct trade-off, wherein gains in flexibility are counterbalanced by increases in both the mass and, occasionally, the thickness of the plates. This also shows the importance of human factor testing to assess how well this compromise performs in practice.

2.2 Radiography & Break-in

Radiography was used to determine if bending the plates might create vulnerabilities at the intersection between tiles. These vulnerabilities would then be exploited during the ballistic test. For this purpose, a low-density foam shaped in accordance with NIJ 0101.07 C-5 standards [1] was used to bend the flexible plates in the same manner as they would during ballistic testing. An example of a plate positioned in front of the X-ray tube is shown in Figure 4. In most cases, bending the plate separated the tiles over one or two rows near the breast region, as shown in Figure 5.

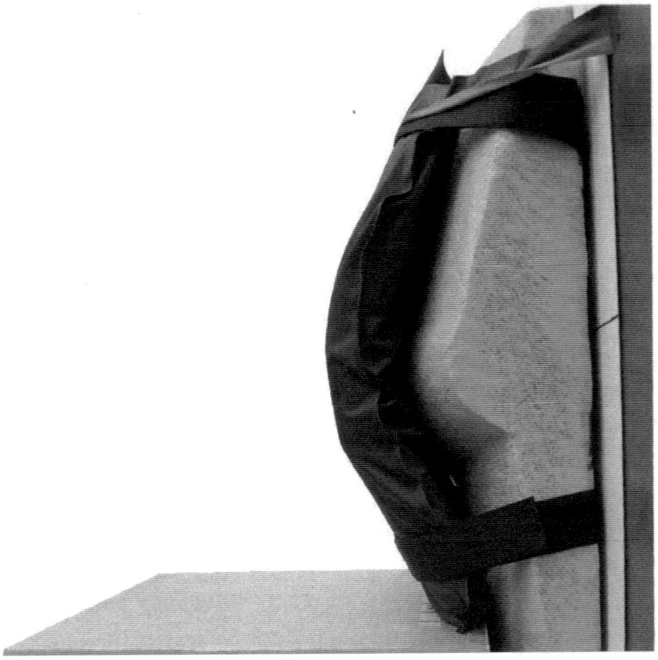

Figure 4 : NIJ 0101.07 C-5 Shape foam for radiography

https://doi.org/10.52202/081621-0026

Figure 6 shows the significant differences in curvature between the NIJ C-5 female applique and the standard SAPI profile.

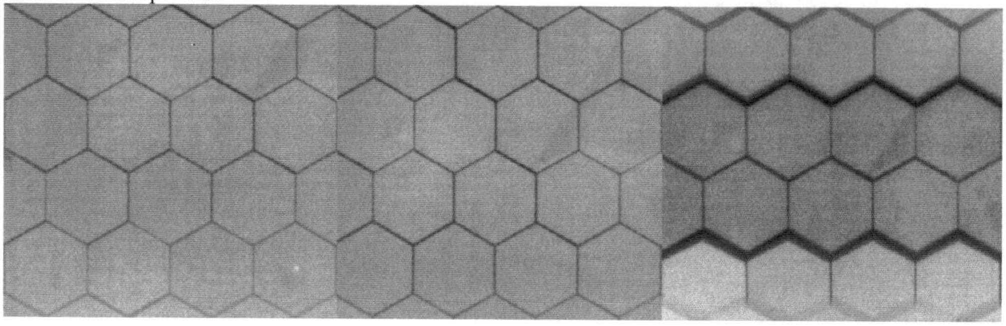

Figure 5 : Tiles before break-in (left), after break-in (middle) and bent on the C-5 applique (Right).

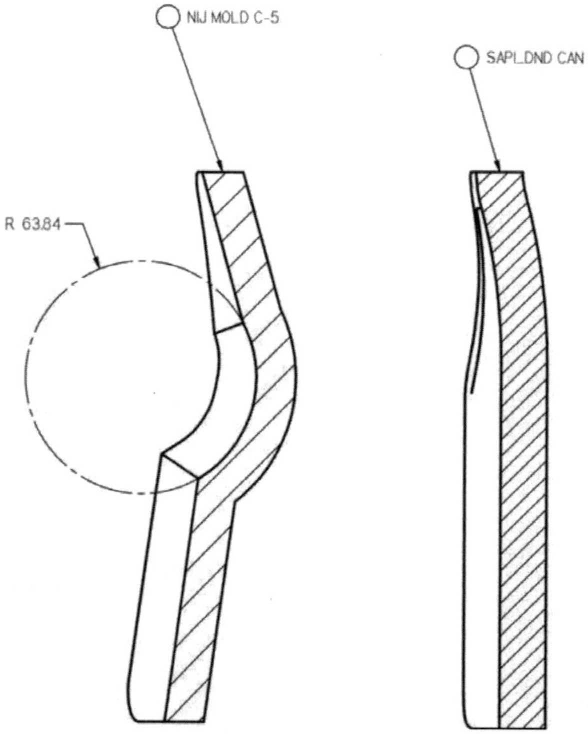

Figure 6 : Curvature difference between the NIJ C-5 and the SAPI plate

The construction and the material used to assemble these plates also affect their flexibility. Some models offer flexibility in both directions simultaneously, while others can only be bent in one direction. It also became apparent that the initial bending resistance decreased significantly with repeated flexion. Therefore, before proceeding with a ballistic test, all plates underwent a break-in process where they were manually bent in the vertical plane several times until the force required decreased and stabilized. Approximately 30 cycles were necessary for each plate to get to this point.

Radiographies were taken before and after this break-in procedure to determine if it had any adverse effect on the position and bonding of the tiles. Minimal to no differences were observed due to the break-in process, as seen in Figure 5.

2.3 Ballistic Test

Following the break-in process, the plates were directly tested without performing any additional conditioning (e.g., drop, temperature cycling, humidity) and following the NIJ test procedures [1]. Prior to testing, plates were conditioned for at least 12 hours at an ambient room temperature of 20±3°C.

ROMA Plastilina No.1 modelling clay was used as the backing material for all tests. The clay was calibrated using the standardized drop-weight technique. A 457 x 406 x 102 mm rigid metallic clay box was used for all tests. For the medium-size plates, the clay backing was shaped in accordance with the SAPI shape, and the NIJ C-5 female applique since the majority of the acquired plates were medium sizes (e.g. 25x 30 cm) in a SAPI shooter's cut profile. For the plate of a smaller size (20x25 cm), the NIJ C1 female applique was used to shape the clay backing. During testing, the plates were successively mounted and bent on the SAPI-shaped clay backing and on the higher curvature NIJ C-1 /C-5 shaped backings (Figure 7), which is referred to as the flex shape designed for testing female soft armours. The plates were all tested in the standalone configuration (e.g. without any soft armour insert behind) except for one plate, which was tested in conjunction with an aramid soft-body armour insert with an areal density of 3.3 kg/m^2. The detailed test matrix is provided in Table 1.

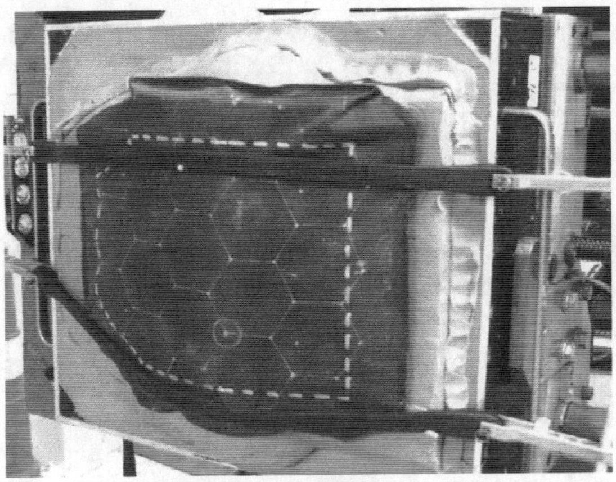

Figure 7 : Flexible Plate mounted on NIJ C-5 clay applique.

In contrast to NIJ 0101-07 [1], the target support for each plate was adjusted to produce normal impacts with a zero-degree obliquity angle. Velocity measurement was done using two independent Doppler radar. Yaw was measured within 150 mm of the impact point using two high-speed orthogonal cameras. To be considered fair, the maximum velocity deviation was ±10m/s, and the maximum acceptable yaw was 5°. Two custom shot patterns were defined for the ballistic tests. For the 7.62 mm rounds, 4 rounds were fired in a 120mm ±15mm square/diamond shooting pattern, and for the 5.56mm rounds, 6 rounds were fired in two triangle configurations, as shown in Figure 8. No shots were made within a 25 mm wide exclusion zone around the plate to minimize edge effects. Triple points and weak areas were targeted as frequently as possible for the SAPI-shaped configuration. For the flex configuration, the shot patterns were adjusted to incorporate as many shots as possible in the areas with the largest inter-tile separations due to bending.

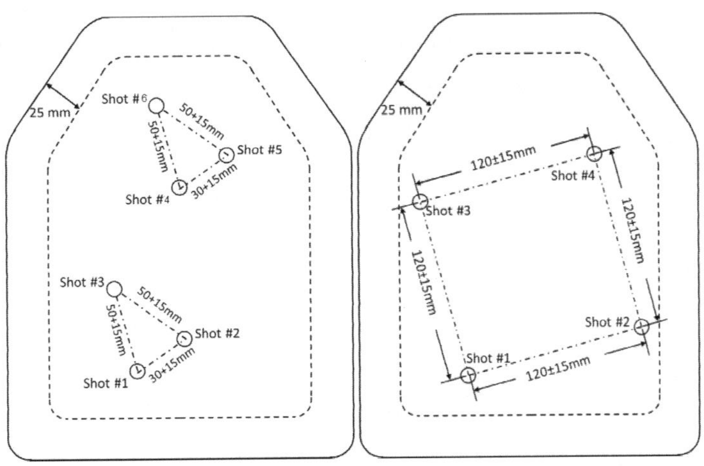

Figure 8: Shooting patterns for the 5.56 mm on the left and the 7.62 mm on the right.

 https://doi.org/10.52202/081621-0026

2.4 Threats

Three different threats were used to test the acquired plates, as described below and illustrated in Figure 9.

2.4.1- 5.56mm C77

This is the Canadian version of the M855 or SS109. This is a 4.0±0.1 grams (61.8±1.5 grain) full metal jacket bullet composed of a lead and steel core. Its test velocity is 950 m/s. This threat is equivalent to the RF2 level as per NIJ 0101-07 [1].

2.4.2 – 7.62mm C21

The C21 is the Canadian version of the M80. This is a 9.5±0.2 grams (147±3 grain) full metal jacket ball round with a lead core and a copper jacket. Its test velocity is 847 m/s. This threat is equivalent to the RF1 level as per NIJ 0101-07 [1] or level III as per NIJ 0101-06 [5].

2.4.3– 7.62mm API-BZ Surrogate

The API-BZ projectile, used for this test series, is a Canadian surrogate developed by DRDC to match the ballistic penetration performance of the original Russian round as per VPAM level 8 [6]. It is a 7.70 grams (118.8 grain) bullet with a test velocity of 740 m/s. Only plates that were rated for this level were tested against it.

Figure 9 : 5.56mm Canadian C77 (Left), 7.62mm Canadian C21 (middle) and 7.62mm API-BZ Surrogate (Right).

3. RESULTS

A total of 116 shots were fired against the 25 flexible plates acquired for this project. Among those 25 plates, six (6) different systems were tested against the three threats from section 2.3.4 in both the SAPI and the flex configuration, as shown in Table 1 below.

Table 1 : Test Matrix

Plate model	AD (kg/m^2)	Level	Construction	Config	Backing	Threat	Nb plates	Shots on tile	Shots on gap
M1	47.9	NJI 06 III+	1.2 in^2 hexagonal SiC mosaic tiles	Flex	C5	C21	1	2	2
						C77	1	1	5
				SAPI	SAPI	C21	1	4	0
						C77	1	3	3
						API-BZ	1	1	3
M2	45.9	NIJ 06 III	4.8 in^2 hexagonal Titanium mosaic tiles	Flex	C5	C21	1	2	2
						C77	1	3	3
				SAPI	SAPI	C21	1	3	1
						C77	1	5	1
M3	35.4	NIJ 06 III	3.9 in^2 hexagonal B4C overlapping mosaic tiles	Flex	C5	C21	1	3	1
						C77	1	3	3
				SAPI	SAPI	C21	1	4	0
						C77	1	6	0
M4	32.1	NIJ 06 III	1.37 in^2 hexagonal Alumina mosaic tiles	Flex	C1	C21	1	1	2
						C77	1	1	3
				SAPI	SAPI	C21	1	2	1
						C77	1	4	2
M5	36.7	NIJ 06 III	1.2 in^2 hexagonal Alumina mosaic tiles	Flex	C5	C21	1	2	2
						C77	1	2	4
				SAPI	SAPI	C21	1	3	1
M6	40.6	NJI 06 III+	1.2 in^2 hexagonal Alumina mosaic tiles	Flex	C5	C21	1	2	2
				SAPI	SAPI	C21	1	3	1
						C77	1	4	2
						API-BZ	1	2	2
M6 In conjunction				SAPI	SAPI	C21	1	2	2

Of the 116 shots, 11 were Complete Penetrations (CP), meaning that the bullet was able to defeat the plate. While the data is limited, there was no indication that shooting in a gap/weak area increased the probability of perforation, as shown in Table 2. From the two plate models (M1 and M6) rated to stop the API-BZ (i.e., VPAM level 8), only the flexible plate model M1 stopped the four projectile impacts. All three CPs with this threat were on the M6 plate model, as shown in Table 2.

https://doi.org/10.52202/081621-0026

Table 2 : **Number of complete penetrations per configuration**

Threat	Plate Model	Flex		SAPI	
		Gap	Tile	Gap	Tile
7.62 mm API/BZ	M1				
	M6			1	2
7.62 mm C21 (M80)	M1				
	M2	1	1		1
	M3				
	M4				
	M5				
	M6				
	M6 in conjunction				
5.56 mm C77 (SS109)	M1			1	
	M2	1			2
	M3				1
	M4				
	M5				
	M6				

The multi-hit ballistic resistance of the acquired flexible plate systems was quite different between the six (6) plate models tested. As mentioned previously, the construction of the different systems is very different from one to the next. Some models use smaller hexagonal ceramic tiles arranged in a mosaic pattern. Tests showed that a single impact on one tile can cause surrounding tiles to shatter into dust, leaving a larger area vulnerable to subsequent impacts. Conversely, other models that used larger metallic tiles, confined impact damage closer to the strike zone while preserving the rest of the plate.

In terms of backface signature (BFS), the C21 (M80), with its high mass and lead core, caused a significant deficiency, with most models having BFS above the 44mm threshold. These failures were unexpected since all plates were claimed to be NIJ-compliant. Table 3 shows similar BFS values in the SAPI and flex configurations, suggesting that the level of plate bending is not a factor that impacts the BFS mitigation performance.

Table 3 : **Average backface signature (mm) per model against the 7.62mm C21.**

Plate Model	Flex		SAPI	
	Gap	Tile	Gap	Tile
M1	43.7	43.1		43.8
M2	47.5	49.8	48.7	51.0
M5	41.7	42.7		43.9
M4	66.9	61.0	62.6	57.1
M3	56.0	51.8	53.2	52.6
M6	51.6	47.5	47.8	50.9
M6 in conjunction			42.7	44.5

Given that the BFS values obtained were above the 44mm threshold, it was decided to assess the performance of a given system in conjunction with a soft body armour insert, which is representative of realistic operational scenario. To this end, the M6 plate model was also tested in conjunction with an

NIJ level 2 soft body armour insert. This arrangement resulted in an average BFS reduction of 13%, as shown in Figure 10, making it almost NIJ compliant.

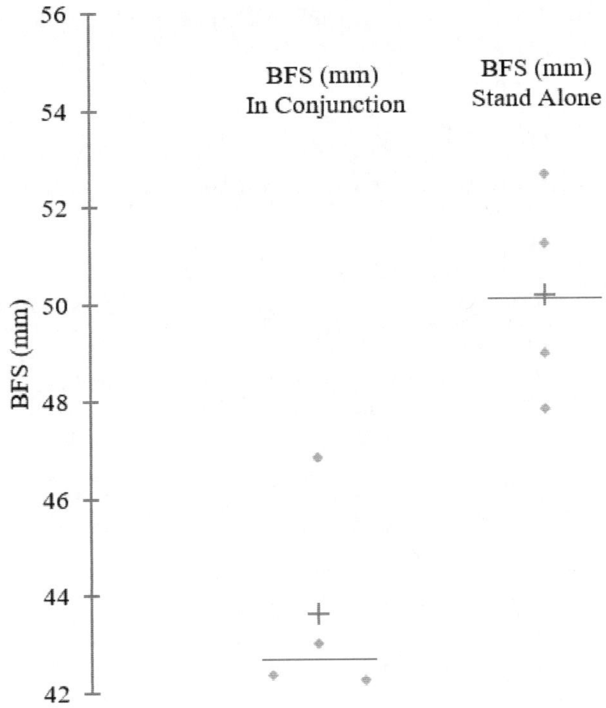

Figure 10 : BFS comparison with and without soft body armour

4. CONCLUSION

One hundred sixteen (116) individual tests were performed on six (6) different flexible systems, offering only a broad assessment of COTS flexible plate ballistic performance. Additional sampling and testing would enhance the reliability of the observations made in this paper. The results indicate that plate conformability, which should increase user comfort and broader morphology accommodation, might come at the cost of thicker and heavier designs and reduced ballistic performance, particularly regarding BFS. Nonetheless, this added flexibility does not seem to create weakness between joints. Multi-hit ballistic performance requirements for flexible systems will need to be defined in more detail in future purchase descriptions and body armour standards, given the large, damaged areas around the impact points on some of the systems tested. Additional testing could demonstrate if the break in procedure impacts performance which could lead to the development of a standardized break in procedure. A tensile machine could potentially be used to perform the break-in a uniform and repeatable manner.

Future testing would benefit from evaluating Vproof and BFS separately. BFS measurements should be conducted with a plate well supported by clay, while Vproof testing could be performed on a ballistic gel torso model that better represents the male and female human body shapes. This approach would provide more realistic boundary conditions where an air gap between the plate and the human body might exist, especially for female users. Additionally, it could be valuable to scan through X-ray or Tomography the plate following each impact. This would provide crucial information on the point of impact and damage to adjacent tiles which in turn, could provide a better understanding of the failure mechanisms. This knowledge could contribute to the development of more robust and reliable flexible armour solutions.

https://doi.org/10.52202/081621-0026

Acknowledgements

The author would like to thank METC's radiography and ballistic lab for their effort on this project and DRDC for their contribution.

References

[1] National Institute of Justice Standard Body Armour, Ballistic Resistance of Body armour, NIJ standard 0101.07
[2] https://www.armschoice.com/female
[3] https://nfm.no/nfm-group-awarded-pioneering-rd-contract-with-cansofcom
[4] https://en.wikipedia.org/wiki/Dragon_Skin
[5] National Institute of Justice Standard Body Armour, Ballistic Resistance of Body armour, NIJ standard 0101.06
[6] VPAM APR2006, General basis for ballistic material, construction and product testing, 2014-11-30
[7] Testing of body armour materials Phase III, National Research Council of the National Academies, 2012

Predicting Whole Helmet Blunt Impact Performance from Isolated Pad Testing Using Machine Learning

J. Zhang[1] and D. Midget[1], M. Yates[1], C. Bradfield[1], J. Dunn[1], J. Leon[1], E. Cawi[1], C. Pyles[1], Q. Luong[1], L. Ruiz[2], C. Trageser[2], J. Hoppings[3]

[1]*The Johns Hopkins University Applied Physics Laboratory, 11100 Johns Hopkins Road, Laurel, Maryland 20723, USA*
Jiangyue.Zhang@jhuapl.edu
[2]*Aberdeen Test Center, 400 Colleran Rd, Aberdeen Proving Ground, MD 21005, USA*
[3]*Program Executive Office Soldier, 10170 Beach Road, Fort Belvoir, VA, 22060, USA*

Abstract. Helmet blunt impact performance evaluation has traditionally relied on whole helmet impact testing. Assessing the performance of helmet pads using this process can be costly, given the destructive nature of the tests and need to consume helmet shells. Additionally, the presence of the helmet shell introduces added complexity and variations, masking differences in pad performance. To overcome these limitations, this study employed machine learning (ML) to establish the intrinsic link between helmet pad properties, from isolated pad impact testing, and their performance in the helmet. This novel approach to rapidly estimate helmet performance eliminates the need for extensive helmet testing or computational analysis. The proposed methodology, combining isolated pad testing and ML prediction, significantly reduces the time and effort required for pad evaluation. This approach enables the efficient screening of helmet pads, promoting accelerated innovation in pad development and facilitating faster, more efficient testing and evaluation processes for helmets. High-rate material property and structural impact response tests were conducted on commercial off-the-shelf helmet pads. Finite element models of whole helmet and individual pad tests were constructed and validated using experimental data. Parametric simulations generated synthetic training data for the ML model, which predicted headform acceleration from isolated pad testing metrics. The approach was applied to 4 combat helmets (ACH and ECH, size large and medium) at 3 temperatures (hot, ambient and cold) and 3 impact velocities (10, 14 and 17 ft/s). Results showed that a gradient boost tree ML model accurately predicted headform acceleration using maximum acceleration, compressed thickness, and impact velocity from isolated pad testing. Validation studies confirmed that predicted accelerations were less than <15g of measured values across all conditions and helmets. However, the study's findings are limited to the specific helmets and pads investigated, and different models may be required for other helmet types and pad designs.

1. INTRODUCTION

Combat helmets rely on both a rigid outer shell and a suspension system to absorb impact forces and protect the head from direct trauma [1, 2, 3]. The current method for evaluating US Advanced Combat Helmets (ACH) and Enhanced Combat Helmets (ECH) performance involves whole-helmet impact testing using a drop tower setup [1, 4, 5], which assesses the effectiveness of helmet pads by measuring headform acceleration under controlled impact conditions. Specifically, this testing ensures that head accelerations remain below 150 g after two consecutive impacts at seven designated locations (crown, front, rear, left/right side, left/right nape) and under three temperature conditions (hot, ambient, cold) at an impact velocity of 10 ft/s [1, 4, 5].

Although effective, whole-helmet testing is both costly and time-consuming. The need to sacrifice helmet shells and complete pad systems limits the rapid development of innovative pad designs. Additionally, testing the helmet and pads together introduces variability and uncertainties, such as pad positioning, chin strap tightness, and helmet shell properties, which can obscure the intrinsic effects of pad material modifications. The increasing demand for enhanced protection requiring performance at higher impact velocities (14.1 ft/s and 17.3 ft/s) while maintaining the same acceleration threshold [2, 3] further underscores the need for a more efficient and cost-effective approach to evaluating new pad designs.

Since headform acceleration for a specific helmet in whole-helmet testing for pad-based liner systems is primarily dictated by pad properties as other conditions remained constant, it is possible to predict helmet performance based on pad behavior. A machine learning (ML) model can establish the relationship between input and outcome in complex impact events [6, 7], offering an efficient predictive framework for evaluating pad designs. However, obtaining both pad impact data and whole-helmet acceleration data through experimental methods alone is prohibitively expensive and time-consuming, making the experimental approach impractical.

https://doi.org/10.52202/081621-0027

Finite element models (FEMs) provide powerful virtual representations of physical testing setups [2, 8, 9], enabling systematic variation of pad properties to generate synthetic data for ML model training. However, FEMs are computationally demanding and require specialized expertise, limiting their accessibility for rapid evaluations. To address these challenges, a fast-running ML model can be trained on synthetic FEM-generated data to correlate pad response to whole-helmet acceleration, offering near-instant predictive capabilities.

Consequently, the Johns Hopkins University Applied Physics Laboratory (APL), in collaboration with U.S. Army Aberdeen Test Center (ATC) conducted the current study to develop a cost-effective ML-driven screening method. This approach aims to identify underperforming pads before committing to full system-level helmet drop tower testing, thereby accelerating development cycles and reducing costs.

2. METHODS

To develop a fast-turnaround helmet pad assessment approach, the Team Wendy Zorbium Action Pad™ 7-Pad NSN Liner System was selected as the baseline. The material properties of these pads were characterized for finite element modeling (FEM). Additionally, a standalone pad testing system was developed at the U.S. Army Aberdeen Test Center (ATC) to evaluate individual pad impact responses under conditions similar to whole-helmet tests. Whole-helmet impact data was obtained from prior testing.

Finite element models of both pad and helmet testing were developed at APL for large and medium-sized ACH and ECH. The pad properties in these FEMs were systematically varied to generate synthetic paired data correlating pad impact response and helmet accelerations. Machine learning (ML) models were then trained using this paired dataset and validated against experimental pad and helmet testing data. Results from large ACH were used to demonstrate the development process.

2.1 Material property testing of Team Wendy pads

The rate-dependent material properties of the Team Wendy pads were characterized by compressing crown, trapezoid, and oblong pads between two rigid platens using Instron machines (Figure 1, left) at four different compression velocities (0.002, 0.02, 0.2, and 2 m/s) under hot (54±3.0°C), ambient (20±5.6°C), and cold (-10±3.0°C) temperature conditions. Force-compression data was converted into engineering stress-strain curves (Figure 1, right) by dividing the pad nominal cross-sectional area and initial thickness, assuming homogeneity and disregarding the multilayer construction of the pads.

To validate the material response, FEMs of the pad compression tests were developed. Pads were meshed with hexahedral elements and modeled using the Fu-Chang foam material model [10], incorporating rate-dependent stress-strain curves obtained from experiments for each temperature condition. The compression plates were modeled as rigid bodies, with the bottom plate fixed and the top plate constrained to allow only vertical motion, simulating the experimental compression test conditions. A friction coefficient of 0.5 was applied at the plate-pad interface in the pad compression FEM to represent interaction between the steel plate and pad fabric. Sensitivity studies with lower and higher coefficients showed negligible effect on the simulated force-deflection responses.. These validated pad models were consistently applied across all FEMs used in the study.

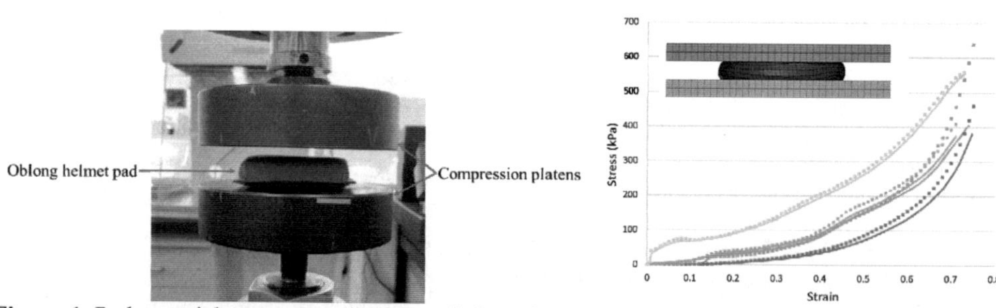

Figure 1. Pad material property testing setup (left) and comparison of experimental engineering stress-strain curve (right, symbols) with verified FEM response (right, solid lines)

2.2 Stand-alone pad impact testing setup, FEM development, and validation

The impact response of individual Team Wendy pads was evaluated using a custom-designed standalone pad drop tower test fixture at ATC as shown in Figure 2. This setup replicates pad impact conditions in whole-helmet testing, utilizing a monorail-guided concave drop mass and a rigid convex bottom anvil. Pads were attached to the inside of the drop mass, and two consecutive impacts were conducted within a one-minute interval. Tests were performed at 10 ft/s under hot, ambient, and cold conditions, with three pads tested per condition. Impact accelerations were recorded using PCB353B18 accelerometers (PCB Piezotronic, Depew NY, USA), while a Keyence G157 (Keyence Corportation of America, Itasca, IL, USA) laser measured the distance between the drop mass and anvil. Acceleration time history and displacement data were used for FEM and ML model validation.

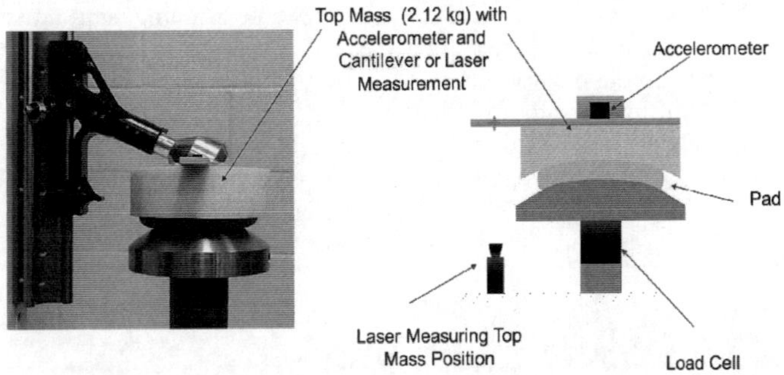

Figure 2. Standalone pad drop tower test fixture (left) and schematics of the test setup (right)

The FEM of the pad test setup was constructed using CAD geometries of the physical fixture (Figure 3 left). The test assembly was auto-meshed with tetrahedral elements and treated as a rigid body. Care was taken to ensure the total mass of drop mass and each component to match exactly to experimental setup. The same pad models from the pad compression tests were incorporated, with pads settled and positioned to conform with bottom face of the concave drop mass. The FEM was constrained to allow only vertical motion of the drop mass while keeping the anvil fixed, ensuring consistency with the experimental setup.

Impact simulations were conducted using LS-DYNA (R9.1.0 11398, Ansys, Inc. Canonsburg, PA, USA) by assigning an initial velocity of 10 ft/s to the drop mass and pad, along with a constant 1 g acceleration to simulate gravity. The nodal acceleration history at the center of the drop mass corresponding to the physical pad test was extracted and filtered at CFC1000 for comparison with experimental results (Figure 3 right).

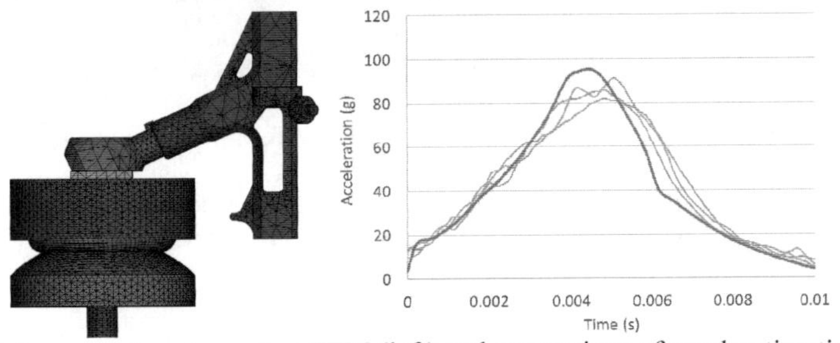

Figure 3. Standalone pad drop tower test FEM (left) and comparison of acceleration time history from experimental testing (right, gray) with FEM (right, red)

2.3 Whole helmet testing, FEM development, and validation

Whole-helmet test data were obtained from prior helmet impact tests conducted at ATC and the Combat Capabilities Development Command Soldier Center (Soldier Center). The combat helmet test protocol, originally described by McEntire and Whitley, was adapted from the Federal Motor Vehicle Safety Standard (FMVSS) 218 for motorcycle helmets [1, 4, 5]. A Team Wendy 7 pads system was installed in the layout dictated by the test protocol as shown in Figure 4a. Testing involved securing the helmet to a Department of Transportation (DOT) headform mounted on a uniaxial monorail drop tower (Figure 4b) and releasing the assembly from a height that resulted in an impact velocity of 10.0 ft/s. A

https://doi.org/10.52202/081621-0027

laser gate positioned near the bottom of the monorail measured the drop mass velocity immediately before impact. Prior to each test, the headform was rotated to one of seven orientations to evaluate impacts at different helmet locations: crown, front, rear, left side, left nape, right side, and right nape. Figure 4b illustrates a crown impact test, where the apex of the helmet makes contact with the hemispherical steel anvil at the base of the drop tower. A uniaxial accelerometer was used to measure headform acceleration during impact.

The finite element model (FEM) for whole-helmet testing was developed using a methodology similar to the pad test FEM. The drop tower, DOT headform, chin strap support, and impact anvil were meshed with tetrahedral elements derived from CAD geometries and modeled as rigid bodies (Figure 4c). The helmet shell (for both large and medium ACH and ECH helmets) was meshed using fully hexahedral elements. The same pad models from the pad testing FEM were incorporated, with pads positioned according to test protocol [4]. To ensure proper pad positioning, settling simulations were conducted to allow the pads to conform to the helmet shell inner surface and headform outer surface, as shown in Figure 4d. Settling simulations were performed by first aligning the undeformed pads to their CT-derived locations. A distributed load was then applied to the inner pad surfaces to press them against the helmet shell, replicating experimental conditions. The resulting nodal positions were saved for subsequent simulations. The helmet-headform assembly was then rotated around the ball joint to model impacts at other locations, following the experimental procedure (Figure 5). Due to the symmetry of the helmet FEMs, left-side and left-nape impacts were not explicitly simulated, as they would yield identical results to their right-side counterparts.

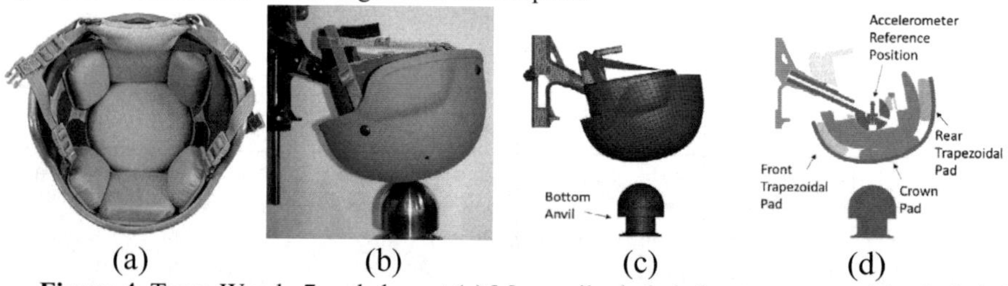

Figure 4. Team Wendy 7 pads layout (a) Monorail whole helmet test setup (b) whole helmet FEM overview (c) and mid-plane cross-section showing settled pads (d)

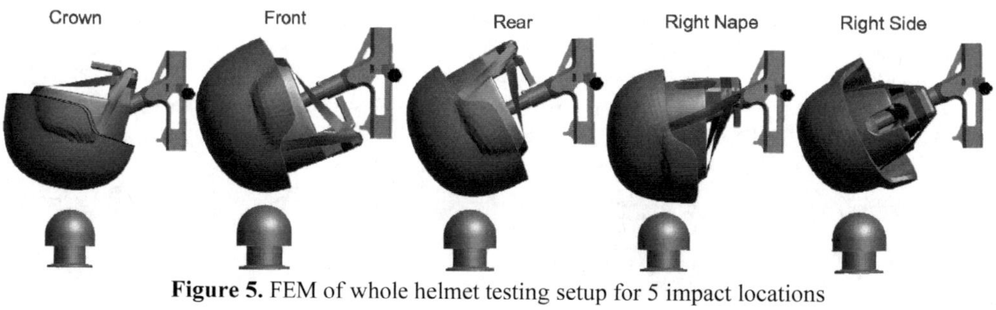

Figure 5. FEM of whole helmet testing setup for 5 impact locations

Impact simulations were conducted using LS-DYNA by assigning initial velocities of 10, 14, and 17 ft/s to the headform and helmet assembly. Acceleration histories (Figure 6 top) and peak accelerations (Figure 6 bottom) from the FEMs were extracted and compared to experimental results to validate the helmet FEMs.

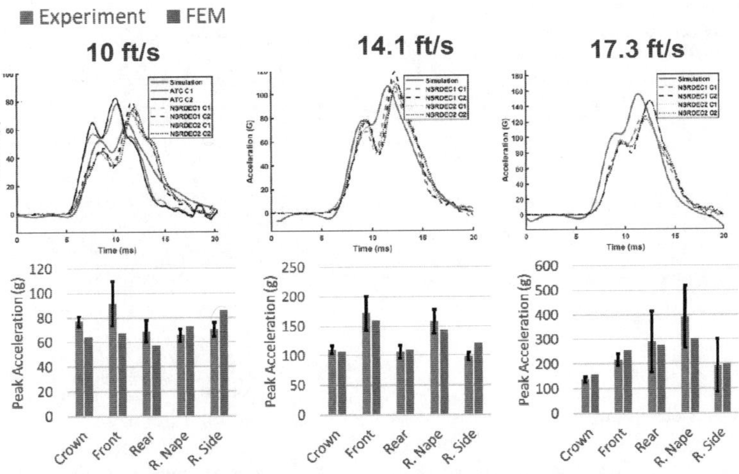

Figure 6. FEM of whole helmet testing setup for 5 impact locations

2.4 Parametric FEM simulation to generate synthetic data for ML training

In principle, a transfer function exists between pad impact responses and the corresponding headform acceleration observed in helmet testing. Machine learning (ML) can be used to establish this transfer function, provided that sufficient data is available from helmet pads with varying material properties. However, obtaining such a dataset experimentally would be prohibitively expensive and time-consuming, making it impractical. To address this challenge, synthetic data can be generated at a reasonable cost using finite element models (FEMs) for both pad impact testing and helmet testing, with systematic variation of pad material properties. Pad density and the ordinate and abscissa of the stress-strain curves were scaled to create helmet pads with varying stiffness. A design of experiments (DOE) approach using Latin Hypercube Sampling (LHS) was implemented to explore the parameter space [11]. The design variables included, pad density, varied from 50 to 500 kg/m^3 [12]; Stress-strain curve scaling factors: Ordinate and abscissa scaling factor ranged from 0.1 to 5, with 1 representing no scaling. A total of 50 unique pad property combinations were generated, and simulations were conducted using LS-DYNA at three impact velocities (10, 14, and 17 ft/s) for all three pad shapes and across five helmet impact locations.

2.5 ML model development and validation

Fast-running machine learning (ML) models were developed and optimized using the synthetic dataset generated from the paired pad and helmet impact FEMs, where pad material properties were systematically varied. Several variables with potential predictive power for helmet acceleration were evaluated, including: Maximum pad compression (percentage and absolute thickness); Maximum compressed thickness; Maximum internal energy; Maximum acceleration; Impact velocity; Time to peak acceleration. These variables were used as input features for ML models developed in Python 3 using the scikit-learn library [13], with the goal of predicting helmet acceleration from pad impact characteristics.

To evaluate model performance, a leave-one-out (LOO) cross-validation approach (Figure 7) was employed. In this method, the ML model was trained on the entire paired dataset, excluding one data point. The trained model was then used to predict the helmet acceleration for the omitted data point. The difference between the predicted and FEM values was recorded as the prediction error. This process was repeated for all data points in the dataset to compute the total prediction error. The inclusion or exclusion of a predictor variable was determined based on whether it significantly reduced the total prediction error in the LOO process. Similarly, selection of a better ML model was also based on minimizing the total prediction error. The final selected and training ML model based on the whole dataset along with the predictor variables were validated using experimental test data from pad impact and whole-helmet testing experiments.

https://doi.org/10.52202/081621-0027

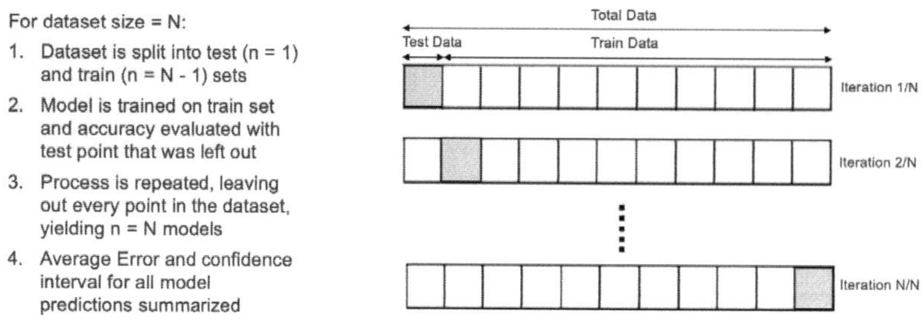

Figure 7. Illustration of leaving one out (LOO) cross-validation approach

3. RESULTS

Accelerations vs. pad compression from crown pad simulation and absorbed energy vs. helmet acceleration for ACH large is shown in Figure 8 below. The LHS design of experiments (DOE) generated a total of 50 cases, exploring a three-parameter design space with ordinate and abscissa scaling factors ranging from 0.1 to 5 and pad density varying from 50 to 500 kg/m³. Paired simulations of pad and helmet impact tests were conducted for three pad shapes and five impact locations at impact velocities of 10, 14, and 17 ft/s. This resulted in a total of 450 pad impact simulations; 750 helmet impact simulations per helmet type for each of the 4 helmets.

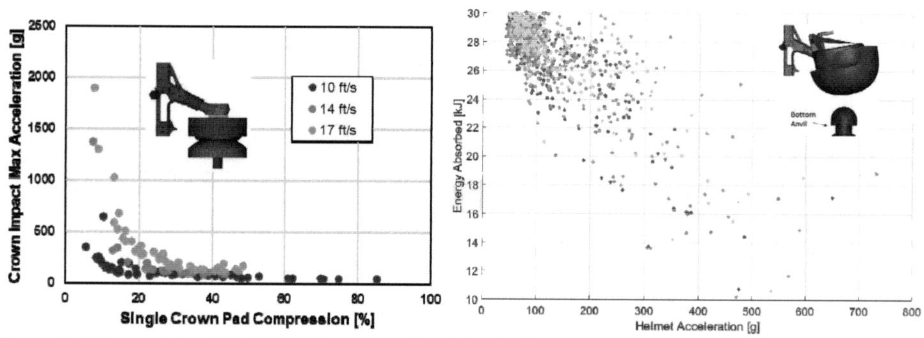

Figure 8. Illustration of DoE FEM generated synthetic data from standalone pad testing crown pad model and ACH large whole helmet impact

A total of 11 machine learning (ML) models from Python 3 scikit-learn were evaluated using the leave-one-out (LOO) cross-validation approach. The analysis identified maximum acceleration (A_{max}), maximum compressed thickness (C_{max}), and impact velocity (V) as the most effective input predictors for ML models. For the ACH large helmet, the Gradient Boosted Tree ML model demonstrated the best predictive performance for most impact locations, except for the front impact location, where the K-Nearest Neighbors (KNN) model was found to be the most accurate. Table 1 summarizes the selected ML models for each impact location, along with their root mean square error (RMSE), relative percentage error, and 95% prediction confidence interval.

Table 1. Final selected ML model and model performance for ACH large helmet

Impact Loc.	Model Type	Predictor Vars.	RMSE (g)	Error (%)	CI 95% (g)
Crown	GBT	A_{max}, C_{max}, V	4.38	5%	2.47
Side	GBT	A_{max}, C_{max}, V	7.66	8%	4.72
Rear	GBT	A_{max}, C_{max}, V	8.92	11%	7.51
Front	KNN	A_{max}, C_{max}, V	6.17	6%	5.24
Nape	GBT	A_{max}, C_{max}, V	7.41	8%	6.87

Figure 9 compares ML predictions with finite element model (FEM) acceleration outputs and experimental helmet testing results. The ML-predicted helmet accelerations for each impact location under hot, ambient, and cold conditions, using experimental pad test data, closely matched the experimental helmet testing data.

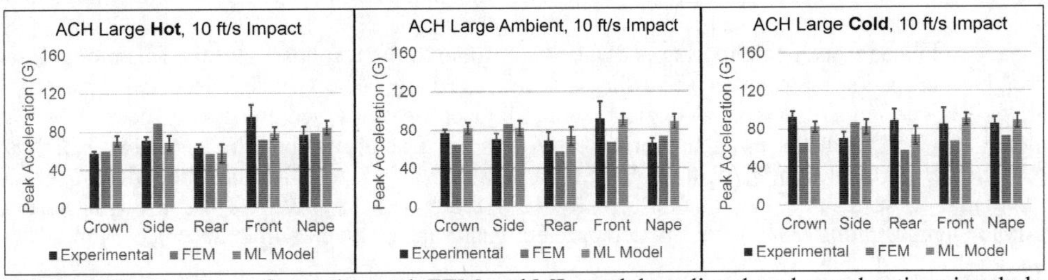

Figure 9. Comparison of experimental, FEM and ML model predicted peak accelerations in whole helmet testing at crown, side, rear, front and nape impacts

4. DISSCUSSION

This study developed a finite element modeling (FEM) and machine learning (ML)-based approach to enable cost-effective evaluation of helmet pad performance in large and medium-sized ACH and ECH helmets without the need for extensive, expensive whole-helmet testing.

The Team Wendy pad system was selected as the baseline due to its widespread use in ACH and ECH helmets in a 7-pad configuration, consisting of crown, trapezoidal, and oblong pads. These pads are polyurethane-based multilayer structures, composed of a comfort layer, an energy absorption layer, a moisture barrier, and an outer fabric cover. To simplify the FEM, the pads were treated as homogeneous materials, as their overall compression behavior predominantly dictates helmet response and the layered deformation mechanics of the pads were not a focus of this study. The strong agreement between experimental stress-strain curves and FEM predictions across all compression velocities confirmed that this approach effectively captured the pads' overall impact compression response.

A custom-designed standalone pad impact test setup was developed to replicate impact compression conditions similar to whole-helmet testing. The test setup used the same accelerometers and data processing methods as the helmet tests to ensure consistency. The drop mass (2.12 kg) was carefully selected to achieve at least 75% compression of the pad, ensuring activation of the energy absorption layer. Similar to helmet testing, each pad underwent two consecutive impacts within one minute to assess pad recovery. Higher acceleration and pad compression were typically observed in the second impact like the whole helmet tests.

The FEM of the pad test setup was developed using CAD-based geometries, ensuring an exact replication of the test fixture. Similarly, whole-helmet FEMs were constructed following the same methodology. These virtual models were used to generate paired pad impact and helmet acceleration data, eliminating the need for costly experimental testing. Since the test fixture and headform are significantly stiffer than the helmet pads, they were treated as rigid materials in the FEMs. The helmets were modeled as elastic materials, based on the observation that ACH and ECH helmets generally withstand 10 ft/s impacts without significant deformation. However, permanent deformation can be observed at higher impact velocities (14 and 17 ft/s). Since this study primarily focused on pad-induced acceleration changes, it was justified to model the helmet shell as elastic without considering its post-yield behavior.

As shown in Figure 3 and Figure 6, the FEM predictions for both pad and helmet testing closely matched experimental results in both waveform shape and amplitude, confirming the validity of these models as accurate virtual test replicates. The helmet acceleration predictions from FEMs fell within or closely matched experimental averages and variations for impact velocities up to 17 ft/s.

The validated FEMs were used to generate paired pad response and helmet acceleration data, mitigating the prohibitive costs of experimental data collection. The Team Wendy pad stress-strain curves and pad density were systematically scaled to generate softer and stiffer pads. The scaling factors ranged from 0.1 to 5, and density varied between 50 and 500 kg/m^3, covering a wide range of material properties. The LHS approach ensured thorough exploration of the design space while limiting the number of required simulations. The resulting helmet acceleration values ranged from ~50 to 750 g, providing an adequate dataset that included both accelerations below the 150 g acceptance threshold and those exceeding the limit, ensuring a balanced dataset for ML training.

https://doi.org/10.52202/081621-0027

A comprehensive evaluation of predictive input variables and ML models was conducted using the leave-one-out (LOO) cross-validation approach. ML models for each impact location were trained using data from all three pad types (crown, trapezoid, and oblong) as input features, with whole-helmet impact accelerations at the corresponding location as the output response. Although each impact site may have a dominant pad, neighboring pads also contribute to the response, and the ML model learns to weight each pad's contribution accordingly through training. The Gradient Boosted Tree (GBT) model was identified as the most accurate for most impact locations, achieving an average RMSE of less than 10 g and a relative prediction error of 5–11%. For frontal impacts, the K-Nearest Neighbors (KNN) model performed best, with an RMSE of 6 g and a 95% confidence interval of 5 g. To contextualize these results, a prediction error of 10 g represents only ~7% of the 150 g acceptance threshold, demonstrating that the ML model's predictive accuracy is well within acceptable limits.

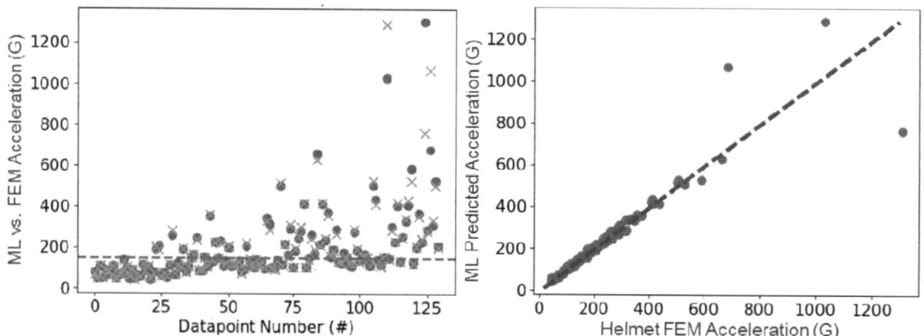

Figure 10. Overlay (left) and cross-plot (right) comparison of FEM and ML model predicted peak helmet accelerations for ACH

A comparison of ML-predicted helmet accelerations with FEM predictions is shown in Figure 10. In an ideal scenario, ML and FEM predictions should align perfectly, with data points overlapping in an overplot or lying on the x = y dashed line in a cross-plot. As observed in Figure 10, the ML models closely replicate FEM predictions, particularly for accelerations below 400 g. However, discrepancies emerge at higher acceleration levels (>600 g), where ML predictions tend to significantly deviate from FEM acceleration. This deviation likely results from nonlinear effects in the FEMs, particularly when the helmet pad bottoms out and acceleration spikes sharply. However, in these cases, both ML and FEM predict accelerations exceeding the 150 g acceptance threshold, meaning both methods would correctly identify the pad as failing, despite differences in absolute values. Accelerations exceeding 400 g were retained to ensure the ML model remained stable at extreme conditions and did not produce false pass predictions. Accuracy at these high acceleration levels was not prioritized, as they far exceed the 150 g pass/fail threshold.

The ML models were further validated using experimental data from standalone pad impact tests conducted at ATC. The maximum acceleration (A_{max}), maximum compressed thickness (C_{max}), and impact velocity (V) were extracted from pad impact test results and used as inputs to the ML models to predict helmet acceleration. As shown in Figure 9, despite differences in batch production and significant time gaps between the pad and helmet test data, the ML-predicted helmet accelerations closely matched experimental helmet test results or fell well within experimental variations. These results further confirm that the ML approach accurately predicts helmet acceleration based on pad response, highlighting the intrinsic link between the two testing methods.

To further simplify the process of predicting helmet acceleration from pad response, a graphical user interface (GUI)-based software tool was developed. As shown in Figure 11, the software allows users to specify test conditions and import pad impact test data via the "Simulation Inputs" section (top left); visually verify imported data via plotted graphs (bottom left); automatically compute key metrics (A_{max}, C_{max}) in the background; select a helmet type and size, then run simulations to instantly obtain predicted accelerations (displayed on the right); and export results for further analysis as needed.

https://doi.org/10.52202/081621-0027

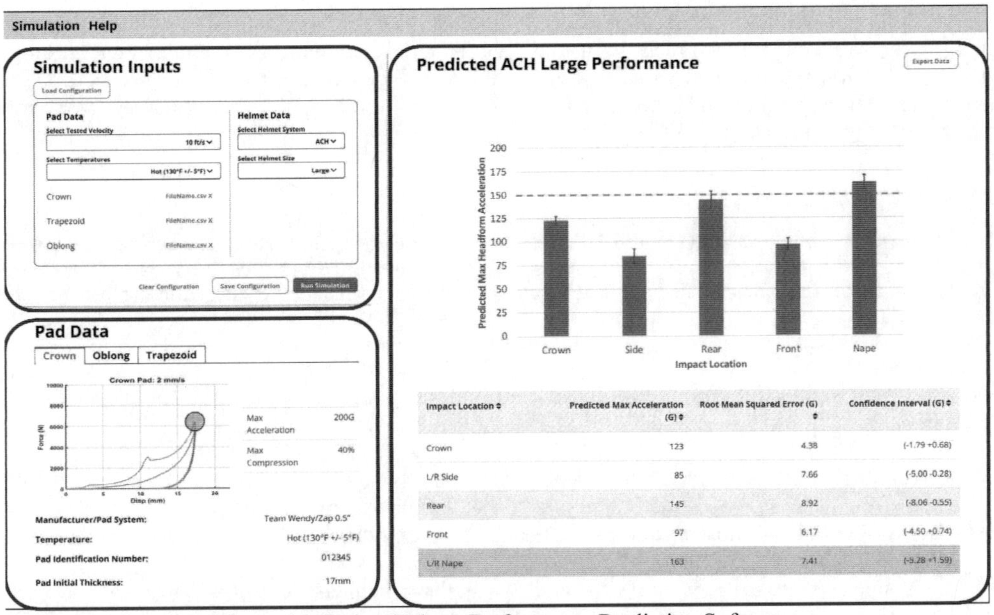

Figure 11. Pad to Helmet Performance Prediction Software

Limitation of the current study include the exclusive use of Team Wendy pads as the baseline pad system. Although scaling factors and density variations were explored to simulate a range of stiffness levels, all synthetically generated pads retained stress-strain characteristics similar to open-cell foam helmet pads (Figure 1). Other novel materials may exhibit significantly different stress-strain behaviors, potentially reducing ML model accuracy when apply to new material classes. Additionally, this study used Team Wendy ZAP pads in a 7-pad layout, following ACH/ECH testing standards. Since the ML models were trained on this specific configuration, changes in pad shape, size, or placement may alter the relative contributions of each pad to helmet acceleration, affecting ML prediction accuracy. Finally, the helmet itself significantly influences blunt impact protection. Different ML models were required for various helmet sizes and types, suggesting that future studies should expand ML model training to include non-standard pad materials and configurations, further enhancing the model's predictive capability and the software's applicability.

5. CONCLUSION AND FUTURE WORK

This study developed a cost-effective standalone pad testing system and a machine learning-based software tool to predict peak accelerations in helmet impact tests for medium and large-sized ACH and ECH helmets. Finite element models (FEMs) of both pad and helmet testing systems were generated and validated throughout the study. Machine learning (ML) models were trained using synthetic paired data, obtained by systematically varying pad stiffness within these validated FEMs.

Among the evaluated features, peak acceleration (A_{max}), maximum compressed pad thickness (C_{max}), and impact velocity (V) were identified as the most indicative predictors of helmet acceleration. The ML model predictions closely matched experimental helmet test results for ACH and ECH helmets, achieving a root mean square error (RMSE) of less than 10 g across all temperature conditions and impact locations at 10 ft/s. These results confirm the validity of using ML models to predict helmet acceleration responses based on standalone pad impact testing, eliminating the need for costly whole-helmet testing in early-stage evaluations. Limitations of this study is the exclusive use of Team Wendy pads as the baseline pad system. As discussed, since these pads are polyurethane-based open cell foam with specific size shape and pad layout dictated by the testing protocol [4], the ML prediction error will likely increase when applied to pads with significantly different mechanical behaviors, or different shape, size and layout. Additionally, the helmet shell plays a critical role in whole-helmet impact response. Specific ML models were developed for each size of ACH and ECH helmets, meaning that new models must be trained for other helmet types following the same correlation process between pad impact response and whole-helmet acceleration. These limitations can be addresses in future studies to further enhance the capability of the ML models.

https://doi.org/10.52202/081621-0027

Overall, the instantaneous results provided by the ML models, based on low-cost standalone pad testing, offer a valuable tool for reducing the cost and material requirements of whole-helmet testing. This approach has the potential to significantly shorten the development cycle of next-generation helmet pad systems, accelerating innovation in impact protection technology and ultimately improving soldier safety and survivability.

Acknowledgments

The authors would like to thank PEO Soldier PdM SPE for sponsoring this effort. This material is based on work supported by PdM SPE under Naval Sea Systems Command Contract No. N00024-13-D-640, Task Order N00024-22-F-8069. Any opinions, findings, and conclusions or recommendations expressed in this material are those of the author(s) and do not necessarily reflect the views of PdM SPE or the NAVAL SEA SYSTEMS COMMAND (NAVSEA).

References

[1] J. McEntire and P. Whitley, "Blunt impact performance characteristics of the Advanced Combat Helmet and the Paratrooper and Infantry Personnel Armor System," 2005.

[2] Moure-Guardiola, Carlos, et al. "Evaluation of combat helmet behavior under blunt impact." Applied Sciences 10.23 (2020): 8470.

[3] Li, Yongqiang, Hualin Fan, and Xin-Lin Gao. "Ballistic helmets: Recent advances in materials, protection mechanisms, performance, and head injury mitigation." Composites Part B: Engineering 238 (2022): 109890.

[4] Aberdean Test Center, "Internal Operating Procedure, ATC_MMTB-IOP-029 REV 13, Blunt impact of Combat Helmets," 2022.

[5] National Research Council, et al. "Review of Department of Defense test protocols for combat helmets." (2014).

[6] Shim, Vickie B., et al. "Rapid prediction of brain injury pattern in mTBI by combining FE analysis with a machine-learning based approach." IEEE Access 8 (2020): 179457-179465.

[7] Zhan, Xianghao, et al. "Deep learning head model for real-time estimation of entire brain deformation in concussion." arXiv preprint arXiv:2010.08527 (2020).

[8] Li, Y. Q., X. G. Li, and X-L. Gao. "Modeling of advanced combat helmet under ballistic impact." Journal of applied mechanics 82.11 (2015): 111004.

[9] Palta, Emre, Hongbing Fang, and David C. Weggel. "Finite element analysis of the Advanced Combat Helmet under various ballistic impacts." International Journal of Impact Engineering 112 (2018): 125-143.

[10] Livermore Software Technology Corporation, LS-DYNA Keyword User Manual, Livermore, California, 2017.

[11] M. Shields and J. Zhang, "The generalization of Latin hypercube sampling, Reliability Engineering & System Safety," 2016.

[12] M. F. Ashby, Material Selection in Mechanical Design (3rd Edition), Butterworth-Heinemann, 2010.

[13] Pedregosa, F. et al., "Scikit-learn: Machine Learning in Python" Jurnal of Machine Learning Research (JMLR), Volume 12, pages 2825-2830, 2011.

Ballistic, mechanical and BABT related properties of UHMWPE helmet shells based on novel matrix systems

R. Proost[1], J. van Elburg[1], W. Roovers[1], J. Maassen[1], U. Heisserer[1], M. van der Kamp[1], D. Louwers[1], J. Stolk[1]

[1]*Avient Protective Materials (APM), PO Box 1163, 6160 BD Geleen, the Netherlands*
Roy.Proost@avient.com

Abstract
It is generally known that Ultra High Molecular Weight Polyethylene (UHMWPE) materials exhibit better ballistic stopping power than aramid counterparts on a weight basis. Besides the higher ballistic properties in protective helmets, however, most UHMWPE materials in earlier studies were reported to demonstrate higher back face deformations and lower rigidity of the ballistic shell, e.g. in a side-to-side compression test. This raised concerns about Behind Armour Blunt Trauma (BABT) for UHMWPE helmets.

With the continued development in the field of UHMWPE laminates, new matrix systems and processing technologies have resulted in strongly improved mechanical properties. A contribution at the PASS 2023 showed that with the right design and processing an UHMWPE helmet shell at attractive areal density can stop 7.62x39mm AK47 Mild Steel Core projectiles at muzzle velocity.

This year's paper provides a performance comparison between helmet shells from European NATO fielded aramid helmet systems on the one hand, and Avient in-house UHMWPE helmet shells manufactured with deep draw technology on the other. The UHMWPE unidirectional material uses a novel matrix system for enhanced mechanical performance. Both helmet systems make use of identical suspension systems. Ballistic testing is executed against 1.1 g Fragment Simulating Projectiles (FSP). Considering BABT, side-to-side compression tests and the effect of 9 mm FMJ DM41 projectile impact on dynamic back face deflection, back face signature (in a witness medium) and transmitted impact forces to a Biokinetics Ballistic Load Sensing Headform are evaluated.

Ballistic testing shows that the UHMWPE helmet shells exhibit at least 20% higher V_{50} against 1.1 g FSP, while being 20% lighter than the tested aramid helmet shells. In addition, the results show that the lighter helmet shell made with UHMWPE material and the novel matrix system for enhanced mechanical performance is on par with the fielded aramid helmet shell on all the above mentioned BABT criteria.

1. INTRODUCTION

Since ages, combat helmets are a crucial part of personal protection equipment. The requirements have changed over time in line with methods of warfare and threats and evolved with development of manufacturing methods and materials. What started as a means of protection to sword strokes, arrows and musketry assault threats changed during World War 1. Combat helmets evolved to protect against shrapnel, debris and small calibre ammunition, as it proved that many fatal head wounds were not caused by impact of bullets or direct blows but by small, limited energy containing fragments. Most recently, protection against rifle threats has been added to the possibilities that modern materials like Ultra High Molecular Weight Polyethylene (UHMWPE) and state of art production processes can provide.

The helmet of today has developed to a system that provides several functions:

- mounting for accessories like communication devices and night-vision goggles
- protection to regular blunt impact
- stopping various threats such as fragments, small calibre and rifle threats.
- Behind Armour Blunt Trauma (BABT) protection

The main function of combat helmets today is to provide protection against fragments. It is generally known that UHMWPE materials exhibit better ballistic stopping power than their aramid counterparts [1,2] on a weight basis. Next to the pure stopping power, also the resistance against deformation is important. In the past, UHMWPE materials were reported to demonstrate higher back face deformations and lower rigidity of the ballistic shell, e.g. in a side-to-side compression test compared to alternative materials. These observations gave rise to concerns and speculations about BABT for helmets made from UHMWPE.

 https://doi.org/10.52202/081621-0028

This paper aims to evaluate and compare helmet shells from European NATO fielded aramid helmet systems on the one hand, and Avient in-house moulded UHMWPE helmet shells on the other. The UHMWPE helmet shells are manufactured with a unidirectional (UD) material, based on a novel matrix system for enhanced mechanical performance.

The same test and evaluation methods are applied to both the UHMWPE and the aramid helmet shells for comparison purposes.

2. HELMET SELECTION

For the various tests, two series of helmet shells are selected:

- The Series #1-aramid is a military combat helmet shell made from plain weave aramid prepregs with a mass of 455 g/m^2, single side coated with 12% toughened phenolic resin. The fibre used is 3140 dtex Kevlar® 129. The shells are moulded into an Advanced Combat Helmet (ACH) shape using a flower cut process. Moulding conditions are in line with guidelines provided by the raw material supplier.
- The Series #2-PE is a helmet shell made from Dyneema® HB30 unidirectional composite (APM). The composite consists of first generation 1760 dtex UHMWPE yarns, embedded in a novel thermoplastic matrix system. The mass of the UD composite material is 266 g/m^2. The proprietary matrix system can be categorized as a polyolefin matrix system with tailored mechanical and ballistic properties, providing a stiffer composite than legacy first generation PE UD composites. The helmet shells are manufactured in an R&D facility in the Netherlands, making use of the available deep draw helmet manufacturing process [3].

These two series of helmets are selected to make an objective comparison between two materials, targeting similar protection levels with similar designs. Admittedly, aramid helmets of this performance level (mass, areal density, V50, mechanical properties) are around for more than 10 years, but still represent the state of the art in ballistic protection and mechanical performance for aramid helmets. This helmet type is still being manufactured commercially and supplied to NATO end-users today. Likewise, the PE fibres are of the first generation, while to date Dyneema® composite materials are available with second and third generation fibres, with increased strength and protection level. The HB30 composite is designed for applications like a standard military combat or police helmet (fragment, handgun protection), whereas later generations target higher threat levels, such rifle threats [3], or lower weight shells. Helmets made with HB30 are also being manufactured and supplied to NATO end-users today.

Table 1 shows some characteristics of the helmet series selected for this study. The helmet shells of both series exhibit an overall thickness of approximately 8 mm. At roughly similar thickness, the Series #2-PE helmet shell is 23% lighter compared to the helmet shell of Series #1-aramid.

Table 1: Key parameters of the helmet shells used in the study

Type	Base material	Shell thickness [mm]	Total area [cm²]	Helmet shell mass[i] [g]	Shell AD[ii] [kg/m²]
Series #1-aramid	Aramid, woven Kevlar® 129	7.5	1182 (100%)	1130 (100%)	9.6 (100%)
Series #2-PE	UHMWPE composite HB30	7.8	1216 (103%)	872 (77%)	7.2 (75%)

Both helmet shells resemble a size L helmet. The geometry of the helmet shells of both series is not fully identical though; small differences in the shape do exist. Figure 1 shows both helmet shells.

Both helmet systems were tested with identical suspension systems. Where applicable for the tests, the helmet shells of both series are equipped with a pad system: Team Wendy® ZORBIUM® Action Pad (ZAP™) 7-Pad NSN Liner System (3/4inch) [8]. The typical pad configuration is shown in Figure 2. A helmet strap system is used to fixate the helmet shells for ballistic testing where applicable.

i Shell mass excluding padding and suspension system
ii Areal Density

https://doi.org/10.52202/081621-0028

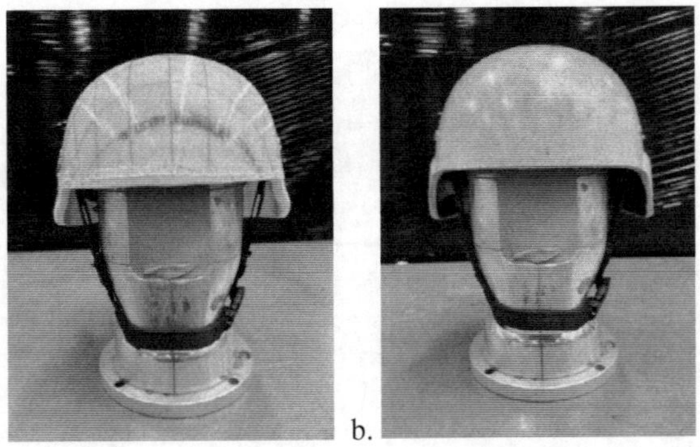

a. b.

Figure 1: Helmet shell including strap and padding system mounted on NIJ head form
(a. UHMWPE helmet shell. b. Aramid helmet shell)

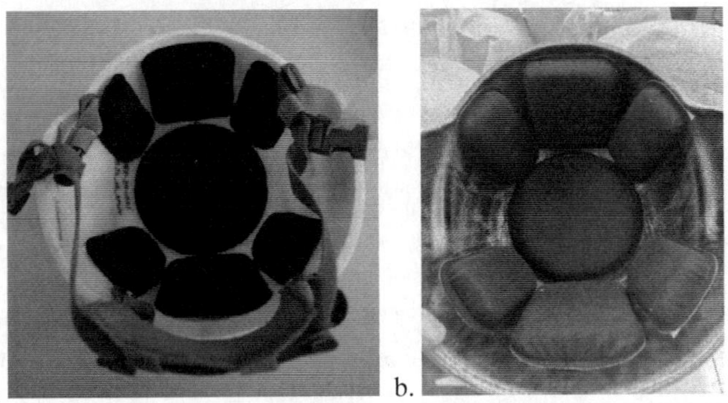

a. b.

Figure 2: Typical padding configuration
(a. UHMWPE helmet shell. b. Aramid helmet shell)

3. TEST METHODS

To evaluate helmet shells of both series, ballistic testing is executed against 1.1 g Fragment Simulating Projectiles (FSP). Considering BABT [4,5], testing is performed with 9 mm Full Metal Jacket (FMJ) DM41 (Ruag Ammotec) projectiles. In summary, five different test methods are used to evaluate the helmet shells:

a. Fragment V_{50} performance executed against 1.1 g FSP.
b. Back Face Signature (BFS) measurement in a witness medium (against clay) with 9 mm FMJ DM41 projectiles
c. Side-to-side compression [7]
d. Dynamic Back Face Deformation (BFD) measurements with Digital Image Correlation (DIC) with 9 mm FMJ DM41 projectiles
e. Measurements of the transmitted impact forces to a Biokinetics Ballistic Load Sensing Headform (BLSH) [6] with 9 mm FMJ DM41 projectiles

The fragment V_{50} performance measurement (a) and BFS measurement (b) are executed at Avient's ballistic shooting range in The Netherlands. All other tests (c, d and e) are carried out at TNO Laboratory for Ballistics Research in The Hague, The Netherlands.

Table 2 presents the number of helmets tested in each test and the amount and type of fired projectiles. In total 24 helmets shells are tested and 116 projectiles are fired.

https://doi.org/10.52202/081621-0028

Table 2: Overview of the number of helmets tested and projectiles fired per test

Test	Helmets		Projectile	
	Series #1-aramid	Series #2-PE	1.1 g FSP	9 mm FMJ DM41
Fragment V_{50} performance	4	4	72	-
BFS in a witness medium	2	2	-	16
Side-to-side compression	2	2	-	-
Dynamic BFD with DIC	2	2	-	16
Transmitted impact forces to a BLSH	2	2	-	12

3.1 Fragment V_{50} performance

To study the 1.1 g FSP V_{50} performance a method derived from STANAG 2920 test procedure [9,10] was used.

The helmet shell without suspension system is mounted in a three-point fixation platform and fixated by clamping it at three points.
The projectile used for the test is a 1.1 g FSP. Each helmet shell is tested according to a standard shot pattern (according to Avient's internal test protocol) on nine locations per helmet. Table 3 presents the typical shot sequence and the distance to the edge at the shot location. The shot-to-shot distance is 8-10 cm. Four helmet shells are tested for both series. The test is executed without backing material; precautions are taken to prevent any complete penetration from damaging the opposite side.

Table 3: Overview of shot sequence and distance to edge

Shot #	1	2	3	4	5	6	7	8	9
Location	Front	Left	Back	Right	Front/ left	Back/ left	Back/ right	Front/ right	Crown
Height to edge [cm]	6	8	6	8	11	13	13	11	-

In this specific study the STANAG 2920 is the method used for the determination of the V_{50}. The definition of V_{50} is the velocity at which the probability of perforation of the chosen projectile is 50%. The V_{50} is calculated by using a probit analysis over all measurements on all shot locations.

3.2 BFS in a witness medium

For the BFS in a witness medium the following method is applied (derived from the "Bullet Resistance Helmet test procedure" - HPW-TP-0401.01B [11]).

For this test, a head form (referred to as the National Institute for Justice (NIJ) head form) is used. The metallic head form is configured for the deformation test and has sagittal and coronal slots or cavities in the head form. Prior to the test, the slots or cavities of the head form are filled with Roma Plastilina nr.1 (calibrated) as described in the NIJ-STD-0101.06 standard [12] and shaped to the contour of the metallic head form. The impact test headform is specified as a size 7 ¼ for testing. The corresponding head circumference of 57.9 cm corresponds to the size L helmets used.

The complete helmet, including strap and pad system, is attached to the head form as it would be attached to the head of the user in a normal operational scenario. After each shot, the indentation or deformation of the clay is measured and recorded.

The projectile used for the test is a 9 mm FMJ DM41 at a speed of 430 ± 9 m/s. Each helmet is tested according to the same shot order before another helmet was tested; the test order is front, left, back, right. The shot locations are indicated on the helmets and have a similar distance to the edge of the helmet shell for both series (front: 6.0 cm, left/ right: 9.5 cm, back: 6.5 cm). Orientation of the impact is perpendicular to the helmet shell surface at the shot location.

3.3 Side-to-side compression

The test method used to measure the rigidity of the helmet is a compression test [7] performed on a Constant Rate of Extension machine. An example of the test setup is shown in Figure 3.

Figure 3: Side-to-side compression test setup with UHMWPE helmet shell sample.

The helmet shell under test is placed in the equipment used for the compression test. Placement is such that compression is executed at the widest point of the helmet. The helmet shell is supported by a plate on the crown. The compression test starts by compressing at a prescribed rate from side-to-side (ear-to-ear) until a pre-defined force is reached (1334 N or 300 lbf). As soon as the pre-defined force is reached, the equipment reduces the force to 22 N (or 5 lbf). This compression cycle is repeated 24 times. The load levels and number of cycles are chosen as such for reasons of comparison to earlier measurements.

Within five minutes after the completion of these 24 repetitions, the distance from the left-hand side to the right-hand side (ear-to-ear) of the helmet is measured (measurement C) and compared to the reference distance measured prior to the test sequence (measurement A). Subsequently, no force is applied on the helmet under test for 24 hours. After this period, the distance from the left-hand side to the right-hand side of the helmet is measured again (measurement D), compared to the reference and reported. A value C–A can now be considered as the remaining deformation directly after the test. The value D-A can be considered as the remaining deformation after 24 hrs.

3.4 Dynamic BFD with DIC

To demonstrate how the deformation develops during the impact of a bullet, a dynamic BFD measurement including DIC is performed. The deformations during the dynamic BFD test will, in contrary to the measurement of the BFS in a witness medium as described in section 3.2, not be affected by any contact between the deforming helmet surface and the witness medium itself

To study the dynamic BFD with DIC, two helmet shells of each series are prepared for the test to perform a duplicate measurement. The helmet shells do not contain a suspension system. The areas surrounding the shot locations are covered with a random speckle pattern at the inside of the helmet (see Figure 4a). The pattern is manually applied by using a permanent marker; this provides a pattern that adheres well enough to the helmet inner surface and the colour of the pattern is chosen to provide sufficient contrast with the inside surface of the helmet shells.

a.  b.

Figure 4: Dynamic BFD test
(a. Speckle pattern on UHMWPE helmet. b. Typical dynamic BFD test set-up)

https://doi.org/10.52202/081621-0028

The prepared helmet shells are placed on a fixture with the crown facing downwards. Two cameras are oriented towards the inside of the helmet shell and trace the bulging of the helmet shell during impact. In Figure 4b the dynamic BFD setup can be seen. The frame rate of the cameras recording the event is 30,000 frames per second (fps). DIC software is used to extract the data from the test.

The projectile used for the test is a 9 mm FMJ DM41 at a speed of approximately 430 m/s. Each helmet is tested according to the same shot order before another helmet was tested; the sequence is front, left, back, right. The shot locations are indicated on the helmets and have a similar distance to the edge of the helmet shells for both series (front: 6.5 cm, left/ right: 11.0 cm, back: 7.0 cm). Orientation of the impact is perpendicular to the helmet shell surface at the shot location.

3.5 Transmitted impact forces to a BLSH

Next to BFS/ BFD, also the force exerted on the head during projectile impact is an important criterium for manufacturers and end users, as it is linked to Behind Armour Blunt Trauma. This paper does not investigate the link between BFS/BFD and BABT, nor aims at estimating which severity of the force is acceptable. However, this paper does compare any differences in the transmitted forces to the head during back face deflection by using the BLSH [6].

To perform a duplicate measurement, two helmets of each series are prepared for the test. All helmet shells are equipped with a full suspension system. Each helmet is shot on three locations: front, back and right side. For the latter location, a second BLSH with sensors mounted for side impact measurement is used. The shot locations are determined based on the position of the central sensor on the BLSH.

Close attention is paid to:

- using the same pad system and it's positioning.
- similar positioning of the helmet shell on the BLSH.

The projectile used for the test is a 9 mm FMJ DM41 at a speed of approximately 430 m/s. Each helmet is tested on the front side first. After execution of all front tests, all helmet shells are tested on the back side and subsequently on the right side of the helmet shell. Orientation of the impact is perpendicular to the helmet shell surface at the shot location. Per test, the individual sensor readings are added to give the total force.

4 TEST RESULTS

In this section, the results of the individual investigations are presented and discussed. No pass or fail criteria are applied on the results, as the tests serve for quantitative comparison only.

4.1 Results on Fragment V_{50} performance and BFS in a witness medium

The fragment V_{50} performance measurement and the BFS measurement in a witness medium are tested according to the methods described above. In Table 4 the V_{50} is reported for both series as well as the average indentation on the different shot locations. All underlying test reports used for this study [13-15] are available from the authors upon reasonable request.

Table 4: Overview of test results

Type	**Helmet shell mass [g]**	**1.1g FSP V_{50} Avg. [m/s]**	**Average BFS [mm] (Min. value/ Max. value)**				
			Front (6.0 cm from edge)	Left (9.5 cm from edge)	Back (6.5 cm from edge)	Right (9.5 cm from edge)	Avg.
Series #1-aramid	1130 (100%)	682 (100%)	20 (20/21)	9 (7/11)	14 (12/15)	10 (8/13)	13.2
Series #2-PE	872 (77%)	841 (123%)	18 (17/19)	10 (9/11)	15 (12/17)	12 (8/15)	13.8

The results in Table 4 show that for the UHMWPE helmet shells, with 23% lower mass, the ballistic stopping power is at least 20% higher compared to the aramid counter parts. These results confirm the better ballistic stopping power of UHMWPE materials.

From the results in Table 4 it also is apparent that the deformation for the sides is smaller than for the front and the back, with the maximum deformation for the front impact. There are multiple potential explanations for that:

- The sides of the helmet experience an additional stiffening from the shape of the ear section leading to lower deformations.
- The curvature is different for the various impact locations. The curvature is highest on the front and back side. The material on the front and back experiences more free length before the material is tensioned during the bulge formation.
- The front and back section are shot more closely to the edge, when compared to the locations on the sides. This would typically mean that less material is available to carry the load.
- The front location is located more towards the crown compared to the back side and as such can experience a negative influence from the additional curvature towards the crown.

Which of these effects is the most dominant with respect to the observed deformation behaviour is not investigated in detail.

The results in Table 4 also demonstrate that the BFS of both Dyneema® HB30 based UHMWPE helmet shells show similar deformation values as the aramid helmet shells, despite the lower mass of the UHMWPE shell.

4.2 Results on side-to-side compression

In Table 5 the average results of the side-to-side compression measurement are reported for both series.

Table 5: Overview of side-to-side compression results, average values per series

Type	Remaining deformation after load cycle completion (C – A) [mm]	Remaining deformation after a 24-hour recovery period (D – A) [mm]
Series #1-aramid	1.0	1.0
Series #2-PE	0.0	0.0

The results in Table 5 show that the UHMWPE helmet shells based on the novel matrix system perform similar compared to their aramid counterparts in the side-to-side compression test. Helmet shells of both series exhibit almost absent remaining deformation levels.

4.3 Results on Dynamic BFD with DIC

A typical result of the DIC measurement is shown in Figure 5.

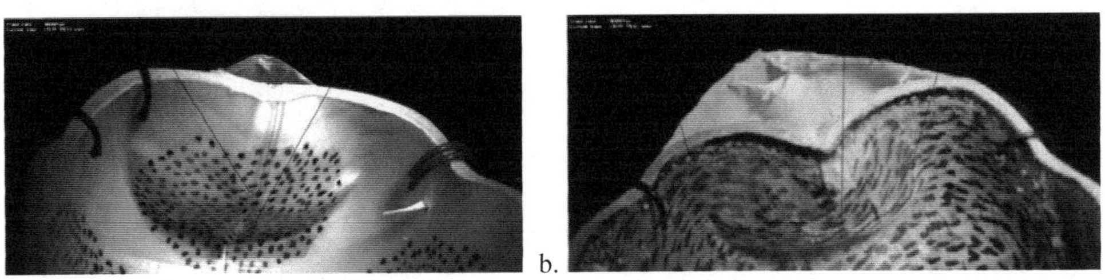

a. b.

Figure 5: Typical deformation pattern result of the DIC measurement in front impact (a. UHMWPE helmet shell. b. Aramid helmet shell)

The peak deformation values from the DIC analysis as obtained from the recorded videos are shown in Figure 6 below. Values are normalized with respect to the maximum occurring value of all tests (i.e. the

https://doi.org/10.52202/081621-0028

displacement of helmet 1 of Series #1-aramid in a frontal impact). In case no value is plotted, no measurement is available from the DIC analysis.

The deformations occurring during the dynamic BFD measurement are larger than the deformations during the BFS measurement in a witness medium. An explanation for this is that the helmet shells can deform without any counter pressure from the witness medium during the dynamic BFD measurement.

For the dynamic BFD a similar trend can be seen as for the results presented for the BFS measurement in a witness medium in section 4.1. For the dynamic BFD measurements, the deformation for the sides is smaller than for the front and the back, with the maximum deformation for the front impact. Similar potential explanations for this difference in deformation can be given as in section 4.1.

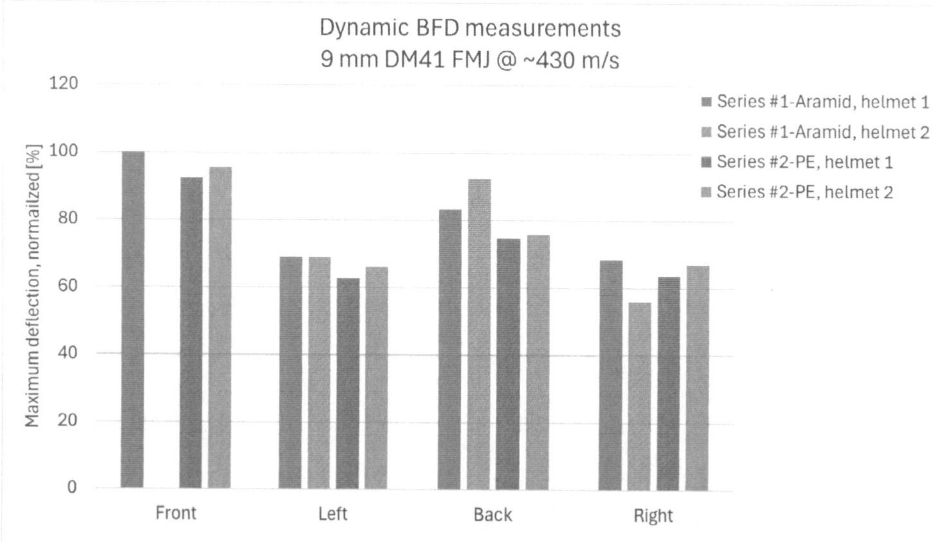

Figure 6: Overview of peak deformations for the dynamic BFD tests at 430 m/s projectile velocity

The relative comparison of the dynamic BFD measurement for both series shows that the deformations for the UHMWPE helmet shells are similar or less compared to their aramid counterparts for all four test locations.

4.4 Results on Transmitted impact forces to a BLSH

The transmitted impact forces to a BLSH are tested as described above. A typical result of the BLSH measurement is shown in Figure 7.

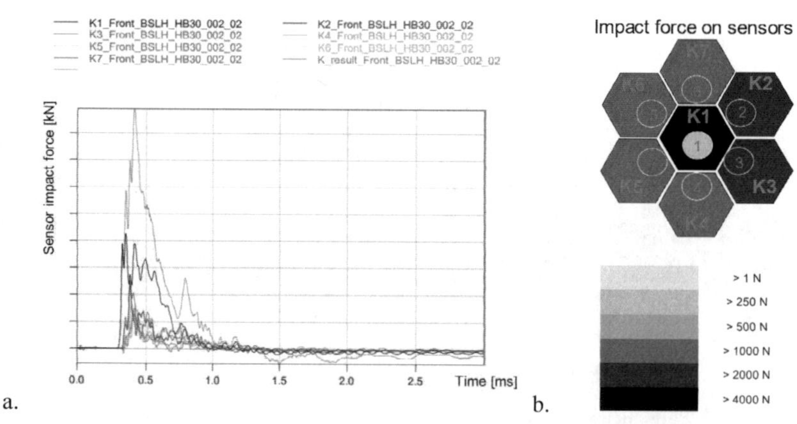

Figure 7: Typical result of a BLSH measurement (front impact on UHMWPE helmet shell)
(a. sensor impact force data. b. sensor pad orientation)

From the unfiltered measurements, the maximum occurring (peak) total force is derived and plotted for each test in Figure 8. Values are normalized with respect to the maximum occurring value in all measurements (i.e. the maximum occurring force of helmet 1 of Series #1-aramid in a frontal impact). Measurements for the left side are not available as deformation of the helmet on the other locations prevents correct positioning of the helmet on the BLSH.

Figure 8 shows that the forces transmitted from the helmet shell to the sensors on the head form are similar, except for the front side where the force exerted by the UHMWPE helmet shell is less compared to the aramid counterparts.

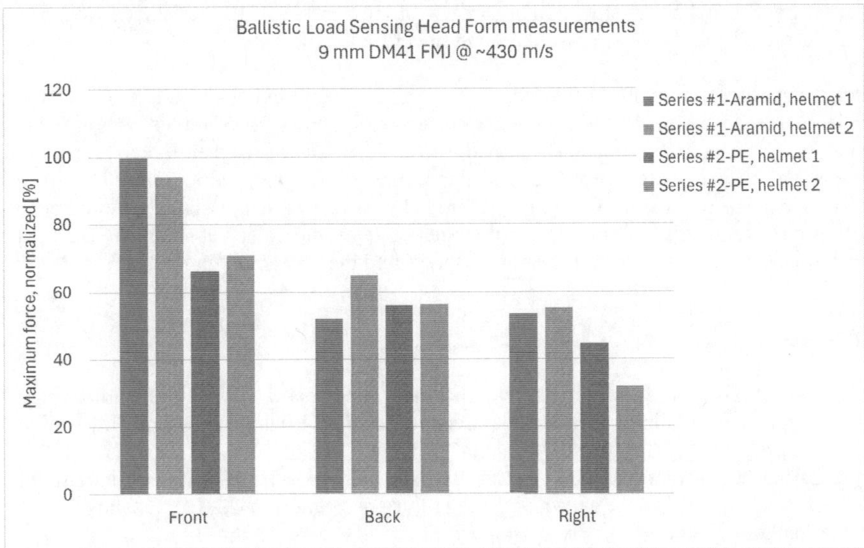

Figure 8: Overview of maximum force for the BLSH tests at 430 m/s projectile velocity.

Differences in performance on the BLSH can not be fully attributed to the material performance per definition. Variation in size can have an influence on the transmitted force. The distance between the BLSH and the helmet inside surface at the actual shot location is not recorded prior to the test. As the aramid helmet shells are slightly wider, the aramid helmet series may exhibit a larger distance between shell and sensor and can benefit from this to generate lower impact forces on the side. Front-to-back, the UHMWPE helmets are slightly larger which can have an influence on the measured forces. As the differences in size are small compared to the deformations reported in section 4.3, the effect of variations in size on the transmitted forces is expected to be small.

On the front and back side, the helmet pads are located beneath the impact location, which is not the case for the side location. Whether variations in transmitted force on the front side can be attributed to different contact area between the pads and the sensor due to a different local curvature is not investigated in this study.

5. CONCLUSIONS

Based on this study, it can be concluded that the ballistic resistance of helmets using UHMWPE based material for fragments can be more than 20% higher, while being more than 20% lighter than the tested aramid helmet shells. At this weight benefit, UHMWPE helmet shells making use of novel matrix systems exhibit at least similar BFS compared to their aramid counter parts. This is confirmed with test results from measurements of the BFS in a witness medium as well as from measurements of the dynamic BFD measured with DIC. The results from the side-to-side compression test complete the picture that UHMWPE helmet shells based on novel matrix systems perform similar compared to their aramid counterparts and exhibit very low remaining deformation levels after the test. From this investigation, it is also concluded that the UHMWPE helmet shells overall demonstrate similar or slightly lower forces on the head sensors in the instrumented head form.

 https://doi.org/10.52202/081621-0028

In summary, compared to their aramid counterparts, at similar thickness, but at 23% lower mass, the helmet shells made from Dyneema® HB30 demonstrate higher resistance to penetration while providing similar BFS/ BFD, similar transmitted force levels and similar permanent deformation after side-to-side compression.

Because of the presented results, procurers, manufacturers and end users are provided with additional possibilities compared to their existing aramid solutions:

- a lighter UHMWPE helmet with increased resistance to penetration at similar BFS and transmitted force levels.
- a UHMWPE helmet at similar mass, but with an increase of the resistance to penetration at increased stiffness levels, lower BFS and lower transmitted force levels.

For future research it is important to note that with reaching increased protection levels at lower mass and lower helmet thickness, the levels of BFS and corresponding transmitted force levels are of increased importance. With the onset of the wish and the possibility to stop higher threat levels, additional investigations and research are needed to understand and define acceptable levels for the BFS and transmitted force levels to prevent BABT. For future helmet systems, manufacturers would benefit from a clear definition of projectile calibres, projectile speeds and test methods that should be used to evaluate helmet performance in terms of BFS, transmitted forces and the corresponding BABT.

REFERENCES

[1] Zheng, J. Q. / Walsh, S. M. Materials, manufacturing, and enablers for future soldier protection, 2016, in Lightweight Ballistic Composites (Second Edition), Bhatnagar, Ashok (Ed.), Woodhead http://dx.doi.org/10.1016/B978-0-08-100406-7.00014-3

[2] Li, Yongqiang / Fan, Hualin / Gao, Xin-Lin, Ballistic helmets: Recent advances in materials, protection mechanisms, performance, and head injury mitigation, 2022, Composites Part B: Engineering https://doi.org/10.1016/j.compositesb.2022.109890

[3] van der Kamp, Mirre / Heisserer, Ulrich / van Elburg, Jos / Louwers, Dirk / Stolk, Jan. Development and performance of an UHMWPE rifle helmet shell, 2023, Personal Armour Systems Symposium, Dresden https://doi.org/10.52202/080042-0027

[4] Nsiampa, Nestor / Coghe, F. Review of Literature: Behind Helmet Blunt Trauma Mechanisms 2023, Human Factors and Mechanical Engineering for Defense and Safety https://doi.org/10.1007/s41314-023-00063-6

[5] Freitas, Christopher J. / Mathis, James T. / Scott, Nikki / Bigger, Rory P. / MacKiewicz, James. Dynamic Response Due to Behind Helmet Blunt Trauma Measured with a Human Head Surrogate, 2014, Int J Med Sci https://dx.doi.org/10.7150/ijms.8079

[6] Anctil, B. / M. Keown / Pageau, G. / D. Bourget, A Novel Test Methodology to Assess the Performance of Ballistic Helmets, 2005, 22nd International Symposium on Ballistics. Vancouver BC, Canada. 14-18 November 2005.

[7] Anctil B. Bayne, T. Williams K. Fourny P. Craigie I. Savard A. Compression Resistance Testing of Combat Helmets and the Effects on Ballistic Performance, 2014, Technical Report Defence R & D Canada -Valcartier Quebec https://apps.dtic.mil/sti/citations/AD1004167, also Personal Armour Systems Symposium 2014. Cambridge, UK.

[8] https://www.teamwendy.com/products/aftermarket-systems/liner-systems/zap-zorbium-action-pad-systems/zorbium-action-pad-zap-7-pad-nsn-system

[9] NATO Standard – Standard Agreement STANAG 2920 Edition 3 - Classification of Personal Armour – June 2015.

[10] NATO Standard - Allied Engineering Publication AEP 2920 Edition A V2 - Procedures for the Evaluation and Classification of Personal Armour - Bullet and Fragmentation Threats- September 2016.

[11] "Bullet Resistance Helmet test procedure" - HPW-TP-0401.01B, October 1995 from the H.P. WHITE LABORATORY, INC.

[12] Mukasey M.B. et al., "Ballistic resistance of body armor", NIJ 0101.06, US Department of Justice, Office of Justice Programs, Washington, DC, USA, July 2008.

[13] Internal test report TOR24460, Avient, 2024, unpublished material.

[14] Internal test report TOR24431, Avient, 2024, unpublished material.

[15] Internal test report TOR24322, Avient, 2024, unpublished material.

Establishing and Monitoring the Quality of Computed Radiography Imaging for Ceramic Body Armor Personal Protective Equipment

M. Khan[1], M. Brothers[1] and T. Gillis[1]

[1]*Non-Destructive Evaluation, Aerospace Research Centre, National Research Council Canada, 1200 Montreal Road, Ottawa, Ontario, Canada* muzibur.khan@nrc-cnrc.gc.ca

Abstract. Ceramic body armor is designed to protect against high-velocity projectiles by leveraging the unique properties of ceramic materials. Therefore, they have been widely employed for lightweight armor protection. However, ceramic body armor personal protective equipment (PPE) plate and inserts can develop defects from several sources: (i) manufacturing, (ii) mishandling, and (iii) projectile impacts during operations. The ballistic performance of armor plate degrades due to the presence of damage and the required protection cannot be guaranteed. Due to their high manufacturing and raw material costs, the end user expects ceramic armor to have a prolonged service life while maintaining their integrity and performance. In order to promptly and nondestructively assess the operational readiness of ceramic protective plate, nondestructive evaluation (NDE) techniques are necessary. A successful body armor inspection guarantees that the PPE maintains its original characteristics.

Radiographic NDE can visualize and quantify damage or anomalies, thereby verifying the integrity of the armor plate. Computed radiography (CR) uses a reusable phosphor imaging plate instead of film and chemicals, making it easier to take faster pictures, analyze them better, store them digitally, while also lowering costs. This comprehensive experimental study of over several thousand plates proved that high-resolution CR effectively reveals typical damage in PPE armor plate. This paper establishes and validates a CR-based PPE inspection procedure, meant to monitor the quality of the CR imaging process, and the control mechanisms that ensure consistent optimal picture quality, high repeatability, and minimal false positives.

1. INTRODUCTION

Body armor PPE, commonly used in bulletproof vests, employs composite materials combining distinct constituents to achieve superior protective performance. Typically, a ceramic strike face fractures incoming projectiles, dispersing impact energy, while a polymer backing, such as aramid or polyethylene fibers, absorbs residual energy and retains fragments. Examples include Kevlar® vests, which may incorporate ceramic inserts for enhanced ballistic resistance.

Ceramic plates and inserts may develop defects during manufacturing or incur damage from mishandling, wear, or operational impacts. While ceramics are favored in armor systems for their low density, high hardness, and compressive strength, their brittleness and poor tensile properties make them prone to cracking, compromising ballistic performance [1]. Given the high cost of manufacturing and raw materials, ceramic armor is expected to have a long and effective service life. However, there is currently no standardized method or reliable mechanism for determining their lifespan, which can vary greatly. Maintaining the structural integrity of armor throughout its service life requires regular inspections. Non-destructive evaluation (NDE) techniques are essential for quickly and effectively assessing the combat readiness of ceramic plate. Properly conducted body armor inspections ensure that the armor worn by personnel remains effective.

Ultrasonic testing and radiographic imaging are key nondestructive methods for inspecting armor plates. However, the complex structure and multiple interfaces of composite materials hinder ultrasonic testing. In contrast, radiography, including film and computed radiography (CR), effectively detects internal defects. CR, which uses reusable phosphor imaging plates, offers faster acquisition, enhanced image analysis, digital storage, and greater efficiency, reducing reliance on traditional film-based methods.

2. COMPUTED RADIOGRAPHY

Industrial radiography, a well-established NDE technique for volumetric analysis, has been widely used in industrial and medical imaging since the discovery of X-rays in 1895 [2-3]. Conventional radiography utilizes penetrating electromagnetic radiation to evaluate the internal structure of solid objects. Common detection media include

https://doi.org/10.52202/081621-0029

photographic films (for film radiography), phosphor plate (for computed radiography), fluorescent screens (fluoroscopy), and solid-state flat panel detectors (digital radiography) [4].

Computed Radiography (CR), introduced in 1980, became a leading eco-friendly digital method for producing radiographic images, following practices similar to traditional film radiography [5-6]. Unlike film radiography, CR uses a reusable photostimulable phosphor imaging plate (IP) with a coated detector layer composed of photostimulated crystals containing halogenides such as bromide, chlorine, or iodine. As illustrated in Figure 1, the imaging plate, loaded into a cassette and positioned beneath the test object (e.g., armor plate), is first exposed to radiation, such as X-rays or gamma rays, depending on the application. While undergoing radiation exposure, a portion of the radiation photon is absorbed by the test object, while the remaining transmitted photon passes through and is captured by the IP plate. The transmitted radiation photon energy excites the electrons in the phosphor material and causes some of them to move to a higher metastable energy state (i.e., electrons trapped in a high energy state) in the crystal lattice layer of the phosphor plate, which in turn creates an invisible latent image that stores the information from the X-ray exposure. The phosphor traps energy proportional to the incoming radiation intensity.

To retrieve the latent image, the exposed imaging plate undergoes scanning by a laser reading device, which uses a helium-neon laser to scan in a raster pattern. The red laser provides sufficient energy to stimulate the trapped electrons from prior exposure, causing them to return to a low-energy state and emit blue light (a process known as photostimulated luminescence, or PSL) which is proportional to the stored latent image. This blue light is captured by a light guide and directed to a photomultiplier tube (PMT). The PMT amplifies the collected light signal, samples it spatially, and converts it into a digital signal through analog-to-digital conversion, resulting in a digital image displayed on a monitor for evaluation. After laser scanning, the phosphor screen is exposed to a high-intensity white light beam to erase any residual latent image (reset), preparing it for reuse. The digitization process typically takes one to five minutes, depending on the image size and resolution. The resulting digital image can be processed, analyzed, stored electronically, and archived with ease, offering the advantage of being easily copied, edited, and shared [4].

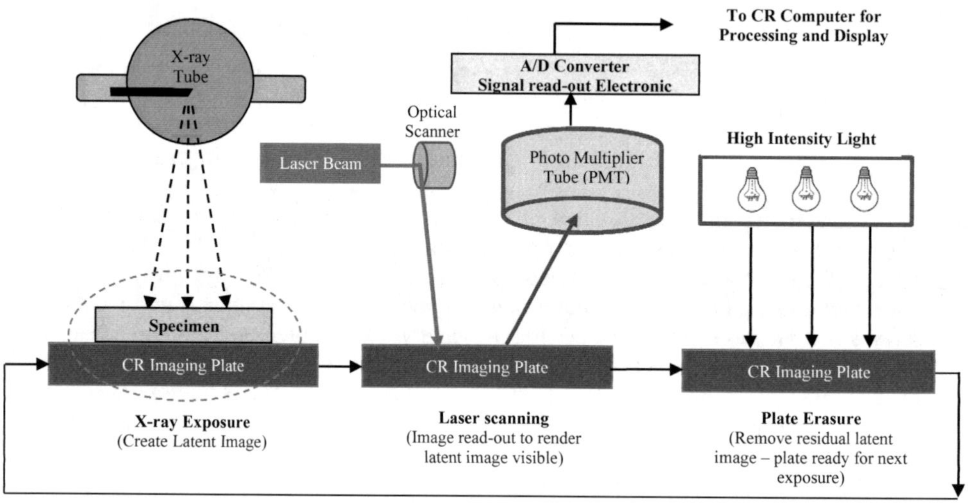

Figure 1. Computed Radiography Principles Based on a Storage Phosphor Plate

2.1 Characteristic Parameters of CR Imaging

Computed radiographic (CR) imaging is governed by a set of critical characteristic parameters that determine image quality, efficiency, and the reliability of defect detection. With the advent of digital radiography technologies such as CR, radiographic images are captured and stored in a digital format, enabling the application of advanced image processing techniques and computational algorithms. These advancements have led to a significant transition from the qualitative and subjective assessment of imaging performance, typical of traditional film radiography, to an objective, separate quantitative evaluation based on critical characteristic metrics such as signal-to-noise ratio (SNR), basic spatial resolution (BSR), contrast-to-noise Ratio (CNR), dynamic range, and others [7].

During the traditional film radiography era "radiographic sensitivity" (often called "sensitivity") is a global and subjective image quality term used that to describe the overall ability of a radiographic imaging procedure when processes were purely analog and lacked digital acquisition, image processing, filtering, and computational evaluation capabilities. As a result, the separate quantitative measurement of individual characteristic parameters (BSR, SNR, CNR etc.) was not possible. Therefore, it can be stated as

SNR, BSR, CNR, Dynamic Range (digital radiography CR, DR) ≡ sensitivity (film radiography)

Additional influencing factors include the characteristics of the inspected object (such as material and thickness), geometric aspects (including X-ray focal spot size, source-to-object distance, properties of the radiation source, and exposure sensitivity or speed class), the type of CR imaging plate (standard vs. high-resolution), and defect orientation relative to the X-ray beam. Exposure conditions (such as kilovoltage or kV, milliampere or mA settings, and exposure time), as well as the defect's size, shape, position, and absorption coefficient, the specific settings of the CR system, and the visual perception of the operator, also play roles in determining CR image quality [8]. All of these parameters collectively affect the clarity, accuracy, and reliability of radiographic evaluations, particularly in industrial applications where precision is critical. While each of these parameters contributes to image quality in distinct ways, they are interrelated, and their quantitative interactions are complex. No single universal equation exists that links these factors across all CR systems, as their relationships depend on specific system characteristics, acquisition parameters, and inspection conditions. A thorough understanding and optimization of these parameters are essential to ensure accurate, repeatable results and to maintain quality assurance in computed radiography.

2.1.1 Contrast Sensitivity

Contrast sensitivity, also known as contrast resolution or simply contrast, refers to the capability of a CR imaging system to identify subtle differences in thickness or density of a specimen due to anomalies, which are represented by changes in pixel or gray values between adjacent areas in the radiographic image [9]. Optimizing the contrast sensitivity enhances defect visibility, ensuring subtle imperfections can be detected effectively. In CR contrast sensitivity is usually expressed or evaluated by contrast-to-noise ratio (CNR), which defines how well two areas (interest vs. background) can be distinguished from each other in the presence of noise.

CNR is commonly evaluated using the ASTM E1647 standard 4-recessed-step gauge, a standardized tool comprising multiple steps of varying depth to assess the system's ability to detect subtle low-contrast details (1%, 2%, 3% and 4% thickness change) [10]. The contrast sensitivity in CR is primarily influenced by factors such as signal-to-noise ratio (SNR) and basic spatial resolution (BSR). Engineering codes and film/CR process control guidelines often mandate achieving at least 2% contrast sensitivity.

2.1.2 Spatial Resolution

Basic spatial resolution (BSR, also known as sharpness or resolution or detail) refers to the capability to differentiate between closely spaced small objects in a digital image or to distinctly identify abrupt changes within an object [11]. The spatial resolution is a critical characteristic for evaluating the efficacy of computed radiography systems in detecting crack-like discontinuities and measuring imaging performance levels. Spatial resolution is typically measured using dedicated gauges known as the duplex wire gauge or the line pair gauge. There is no industry standard guidance or universally defined minimum requirement for spatial resolution; however, higher spatial resolution (represented as a smaller value) is generally preferred because it allows for better visualization of fine details in the image, reduces blurring and yields sharper edges. Spatial resolution tends to be more of a qualitative goal: the smaller the details the system can resolve, the better.

Optimizing spatial resolution is important for detecting crack-like subtle indication. During PPE armor plate inspections, crack indications on radiographs appear as localized intensity variations, represented by pixel value changes, resulting from cracks within the test object. Owing to the characteristics of the radiographic technique, these crack indications typically exhibit low contrast and sharpness and are superimposed on a non-uniform and noisy background.

https://doi.org/10.52202/081621-0029

2.1.3 Signal-to-Noise Ratio (SNR)

Noise is unwanted random variations in the image signal intensity (variations in gray values or pixel values) that do not correspond to actual differences in the X-ray attenuation of the specimen or object. Quantum noise (photon noise), electronic noise, scatter noise, and image processing noise are types of inherent noise in CR system that may be present during CR imaging. CR systems convert X-ray signals into digital images, and during this process, noise can degrade image quality. Any irrelevant signal that disrupts the detection or processing of the desired defect signal is considered noise, which significantly impacts the overall image quality. Noise is typically described using the signal-to-noise ratio (SNR), which serves as an indicator of image quality; the higher the SNR, the less distracting the background noise. SNR reflects the signal intensity in relation to the CR system's inherent noise.

No specific hardware tool is used for SNR measurement. In the output radiographic image, SNR is determined mathematically by analyzing pixel intensity values, comparing signal strength in a region of interest to the standard deviation of background noise. No universally accepted minimum signal-to-noise ratio (SNR) requirement has been established across all industries or regulatory authorities. However, guidelines are often provided to minimize the impact of noise. In general, a higher SNR results in better image clarity, as a stronger signal relative to noise enhances the visibility of defect and improves inspection accuracy.

2.1.4 Bit Depth and Dynamic Range

Bit depth refers to the number of bits used to represent the gray levels (shades of gray) in each pixel of a digital image. The term "bit depth" is primarily associated with hardware devices (not software) such as computers, display monitors, and CR system, representing the maximum range of pixel values that a computer or other hardware can store. Displaying images at full bit depth is unnecessary, as it slows processing. Instead, system software adjusts brightness levels based on the imaging machine's application. If a monitor's pixels have an 8-bit depth (2^8 or 256 pixel values), while the human eye discerns only about 5 bits (2^5 or 32 perceivable brightness levels), the imaging equipment's bit depth exceeds human capacity, so images often appear identical to analog ones. Thus, bit depth generally describes the capabilities of the hardware equipment being used to capture process and display an image.

However, the dynamic range (a subset of bit depth) of an imaging system represents the range of X-ray exposure levels or the range of pixel values, brightness, or gray levels made available by the combined hardware and software of an imaging system to build up a final image on the display monitor. A broad dynamic range system improves overall image quality, allowing for detailed visualization of components with varying material densities/thicknesses from very low-density to high-density area in a single exposure. In film-based radiography systems, this window is narrow; whereas CR systems have a much wider dynamic range, allowing more flexibility in exposure and reducing the need for repeat images. Images can suffer from overexposure or underexposure that falls outside the dynamic range, thus affecting imaging performance (Figure 2).

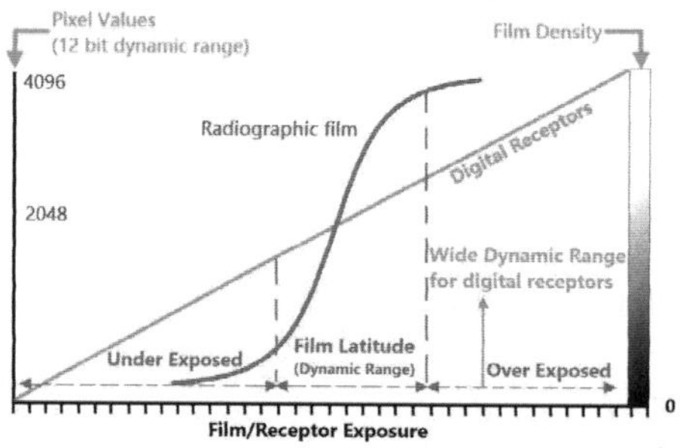

Figure 2: Dynamic Range: Film vs Digital System [12]

The dynamic range of a CR system is influenced by multiple factors, which are summarized in Table 1. These interdependencies emphasize the need for precise calibration, consistent hardware performance, and controlled environmental conditions to maintain the accuracy and reliability of the CR imaging results.

Table 1: Key Parameters Influencing Dynamic Range

Influencing Factors	Effect on Dynamic Range
Imaging Plate Type/Properties and Sensitivity	Higher sensitivity plates detect lower exposures; plate storage capacity affects saturation point (upper limit of dynamic range).
Laser Scanning System (Resolution and Stability)	High laser resolution and stable scanning improve the system's ability to capture small exposure differences; poor resolution or instability reduces the ability to distinguish subtle differences, especially at low exposures.
Photomultiplier Tube (PMT)	PMT consistency and sensitivity impact the accurate detection of emitted light during scanning, influencing dynamic range fidelity.
Analog-to-Digital Converter (ADC) Bit Depth	Higher bit depth (e.g., 12-bit or 16-bit) allows finer distinction between intensity levels, expanding the usable dynamic range.
Exposure Technique	Incorrect exposure settings (too high or too low) result in saturation or loss of data below detection threshold, narrowing the effective dynamic range.
Scatter Radiation	Increases noise, lowers signal-to-noise ratio (SNR) and contrast-to-noise ratio (CNR), effectively reducing usable dynamic range.
Image Processing Algorithm	Does not alter the physical dynamic range but affects perceived and diagnostically useful dynamic range through contrast enhancement, noise reduction, and windowing.
Quality of Radiation (kV)	Higher kV lowers contrast and narrows dynamic range; lower kV increases contrast and broadens dynamic range, requiring greater system capacity to accommodate intensity variations.
X-ray Focal Spot Size	Smaller focal spots improve image detail and contrast but may limit dynamic range; larger focal spots reduce detail but broaden dynamic range by accommodating wider intensity variations.
Change in Hardware, System components or Calibration	Modifications to X-ray tube characteristics, such as focal spot size or radiation quality, can influence the system's ability to manage varying intensity levels
Environmental conditions	temperature and humidity can indirectly influence system performance

In CR, a step wedge of varying thickness is often used to evaluate the dynamic range of an imaging system by counting the number of steps (exposure levels) that are visible, which provides an estimate of the usable dynamic range. There is no strict minimum dynamic range requirement, but a wider dynamic range is preferred as it enhances image detail and contrast across varying exposure levels.

2.1.5 Radiographic Sensitivity

The term *radiographic sensitivity*, originating from the traditional film-based approach, continues to be used in the context of digital radiography to describe a system's imaging performance or its capability to detect small or low-contrast features. In practice, it is often expressed as the ability to resolve a specified percentage of thickness variation within the specimen; however, it does not directly correspond to the actual defect size. Instead, it functions as an indirect indicator of the overall performance of the radiographic system, reflecting both process parameters and the system's ability to detect defect or thickness variations within the specimen. Lower percentage values indicate higher sensitivity.

The image quality indicator (IQI) is a well-established method for assessing image quality (specifically for measuring sensitivity), originally developed for film-based radiography but still widely utilized in digital radiography systems. Many industry standards (e.g., ISO 17636-2, ASME BPVC Section V) continue to require the use of IQIs, which serve as standardized gauges placed on or near the test specimen to measure and maintain radiographic sensitivity, thereby ensuring consistent and reliable image evaluation. The three primary types of IQIs (illustrated in Figure 3) commonly employed are: (i) wire type, predominantly used in Europe; (ii) step hole type, commonly utilized in France; and (iii) plaque hole type, frequently referred to as penetrameters in the United States.

The wire-type IQI consists of a set of thin, parallel wires (typically 16) of varying diameters, usually made

https://doi.org/10.52202/081621-0029

of the same material as the test object or a material with similar attenuation properties. Sensitivity is directly indicated by the smallest visible wire diameter or corresponding wire number. The French step-hole IQI consists of a single flat plaque with several steps (commonly 4, 5, or 6) of increasing thickness, without any holes. Sensitivity is measured based on the smallest thickness difference (step height) that can be detected on the radiograph.

However, only for the plaque hole type IQI (Figure 3C), the term Equivalent Penetrameter Sensitivity (EPS) is used instead of the general term "sensitivity" (as applied to wire or step hole IQIs), and the EPS value is defined in ASTM E1025-11 as follows:

$$EPS\ (\%) = \frac{100}{t_{material}} \sqrt{\frac{T_{IQI} d_{hole}}{2}} \quad (1)$$

Where: EPS - Equivalent Penetrameter Sensitivity
$t_{material}$ - Thickness of penetrated material
T_{IQI} - Thickness of IQI
d_{hole} - Diameter of IQI hole

These IQIs consist of multiple plaques of uniform thickness, each containing three holes of known diameters drilled perpendicularly through the material. The diameters of these holes correspond to one, two, and four times the plaque thickness (**T**), and are designated as 1T, 2T, and 4T holes, respectively. The EPS directly relates to the percentage of thickness change represented by the visibility of the holes in the plaque radiograph (e.g., 2-2T corresponds to 2% sensitivity). A sensitivity level of 2-2T, indicating 2% of the material thickness with a hole diameter equal to twice the plaque thickness (2T), is considered the threshold of detectability on the radiograph [13]. The detectability of these holes on the resulting radiograph is influenced by several factors, including hole diameter, plaque thickness, image contrast, and signal-to-noise ratio. Plaque-type penetrameters provide a binary assessment (pass/fail) to verify whether the achieved sensitivity meets the specified acceptance criteria (e.g., 1-2T or 2-2T); however, they do not yield a precise quantitative measurement of the system's sensitivity.

For computed radiographic imaging of armor plates, the plaque hole type IQI (Figure 3C) was employed, which enables radiographers to detect fine variations in image quality as radiographic parameters are optimized. Defect in armor plates such as clusters of porosity, chipped ceramic dimples, extra ceramic material fragments, micro-cracks, micro-tears, voids, or missing dimples, depend on the CR system's sufficient contrast sensitivity. In compliance with established engineering codes and film/CR process control guidelines, a minimum contrast sensitivity of 2% (i.e., visibility of the 2-2T hole) is required in radiographic imaging, which is also specified in the procedure as the target for imaging armor plates [14]. A 2% sensitivity refers to the imaging system's ability to distinguish at least 2% in material thickness or density, serving as a key measure of its capability to resolve subtle variations in material properties or low-contrast features, such as minor density variations or small defects, which is essential for accurate assessment in complex materials.

(a) Wire type IQI (b) Step hole type IQI (c) Plaque hole type IQI

Figure 3: Three Conventional Types of Image Quality Indicators (IQIs) [15-16]

3. IMAGING PROCEDURE OF ARMOR PLATE AND QUALITY MONITORING

The schematic of the experimental setup for computed radiographic inspection of armor plate is illustrated in Figure 4. The armor chest plate is positioned on the CR standard resolution phosphor imaging plate with the strike face facing downward. The chest plate is aligned with the phosphor imaging plate, ensuring that its long axis runs parallel to the long axis of the imaging plate. The ceramic plates inspected in this study measure approximately 254 mm (10 inches) by 304.8 mm (12 inches), consisting of a hard ceramic dimpled front face and a backing layer made of high-performance polymer composite. The body armor chest plate is considered sensitive material in terms of their fabrication and intellectual property; therefore, preventing disclosure of additional information.

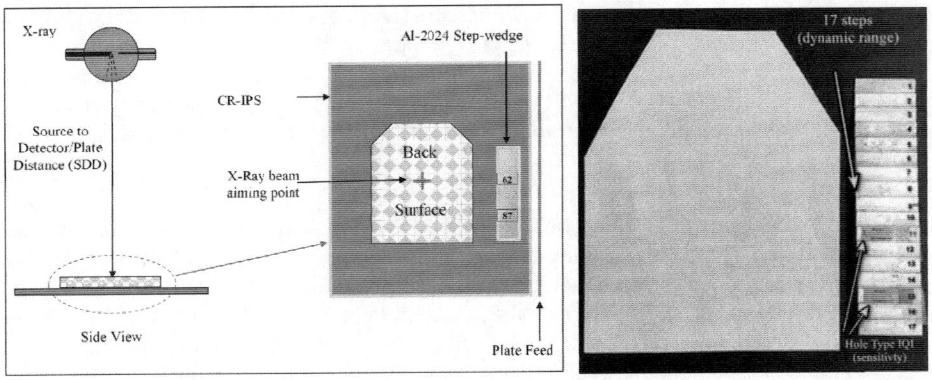

Figure 4: Experimental Setup for Armor Plate Inspection

The procedure for armor plate inspection was initially developed using an X-ray tube paired with a computed radiography (CR) system. During the initial setup, no specific image quality indicator was employed. Nevertheless, after optimizing inspection parameters such as kilovoltage (kV), milliampere (mA), source-to-object distance, and other critical variables, the system produced satisfactory CR images. These optimized images effectively revealed typical defect in the armor plate, demonstrating the setup's initial effectiveness.

However, complications arose when the X-ray tube and CR system were later replaced with new equipment. Although the same inspection parameters, including kV, mA, and exposure settings, were retained, the resulting images showed inconsistencies. The system's reliability was compromised as the new setup frequently generated inaccurate results, such as over-saturation at the edges of the armor plate, and occasional false-positive findings.

Investigations revealed that hardware changes altered the dynamic range of the imaging system, despite identical inspection parameters being maintained, which resulted in over-exposure of the ceramic material at the edge of the chest plate. This finding highlights the sensitivity of CR systems to hardware variations and underscores the necessity of maintaining repeatability and calibration to ensure consistent image quality. This case underscores the necessity of implementing a robust protocol to account for hardware variations and highlights the potential utility of incorporating image quality indicators (IQIs) to ensure consistency and accuracy in inspection results.

3.1 Revised Procedure with IQI and Step Wedge

To address inconsistencies in imaging results, both the imaging procedure and the setup were carefully revised. As part of the updated procedure, and to ensure both optimal image quality and repeatability, a seventeen-step Al 2024 (aluminum alloy 2024) step wedge was placed adjacent to the specimen on the imaging plate to assess the system's dynamic range (as shown in Figure 4). Accurate exposure settings were validated by ensuring that all steps on the Al-2024 step wedge (steps 3 through 17), except for the two thinnest steps, were clearly visible and readily distinguishable in the radiographic image. Additionally, approximate target gray-scale values for each step were specified and had to be achieved by varying the exposure settings in order to maintain consistency and precision in the imaging process. Similarly, the target gray scale value range for both the central and edge portions of the armor plate was also defined. X-ray parameters, including voltage, current, and exposure duration, were systematically adjusted to achieve target grayscale values at both the central dimples and the plate edges while preventing overexposure that could potentially produce false positives. The thickness of the ceramic material may vary slightly, potentially causing gray-scale values to deviate marginally from the specified range. In such instances, verification of the Al step wedge gray-scale values to ensure alignment with the specified range, as well as evaluation of the chest plate, is required.

Furthermore, two ASTM E1025-specified penetrameters were strategically positioned on two distinct steps of the wedge. These penetrameters serve as image quality indicators (IQIs), which were instrumental in verifying that the resulting CR image achieves the required sensitivity of 2% [Equation 1]. By integrating the step wedge (for

https://doi.org/10.52202/081621-0029

dynamic range assessment) and ASTM IQIs (for sensitivity verification), the procedure provides a standardized and robust approach for maintaining high-quality radiographic results.

In CR, grayscale values are represented as digital numbers (DN), or pixel intensity values, corresponding to the detected X-ray attenuation at each pixel. The CR system operates with a 16-bit processor, producing DN values ranging from 0 to 65,535 (2^{16} levels). Exposure parameters were carefully calibrated to achieve a dynamic range spanning from 10% (DN 6,553) to 90% (DN 58,981) of the system's maximum pixel intensity. This range minimizes the effects of low-end noise and high-end saturation, ensuring sufficient contrast and detail for reliable image interpretation.

The aluminum step wedge provides a controlled range of X-ray attenuation through its graduated thickness steps. Dynamic range is evaluated by identifying the lowest thickness step where the signal exceeds the noise floor (the lower limit) and the highest step before signal saturation (the upper limit). A characteristic response curve can be plotted with step thickness or logarithmic exposure on the x-axis and pixel value (DN) on the y-axis. The linear portion of this curve defines the usable dynamic range, where pixel values increase proportionally with exposure. The total dynamic range is determined by the span between the minimum detectable signal and the saturation point, reflecting the system's capability to capture varying levels of X-ray attenuation.

This also reflects the system's ability to distinguish minimal variations in attenuation due to small differences in material thickness, thereby assessing contrast resolution. It ensures that exposure parameters (kV, mA, exposure time) are optimized for the specific material and imaging setup. Additionally, the step wedge plays a critical role in periodic quality assurance by monitoring hardware or inspection parameter changes, detecting any decline in imaging performance, and maintaining consistent accuracy and reliability.

The incorporation of image quality indicators (IQIs) ensured that the acquired images met specific quality standards by providing a consistent reference for contrast sensitivity and defect detectability. The combined use of the step wedge and IQIs standardized the inspection process, enabling reliable results despite minor variations in hardware or operational parameters. This methodological refinement enhanced both the accuracy and repeatability of the computed radiography system, substantially reducing false positive indications and increasing confidence in defect detection. A comprehensive inspection of several thousand armor plates was conducted, with each inspection requiring less than five minutes. Interpretation of the images was performed by certified personnel specialized in radiographic nondestructive testing.

4. INSPECTION RESULT

The CR inspection primarily aimed to identify and evaluate defect such as cracks and voids that could adversely affect ballistic performance of ceramic PPE. Cracks of varying lengths could appear anywhere on the plate, whereas voids were typically located along the flat outer edges, often resulting from wear-related chipping or breakage. During the inspection, anomalies or discontinuities associated with the manufacturing process such as clusters of porosity, chipped ceramic dimples, fragments of additional ceramic material, micro-cracks, or micro-tears were also observed. These manufacturing-related defect did not constitute grounds for rejection, as the primary objective was to detect and assess service-induced defect. Acceptance or rejection of the plate was determined according to pass/fail criteria established from prior ballistic performance testing. Specifically, cracks (excluding craters or micro-cracks), missing dimples in the central region of the chest plate, and defect exceeding 10 millimeters along the edges were classified as reasons for rejection due to their detrimental effects on the overall structural integrity of the plate.

Cracks and voids resulting from service conditions constituted the primary defect that led to the rejection of plates. The crack profiles displayed a wide range of sizes, extending from a few millimeters to the full length or width of the armor plate. These cracks exhibited characteristics similar to conventional linear defect typically observed in both brittle and ductile materials. Potential causes for the formation of these cracks included pressure exerted during the pressure-casting stage of manufacturing, the release of localized stresses within the material, or external mechanical forces such as impacts caused by dropping or striking the plate.

Certain objects detected at the dimples were identified as unwanted inclusions, such as ceramic chips, which were likely introduced during packaging from adjacent layers. Edge void defects were primarily attributed to chipping along the ceramic edges, occurring either during handling or manufacturing processes. Localized discontinuities in ceramic thickness were attributed to deviations in the molding and casting processes, which resulted in the incorporation of multiple small fragments of excess ceramic material.

The observed defects, regardless of whether their origin was manufacturing or service-related, encompassed porosity, micro-tears or micro-cracks, missing dimples at the edges, trimmed dimples along the edges, small voids with diameters less than 10 mm situated along the edges, unwanted ceramic particles dispersed between dimples, and substantial thinning voids at the edges. These defects were discernible in the radiographic images. During the evaluation process, all indications, whether deemed acceptable or constituting grounds for rejection, were systematically annotated within the CR images using specialized software tools. A selection of representative X-ray images depicting various defect types is provided in Figure 5 to Figure 7 and in Table 2.

(a) Small Crack at Edge (b) Through-Plate Crack (c) Void at Edge

Figure 5: Crack and Void Defect (with Step-Wedge Gauge)

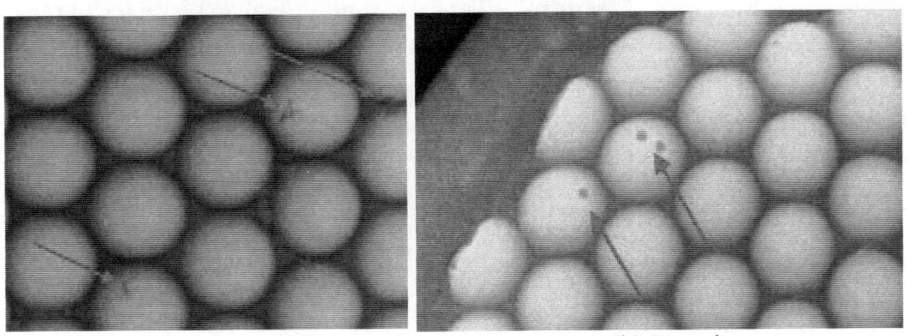

(a) Miro-tears /Micro crack (b) Porosity

Figure 6: Manufacturing Defect

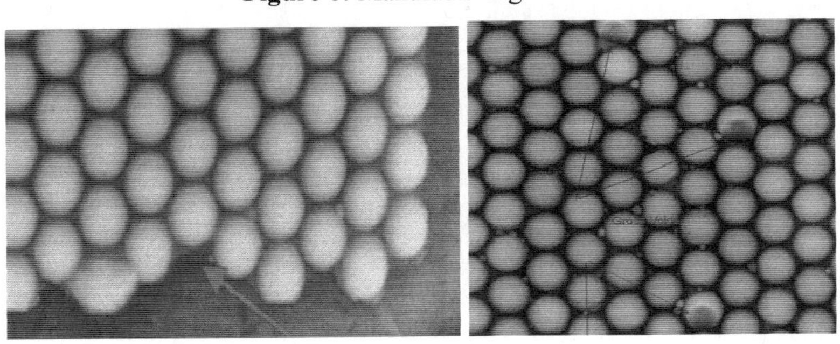

Figure 7: Manufacturing Process Defects (missing dimple at the edge in left image)

https://doi.org/10.52202/081621-0029

Table 2: Defects Detected in Radiographic Images of Ceramic Armor Plates

Cracks: Varying length, presence anywhere in plate Few millimeters to the through length or width of the ceramic plates	Figure 5a,b
Void at Edge	Figure 5c
Micro-tears / Micro crack	Figure 6a
Porosity	Figure 6b
Missing dimple at edge	Figure 7a
Gross Voids	Figure 7b

5. CONCLUSION

Over several thousand PPE armor plate were inspected, with damage types categorized for acceptance or rejection. Cracks and void-like discontinuities, as per inspection criteria, were the main reasons for rejection. Using standard-resolution CR imaging plate, the system was able to identify non-critical manufacturing and service-induced defects such as porosity, micro-tears, micro-cracks, edge-related dimple defects, small edge voids (<10 mm in diameter), dispersed ceramic particles, and significant edge thinning voids. The CR inspection procedure, integrated with a comprehensive quality monitoring setup, demonstrated the capability of CR to facilitate efficient, high-throughput nondestructive evaluations. It demonstrated resilience against the considerable variations in the design, size, and condition of armor inserts and their coverings. It successfully detected various types of defects and provided clear, digital image records of inspected plate for verification purposes.

6. REFERENCES

[1] Wang et al; “Design and Characteristics of Hybrid Composite Armor Subjected to Projectile Impact” Journal of Materials and Design 46 (2013) pp. 634–639.

[2] Eastman Kodak Company, “Radiography in Modern Industry” [www.kodak.com accessed on July 2010].

[3] Bossi, R.H., Iddings, F.A., Wheeler, G.C., and Moore, P.O, “Radiographic Testing, Nondestructive Testing Handbook” (3rd. Ed.), Volume 4, American Society for Nondestructive Testing, Columbus, OH, 2002.

[4] Khan, M., “Film and Digital Radiography System” NRC Technical Report LM-SMPL-2010-0032, Feb 2010

[5] Korner, M., et al., “*Advances in digital radiography: Physical principles and system overview*”, Radiographics 27:675–686, 2007.

[6] Nondestructive Inspection Methods, Basic Theory, NAVAIR Technical Manual T.O.33B-1-1, May 2014.

[7] LaCivita, K., Ewert, U., "Equivalent Penetrameter Sensitivity as Applied to Computed Radiography" Materials Evaluation, Volume 70, Issue 8, August 2012.

[8] Ewert, U. et al., “Performance Control, Nondestructive Testing and Reliability Evaluation, in “Springer Handbook of Metrology and Testing”, (Ed. Czichos, H., Saito), pp-904, Springer, Heidelberg, 2006.

[9] Soltani, P., Neary, M., “Filmless Radiography for Aerospace Application”, United States Airforce Research Laboratory Technical Report No. AFRL-ML-WP-TR-1999-4144, July 1999.

[10] ASTM International. ASTM E1647-17: Standard Practice for Measuring Contrast Sensitivity of Radiographic Systems. ASTM International, 2017.

[11] Hussein, E.M.A., “Handbook on Radiation Probing, Gauging, Imaging and Analysis, Volume I: Basics and Techniques”, Kluwer Academic Publishers, May 2003, pp: 276.

[12] Carestream, AINDT, “Digital Imaging Basics” Radiography Equipment Training Manual.

[13] Ewert, U., Jaenisch, G-R., Zscherpel, U., Osterloh, K., Redmer, B., “Industrial Radiology”, in Czichos, H. (Ed.), Handbook of Technical Diagnostics (Fundamentals and Application to Structures and Systems, pp 229, Springer, Heidelberg 2013.

[14] Soltani, P., Neary, M., “Filmless Radiography for Aerospace Application”, United States Airforce Research Laboratory Technical Report No. AFRL-ML-WP-TR-1999-4144, July 1999.

[15] www.ndt-consultants.co.uk [Accessed November 2017]

[16] Individual Penetrameters ASTM-E1025, www.ndtsupply.com [Accessed January 2016]

Case Study of IED Incident in Cauca, Colombia – A Review of the Blast Threats and Protection Provided to the EOD Technician

A. Makris, Jeff Levine and J.P. Dionne

Med-Eng Holdings ULC, 2400 St. Laurent Blvd, Ottawa, Ontario, Canada, K1G 6C4
aris.makris@safariland.com

Abstract. In June 2022, an Explosives Technician of the Colombia Dirección Central de Policía Judicial e Inteligencia (DIJIN) responded to a suspect Improvised Explosive Device (IED) mounted on a motorcycle, in Cauca, Suarez, Colombia. The local unit of the Policia, responsible for responding to explosives incidents arrived at the scene to investigate. The explosives technician conducted a search and render safe procedure of the suspect device. After a canine unit and a robot could not definitively confirm an explosive device, the decision was made to conduct a manual approach of the motorcycle with suspect device using a Med-Eng EOD 9 bomb suit. During the course of procedures, the IED detonated. The ensuing blast posed a fatal threat that was mitigated by the bomb suit the technician was using. He was launched several meters away and suffered non life-threatening injuries to the extremities. This paper is a case study that analyses the four blast threats (overpressure, fragmentation, impact, heat) related to the particular device and circumstances. Through a detailed analysis of the Explosive Ordnance Disposal (EOD) suit including a numerical simulation, one can ascertain the protection provided across the entire body and some inherent limitations. Injury outcome is compared with expectations based on the blast conditions that confronted the victim. Lessons learned from the response and outcome of this specific incident can be beneficial in mitigating injury risk to EOD technicians responding to similar situations.

1. DESCRIPTION OF THE EVENT AND OUTCOME

According to Cardash [1], motorcycle VBIEDs (Vehicle-Borne Improvised Explosive Devices) have become a favoured tactic, technique, and procedure (TTP) employed by FARC (Fuerzas Armadas Revolucionarias de Colombia) dissidents in recent years, particularly in Colombia's Cauca department. Historically, these attacks involved motorcycles carrying large explosive devices, often targeting police stations and government buildings. However, Cardash highlights a recent shift toward smaller explosive payloads aimed at moving police vehicles or officers, including bomb disposal technicians responding to investigate these motorcycles. The present study includes a comprehensive analysis of the four main threats posed by the blast in that specific incident and a focus on how the bomb suit effectively protected the bomb technician exposed to the explosion. Details included in the study were informed by an interview with the bomb technician involved in the incident, as well as by direct examination of the bomb suit he was wearing at the time of the explosion.

1.1 The blast event

On the morning of June 16, 2022, in Suárez, located in Colombia's Cauca region, a local bomb technician was dispatched to investigate a suspected explosive device possibly attached to a motorcycle parked near the rear garage door of the Suárez police station (Figure 1). This incident followed two similar cases in the preceding weeks in the same area, both of which had turned out to be hoaxes. Despite this precedent, the bomb technician approached the situation with appropriate caution.

During the initial assessment, a canine unit was deployed to inspect the area; however, no suspicious indicators were detected. Next, a small EOD robot was sent to examine the motorcycle, but it could not confirm the presence of an explosive device (Figure 2). Wearing the Med-Eng EOD 9 bomb suit ensemble, the bomb technician proceeded with a manual approach to determine whether there was indeed a suspect device located beneath the motorcycle seat. While attempting to dislodge a possible suspected package wrapped in fabric and appearing stuck under the seat, the technician braced his left foot against the motorcycle and pulled on the package. At 9:17 AM, the suspected IED detonated while the technician was in this position (Figure 3, centre). The force of the blast propelled the technician approximately three meters from the motorcycle.

The entire event, up to the moment of the explosion, was captured on video by bystanders and later retrieved online. No visual records exist of the aftermath, except for photos provided by Colombian authorities. The photo on the right of Figure 3 identifies the approximate landing location of the technician, marked by where the bomb suit and helmet were left after first responders rescued

 https://doi.org/10.52202/081621-0030

and evacuated the injured operator. Figure 3 also illustrates the severe damage sustained by the motorcycle, as well as minor damage to the garage door and surrounding area.

A forensic report from the Colombian authorities (DIJIN), later provided to Med-Eng, revealed that the IED contained approximately 2 kg of explosives. According to the DJIN official technical report, the explosive was identified as a high explosive, although its exact composition could not be determined at the time of investigation. It was remotely detonated using a radio frequency device modified from a vehicle alarm system, enabling activation from a distance of up to 400 meters. From the videos and photos, it is estimated that the bomb technician's chest was located at a standoff of approximately 80 cm from the IED at the time of the explosion.

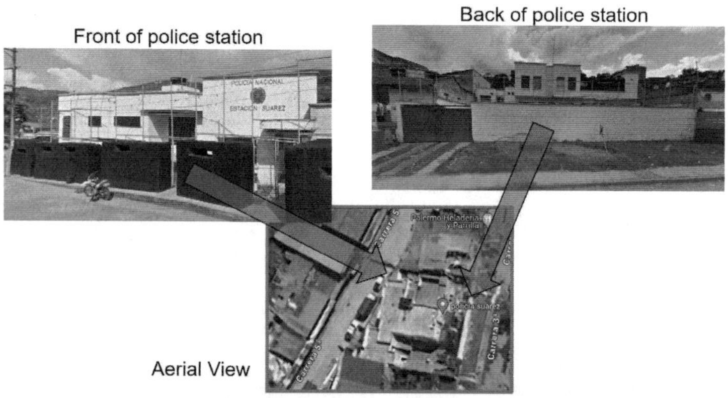

Figure 1. Photos from the area where the blast even took place, extracted from Google Maps (©2025 Google, Image ©2025 Maxar Technologies).

Figure 2. Pre-explosion images. Left: The suspected motorcycle VBIED positioned near the garage door. Centre: The bomb technician being suited up by an assistant. Right: The technician advancing toward the scene for manual inspection. All images courtesy of Colombia's DIJIN

Figure 3. Left and Centre: Still images extracted from a video of the incident obtained from the Internet, capturing the moment of the explosion. In the centre image, the technician's left foot is resting against the motorcycle while handling the explosive package with no standoff. Right: A post-incident photograph of the scene, showing the aftermath after first responders had removed the injured technician from the bomb suit. The image on the right is provided courtesy of Colombia's DIJIN

1.2 The medical outcome

Following the blast, the bomb technician sustained various injuries, including the amputation of his left forefoot, which had been in direct contact with the motorcycle (and underlying explosive), and a fracture of his left tibia. Since he was holding the fabric wrapping the explosive at the moment of detonation, both of his hands were injured, with the right hand being more severely affected, including multiple finger fractures. He also suffered minor burn injuries to his forearms due to flames propagating up the bomb suit sleeves.

Notably, the technician did not sustain any penetrating wounds, blast overpressure injuries (including to the ears), or head trauma. He remained conscious and alert upon arrival at the hospital. Figure 4 shows photos of the technician taken a few days after the incident, highlighting his primary injuries.

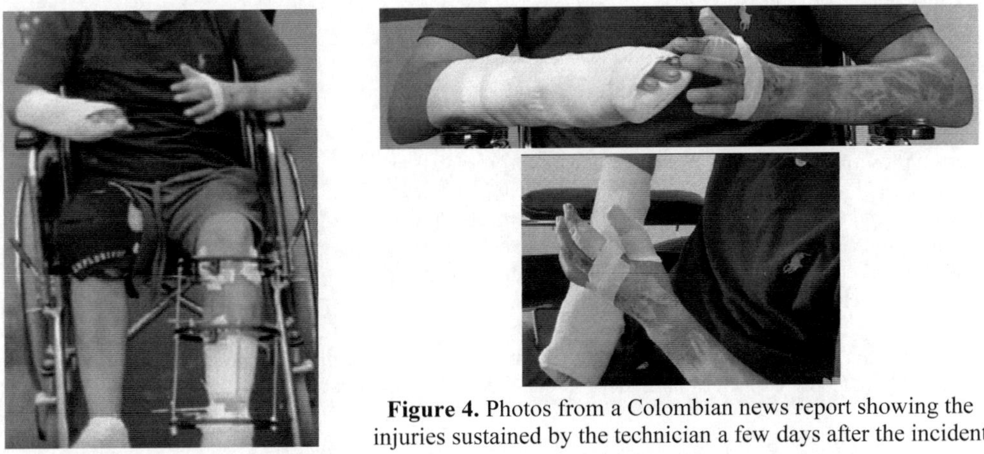

Figure 4. Photos from a Colombian news report showing the injuries sustained by the technician a few days after the incident

1.3 Personal Protective Equipment after the blast

Due to the bomb technician's close proximity to the blast source, the left side of the bomb suit trousers sustained significant fragmentation damage, particularly in the lower exterior portion. This damage exposed the soft armour aramid layers and shattered the shin plate. In contrast, the right side of the trousers, positioned slightly farther from the blast, experienced only minor fragmentation damage. Importantly, no fragments fully penetrated the stack of materials comprising the trousers worn on either leg. See images on Figure 5.

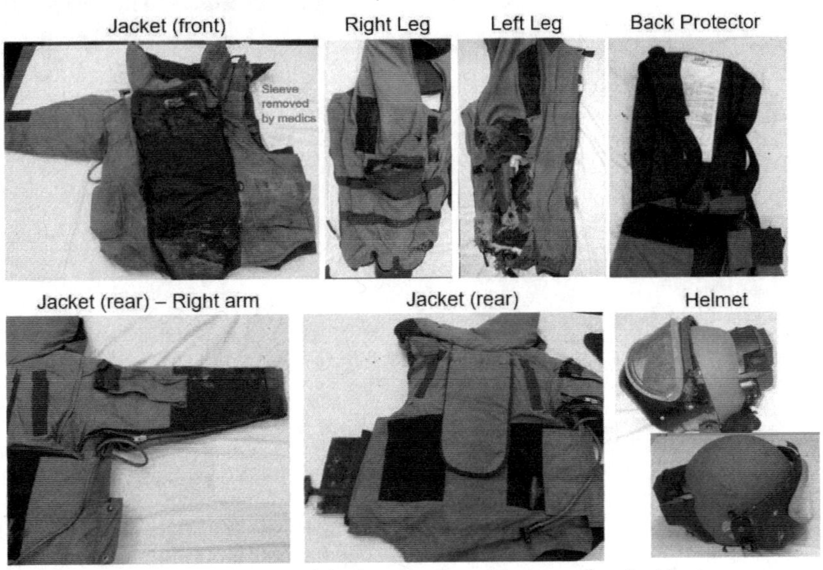

Figure 5. Photos of the bomb suit and helmet after the blast

https://doi.org/10.52202/081621-0030

Similarly, the upper portion of the bomb suit jacket, which directly faced the blast, exhibited concentrated fragmentation damage. Although fragments penetrated the soft armour and some of the rigid armour at the chest and groin levels, none managed to breach the innermost protective layers, thereby ensuring the wearer's safety. Some deformation of the rigid protective plates was observed. The rear portions of the trousers and jacket, which were shielded from the blast, remained largely unaffected. While the left sleeve was removed by medical personnel to facilitate the victim's evacuation, the remaining sleeve showed no significant damage.

Regarding the helmet, a single fragment penetrated the surface of the outer acrylic layer of the visor but failed to breach the innermost polycarbonate layer. The visor primarily exhibited cosmetic fragmentation damage, and the rubber trim housing the LED lights was dislodged during the blast. Despite this, the retention system remained fully functional, and the helmet shell sustained no penetrations. Figure 5 presents photos of the bomb suit ensemble, highlighting the damaged areas.

2. ANALYSIS OF THE MAIN BLAST THREATS

2.1 Primary blast threat: overpressure

An estimation of survivability based on blast overpressure can be derived from the work of Van der Voort et al. [2], which updated the original survivability curves developed by Bowen et al. [3]. These curves, referred to here as the "NATO curves", presented in terms of peak overpressure and positive phase duration (Figure 6), estimate the probability of survival for an unprotected individual. Similarly, Figure 7, inspired from previous work [4] translates these survivability predictions into a format based on explosive charge and standoff distance, using free-field blast calculations. For the current scenario—2 kg of TNT detonated at an approximate standoff distance of 80 cm—the corresponding data points are marked on both figures. The predicted survivability level (<1%) highlights the unfavourable outcome expected for an unprotected individual *not* wearing a bomb suit.

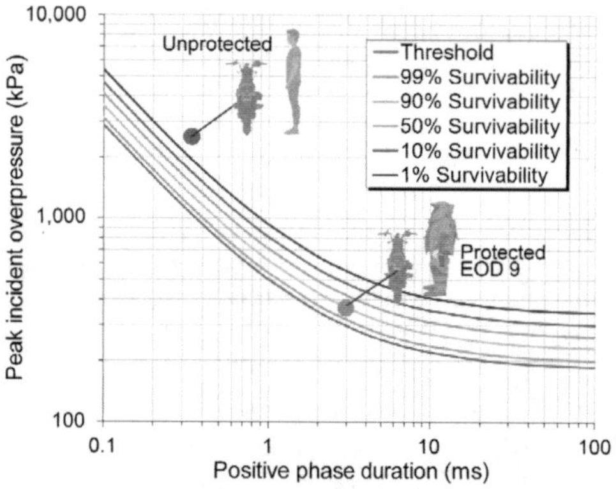

Figure 6. NATO blast injury prediction curves [2] along with data points corresponding to the motorcycle VBIED scenario

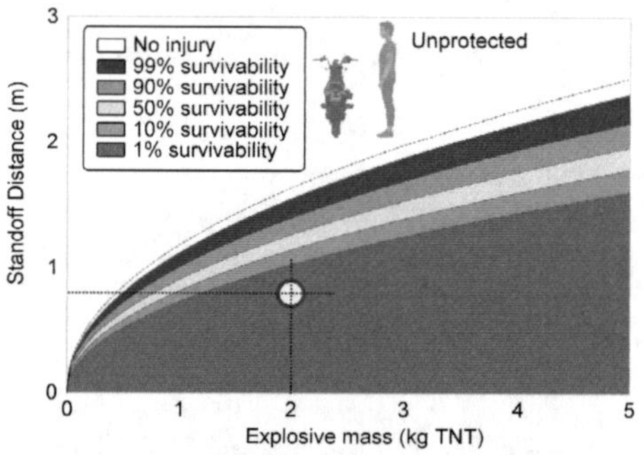

Figure 7. NATO curves expressed in terms of explosive charge mass and standoff distance, with the motorcycle VBIED scenario data point for an unprotected scenario

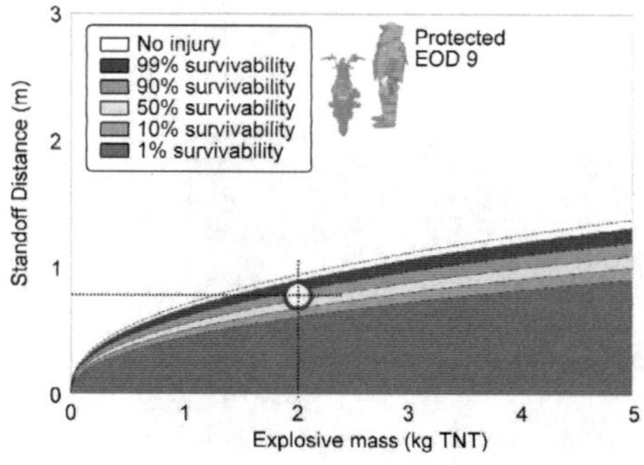

Figure 8. NATO blast injury curves from Figure 7 modified to account for the protection provided by the EOD 9 bomb suit. Data shown for the motorcycle VBIED scenario

Figure 8 presents the NATO injury curves, adjusted to reflect the protective capabilities of the EOD 9 bomb suit (assuming a conservative 85% reduction in overpressure and a fourfold increase in positive phase duration, see [5]). Based on Figure 8, the bomb technician was predicted to have a 90% chance of survival from the blast overpressure threat alone, consistent with the actual survival outcome of this incident. However, it is important to note that the charts in Figures 6 to 8 are intended as general guidance for free-field blasts, and do not provide definitive or fully validated injury predictions.

The threshold for eardrum perforation is as low as 35 kPa. At 100 kPa, there is a 50% probability of perforation, increasing to 95% at 200 kPa [6]. Figure 9 illustrates the potential for eardrum perforation based on these thresholds, shown in terms of explosive charge mass and standoff. Figure 10 presents the same data for an individual wearing the EOD 9 Helmet, which provides a conservative 90% reduction in overpressure at the ear. In the current scenario, the unprotected probability of eardrum perforation exceeds 95%, while the helmet reduces it to 50%. Notably, the bomb technician did not sustain any ear injuries during this incident.

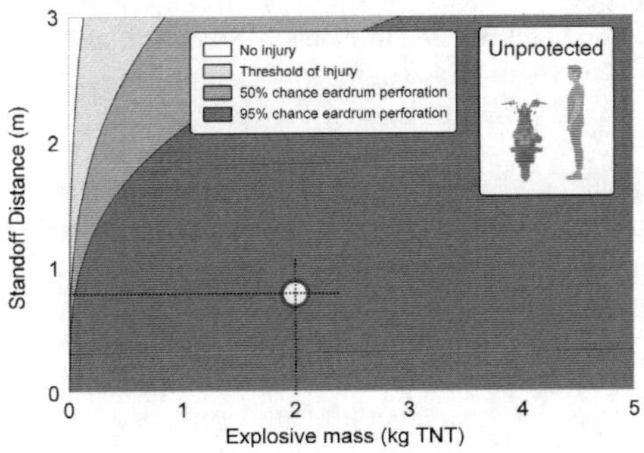

Figure 9. Ear overpressure injury chart (unprotected case)

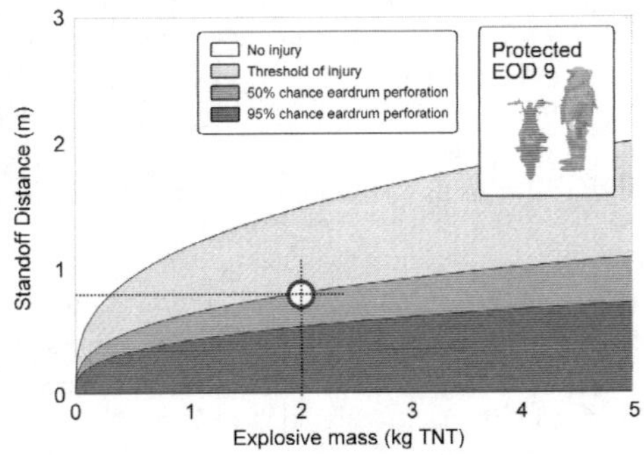

Figure 10. Ear overpressure injury chart (protected case – EOD 9 Helmet)

2.2 Secondary blast threat: fragmentation

Cardash [1] observed that adding fragmentation to small motorcycle VBIEDs significantly enhances their impact and effective range, effectively offsetting the limitations of smaller explosive payloads typically used in targeted attacks. For the incident under study, however, Colombian authorities did not provide specific details regarding the fragmentation threat posed by the IED itself. Nevertheless, even in the absence of engineered fragmentation, the motorcycle itself can act as a source of secondary fragmentation upon detonation, generating debris of varying sizes capable of causing severe penetrating injuries in the absence of proper protective measures.

The images in Figure 5 highlight the significant fragmentation and peppering damage sustained by the bomb suit due to the blast. This resulted in considerable cosmetic damage, along with a few penetrations through the suit's outermost layers and protective plates. For instance, Figure 11 illustrates fragmentation damage to the groin protective plates. However, none of the fragments fully penetrated the innermost composite rigid layer. Similarly, only one fragment damaged the helmet visor without completely penetrating its material layers. The bomb suit ensemble thus provided essential fragmentation protection, effectively preventing fragment penetration to critical areas.

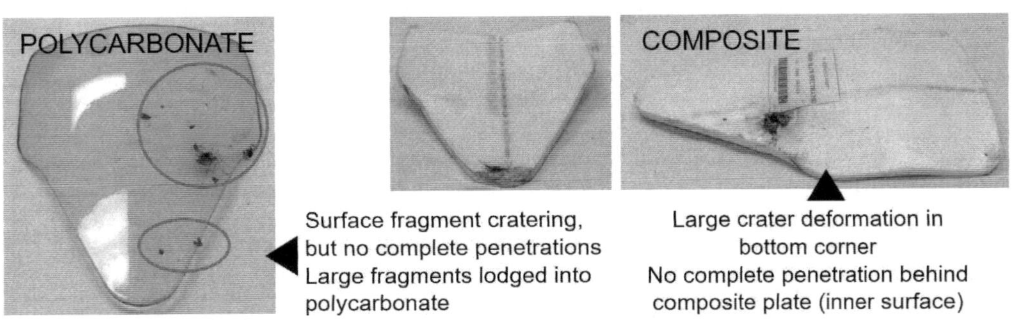

Figure 11. Photos of the innermost groin protective layer showing no complete penetration. No complete penetrations were observed anywhere on the bomb suit

https://doi.org/10.52202/081621-0030

2.3 Tertiary blast threat: impact

Blast exposure can cause rapid head acceleration, leading to back-and-forth brain movement within the skull, which is a potential "impact" mechanism for blast-related brain injury. Although the Head Injury Criterion (HIC, [7]) has not been specifically validated for blast scenarios, it can still be used to correlate head acceleration measurements with the likelihood of brain injury through applying correlations to the Abbreviated Injury Scale [8]. Leveraging HIC, Dionne et al. [9] developed predictive tools that relate AIS injury levels to explosive charge and standoff distance for free-field blasts, based on data from multiple blast tests using instrumented mannequins. Figure 12 illustrates the chart developed for the EOD 9 bomb suit, showing the data point corresponding to the motorcycle VBIED incident under study. The chart predicts "no injury," consistent with the medical report provided by Colombian authorities, which noted no brain injuries, or concussion symptoms. Furthermore, the technician was reportedly able to converse shortly after the incident. These findings suggest that the EOD 9 Helmet provided effective protection, preventing brain injury in this case.

Beyond the direct impact of the blast wave on the head, the technician's head could have also impacted the ground as the body was propelled by the blast force. To explore this further, an LS Dyna (Ansys Inc., Canonsburg, PA) computer simulation was performed to model the technician's motion following the blast, inspired from earlier work [10]. A Hybrid III model was adapted to account for the additional mass of the bomb suit. The mannequin was positioned at the representative distance and in a posture similar to that of the technician during the motorcycle VBIED incident and subjected to a 2 kg TNT explosion. The simulation results (see Figure 13) were successfully validated against the technician's final landing position. While the figure includes the motorcycle for reference, it was not part of the simulation itself. The simulation suggests that the technician's head may have impacted the ground, potentially with a frontal impact. However, it must be noted once again that no head injuries were reported for the technician following the incident, based on the helmet protection provided.

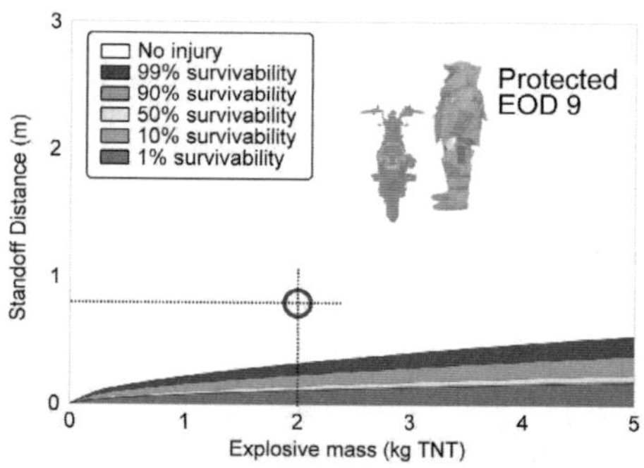

Figure 12. Blast induced traumatic brain injury chart for the EOD 9 protected case, from [9]

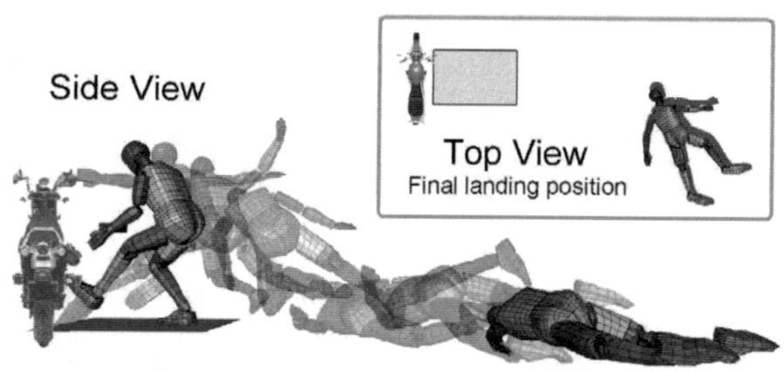

Figure 13. LS Dyna simulation of the blast event using a Hybrid III mannequin whose weight was adjusted to account for the weight of the bomb suit

2.4 Quaternary blast threat: heat (fireball)

When an explosive detonates, it generates a fireball composed of extremely high-temperature gases. Although this intense thermal hazard only persists for a few milliseconds, it can cause severe burns to unprotected skin if an individual is within the fireball's radius. Figure 14 illustrates the estimated size of the fireball in free-field experiments as a function of the explosive charge, based on high-speed video analysis of blast events [11]. For the 2 kg high explosive charge used in this motorcycle VBIED incident, the graph suggests that the bomb technician—whose foot was in contact with the motorcycle and whose chest was approximately 80 cm from the explosive centre—would have been positioned within the fireball.

Despite the protection provided by the bomb suit, the technician sustained burn injuries to the forearms (Figure 15) due to the fireball entering through the uncinched sleeve openings around the hands. This occurred because the arms were directly facing the IED at the moment of detonation. Medical documentation provided by DJIN classified these injuries as second-degree (superficial partial-thickness) burns, specifically affecting the anterior and lateral aspects of the right forearm. It is important to note that bomb technicians prefer to work with their bare hands during render-safe procedures, which leaves them vulnerable to such risks. Although the burn damage to the technician's forearms was not extensive, it underscores the inherent dangers faced in such situations. Some clues of burning damage were also visible on the bomb suit itself, visible on the photos from Figure 5. The sleeves feature an adjustable strap that permits tightening of the opening at the wrist and minimizing flame ingress; this feature was not used by the bomb technician in this incident.

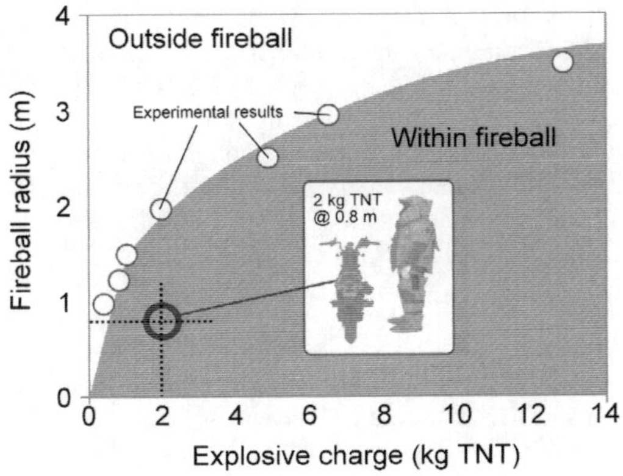

Figure 14. Estimated fireball size as a function of the explosive charge, indicating that the bomb technician was within the radius of the fireball [11]

Figure 15. Second degree burn injuries suffered by the technician on the forearms

3. DISCUSSION

Incidents involving bomb technicians are rare due to their expertise, the remote tools they use, as well as the sensitive nature of information related to such events, which is often guarded by authorities to prevent its misuse by terrorists or perpetrators. When such information is available, it is frequently limited, incomplete, or ambiguous. However, the incident discussed in this case study provided a wealth of data, making it particularly valuable for analysis. The availability of bystander video footage captured the technician's movements and exact posture just before the explosion, offering critical insights into standoff distance and positioning at the time of detonation. Additionally, the Colombian authorities (DIJIN) shared extensive materials, including original medical and forensic reports, as well as the damaged bomb suit and helmet for direct inspection. A meeting with the injured technician in Colombia a few months after the event (see Figure 16) provided additional information for the study, allowing for a comprehensive reconstruction of the events.

Figure 16. Injured bomb technician (right) in November 2022 (5 months after the explosion)

https://doi.org/10.52202/081621-0030

This incident highlights the importance of studying bomb suit protection under such severe threat conditions. The blast involved was lethal to an unprotected individual and posed a significant risk even to a technician wearing a bomb suit, resulting in serious injuries despite the suit's protective capabilities. This case study analysed the four primary blast threats—overpressure, fragmentation, impact, and heat—and evaluated the performance of the bomb suit using numerical simulations and forensic analysis. A comparison between the injury outcomes and the expected effects of the blast provided valuable insights into the protective capacity of the bomb suit and its limitations. Lessons learned from this incident are crucial for improving the safety of EOD technicians and mitigating injury risks in similar high-threat scenarios. Furthermore, the findings underscore the life-saving potential of bomb suits while providing a realistic understanding of their limitations as well as the injuries that may still occur under extreme conditions.

Acknowledgments

Special thanks to the Colombia Dirección Central de Policía Judicial e Inteligencia (DIJIN) for providing access to the blasted suit following the incident, as well as for sharing intelligence on the event to benefit the EOD community and its end users.

References

[1] Cardash, M., Motorcycle VBIED attacks targeting security forces, Colombia, (2023) Counter-IED Report, Winter 2022-23, pp 53-66

[2] Van der Voort, M. M., Holm, K. B., Kummer, P. O., Teland, J. A., Van Doormaal, J. C. A. M., & Dijkers, H. P. A. (2016). A new standard for predicting lung injury inflicted by Friedlander blast waves. Journal of Loss Prevention in the Process Industries, 40, 396-405

[3] Bowen, I.G., Fletcher, E.R., Richmond, D.R., Estimate of Man's Tolerance to the Direct Effects of Air Blast, Lovelace Foundation for Medical Education and Research, Albuquerque, New Mexico, DASA 2113, DA-49-146-XZ-372, 1968

[4] El Maach, I., Dionne, J.P., & Makris, A., New model for survivability predictions for simulated free field blast, Proceedings of the International Symposium on Military Aspects of Blast and Shock (MABS 19), Calgary, Canada, 2006

[5] Nerenberg., J., Dionne, J.P., Makris, A., Fisher, G., Evaluation of the ABS-LPU Ensemble for Compliance with US Army Advanced Bomb Suit Program, (2002) UXO Countermine Forum, Orlando, FL

[6] Garth RJN, Blast Injury of the Ear, Scientific Foundations of Trauma, GJ Cooper: editor, Oxford: Butterworth-Heinemann Publisher (1997) pp. 225-235

[7] Versace, J., A Review of the Severity Index, Proceedings of the Fifteenth Stapp Car Crash Conf., pp. 771-796, 1971

[8] Prasad, P., Mertz, H.J., The Position of the United States Delegates to the ISO Working Group 6 on the Use of HIC in the Automotive Environment, SAE Paper Number 85-1246, Society of Automobile Engineers, Warrendale, PA, 1985

[9] Levine, J., Dionne, J.P., & Makris, A., Blast injury charts for EOD scenarios based on scaled explosive parameters, Proceedings of the International Symposium on Military Aspects of Blast and Shock (MABS 21), Tel Aviv, Israel, 2010

[10] Dionne, J.P., Cheng, M., Levine, J. & Makris, A. (2016), Tertiary Blast Injury and the Need for Impact Protection, Proceedings from the 13th Personal Armour Systems Symposium (PASS), Amsterdam, Netherlands

[11] Dionne, J.P., Makris, A., & Levine, J., Safety through Standoff: Remote Manipulator Tools, The Detonator, May/June 2014, International Association of Bomb Technicians and Investigators (IABTI), pp. 26–31

Improving blunt impact performance of combat helmets

Y.S. Khoe[1] and C.Iványi[1]

[1]*TNO, Ypenburgse Boslaan 2, 2496 ZA Den Haag*
yoeng_sin.khoe@tno.nl

Abstract. TNO performs research on soldier survivability to support the Ministry of Defence (MoD) in the procurement of materiel. The in service combat helmet is primarily designed to protect against ballistic threats, which does not b definition protects against other types of relevant injury risks for the soldier. Recently, protection against blunt impact to the head in the military context has gained attention , because blunt impact injury accounts for a substantial portion of incidents. The loading as a result of blunt impact, either from falling debris or due to accidents during activities such as parachuting, mountaineering or operating open motorized vehicles, differs from ballistic loading and as such protection requirements differ. The search for a solution that improves the blunt impact performance adds to the traditional trade-off for a solution between minimal burden for the soldier and maximal protection. In addition, the constraints on this trade-off due to legislation may differ between peacetime and wartime. This paper provides an overview of the mechanical loadings that are relevant to military activities. A novel helmet add-on prototype was evaluated, showing that a combat helmet's blunt impact performance can be substantially upgraded, reducing the risk of injury. The add-on concept allows soldiers to 'train-as-you-fight' during peacetime scenarios, whilst benefiting from increased blunt protection.

1. INTRODUCTION

The soldier presents probably the most versatile fighting system of any army. Soldiers are tasked with a wide variety of objectives and are given an equally wide variety of tools to achieve those. Whilst the 'job' of a soldier inherently carries a higher risk of injury, the MoD has the obligation to provide the tools and training to minimize those risks where possible. In this context, consequences of investigate blunt impact loadings on the combat helmet were investigated. TNO aids the MoD in this goal by performing research and analyses and has recently started to.

Soldiers are taught to 'train-as-you-fight' to replicate a scenario as realistically as possible. However, training, during peace-time, requires that soldiers must also adhere to civil law (unless exempted). For example, as a combat helmet is not certified for use on motorized vehicles, switching to wearing a certified motor helmet is mandatory when they operate a motor vehicle (such as a quad, see Figure 1) during the training. This is neither 'train-as-you-fight', nor a practical solution. However, the adherence to these regulations provides essential worker safety during peace-time activities. For reference, in the Iraq and Afghanistan campaign, motor vehicle crashes were the second leading cause (Afghanistan: 26%; Iraq: almost 18%) for Type 2 Traumatic Brain Injury hospitalizations [1].

Figure 1. NLD KCT operating a quad (Mali, Africa) wearing a combat helmet with night vision goggles.

The aim of the present research was to compare the performance of the Team Wendy combat helmet and improvements in situations posing risks of blunt impact, to that achieved by helmets designed for use in such contexts. This approach intends to provide input for certification of the combat helmet, or support a request for exemption to the relevant law.

https://doi.org/10.52202/081621-0031

The results highlight the requirements for protection of the head against blunt impacts that are compatible with the military's motto to train-as-you fight, and the evaluation of a blunt impact add-on for the combat helmet.

2. OVERVIEW OF BLUNT IMPACT RELATED RELEVANT STANDARDS

Several civil standards, regarding blunt impact to the head, were compared to assess the variation in loading and thresholds employed by the different standards. Table 1 shows the selected helmet certification standards based on the potential relevance to military applications.

Table 1. Military activity and corresponding selected (helmet) standard

Military activity	Relevant Standard
Operating a motor vehicle	Motor cycle (ECE 22.06 [2]) Racing helmet (FRGPhe-01 [3])
Riot control	PSDB 2004 [4]
Climbing	Mountaineering EN 12492 [5]
Parachuting	Airborne sports EN 966 [6]
Construction	Industrial Helmets EN 397 [7] Protective Helmets EN 13087 [8] High Performance Industrial Helmet EN 14052 [9]
Miscellaneous	Equestrian (ASTM F1163-23) [10] Bike (EN 1078) [11] Speedelec (NTA 8776) [12]

The blunt impact loading in helmet standards is generated in two ways; either by dropping the helmet with headform onto an anvil, or by dropping an impactor onto the helmet. The variables measured in the tests are compared to the thresholds defined in the respective standards (see Table 2)

Table 2. Overview of thresholds used in blunt impact standards

Threshold	Loading type	Standard
Peak Linear acceleration (PLA)	Linear	ECE22.06 [2], PSDB(2004) [4], FRHPhe-01 [3]
Maximum transmitted force	Linear	EN 397 [7], EN 12492 [5] , EN14052 [9]
Severity index (SI)	Linear	NOCSAE ND-001-17m23 [13]
Head Injury criterion (HIC)	Linear	ECE22.06 [2], FRHPhe-01 [3]
Peak rotational acceleration (PRA)	Oblique	ECE22.06 [2], NOCSAE ND-001-17m23 [13]
Brain Injury Criterion (BrIC)	Oblique	ECE22.06 [2]

In order to get an initial impression of similarities between standards, the linear impact velocity and corresponding thresholds (PLA) were compared. Where no PLA was specified, the threshold was converted to a PLA using assumptions as specified in the caption of Figure 2. The resulting overview is presented in Figure 2.

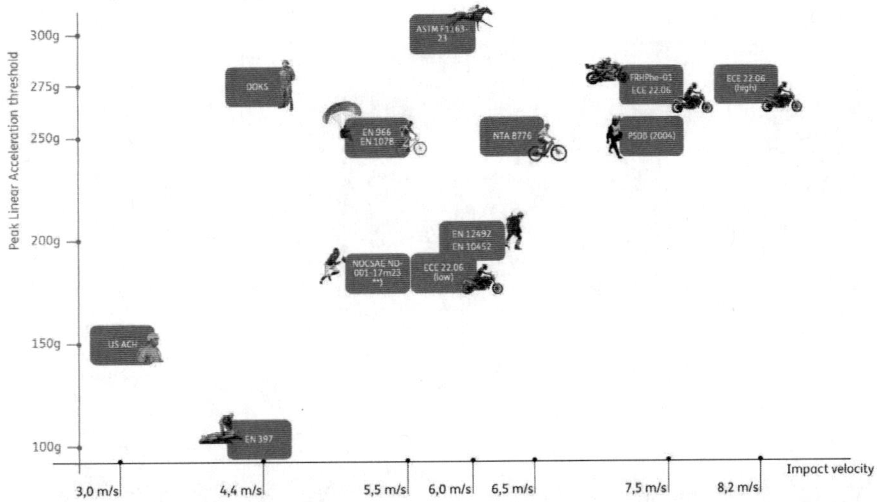

Figure 2. Overview of head blunt impact standards and corresponding PLA or force threshold.
*) Conversion from force threshold to PLA is done using $F = ma$, using a mass of 5 kg
**) Conversion from Severity Index to PLA assumes triangular pulse of 10ms

Current military helmets do include blunt impact requirements. For example, the Netherlands DOKS helmet specifies a maximum PLA of 275g for an impact at 4,4m/s and the US ACH helmet [14] specifies an impact velocity of 10 ft/s (3 m/s) with a maximum PLA of 150g.

This overview of activities, standards and thresholds provides a reference frame that can be used to judge blunt impact loadings and thresholds. Notable observations in this overview are that higher loadings generally coincide with a higher level of accepted risk. A risk associated to occupational hazards (e.g. EN397 industrial helmets) has a lower threshold (and thus a lower accepted risk) than a voluntary sport (e.g. EN966 paragliding).

A choice of loading and threshold depends on the institution that issues the standard and it will reflect the risk acceptance regarding the intended activities, as well as a maturity of the products at hand. Sometimes new requirements (or standards), opt for relatively easier to reach thresholds so that a minimum level of protection is achieved faster whilst also giving producers a chance to adapt their products. As the requirement matures, a more critical threshold can be chosen to incentivize producers to improve the performance of their products.

3. EVALUATION OF BLUNT IMPACT ADD-ON

3.1 Methodology

3.1.1 Helmet and add-on description

An add-on prototype for the Team Wendy helmet was evaluated that was designed to improve the blunt impact protection (see Figure 2). The add-on consists of an additional shell that attaches to the helmet using Velcro and has a separate element to cover the front-facing mount. The side rails are not yet covered by the current prototype. The intended goal of the add-on is to improve blunt impact protection by redistributing the impact force over a larger area and absorbing and dissipating impact energy.

https://doi.org/10.52202/081621-0031

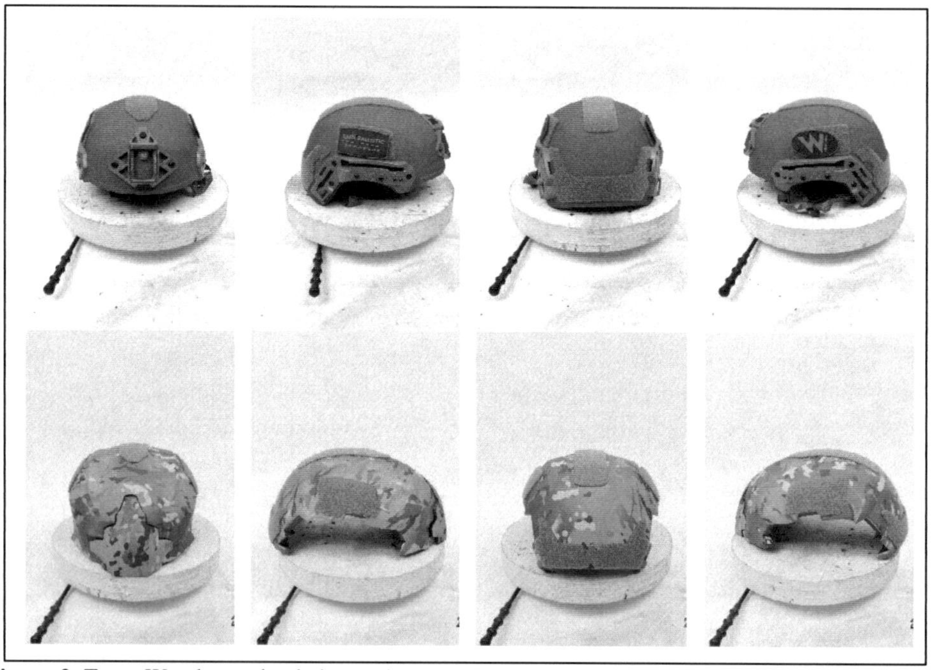

Figure 3. Team Wendy combat helmet (above) and blunt impact add-on (below), from left to right: front-, right-, back- and left side.

3.1.2 ECE22.06 drop test

An analysis was performed to assess the blunt impact performance of this prototype, evaluate the added protected, and identify necessary improvements. The ECE22.06 drop test was selected as the test methodology due to its relevance for the operation of motorized vehicles.

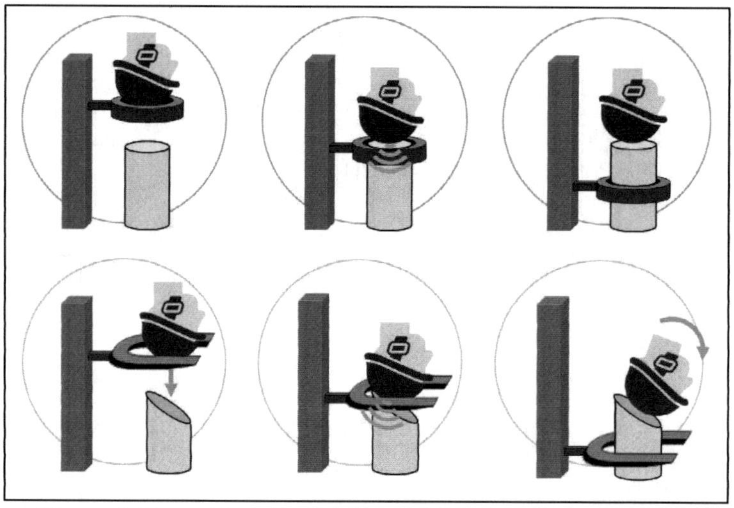

Figure 4. Linear (above) and oblique (below) impact test procedure

The ECE22.06 drop test includes both linear and oblique impacts (see Figure 3), where accelerations are measured in the headform. From these accelerations, the Head Injury Criterion (HIC) for linear impacts and the Brain Injury Criterion (BrIC) for oblique impacts are determined.

The HIC and BrIC are metrics used to assess the severity of head injuries in impact scenarios. By linking them to established injury risk scales, they help estimate the probability of specific types of brain injuries.

https://doi.org/10.52202/081621-0031

Unlike PLA, which only considers the instantaneous maximum acceleration, HIC accounts for both the magnitude and duration of linear acceleration. A higher HIC value indicates a greater risk of head injury. The calculation integrates acceleration over a specific time window (up to a maximum of 36 ms) to capture the most critical phase of the impact.

BrIC is used for oblique impacts and focuses on rotational kinematics, particularly angular velocity. Since brain injuries such as concussions and diffuse axonal injury are strongly linked to rotational forces, BrIC complements HIC by providing a broader assessment of head injury risk. However, as the inclusion of oblique impact testing is relatively new in ECE 22.06 [2], the current thresholds for rotational acceleration and BrIC are generally considered less constraining compared to those for linear impacts, which is not reflected in epidemiology studies [15].

The ECE22.06 specifies three impact velocities and corresponding thresholds for the Peak Linear Acceleration (PLA) and HIC, and one impact velocity for the Peak Rotational Acceleration (PRA) and BrIC. These thresholds are provided in Table 1.

For the linear impact test, four impact points are defined, that are all tested one helmet: Frontal area (B), Left or right lateral area (X), Top (P) and Rear (R). For the oblique impact test, the ECE22.06 prescribes five impact points to be tested on two helmets, out of which the following three impact points were selected for this study: Rear (180°), Lateral left (270°) and Front (0°). All impact locations must meet the thresholds defined according to ECE 22.06.

Table 1. Thresholds prescribed by ECE22.06 for linear (top table) and oblique (bottom table) tests.

Test	**Impact velocity**	**PLA threshold**	**HIC threshold**
Linear blunt impact	6,0 m/s	≤ 180 g	≤ 1300
Linear blunt impact	7,5 m/s	≤ 275 g	≤ 2400
Linear blunt impact	8,2 m/s	≤ 275 g	≤ 2880

Test	**Impact velocity**	**PRA threshold**	**BrIC threshold**
Oblique blunt impact	8,0 m/s	≤ 10400 rad/s^2	≤ 0,78

The tests were performed by TASS International, where a total of five helmets was tested. The helmet without the add-on was only subjected to the linear impact test, as this represents the most critical impact loading condition. The test was conducted at the standard velocity of 7.5 m/s.
Helmets with the add-on were tested at all three linear impact velocities (6.0 m/s, 7.5 m/s, and 8.2 m/s) as well as in the oblique impact test at a velocity of 8.0 m/s.

3.2 Results

Figure 5 and Figure 6 show the PLA and HIC results for linear impacts on the helmet with and without add-on. The oblique drop tests were also performed, but since the results with the add-on were well below the ECE22.06 thresholds those results have not been included in this publication. In contrast, the ECE22.06 tresholds for the linear impact tests appear to be harder to satisfy which is in line with other results [15],.

Note that (for the tests without the add-on) TASS indicated that some of the measured accelerations exceeded their equipment specifications. In accordance with ECE22.06 the raw accelerometer signals are filtered, leading to the results presented here. However, the unfiltered signals may contain peaks that exceed sensor limits. Where applicable, those results have been indicated in the results.

https://doi.org/10.52202/081621-0031

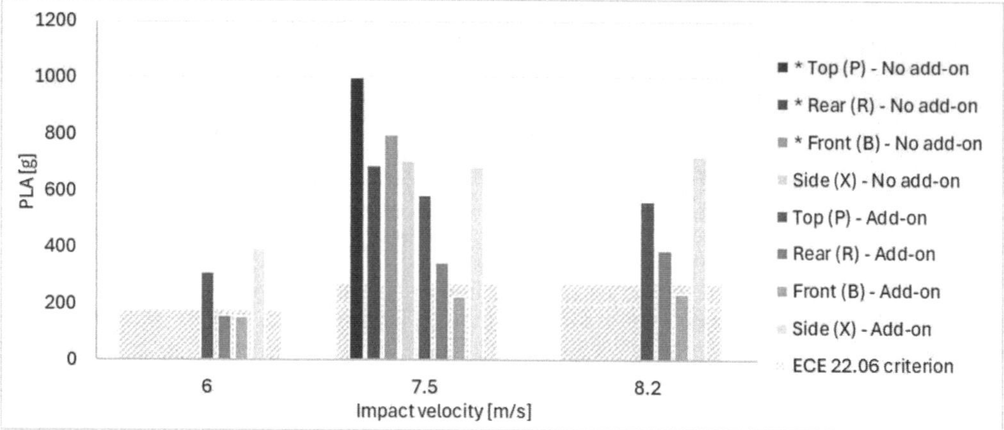

Figure 5. Peak Linear Accelerations (PLA) measured during drop tests with and with-out the blunt impact addon. * indicate measurements exceeding test equipment specifications

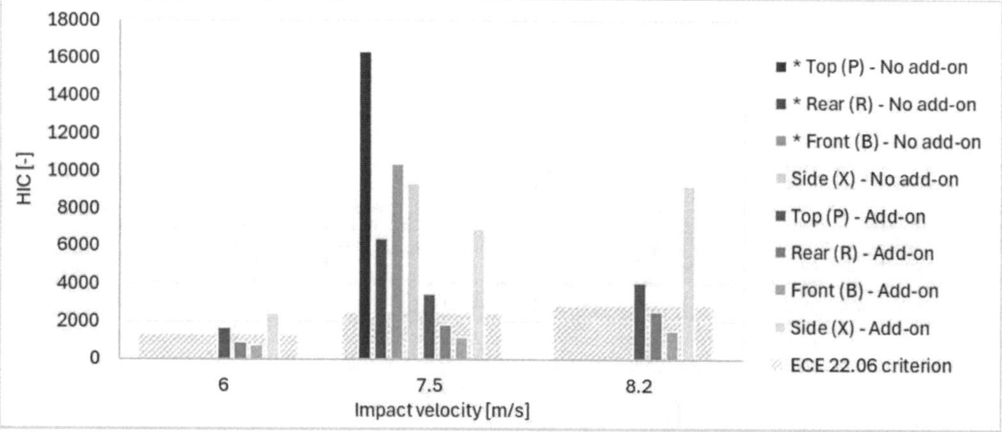

Figure 6. Head Injury Criterian (HIC) values based on accelerations measured during drop tests with and with-out the blunt impact addon. * indicated measurements exceeding test equipment specifications.

3.2.1 Helmet without add-on

The helmet without the add-on, serving as a reference for comparison of the add-on performance, was tested in a linear impact at the standard velocity of 7.5 m/s. The measured PLA and HIC values can be found in Figure 5 and Figure 6 respectively. In 3 out of the 4 tests, the measured acceleration exceeds the capacity limit of the accelerometers. The capacity limit of the accelerometers used is 500 g per direction, with a theoretical limit of 866 g, and any values reported above this limit are unreliable and should be treated with caution. For the top, rear and front impact locations, the measured accelerations exceed the equipment limit and cannot be used for calculations. These measurements are indicated in the label of the figures with an asterisk.

The ECE22.06 PLA threshold is well below the capacity limit. The exceeding of this limit during the helmet test with the available equipment strongly indicates the poor blunt impact performance of the helmet without the add-on, reinforcing the need for additional protective measures.

Only the acceleration for the side impact location could be reliably measured. However, with a PLA of 709 g and a HIC of 9355, the performance still exceeds the threshold values provided in ECE 22.06 (indicated by the hatched areas in the figures).

3.2.2 Helmet with add-on

Linear impact tests with the add-on were conducted at three different velocities (as specified in ECE22.06), with the results presented Figure 5 and Figure 6. The performance of the helmet add-on varied depending on the impact location.

The top location demonstrated the worst performance. At all tested velocities, the helmet did not meet the thresholds set by the ECE 22.06 standard. The measurement of acceleration over time, presented in Figure 7 for the impact velocity of 6 m/s, shows two peaks in the acceleration experienced by the headform. The first peak is likely attributed to the outer shell of the add-on that deforms, and dampens the impact. Damage on the outer shell of the add-on is observed after testing (see Figure 8). The second peak is hypothesized to originate from either the contact of the outer shell with the inner foam of the add-on, or the contact of the add-on with the helmet.

Similar behaviour with double peaks in acceleration is observed for the rear impact location. However, the acceleration is significantly lower at this location, and it meets the ECE 22.06 standard threshold for HIC at all impact velocities and the PLA threshold for the 6 m/s impact.

In contrast, the front location demonstrated the best performance with a single, spread-out peak. The helmet with the add-on met the ECE 22.06 standard thresholds at all tested velocities. Plastic deformation of the add-on was observed at the front location after the tests, which likely contributed to its superior performance due to additional energy absorption through crushing. The front is also the only location that meets the PLA requirements set in the ECE 22.06 standard at all impact velocities.

The side impact location was not covered by the add-on. Results are (as expected) similar to the results without the add-on. The results showed a single peak in the headform acceleration.

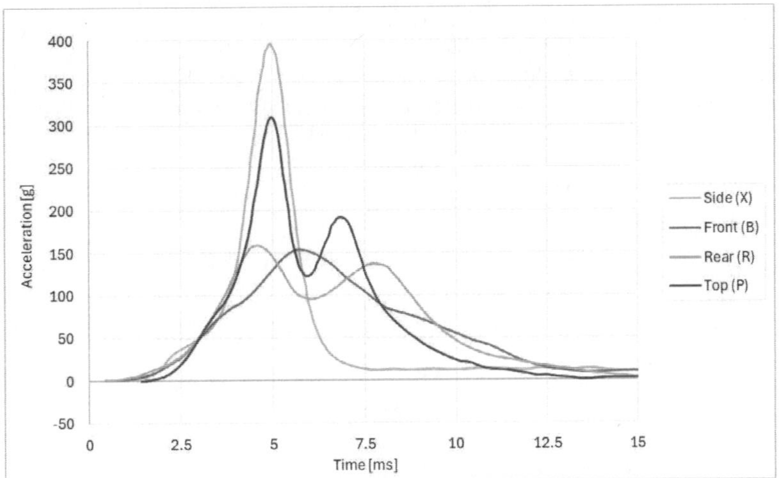

Figure 7. Acceleration of the headform over during the linear impact test at 6 m/s with the add-on

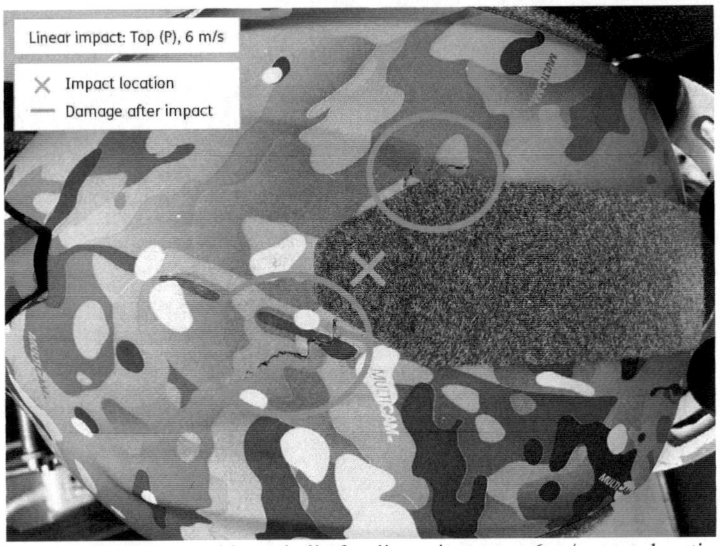

Figure 8. Damage to add-on shell after linear impact at 6 m/s at top location.

A comparison of the helmet's performance with and without the add-on at an impact velocity of 7.5 m/s clearly demonstrates an improvement in blunt impact protection. The add-on dampens the impact

https://doi.org/10.52202/081621-0031

and spreads out the acceleration to enhance protection. The most substantial improvement was observed at the front location. For the top and rear, there is still a substantial improvement, but the acceleration peak is less spread out and instead a double peak behaviour is obtained, indicating room for improvement.

4. CONCLUSIONS

This paper presents a pilot evaluation of the blunt impact performance of a prototype for a blunt impact add-on for a military combat helmet. Without the add-on, drop tests (according to the ECE 22.06) with the military helmet result in PLA and HIC values that far exceed the ECE22.06 criteria. With the add-on there is a significant improvement and shows some values that meet the ECE22.06 criteria.

Thus, the current add-on improves the blunt impact protection of the helmet and thereby reduces risks of blunt impact injuries. It is not sufficient yet to pass the requirements outlined in the ECE 22.06, but it is expected that further engineering can result in a next iteration that meets the ECE22.06 criteria. At this moment critical points to address are impact performance at the top (P) location and a solution that covers the side (X) locations without significantly impeding the functionality of the side rails.

This study shows that a fairly simple add-on construction can substantially improve a military helmet's performance against blunt impact. Improving the blunt impact performance can potentially reduce a substantial number of TBI related injuries that occur through e.g. motor vehicle accidents. This is especially relevant in training scenario's where military personnel is expected to adhere to civil laws and regulations. In peacetime conditions, balancing the motto to train-as-you-fight, with the safety of personnel might lead to different decisions regarding the utilized personal equipment. At the moment this leads to scenarios in which military personnel, operating motor vehicles, are expected to use a separate combat helmet and a motor helmet during training. An add-on, such as the one evaluated in this paper, is a solution that removes the need for two separate helmets.

Acknowledgments

The authors would like to acknowledge the Netherlands Ministry of Defence for funding this research, especially Adjudant D. Brouwer.

References

[1] B. E. Wojcik, C. R. Stein, K. Bagg, R. J. Humphrey en J. Orosco, „Traumatic Brain Injury Hospitalizations of U.S. Army Soldiers Deployed to Afghanistan and Iraq," *American Journal of Preventive Medicine,* vol. 38, nr. 1, pp. S108-S116, 2010.

[2] „ECE 22-06 Uniform provisions concerning the approval of protective helmets and of their visors for drivers and passengers of motor cycles and mopeds," United Nations, 2022.

[3] „FRHPhe-01 FIM RACING HOMOLOGATION PROGRAMME FOR HELMETS," Fédération Internationale de Motocyclisme, 2019.

[4] C. Malbon en J. Croft, „PSDB Protective Headwear Standard for UK Police 21/04," POLICE SCIENTIFIC DEVELOPMENT BRANCH, Sandridge, 2004.

[5] „EN 12492 - Bergbeklimmersuitrusting - Helmen voor bergbeklimmers - Veiligheidseisen en beproevingsmethoden," Nederlands Normalisatie-instituut, Delft, 2012.

[6] „EN 966 - Helmets for airborne sports," Nederlands Normalisatie-instituut, Delft, 2012.

[7] „EN397 Industrial safety helmets," Nederlands Normalisatie-instituut, Delft, 2012.

[8] „EN 13087 - Protective helmets," Nederlands Normalisatie-instituut, Delft, 2000.

[9] „EN14052 - Industriële helmen met een hoog beschermingsniveau," Nederlands Normalisatie-instituut, Delft, 2005.

[10] „ASTM F1163-23 Standard Specification for Protective Headgear Used in Horse Sports and Horseback Riding," American Society for Testing and Materials, West Conshohocken, Pennsylvania, U.S, 2023.

[11] „EN 1078 - Helmen voor fietsers en gebruikers van skateboards en rolschaatsen," Nederlands Normalisatie-instituut, Delft, 2012.

[12] „NTA 8776 - Helmets for S-EPAC riders," Nederlands Normalisatie-instituut, Delft, 2016.

[13] „STANDARD PERFORMANCE SPECIFICATION FOR NEWLY MANUFACTURED FOOTBALL HELMETS - NOCSAE DOC (ND)002-17m21," NOCSAE, 2021.

[14] „CO/PD-05-04 PURCHASE DESCRIPTION HELMET, ADVANCED COMBAT (ACH)," US Army Natick Soldier Centre, 2007.

[15] M. Fahlstedt, J. Bergström, D. Lanner en P. Halldin, „How Efficient are the Rotational Impact Tests in ECE R22.06 Motorcycle Helmet Test Standard to Decrease the Rotational-Induced Brain Injuries?," in *IRCOBI*, Porto, Portugal, 2022.

https://doi.org/10.52202/081621-0031

Design and Development of Ultra-Lightweight Personal Body Armor for NIJ Level III++ Protection

H.H.Türkmen[1], M.B.Ünal[2], K.Kaya[3], M.C. Şimşek[4]

[1]Nurol Teknoloji Ltd, Oğulbey District. 3060 Street. No.5 06830 Gölbaşı Ankara, Türkiye, huseyin.turkmen@nurolteknoloji.com
[2]Nurol Teknoloji Ltd, Oğulbey District. 3060 Street. No.5 06830 Gölbaşı Ankara, Türkiye, bartu.unal@nurolteknoloji.com
[3]Nurol Teknoloji Ltd, Oğulbey District. 3060 Street. No.5 06830 Gölbaşı Ankara, Türkiye, koray.kaya@nurolteknoloji.com
[4]Nurol Teknoloji Ltd, Oğulbey District. 3060 Street. No.5 06830 Gölbaşı Ankara, Türkiye, mert.simsek@nurolteknoloji.com

Abstract. Ceramic body armor has been widely used in military and law enforcement applications due to its excellent ballistic performance. The ballistic performance of a protective armor system comprising Boron carbide – Silicon carbide composite ceramics (Nurol Technology's special ceramic product) coupled with ultra-high molecular weight polyethylene (UHMWPE) with varying protective levels, crucial research to design the lightest weight of the personal body armor. The ceramic thickness ranges from 2 mm to 3 mm, aiming to optimize impact resistance and energy absorption in personal armor applications. Experimental ballistic tests are conducted to measure impact resistance under controlled conditions, providing insights into the penetration resistance, energy dissipation, and structural integrity of the ceramic-UHMWPE composite when subjected to high-velocity projectiles. To support and be validated with empirical results, LS-DYNA© simulations are employed, enabling detailed analysis of the dynamic response of the composite materials under ballistic impact. LS-DYNA©, a widely used finite element analysis (FEA) software in impact simulations, allows for detailed modeling of material behaviors, including the brittle nature of ceramics and the high tensile strength and toughness of UHMWPE. The simulations account for material heterogeneity, varying ceramic thicknesses, and hardness differences, replicating realistic ballistic impact scenarios.
This research aims to enhance the protection level of ballistic armor while minimizing its weight. It includes an assessment of the correlation between simulated and experimental outcomes, identifying any inconsistencies, and exploring potential causes, such as assumptions or simplifications in the finite element analysis (FEA) model. Key performance metrics, such as minimum backface deformation, erosion of the ammunition, and overall material integrity, are compared across both methods. This research ultimately seeks to establish the ultimate personal armor solution. By comparing ballistic testing with LS-DYNA© simulations, the study aims to refine simulation methodologies and enhance material configurations, contributing valuable insights to the field of ballistic protection and impact resistance.

1. INTRODUCTION

Body armors are specialized equipment designed to protect individuals from various threats. Offering different features such as protection against ballistic, cutting, piercing, and impact threats, body armors vary based on their applications and specific needs [1].
One specific focus in recent studies on body armors involves enhancing ballistic performance using advanced materials and design methodologies. This is particularly significant in addressing the growing demand for lightweight, high-performance protective solutions in various fields, including military and law enforcement applications. The ballistic performance of a protective armor system made of Boron carbide–Silicon carbide ceramic composite materials combined with ultra-high molecular weight polyethylene (UHMWPE) at different protection levels is a challenging study in terminal ballistic literature.

Ballistic ceramics should exhibit high hardness. However, to enhance multi-shot performance, they must also possess improved flexural strength and fracture toughness. In the industry, the reaction-bonding process, which involves metal infiltration into Boron carbide ceramics, is a commonly used method to achieve these properties [2]. However, Nurol Teknoloji's specialized product utilizes a Boron carbide – Silicon carbide composite ceramic instead of metal infiltration. This method notably preserves the high hardness of B4C while significantly enhancing its flexural strength and fracture toughness. At high-velocity impacts, multiple energy absorption mechanisms are involved. These include the

https://doi.org/10.52202/081621-0032

formation of macro-damage areas (e.g., cone cracks and radial cracks), micro-fragmentation zones, and mixed fracture modes [3].

This research prioritizes the development of lightweight ballistic armor with high protection levels. It leverages the finite element modeling approach to achieve this goal. In high-velocity impact scenarios, conducting real-life tests to measure the response of armor materials and predict their behavior is highly challenging, costly, and time-consuming. Thanks to the finite element method, numerous armor solutions can be tested, allowing the structure to be optimized and made more durable [4]. Finite element calculations in this work were conducted using the LS-DYNA© commercial software, which provides robust tools for simulating complex ballistic impact scenarios [5].

In this study, key performance indicators such as back-face signature, and overall material performance were evaluated both experimentally and through the finite element model. This study aims to bridge the gap by combining Boron carbide and Silicon carbide properties into a composite armor system, providing both high protection and reduced weight, addressing the challenges in modern armor design.

2. BALLISTIC EXPERIMENT PROCEDURE

2.1.Test Facility and Setup

All ballistic tests were conducted at the certified indoor ballistic shooting range of **Nurol Teknoloji**, by standardized protocols for evaluating the back-face signature (BFS) and penetration performance of armor systems. The shooting experiments were performed under controlled environmental conditions to ensure the repeatability and reliability of the results.

Figure 1: Nurol Teknoloji's Shooting Range.

The final tested armor as shown in Figure 2 assembly comprised a **3 mm thick proprietary NT ceramic** strike-face layer, backed by a **12.5 mm thick high-pressure consolidated ultra-high-molecular-weight polyethylene (UHMWPE)** layer. **The total weight of the final test armor is 1705 g with having areal denstiy 0.25 kg/m2 which is significantly lower than comparable commercial products.** The armor plates were produced using standard pressing and sintering parameters provided by the manufacturer to ensure uniformity and quality across all test samples.

https://doi.org/10.52202/081621-0032

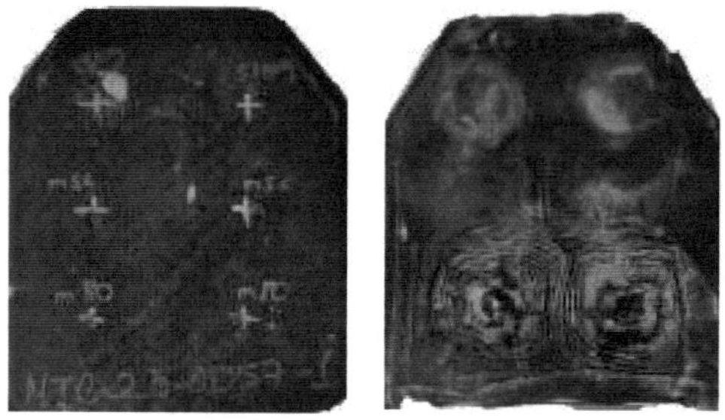

Figure 2*:* Body Armor Plate after the experiment.

The armor samples were rigidly clamped in a vertical position and backed with a calibrated Roma Plastilina No.1 clay block as shown in Figure 3, conditioned to the specifications outlined in NIJ Standard 0101.07, for measuring back-face signature (trauma). Each sample was subjected to a sequence of six shots, conducted at a nominal distance of **15 meters**, which is a standard testing range for personal ballistic protection materials.

Figure 3: BFS measurement setup with P1 plastina.

2.2. Ammunition and Firing Sequence

A total of six ballistic impacts were performed on each armor sample. Three types of ammunition, representative of common ballistic threats encountered in both military and law enforcement scenarios, were used. The firing sequence was standardized across all test samples to ensure comparability:

1. **Two rounds of 5.56×45 mm NATO SS109 (also known as M855)** — a standard military ball round featuring a steel penetrator with striking velocity of 947 - 953 m/s.
2. **Two rounds of 7.62×39 mm MSC (Mild Steel Core)** — a widely used intermediate cartridge associated with AK-pattern rifles with striking velocity of 732 -724 m/s.
3. **Two rounds of 7.62×51 mm NATO M80 Ball** — a full metal jacket round with a lead core, representing a high-energy rifle threat with a striking velocity of 841-855 m/s.

The shots were fired consecutively, with careful targeting to avoid direct overlap while maintaining sufficient spacing to eliminate edge effects and mutual influence between adjacent impact points. Following each ballistic test, X-ray imaging of the body armor plate was conducted to monitor the crack patterns formed on the ceramic layer.

3. NUMERICAL MODELLING

Ballistic impact simulation is a highly complex and non-linear problem. Non-linearity can come from both geometry and material non-linearity. In addition to that, there is excessive large deformation in the materials. All these models and techniques are used in the commercial finite element solver software LS-Dyna© for this study.

3.1. Integration-Explicit

In the ballistic impact simulations due to the high number of elements, complex contacts, high-speed dynamics, and very large deformations, explicit time integration is more suitable. In addition to that in terms of computational efficiency explicit time integration calculates the system's future state directly from its current state, making it simple and efficient but conditionally stable, requiring small time steps for accuracy. An explicit scheme can be investigated from Equation 1 for an update of the displacement.

$$u_{t+\Delta t} = u_t + \Delta t \dot{u}_{t+\Delta t} \tag{1}$$

3.2. Contact

Contact algorithm and contact type are other important points in ballistic impact modeling. Although there are several challenges in the implicit algorithm, the explicit scheme is very powerful for the contact algorithms. There are two types of contact that are used in LS-DYNA© such as Automatic Surface to Surface Tiebreak and Eroding Single Surface in this paper.

Automatic Surface-to-Surface Tiebreak Contacts are a specialized type of contact used to model the behavior of interfaces that can initially hold two surfaces together but may separate or fail under certain conditions. Its ability to simulate interface failure between similar or different layers under both tensile and shear loading. After the given shear and tensile traction stress limit is exceeded the layers start to separate gradually. Thus, it can be used within the composite structure between the layers of UHMWPE.

Eroding Single Surface Contact refers to a contact definition that allows for interaction between different parts of the same body or mesh (self-contact) and accounts for erosion of elements due to failure criteria. This type of contact is particularly useful in simulations involving severe deformation, fragmentation, or material failure where self-interaction and element deletion are critical aspects. During the definition of this contact type at LS-DYNA© choice of the critical erosion, type and its value are too important. After this value reaches the critical limit that element will be deleted.

3.3. Material Models

3.3.1. Johnson-Cook

The Johnson-Cook material model (**MAT_015-JOHNSON_COOK**) in LS-DYNA© is widely used for simulating strain- and temperature-dependent plasticity in materials subjected to high strain rates. This model accounts for material softening due to plastic heating and defines flow stress as a function of effective plastic strain, strain rate, and temperature, expressed as:

$$\sigma_y = [A + B\varepsilon_p{}^n][1 + C\ln(\dot{\varepsilon}^*)][1 - (T^*)^m] \tag{2}$$

where;

$$T* = \frac{T - T_{ref}}{T_m - T_{ref}}, and, \ \dot{\varepsilon}* = \frac{\dot{\varepsilon}_p}{\dot{\varepsilon}_o} \tag{3}$$

The coefficients in the equation 2 and equation 3 are stated at the below Table 1.

Table 1: Johnson-Cook Parameter Identification

https://doi.org/10.52202/081621-0032

A	Yield Stress
B	Strain Hardening Coefficient
n	Strain hardening Exponent
ε_p	Effective Plastic Strain
$\dot{\varepsilon}_p$	Effective Plastic Strain Rate
$\dot{\varepsilon}_0$	Referance Strain Rate
C	Strain Rate Coefficient
T_m	Melting Tempereature
T_{ref}	Referance Temperature

In terminal ballistic analysis, the Johnson-Cook model is commonly employed to study failure mechanisms in materials like hardened steel cores [6]. The hydrostatic response of materials under these conditions is often modeled using the Mie-Gruneisen equation of state, which relates pressure (p), density (ρ), and internal energy (e), with the Gruneisen parameter (Γ) defined in Equation 4.

$$\Gamma = \frac{1}{\rho}\left(\frac{\delta p}{\delta e}\right)_V \quad (4)$$

The Johnson-Cook Material model are characterized based on the terminal ballistic studies. The parameters of the material model are provided by Table 2.

Table 2: Johnson-Cook Material Parameters for Hardened Steel Core

Part	**Material**	**Model**	**Source**
Core - SS109	Hardened Steel	MAT_015	Weiss et al. (2012) [6]
Core - MSC	Mild Steel	MAT_015	Hála et al. (2024) [7]
Roma Plastina	Clay	MAT_015	Eckhoff et al. (2012) [8]

Analyzing ammunition failure mechanisms is challenging due to the complex interplay of extreme deformation, high strain rates, and elevated temperatures and pressures. Failure criteria are typically based on the maximum effective plastic strain that material elements can endure. When elements reach their critical strain, they are removed from the simulation, providing an efficient computational approach to model material failure. However, this method has limitations, such as ignoring strain rate effects and continuum damage evolution.

3.3.2. Enhanced Composite Damage(MAT_054)

Enhanced Composite Damage Model **(MAT_054-Enhanced_Composite_Damage)** is designed to model laminated composite materials under impact, crash, and dynamic loading conditions. It incorporates an anisotropic constitutive model, progressive failure mechanisms, and element deletion capabilities to capture the behavior of fiber-reinforced composites. The material follows orthotropic elasticity, meaning the stiffness matrix evolves based on damage accumulation. The constitutive relation is depicted in Equation 5.

$$\sigma = D \cdot \varepsilon \quad (5)$$

where σ is stress tensor, D is stiffness matrix (updated with damage) and ε is strain tensor

The elastic response is defined by Young's moduli (E_A, E_B, E_C), shear moduli (G_{AB}, G_{BC}, G_{CA}), Poisson's ratios ($\nu_{AB}, \nu_{BC}, \nu_{CA}$)

Failure Criteria (Chang-Chang Model): This model differentiates between fiber and matrix failure under tension and compression.
Fiber Failure

1. Tensile failure in fiber direction (DFAILT) in Equation 6:

$$\sigma_{aa} > 0 \Rightarrow e_f^2 = \left(\frac{\sigma_{aa}}{X_t}\right)^2 + \beta\left(\frac{\sigma_{ab}}{S_c}\right)^2 - 1, \quad \begin{array}{c} e_f^2 \geq 0 \Rightarrow \text{ failed} \\ e_f^2 < 0 \Rightarrow \text{ elastic} \\ E_a = E_b = G_{ab} = v_{ba} = v_{ab} = 0 \end{array} \tag{6}$$

2. Compressive failure in fiber direction (DFAILC) in Equation 7:

$$\sigma_{aa} < 0 \Rightarrow e_c^2 = \left(\frac{\sigma_{aa}}{X_c}\right)^2 - 1, \quad \begin{array}{c} e_c^2 \geq 0 \Rightarrow \text{ failed} \\ e_c^2 < 0 \Rightarrow \text{ elastic} \\ E_a = v_{ba} = v_{ab} = 0 \end{array} \tag{7}$$

Matrix Failure

1. Tensile failure in matrix direction (DFAILM) in Equation 8:

$$\sigma_{bb} > 0 \Rightarrow e_m^2 = \left(\frac{\sigma_{bb}}{Y_t}\right)^2 + \left(\frac{\sigma_{ab}}{S_c}\right)^2 - 1, \quad \begin{array}{c} e_m^2 \geq 0 \Rightarrow \text{ failed} \\ e_m^2 < 0 \Rightarrow \text{ elastic} \\ E_b = v_{ba} = 0 \Rightarrow G_{ab} = 0 \end{array} \tag{8}$$

2. Compressive failure in matrix direction (DFAILM) in Equation 9:

$$\begin{array}{l} \sigma_{bb} < 0 \Rightarrow e_d^2 = \left(\frac{\sigma_{bb}}{2S_c}\right)^2 + \left[\left(\frac{Y_c}{2S_c}\right)^2 - 1\right]\frac{\sigma_{bb}}{Y_c} + \left(\frac{\sigma_{ab}}{S_c}\right)^2 - 1, \quad e_d^2 \geq 0 \Rightarrow \text{ failed} \\ e_d^2 < 0 \Rightarrow \text{ elastic} \\ E_b = v_{ba} = v_{ab} = 0 \Rightarrow G_{ab} = 0 \\ X_c = 2Y_c, \text{ for } 50\% \text{ fiber volume} \end{array} \tag{9}$$

Although the material model defines separate failure criteria, it uses them simultaneously. Since by nature matrix fails initially, the model progressively reduces the effective fiber strength in proportion to the matrix damage accumulates.

3.3.3. Modified Johnson-Cook

The Modified Johnson-Cook model incorporates adiabatic heating into the standard Johnson-Cook material formulation. The stress response for this model is given by Equation 10.

$$\sigma_{eq} = \left(A + B\varepsilon_{neq}\right)\left(1 + \frac{\dot{\varepsilon}_p}{\dot{\varepsilon}_0}\right)(1 - T^m) \tag{10}$$

where A, B, C, n, and m are material parameters as defined in the Johnson-Cook model. Here, ε_p represents the equivalent plastic strain, $\dot{\varepsilon}_p/\dot{\varepsilon}_0$ while denotes the normalized plastic strain rate. The material model used in the analysis are listed in Tables 3 and 4.

Table 3: Modified Johnson-Cook Parameters for Brass

Part	**Material**	**Model**	**Source**
Jacket	Brass	MAT_107	Manes et al. (2020) [9]

Table 4: Modified Johnson-Cook Parameters for Lead

Part	**Material**	**Model**	**Source**
Lead	Lead	MAT_107	Choudary et al.(2020) [10]
Core – M80	Lead	MAT_107	Choudary et al.(2020) [10]

The failure model used in this formulation is based on the Cockroft-Latham failure criterion, which considers both stress and strain. It is defined by Equation 11:

https://doi.org/10.52202/081621-0032

$$D = \frac{1}{W_{cr}} \int_0^{\varepsilon_{eq}} max(\sigma_1, 0) d\varepsilon_{peq} \tag{11}$$

where σ_1 is the maximum principal stress, and W_{cr} represents the Cockroft-Latham parameter, indicating the total plastic work. According to this failure criterion, material failure occurs when D reaches a value of one. The key differences between the Modified Johnson-Cook model (**MAT_107-MODIFIED MAT_015-JOHNSON_COOK**) and the standard Johnson-Cook model (**MAT_015-JOHNSON_COOK**) lie in strain rate dependency and failure criteria. The Cockroft-Latham parameter can be determined through a simple uniaxial tensile test, where the area under the stress-strain curve represents the total plastic work.

3.3.4. Johnson-Holmquist-2

The Johnson-Holmquist model (**MAT_110-JOHNSON_HOLMQUIST_CERAMICS)** is designed for brittle materials and incorporates a transition regime between intact and fractured states. It specifically accounts for material strength in both conditions using a pressure-volume relationship that includes the effects of bulging. This model considers damage, pressure, and strength as key factors in material behavior. To simplify the equations, normalization is applied, and the normalized equivalent stress is given in Equation 12 for the intact material [5]:

$$\sigma_i^* = A(P^* + T^*)^N (1 + C \ln \dot{\varepsilon}^*) \tag{12}$$

For damaged material, the stress is adjusted based on the level of accumulated damage D, as shown in Equation 13 [5]:

$$\sigma_i^* = \sigma_i^* - D(\sigma_i^* - \sigma_f^*) \tag{13}$$

where σ_i^* is the normalized intact equivalent stress, σ_f^* is the normalized fractured stress, and D represents the damage variable, ranging from 0 to 1 [5].

For fully fractured material, the stress follows Equation 14 [5]:

$$\sigma_f^* = B(P^*)^M (1 + C \ln \dot{\varepsilon}^*) \tag{14}$$

The damage accumulation leading to fracture is modeled as the sum of incremental equivalent plastic strain over each integration cycle, as given in Equation 16 [5]:

$$D = \sum \frac{\Delta\varepsilon_p}{\varepsilon_{pf}} \tag{15}$$

where the fracture strain ε_{pf} is expressed in Equation 17 as [5]:

$$\varepsilon_{pf} = D_1 (P^* + T^*)^{D_2} \tag{15}$$

Here, D_1 and D_2 are material constants, while $P^*$and T^* represent the normalized pressure and the normalized maximum tensile hydrostatic pressure, respectively.[5]

As plastic deformation progresses, the material undergoes fragmentation, reducing the particle size and softening the structure. When the accumulated damage reaches one, the material fully fractures. Equation 18 describes the bulking phenomenon, where hydrostatic pressure increases due to volumetric strain after fracture. The added internal energy, represented by ΔP, contributes to potential internal energy [5].

$$P = K_1\mu + K_2\mu^2 + K_3\mu^3 + \Delta P \tag{17}$$

3.4. Mesh Structure

In this research, there are three types of ammunitions. These are namely;

- 5.56x45 mm M855/SS109
- 7.62x39mm MSC
- 7.62X51mm M80

The cores for all ammunition types are modeled with a hexahedral mesh. However, depending on the shape of the cover or backing (lead), the mesh structure could be either hexahedral or triangular. In addition, the cover always has a triangular mesh structure. The ceramic and UHMWPE are modeled with a hexahedral mesh structure with a size of 0.5 mm.

The size of the mesh varies, the reason for this is based on the convergence test. Thus, these ammunitions were characterized and their verifications were accurate at these mesh sizes.

4. RESULTS

Numerical models play a crucial role in the design of hard body armor by allowing the evaluation of key parameters such as weight, thickness, multi-hit resistance, and penetration status—where partial penetration (PP) indicates successful stoppage of the projectile and complete penetration (CP) indicates full perforation—and back-face signature (BFS) performance. This section aims to examine the effect of the different ammunition threads, and the ceramic composite material (Nurol Technology's specialized product) on BFS through numerical examples. By studying the state-of-the-art computational techniques, the behavior of different ammunition such as 5.56x54 mm SS109, 7.62x39 mm MSC, and 7.62x51 mm M80, ceramic composite material is investigated by comparing real test campaigns. The deformations caused by the 5.56×45 mm SS109 ammunition, which features a hardened steel core, were investigated by examining the BFS. The analysis of the resulting BFS indicates that the performance of such a lightweight plate at this protection level was successfully simulated with a very low margin of error. The BFS values are listed in Table 5 for both numerical and experimental results. The ballistic performance of the armor plate was investigated against 7.62x39 mm MSC (Mild Steel Core) ammunition through the analysis of BFS. The results presented in Table 5, demonstrate the plate's effective performance. These tests were successfully replicated through numerical modeling, confirming the accuracy of the simulation approach. 7.62x51 mm M80 ammunition poses a significant threat to body armor plates, primarily due to the substantial trauma it can inflict. The impact of this threat on the developed product was analyzed with the BFS. According to the results, the M80 threat caused greater trauma compared to other ammunition types; however, in parallel with ballistic test outcomes, the measured BFS values as depicted in Table 5 remained well below the 44 mm threshold based on NIJ 0107.01.

Table 5 summarizes the simulated and experimental BFS depths for each shot. All rounds resulted in partial penetration, defined as the projectile not perforating completely through the armor, yet causing measurable back-face trauma in the backing material. The BFS was measured at the point of maximum clay deformation behind the armor panel immediately after each shot as shown Figure 4.

Table 5: Back-face signature (BFS) depths recorded after each shot.

Shot No.	Ammunition Type	Simulation-BFS (mm)	Experiment-BFS (mm)	Penetration Type
1	5.56×45 mm NATO SS109 (M855)	12 mm	13 mm	Partial Penetration
2	5.56×45 mm NATO SS109 (M855)	11 mm	14 mm	Partial Penetration
3	7.62×39 mm MSC	16 mm	10 mm	Partial Penetration
4	7.62×39 mm MSC	18 mm	14 mm	Partial Penetration

https://doi.org/10.52202/081621-0032

5	7.62×51 mm NATO M80 Ball	28 mm	32 mm	Partial Penetration
6	7.62×51 mm NATO M80 Ball	34 mm	34 mm	Partial Penetration

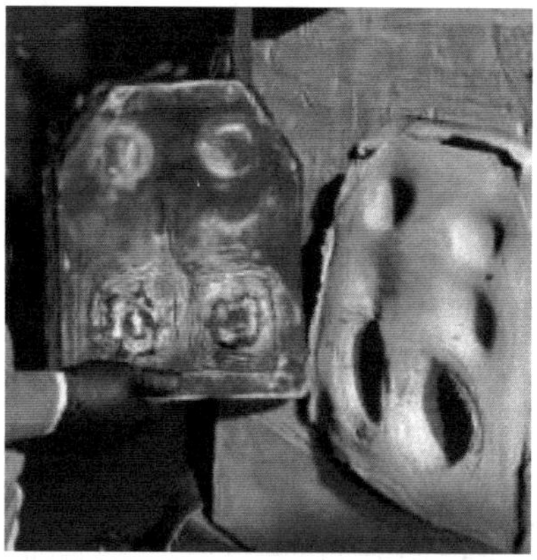

Figure 4: BFS Marks after the shoot for the proper measurement.

BFS depths for the lighter rounds (5.56x45 mm SS109 and 7.62x39 mm MSC) remained well below the critical threshold of 44 mm based on NIJ 0107.01, suggesting an effective dissipation of kinetic energy through the ceramic and UHMWPE layers. However, as expected, 7.62×51 mm M80 projectiles resulted in significantly higher deformation values, though still within survivable limits. The simulated and experimental BFS results in Table 5 show good overall agreement. For 5.56×45 mm SS109 rounds, differences are minimal (1–3 mm). Simulations slightly overpredict BFS for 7.62×39 mm MSC rounds, with up to 6 mm deviation. For 7.62×51 mm M80 rounds, the results are close, with one exact match. Overall, the simulation reliably reflects experimental trends for partial penetrations. Upon examining the layers of UHMWPE as shown in Figure 5, it was observed that the first 5.56x45 mm SS109 projectile did not penetrate any layers, while the second one was stopped at the sixth layer. The 7.62x39 mm MSC projectile was found to be halted at the thirth and fifth layers following the layer separation process. The 7.62x51 mm M80 projectile was determined to have been stopped at the seventy seventh and eighty second layers.

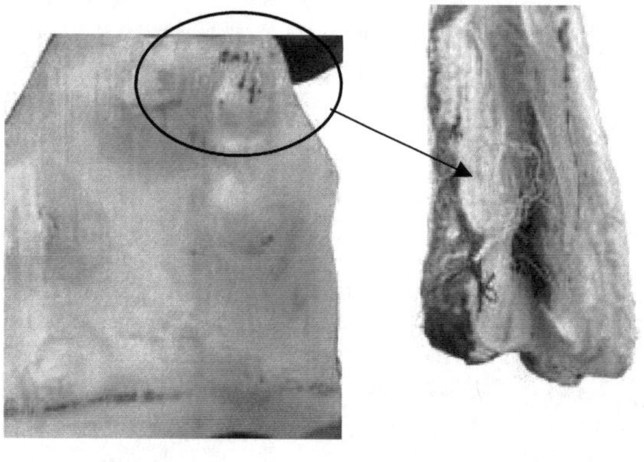

Figure 5: The layer of UHMWPE after test

Due to the complete fragmentation and deformation of the 5.56x45 mm SS109 and 7.62x51 mm M80 projectiles after testing, it was not possible to make a quantitative assessment. Only the MSC projectile was recovered from within the plate in a mushroomed shape, as shown in Figure 6. A similar geometry was also observed in the analysis results.

(a) (b)

Figure 6: The deformation on the 7.62x39 mm MSC for experiment(a) and simulation(b)

5. CONCLUSION

In this study, the ballistic performance of an ultra-lightweight personal body armor system developed by Nurol Teknoloji was successfully evaluated through both experimental testing and numerical simulations. The results demonstrated that the specialized NT ceramic-UHMWPE composite armor system effectively withstood multiple ballistic threats while consistently maintaining BFS values well below the critical NIJ 0101.07 threshold of 44 mm. A key achievement of this research was the armor's ability to provide high-level ballistic protection without any instance of complete penetration (CP) across all tested ammunition types and impact sequences. The correlation between experimental and simulation results confirms the reliability of the computational models employed and validates their use in future armor development projects. Overall, this research reaffirms NT's commitment to delivering premium, lightweight personal body armor solutions characterized by exceptional ballistic protection, and minimal backface trauma under demanding threat conditions.

6. REFERENCES

[1] Nayak, R., Crouch, I., Kanesalingam, S., Ding, J., Tan, P., Lee, B., Miao, M., Ganga, D., Wang, L., (2017). Body armor for stab and spike protection part 1: scientific literature review, Textile Research Journal, 88(7).

[2] Karandikar, P. G., Evans, G., Wong, S., Aghajanian, M. K., & Sennett, M. (2009). A review of ceramics for armor applications. Advances in Ceramic Armor IV, 29, 163-175.

[3] Wilshaw, T.R. (1971) 'The Hertzian fracture test', Journal of Physilicon carbides D: Applied Physilicon carbides, 4(10), p. 1567.

[4] Crouch IG, Eu B (2017) Ballistic testing methodologies In IG Crouch (ed.) The Science of Armour Materials, Woodhead Publishing.

[5] Livermore Software Technology Corporation (LSTC), 2012. LS-DYNA Keyword User's Manual, Volume I - Keywords, Version 971 R6.1.0. Livermore, CA: Livermore Software Technology Corporation.

[6] Paris, V., Weiss, A., Vizel, A., Ran, E. & Aizik, F. (2012). Fragmentation of armor-piercing steel projectiles upon oblique perforation of steel plates. EPJ Web of Conferences, 26, 04032. DOI: 10.1051/epjconf/20122604032

[7] Hála, P., Hurtig, K., Řídký, R., Sovják, R., & Konvalinka, P. (2024). Resistance of high-performance concrete panels with dispersed fiber reinforcement to oblique-angle projectile impact. Journal of Materials in Civil Engineering, Advance online publication. https://doi.org/10.1061/JMCEE7.MTENG-18522

[8] Eckhoff, D., Thomesen, S., Heisserer, U., & Børvik, T. (2025). Ballistic response behaviour of Dyneema® HB210 curved armour plates: An experimental and numerical study. Thin-Walled Structures, 213, 113265. https://doi.org/10.1016/j.tws.2025.113265

[9] Scazzosi, R., Giglio, M., & Manes, A. (2020). Fe coupled to SPH numerical model for the simulation of high-velocity impact on ceramic based Ballistic Shields. Ceramics International, 46(15), 23760–23772. https://doi.org/10.1016/j.ceramint.2020.06.151

[10] Choudhary, S., Singh, P. K., Khare, S., Kumar, K., Mahajan, P., & Verma, R. K. (2020). Ballistic impact behaviour of newly developed armour grade steel: An experimental and numerical study. International Journal of Impact Engineering, 140, 103557. https://doi.org/10.1016/j.ijimpeng.2020.103557

 https://doi.org/10.52202/081621-0032

Effect of Personal Body Armor's Ceramic Geometry and Shot Location in Defeating 7.62 mm Hardened Projectiles

Kartikeya Kartikeya[1], Mohit Garg[1], Hemant Chouhan[1], Makhan Singh[1] and **Naresh Bhatnagar**[1]

[1]*Department of Mechanical Engineering, Indian Institute of Technology Delhi, Hauz Khas, New Delhi – 110016, India*
nareshb@mech.iitd.ac.in

Abstract. Modern state-of-the-art B_4C/UHMWPE composite armour is used to defeat 7.62 × 39 mm hardened projectiles. Nowadays, B_4C tiles are preferred over monolithic plates. The tiles are available in many shapes. In this preliminary work the effect of the two most popular shapes of the tiles i.e., hexagonal and quadrilateral, on ballistic resistance against 7.62 × 39 mm projectiles is studied. Test specimens were prepared by bonding tiles to thick aluminium and polycarbonate blocks of known mechanical properties. The test specimens were subjected to impact from these hardened steel core (HSC) projectiles. The impact locations were: the centre of tiles, edges, and common vertices (of two or more adjacent tiles) for both the shapes. The depth-of-penetration (DOP) and residual core length (RCL) were recorded after each test to understand the effect of the shape of the tile and shot location. It was observed that both tile shape and shot location affect DOP for the B_4C tiles.

1. INTRODUCTION

Modern personal body armours are fabricated from ceramic tiles backed with fibre-reinforced composites to defeat hardened core projectiles [1–3]. The role of ceramic is to erode the projectile's hardened core, and the role of composite backing is to catch the eroded core [4,5]. Boron carbide ceramic is most popular among alternatives like alumina or silicon carbide. Thus, boron carbide is extensively studied in the literature [6–9].

Woosley et al. [10] introduced the depth of penetration (DoP) method for the characterization of ceramic materials under high-velocity impact. The authors identified a region between a case of ceramic overmatched and a case of projectile overmatched on a DoP and areal density plot, where the relationship between areal density and DoP was almost linear with negative slope. Moynihan et al. [11] performed DoP tests using 0.30 calibre AP M2 projectile on ceramic materials viz. aluminium oxide, silicon carbide and boron carbide. The authors noted that boron carbide (B_4C) has the highest ballistic efficiency, followed by silicon carbide (SiC) and then aluminium oxide. Similarly, Kauffman et al. [12] determined the ballistic performance of SiC, alumina and B_4C by conducting DoP tests with 0.50 calibre AP projectiles. The authors measured DoP in an aluminium block using X-rays. In their tests, especially at velocities greater than 750 m s^{-1}, SiC and B_4C provided better protection than alumina.

Savio et al. [13,14] conducted depth of penetration tests on reaction-bonded and hot-pressed B_4C tiles backed with aluminium alloy blocks using 7.62 × 54R AP projectiles. The hot-pressed B_4C had higher ballistic efficiency than reaction-bonded B_4C. Also, hot-pressed B_4C had a peak efficiency range at tile thickness to projectile diameter ratios in the range of 0.85 to 1.16. Carton et al. [15] showed that as the areal density (AD) of a ceramic tile increased eroded length of the projectile's core also increased. Zhai et al. [16] studied the ballistic impact of long-rod projectiles on ceramic targets. The authors determined that higher flow stress of the projectile results in higher transition velocity. A transition velocity is a range between complete interface defeat and penetration. The authors also noted that flat nose-shaped projectiles induced extensive damage to the ceramic.

The measurement of DoP in an aluminium block is the most popular method. However, for comparatively softer projectile cores, there is marginal penetration into the block. Thus, researchers shifted to studying DoP in polycarbonate blocks [17,18]. Hazell et al. [19] evaluated the performance of 7.5 mm thick SiC tiles of various sizes. The SiC tiles backed with semi-infinite polycarbonate block were subjected to the impact of 7.62 mm WC-Co cored projectiles at 838 m s^{-1}. The authors prescribed the use of a minimum of 70 mm square tile against a 7.62 mm WC-Co cored projectile as this ballistic configuration had the least edge proximity effect. Seifert et al. [20] tested the effect of adhesives, inter-tile gap width and impact location on the ballistic performance of a ceramic/metal target when impacted with a tungsten carbide AP projectile. The authors noted that ballistic limit velocity was higher in the case of an epoxy-based adhesive than in the case of a modified polyurethane-based adhesive. In the case of both adhesives, an increase in the inter-tile gap resulted in higher residual velocity and lower damage

https://doi.org/10.52202/081621-0033

to the projectile. It was also observed that shots on the tile's edges resulted in higher residual velocity than shots in the middle of the inter-tile gap.

It is clear from the literature that the role of ceramic tile shape and shot location has not been studied extensively. Thus, in this preliminary work, the effect of tile shape and shot location on the ballistic resistance of ceramic is explored. First, targets were fabricated from B_4C tiles of two different shapes. The tiles were backed by aluminium blocks or polycarbonate bars. The targets were then subjected to high-velocity impact (HVI) tests using 7.62 × 39 mm hardened steel core (HSC) projectiles in a single-stage gas gun. Residual core length (RCL) was measured after each test along with the depth of penetration of the projectile in the backing material. The study shows the interesting effect of backing material stiffness, tile shape and shot location on the ballistic performance of B_4C tiles.

2. MATERIAL AND METHODS

This section contains details of materials used for the study. It also explains the experimental methods used in the study.

2.1. B_4C Targets

The hot-pressed B_4C tiles of 6.5 mm thickness were procured in two shapes i.e., hexagonal and quadrilateral as shown in Figure 1. The hexagons had an edge length of 17 mm and an edge-to-edge distance of 30 mm. The quadrilaterals were squares of 25 mm edge length. The B_4C tiles were compliant with MIL-P-64153 as per the supplier's report. Two different types of backing material were also procured i.e., aluminium and polycarbonate. A 100 mm × 100 mm square rod of Al-6061-T6 was procured and similarly Ø100 mm polycarbonate rod was also procured.

The targets were set in a way to allow impact at the centre of the tile, at two adjacent edges and at a vertex of three edges as shown in Figure 2. The targets, as shown schematically in Figure 2, were fabricated in an autoclave. The B_4C tiles were bonded with a polycarbonate rod or Al 6061-T6 block using two-part epoxy adhesive as per literature [20]. The thickness of the adhesive was kept minimal as per the literature [21]. After fabrication, the top surface of the target was confined with glass-fibre tape which ensures minimal material ejection during impact. The depth of the polycarbonate rod or aluminium block was either 150 mm or 75 mm.

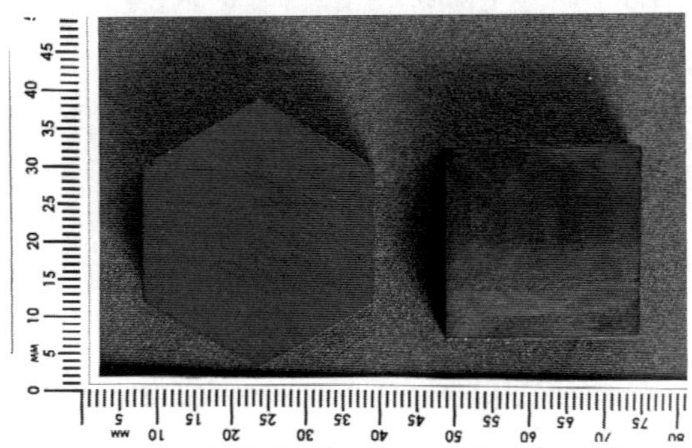

Figure 1. B_4C tile shapes used in this study

https://doi.org/10.52202/081621-0033

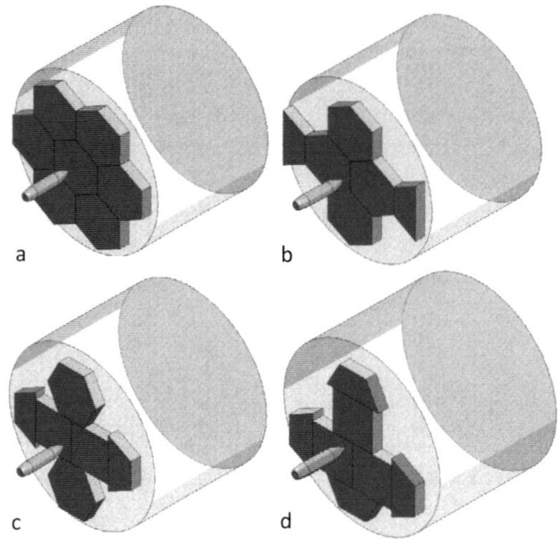

Figure 2. Impact location on B_4C tiles, a. Centre of hexagonal tile, b. Edge of two hexagonal tiles, c. Edge of two quadrilateral tiles and d. Vertex of three quadrilateral tiles

2.2. 7.62 × 39 mm HSC Projectile

The projectiles used in this study were procured from Ordnance Factory Varangaon, India. The 7.62 × 39 mm HSC projectile contains a hardened steel core and lead filler enclosed in a copper jacket. The hardness of the core varies between 46 to 55 HRC and it weighs 3.52 g. The variation in hardness of core may affect DoP values obtained from experiments. The height of the core is 17.82 mm. The 7.62 × 39 mm HSC projectile along with its core is shown in Figure 3.

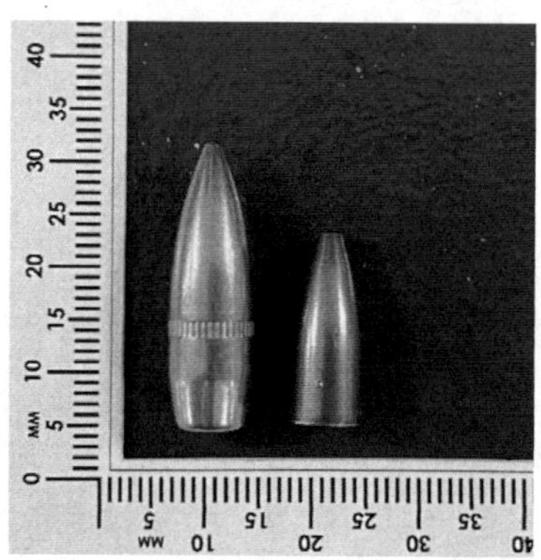

Figure 3. 7.62 mm HSC projectile and its hardened steel core

2.3. High-velocity Impact Test

The HVI tests were conducted using a single-stage gas gun at IIT Delhi. A detailed account of the single-stage gas gun is present in literature [4]. A schematic of the single-stage gas gun is shown in Figure 4. The single-stage gas gun was aligned using a laser. First, the chamber and the barrel were evacuated up to 500 mbar. Helium gas was then filled in the reservoir up to a precalculated pressure level (72 bar). The sudden release of pressure from the reservoir launched a sabot which houses a 7.62 × 39 mm HSC projectile. The sabot-projectile assembly travelled in the barrel while accelerating and gaining velocity. At the entry of the chamber, a sabot trapper assembly is installed which destroys the sabot and allows only the projectile to pass through to impact the target. The target was mounted in the chamber using flexible tapes and backed with softwood. The event was recorded using a high-speed camera to ensure normal impact on the target. The velocity was measured using IR-based velocity screens in the chamber. The residual core was recovered by vacuuming the chamber.

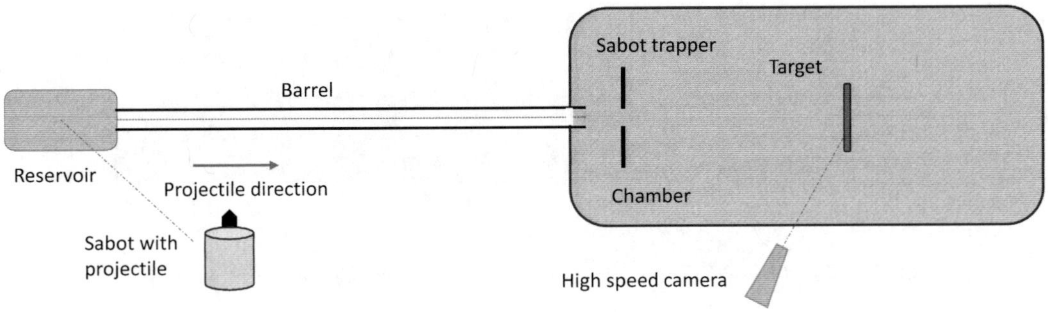

Figure 4. Schematic of single-stage gas gun installed at IIT Delhi

3. RESULTS AND DISCUSSION

The results from impact tests are summarized in Table 1. Initially, HVI tests were performed on adjoining edges of hexagonal B_4C tiles backed with aluminium blocks. However, it was noted that at 150 mm backing thickness, there was no DoP in the block. As can be seen from Table 1 and Figure 5 the projectile's hardened core got eroded and broke into a number of pieces. The 150 mm aluminium block has very high stiffness and thus increases the dwell time of projectiles on B_4C tiles which translates to the core's higher erosion as corroborated by other researchers also [22,23]. Thus, the next HVI test was performed on B_4C tiles with an aluminium backing of 75 mm. On impact with B_4C tiles, a compressive stress wave is generated beneath the projectile. This wave travels towards the backing material and gets reflected as a tensile wave [24]. The tensile wave causes cracks in the B_4C tiles. Initially, a fracture conoid is generated, followed by radial cracks at the bottom of a tile and ultimately circumferential cracks. The fracture conoid contained comminuted B_4C particles which remained confined for a certain significant duration by adjacent material laterally. However, some of the comminuted particles get ejected from the projectile's entry hole, which was reduced by the use of glass-fibre tape in this study. The confined yet comminuted B_4C particles eroded a projectile's core during an impact event [1]. It can be seen from Figure 6 that although the impact location was between two edges, all the tiles separated from the block. In this test, DoP was recordable i.e. 1.1 mm, however, it was still low as shown in Figure 6. It can also be noted that the RCL was considerably higher in this case which can be attributed to lower erosion by the B_4C tiles backed with only 75 mm aluminium block. Thus, the stiffness of backing plays an important role in RCL values of the projectile's core is now experimentally established.

Table 1. Summary of HVI tests on B_4C tiles backed by aluminium or polycarbonate

Test	Backing Material	Backing Material Size	Backing Thickness (mm)	Tile Shape	Impact Location	Velocity (m/s)	DoP (mm)	RCL (mm)
a	Al 6061 - T6	100 mm square	150	Hex	Edge	734.57	No DoP	4.72
b	Al 6061 - T6	100 mm square	75	Hex	Edge	705.34	1.1	8.1
c	PC	Ø100 mm rod	75	Hex	Edge	748.57	4.9	8.62
d	PC	Ø100 mm rod	75	Hex	Centre	634.55	1.8	10.83
e	PC	Ø100 mm rod	75	Quad	Vertex	732.13	6.6	13.91
f	PC	Ø100 mm rod	75	Quad	Edge	715.16	4.6	12.72

https://doi.org/10.52202/081621-0033

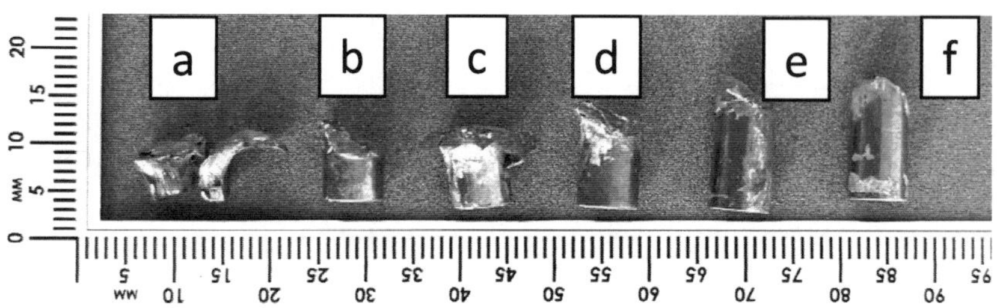

Figure 5. Residual core length (RCL) after HVI tests, designations as per Table 1

Since DoP was not considerable with aluminium block backing, further HVI tests were conducted with B_4C tiles backed with polycarbonate round block cut from bars. In these tests, tile thickness was fixed at 6.5 mm due to material limitations. It can be seen from Table 1 that this thickness does not allow for higher DoP values, even with polycarbonate backing, which is desirable for distinguishing roles of various parameters. The HVI tests were conducted at the centre of the tile, adjoining edge of two tiles and at adjoining vertex of three tiles. The results are shown in Figure 7 and designations are as per Table 1. It can be seen from Figure 7c that the arrangement of tiles was different from that shown in Figure 2. However, it was ensured during the test that the impact occurred at adjoining edges of two hexagonal tiles. It can be noted that impact tests were not conducted on centre of square and hexagonal tiles due to material limitations. The reasoning behind this decision was that both tiles were of same material, same thickness and were confined in similar way and thus should provide similar ballistic resistance.

Figure 6. Aluminium block after HVI test

Figures 7c and 7e show an interesting phenomenon that along with local DoP the polycarbonate also exhibited global cracking and failure for certain tests. The global cracking and failure rendered generally translucent polycarbonate opaque for certain viewing angles. The global response implies that force from impact was distributed to a larger area in those cases and was not concentrated just beneath the projectile. A concentrated force is desirable to get proper DoP values which may be achieved with higher thickness of polycarbonate bar. However, a higher thickness of polycarbonate bar will increase its stiffness and may reduce DoP values. Thus, DoP values in an HVI test become a function as shown in Equation 1 and can be optimized through numerical studies for better experimental results.

$$DoP = f\,(E_{backing}, E_{ceramic}, V_{projectile}, E_{projectile}) \quad (1)$$

In Equation 1, $E_{backing}$ represents the stiffness of a backing material i.e. thickness and material; $E_{ceramic}$ represents hardness and thickness of a ceramic; $V_{projectile}$ is the impact velocity of a projectile; and $E_{projectile}$ is flow stress of projectile i.e. hardness and yield stress.

Figure 7. Polycarbonate block after HVI tests, designations as per Table 1

The HVI test designated as 'c' and 'f' in Table 1 represents impact at the adjoining edge of two B_4C tiles for hexagonal and quadrilateral shapes respectively. It can be noted that both tests resulted in similar DoP values even with different impact velocities. However, there is a considerable difference in RCL values. The RCL values were higher for quadrilateral tiles than for hexagonal tiles which shows that hexagonal tiles eroded projectile's core better than quadrilateral tiles. In the case of quadrilateral tiles, an impact at the vertex of three adjoining tiles will result in higher DoP and RCL values than an impact at the edge of two adjoining tiles. Thus, it can also be stated that a vertex is weaker than an edge.

4. CONCLUSIONS

In this study, targets were fabricated by bonding B_4C tiles to aluminium and polycarbonate blocks (square and round) using an autoclave. The targets were then subjected to an HVI test using 7.62 × 39 mm HSC projectiles. After each HVI test, DoP and RCL were measured. Based on results and discussion following conclusions can be drawn from the study.

1. The backing material plays an important role in DoP values determined through HVI tests. A high stiffness backing will allow greater dwell of projectile at tile and thus lower DoP values. The DoP values were higher for targets backed with a polycarbonate bar than those backed with an aluminium block due to its lower density and deformability.
2. The DoP values are a function of parameters of ceramic, projectile and impact velocity and can be optimized for better experimental results.
3. In this preliminary work, the difference in DoP values between different tile shapes was not apparent. However, the RCL values for quadrilateral tiles were higher than for hexagonal tiles.
4. It was also noted that an impact at the vertex of three adjoining tiles will result in higher DoP and RCL values than an impact at the edge of two adjoining tiles. Thus, a vertex is weaker than an edge.
5. A similar study with lower tile thickness will allow higher DoP values is also planned in the future. The study will also include more replicates of each test.

https://doi.org/10.52202/081621-0033

Acknowledgements

The authors acknowledge the use of manufacturing and testing facilities at the Centre of Excellence in Personal Body Armour at the Indian Institute of Technology Delhi.

References

[1] Crouch IG. The Science of Armour Materials. 1st ed. Woodhead Publishing Limited; 2017.
[2] Medvedovski E. Ballistic performance of armour ceramics: Influence of design and structure. Part 1. Ceram Int 2010;36:2103–15. https://doi.org/10.1016/j.ceramint.2010.05.021.
[3] Hong Y, Xie F, Xiong L, Yu M, Cheng X, Qi M, et al. The ballistic performance of laminated sic ceramics for body armour and the effect of layer structure on it. Applied Sciences (Switzerland) 2021;11. https://doi.org/10.3390/app11136145.
[4] Kartikeya K, Chouhan H, Ram K, Prasad S, Bhatnagar N. Ballistic evaluation of steel/UHMWPE composite armour system against hardened steel core projectiles. Int J Impact Eng 2022;164:104211. https://doi.org/10.1016/j.ijimpeng.2022.104211.
[5] Ong CW, Boey CW, Hixson RS, Sinibaldi JO. Advanced layered personnel armour. Int J Impact Eng 2011;38:369–83. https://doi.org/10.1016/j.ijimpeng.2010.12.003.
[6] Holmquist TJ, Johnson GR. Response of boron carbide subjected to high-velocity impact. Int J Impact Eng 2008;35:742–52. https://doi.org/10.1016/j.ijimpeng.2007.08.003.
[7] Holmquist TJ, Johnson GR. Characterization and evaluation of boron carbide for plate-impact conditions. J Appl Phys 2006;100. https://doi.org/10.1063/1.2362979.
[8] Crouch IG, Appleby-Thomas G, Hazell PJ. A study of the penetration behaviour of mild-steel-cored ammunition against boron carbide ceramic armours. Int J Impact Eng 2015;80:203–11. https://doi.org/10.1016/j.ijimpeng.2015.03.002.
[9] Vozarova M, Neubauer E, Carton E, van Zeijl E, Olevano D, Lionetti S. Enhancing Armor Performance: Investigation into Lightweight Ceramic Composites for Defence and Security. Light-Weight Armour for Defence and Security, 2024.
[10] Woolsey P, Kokidko D, Mariano SA. Alternative Test Methodology for Ballistic Performance Ranking of Armor Ceramics. 1989.
[11] Moynihan, Thomas J.; Chou, Shun-Chin; Mihalcin AL. Application of depth-of-penetration test methodology to characterize ceramics for personnel protection. 2000.
[12] Kaufmann C, Cronin D, Worswick M, Pageau G, Beth A. Influence of material properties on the ballistic performance of ceramics for personal body armour. Shock and Vibration 2003;10:51–8. https://doi.org/10.1155/2003/357637.
[13] Savio SG, Ramanjaneyulu K, Madhu V, Bhat TB. An experimental study on ballistic performance of boron carbide tiles. Int J Impact Eng 2011;38:535–41. https://doi.org/10.1016/j.ijimpeng.2011.01.006.
[14] Savio SG, Rao AS, Reddy PRS, Madhu V. Microstructure and ballistic performance of hot pressed & reaction bonded boron carbides against an armour piercing projectile. Advances in Applied Ceramics 2019;118:264–73. https://doi.org/10.1080/17436753.2018.1564416.
[15] Carton E, Roebroeks G, Weerheijm J, Diederen A. Inertia as main working mechanism for ceramic based armour Inertia as main working mechanism for ceramic based armour 2019:1–14. https://doi.org/https://doi.org/10.6084/m9.figshare.9923807.v1.
[16] Zhai YX, Wu H, Fang Q. Interface defeat studies of long-rod projectile impacting on ceramic targets. Defence Technology 2020;16:50–68. https://doi.org/10.1016/j.dt.2019.05.021.
[17] Dorogoy A, Rittel D, Brill A. Experimentation and modelling of inclined ballistic impact in thick polycarbonate plates. Int J Impact Eng 2011;38:804–14. https://doi.org/10.1016/j.ijimpeng.2011.05.001.
[18] Weiss A, Vizel A, Durban D. AN EXPERIMENTAL INVESTIGATION OF DEEP PENETRATION INTO POLYCARBONATE TARGETS. 27th International Symposium on Ballistics, Freiburg: 2013.
[19] Hazell PJ, Roberson CJ, Moutinho M. The design of mosaic armour : The influence of tile size on ballistic performance 2008;29:1497–503. https://doi.org/10.1016/j.matdes.2008.03.003.
[20] Seifert W, Strassburger E, Dolak M, Schaare S. Experimental study on the dependency of the ballistic performance of tiled ceramic/metal targets on inter tile gap width and projectile impact position. Int J Impact Eng 2018;122:50–9. https://doi.org/10.1016/j.ijimpeng.2018.08.006.

[21] López-Puente J, Arias A, Zaera R, Navarro C. The effect of the thickness of the adhesive layer on the ballistic limit of ceramic/metal armours. An experimental and numerical study. Int J Impact Eng 2005;32:321–36. https://doi.org/10.1016/j.ijimpeng.2005.07.014.
[22] Wilkins ML. Third Progress Report of Light Armor Program. Livermore, California: 1968.
[23] Bhatnagar N, Kartikeya K, Narayan D, Chouhan H, Ram K, Ahmed A. Effect of Backing Material Stiffness on Ballistic Performance of Ceramic/UHMWPE Personal Body Armour. PERSONAL ARMOUR SYSTEMS SYMPOSIUM 2023 Proceedings, 2023.
[24] Li-li W. Foundations of Stress Waves. First. Elsevier; 2007.

 https://doi.org/10.52202/081621-0033

Back-face Dynamic Profiling System: A Tool for Behind-Armour Blunt Trauma

A. Goertz[1], A. Dannaoui[2], A. Brown[3], D. Sherman[2], C. Bir[2], and K. Rafaels[4]

[1]*SURVICE Engineering Company, 4695 Millennium Drive, Belcamp, MD 21017, USA*
alan.goertz@survice.com
[2]*Wayne State University, 818 W. Hancock, Detroit, MI 48201, USA*
[3] *U.S. Army Combat Capabilities Development Command Army Research Laboratory, 800 Park Office Dr, Research Triangle Park, NC 27709, USA*
[4]*U.S. Army Combat Capabilities Development Command Army Research Laboratory, Aberdeen Proving Ground, MD 21005, USA*

Abstract. Armour back-face dynamics are responsible for injuries sustained by the wearer during non-penetrating ballistic impacts. Capturing armour back-face dynamics has been an ongoing challenge for researchers. Body armour is intended to be worn and tested with a tight fit, which visually obscures the armour back-face, and the visible armour front-face dynamics may not reflect the armour back-face dynamics. A novel Back-face Dynamic Profiling System (BFDPS) for measuring three-dimensional (3-D) dynamic armour deformation has been developed that consists of a slim fixture that is positioned between the armour back-face and surrogate or test device. The BFDPS allows for calculation of the dynamic deformation of a visually obscured rectangular fibre grid using high-speed video motion tracking of the fibre ends, which extend beyond the armour. Considering all the fibre displacements simultaneously provides constraints for an optimisation that solves for the out-of-plane displacements of the intersecting points of the fibre grid for each frame of the video. Optimisations of sequential video frames are combined to provide back-face dynamics. To assess the accuracy of the BFDPS, a ballistic experiment was conducted using a .45 ACP (Automatic Colt Pistol) bullet striking a soft armour panel backed by a transparent ballistic gelatine block. The experimental back-face dynamics were captured using motion-tracking of the high-speed video of the armour back-face profile intruding into the gelatine block. Similarly, the BFDPS fibre-grid tip displacements were tracked with high-speed video motion tracking and underwent optimisation to calculate the 3-D back-face dynamics. Superposition of the back-face profiles obtained from both methods indicate good agreement throughout the time history of the event. The promising results suggest the BFDPS has utility for tracking back-face dynamics under conditions where the back-face cannot be visually observed or measured by other means.

1. BACKGROUND

Characterising dynamic surface deformation of objects that are visually obscured by surrounding opaque surfaces is an ongoing challenge. A prime example being the measurement of the back-face dynamics of personal protective body armour, backed by clay, subjected to ballistic impact loading. Body armour is generally layered such that the dynamics of the front-facing outer armour layer are not representative of the rear back-face layer of the armour due to separation of interior armour layers. The back-face surface of the armour is normally in intimate contact with the clay so it cannot be directly observed.

Various schemes have been developed to measure 3-D shapes using thin planar sensing devices. A planar mesh using a two-layer orthogonal mesh of fibre Bragg grating (FBG) strands embedded in a 5 mm thick sheet of silicon was designed to measure 3-D shapes [1]. FBG senses strain by measuring light reflection in the fibre optic strands. This device is not well suited to ballistic armour application since the 5 mm thick gelatine would almost certainly alter the interaction between the armour back-face and the backing object. In addition, it is unlikely the fibres could withstand or accommodate large deformations that often occur in armour back-face dynamics. An FBG strand was woven into a Kevlar® body armour panel, backed by gelatine, which was impacted by a 12.59 mm steel ball with a velocity of 425 m/s [2]. The FBG slipped in the fabric so the FBG strains could not be used to determine back-face dynamics, but the strains were correlated to observed back-face dynamics determined by video through the gelatine. In follow-up studies, a method was devised to account for fibre slip by observing drops in FBG strain that allowed for estimating the back-face displacement[3, 4]. The methodology requires knowing or estimating the initial velocity of the deformation, which significantly limits the technique to environments where that is possible.

A novel fibre (wire) grid was developed, the Back-face Dynamic Profiling System (BFDPS), that uses changes in displacements of the ends of the fibres to determine changes in the shape of the grid. The grid is positioned between body armour and the body armour backing material (e.g., clay) with the fibre ends visible outside the perimeter of the armour. The fibre grid consists of flexible, low-friction, high

Distribution Statement A. Approved for public release: distribution is unlimited.

https://doi.org/10.52202/081621-0034

tensile strength Nitinol wires. The grid has a low resistance to out-of-plane deformation and therefore should have a minimal influence on the back-face dynamics. As the armour back-face deforms under impact, the free ends of the grid wires (outside the grid) are pulled into the grid to accommodate the geometric expansion of the back-face in deformation. The shape change of the grid is determined by tracking, currently with high-speed video, the linear dynamics of the free ends of the wires, which reside outside the periphery of the armour and are visible to the high-speed camera(s). The shape of the obscured grid is calculated by solving a constrained optimisation problem where the unknown Z-displacement (deformation) of each interior wire grid intersection is shared between and defined by four adjacent trapezoids, two in the X-direction and two in the Y-direction, all with known grid base lengths, which are the length of the initial grid (Figure 1). A proprietary Python™ optimisation program solves for a best-fit deformation contour that matches the observed free-end wire displacements at each camera frame, which provides the back-face dynamics when the camera frames are concatenated. The optimisation algorithm was verified by calculating a theoretically fibre pull-in history from a simulated BABT event and confirming the optimisation program's reconstruction matched the simulation results.

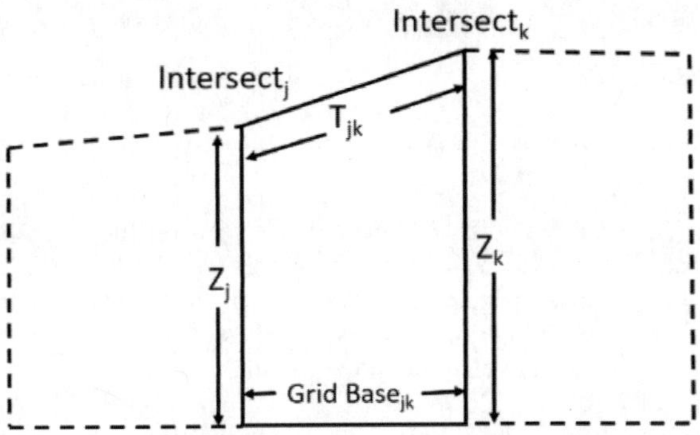

Figure 1. Wire intersection right trapezoid. Intersect_j and Intersect_k: wire intersections; T_{jk}: deformed spanning oblique segment; Grid Base_{jk}: plane projected undeformed spanning segment; Z_j and Z_k: deformed wire intersection heights.

2. METHODOLOGY

A ballistic experiment with a BFDPS fixture positioned between a soft armour panel and a clear gelatine block was conducted to assess the predictive accuracy of the BFDPS compared to the armour back-face deformation observed through gelatine. Two high-speed video cameras recorded a .45 ACP bullet striking a soft armour panel, one view facing the striking surface of the armour to capture the BFDPS wire dynamics and a second side view capturing the armour back-face intrusion into the gelatine.

2.1 BFDPS Fixture

A 71 cm × 56 cm muslin fabric (Ticking Solid Natural Colour Cotton Fabric, Walmart) was used as the base material. The warp and weft of the fabric are unknown, but it has a weight of 0.23 kg/m^2. A 22 × 22 grid of 1 cm squares was drawn in the centre of the fabric. A 1 mm leather hole punch was used to create weaving holes at the grid line intersections. Non-intersecting diagonal stress relief cuts of approximately 2 cm in length were introduced into the fabric to reduce tension in the fabric during deformation, which also diminish the influence of the fabric properties on the BFDPS response. Round 0.04 cm diameter super elastic Nitinol wire (Kellogg's Research Labs, New Boston, New Hampshire) was cut into twenty-three, 51 cm sections for vertically oriented wires and another twenty-three sections of 61 cm in length for horizontally oriented wires. The wires were woven in alternating over-under patterns, skipping every other hole, through the weaving holes with the grid centred in the length of the wires for both orientations. To maintain wire alignment outside the grid, the wires were anchored to the fabric approximately 9 cm from the free end using a combination of fabric weaves and 32 cm Nitinol wires acting as straps. The wires were inserted through sections of 0.15 cm inner diameter silicon tubing between the grid and the outer alignment anchors. The silicon tubes were fastened together using gaffer's tape to keep the tubing and wires aligned during ballistic loading (Figure 2).

https://doi.org/10.52202/081621-0034

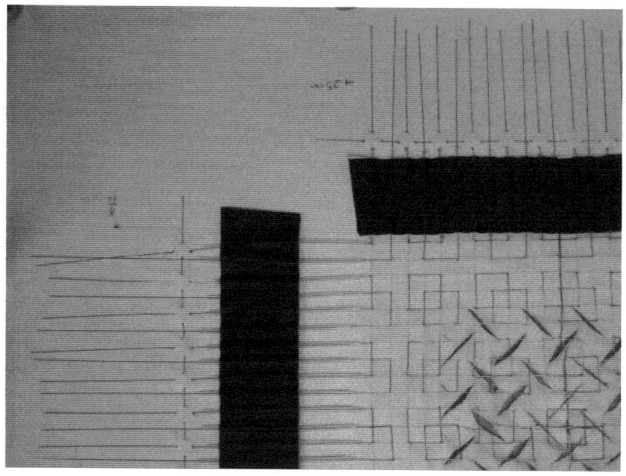

Figure 2. Quarter view (upper-left quadrant) of the BFDPS fixture.

2.2 Experimental Protocol

The chamber and barrel (.45 ACP Velocity, M34917, H-S Precision Inc) were securely mounted to a universal receiver. The barrel's end was positioned 152.4 cm from the BFDPS. The universal receiver, equipped with a remote firing mechanism, used compressed air to activate the firing pin. A full metal jacket (FMJ) round (0.45 ACP FMJ 230 grain, Q4170, Winchester) was used for this test. Three sight-in shots were fired to warm the barrel and verify the point of aim.

A 20% Clear Ballistics® gelatine block (20% NATO block, Fort Smith, Arkansas) served as the witness material. Rafaels et al. previously described the mounting fixture for the gelatine block [5].

Two high-speed cameras were employed for video capture: a Phantom® V1212 (Vision Research Inc., Wayne, New Jersey) for the lane view to visualize the BFDPS and an HX-3 (NAC Image Technology, Yokohama, Japan) for the perpendicular side view to visualize the bullet pre-impact and the back-face signature in the gelatine. Both cameras were equipped with 2.4–8.5 cm Nikon lenses (Nikon, AF Nikkor, Melville, New York) and recorded at 20,000 frames per second (FPS). The software for the V1212 and HX-3 are PCC 3.6 (PCC 3.6, Vision Research Inc, Wayne, New Jersey) and MLink (MLink x64, NAC Image Technology, Yokohama, Japan), respectively. Yellow Jacket LED lights (YJ5X, Crash Lab Support, Howell, Michigan) were positioned to backlight the gelatine block and illuminate the BFDPS, ensuring both were fully visible in the recorded videos. A 15 cm × 15 cm woven sheet of Kevlar® was positioned on the face of the gelatine block to protect it from damage from the wires.

The BFDPS fabric was clamped tautly onto a wooden frame. The clamping process involved aligning the long and short edges of the fabric and securing it with 12 small ratcheting bar clamps. The BFDPS and wooden frame were then mounted and secured using straps, with careful alignment to ensure the system's centre corresponded with the block's centre. The soft armour shoot pack was mounted onto the BFDPS (Figure 3), and the entire setup was centred using a laser. Before testing, pre-test images with a photo scale were taken. These images included the gelatine block's face, the overall setup from the lane view, close-ups of the mounted fixture, and a perpendicular view of the system. A final review of the setup ensured alignment and lighting. The V1212 camera recorded the lane-view from the point the bullet exited the barrel until the wires stopped moving post-impact, typically spanning 0.015–0.02 s. The HX-3 camera captured the side-view event from three frames before impact until the back-face deformation into the gelatine reached its maximum.

Figure 3. Assembled BFDPS with 12 ratcheting bar clamps keeping the fabric taut and hook-and-loop fastener straps holding the armour up to the fabric.

2.3 Post-processing Protocol

The high-speed video motion capture analyses for the lane and the perpendicular side camera views utilised Tracker software (Tracker 6.0.8, Open-Source Physics Project). The analysis of the lane camera view provided the positional X and Y coordinates of the BFDPS wire tips for each video frame of the analysis (Figure 4), which were subsequently used to calculate the wire tip "pull-in" displacements. An in-house custom Python™ program was used to read the Tracker motion capture analysis and solve a series of optimisation problems, one per video frame, that predicts the armour back-face contour based on the amount of wire pull-in for each wire in the BFDPS grid. To assess the profile's accuracy, the armour back-face deformation into the gelatine was also tracked.

The perpendicular side camera view provided the Y and Z coordinates of the armour back-face deformation profile (Figure 5). The armour deformation side profile history was measured along 11 lines parallel to the shot line, spaced 10 mm apart, resulting in a 100 mm profile height centred about the shot line, which was adequate for capturing the height of the back-face deformation profile. For frames where the margin of the gelatine deformation, or the wire tips for the front view, were excessively blurred or there was no detectable movement, point measurements were skipped and the coordinates were computed using the Tracker software's linear interpolation. In addition, the side camera view also provided images of the threat, prior to impacting the armour, which were used to estimate the initial threat velocity.

Figure 4. Tracker motion capture analysis of the lane camera view.

https://doi.org/10.52202/081621-0034

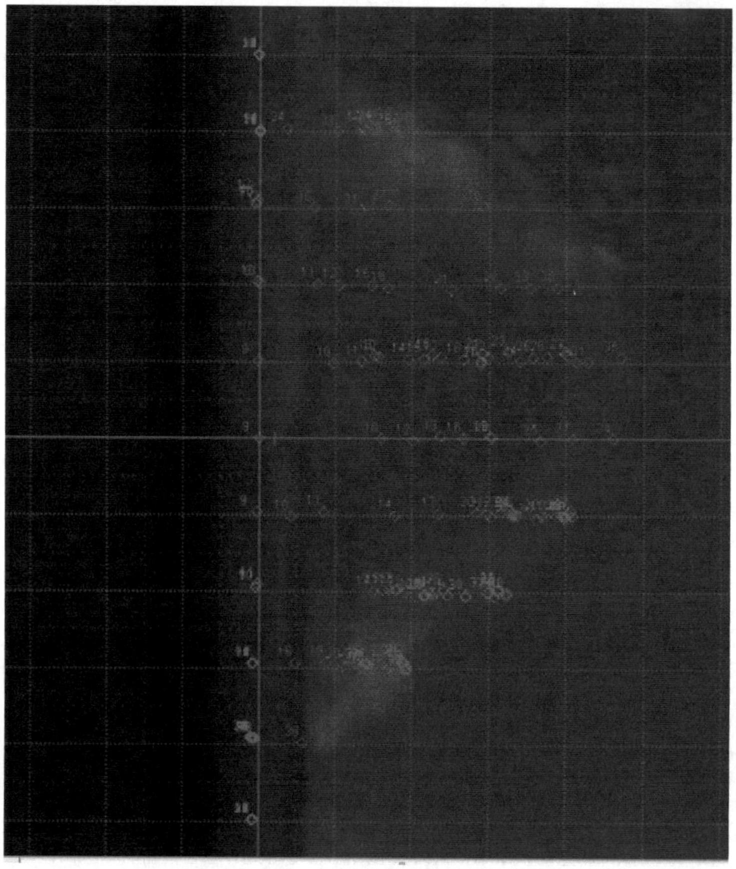

Figure 5. Tracker motion capture analysis of the perpendicular side camera view. Purple axes identify shot line and initial back-face plane.

3. RESULTS AND DISCUSSION

The motion capture analysis of the threat prior to striking the armour indicated its velocity was approximately 228 m/s at impact. The motion capture and contour optimisation analyses for the lane view video were used to generate a peak displacement history, defined by the maximum displacement of the 3-D contours of the back-face deformation for each frame of the video (Figure 6). The side view motion capture analysis generated a displacement history, also defined by the peak displacement history and a profile history. The peak displacement comparison (Figure 7) for this study was plotted through 1.4 ms, when the BFDPS analysis indicated that the armour back-face was no longer demonstrating meaningful displacements per time step. The velocity histories for both analyses were calculated using the derivative of the displacement histories and plotted (Figure 8).

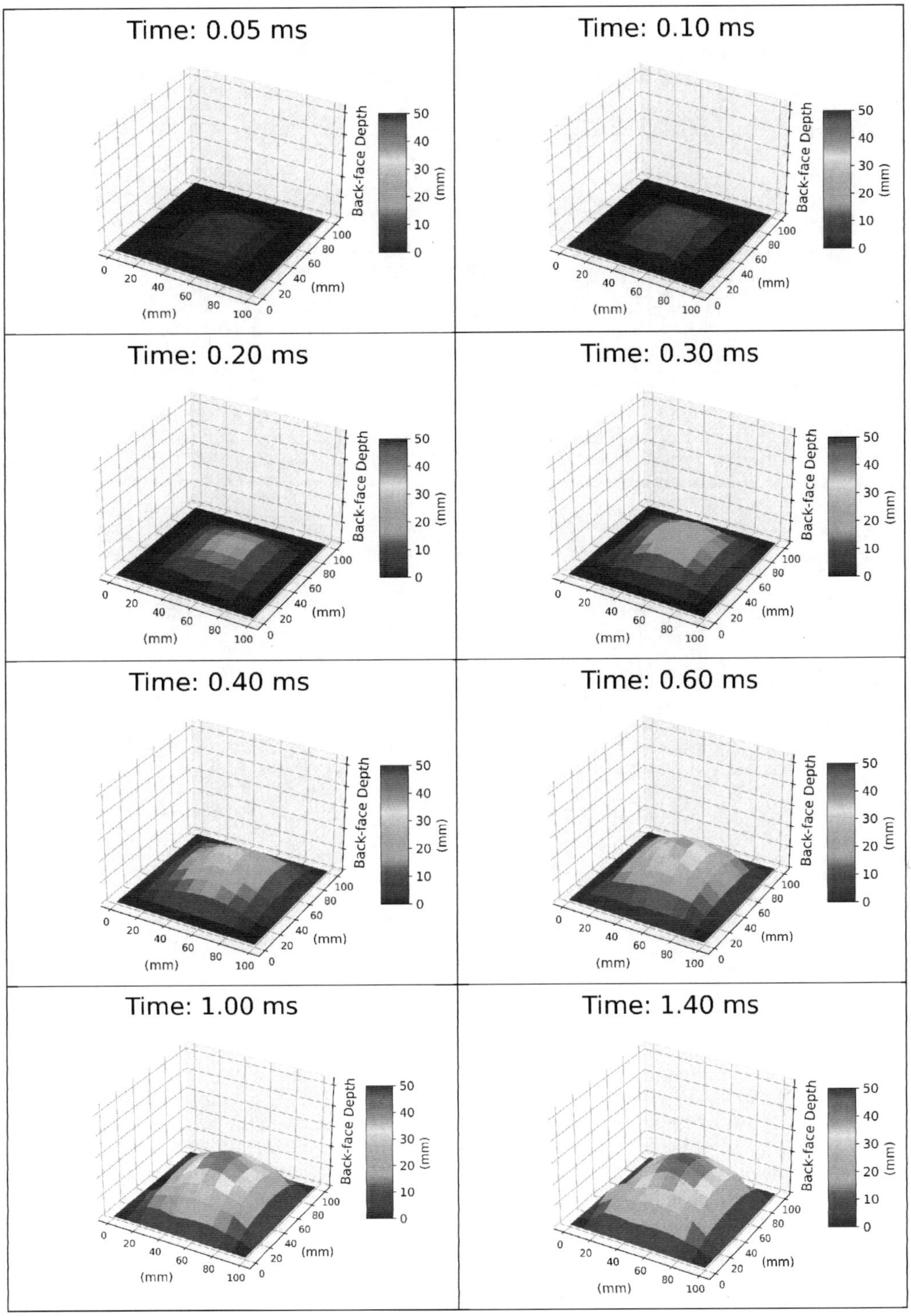

Figure 6. BFDPS-predicted back-face deformation progression.

https://doi.org/10.52202/081621-0034

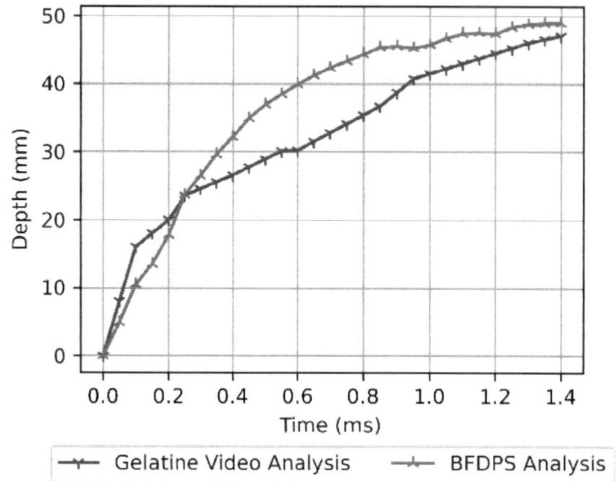

Figure 7. Peak back-face displacement histories for the gelatine video analysis and the BFDPS analysis.

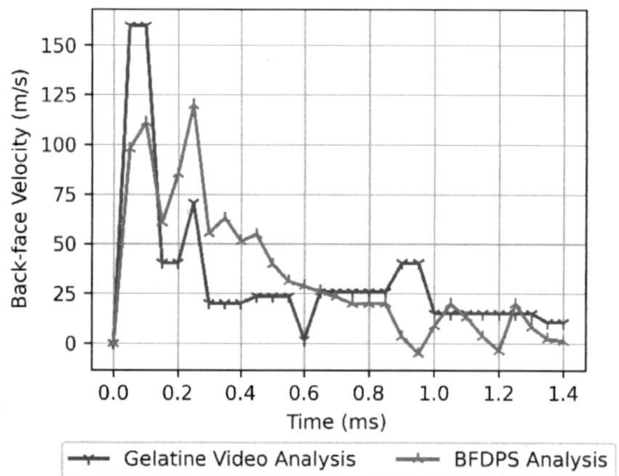

Figure 8. Peak back-face velocity histories for the gelatine video analysis and the BFDPS analysis.

The predicted BFDPS displacement history was in reasonable agreement with the gelatine displacement history through 225 μs, when the BFDPS displacements diverged from the gelatine displacements. In-house simulations of ballistic impacts to biological specimens indicate the armour back-face is largely disengaged with the surrogate at 200 μs, which is before the BFDPS displacements diverge from the backing in this study. Therefore, from a biomechanics standpoint, it appears early armour back-face dynamics are critical with respect to injury outcome since armour dynamics after separation from the backing are largely irrelevant.

In the authors' experience, the backing used in ballistic experiments, in this case gelatine, commonly separates from armour backing as the retarding mechanisms built into protective armour slow the armour back-face, while the inertia imparted to the backing during initial back-face loading continues to drive the backing material. The shape of the displacement curves in this analysis reflects those characteristics (i.e., a decelerating armour back-face and a relatively constant post-impact velocity for the gelatine backing). However, at later times, the BFDPS estimates greater displacements than the gelatine, which is contrary to what would be expected. Some of the error in calculating the displacement in the gelatine could result from optical distortion of the gelatine as it is deforming. Another reason the BFDPS could predict larger displacements than witnessed in the gelatine may be due to extraneous deformation in the wires that artificially increase the observed wire pull-in. Video analysis of other experiments has indicated that the armour may separate from not only the gelatine but also the BFDPS at later times, such that the wires can exhibit inertially driven displacements as well (Figure 9).

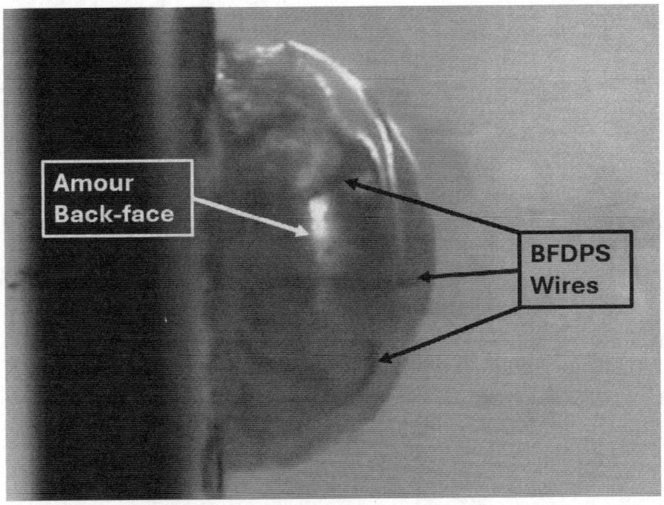

Figure 9. Example of armour and BFDPS wires separating from a hard armour plate.

The comparison of the back-face velocity predictions from the analyses confirms that derivations of dynamic histories are notoriously noisy [6-8]. The back-face velocity history for the gelatine analysis suggests approximately 160 m/s peak back-face velocity, which is reasonable given the estimated 228 m/s threat impact velocity and the resolution of the analysis (25 μs). The BFDPS analysis peak velocity occurred at 225 μs, which is unreasonable. This suggests that the early displacement estimates for the BFDPS were low. An adjustment manually overriding the optimisation analysis while maintaining a constant area under the velocity trace to preserve the later BFDPS displacement estimates might improve the consistency between the gelatine and BFDPS displacement histories prior to 225 μs.

https://doi.org/10.52202/081621-0034

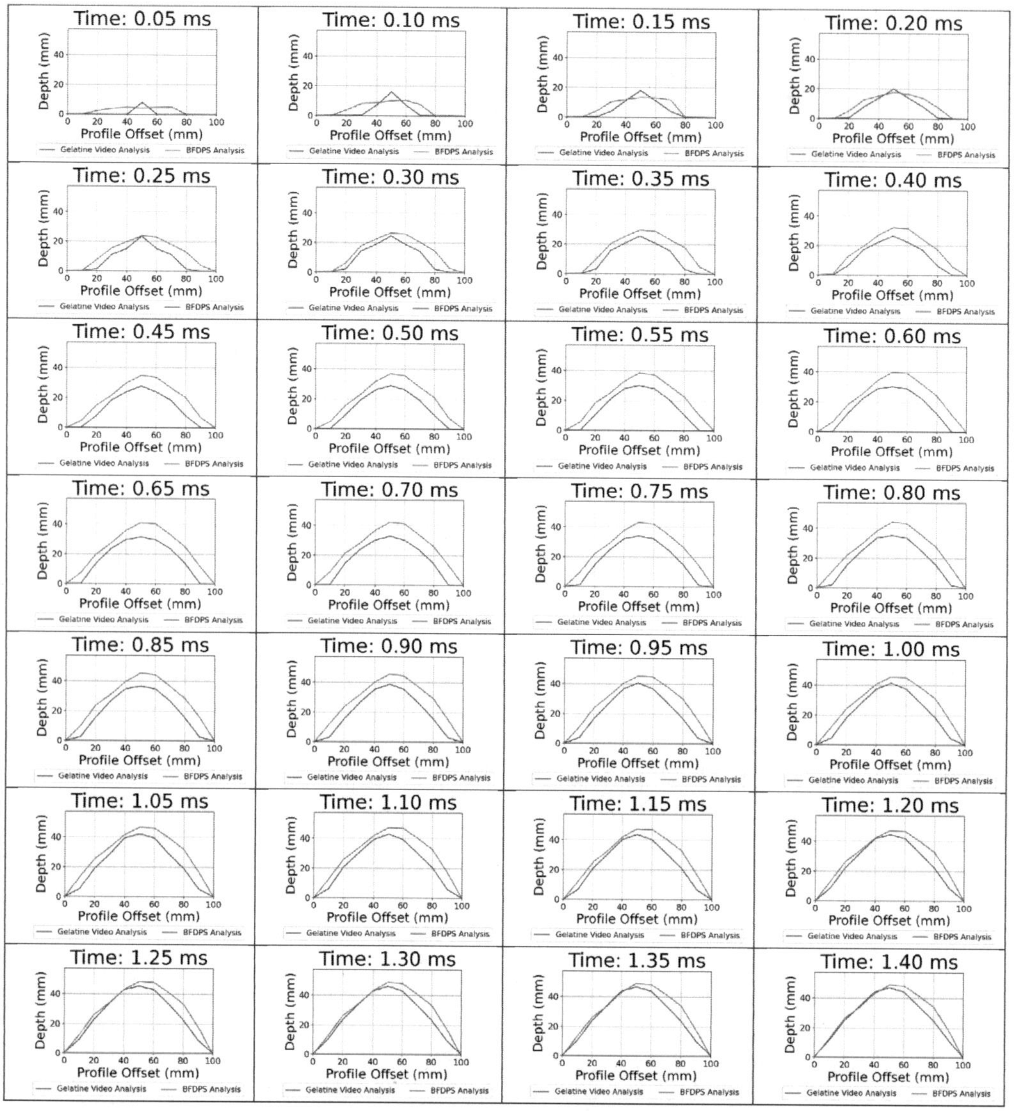

Figure 10. Profile histories comparing the gelatine video (blue) and the BFDPS (red) analyses.

The profile history comparisons between the gelatine and BFDPS analyses are shown in Figure 10. The BFDPS optimisation depicts an initially wider and shallower deformation profile than the gelatine analysis. It is noteworthy that the camera view of the BFDPS showed movement over a 70 mm span (a span containing six wires), while the gelatine analysis showed movement over a 20 mm span (a span containing the equivalent of only one wire). Although strain relief cuts were introduced into the BFDPS weaving fabric, it is possible that once a single wire of the BFDPS was loaded by the armour, ancillary loading paths through the BFDPS fabric and grid could induce anomalous wire movement of neighbouring wires not yet directly deformed by the armour back-face. Beyond the relatively coarse initial profile shapes, the BFDPS profile shape tracked the gelatine well.

The BFDPS does have limitations. The BFDPS cannot detect or represent deformation that is undercut (e.g., mushroom capping) or purely vertical. The resolution of the BFDPS is limited to the spacing of the grid fibres, which itself is practically constrained by frictional binding when fibres are too closely spaced, and the temporal resolution of the high-speed cameras. Frictional binding can be reduced by increasing the spacing between over-under weaves, if implemented. The BFDPS wire thickness may need to be adjusted to accommodate changes in back-face loading profiles. For example, the researchers in this study found they had to increase the wire sectional size as the threat level and anticipated response increased to prevent damaging the wires. In cases where there are substantial strains in the back-face surface, the fibre intersections may deviate laterally during back-face deformation, particularly when large back-face strains are present. The optimization algorithm does specifically not account for potential lateral offsets, which is an area for future consideration.

4. SUMMARY

This experiment and analysis demonstrated a novel method for capturing 3-D contour histories of surfaces that may be completely visually obscured. Capturing the dynamic profile is important for understanding the potential for injury behind armour systems [9]. This method fills a current gap for behind-armour blunt impact injury researchers who are essentially blind to the insulting dynamics of the armour back-face during a ballistic impact that is backed by an opaque backing material. It uses readily available, low-cost materials, other than a high-speed camera, which many ballistic facilities already possess. This study demonstrated the concept using a handgun threat impacting soft body armour. Subsequent testing by the authors has shown the system has utility with rifle threats impacting more robust armour systems. Those findings are beyond the scope of the present study and warrant a separate report.

Acknowledgments

The authors would like to thank Robert MacDonald for his testing expertise and support throughout the development of the BFDPS.

The research reported in this document was performed in connection with contract/instrument W15P7T-19-D-0126 with the Combat Capabilities Development Command (DEVCOM) Army Research Laboratory.

The findings in this report are not to be construed as an official Department of the Army position unless so designated by other authorized documents. The views and conclusions contained in this document are those of SURVICE Engineering Company, LLC and the DEVCOM Army Research Laboratory.

Conflicts of interest: The BFDPS is under patent pending by SURVICE Engineering Company, LLC with authors A. Goertz, A. Brown, and K. Rafaels as co-inventors.

References

[1] L. Xu, J. Ge, J. H. Patel, and M. P. Fok, Dual-layer orthogonal fiber Bragg grating mesh based soft sensor for 3-dimensional shape sensing, *Optics Express,* vol. 25, no. 20, pp. 24727-34, 2017.

[2] D. A. Hackney, T. Goode, F. Seng, S. Schultz, M. Pankow, and K. Peters, In-situ strain measurement of ballistic fabrics during impact using fiber Bragg gratings, *Optical Fiber Technology,* vol. 59, pp. 1-9, 2020.

[3] F. Seng *et al.*, Dynamic back face deformation measurement with a single optical fibre, *Int J of Impact Engineering,* vol. 150, pp. 1-9, 2021.

[4] I. Velasco, P. Spackman, and S. Schultz, Body armor shape sensing using fiber Bragg gratings, presented at the Intermountain Engineering, Technology and Computing (IETC), 2020.

[5] K. A. Rafaels, K. Choi, G. Glasser, and C. Bir, "Estimation of Armour Backface Velocity," in *Proceedings of the Personal Armour Systems Symposium 2020*, Copenhagen, Denmark, Oct 11-15, 2021.

[6] D. A. Murio, Automatic numerical differentiation by discrete mollification, *Computers & Mathematics with Applications,* vol. 13, no. 4, pp. 381-386, 1987.

[7] F.-F. Dou, C.-L. Fu, and Y.-J. Ma, A wavelet-Galerkin method for high order numerical differentiation, *Applied Mathematics and Computation,* vol. 215, no. 10, pp. 3702-3712, 2010.

[8] T. Wei, Y. C. Hon, and Y. B. Wang, Reconstruction of numerical derivatives from scattered noisy data, *Inverse Problems,* vol. 21, no. 2, p. 657, 2005.

[9] K. A. Rafaels, "Importance of Areal Measurements for Thoracic BABT Test Devices," in *Proceedings of the Personal Armour Systems Symposium 2016*, Amsterdam, Netherlands, Sep 19-23, 2016.

https://doi.org/10.52202/081621-0034

Comparing Laboratory Blast Simulators and Exploring Influence on Helmet Test and Evaluation

A. Lurski[1], J. Clark[1], G. Holt[1], L. Reider[1], B. Arnold[1], V. Alphonse[1], P. VandeVord[2], V. Sujith Sajja[3], A. Iwaskiw[1]

[1]*The Johns Hopkins University Applied Physics Laboratory (JHU/APL), 11100 Johns Hopkins Road, Laurel, MD 20723, alex.lurski@jhuapl.edu*
[2]*Virginia Polytechnic Institute and State University (VT), 2280 Kraft Ave, Blacksburg, VA 24061*
[3]*Walter Reed Army Institute of Research (WRAIR), 503 Robert Grant Ave, Silver Spring, MD 20910*

Abstract. Preventing warfighter exposure to primary blast waves (BW) is important to mitigate injuries. Experimentally recreating BW is necessary for test and evaluation (T&E) of head protection systems (HPS) [1,2]. Advantages of laboratory blast simulators (LBS) over live-fire blast testing include reliability, repeatability, safety, and cost. However, there exists a paucity of cross-comparisons between different LBSs [3]. The goal of this study is to investigate if different LBSs can achieve comparable T&E results for the purposes of HPS performance by targeting the same peak incident overpressure at the test subject location. Two target peak incident overpressure conditions were tested with four different LBSs. Differences in these LBSs include geometry and shockwave generation method (diaphragm or gas-detonation) and each LBS was characterised with the same pressure sensor array. A head surrogate was added to understand differences between LBSs when evaluating HPSs. Measured incident peak overpressures across LBSs neared the nominal target; however, statistical differences were observed in a portion of sensor locations when cross-comparing LBSs. Positive phase duration and impulse varied significantly across LBSs. Achieving matching peak incident pressure and positive phase impulse simultaneously was not reconciled due to the need for costly permanent changes to the LBSs. A scaling methodology to predict the effect of two different helmets at an LBS based on results at another LBS was explored, but was not able to overcome differences. It was found that the differences in the generated shockwave characteristics by the LBSs, not differences in helmet design, drove discrepancies in helmet evaluation. While no agreed-upon injury risk curves (IRCs) currently exist for blast-induced traumatic brain injury (bTBI), standardised test metrics would help motivate the different research teams to generate similar shockwaves from different LBSs to achieve reproducible test results to assess HPSs.

1. INTRODUCTION

Blast injuries are a concern for the warfighter, and the study of blast-induced traumatic brain injury (bTBI) is becoming increasingly important to the design and evaluation of head protection systems (HPS) [1,2]. Live-fire blast events using high energy explosive materials and high-fidelity measurement systems are essential to evaluate HPS performance; however, there are many challenges associated with live-fire blast testing including repeatability, reproducibility, cost, safety, skilled personnel and environmental control. Laboratory testing has historically been able to create complex loading events in a repeatable and reproducible way that is well suited for HPS testing. Therefore, the ability to replicate realistic blast overpressure conditions in the laboratory through the use of laboratory blast simulators (LBS) is of great interest to support efficient and effective testing and evaluation of HPSs.

Several organizations have conducted LBS testing to evaluate the performance of HPSs using different LBS designs, instrumentation, and data analysis methods [2,3,4]. While these studies have done much to characterise the individual LBS, little work has been done comparing different LBSs using near-identical data collection methods or test articles. Studies comparing various LBSs have been limited due to the complex nature of shockwaves as well as challenges associated with coordination across different institutions. There is a desire for a standardised test methodology to assess blast overpressure attenuation performance of HPSs, and the feasibility of comparing test results across different LBSs is unknown. In an effort to investigate if different LBSs can achieve comparable HPS test results, experiments were conducted to determine if HPS testing could be conducted with consistent inputs (e.g. consistent shockwave metrics) and consistent outputs (e.g. HPS blast attenuating performance metrics) across four disperate LBSs.

https://doi.org/10.52202/081621-0035

2. METHODS

The Johns Hopkins University Applied Physics Laboratory (JHU/APL) led this effort to evaluate LBSs at four different partner institutions; JHU/APL [2,4,5], Aberdeen Test Center (ATC), Virginia Polytechnic Institute and State University (VT) [6], and Walter Reed Army Institute of Research (WRAIR) [7]. Table 1 summarises key differences in the LBS design, shockwave generation, and shockwave expansion.

Table 1. LBS Design Metrics

	JHU/APL	ATC	VT	WRAIR
Cross-sectional dimensions	91 x 91cm	71 x 71cm	119 ×122cm	60 x 60cm
Driver gas used for this study	Air	Helium	Oxyacetylene	Air
Shockwave generation method	Diaphragm-rupture	Diaphragm-rupture	Ignition- gas combustion	Diaphragm-rupture
Driven chamber expansion direction	Horizontal only	Horizontal and vertical	Horizontal only	Horizontal only
Design paradigm	Advanced Blast Simulator [8]	Shock Tube	Advanced Blast Simulator [8]	Advanced Blast Simulator [8]

In the first phase of the study, tests were conducted with indentical experimental and analytical methods in order to characterize effects of different LBS designs and assess the consistency of the shockwaves generated at two different pressure conditions, as outlined in Table 2. This approach enables a direct comparison of shockwave variations across LBSs to inform future test standardization. Each LBS was tuned to produce peak incident pressures of 69 and 103 kPa measured at the center of the LBS. These two pressure values were chosen because each LBS in the study was already capable of achieving these conditions and because the conditions were deemed relevant to blast dosages where HPSs could potentially make a difference in health outcomes as these levels are below the threshold for lung injury [9]. Limited repeats were conducted at VT at 69 kPa due to the original test plan prioritizing an alternative test condition.

A custom designed array of pressure sensors was mounted to the test region of each LBS and used to quantify the shockwave characteristics of these four LBSs. Tests at JHU/APL and VT were conducted using the nine-probe-array configuration, as shown in Figure 1, whereas the five-probe-array in Figure 1 (outlined in the red box) was used at ATC and WRAIR due to the smaller cross-sectional areas of these LBSs. The probe-array consists of free-standing aerodynamic pitot-static probes [10] held in place by aerodynamic rigid stands of specified lengths. Pitot-static probes measure incident and stagnation pressure time-histories by utilizing two perpendicular pressure sensors. The stands were mounted to steel plates on the floor and ceiling of the LBSs. A coordinate measurement machine (ROMER Absolute 7535, Hexagon Metrology) was used to check sensor locations after installation. Endevco (END) 8530C-100 (piezoresistive) and PCB Piezotronics 113B27/113B21 (piezoelectric) sensors were utilised in different pitot-static probe locations in the custom pressure sensor array. Tests were executed in sets of 5, which was informed by a design of experiments to increase testing efficiency and to reduce the effects of confounding variables such as test day, time, humidity, and temperature. If an LBS misfire, data loss, or other anomaly was detected on the test day, the shot condition was repeated. However, if the issue was not identified within the test period, the test was not repeated. Pressure measurements from the sensors were sampled at 1MHz and then filtered using a 40 kHz low-pass Butterworth filter.

In the second phase of the study, the Wireless Blast Anthropomorphic Test Device (WBATD) Headform (Figure 2) [11] was used, in lieu of the pressure probe array, to assess the consistency of the headform response and HPS performance metrics across different LBSs. The WBATD intracranial pressure (ICP) measurements were the primary metrics assessed in both barehead and helmeted configurations. Two different combat helmet systems were used (Helmet A and Helmet B). The WBATD Headform was mounted to a Hybrid-III neck in the forward-facing orientation in the test region of the LBSs [4]. A pitot probe was also positioned in the inner bottom left of the LBS for the headform characterization tests at JHU/APL, and in the inner bottom left and inner top left at WRAIR. Due to resource constraints, WBATD tests were only conducted at JHU/APL and WRAIR, which were downselected from the four LBSs due to similarities in their design and performance. Tests were performed as outlined in Table 3.

https://doi.org/10.52202/081621-0035

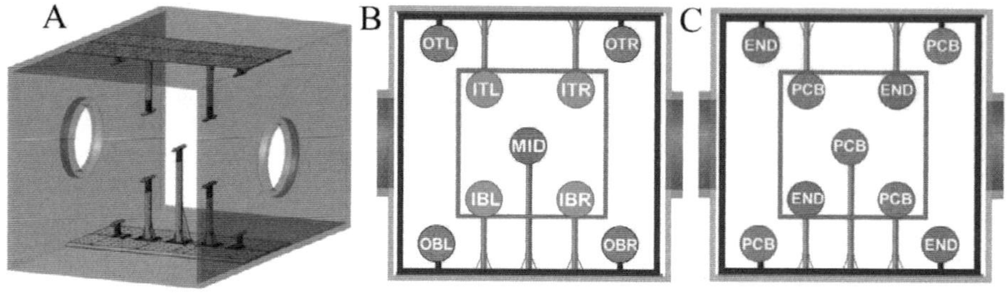

Figure 1. Nine Probe-Array Configuration (A), Pressure Sensor Naming Convention (I/O – Inner/Outer, T/B – Top/Bottom, L/R – Left/Right, and MID – Middle) (B) and Pressure Sensor Layout (C), five probe array configuration indicated by red box

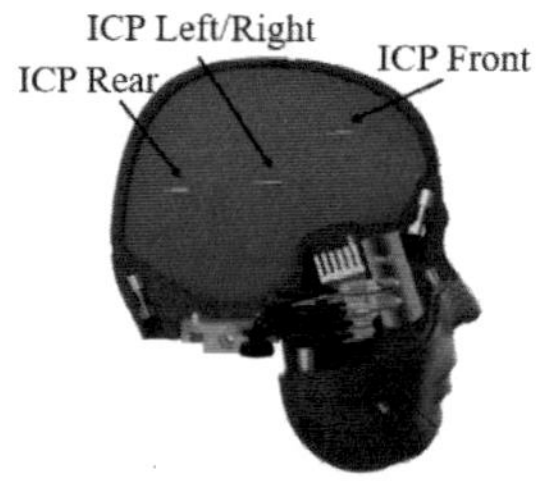

Figure 2. CAD Rendering of WBATD Headform showing the ICP sensor locations. The ICP Left and ICP Right sensors are symmetrical with respect to the mid-sagittal plane

Table 2. Probe-Array Test Matrix

Target Incident Pressure (kPa)	JHU/APL	ATC	VT	WRAIR
69	15	15	4	15
103	25	23	25	25

Table 3. Headform Test Matrix

Target Incident Pressure (kPa)	HPS Configuration	JHU/APL	ATC	VT	WRAIR
69	Barehead	5	0	0	5
103	Barehead	5	0	0	5
69	Helmet A	5	0	0	5
103	Helmet A	5	0	0	5
69	Helmet B	5	0	0	5
103	Helmet B	5	0	0	5

These experimental methods allow for a quantitative evaluation of features of the shockwaves generated in each LBS and establish a baseline for use in the development of LBS helmet testing standards. Additionally, for the first time, the same experimental (i.e sensor type, placement, headform surrogate) and analytical (i.e. data processing and feature extraction) methods were used across these disparate LBSs, enabling direct comparison and a better understanding of the differences between the shock waves simulated by each system.

3. RESULTS

The time history plots in Figure 3 and Figure 4 demonstrate the propagation of the shockwave across the middle probe of the probe-array at the two test conditions. The middle probe was selected as a representative sensing location as it is the farthest from walls and corners (and associated predicted reflections) of the tube. Side-facing sensors were selected as they measure incident overpressure, which the laboratories in this study utilise to drive their nominal pressure conditions. The comparison corridors for each LBS represent the average time history trace plus or minus one standard deviation. Each laboratory demonstrates a Friedlander wave shape, with a sharp rise to the peak overpressure and a decay into a negative phase before the pressure returns to zero.

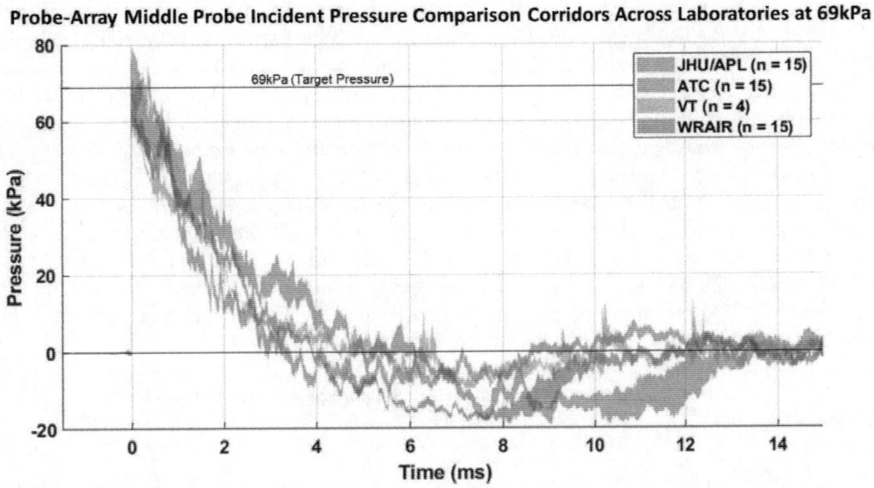

Figure 3. Probe-array Overpressure Comparison Corridors for JHU/APL, ATC, VT, and WRAIR, 69-kPa Nominal Target Overpressure Condition, Middle Probe, Side-Facing Sensor

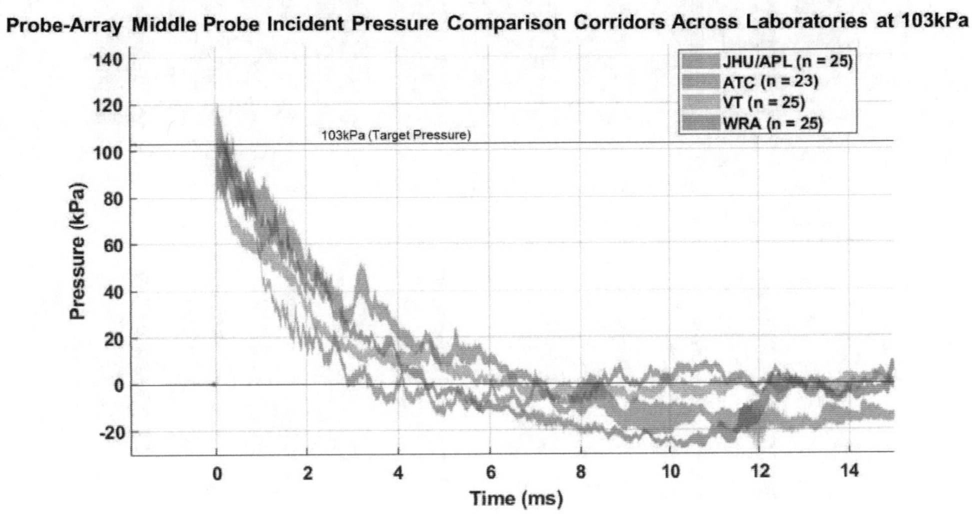

Figure 4. Probe-array Overpressure Comparison Corridors for JHU/APL, ATC, VT, and WRAIR, 103-kPa Nominal Target Overpressure Condition, Middle Probe, Side-Facing Sensor

Peak overpressure, positive phase duration, and positive phase pressure impulse features were extracted from the 40 kHz filtered pressure sensor time histories of the side-facing sensors in the pitot probe-array (Figure 5). Peak pressure was defined as the signal's maximum value. The start of the positive phase duration (or time of arrival) was defined by over-filtering (10 kHz low-pass Butterworth filter) the signal and determining the time point at which the signal reached 10% of the peak value. The end of the duration was the time point at which the over-filtered signal crosses zero for the first time after the time of arrival. The pressure impulse was then computed by integrating the 40 kHz filtered pressure time history across the positive phase duration. Cases in which the automated featured extraction did not capture the actual peak as a result of some anomaly in the data were corrected manually and the duration and pressure impulse were then recalculated.

Peak incident pressure measurements were used to evaluate the repeatability of a given LBS, the shockwave uniformity of a given LBS, and the ability of each LBS to achieve the target peak incident pressure conidtion. Of the four LBSs evaluated, ATC had the most repeatable peak incident pressure measurements, with a mean coefficient of variation (CV) of 3% across sensor locations and across test

 https://doi.org/10.52202/081621-0035

conditions. JHU/APL's LBS had the least repeatable peak incident pressure, with mean CVs across sensor locations of 13% at 69 kPa and 9% at 103 kPa. ATC's LBS was the least uniform, with a large difference in peak incident pressure between the middle probe and the peripheral probes (25 kPa mean difference at the 69kPa test condition and 39 kPa mean difference at the 103kPa test condition). The JHU/APL, VT, and WRAIR shockwaves were generally more uniform in comparison. All four LBSs achieved mean middle probe peak incident pressure measurements within 10% of the target values.

The positive phase duration and impulse varied greatly between the four LBSs. Positive phase impulse is affected by both the magnitude and duration of the pressure wave. Since the peak pressure was targeted to be the same in the test conditions across the different laboratories, positive-phase duration was the main driver of impulse differences between the different LBSs. As shown in Figure 5, JHU/APL had the longest duration shockwaves and, as a result, it had the higest pressure impulse values across all sensor locations and test conditions. In contrast, ATC had the shortest duration shockwaves and lowest pressure impulse values across all sensor locations and test conditions. The exception to this trend is the VT LBS, which ranks the second highest for duration but third highest for impulse. The differing wave shape generated by VT drives this phenomenon, primarily due to VT's shockwave generation method which utilizes a gas detonation system, unlike JHU/APL, WRAIR, and ATC.

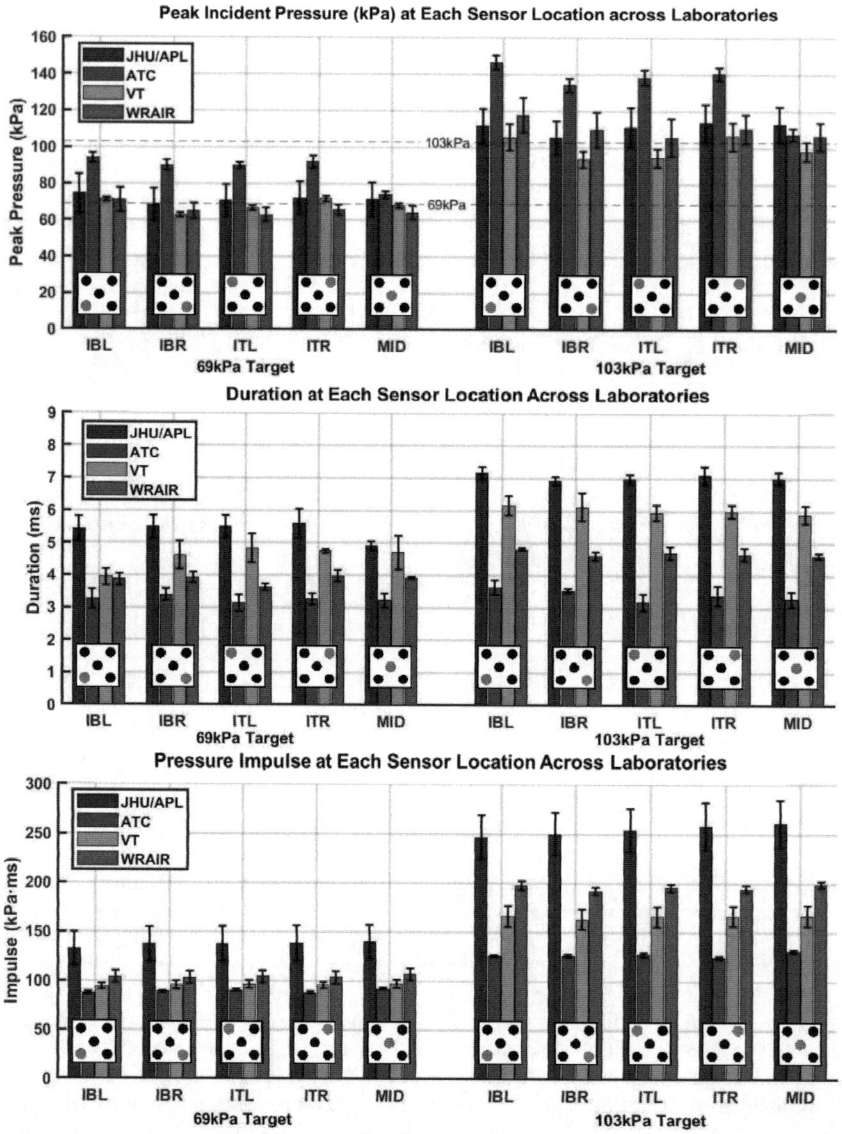

Figure 5. Probe-array Incident Pressure Metrics Comparison Across Sensor Locations for JHU/APL, ATC, VT, and WRAIR

Analysis of Variance (ANOVA) tests were conducted with both stagnation and incident peak pressures to determine if there were statistical differences between the LBSs. A Tukey's honestly significant difference (HSD) procedure, with alpha (**α**) set to 0.05, was conducted to determine which specific groups were statistically different from one another. To control for family-wise error rate, a Bonferroni p-value correction (*m*) was implemented. The null hypothesis is the peak incident pressure is statistically the same across laboratories. If the null hypothesis was not rejected, the box is grey. If the null hypothesis was rejected and the lab listed first had a greater metric than the lab listed second, the box is orange. If the lab listed second had a greater metric than the lab listed first, the box is green. The VT 69-kPa condition was not included due to only 4 repeats completed, noted in black.

Figure 6 and Table 4 show the results of the ANOVA tests. In the 69-kPa condition, JHU/APL, ATC, and WRAIR were found to be statistically similar at the middle pressure probe. At three of the four inner probes, JHU/APL and WRAIR were statistically similar. As expected, ATC's inner probe measurements were statistically different than JHU/APL and WRAIR. At 103-kPa, no laboratories were statistically similar across all five sensors. At the middle sensor, the null hypothesis was only rejected when comparing ATC and VT's peak incident pressures (Table 4).

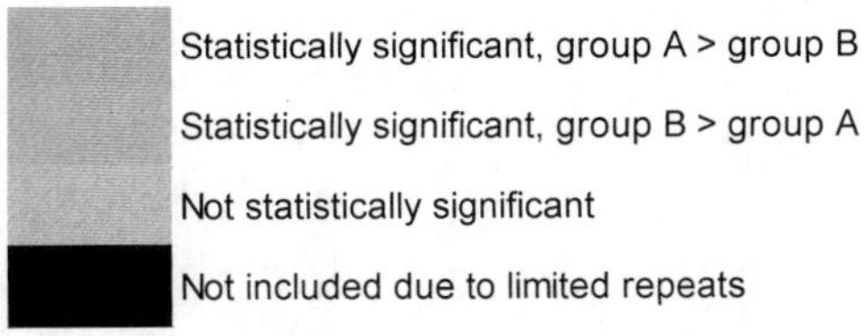

Figure 6. ANOVA and Tukey's HSD Table Legend

Table 4. Wave Metric ANOVA Tukey's HSD Results

	Incident Overpressure Filtered Peak (69-kPa)					
IBL						
IBR						
ITL						
ITR						
MID						
	Incident Overpressure Filtered Peak (103-kPa)					
IBL						
IBR						
ITL						
ITR						
MID						
	APL vs. ATC	APL vs. VT	APL vs. WRAIR	ATC vs. VT	ATC vs. WRAIR	VT vs. WRAIR

The second phase of the study (Table 3) was conducted to assess the effect of different LBSs and different helmets on the WBATD headform's intracranial pressure (ICP) measurements. JHU/APL and WRAIR LBSs were selected for the study due to the similarity in their design, other than cross-sectional area, and the statistical similarity observed at seven of ten peak incident pressure readings across all probes. The WBATD headform ICP peak pressure measurements and presure impulse measurements are shown in Figure 7 and Figure 8, respectively. For the peak ICPs, JHU/APL was greater than WRAIR at seven of the eight test conditions and sensor locations, with the lone exception being the right sensor at 69kPa. For comparison, the mean difference across the four sensors in peak ICP while targeting 69kPa was 16 kPa, and while targeting 103kPa was 8 kPa. The mean difference in the pressure probe peak pressure measurements was 7 kPA at the middle sensor while targeting both 69kPa and 103kPa.. For the positive impulse of the ICPs, JHU/APL was greater than WRAIR at all test conditions and sensor locations. This is driven by the higher positive phase duration and impulse at the JHU/APL LBS. The mean difference across the four sensors in impulse was 26 kPa*ms at 69kPa and 35 kPa*ms at 103kPa, whereas the mean difference in the pressure probe measurements was kPA*ms at the middle sensor was 32 kPA*ms for 69kPa and 62 kPA*ms for 103kPa.

https://doi.org/10.52202/081621-0035

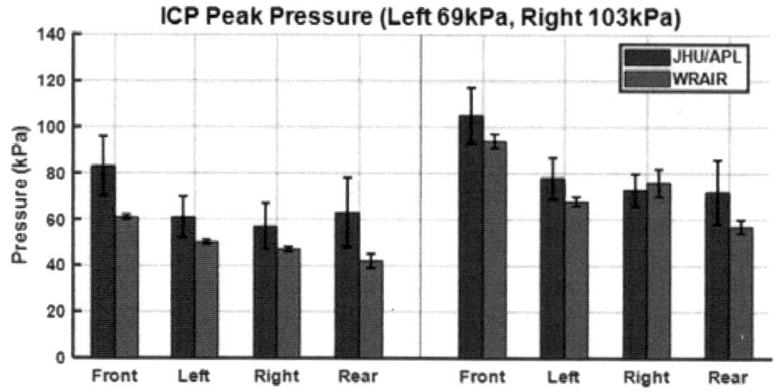

Figure 7. WBATD Peak ICP at 69kPa (left) and 103kPa (right) target peak incident pressure test condition for Barehead at JHU/APL and WRAIR

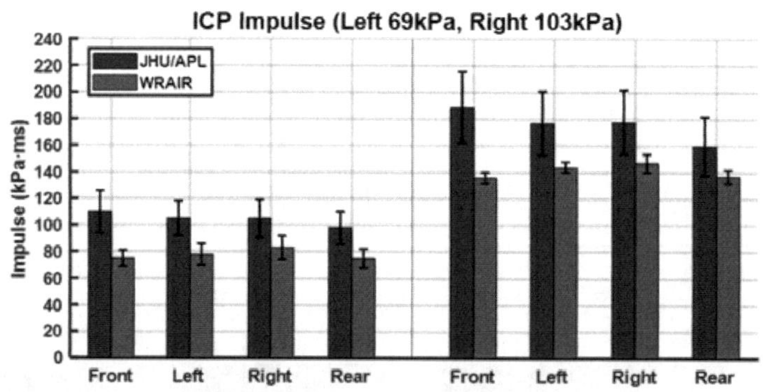

Figure 8. WBATD ICP Impulse at 69kPa (left) and 103kPa (right) target incident pressure test condition for Barehead at JHU/APL and WRAIR.

The helmeted headform conditions exhibited varying responses across the JHU/APL LBS and the WRAIR LBS. To separate the effects of different HPSs, a scaling methodology was applied. Specifically, the mean ICP metrics in the helmeted conditions were normalise against (divided by) the corresponding mean ICP metric measured on a bare headform under the same conditions. A two-tailed independent t-test with alpha (**α**) set to 0.05 was conducted to determine if the normalised ICP measurements were statistically different between the two helmets. The null hypothesis is the normalised ICP measurements is statistically the same across Helmet A and Helmet B. If the null hypothesis was not rejected, the box is grey. If the null hypothesis was rejected and Helmet A outperformed (i.e. resulted in lower normalised ICP measurements than) Helmet B, the box is orange. If the null hypothesis was rejected and Helmet B outperformed Helmet A, the box is green (Figure 9). In the 69-kPa condition, Helmet B showed better performance in six of the eight normalised ICP metrics using the JHU/APL LBS, while Helmet A showed better performance in seven of the eight normalised ICP metrics using the WRAIR LBS. Evaluated at 103-kPa nominal dose, the performance is split with seven of the eight normalised ICP metrics across laboratories not having statistically significant differences using the JHU/APL LBS. Four of the eight normalised ICP metrics are statistically different at WRAIR, with Helmet A having better performance at the left and right normalised peak metrics but Helmet B having better normalised performance at both rear metrics (Table 5).

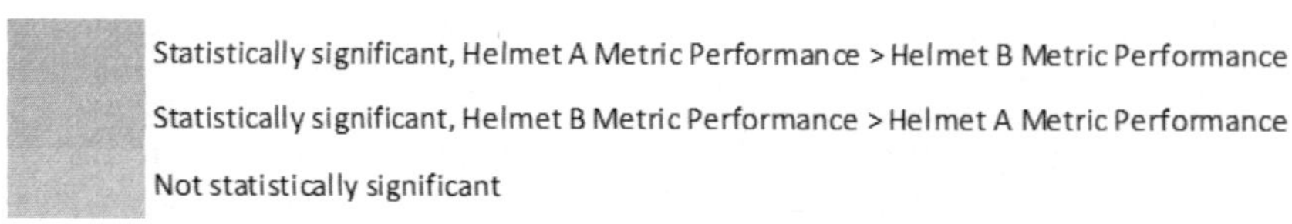

Figure 9. Helmet Comparison T-test Legend

Table 5. Helmet Comparison T-test Results

		Front Peak	Left Peak	Right Peak	Rear Peak	Front Imp	Left Imp	Right Imp	Rear Imp
69 kPa	JHU/APL								
	WRAIR								
103 kPa	JHU/APL								
	WRAIR								
		Helmet A ICP Metric Performance vs. Helmet B ICP Metric Performance							

4. DISCUSSION

4.1 Shockwave Characterization with Probe-Array and Headform

In an attempt to standardise across different LBSs, the purpose of the study was to determine if response characteristics of four disperate LBSs could be considered similar enough for consistent evaluation of HPSs. It is evident from the results that each LBS can achieve defined target peak overpressures at the centre sensor location despite the unique geometry and shockwave generation characteristics of each system. While each institution had demonstrated this prior to the current study, different data collection methods and shockwave initiation methods show a clear affect to the creation of this peak pressure as shown by the variation in responses across the tube. Furthermore, it was found that positive phase duration and positive phase impulse varied significantly across laboratories, as shown in Figure 5.

The ANOVA tests comparing peak incident pressure measurements across laboratories showed that while each laboratory was close to the target peak incident overpressure at the middle of the test region, statistical differences are prevalent between the different systems. This study made clear that standardised experimental and analytical methods must be used to properly characterise lab-generated shockwaves to compare their similarity, and comparing response at a single point in these systems is insufficient. Additionally, multiple features of the lab-generated shockwave must be considered when evaluating the similarity of LBSs. As shown later by the differences in headform response, targeting a single feature such as peak incident pressure is not sufficient for comparing systems. LBSs in this study acheived specific peak overpressures, but key differences in duration (and to a lesser extent, wave shape) drove differences in impulse. In the future, multiple metrics must be reported in research publications as well as test standards to fully communicate the simulated blast exposure achieved by LBSs.

Repeatability is an essential determinant for the utility of a test device. At both WRAIR and JHU/APL, the WBATD bare headform response was repeatable, with peak pressure and impulse metrics having standard deviations similar to those of the probe-array sensor metrics, indicating that shot-to-shot variability in the WBATD response was driven in large part by shot-to-shot variability in the LBSs. Based on these results, if it is determined that ICP is correlated with blast injury, the WBATD or similar headform composed of ICP sensors could be a useful test equipment for standardized HPS testing. Significant challenges arose during this effort when assessing helmet performance between LBSs based on ICP metrics. These challenges are not yet well understood.

4.2 Observed Helmet Performance

A successful scaling methodology could allow for the comparison of helmet performance from one LBS to another where the blast overpressure signature or the headform response are different. Additionally, a valid scaling methodology would demonstrate evidence that the differences observed in pressure metrics or helmet performance between LBSs were due to predictable physical phenomenon. A hypothetical ideal scaling methodology would look something like "HPS tend to show 20% greater blast attenuation at 'Lab B' compared with 'Lab A' due to the positive phase impulse of the shockwave generated at 'Lab B' being 10% greater than that of 'Lab A'. For a standardised test methodology utilised across multiple LBSs to be useful, the results must forecast— either directly or through a transfer function— how HPSs would perform at 'Lab B' given only performance data measured at 'Lab A'.

Of the LBSs investigated in this effort, WRAIR and JHU/APL had the most a similar design, with the driving difference being the larger cross-sectional area at JHU/APL. JHU/APL and WRAIR also had statistically similar nominal peak incident pressures when targeting 69-kPa and 103-kPa at seven of the ten possible sensor locations. However, JHU/APL had longer duration and higher impulse. In addition, all the data has used incident pressure as a basis for analysis, it is yet unclear the role of stagnation pressure in affecting these variables. When assessing pressure transmission to the helmeted WBATD brain surrogate across WRAIR and JHU/APL using ICP sensor measurements, convoluting

https://doi.org/10.52202/081621-0035

factors between LBS characteristics and helmet performance became apparent. No trends that could be leveraged for a scaling methodology were observed across LBSs or blast doses. The underlying physics of the headform, brain simulant, and sensor response are complex and not yet well understood. The differences in the LBS design metrics (Table 1) were more significant relative to the differences in ICP metric performance of Helmet A and Helmet B. This demonstrates that simply defining a target peak incident overpressure is not sufficient in defining standard testing conditions for consistent evaluation of HPSs across LBS that differ in nominal test conditions to the degree of the LBSs in this study.

4.3 Limitations and Future Work

Many wave characteristics (e.g., stagnation overpressure metrics, negative phase waveform metrics) are not currently captured by this shockwave characterization and may be an important component to describing the shockwaves at each lab. The significance of wave non-uniformity (differing wave metrics across the plane of the test region) or wave non-planarity (differing time of arrival across the plane of the test region) on headform response and HPS performance is not yet understood and requires further investigation. In addition, the difference in stagnation pressure to account for the effect of "blast wind" on these variables needs to be taken into account. Future work for test standardization may necessitate developing requirements for an acceptable level of wave uniformity and planarity. A future investigation could conduct statistical tests to compare the outer probe shockwave features to the middle probe shockwave features to better assess the uniformity of the shockwaves. Time of arrival analysis across the probe locations would allow for a better understanding of the planarity across LBSs. This study utilised two different sensing modalities—piezoelectric and piezoresistive pressure transducers—installed in the probes. While the data collected presents an opportunity to further understand different sensing modalities for blast pressure measurements, this study has not looked at differences between these sensors, and data from both sensor types are currently being treated as interchangeable and comparable.

A key limitation is the scale of the headform test series. Larger sample sizes across more LBSs with additional test conditions could provide further evidence that simply defining a target peak incident overpressure is not sufficient in defining standard test conditions for consistent evaluation of HPSs. Helmet A and Helmet B were not designed for blast attenuation. Tested in the forward orientation, they have relatively small blast attenuation effect sizes when compared to other HPS conditions (e.g., rear orientation, helmet with visor), as they do not directly interfere the blast wave prior to it impacting the face of the headform. It may be the case that HPSs exhibiting a larger effect size would show clearer trends agnostic of the LBS. A future study could test additional headform orientations or configurations with larger effect sizes to better determine scalability or comparison of helmet performance between labs. Furthermore, normalizing headform data was considered to address shot-to-shot variance, but differences in pitot probe measurements during headform testing stemmed from reflections off of the test article rather than blast overpressure variations. Further studies could optimize probe placement or explore alternative measurement methods to improve shot-to-shot normalization of headform data.

At this time, tuning two LBSs to better match both peak pressure and impulse is a significant challenge. An encouraging study demonstrated the ability of tuning both peak pressure and positive phase impulse, however, this methodology of tuning would require permanent modifications to the LBS, and these modifications may negatively affect wave uniformity and planarity [12]. It is also unclear to what extent a LBS shockwave needs to perform similarly to a live-fire wave to provide operationally relevant simulations of blast in the context of HPS performance. Complimentary live-fire testing would provide insight in deconvoluting the effect of the helmet and the LBS. A future effort could aim to tune a LBS to closely match different live-fire test conditions, and then directly compare live fire and LBS HPS results.

A 2D computational fluid dynamics (CFD) model was completed to asses LBS and test article size influence on overpressure loading [13]. Building upon that study, 3D CFD models of LBSs using realistic headform and HPS geometries could assist in deconvoluting variables influencing 'helmet effects' observed in this study, while also offering the opportunity to be validated experimentally. These higher fidelity models could also help differentiate shockwave loading effects purely from geometric differences and effects from difference in the wave generated. This may provide valuable information regarding a future tuning effort, or development of future LBSs.

Without clear metrics for standardised testing in LBSs, there has been less motivation to fully explore extensive modifications to tune LBSs. This problem is exacerbated by the current lack of blast injury risk curves (IRCs). Blast IRCs would inform and motivate efforts to tune multiple LBSs to a point where they are similar enough to be used in conjunction with one another. Some key performance issues, such as the lengthy blast duration generated by JHU/APL's LBS, or the non-uniform wave generated at ATC, could be the focus of future tuning efforts; however, without IRCs it is unclear how important these differences are in the context of the evaluation of HPSs regarding blast injury. The data collected in this effort could be revisited after the development of an IRC and provide meaningful context for future blast standardization methods. This data could also influence the design of future LBSs, head surrogates, and HPSs as it is the first investigation of its kind comparing probe-array pitot probe data and headform characterization data across a variety of LBSs.

5. CONCLUSIONS

The first phase of this study demonstrated that the four disparate LBSs varied drastically in duration, impulse, and other characteristics despite having simlar peak pressure values. Measures of repeatability, reproducibility, and shockwave planarity varied between the different LBSs. Headform and helmet testing with the two more similar LBSs resulted in consistent results within a LBS but inconsistent results between LBSs, and no trends could be leveraged to develop a scaling method to compare test results between LBSs. Simply defining a target peak overpressure is not sufficient in defining standard test conditions across different LBSs. Tuning the disparate LBSs to match in more than one metric was challenging and not explored fully due to resource constraints. While this study does not rule out the feasibility of using a combination of tuning and scaling techniques to enable HPS testing across LBSs, it does provide evidence of many challenges with that approach. This study also does not challenge the ability of LBSs to be used for standardized HPS testing, especially if a single LBS or very similar LBS designs are used. One major outcome of this study is the development of a reproducible test method and pressure probe array that can be used as a standarized method to characterize different LBSs and to compare the characteristic responses of other LBS to those of the four LBSs evaluated in this study.

Acknowledgments

The authors would like to thank the PEO Soldier Product Manager Soldier Protective Equipment for sponsoring this effort and Aberdeen Test Center for participating in the study. Any opinions, findings and conclusions or recommendations expressed in this material are those of the author(s) and do not necessarily reflect the views of PEO Soldier, Aberdeen Test Center, or the Naval Sea Systems Command(NAVSEA).

References

[1] P. Hayward, "Traumatic brain injury: the signature of modern conflict," The Lancet Neurology, 7(3), pp. 200–201, 2008

[2] V. Alphonse, "Helmet blast attenuation performance," Personal Armour Systems Symposium, 2021

[3] E. Needham et al., "Blast testing issues and TBI: experimental models that lead to wrong conclusions," Frontiers in Neurology, 6(72), 2015

[4] V. Alphonse, "Effect of helmet and eyewear on headform kinematic response to primary blast overpressure exposure," Personal Armour Systems Symposium, 2018

[5] C. Carneal, "Development of a laboratory shock tube system for helmet blast overpressure performance assessment," Personal Armour Systems Symposium, 2016

[6] A. Nelson et al., "Characterization of an advanced blast simulator for investigation of large scale blast traumatic brain injury studies," Annals of Biomedical Engineering, 53(1), 2024

[7] V. Sajja et al., "Rodent model of primary blast induced traumatic brain injury: guidelines to blast methodology," Pre-Clinical and clinical methods in brain trauma research. Neuromethods, Volume 139, Humana Press, New York, NY (2018)

[8] D. Ritzel, S. Parks, "Shock tube apparatus for blast wave simulation," United States Patent US20130042665A1. United States Patent and Trademark Office. 2015

[9] A. Courtney and M. Courtney, "The complexity of biomechanics causing primary blast-induced traumatic brain injury: a review of potential mechanisms," Frontiers in Neurology, 6, 2015

[10] T. D. Josey, "Development of a miniature double Pitot-static probe and its application to calibrating blat flow conditions," 24th International Symposium on Military Aspect of Blast and Shock, Halifax, Canada, 2016

[11] J. Clark, "A novel repairable wireless blast anthropomorphic test device," Personal Armour Systems Symposium, 2025

[12] E. C. J. Gan et al., "Blast waveform tailoring using controlled venting in blast simulators and shock tubes," Defence Technology, 3, pp.14-16, 2024

[13] M. Baker, "Shock tube size considerations for headborne personal potective equipment: A computational sensitivity study," Personal Armour Systems Symposium, 2023

 https://doi.org/10.52202/081621-0035

The Combined effects of Blast and Fragment impacts on Textiles

G. Kechagiadakis[1], D. Lecompte[1] and W. Paepegem[2]

[1]Royal Military Academy, Avenue de la Renaissance 30, 1000, Brussels, Belgium
georgios.kechagiadakis@mil.be
[2]Department of Materials, Textiles and Chemical Engineering, Ghent University, Technologiepark–Zwijnaarde 46, B-9502 Zwijnaarde, Belgium

Abstract. Explosion protection of Personal Protective Equipment is a widely discussed topic in the Defence sector albeit largely misrepresented experimentally. Oversimplified STANAG 2920 single-impact tests and unregulated field blast tests with limited repeatability and ambiguous results, hinder a comprehensive understanding of the problem: the synchronised impact of fragments and a blast wave on a given target. A laboratory experimental setup capable of simulating the combined effects of blast and fragment impacts of a wide range of explosive devices at a defined stand-off distance from a target is developed. A single 1.1g FSP is ejected using a 5.56mm weapon system; tautochronously, a blast wave is generated in an explosives driven shock tube. The two loads are synchronised based on a predetermined simulated stand-off distance. Aramid dry fabrics are tested in different combinations of fragment impacting velocity, blast overpressure and time interval between the two loads, to evaluate their fragment impact resistance in the presence of blast. The tests reveal a time-dependent material ballistic resistance, influenced positively or negatively by the material's oscillating phase. The blast overpressure/impulse acts as an amplifying factor, exacerbating the synergistic effects observed at different time intervals.

1. INTRODUCTION

Protection against the effects of explosions is a topic that has received extensive attention from the scientific community over the past decades. Many military or civilian applications involve the potential exposure to explosive loads of people, vehicles, or infrastructure. Certain test procedures were developed in order to ensure a basic level of protection can be achieved. Standardised ballistic impact tests incorporating fragment simulants have been at the forefront of fragment protection for personal protective equipment. These standards provide practical and clear instructions and can produce repetitive and comparable results. However, these methods often present an oversimplified view of reality. Explosions are complex multi-physics threats that can affect the surrounding space in many ways. The blast is a rapidly expanding shockwave that exerts immense crushing pressure, transmitting high impulse loads. The shattering of the casing can produce fragments, debris, pebbles and dirt from the surrounding area can be propelled at lethal speeds causing damage. Explosions are exothermic reactions that generate high heat that can ignite flammable materials and cause burns. These phenomena manifest in a successive order that can span from microseconds to minutes and have synergistic effects. This means that by limiting the analysis to one of these aspects the conclusions may have little connection to the initial physical problem.

In literature, there have been a few articles describing such synergistic effects between loads that result from explosions. Combined blast and fragment loading can cause more severe damage than the sum of individual blast and fragment damage [1, 2, 3]. Many researchers consistently highlight the necessity of moving beyond single-threat assessments to integrated approaches that account for the combined effects. [4, 5, 6]. Numerical simulations are crucial tools for analysing combined blast and fragment loading scenarios [2, 7]. 3D Finite Element simulations present improved accuracy over 2D F.E. simulations, as they capture the complex interactions between blast waves, fragments and target structures [7]. Experimental studies were also used as essential tools for validating numerical models and for evaluating the material and structural responses to combined loads [8, 9, 10]. Many of the published articles propose tests in which the combined blast and fragment impact loads are generated using high explosives and precut fragments [8, 9 ,10, 11, 12, 13, 14, 15, 16, 17].

Regarding the post-treatment of the evaluated material and structural response, this includes studying the behaviour of S-2 glass/epoxy composite laminates [11], sandwich panels with different core configurations [9, 12], polyurea-coated steel plates [10, 13], reinforced concrete [2, 3], and UHMWPE composites [14, 18]. Failure mechanisms and damage patterns under combined loads are analysed to promote design and protection strategies [3, 9, 12, 15].

Aramid fabrics have been the most popular textiles for PPE in the past decades. High toughness-to-density ratio of aramid fibres give aramid dry fabrics increased impact absorption capacity. The

https://doi.org/10.52202/081621-0036

woven structure of aramid fabrics is crucial to enhancing their impact resistance. The interlaced yarns create an entangled network that disperses the energy from projectile impacts through a frictional yarn-to-yarn motion. Aramid fabrics are used extensively in personal protective equipment against explosions. Even though their multi-hit fragmentation resistance has been analysed extensively [19, 20], their impact performance in the presence of blast loads hasn't been documented. Their near instantaneous reaction times make them difficult to hold in place while high impulse testing. What is more, until recently the problem of applying the combination of a blast wave in conjunction with a single or multiple fragments, consistently, has been largely an unsolved problem. In this article, a consistent, highly controllable, synchronised blast wave and single fragment impact loading test is presented.

2. METHODS

An experimental setup that can simulate a range of explosive events in a consistent and controllable manner is developed in the Department of Structures and effects of Explosions of the Royal Military Academy of Belgium. The setup ensures the independent control of specific characteristics that define the conditions of an explosion:

- The explosive device:
    - The blast wave has a specific overpressure/impulse at a given stand-off distance
    - The fragment generation yields specific fragment masses/velocities
- The Stand-off Distance:
    - The blast and the fragments travel at different speeds, hence the time interval between these loads correspond to a specific stand-off distance.

These three constituents can characterise an explosive event.

The Blast: The blast wave is generated using an Explosive Driven Shock Tube (EDST). This ensures that the blast wave has consistently repeatable pressure profiles with only a small charge as an input. The applied blast load is also near-planar, free of reflections from near surfaces as it is guided onto the target through the tube. The clamping system is places 20mm away from the tube's end.

The Fragments: A fragment simulant can be propelled using a pyrotechnic launcher. The universal receiver supports barrels from 5.56mm to 12.7mm. This enables the use of a wide range of fragment simulants including NATO's Fragment Simulating Projectiles (except the 20mm version) and other commercial or custom-made projectiles. What is more, with the use of a velocity base light gate system, the projectile velocity can be precisely measured and the impact point can be accurately defined.

The third parameter is the time interval between the loads, which is calculated using the projectile and the blast wave velocities.

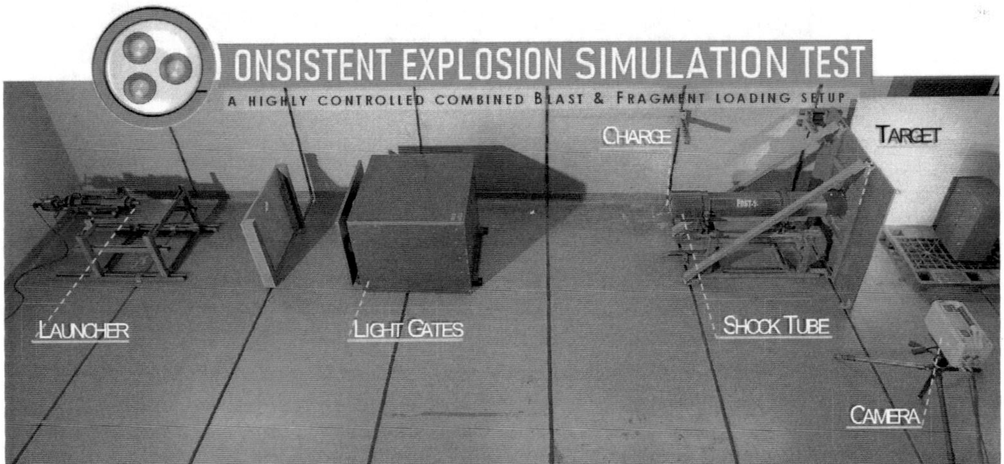

Figure 1. The Combined Blast and Fragment Loading setup is presented: A weapon system is used to launch the FSP through the Blast Tube on to the target. An explosive charge is activated at the front end of the Blast Tube generating a planar blast wave at the target. The two systems are timed for a synchronised arrival of combined loads with specific, impacting velocity, blast overpressure and time interval between the loads.

https://doi.org/10.52202/081621-0036

2.1 Materials

An extensive experimental campaign of 55 combined blast loading and fragment impact tests is conducted on samples comprising 15 layers of plain weave K29 Kevlar fabrics with a total areal density of 2.9kg/m^2. This specific dry fabric type was used in previous studies and it was selected for comparison purposes. The fabric is supported using a Ø135mm circular clamping frame which it comprises a neck flange, a support frame and an outer ring. The fabric is pushed through the outer ring by the neck flange as the outer ring is tightened using 10x M10 screws, and the fabric is slightly stretched and auto-aligned with the blast tube.

2.2 Loads

The proposed setup is capable of a broad range of output loads. The test case proposed by the International Mine Action Standards (IMAS) 10.30 was selected as it provides detailed requirements including the explosive device, the equivalent fragmentation resistance when considering the 1.102g FSP and the stand-off distance from the target [21]. In the IMAS 10.30, a test case of an accidental detonation of a PMN anti-personnel landmine is described, placed 60cm away from the torso of a deminer. The standard also determines also the minimum threshold of the ballistic resistance of the deminer's vest of 450m/s, tested according to NATO's STANAG 2920. The PMN is the largest antipersonnel blast landmine, containing 240g of TNT. Based on the 240g TNT charge and the 60cm stand-off distance, a blast wave with an incident/reflected overpressure in the proximity of 1.4/8.9MPa was calculated using the Kingery-Bulmash equations [22].

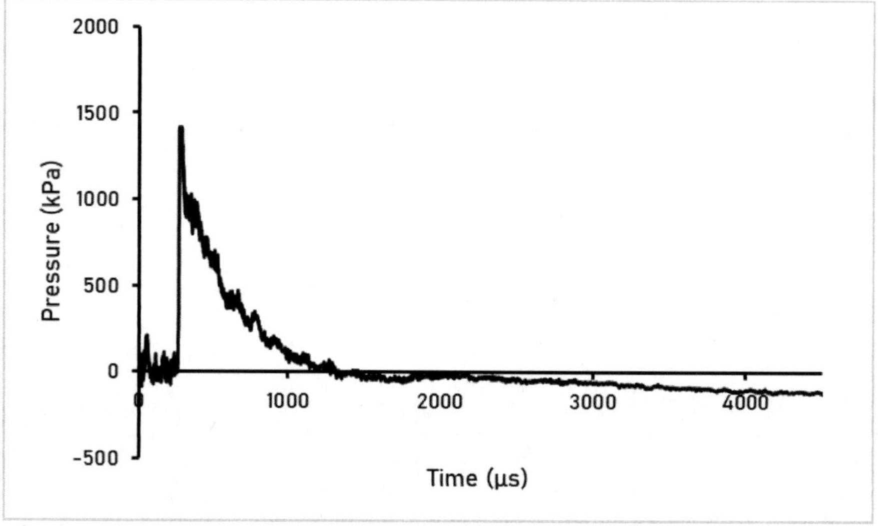

Figure 2. The incident pressure of the blast wave travelling through the Blast Tube is measured 10cm before the end of the tube. No target was in place for this measurement.

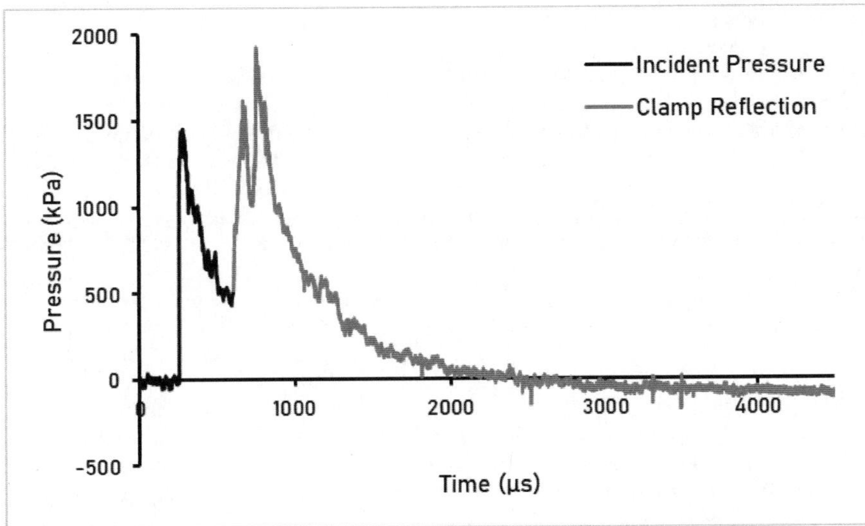

Figure 3. The incident pressure of the blast wave travelling through the Blast Tube is measured 10cm before the end of the tube. In this case, the target is causing reflections, which are distinguishable in the pressure measurements with red line.

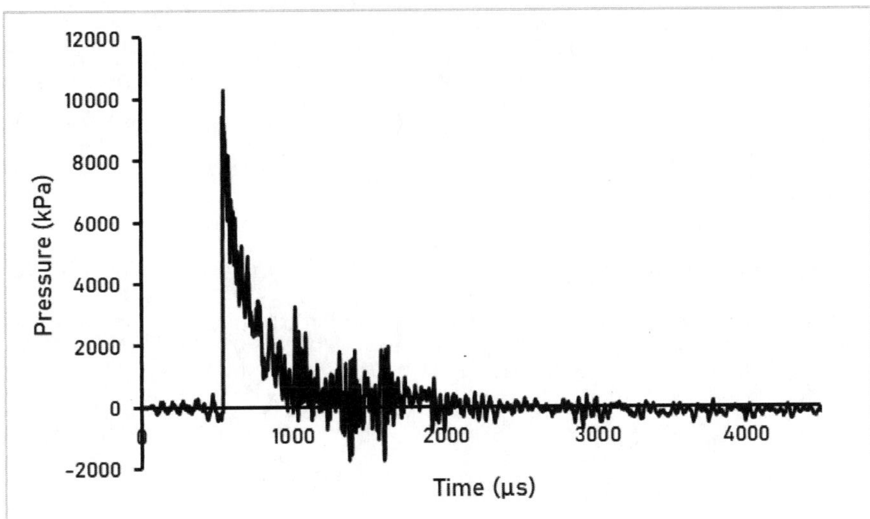

Figure 4. The reflected pressure measured at the exit of the Blast Tube.

3. RESULTS

The response of the aramid fabric under blast loads goes through phases. Initially, the fabric reacts almost instantaneously with the arrival of the blast wave. Longitudinal and transverse waves are generated from the edges of the free surface adjacent to the clamp travelling towards the centre. A truncated cone gradually forms, with the central part of the fabric remaining flat. As the stress waves continue to propagate, the fabric assumes a fully conical shape and its out-of-plane velocity maximises at about 250m/s. At this moment, the clamping forces are insufficient to retain the accelerated fabric, resulting in significant slippage with a magnitude of 16-17mm in the warp/weft directions and about 6-8mm at 45°, 135°, 225° and 335° [23]. The tip of the cone backlashes as the rest of the fabric continues to move forward, slipping through the clamping system. This results in the formation of a hemi-ellipsoid which remains until the end of the blast's positive phase. Beyond that point the fabric gets sucked backwards during the negative phase.

Aramid fabric present a phase dependent modulation in their ballistic resistance.

https://doi.org/10.52202/081621-0036

The time interval between the two loads appears to be a critical factor when it comes to the outcome of a combined loading event. In this study, the time interval is considered using the arrival of the blast wave as a reference. With this in mind, when the blast wave arrives first this results in positive time intervals. When the fragment simulant arrives first, the time interval assumes negative values. Of the 55 total combined blast loading and fragment impact tests, 9 have a negative time interval. As ballistic impact is a phenomenon that takes place within tens of microseconds, the fragment impact concludes before the blast wave may have any effect on the fabric. Furthermore, the blast wave at the levels examined (4.6MPa-10MPa) the fabric is able to withstand the loads with only a mild tearing of the first of the 15 layers at the boundaries. Even after a full perforation, the fabric remained unaffected by the subsequent blast load.

In the case of positive time intervals, two very distinct material behaviours are observed. Combined loading tests with short time intervals, below 150μs, result in elevated ballistic performance. The ballistic limit shows an increase of 18% from a base value of 483m/s without the blast, to 570m/s. On the other hand, time intervals over 150μs up to 516μs, exhibit a critical decrease in the ballistic performance of the aramid fabric by 17%, from 483m/s to 413m/s. The results can be seen in the table 1. This is a major security concern as in many cases textiles are being used as a means of protection against fragmentation resulting from explosions. Given that this critical time interval corresponds to a range of stand-off distances, this means that at certain distances the ballistic resistance may demonstrate critically low performance.

The fabric's variable ballistic performance is accounted for three material changes:

- The damage accumulation that resulting from the applied loads exceeding locally the strength limit of the aramid.
- The superposition of stress energy deposit due to the application of a near-uniform pressure over the fabric's surface during fragment impact.
- The dynamic material response from two consecutive loads that can result in an increased or decreased relative impacting velocity between the fragment and the fabric, accelerated previously by the blast. Previous studies indicated that this mechanism is very impactful when it comes to flexible materials under multiple impulsive loads. Additionally, multiple impacts aligned to the material's natural frequencies become resonant impacts and have increased potential in resulting in catastrophic failure [20].

Figure 5. An aramid fabric sample before and after being tested with the combined blast and fragment impact loading test.

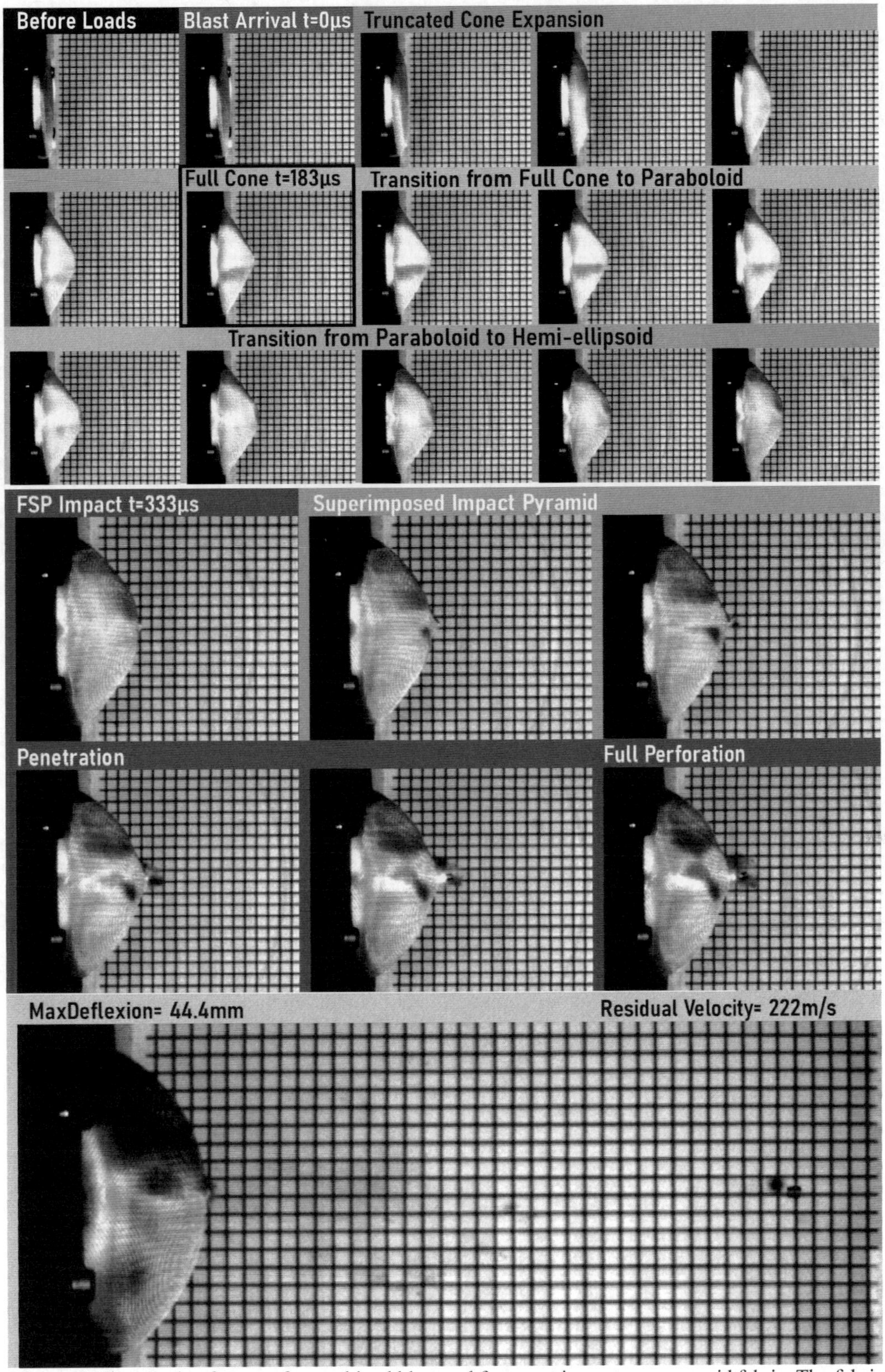

Figure 6. Consecutive frames of a combined blast and fragment impact on an aramid fabric. The fabric undergoes various phases. At t=333µs after the arrival of the blast wave, the FSP arrives on the target.

https://doi.org/10.52202/081621-0036

Table 1. Test results from combined blast and fragment impact tests at different time intervals; Negative: [-210μs,-17μs], Positive:[0,150μs] and Positive:[150μs, 516μs].

(-210) - (-17) μs		0-150μs		150-516μs	
Impacting velocity (m/s)	Result	Impacting velocity (m/s)	Result	Impacting velocity (m/s)	Result
571	P	580	NP	399	NP
562	P	-	P	447	P
600	P	574	P	449	NP
577	P	637	P	438	P
565	P	441	NP	449	P
580	P	472	NP	454	P
569	P	526	NP	449	P
534	P	524	NP	502	P
434	NP	561	NP	413	NP
		572	P	473	P
		566	NP	470	P
		574	NP	396	NP
		586	P	434	P
		580	P	413	NP
		493	NP	417	P
		451	NP	437	P
		430	NP	449	P
		445	NP		
		492	NP		
		507	NP		
		563	P		
		437	NP		

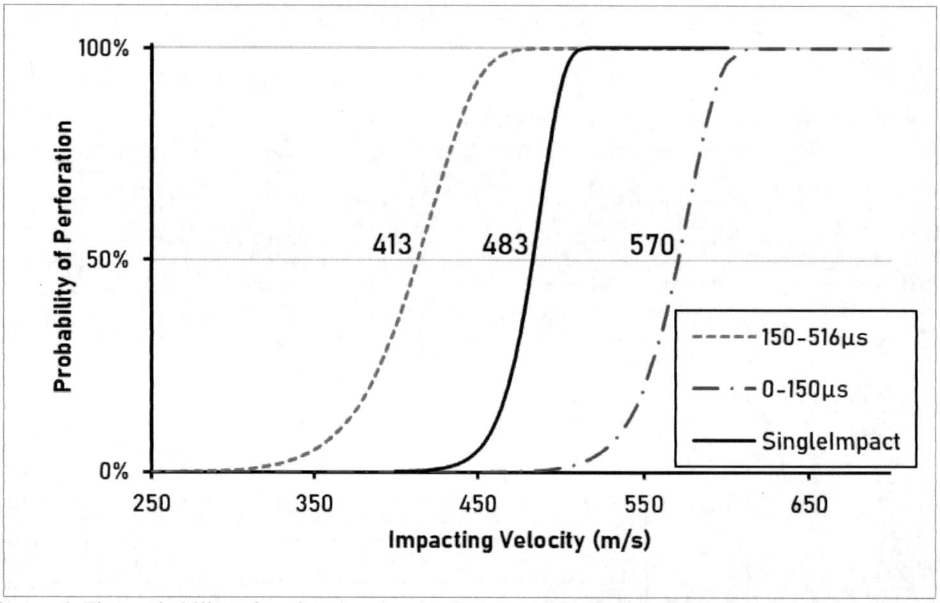

Figure 6. The probability of perforation density functions (Weibull) are fitted on the results from the Combined Blast and Fragment impact tests. The results are split into two distinct groups based on the time interval between the blast and the fragment impact. The results are compared with the base case in which only a single FSP is considered (without blast).

4. CONCLUSIONS

An experimental setup which allows the simulation of an explosive event in a laboratory scale is presented. The proposed setup is capable of generating a synchronised arrival of a planar blast wave and a fragment simulant impact. The three main parameters of the test - the impacting fragment velocity, the blast wave overpressure and the time of relative arrival of the two loads - can be controlled independently with millimetre and microsecond-level of accuracy.

The setup is used to test the fragment impact resistance of aramid fabrics in the presence of blast within a range of impacting velocities and time intervals, following the threat levels described by IMAS 10.30. The aramid fabrics reveal a phase-dependent ballistic resistance under the combined loads. The time interval between the loads plays a crucial role in the final outcome of a combined loading test. At short time intervals the ballistic resistance appears to increase greatly. At time intervals over 150μs, a significant drop in the ballistic resistance is observed.

Three major mechanisms of synergistic interactions between the blast and the fragment are identified: firstly, the damage accumulation from the blast wave influencing the subsequent fragment impact; secondly, the elevated stress-strain energy density due to the continuous application of the blast wave during the fragment impact; and finally, the change of the kinetic state of the fabric after the arrival of the blast wave can reduce or increase the relative impacting velocity of the projectile with respect to the moving fabric resulting in stronger or weaker impact.

References

[1] Li, J., Huang, C., Ma, T., Huang, X., Li, W., Liu, M. Numerical investigation of composite laminate subjected to combined loadings with blast and fragments. Composite Structures 214 (2019), 335–347.
[2] Nyström, U., Gylltoft, K. Numerical studies of the combined effects of blast and fragment loading. International Journal of Impact Engineering 36 (2009), 995–1005.
[3] Goswami, A., Ganesh, T., Adhikary, S.D., 2022. RC structures subjected to combined blast and fragment impact loading: A state-of-the-art review on the present and the future outlook. International Journal of Impact Engineering 170 (2022), 104355.
[4] Grisaro, H.Y., Dancygier, A.N., 2018. Characteristics of combined blast and fragments loading. International Journal of Impact Engineering 116 (2018), 51–64.
[5] Grisaro, H.Y., Benamou, D., Dancygier, A.N. Investigation of blast and fragmentation loading characteristics – Field tests. Engineering Structures 167 (2018), 363–375.
[6] Mellen, P., Shanahan, C., Bennett, T. Blast and fragmentation loading indicative of a VBIED surrogate for structural panel response analysis. International Journal of Impact Engineering 126 (2019), 172–184.
[7] Grisaro, H., Dancygier, A.N. Numerical study of velocity distribution of fragments caused by explosion of a cylindrical cased charge. International Journal of Impact Engineering 86 (2015), 1–12.
[8] Li, D., Hou, H., Chen, C., Zhu, X., Li, M., Yi, Q. Experimental study on the combined damage of multi-layered composite structures subjected to close-range explosion of simulated warheads. International Journal of Impact Engineering 114 (2018), 133–146.
[9] Cai, S., Liu, J., Zhang, P., Li, C., Cheng, Y., Chen, C. Experimental study on failure mechanisms of sandwich panels with multi-layered aluminum foam/UHMWPE laminate core under combined blast and fragments loading. Thin-Walled Structures 159 (2021), 107227.
[10] Wu, G., Wang, X., Ji, C., Liu, Q., Xie, X., Zhao, C., Liu, P. Damage response of polyurea-coated steel plates under combined blast and fragments loading. Journal of Constructional Steel Research 190 (2022), 107126.
[11] Li, J., Huang, C., Ma, T., Huang, X., Li, W., Liu, M. Numerical investigation of composite laminate subjected to combined loadings with blast and fragments. Composite Structures 214 (2019), 335–347.
[12] Zhang, P., Mo, D., Ge, X., Wang, H., Zhang, C., Cheng, Y., Liu, J. Experimental investigation into the synergetic damage of foam-filled and unfilled corrugated core hybrid sandwich panels under combined blast and fragment loading. Composite Structures 299 (2022), 116089.
[13] Zhang, L., Ji, C., Wang, X., Wang, Y., Wu, G., Zhu, H., Han, Z. Strengthening and converse strengthening effects of polyurea layer on polyurea–steel composite structure subjected to combined actions of blast and fragments. Thin-Walled Structures 178 (2022), 109527.
[14] Liang, M., Zhou, M., Li, X., Lin, Y., Lu, F. Synergistic effect of combined blast loads on UHMWPE fiber mesh reinforced polyurea composites. International Journal of Impact Engineering 183 (2024), 104804.

https://doi.org/10.52202/081621-0036

[15] Cai, S., Liu, J., Zhang, P., Li, C., Cheng, Y., Chen, C. Experimental study on failure mechanisms of sandwich panels with multi-layered aluminum foam/UHMWPE laminate core under combined blast and fragments loading. Thin-Walled Structures 159 (2021), 107227.
[16] Zhang, P., Wang, K., Hu, P., Li, Z., Chen, G., Cheng, Y. Experimental study on the synergetic damage of sandwich panel with armored hybrid core under combined blast and fragment loading. Thin-Walled Structures 199 (2024), 111814.
[17] Wu, G., Wang, X., Ji, C., Liu, Q., Gao, Z., Zhang, K., Zhao, C. Experimental and numerical simulation study on polyurea-coated fuel tank subjected to combined action of blast shock waves and fragments. Thin-Walled Structures 169 (2021), 108436.
[18] Zhao, C.-Z., Qiang, L.-S., Zhang, R., Zhang, Q.-C., Zhong, J.-Y., Zhao, Z.-Y., Lu, T.J. Dynamic response of UHMWPE plates under combined shock and fragment loading. Defence Technology 27 (2023), 9–23.
[19] Kechagiadakis, G., Lecompte, D., Van Paepegem, W., Coghe, F., Pirlot, M. Experimental evaluation of the ballistic resistance of aramid fabrics under near simultaneous multiple fragment impacts. International Journal of Impact Engineering 180 (2023), 104675.
[20] Kechagiadakis, G., Lecompte, D., Van Paepegem, W., Coghe, F., Pirlot, M. Near simultaneous multiple fragment impacts on aramid fabrics: Effects of velocity, dispersion and time intervals. International Journal of Impact Engineering 185 (2024), 104830.
[21] International Mine Action Standards 10.30. 2nd Edition, Amendment 5 (2023)
[22] Kingery, C., Bulmash, G. Airblast parameters from TNT spherical air burst and hemispherical surface burst. Technical report ARBL-TR-02555 (1984).
[23] Kechagiadakis, G., Lecompte, D., Van Paepegem. Sliding Boundaries in Finite Element modelling of soft materials under blast loads.

Sensor System to Capture Spatial & Temporal Pressure Distribution from Ballistic Impact Helmet Backface Deformation

L. Voo[1], A. Lurski[1], A. Veith[1], E. Butler[1], J. Marcus[1], Q. Luong[1], and M. Kleinberger[2]

[1] *The Johns Hopkins Applied Physics Laboratory (JHU/APL), 11100 Johns Hopkins Rd, Laurel, MD, USA*
liming.voo@jhuapl.edu
[2] *U.S. Army Combat Capabilities Development Command Army Research Laboratory, Aberdeen Proving Ground, MD, USA*

Abstract. Limited knowledge of the spatial and temporal pressure distribution over the course of body armour and helmet back-face deformation during a ballistic impact is a significant impediment in developing effective injury metrics and better protective systems for countering behind helmet blunt trauma (BHBT). This knowledge gap can largely be attributed to the lack of viable solutions to measure the contact pressure distributions with minimal interference at the interface between the human surrogate and the armour system. JHU/APL has recently developed a spatially integrated thin-film stress sensor array system to address this experimental measurement gap. The packaged system prototype is only 0.20mm (0.008inch) thick and could easily fit between the human surrogate and armour. A version of the system with 13 stress sensors has been tested on and below the impact pad of the Adaptable Testing and Load Assessment System (ATLAS) Headform using an air-cannon propelled projectile that simulated the impact momentum and shape of a helmet back-face deformation during a non-penetrating ballistic impact. The prototype sensor system proved to be capable of surviving, sensing high-speed pressure response, and producing quality signals under severe mechanical insults. The spatial and temporal distributions of the resulting impact pressure were captured during high-speed loading conditions. The sensor spatial distribution density of the current prototype can be increased by including more individual stress gauges or custom-designing more stress-sensing elements. This new sensing capability could enable more precise measurement of the injurious loading conditions behind the helmet and body armour, and thereby improve performance evaluation of personal protective equipment. Further development of this technology will include exploration of its capabilities, spatial resolution, and pressure measurement accuracy, under ballistic BHBT conditions.

1. INTRODUCTION

Combat helmets are designed to stop the bullets from penetrating the helmet shell, and therefore mitigating the risk of ballistic penetrating injuries to the head. It has been well documented that the rapid deformation of the helmet back-face, without complete ballistic penetration, can still cause significant injuries to the head, including severe skull fractures [1-3]. This phenomenon is commonly referred to as Behind-Helmet Blunt Trauma (BHBT). Several studies have attempted to investigate different injury mechanisms and injury risk curves using simulated back-face deformation impactors [4-5]. The apparent differences in the resulting skull fracture patterns in these studies (with and without helmet), however, may be a result of different contact pressure distributions, which have not been measured experimentally [6].

Several studies have attempted to investigate stress conditions under BHBT conditions [2, 7-9]. Ommaya et al. estimated impact stress values for adult human skull fracture to be 43 MPa [7]. Other studies calculated mean stress at 50% risk of skull fracture to be 15 MPa to 51 MPa [2] based on analysis of strains from cadaveric skull fractures and match-paired anthropomorphic test dummy tests. On the other hand, computational modelling and simulation of the BHBT mechanisms also reported calculated stresses on the skull tissues [8-9]. Although such numerical results are useful for parametric assessments and qualitative understanding of the effect various variables on BHBT related stresses in the head, the simplified bone geometry and lack of tissue level model validation make quantitative prediction of tissue stresses uncertain. The validity of such calculated estimates remains a gap due to lack of a suitable sensing system to measure the contact stress experimentally.

The current method of assessing BHBT injury risk of various combat helmets is based on residual plastic deformation left in clay [10-11], with the assumption that the deformation depth correlates to skull fracture risk. The existing biomechanical data on skull fracture are often reported in terms of peak impact force magnitude depending on the impactor profile [12]. An alternative assessment method for skull fracture that could utilize force measurement would provide a more scientifically supported injury

 https://doi.org/10.52202/081621-0037

correlate for BHBT risk assessment. A multi-location test head surrogate, the Adaptable Testing and Load Assessment System (ATLAS) Headform, was developed by the Johns Hopkins Applied Physics Laboratory (JHU/APL) to collect behind helmet impact force responses for helmet BHBT performance assessment [13,14]. Other impact force-based head surrogates include multiple load cells, such as the seven-cell Ballistic Load Sensing Headform (BLSH) [15-18] and the 17-cell Pk17dynA [19,20] that enable measurement of force distribution at a fixed impact location [19,20] or locations [13-18]. While excellent repeatability and reproducibility is a great advantage of such load-cell based measurement systems [21], they are limited by the lack of deformation in the direction of impact, which a human skull possess, and fixed measurement locations. It is, therefore, desirable to have a dynamic surface contact stress sensing system for BHBT testing with minimum deformation resistance in the direction of impact and ability to measure at any location on a head surrogate.

This paper describes a prototype thin-film stress sensing system custom-designed and fabricated to measure dynamic contact pressure spatial and temporal distribution under BHBT environment. Its experimental usability, ability to survive such harsh mechanical conditions, and sensing capabilities are demonstrated through simulated BHBT testing in the laboratory.

2. MATERIALS AND METHODS

The ATLAS Headform [13] was employed as the human surrogate in this study (Figure 1). Its surface represents that of human head anatomy. Response is measured by a high-capacity load cell (Model 224C, PCB Piezotronics, Depew, New York, USA) supported by a metal shaft on the back and covered by a steel cap (Impact Cap) surface (Figure 1a). The Impact Cap is also covered by a polymer Impact Pad (Shore 40A hardness neoprene rubber). A critical challenge of developing the thin-film stress sensing system is its form factors that allows it to fit between the helmet and the test headform, as well as between the headform Impact Pad and Impact Cap (Figure 1). There is a very limited space on the Impact Cap area for the packaged sensing system to fit (Figure 1b). The total thickness of the sensing system package, besides the wires, needs to be less than 0.254mm (0.01inch).

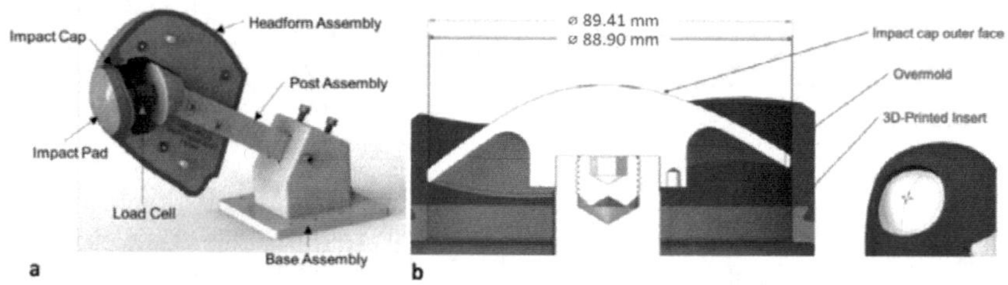

Figure 1: **(a)** Cross-sectional view of the ATLAS Headform (frontal impact module); (b) Cross-sectional view of the Impact Cap and its surrounding

The individual thin-film piezoelectric polyvinylidene fluoride (PVDF) stress sensor from Dynasen (PVF2 11-0.040-EK, Dynasen Inc., Goleta, CA, USA) was selected to construct the thin-film sensing system for BHBT contact pressure spatial and temporal distribution measurement. It has high stress sensing capacity range (10-10,000MPa), rapid response (nanosecond), very thin, and a long-thin form factor (Figure 2).

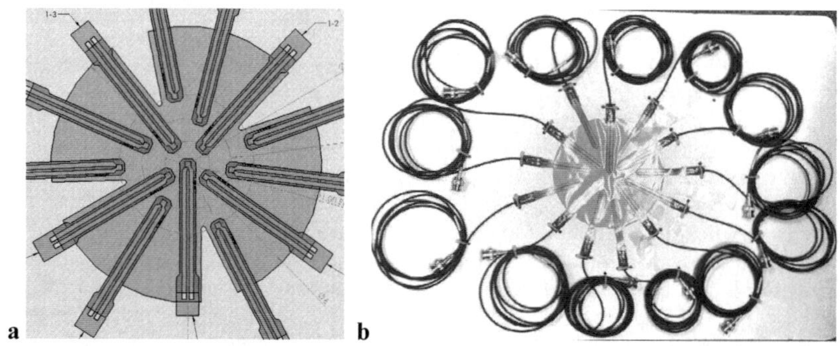

Figure 2: Packaged pressure sensing system with fixed sensor layout (a) and cables (b)

The PVDF sensor used in this study produces an electrical potential proportional to the mechanical stress on its material surface. It can be used as a force sensor or stress sensor depending on how the electrical signals are converted into the values in an engineering unit and how the contact area is accounted. In this study, it is used as a stress sensor with a very small sensing element and assumed uniform stress distribution within the sensing element area. Each individual stress sensor is supplied with a static calibration value while its dynamic calibration data is obtained using an equation provided by Dynasen Inc. Similar PVDF sensors have a track record of being used in blast and ballistic impact conditions [22,23]. The sensing element sizes (PVF2 11-0.040-EK) adopted for this project were 1.016mm×1.016mm (0.04inch x 0.04inch). The placement design of the stress sensors considered the shape and size of the helmet back-face deformation as well as potential spatial distribution gradient: the contact pressure is expected to be greatest at the impact centre and decreasing radially from the centre. In this initial design, the sensor placement covered the total potential compressive stress area induced by the impact projectile which has a 62mm diameter. The individual stress sensors, therefore, were packaged concentrically starting with one sensor at the expected impact centre and four sensors in each of the three concentric rings at distances of 12.7mm, 25.4mm, and 38.1mm (0.5inch, 1.0inch, and 1.5inch) from the centre (Figure 2a). It was expected that the compressed area behind the Impact Pad would be slightly greater than the projectile diameter due to pad deformation. The sensor locations were fixed with adhesive and sandwiched between two layers of thin films with a total thickness of 0.2032mm (0.008inch). The whole sensor system package can fit into the ATLAS Headform sensing area on top of the Impact Cap and covered by the Impact Pad (Figure 3a). The non-sensing area of the hardware, including the cables (Figure 2b), could be routed through the backside of the ATLAS Headform load cell and exit the side (Figure 3b and 3c).

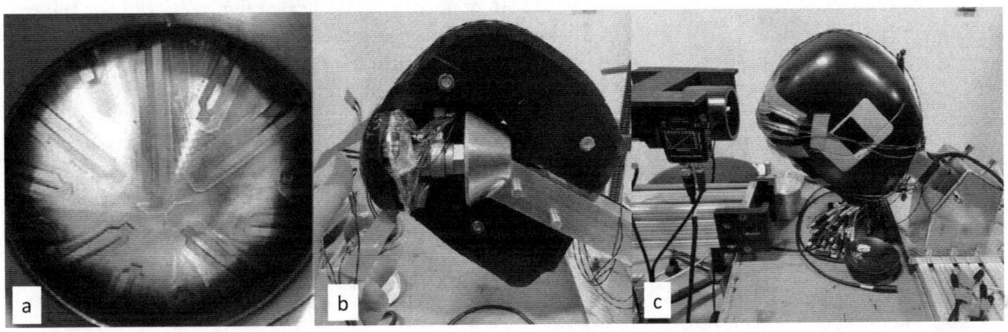

Figure 3: Packaged thin-film pressure sensing system fitted into the ATLAS Headform: (a) on the load cell cap; (b) with cables wrapped around the back; and (c) on the Impact Pad and ready for testing

BHBT impact conditions were simulated using an air-cannon driven projectile impactor representing the bullet-helmet mass and deformation shape. This cylindrical projectile has a spherical front surface, 192g mass, and 62mm (2.45inch) diameter. A pressure-velocity curve was developed to determine the appropriate pressure for the desired projectile velocities. Two lasers at the end of the barrel 10 cm (3.94inch) apart act as a velocity gate to measure the velocity of the projectile immediately before impact. The impact velocities used in the experiments,15, 23, 30, and 45 m/s (50, 75, 100, and 150 ft/s), enabled evaluation of the thin-film sensor system over a BHBT severity range.

Eleven simulated BHBT tests were conducted on the front impact location of the ATLAS Headform, with the thin-film stress sensing system installed below the Impact Pad and another on top of it (Figure 3c). The sequence of the tests was: 3 tests at 15 m/s (50 ft/s), 3 tests at 30 m/s (100 ft/s), 1 test at 15 m/s (50 ft/s), 1 test at 23m/s (75ft/s), and 3 tests at 45 m/s (150 ft/s). The signals from the stress sensors and the ATLAS load cell were recorded at a sampling rate of 200 kHz using a DEWETRON Inc. data acquisition system (DEWE-801, Orion MDAQ-V200, East Greenwich, RI, USA). No post-test data filtering was applied for the data analysis.

3. RESULTS

3.1. Sensor System Conformity to Curve Surface

The sensor system fit condition on the curved surface of the impact cap was assessed at the front impact location of the ATLAS Headform using both visual inspection and computed tomography (CT)

 https://doi.org/10.52202/081621-0037

imaging approaches. CT scans of ATLAS Headform with and without sensor installation were obtained. Although not apparent from the visual inspection, the CT images demonstrated that the sensor system could fit within the curved gap space with no noticeable gap increase between the Impact Pad and the Impact Cap surface. Note, to eliminate artifacts from the CT scanner, a 3D printed polymer Impact Cap was used instead of the steel Impact Cap.

3.2. Sensor System Survivability under BHBT Test Environment

With some iterative design improvements, the pressure sensing system based on PVDF stress sensor proved to be capable of surviving the air-cannon–simulated BHBT impacts. All the PVDF sensors produced expected signals in repeated tests. The earlier tests (prior to the test series reported here) caused damage to the sensor outside the Impact Pad at the impact centre under the most severe impact condition (45 m/s). Covering the sensor package with two additional layers of thin films prevented such damage. In addition, one connector on the outside sensor package was separated from the sensor due to large deformation during the most severe test condition (45 m/s) in this test series.

3.3. Temporal Contact Stress Distribution Measurement

The stress sensing system produced signals that did not require post-test signal filtering. No signal clipping was ever observed. The following event response times of the tests can be identified (Figure 4): initial impact contact with the outside centre stress sensor, load cell response initiation, load cell signal peak, Impact Pad full compression, end of projectile contact, peak Impact Pad expansion, and Impact Pad recovery. Those event-times were identified based on analysis of high-speed video images, load cell response signal, and signals from individual stress sensors. The positive values in the figures represent compressive stress (or positive pressure) while the negative values represent tensile stress (or negative pressure). The projectile contact initiated the response of the outside centre stress sensor response, followed by the signal initiation of the inside centre stress sensor and load cell (Figure 4).

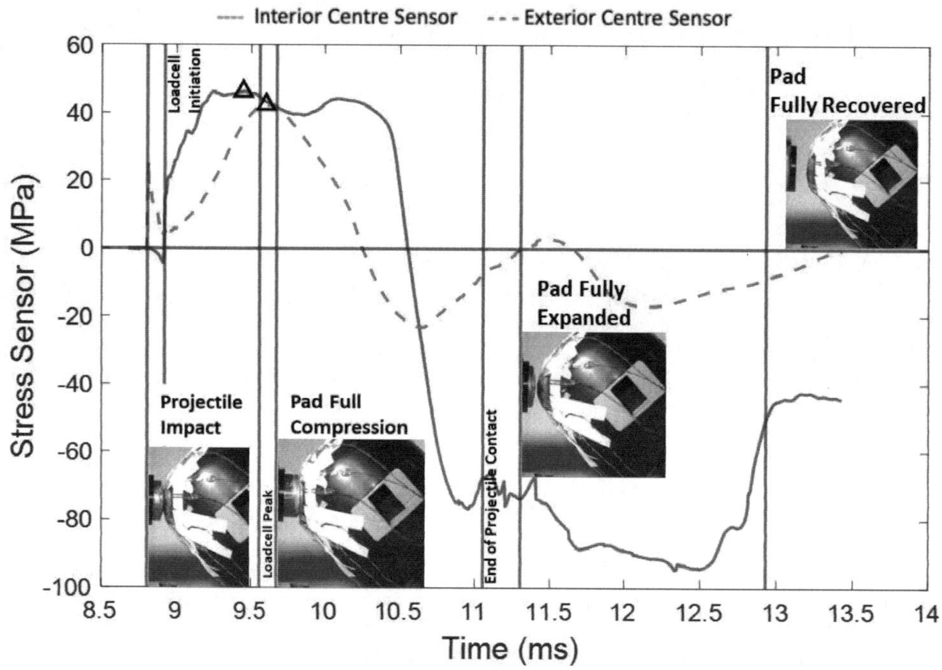

Figure 4: Event times illustrated on centre stress sensor signals inside and outside the Impact Pad (Note: positive values indicate compressive stress while negative values indicate tensile stress)

The initiation of the off-centre signal came after the centre signal and also from the exterior sensors to the interior sensors (Figure 5). The timing of signal peaks varied depending on the impact velocity. A tensile stress was observed inside the Impact Pad during the projectile rebound phase (Figure 4), consistent with what was observed in high-speed video.

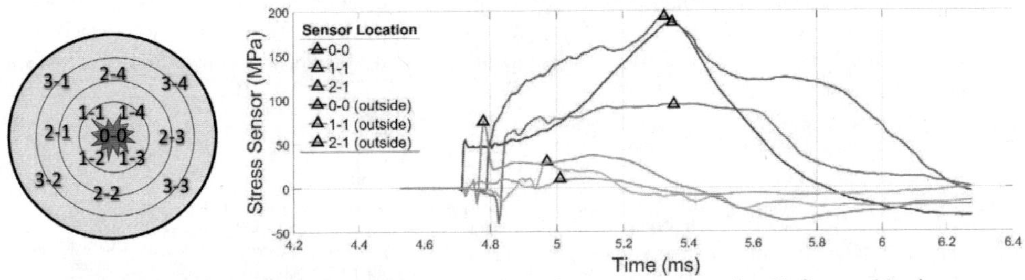

Figure 5: Contact stress responses at locations (impact centre to ring 2) from a 30m/s test

Figure 6 illustrates the force or peak pressure time from when the projectile crossed the laser velocity gate for each test. The mean peak pressure times, averaged from multiple repeat tests, are shown along with their corresponding standard error. The peaks occurred at the same time at the centre and peripheral locations with minimum difference between inside and outside of the Impact Pad and the load cell in any given test. In general, the peak time decreased as the projectile impact velocity increased.

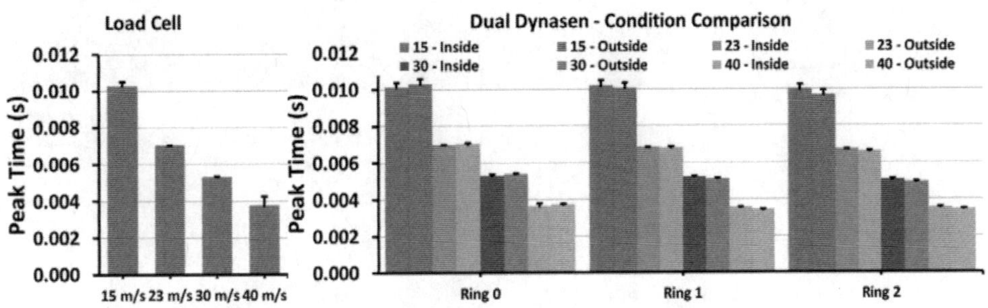

Figure 6: Peak times of force and contact stress signals at various locations (ring 0, ring 1, ring 2) inside and outside the Impact Pad at different velocities

3.4. Spatial Contact Stress Distribution Measurement

The force and pressure response peaks increased as the impact velocity increased (Figure 7). In general, the peak contact pressure was higher inside the Impact Pad than the outside. This difference, however, diminished or even reversed as the impact velocity increased. The peak contact pressure was also greater at the impact centre (Ring 0) and decreased with the distance away from the centre (Ring 1 and Ring 2).

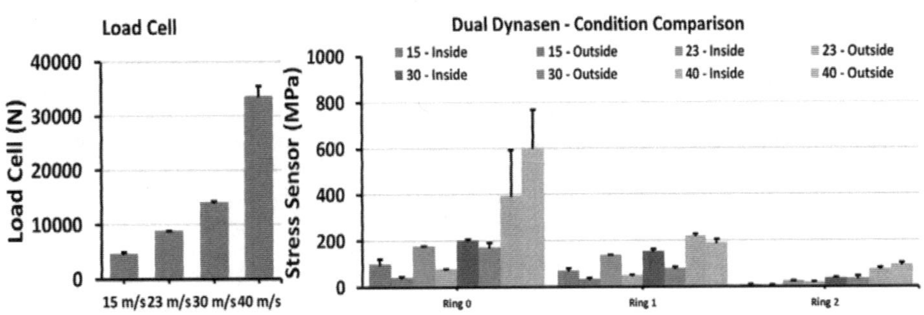

Figure 7: Effect of impact velocity on peak response force and contact stress peak at different distances from the impact centre (Ring 0) to more peripheral locations (Rings 1 and Ring 2)

https://doi.org/10.52202/081621-0037

Heat maps of contact pressure inside and outside of the Impact Pad were created to explore their distribution pattern change over time at different velocities (Figure 8). The greatest amount of contact pressure is concentrated in the centre and the first ring of the sensing area for both the sensor placed on the inside and outside of the pad across all velocities. The pressure maps show little response beyond the projectile frontal contact surface, which is captured by the centre and first two rings of sensors, even though the peripheral sensors in ring 2 did produce low readings because of Impact Pad deformation. Figure 8 also shows the effect of the impact velocity on the pressure pattern inside and outside the impact pad.

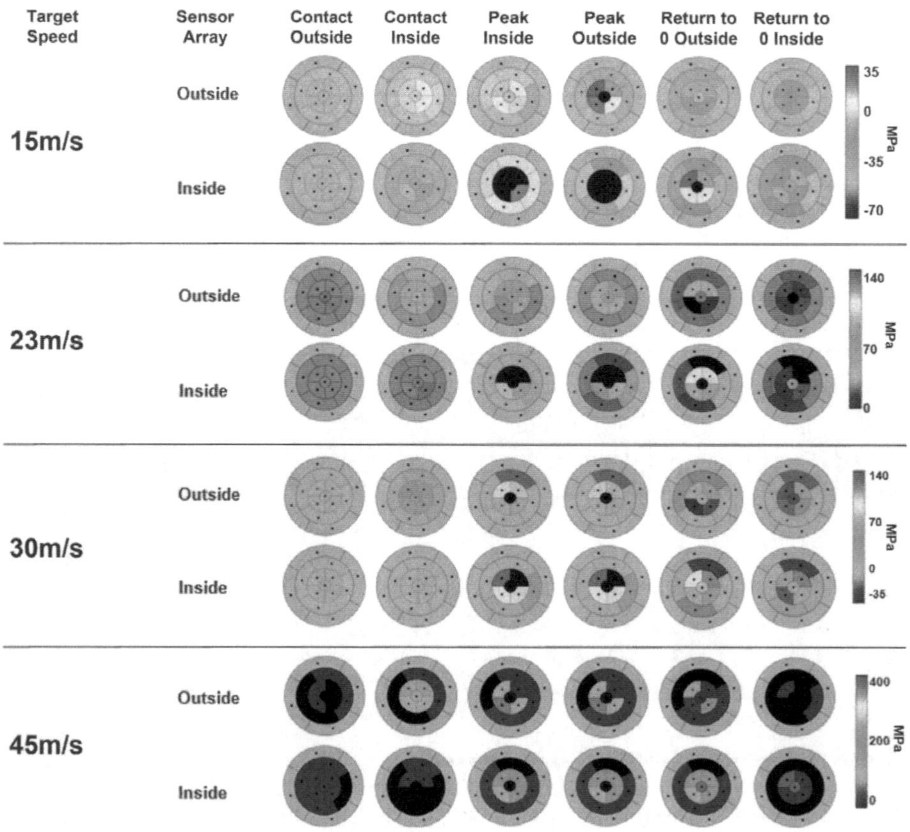

Figure 8: Contact stress patterns at distributed locations, impact progression time, and different impact velocities

3.5. Detecting Impact Pad Damping Capability Change

The thin-film pressure sensing system was able to detect the effect of degradation of the polymer Impact Pad on impact pressure transmission (Figure 9). To compare across different tests of the same target velocity, time zero of the response signal was established to be when the centre outside sensor crossed a threshold that correlated with projectile contact initiation. The centre contact pressure response inside the Impact Pad showed progressively higher peak values from the first, second, third, and seventh test of a test series (figure 9a). The peak transmitted pressure from the 7th test was more than three times that of the first test. The differences in the force responses did not show such a progressive change, instead showing the effect of projectile velocity variation (Figure 9b).

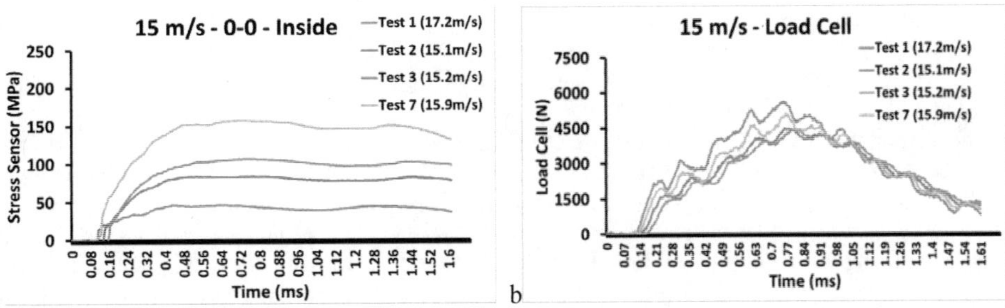

Figure 9: The interior centre stress responses (a) and ATLAS Headform load cell force response (b) from the first, second, third, and seventh test of a test series

A deeper dive analysis of the contact pressure inside the Impact Pad demonstrated more granular pressure measurement capability of this sensing system (Figure 10). The contact pressures at the four locations on ring 1 were similar to each other and the centre pressure location (ring 0) (Figure 10, left) while the pressures at the four more peripherical locations (ring 2) were much lower. The loading damping and distributing effect of the Impact Pad was dramatically reduced after sustaining three tests of six tests (Figure 10 right). Two of the ring-1 gauges sensed much greater pressure than in the first test. The pressure at the impact centre (A00) now increased the greatest and reflected the most prominent surface at that location. The pressures at more peripheral locations (A12-A24) did not show meaningful increase in pressure magnitude.

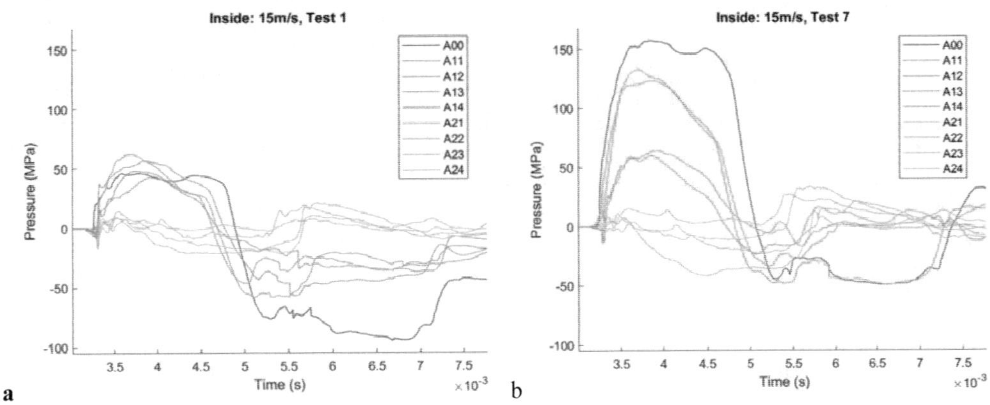

Figure 10: The inside contact pressures at the impact centre (A00), near centre (A11-A14) and more peripheral (A21-A24) from the first test (a) and the 7th test (b) at the same velocity

4. DISCUSSION

The feasibility and capability of measuring contact pressure distribution spatially and temporarily under BHBT impact conditions was investigated using the JHU/APL prototype thin-film stress sensor package. The experimental testing employed the ATLAS Headform as the human head surrogate, while simulated BHBT impacts were generated by an air-cannon propelled projectile. The contact pressure measurement leveraged a custom-designed thin-film stress-sensor system fitted into and over the ATLAS Headform Impact Pad at the front impact location. The experimental results demonstrated that such a sensor system 1) can survive BHBT relevant test conditions; 2) can be fitted into the ATLAS Headform without need for hardware modification; 3) produces low noise stress signals; and 4) allow analysis of contact pressure distribution patterns and magnitude change over the duration of the impact. Additionally, the measurements are able to assess the effect of impact severity and impactor shape.

The packaged thin-film stress sensor system was shown to be able to sustain simulated BHBT impacts repeatedly with minimum damage if sufficient caution and protection were implemented. The sensor package damage around the centre sensor was determined to be the result of the extreme stress. Subsequent tests, which produced the data for the analysis of this study, demonstrated that such damage could be mitigated with additional protective layers of thin-film material applied to the

https://doi.org/10.52202/081621-0037

potentially damaged area. Damage occurred at the connection between the sensor wire strip and the extension wires on sensors installed outside the Impact Pad as a result of the large deformation of the Impact Pad. Such damage could also be prevented with additional cable slack during the sensor system installation.

The response forces measured by the ATLAS load cell are consistent with range of BHBT mechanical environment reported in the literature [4,14,24]. Iwaskiw produced BHBT skull fractures using post-mortem human subject (PMHS) heads in live-fire tests and conducted matched-pair test on the ATLAS Headform which measured forces in the range of 9.6 - 22.6 kN (2.2 – 5.1 klbf) [14] and impulse in the range of 2.62 to 3.81 Ns. Raymond reported mean peak force of 5.8 kN from BHBT skull fracture experimental study using a projectile impact to the lateral side of PMHS head [4]. Daniel reported mean projectile impact force ranges of 12.3, 6.3, and 21.6 kN (2.8, 1.4, and 4.9 klbf) for impacts to the front, side and rear of PMHS head, respectively that produced skull fractures [24]. The current study produced headform response forces in the range of 4.5 - 33.0 kN (1.0 - 7.4 klbf) (Figure 7) which spans those of the BHBT test conditions reported in the BHBT literature.

The response contact pressures measured by the thin-film stress sensor system are consistent with the expected outcomes for the test conditions and sensor locations. The measured contact pressure at each sensor location increased progressively as the impact velocity increased. The measured pressure generally decreases as the sensor location further away from the impact centre. The pressures from the thin-film sensors backed by a rigid steel surface (Impact Cap of the load cell) are mostly higher than those backed by the Impact Pad (Figure 7), consistent with the boundary stiffness effect.

The thin-film stress sensor system has sufficient sensitivity to even detect local padding material degradation due to repeated impact and surface geometry effects (Figures 9 and 10). Such effects cannot be observed visually nor easily analysed using the single load cell signal in ATLAS headform. This capability is particularly useful for BHBT and BABT applications, as injury risk in such scenarios largely depend on the local loading conditions and affected by the surface geometry and local stiffness of the head. The ability to detect changes in the test system that could affect injury prediction would be critical for performance evaluation of personal protective equipment. A spatially distributed sensing system, such as the thin-film stress sensor system in this study, demonstrated such capabilities. Test headforms with multicell rigid load cells such as BLSH (7-cell) [15-18] and Pk17dynA (17-cell) [19,20] may be able to measure such pad damping changes as well, although such result has not been reported in the open literature. On the other hand, the thin-film sensing system also has the advantages of being used in any surface location of a human surrogate while not resisting deformation in the direction of impact loading. Increasing the spatial density of the stress sensor system (more sensing elements in a given area) could further improve its local measurement resolution and accuracy, and is technically feasible.

The contact stresses measured by the thin-film stress sensors on the ATLAS Headform surface and below the Impact Pad are in the same order of magnitude as those reported in the BHBT biomechanics literature. The compact skull bone stress estimated by finite element analysis studies ranged 55 – 80 MPa (7977 – 11603 psi) depending on suspension pad stiffness in BHBT conditions [8]. Similarly, skull bone stress was reported to range 75 – 89 MPa (10877 – 12908 psi) depending on the velocity of the bullet impacting the helmet [9]. The experimental skull fracture stress under impact was reported to be on average 43 MPa (6237 psi) [7] or estimated to be 51 MPa (7397 psi) [2]. The surface stresses in the central area (25.4mm diameter circle) of this study were 34 – 175 MPa (4931 – 25382 psi) outside and 69 – 209 MPa (10008 – 30313 psi) inside the Impact Pad (Figure 7). However, these stress quantities are likely not of the same form or location and therefore should be used for reference of order of magnitude rather than for direct comparison. Nevertheless, this comparison further supports the utility of this thin-film sensor system for BHBT contact pressure measurement application. Additional research could enable quantitative BHBT risk assessment based on stress-related injury metrics.

The impact test conditions in this study adopted a well-established method for producing BHBT skull fractures using air-cannon launched impact projectiles to simulate the impact of the helmet back face on the head during the ballistic impact [4,5,24]. The impact energy and velocity used in such studies represent the helmet back face rather than those of the striking bullet which are expected to be much higher. The projectile impact energy range of 4.4 to 31.3 J in this study is comparable to the range of 5 to 40 J reported in the two experimental studies using cadaveric specimens [5,24]. The projectile impact velocity range of 15 to 45 m/s in this study spans another cadaveric study's 18 to 37 m/s velocity range [4]. However, such experimental test method may not replicate all aspects of the BHBT conditions important for BHBT injury mechanisms and/or helmet performance. For example, the impact velocity range used in projectile impact testing may only represent the lower bound of those in the deforming helmet back face in a helmeted ballistic impact under BHBT conditions. The nonrigid and progressively changing shape of the deforming helmet back face could be critical factors for contact pressure distribution and injury mechanism that are not simulated in the tests using the projectiles. Therefore, the pressure distribution results from this study may not represent the true BHBT conditions of the helmeted human head. Nevertheless, those results do demonstrate the measurement potential of such thin-film stress sensor system to enable better understanding of the biomechanical factors associated with BHBT injuries and developing more effective injury correlation metrics. Future evaluation of the thin-film stress sensor system under helmeted BHBT test conditions would be valuable and more definitive for its intended application.

The current prototype of this thin-film stress sensor system can be further improved. Increasing spatial sensing density could improve measurement of pressure gradients in the critical area and capture more precisely the surface geometry effect between different locations of the head. The number of sensing elements in this study is largely limited by the project scope and budget. Doubling the stress gauge density is feasible without requiring change in the gauge design. Custom-designed stress sensing grid for the targeted BHBT and BABT application could be done to optimize sensing performance, test operation, and cost. The current prototype has only been tested under simulated BHBT conditions using an air-cannon propelled projectile, and needs to be evaluated under live-fire ballistic testing with a helmet system to fully demonstrate its utility. Further development of this technology will include exploration of its capability, spatial resolution, and pressure measurement accuracy under ballistic BABT/BHBT conditions.

5. CONCLUSION

The prototype thin-film stress sensor system, which was designed to fit onto head surrogates and enable additional measurements of the spatial distribution of BHBT loading, proved to be capable of surviving, sensing high-speed pressure response, and producing quality signals under severe simulated BHBT impact conditions. The spatial and temporal distributions of the impact pressure were captured during such high-speed loading conditions and comparable to those expected in BHBT loadings. The spatial sensor density of the current prototype can be increased by including more individual stress sensors or by custom-designing more stress-sensing elements. This new sensing capability could enable more precise measurement of the injurious loading conditions behind the helmet and body armour, and thereby improve performance evaluation of personal protective equipment.

Acknowledgments

This research effort was supported by the U.S. Army Combat Capabilities Development Command Army Research Laboratory (DEVCOM ARL). Any opinions, findings and conclusions or recommendations expressed in this material are those of the author(s) and do not necessarily reflect the views of DEVCOM ARL.

References

[1] A. W. Carroll and C. A. Sonderstrom, "A new nonpenetrating ballistic injury," *Annals of surgery,* vol. 188, no. 6, p. 753, 1978.

[2] Bass, C.R., B. Boggess, B. Bush, M. Davis, R. Harris, M.R. Rountree, S. Campman, J. Ecklund, W. Monacci, G. Ling, G. Holborow, E. Sanderson, and S. Waclawik. "Helmet Behind Armor Blunt Trauma." Paper presented at the RTO Applied Vehicle Technology Panel/Human Factors and Medicine Panel Joint Specialists' Meeting held in Koblenz, Germany, May 19-23, 2003.

[3] K. A. Rafaels, et al. "Injuries of the head from backface deformation of ballistic protective helmets under ballistic impact," *Journal of forensic sciences,* vol. 60, no. 1, pp. 219-225, 2015.

[4] D. Raymond, Van Ee ., et al. "Tolerance of the Skull to Blunt Ballistic Temporo-Parietal Impact," *Journal of biomechanics,* vol. 45, no. 12, pp. 2479-2485, 2009.

[5] C. A. Weisenbach, et al. "Preliminary Investigation of Skull Fracture Patterns Using an Impactor Representative of Helmet Back-Face Deformation," *Military medicine,* vol. 183, pp. 287-293, 2018.

[6] Y. Li,, Adanty K., et al. "Review of Mechanisms and Research Methods for Blunt Ballistic Head Injury," *Journal of biomechanical engineering,* vol. 145, no. 1, 2023.

[7] Ommaya, A.K., W. Goldsmith, L.E. Thibault. "Biomechanics and neuropathology of adult and paediatric head injury." British Journal of Neurosurgery 16(3):220-242, 2002.

[8] Li X., Gao X., Kleiven S. "*Behind helmet blunt trauma induced by ballistic impact: A computational model.*" International Journal of Impact Engineering. Vol 91 (2016) 56–67, 2016.

[9] Cai Z., Huang X., Xia Y., et al. "*Study on Behind Helmet Blunt Trauma Caused by High-Speed Bullet.*" Applied Bionics and Biomechanics, Article ID 2348064, 2020.

 https://doi.org/10.52202/081621-0037

[10] U.S. Department of Justice, National Institute of Justice Standard for Ballistic Helmets National Institute of Justice, NIJ 0107.01, 2024.
[11] National Research Council (U.S.) Committee on Testing Armor Materials, "Testing of Body Armor Materials: Phase III." National Academies Press, Washington, DC., 2012.
[12] Goldsmith W. and Monson K.L., "The State of Head Injury Biomechanics: Past, Present, and Future, Part 2: Physical Experimentation." Crit Rev Biomed Eng, 33 (2005), 105-207.
[13] M. M. Trexler, et al. "Development of a Modular Load-Sensing Headform," in *Personal Armour Systems Symposium*, Washington D.C., 2018.
[14] Iwaskiw A., Howes C., Hingorani R., Bradfield C., Mazuchowski E., Clark M., Luong Q., Drewry D. Loads Associated with Behind Helmet Blunt Trauma: Matched-Pair Load Sensing Headform Tests Correlated with Skull Fracture Severity. Personal Armour Systems Symposium, Dresden, Germany, 2023.
[15] Anctil B., Bourget D., Pageau G. Rice, K. and Lesko J. "Evaluation of Impact Force Measurement Systems for Assessing Behind Armour Blunt Trauma for Undefeated Ballistic Helmets." Personal Armour Systems Symposium, Hague, The Netherlands, 2004.
[16] Bolduc, M, Anctil, B. Improved Test Methods for Better Protection, a BABT Protocol, Proposal for STANAG 2920, Personal Armour Systems Symposium, Quebec, Canada, 2010.
[17] Philippens M.M.G.M., Anctil B., Markwardt K.C., Results of a Round Robin Ballistic Load Sensing Headform Test Series. PASS 2014, London, UK, 8th-12th September 2014, pp.
[18] Philippens M., BABT protection for composite ballistic helmets; Ballistic Load Sensing Headform & Suspension System; TNO report TNO2014 R10151, Technical Sciences, P.O. Box 2280 AA Rijswijk, The Netherlands, 2014.
[19] Grobert S., Gedon H., Peschel O. "Development of a Test Setup for an Integrated Assessment of Ballistic Backface Deflections on Combat Helmets." *Personal Armour Systems Symposium*, Amsterdan, Netherland, 2016.
[20] Grobert S., Gedon H., Peldschus S., Peschel O. "Comparative Tests with Pk17dynA for an Integrated Assessment of Ballistic Backface Deflections on Combat Helmets." *Personal Armour Systems Symposium*, Washington D.C., 2018.
[21] Voo L., Swetz S., Luong Q., Warfield J., Alvarez B., Brinkley K., O'Donnell J., Bevan M., Merkle A. "Improved Repeatability and Reproducibility of the Ballistic Load Sensing Headform." Personal Armour Systems Symposium, Amsterdan, Netherland, 2016.
[22] Tanaka S., Hokamoto K., et al. "High-velocity impact experiment of aluminum foam sample using powder gun." *Measurement* 44, 2185–2189, 2011.
[23] Bartyzak S. and Mock W. "Characterization of viscoelastic materials for low-magnitude blast mitigation." J Physics: Conference Series 500 (2014) 182003.
[24] Daniel R., Rooks T., Gomez J., et al. "Evaluation of Skull Fracture Risk Associated with Helmet Back-Face Deformation." USAARL-Tech-FR—2022-24, 2022.

Detection of defects in ceramic faced armour

M.J. van der Jagt-Deutekom[1], J.P.F. Broos[1], and H. Jager[2]

[1] *TNO, Defence, Safety & Security, PO box 480, 2501 CL Den Haag, The Netherlands*
marjolein.vanderjagt@tno.nl
[2] *Clothing & Personal Equipment Branch, Materiel and IT Command, Ministry of Defence, P.O. Box 90125, 3509 BB, Utrecht, The Netherlands*

Abstract. To protect against high velocity AP rifle threats, ceramic plates are used as body armour. Nowadays mainly multi curved, monolithic plates are used, in order to better conform to the human torso. Ceramic is a brittle material and thus susceptible to cracking after a blunt impact or drop. These cracks have shown to decrease the ballistic performance of the plate. Therefore, regular inspections are necessary to ensure the plate is safe to use. Since the ceramic in the armour plate is covered, it is impossible to observe from the outside whether the ceramic is damaged or not. There are several non-destructive test methods to inspect the condition of the ceramic. X-ray imaging is commonly used and can reveal cracks and manufacturing defects. There are also techniques on the market that merely distinguish plates in good condition from defect ones. These are available as simple, portable systems, for use in the field. This paper gives an overview of different detection methods known to be available and their potential to identify different types of defects. It compares the results obtained by two different techniques on a set of plates.

1. INTRODUCTION

Nowadays, several types of ceramic body armour plates are used to protect against incoming high energy, small calibre, Armour Piercing (AP) bullets. Ceramic body armour plates are made of two basic parts: a hard ceramic top layer and a ductile composite backing glued together. Additionally, in some types an impact absorbing layer is placed in front of the ceramic and around the edge to protect the ceramic against impacts.

Cracks within the ceramic are easily caused by impacts, falls or improper storage. Cracks could have a reducing effect on the ballistic performance [1][2] although the drop in performance depends on the distance between the point of impact of the projectile and the pre-existing crack [3].

To assure the quality of ceramic armour plates during storage and use, regular inspections of the ceramic in the plate are necessary. The methods used need to be reliable, not endangering the condition of the plates and are preferably quick and easy to perform. Since the ceramic top layer of the armour plate is always covered, it is impossible to observe from the outside whether the ceramic is damaged. There are several Non-destructive Inspection (NDI) techniques possible to inspect the condition of the ceramic.

2. NDI TECHNIQUES TO INSPECT CERAMIC PLATES

This paragraph gives a short overview of different NDI techniques applied to ceramic body armour plates. The aim is not to be comprehensive, but to illustrate various techniques and their visualization capabilities.

2.1 X-ray and CT scan

X-ray imaging machines work by using electromagnetic radiation to create images of the internal structure. In a shielded cabinet, the ceramic plate is placed between an X-ray source and a detector, such as photographic film or a digital sensor. The X-ray image is a black and white image that is formed by absorption, transmission and scattering of the incoming X-ray radiation. Cracks and defects in the ceramic plate become visible because they alter the intensity of the radiation passing through the material. Cracks create gaps or changes in density, resulting in varying levels of X-ray absorption. Additionally, ceramic manufacturing defects, such as porosity from incomplete infiltration or unsintered material, will also appear due to density differences, as illustrated in Figure 1.

 https://doi.org/10.52202/081621-0038

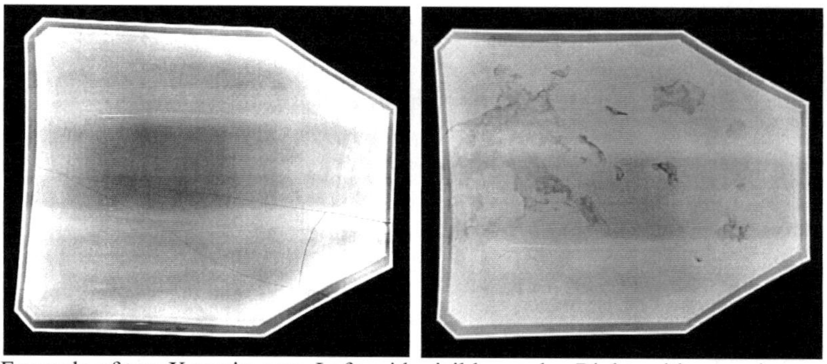

Figure 1. Example of two X-ray images. Left: with visible cracks. Right: with ceramic manufacturing defects.

A CT scan is an advanced imaging technique that combines multiple X-ray images taken from various angles to produce detailed cross-sectional images (slices). CT scans provide more detailed and comprehensive images compared to standard X-rays. They can visualize: cracks parallel to the surface, the through-thickness direction and size of the cracks and delamination of the ceramic from the backing layer. However, this technique is more expensive and time-consuming, and the equipment requires more space and infrastructure, including specialized rooms with shielding to protect against higher radiation levels.

2.2 Infrared

TNO investigated Infrared (IR) photography techniques for crack detection in ceramic armour plates [4]. The IR detection technique involves three steps: heating the plate, cooling down the plate in one direction to start heat flow and a cooling phase during which an IR camera monitors the temperature distribution across the plate. As cracks obstruct the heat flow, discontinuities in the temperature distribution will become visible.

An experimental set-up was constructed using existing components. Once the plate was heated to around 80°C, it is positioned upright between two copper tubes with circulating cold water (5-10°C), which are conductively connected to its left and right side. During the cooling process, the IR camera was able to detect cracks, as illustrated in Figure 2.

The experiments were only a proof of principle and were only conducted on a few plates with clear distinct cracks from edge to edge. These large sharp cracks could be identified as shown in Figure 2 (right).

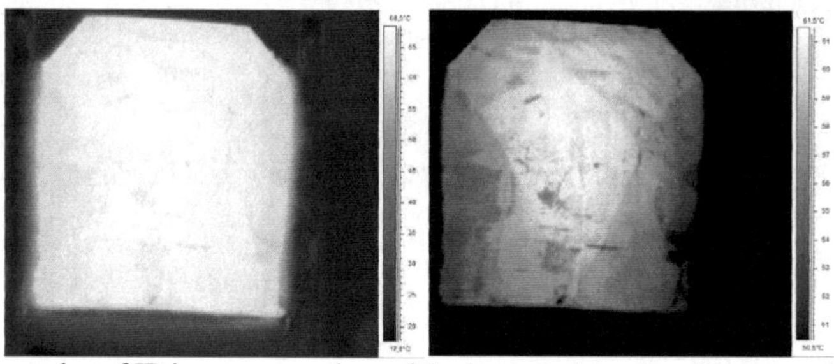

Figure 2. Two examples of IR images over time of the heated plate placed between the cooling tubes: immediately after placement (left) and after a few minutes of cooling (right). [4]

2.3 MEMS-sensor array inspection

A low-cost micro-electro-mechanical system (MEMS) sensor array and dedicated processing system has been developed by TNO for quality assurance of aerospace structures [5]. This inspection approach uses a single piezoelectric actuator in contact with the laminate to excite it with Lamb waves, a type of elastic wave that propagates in solid plates. The leaky Lamb waves lose energy to the surrounding medium through acoustic radiation. As these waves travel through the material, defects alter their behaviour. An array of MEMS sensors records the energy that radiates into the surrounding air. The processing system translates the measured wavefield into a thickness map of the inspected part.

A first attempt has been made to apply this technique to ceramic body armour plates. Six plates were scanned, four contained visible cracks on X-ray and two did not. This was not revealed to the operator beforehand. Because of lack of detailed material information, the processing system displays velocity scans instead of thickness.

Although this technique is not developed for this material composition, first results are promising as illustrated in Figure 3. The two plates with no visible cracks on X-ray, also did also not show a large variation in output with the MEMS technique. So both techniques distinguished the undamaged plates. Of the 4 plates with cracks, the MEMS technique clearly indicated the locations of the cracks in 3 of the plates. However, for the plate with a crack at the top, the MEMS technique did not show a sufficient difference to identify a crack at that location. So, for the damaged plates the success rate was three out of four.

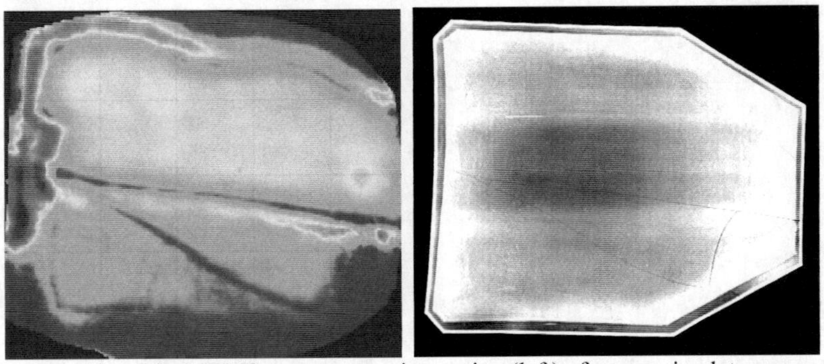

Figure 3. Example of the MEMS-sensor array inspection (left) of a ceramic plate compared (The cracks show up in blue) to the X-ray image (right) of the same plate.

2.4 Ultrasound

Ultrasound uses high-frequency sound waves to inspect materials. A transducer sends these sound waves into the material. When the sound waves hit a boundary (like a crack or defect), they bounce back. The transducer captures these returning echoes. The ultrasound machine processes the echoes into an image, showing the internal structure of the material.

There are two main types of ultrasonic testing: contact and immersion. Contact testing uses a gel or oil to transmit sound waves from the transducer into the material. For immersion testing the sample is submerged in water. The gel or water is used to facilitate the transmission of sound waves between a transducer and the object being inspected in ultrasonic testing. It helps ensure efficient transfer of sound energy by eliminating air gaps, which can interfere with the sound waves. Immersion testing is used in laboratory settings for complex shapes and detailed inspections.

The state of the sample is assessed by analysing the intensity of ultrasound signals and the transit time through the material (known as Time-of-Flight or TOF). Kesharaju conducted research using an ultrasonic immersion testing method on defected ceramic armour components [6]. This research investigated how porosity affects the ultrasonic TOF of reflected signals to establish a correlation between velocity and density across the ceramic component. Figure 4 shows an example of the generated output for the ceramic part of a body armour plate. Do note, that this was a scan of only the ceramic part and not the complete body armour plate.

An accuracy of 96% was obtained by comparing the ultrasonic results against X-ray results. An overall accuracy of 100% was achieved using ultrasonic inspection coupled with artificial intelligence based signal processing and validated against micro-CT scan results. Crouch [6] concluded that the ultrasonic detection method, coupled with intelligent signal processing, has been identified as being a useful, complementary tool for measurement of local variations in thickness and/or bulk density to be used in manufacturing process of the ceramic component.

 https://doi.org/10.52202/081621-0038

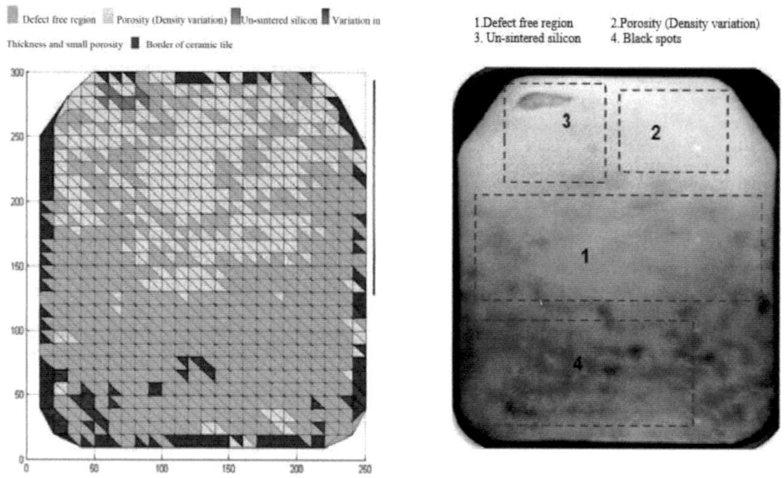

Figure 4. Example of the ultrasonic mapping of a ceramic tile (left) in a plate together with the X-ray image (right) of the plate. [6]

2.5 Ultrasonic resonance

Resonance Ultrasonic Vibrations (RUV) technology measures the ultrasonic resonance response. This involves sending ultrasonic waves selected from a broad frequency range (20 - 250 kHz) through the material and analysing the resonance frequency curve [7].

Cracks and delamination in the ceramic plates cause deviations in the resonance response. These flaws are detected by comparing the resonance response of a flawed plate to that of a standard, non-flawed plate. A crack introduced into the ceramic alters RUV peak parameters: bandwidth and peak position, as illustrated in Figure 5 (left plot). A delamination within the ceramic armour plate components alters RUV peak parameters: amplitude, bandwidth and peak position, as illustrated in Figure 5 (right plot). This method uses specific quantitative criteria, such as changes in amplitude, bandwidth, and shifts in peak position of the resonance frequency, to identify flaws.

The RUV method to inspect delamination is based on a statistical approach. The manufacturer states that in case studies, the capture rate of RUV method approaches 100%. It is not known from the website [7] how many different plate constructions were tested in these case studies.

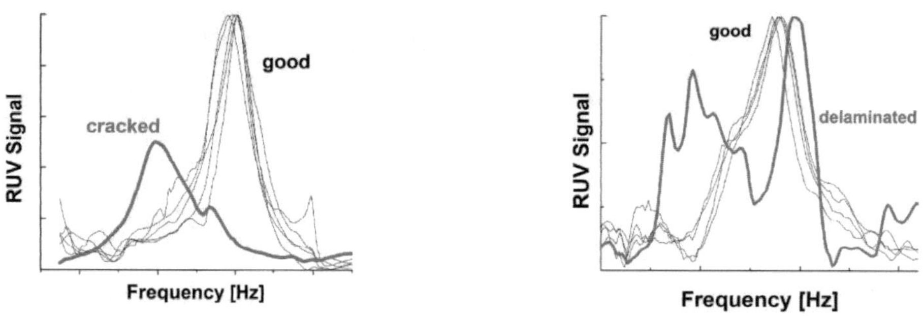

Figure 5. Examples of the frequency response of the RUV technique, showing a typical response for a cracked plate (left – red line) and a delaminated plate (right – red line). [7]

2.6 Impulse excitation

The principle of impulse excitation relies on the fact that every object has natural vibration frequencies tied to its mechanical properties. The specimen is mounted on supports that allow it to vibrate freely in one or more directions. When an object is struck with a small hammer, it vibrates at these natural frequencies. These vibrations can be detected with a transducer, and the collected data can then be used to calculate material properties or to capture the object's resonance signature. Figure 6 gives an illustration of this technique.

https://doi.org/10.52202/081621-0038

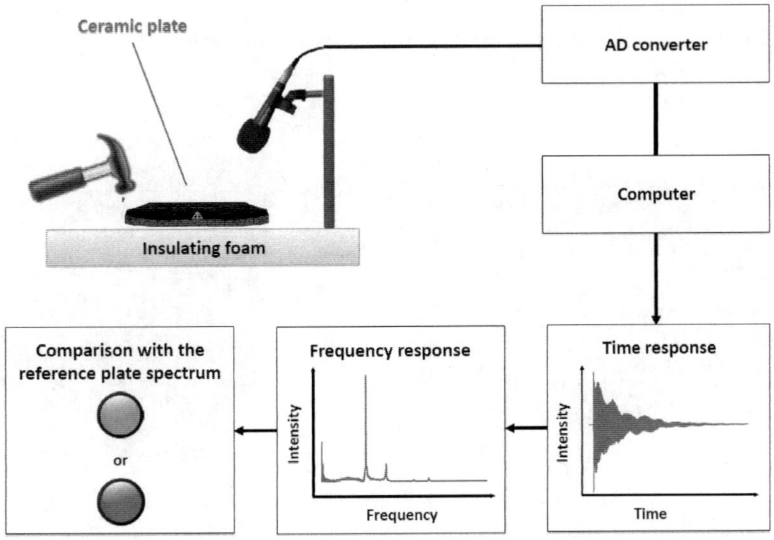

Figure 6. Schematic view an impulse excitation device; the Acoustic Test Device [8].

Impulse excitation is a fast and straightforward, non-destructive testing method for measuring the mechanical properties of materials and is commonly used in quality control, materials science, engineering, and industry. A study using this method will be presented in Chapter 3.

2.7 Summary NDI techniques

Some NDI techniques provide in-depth information about the size and location of cracks, as well as potentially other defects. They can be complex, and often require specialized equipment that could be difficult to transport. These techniques are ideal for laboratory settings, where a more scientific examination can be conducted. Other methods simply determine the presence or absence of cracks, these are classified as go/no-go methods. They are straightforward, and could be suitable for use in portable systems, making them ideal for field applications. Table 1 gives an overview of the different methods, including the complexity of it.

Table 1. Summary of NDI techniques to inspect ceramic plates

	Go-no go	Visualized defects			Complexity
		Cracks	Manufacturing defects	Delamination	
X-ray		✓	✓		+
CT scan		✓	✓	✓	++
Infrared		✓			-
MEMS sensor array		✓			-
Ultrasound			✓		+
Ultrasonic resonance	✓	-		-	--
Impulse excitation	✓	-			--

3. IMPULSE EXCITATION MEASUREMENT CAMPAIGN

3.1 Methodology

For detection of defects in ceramic body armour plates, a research demonstrator based on impulse excitation was developed by the Swiss Engineering School "HE-Arc" in collaboration with Armasuisse [8]. Figure 7 gives an illustration of the first generation Acoustic Test Device (ATD), which had a fixed geometry regarding excitation and response measurement and only one microphone to record the acoustic response.

 https://doi.org/10.52202/081621-0038

The test equipment utilised in the measurement campaign is capable of conducting an evaluation based on the analysis of up to five microphones. These microphones are mounted on magnetic mounts (as illustrated in figure 8). The microphones are freely movable to adapt to different plate geometries. The position of the hammer is fixed. Figure 8 illustrates the positions of the microphones used for experiments on Type 1 plate.

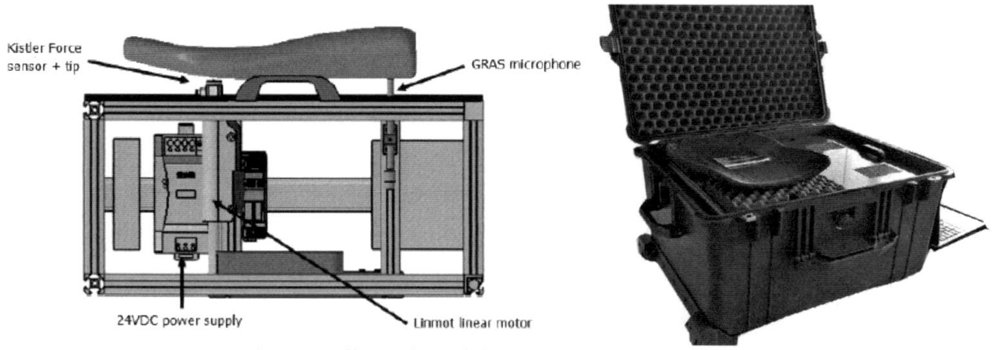

Figure 7. Illustration of the Acoustic Test Device. [8]
Left: Schematic cross section without foam layer (red indicates a plate). Right: Picture of the ATD.

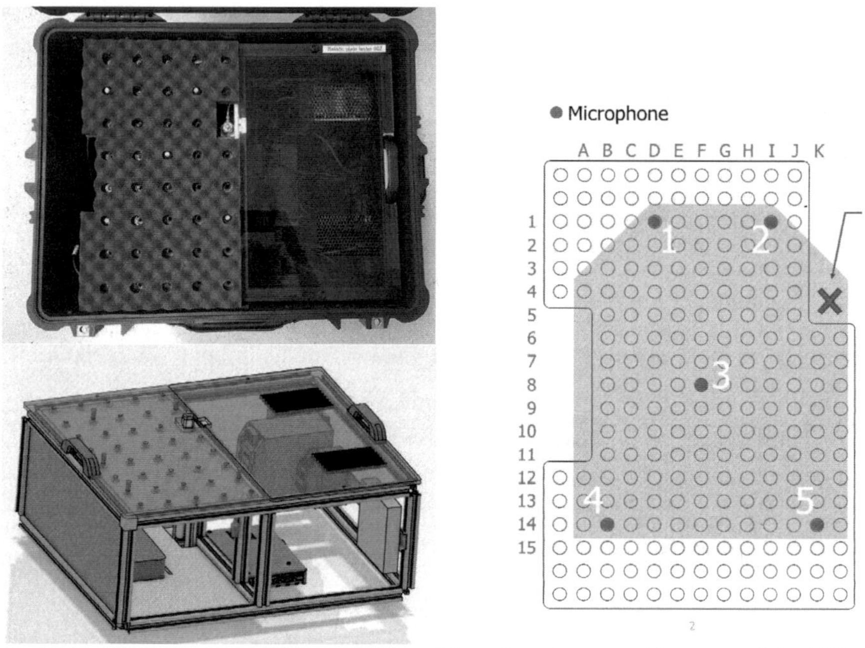

Figure 8. Illustration of the positioning of the microphones and hammer of the ATD used in the measurement campaign. Left-top: top view of the ATD with multiple microphones in the foam layer. Left-bottom: Schematic illustration without the foam layer. Right: positioning of the microphones for the experiments.

This method relies on assessing six criteria, which are derived from an analysis of the acoustic response of the plate in both the time and frequency domains: 1) frequency and 2) the peak width of the first vibrational mode, 3) mean of the Frequency Response Function (FRF), 4) damping time constant, 5) initial amplitude and 6) the damping in the time domain (R^2 coefficient derived from the exponential regression – noted as R2).

The acceptance limits for these criteria must be established on data from a significant number of pristine plates. For a plate to be accepted, all six criteria must fall within the specified limits for its corresponding plate type. If any criterion falls outside the acceptable range, the plate is rejected [9]. Once the criteria are established, this method can effectively identify whether cracks are present in a plate.

3.2 Experimental

Two types of ceramic body armour plates have been used for a measurement campaign with the ATD research demonstrator [9]. These plates were examined with X-ray imaging by TNO for the Clothing & Personal Equipment Branch of the Dutch military of Defence. The condition assessment on X-ray analysis was done by TNO and focussed on both cracks and manufacturing defects. If one or both defects were observed on the X-ray image, the plate was labelled "out of use". It must be noted that the assessment of the plate condition based on X ray analysis is not always trivial as no clear definition is available which size of manufacturing defect should lead to a reject.

3.2.1 Result Type 1 plate

The construction of the Type 1 plate is straightforward as it consists only of a monolithic SiC ceramic glued onto a UHMWPE backing enveloped in a textile cover. First, the preliminary acceptance limits were set, based on the response of seven plates of known condition (four undamaged and three damaged). The signal of microphone number 1 was used for the analysis. Microphone number 1 is positioned in the upper left corner, opposite to the shock location in the upper right corner along the plate's longitudinal axis as shown in figure 8 (right).

Figure 9 shows the Frequency Response Function (FRF) of the seven plates around the first frequency mode. Plots of damaged plates are in bold. As a first observation, the FRFs differ significantly between damaged and undamaged plates. For this plate type, the interval of [200 Hz, 1500 Hz] is set in the acoustic tester software to find the peak corresponding to the first vibrational mode. For each of the seven plates, the values of the six evaluation criteria were extracted from the signals, and preliminary acceptance limits were defined.

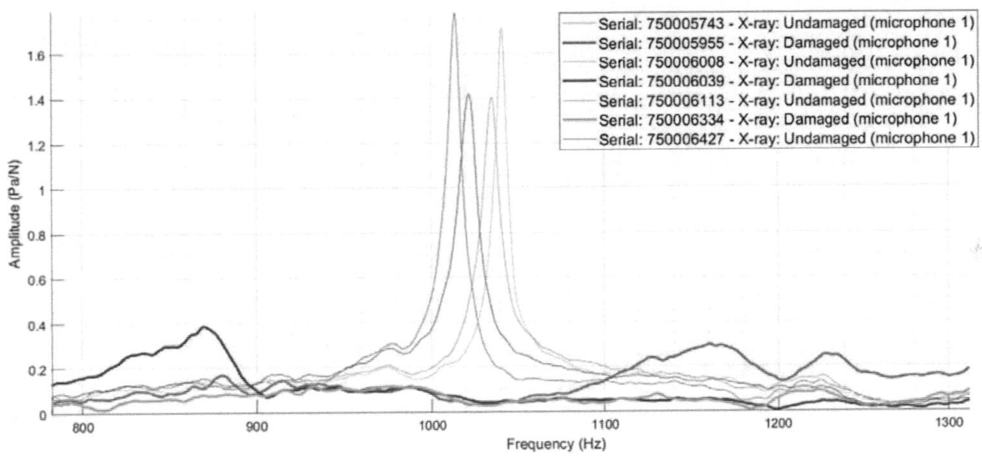

Figure 9. Frequency Response Function (FRF) of the seven plates of type 1 used for pre-tests with zoom on the 1st vibrational mode. Damaged plates are plotted in bold. [9]

Throughout the campaign, a total of 125 type 1 plates were measured using the ATD. At the conclusion of the campaign, the X-ray images of these plates, along with the corresponding assessments by TNO, were provided to Armasuisse.

As explained, for a plate to be accepted with the ATD, all six criteria must lie within the limits defined for the corresponding plate type. As soon as a criterion is out of the acceptance limits, the plate is rejected. Figure 10 shows the preliminary acceptance limits and an overview of the values for each plate criterion. The fact that the response of defect plates may fall within one or more of the criteria for undamaged plates demonstrates that multiple criteria are needed to distinguish the damaged from the undamaged plates. Table 2 compares the results of both methods.

https://doi.org/10.52202/081621-0038

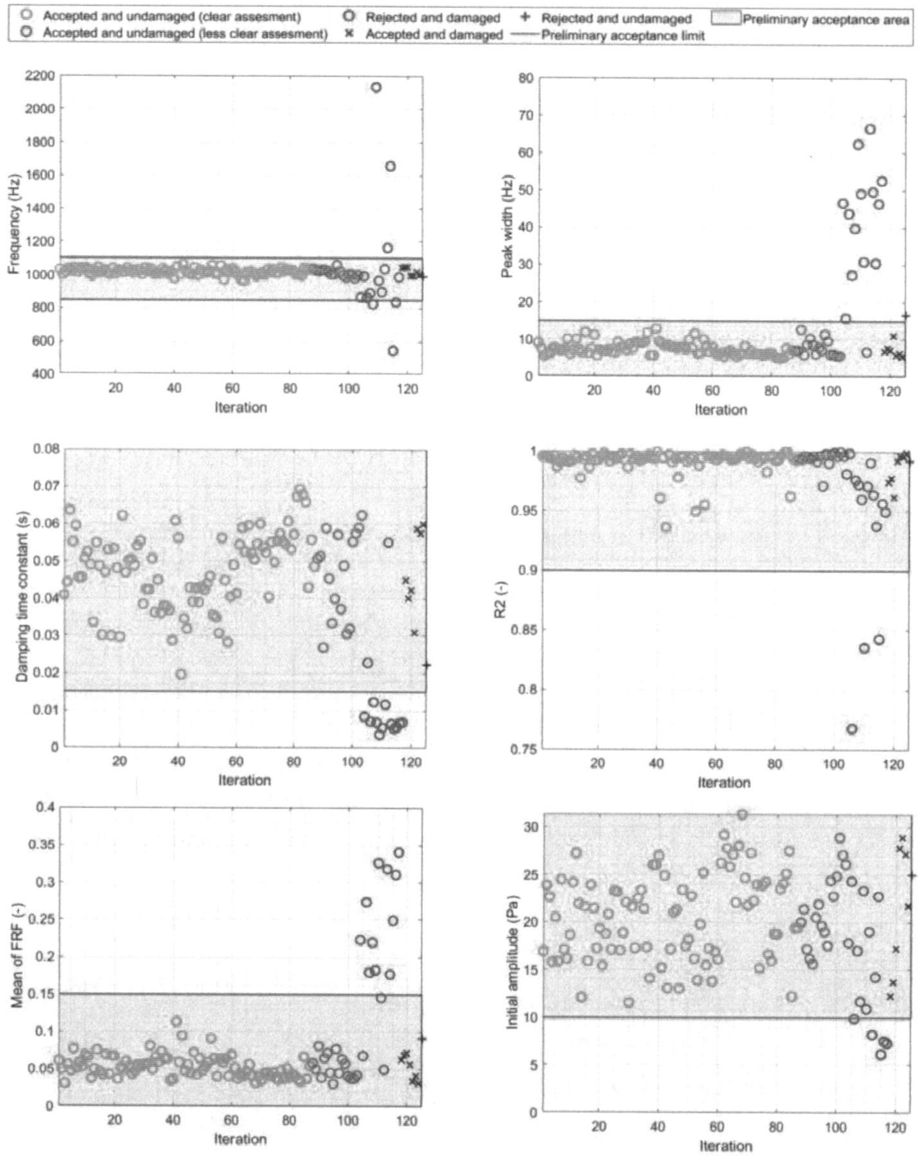

Figure 10. Overview, for each criterion, of the values obtained for each measured plate of type 1. The colour of the markers is associated with the X-ray assessment (damaged or undamaged) and indicated in the legend box. [9]

Table 2. Overview of ATD and X-ray image results for plate Type 1.[9]

	ATD result: accepted	ATD result: rejected
X-ray result : undamaged	103	1
X-ray result : damaged	7	14

Despite the very small number of plates used to determine the acceptance limit, 117 out of 125 acoustic measurements gave the same results ("match") as the X-ray analysis, i.e. 93.6%, and eight acoustic measurements gave different results ("mismatch"), i.e. 6.4%. These promising results, can be explained by the large difference in the acoustic response ("on/off character") between damaged and undamaged plates. The X-ray images revealed that the ATD failed to notice in 7 out of the 21 plates labelled as a mismatch "x", because of the following defects:

- One plate had a small piece of ceramic broken off in a corner. Such defects typically show little influence on the acoustic response.

- Three plates showed manufacturing defects on the X-ray (see Figure 11). The influence of such defect on the acoustic response is not precisely known.
- One plate showed an unusual crack pattern in a serpentine shape, possibly caused by shrinkage.
- One plate showed a crack visible on the upper right corner of the plate. The influence of such a crack on the acoustic response might be limited, as its position lies well off the path between actuator and microphone.
- One plate showed three clear cracks, but the FRF was very similar to a undamaged plate.

This analysis shows that 3 out of 17 plates with visible cracks on X-ray were undetected with the ATD.

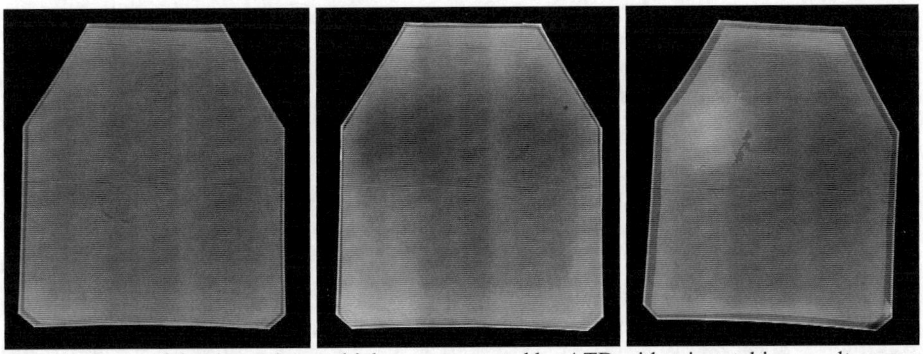

Figure 11. X-ray of the three plates which were accepted by ATD with mismatching result compared to the X-rays analysis (i.e. damaged) due manufacturing defects.

3.2.2 Result Type 2 plate

The construction of the Type 2 plate consists of a thin impact protective foam, a fully wrapped monolithic SiC ceramic bonded onto a UHMWPE backing, all enveloped in a textile cover. Evaluating their condition based on acoustic response is challenging, as in one of the damaged plate even large cracks had a less significant impact on the FRF compared to type 1 plates as shown in Figure 12.

During the campaign, 53 type 2 plates were measured using the ATD. Establishing preliminary acceptance limits was not yet possible. Further studies are taking place to understand the reasons behind this limitation of the ATD, and to improve the method's applicability and sensitivity.

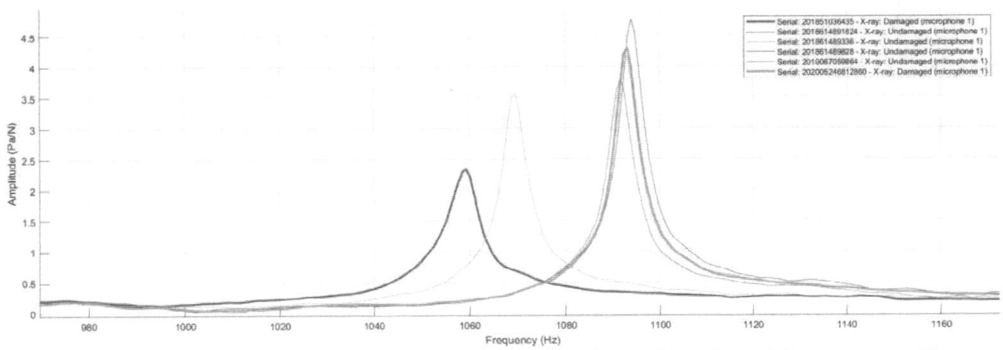

Figure 12. Frequency Response Function (FRF) of the six plates of type 2 used for pre-tests with zoom on the 1st vibrational mode. Damaged plates are plotted in bold. [9]

5. SUMMARY

Different techniques for the detection of defects in ceramic body armour were discussed. Two NDI techniques have been assessed by TNO to detect defects in ceramic body armour plates: infrared inspection during cooling down a heated plate and the MEMS sensor array inspection. Both methods provide information about the size and location of cracks and could be suitable for use in portable systems. Especially the MEMS sensor array system is an interesting technique to further investigate its potential for ceramic body armour plates.

https://doi.org/10.52202/081621-0038

One go/no-go evaluation technique, impulse excitation, was evaluated with two plate types differing in construction. Although this technique is simple to use, setting the conditions for a go/no-go approach is dependent on the plate construction. The ATD shows promising results; a relatively simply constructed plate shows large difference in the acoustic response between damaged and undamaged plates; of the 17 plates with visible cracks on the X-ray, 14 were detected by the ATD. Additionally, only 1 out of 104 plates with no visible defects on the X-ray was incorrectly identified as defective. It becomes challenging when even large cracks have a less significant impact on the acoustic response as appears to be the case with a plate more complex in design. Understanding the reasons behind this is essential for further developing the ATD, to improve the method's applicability and sensitivity.

Acknowledgments

The authors would like to thank ArmaSuisse and the Swiss Engineering School "HE-Arc" for the collaboration in the measurement campaign and technical information on the methodology of the ATD device as provided for this paper.

References

[1] Broos J.P.F. and Gunters R., 'Study on the ballistic performance of monolithic ceramic plates', 23rd International Symposium on Ballistics, Taragona, Spain 16-20 April 2007.

[2] Watson C., Ringrose T., Champion S., Horsfall I. and Mallon B., 'The effect of cracks on the ballistic performance of contoured protective body armour plates', 23rd International Symposium on Ballistics, Taragona, Spain 16-20 April 2007.

[3] Crouch I.G., 'Effect of cladding ceramic and its influence on ballistic', 28th International Symposium on Ballistics, Atlanta, GA, 22-26 September 2014.

[4] Broos J.P.F, Brouwer S.D., Carton E.P. and Frade J., 'Crack detection in ceramic faced armour and the influence of cracks on ballistic protection capacity', TNO report TNO 2014 R11434, January 2015.

[5] Volker A.W.F, Vrolijk J.W., Merks-Swolfs E.J.W., van der Burg D.W., van der Heiden M.S. and Martina Q.E.V.N, 'Non-contact MEMS-sensor array inspection of composites and metallic parts using lamb waves', Journal of Nondestructive Evaluation Diagnostics and Prognostics of Engineering Systems Vol. 6 no. 4, 041002-1 - 041002-8.

[6] Crouch I.G., Kesharaju M. and Nagarajah R., 'Characterisation, significance and detection of manufacturing defects in reaction Sintered Silicon Carbice armour materials', Ceramics International Volume 41(2015), 11581–11591.

[7] Crack and Flaw Detection in Ceramic, https://www.ultrasonictech.com/applications/armor-plates; webpage visited on 3-3-2025.

[8] Thévenaz D., Fatton Q., Drapela P., Moutarlier N., Vuille D., Lampert S., Folly P. and Lüthi J.-D, 'Condition assessment of ballistic plates using modal analysis, 3rd international workshop on ageing effects in protective systems, components and materials', French-German Research Institute of Saint-Louis, France, 15-17 October 2019, poster.

[9] Moutarlier N., and Thévenaz D., 'Acoustic measurement campaign in Soesterberg (NL)', Armasuisse / Panda report 23MICa10, 24-4-2023.

Rigorous Bayesian Improvements to Ballistics Armour Testing Methods

M. McKibben[1], L. Stabile[1], A. Iwaskiw[1] and E. Cawi[1]

[1] *Johns Hopkins Applied Physics Laboratory, 11100 Johns Hopkins Road, Laurel, 20723 MD., USA, Michael.Mckibben@jhuapl.edu*

Abstract. While testing methods such as Neyer and 3POD offer marked improvements over earlier test methods such as Bruceton and the step method in NIJ Standard 0101.07 [1], there is still some susceptibility to the initial user guess (μ_{Guess} and σ_{Guess}). Johnson et al [2] offered a rigorous comparison of all testing methods currently in the field, and noted that a mis specified σ_{Guess} can lead to large bias for all tested methods. Additionally, while in some cases prior data such as manufacturer testing and previous first articles tests may be available to inform μ_{Guess} and σ_{Guess}, there is not an existing framework for how this data should be included in a model. Note that updating σ_{Guess} is also consistent with the movement to incorporate unbiased error measurements in the broader design of experiments community initially spearheaded by Gilmour and Trinca [3]. In this paper, we offer both a rigorous Bayesian framework to include prior data, as well as additional considerations on when to include and how to weight that information. We also include a simulation study demonstrating the implications both of updating σ_{Guess}, as well as of a Bayesian design more generally. The results have implications for a variety of testing methodologies; for example, it is well known that Robbins Monro may not be able to achieve an unbiased estimate of V50 even asymptotically when initial parameters are incorrectly specified (i.e. when σ_{Guess} is too small).

1. INTRODUCTION

Personal protective equipment (PPE), which is designed to stop ballistic threats, is an important tool to protect military and law enforcement personnel. In order to ensure optimal performance of these products, PPE must be tested using live ammunition to evaluate their protective capabilities. This is primarily achieved by employing a scheme of testing where armour performance (scored as a partial or complete penetration) is assessed at various ballistic conditions (ambient versus elevated temperature, etc.) in a limited range of firing velocities. Various testing methodologies [4], [5], [6] describe velocity selection procedures, tools, measurements and techniques to test and assess the ballistic performance of PPE. These methods are used to determine various ballistic performance metrics, such as resistance to penetration, where 0 percent of events is expected to result in complete penetrations (RTP), and the ballistic striking velocity where 50 percent of events will be complete penetrations ($V50$). However, these tests are costly and therefore there is a strong desire to reduce the testing resources needed to calculate these ballistic measures to a specific confidence, as well as to expand the characterization of the PPE ballistic response curve. Using advanced statistical approaches may reduce the resources needed to accurately assess ballistic PPE performance [7]. This would also allow practitioners to calculate additional metrics, such as the ballistic striking velocity where 10 percent of events will be complete penetrations ($V10$).

Most statistical methods used in ballistics testing are Frequentist [7]. Bayesian methods allow the incorporation of information from previously collected data into the selection of tested velocities, through the use of priors. In complicated live-fire destructive testing firing schemes, Bayesian methods could better inform both initial test conditions and the subsequent testing sequence. These methods could also reduce the number of tests needed to achieve a certain level of confidence with respect to ballistic performance.

Studies of existing methods have generally not included the perturbation of target velocity to actual velocity in their simulations, i.e. when the striking velocity differs from the intended velocity. This additional source of variance is essential when considering the efficacy of existing methods in practice. This paper compares commonly used ballistic firing sequences (Neyer, 3POD, Bruceton) via a simulation study in which this variation is captured to assess the ability to reduce the number of tests needed through these approaches. Additionally, we introduce the idea of a Bayesian methodology to incorporate additional data that is currently unused in the test generation of initial conditions estimates (μ_{Guess} and σ_{Guess}). In certain scenarios, it is hypothesized that this will allow the practitioner to find the Zone of Mixed Results (ZMR) faster, and produce more accurate response curves than currently used methods.

https://doi.org/10.52202/081621-0039

2. LITERATURE REVIEW

One of the earliest ballistics testing procedures to gain popularity was Bruceton [8], a binary search procedure published in 1948. While Bruceton is consistently able to find V50, it is generally inefficient unless an optimal starting step size is chosen. The Robbins-Monro and Delayed Robbins-Monro [9], [10] from 1961 and 1985, respectively, improved upon this method by dynamically altering the step size throughout the test. Robbins-Monro Joseph [11] incorporates the notion of V50 estimation following a normal distribution in shot selection, which was built upon by later methods. Langlie [12] combined several of these early ideas in 1962 into a sequential testing methodology. Ultimately, many of these methods are inefficient when compared to the advancements made in the statistical field of modern optimal design methodology [13] and may be overly dependent upon starting parameters such as μ_{Guess}, σ_{Guess}, and the initial shot velocity. For a more substantive review of historical ballistics testing methods, the reader is directed to Johnson et al. [2].

Neyer was the first to propose a systematic method for generating an optimal design in 1994 [14]. The method consists of two phases: (1) an initial series of shots to establish the appropriate stimulus range and break separation (i.e. finding the Zone of Mixed Results). This is necessary as unlike many traditional Design of Experiments (DoE) settings, the bounds are not known a priori. This is followed by stage (2) the use of a D-Optimal design for subsequent shots [14]. Sequential D-Optimal designs have seen substantial use in the broader DoE applications literature [13], however, Neyer was the first to apply it to destructive sensitivity testing. One of Neyer's contributions was the inclusion of the first phase of the experiment to establish the bounds of the experimental range. This differs from many traditional DoE settings where the bounds are known a priori through subject matter expertise.

3POD is a three-phase sequential optimization procedure that aims to find performance metrics of interest, such as V50, more efficiently than Neyer [15]. Similar to Neyer, the first phase establishes the appropriate experimental range. This is followed by finding the Zone of Mixed Results (breaking separation), and adding points to optimize the remaining tests. Phase 1 of 3POD chooses different points from Neyer to break separation more quickly, though the underlying approach is similar. Unlike Neyer, 3POD explicitly adds a third step with n_2 points to optimize for specific quantiles of interest, as specified by the user. This follows a stochastic search procedure proposed by Joseph [11] to estimate specific quantile of interest; while this method has desirable asymptotic properties, there is some question as the how it performs in the presence of both small n and when the target shot velocity is perturbed. Follow on work for 3POD 2.0 [16] offers some technical improvements, while GONOGO [17] offers an implementation of both 3POD, Neyer, and other methods.

This paper also drew inspiration from modern design techniques in other spaces; specifically, the Bayesian D-Optimal designs of DuMouchel and Jones [18], the Bayesian perspective of Dror and Steinberg [19], and work in the clinical trials space [20], [21], [22], [23]. Collectively, the promise of Bayes is two-fold. First, it allows for the logistic regression curve to be estimated before separation has been broken. This can be particularly useful when choosing future points, as current techniques rely on an average of velocities. Second, these methods have repeatedly demonstrated the power of relevant prior information when creating an optimal design and applying the results to a sequential optimization procedure such as 3POD.

3. METHODS

3.1 A Bayesian Approach

In the broader design community, there is a movement to use Bayesian statistics in both the creation of design matrices and in analysis. Note that priors are generally said to be strong when the prior has a large impact on the analysis or test points selected, and weak or vague when the impact is minimal. In some sense, all the methods discussed in the literature review use prior information. Specifically, the μ_{Guess} and σ_{Guess} parameters in 3POD and Neyer are often based on previous step-down experiments conducted before the test of interest. This is then incorporating summary statistics of prior information, while a Bayesian approach would use the actual test information. Other methods, such as Bruceton, require some estimates related to $V50$ range and variability. There is generally minimal issue with using additional information in choosing initial starting points, provided those starting points are not part of an analytical solution once the data has been collected.

The study presents two particular recommendations. The first is the use of a Bayesian framework to generate subsequent test points, i.e. target velocities. The priors allow for the easy and relevant incorporation of prior information into the design matrix. The second is for additional prior information to be included for the sole purpose of optimal points for shot selection. In many cases, manufacturers have conducted extensive internal testing prior to submission to the government for acceptance or rejection (First Article Testing, i.e. FAT). In some cases, this testing is contracted to the same facility that will perform the FAT. It is easy to demonstrate that incorporating data from those manufacturer tests into the detemination to accept or reject a piece of equipment would create various moral hazards; for this reason, many regulations generally state that only information from the official tests may be used for analysis. It is also clear that, in some cases, the manufacturer information could be used to form an unbiased estimate of the response curve that would lead to better testing points. Specifically, the manufacturer data would be used to identify optimal target velocities for an acceptance, but would not be included in the acceptance test analysis. This would allow for a variety of benefits; separation would be broken faster, the response curve would fit more accurately by choosing more optimal test points, and higher confidence and lower variability would be gained in final estimates.

3.2 Method Description

The proposed procedure in this section is named Sequential Bayesian Optimal Procedure (SBOP) and is broadly inspired by 3POD [15]. This method is most suitable when prior infomation can be leveraged to inform estimates of the input parameters. These estimates may be obtained from a previous step-down test as is common for 3POD and Neyer, or from manufacturer data described in the previous section. The method has three phases that closely align with the phases of 3POD:

- Phase I, in which initial shots are added and separation is broken
- Phase II, in which points are added to optimally estimate the logistic response curve
- Phase III, in which points are added to better estimate quantiles of interest

3.2.1 Phase I

The goal of the first phase of the experiment is to establish valid test regions and break separation by varying the velocity of the projectile. This phase can be divided into the following steps: (I) Obtaining at least one partial and complete penetration (II) Breaking separation, i.e. finding the Zone of Mixed Results.

3.2.2 M.1 Obtain partial and complete penetrations

When the priors for location (μ_{Prior}) and scale (σ_{Prior}) are identical to μ_{Guess} and σ_{Guess}, the initial shot velocities for shots 1 and 2 of 3POD and SBOP are the same as shown in Equations 1 and 2:

$$Shot\ 1 = \mu_{Prior} - 2\ \sigma_{Prior} = \frac{3}{4}(\mu_{Guess} - 4\ \sigma_{Guess}) + \frac{1}{4}(\mu_{Guess} + 4\ \sigma_{Guess}) \tag{1}$$

$$Shot\ 2 = \mu_{Prior} + 2\ \sigma_{Prior} = \frac{1}{4}(\mu_{Guess} - 4\ \sigma_{Guess}) + \frac{3}{4}(\mu_{Guess} + 4\ \sigma_{Guess}) \tag{2}$$

Define x_{min} and x_{max} as the lowest and highest actual velocities, respectively. These values are updated after each shot. There are four possible scenarios described below for shots one and two, and each of these has implications for subsequent shot selections. Let y_1 be the outcome of shot one and y_2 be the outcome of shot 2, where 1 indicates a complete penetration (CP) and 0 indicates a partial penetration (PP).

A. [y_1=0, y_2 = 1] This represents the ideal (and also most common) scenario. The practitioner is testing in the appropriate range, and can move to step 1 in M.2

B. [y_1=0, y_2 = 0] Both shots are partial penetrations, and it is inferred that the actual range of interest is further occurs at a higher velocity. The next target velocity is

$$x_{next} = x_{max} + 1.5\ \sigma_{Prior}, \tag{3}$$

and tests are iteratively added until a complete penetration is observed.

C. [y_1=1, y_2 = 1] Both shots are complete penetrations, and it is inferred that the actual range of interest is occurs at a lower velocity. The next target velocity is

$$x_{next} = x_{min} - 1.5\ \sigma_{Prior}, \tag{4}$$

and tests are conducted iteratively until a partial penetration is observed.

D. [y_1=1, y_2 = 0] This scenario represents the case that the actual variance is substantially larger than the specified σ_{Guess}. Two additional shots are added:

$$Shot\ 3 = x_{min} - 3\ \sigma_{Prior}, \tag{5}$$

$$Shot\ 4 = x_{max} + 3\ \sigma_{Prior}, \tag{6}$$

https://doi.org/10.52202/081621-0039

In the event that shot 3 is a PP and shot 4 is a CP, we revise the estimate of σ_{Prior} and move to step 1 in M.2.

For any other result, the authors recommend assessing the input parameters and priors for reasonableness, as this suggests the values have deviated very strongly from both subject matter expert and prior data expectations.

There may be cases where the target velocity is substantially different from the actual velocity; historically, these shots have been omitted from the test matrix and subsequent analysis. The authors recommend retaining these data points in the test matrix if the shot is otherwise within specification for other factors such as yaw and obliquity, even if it is necessary to remove them from subsequent analyses related to pass/fail. Existing methods generally do not comment on this issue.

3.2.3 M.2 Break Separation

Once the initial boundaries are established, the next step is to break separation. The practitioner adds shots until a zone of mixed results is established, i.e. the highest velocity of a partial penetration ($M0$) is greater than the lowest velocity of a complete penetration ($m1$). Similar to 3POD, shots are added based on the size of the separation interval, defined as $sep = M0 - m1$:

- If ($sep < 0$): move to the next stage
- If ($sep > 1.5\sigma_{Prior}$): $x_{next} = \widehat{V50}$
- If ($0 \leq sep < 1.5\sigma_{Prior}$):
    - If (Number of CPs > Number of PPs): $x_{next} = m1 + .3\sigma_{Prior}$ *(7)*
    - Else: $x_{next} = M0 - .3\sigma_{Prior}$

SBOP recommends iteratively adding a minimum of 4 shots near $\widehat{V50}$ until $m1 - M0 < \sigma_{Prior}$. Once the separation interval is sufficiently small, the next points added are at the 40th and 60th quantile estimates on the curve. This is broadly in line with the $mean(X) \pm .3\sigma_{Guess}$ recommended by 3POD. Note that the so-called "separation trap" [15] is not an issue in this methodology or 3POD as the chosen points are outside of the separation interval.

3.2.4 Phase II:

Phase II adds points to optimize estimates for the logistic regression curve parameters. In contrast to methods that use a traditional generalized linear model (GLM) for this step, Bayesian Methods use a Markov Chain Monte Carlo (MCMC) to estimate the logistic response curve. In this implementation, SBOP uses MCMCpack [24]. This has the effect of centering the initial estimates at the prior, while allowing the data to play a larger role as the sample size increases. Optimizing for subsequent points has the form of optimizing the information matrix.

Note that n_1 points are assigned to Phase I and Phase II; the faster separation is broken, the more points will be available to optimize the estimate of the logistic regression curve. Future work will include dynamically altering the distribution of points between Phase I, Phase II, and Phase III, as it is likely that some points should always be assigned to Phase II prior to entering Phase III. To avoid cases of poor estimation, clipped estimates of the mean and scale parameter are used as shown in the equations below.

$$\widetilde{\mu_k} = max\left(\underline{x}, min(\widehat{\mu_k}, \bar{x})\right), \widetilde{\sigma_k} = min(\widehat{\sigma_k}, \bar{x} - \underline{x}), \text{ where } \underline{x} = min(x_1, \ldots, x_k) \text{ and } \bar{x} = max(x_1, \ldots, x_k) \quad (8)$$

3.2.5 Phase III:

Similar to 3POD, SBOP adds n_2 points that optimize for quantiles of interest. 3POD and SBOP use the modified stochastic optimization procedure suggested by Joseph [11]. Previous criticism of Robbins-Monro-Joseph is that it may converge slowly when the algorithm starts far away from the true value. This is less of a concern in the SBOP and 3POD implementations given n_1 runs have already occurred. However, as noted by Joseph, "The optimal choice of a_n in small samples has not been investigated, although in most experiments this is the most interesting case." An area for future work is deriving guarantees for small values of n_2. This is particularly relevant in the face of perturbations between the target velocity and actual velocity. Additionally, it is reasonable to compare the performance of the stochastic optimization procedure against a sequential c-optimal design for the quantities of interest for various n_1 and n_2.

4. SIMULATION STUDY

The following simulation study was designed to answer two essential questions:

- Does the Bayesian methodology have potential and warrant further study?
- What is the impact when the priors are correct, i.e. have perfect information, or incorrect, i.e. are substantially different from the truth?

Broadly, two classes of scenarios were considered: one in which the priors and guesses were centred to be exactly on the true response curve (i.e. a given method is provided with oracle guesses), and a second in which they were set to be substantially far apart from the true curve. The response curve equation is shown in Equation 9, where p is the probability of CP. Note that there is an equivalence between the prior β_0 and β_1 for a given logistic response curve in a single variable and the location-scale parameterization with scale $=\frac{1}{\beta_0}$ and location $= -1 * \frac{\beta_1}{\beta_0}$. Code was run in R version 4.4.2 [25], with the base glm function used to calculate the maximum likelihood estimates (MLE) once separation had been broken, and MCMCpack [24] used when a Bayesian logistic regression estimate was needed. The test matrix is shown in

Table *1*.

$$log(p/(1-p)) = \beta_0 + \beta_1 * Velocity \quad (9)$$

Table 1: Test matrix for simulation study

Scenario	μ_{Guess}	σ_{Guess}	β_0 *True*	β_1 *True*	Velocity $\mu_{Perturb}$	Velocity $\sigma_{Perturb}$	μ_{Prior}	σ_{Prior}	n_1	n_2
Perfect Information, N=40	2000	10	-200	0.1	0	5	2000	10	20	20
Perfect Information, N=10	2000	10	-200	0.1	0	5	2000	10	8	2
Imperfect Information, N=10	1800	50	-200	0.1	0	5	1800	50	8	2

For each method in each scenario, the scenario was run over 1000 iterations using the same true ballistic response curve shown in Figure 1. For each iteration:

1. Each method was followed to select the desired velocity for the starting and subsequent shots until $n_1 + n_2 = n$ shots were recorded
2. The desired shot velocity was perturbed according to a $N(0,5^2)$ distribution and the actual velocity was output
3. The outcome of PP/CP was drawn according to the true ballistic response curve for the actual velocity
4. The outcome was recorded, as were the desired and actual velocities

4.2 Results

The ultimate goal of an analysis of ballistics armour is to understand the behaviour of the protective gear under a variety of threats and conditions. While this is often simplified to the V50 of a given threat or condition (e.g. temperature), understanding the full ballistic response curve allows for more informed decisions to be made regarding warfighter and personnel safety.

To this end, several metrics were studied as part of this analysis:

- Percent of runs where a Zone of Mixed Results is identified (i.e. separation is broken)
- Median shot number where overlap is achieved
- Properties of β_0 and β_1 estimates
- Properties of V50, and V10 estimates
- Width of V50, V10, β_0 and β_1 80% intervals

While $V50$ is an oft-reported metric, there is increasing interest in the performance of protective gear at other quantiles, such as $V10$, which may be important metrics under alternative safety and performance thresholds; there is related interest in better understanding the behaviour of the whole curve as defined by β_0 and β_1. These curves can be expanded to include additional variables such as yaw, which would allow practitioners to model armour performance

https://doi.org/10.52202/081621-0039

under more complex conditions. For experimental efficiency, it is preferable to have a smaller number of runs where the ZMR was not achieved (e.g. break separation faster), and to have a smaller median test number at which the ZMR is found. It is also preferable to have a smaller median absolute deviation (MAD) for estimates such as $V50$ and β_0. There were several cases of simulation runs with outliers for all methods; for this reason, MAD was chosen as a more robust metric than standard deviation. In practical applications, these tests would have been discontinued. Unless otherwise stated, all final results are calculated using GLM to calculate the logistic regression MLE coefficients; the stochastic approximation method of 3POD $V50$ estimation is considered separately. For all tables, the bolded values indicate the best result.

4.2.1 Perfect Information Results

All methods performed fairly well in the case where our starting parameters, σ_{Guess} and μ_{Guess}, match the truth and there is sufficient sample size (N=40). Table 2 shows the MADs for the metrics of interest for N =40 and N = 10. While the regression coefficients for SBOP were much closer to the ground truth of β_0=-200 and β_1=0.1 for N = 10, the MAD for $V10$ and $V50$ estimates was generally slightly better for Neyer and Bruceton. All methods performed similarly for N = 40.

Table 2: Median Absolute Deviations for perfect information scenarios

MAD	N = 40				N = 10			
Method	β_0	β_1	**V50**	**V10**	β_0	β_1	**V50**	**V10**
Bruceton	**54.2**	**0.0272**	**6.09**	**2.3**	157	0.0788	11.8	**4.37**
SBOP	55.2	0.0277	6.42	2.69	**116**	**0.0578**	11.8	5.43
Neyer	57	0.0286	**6.09**	2.48	153	0.0762	**11.2**	5.7
3POD	54.9	0.274	6.19	2.52	187	0.0938	11.8	5.79

With N = 40, all methods broke separation. For N = 10, Bruceton broke separation most often at 91.8%, followed by 3POD, Neyer and SBOP, as shown in **Error! Reference source not found.**. This is not surprising as Bruceton performs well when given an appropriate step size; velocity selection is highly influenced by step size. Additionally, this is expected in the presence of perfect information as the other methods use the first two shots (x_1 = 1960 and x_2 = 2040) to establish the appropriate range while Bruceton is already iterating through finding $V50$ (x_1=2000).

We calculate the difference between the 90th percentile and the 10th percentile for β_0, β_1, V50 and V10 as a measure of how closely these metrics are clustered for 80% of the simulation runs. The results are shown in

Table 3. SBOP has the smallest intervals around β_0 and β_1 for N =40, while Bruceton and Neyer have the smallest intervals around V50 and V10, respectively. For N = 10, Neyer has the smallest intervals around β_0 and β_1, followed closely by SBOP. 3POD performs relatively poorly in this metric for N = 10, but does have the smallest interval compared to the other methods around V10. When given a larger sample (i.e. N = 40), SBOP returns the tightest 80% intervals around β_0 and β_1 and performs comparably to the best method when given a smaller sample (i.e. N = 10).

Table 3: 80% intervals for key values, perfect information

80% Intervals	N = 40				N = 10			
Method	β_0	β_1	**V50**	**V10**	β_0	β_1	**V50**	**V10**
Bruceton	229.33	0.11	**9.18**	20.76	708.10	0.35	**17.80**	32.78
SBOP	**217.96**	**0.11**	10.09	21.35	672.68	0.34	21.35	40.06
Neyer	265.63	0.13	9.31	**19.35**	**652.71**	**0.33**	21.23	32.80
3POD	256.20	0.13	9.53	20.85	1,331.29	0.67	21.70	**32.34**

One area in which results differed from Johnson et al [2] is the performance of 3POD when the stochastic search approximation was used as recommended by 3POD, rather than the MLE estimates. The primary difference is that the simulation study in this paper included a perturbation of the true versus desired velocity, consistent with conversations with field technicians. This implication is that in practice, the prescribed small step sizes may instead be large or in the opposite direction.

This is a challenge as the stochastic approximation algorithm relies on achieving an increasingly smaller step sizes (trending towards zero) between the target velocity and actual velocity, with a guarantee of asymptotically achieving the desired quantile. However, in the presence of perturbations between target and actual velocity, most steps will have sizes substantially different from zero. One option is to use the actual velocity rather than the target velocity; while this is still centered at a $V50 = 2000$, there is substantially more variation than an MLE estimate, as pictured in Figure 1 and described in

Table 3 (Right): Separation Statistics for N = 10

Table *4*.

Table 3 (Right): Separation Statistics for N = 10

	N = 10	
Method	**Percent of Runs with Overlap**	**Median Overlap Test Number**
Bruceton	**91.8%**	**6**
SBOP	83%	7
Neyer	85.2%	7
3POD	85.3%	7

Table 4 (Below): V50 estimates for 3POD

Scenario	MLE	Stochastic Approximation
Perfect Info, N = 40	2.52	7.75
Perfect Info, N = 10	5.86	8.52
Imperfect Info, N = 10	7.79	10.2

Figure 1: Baseline true response curve (Left), distribution of V50 estimates for 3POD (Right). Velocity in ft/sec

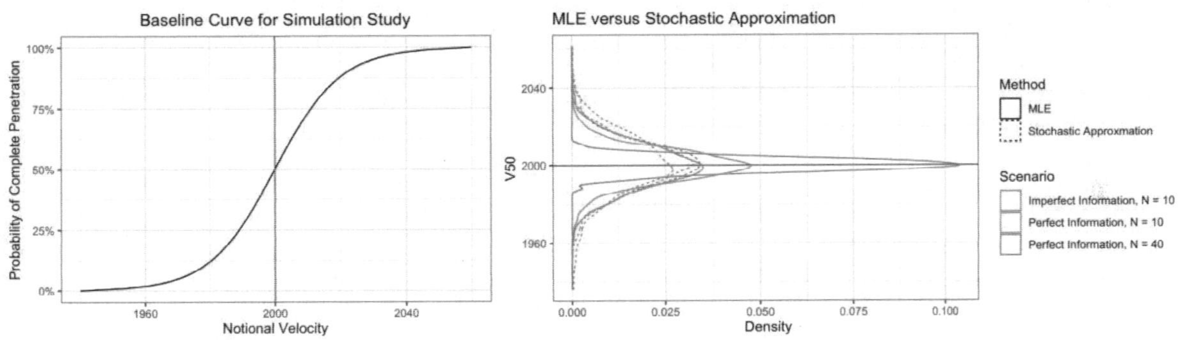

Table 6: Percent of simulation runs with overlap (Left)

Table 7: Median Absolute Deviation for imperfect information, N = 10 (Right)

MAD	N = 10			
Method	β_0	β_1	**V50**	**V10**
Bruceton	**50.1**	**0.248**	**9.86**	**7.37**
SBOP	79.8	0.04	17.6	8.97
Neyer	83.9	0.0418	19.9	12.4
3POD	74.6	0.0373	13.1	7.71

https://doi.org/10.52202/081621-0039

4.2.2 Distribution of V50 estimates for 3POD

The alternative option of using results from actual velocities and assigning them to the desired shot velocities is also problematic; for example, when at $V10$, perturbing the velocity up by 10 unit/second will not have an equal effect on the probability of a complete penetration as perturbing the velocity down by 10 units/second. Additional work is needed comparing the performance of a c-optimal design for various combinations of n_2 and variances.

4.2.3 Imperfect Information Results

In the case of imperfect information, SBOP does not perform well when both the μ_{Prior} or σ_{Prior} are incorrect and there is a small sample size (in our case, N = 10). When both the μ_{Prior} or σ_{Prior} are significantly incorrect, there is a cost in terms of how quickly overlap occurs, if it occurs at all. A prior standard deviation of $\sigma^2=10^2$ was used in MCMCpack function. The percent of runs where separation is broken and overlap is achieved is shown in **Error! Reference source not found.**.

In practice, the prior variance is generally a predefined weight or treated as a tuning parameter; this was not explored in this analysis. When the prior information is relatively weak, the prior variance on the tuning parameters can be increased. More work on the Bayesian method is needed to appropriately identify when the prior parameters appear to be incorrect, as this is valuable information to the experimenter and can be identified relatively early in the experiment.

Bruceton outperforms all other methods on the MAD of the variables under consideration in this scenario as shown in **Error! Reference source not found.**. Neyer is broadly accepted as a superior and more efficient methodology to Bruceton; it is unclear if Bruceton's superior performance with regards to achieving overlap is an artifact of the specific parameters tested, or a robustness to the differences in target and actual velocity. Future research will investigate if Bruceton is in fact more robust to velocity perturbations with different pairs of input parameters. Note that SBOP and 3POD perform similarly for β_0, β_1, and $V50$ while SBOP performs similarly to Neyer for $V10$. SBOP's performance is in line with other standard methods for key metrics when given imperfect starting parameters, which demonstrates the viability of the proposed approach.

Again, we calculate the difference between the 90th percentile and the 10th percentile for β_0, β_1, $V50$ and $V10$ as a measure of how closely these metrics are clustered for 80% of the simulation runs. The results are shown in **Error! Reference source not found.**. With a small sample size and imperfect starting parameters, Neyer and SBOP return β_0 and β_1 values within a smaller range than the other methods. Neyer has the tightest interval followed closely by SBOP for β_0, and β_1. 3POD has the tightest interval around V50 and Neyer has the tightest interval around $V10$.

Method	Percent of Runs with Overlap
Bruceton	**34%**
SBOP	16.3%
Neyer	15.2%
3POD	26%

Table 8: 80% intervals for key values, imperfect information

80% Intervals Method	N = 10			
	β_0	β_1	V50	V10
Bruceton	516.57	0.26	33.18	47.92
SBOP	111.55	0.06	40.33	48.39
Neyer	**107.74**	**0.05**	34.77	**43.43**
3POD	503.03	0.25	**31.83**	45.88

5. CONCLUSIONS AND FUTURE WORK

5.1 Conclusions

This paper demonstrates the potential to use Bayesian methodology in ballistics testing, in order to allow for the incorporation of prior information in the selection of new test points without impacting the final analysis. Additionally, this is one of the first paper to incorporate velocity perturbation into a simulation study [2]; Note that the Institute for

Defense Analyses enforced a maximum bound on perturbation size, while no limit was applied in this study. Additional work should analyse the sensitivity of methods to the size of the perturbation, as conversations with technicians indicated there can be substantial variability in the variance based on factors such as threat, season, and day to day variance. Under some circumstances, this methodology breaks separation faster than Frequentist methods, although further development is needed. Specifically, while the coefficients for both β_0 and β_1 were substantially closer to the true values, this did not translate to a lower MAD for $V50$ and $V10$. This is a promising result, and it is likely that future iterations of SBOP can correct this issue.

5.2 Future Work

There are multiple avenues for future work. For work relating specifically to SBOP, the following are of particular interest:

- In Phase I, initial shots are based on estimates of μ_{Guess} and σ_{Guess}. However, these shots are not defined as optimal by either a design or risk-adjusted method. Given priors for a response curve, it would be possible to estimate the probability that multiple shots are partial penetrations or complete penetrations, and to adjust accordingly.
- In Phase I and Phase II, the priors play a role in determining the selecting subsequent shots. There is robust literature on the impact of tuning various parameters, such as the precision. However, this has not been done specifically in the ballistics testing space. Future work includes implementing this method in live-fire testing to understand the sensitivity and impact of these parameters.
- In Phase III, SBOP follows the stochastic approximation method proposed by [11] and used in 3POD. While this has desirable asymptotic properties, guarantees do not exist for small n_1 and n_2, and it is unclear how a c-optimal design would compare. A separate study would robustly analyze if a method was dominant for particular combinations of n_1 and n_2, particularly when more than one quantile is of interest.

Additionally, more analysis is needed to better understand the robustness of existing methods against the impact of velocity perturbation between the target and actual velocity is needed. Another area stems from discussion with practitioners, who note that for some threat and velocity combinations, the variance in velocity is heterogeneous. Expanding the design to account for cases where it is known that the shot variance is not homogeneous would be quite useful in these cases.

Acknowledgements

The authors would like to thank the PEO Soldier Product Manager Soldier Protective Equipment for sponsoring this effort. Any opinions, findings and conclusions or recommendations expressed in this material are those of the author(s) and do not necessarily reflect the views of PEO Soldier or Naval Sea Systems Command (NAVSEA).

References

[1] *Ballistic Resistance of Body Armour, NIJ Standard 0101.07*, National Institute of Justice, 2023.
[2] T. H. Johnson, L. Freeman, J. Hester, and J. L. Bell, "A Comparison of Ballistic Resistance Testing Techniques in the Department of Defense," *IEEE Access*, vol. 2, 2014, pp. 1442–1455.
[3] S. G. Gilmour and L. A. Trinca, "Optimum Design of Experiments for Statistical Inference," *Journal of the Royal Statistical Society Series C: Applied Statistics*, vol. 61, no. 3, May 2012, pp. 345–401.
[4] *V50 Ballistic Test for Armour*. 1997.
[5] *Ballistic Test Method for Personal Armour Materials and Combat Clothing*. NATO Standardization Agency, 2003.
[6] M. E. et al. Greene, *The Next Revision of the NIJ Performance Standard for Ballistic Resistance of Body Armour, NIJ Standard 0101.07: Changes to Test Methods and Test Threats*. 2018.
[7] S. Magnan, G. Pageau, and A. Bouamoul, "Beyond V50: A More Comprehensive and Efficient Methodology for Assessing Armour Perfomance," Personal Armour Systems Symposium, Dresden, Germany, Sep. 2023.
[8] W. J. Dixon and A. M. Mood, "A Method for Obtaining and Analyzing Sensitivity Data," *Journal of the American Statistical Association*, vol. 43, no. 241, 1948, pp. 109–126.
[9] H. Robbins and S. Monro, "A Stochastic Approximation Method," *The Annals of Mathematical Statistics*, vol. 22, no. 3, Sep. 1951, pp. 400–407.
[10] V. Dupac and U. Herkenrath, "Stochastic Approximation With Delayed Observations."
[11] V. R. Joseph, "Efficient Robbins-Monro procedure for binary data," *Biometrika*, vol. 91, no. 2, Jun. 2004, pp. 461–470.

 https://doi.org/10.52202/081621-0039

[12] H. J. Langlie, "Proceedings of the Eighth Conference on the Design of Experiments in Army Research Development and Testing," in *Technical Report U-1792, Third edition*, 1962, pp. 146–165.
[13] C. F. J. Wu and M. Hamada, *Experiments: Planning, Analysis, and Optimization*, 1st ed. John Wiley & Sons, Inc., 2021.
[14] B. T. Neyer, "A D-Optimality-Based Sensitivity Test," *Technometrics*, vol. 36, no. 1, 1994, pp. 61–70.
[15] C. F. J. Wu and Y. Tian, "Three-phase optimal design of sensitivity experiments," *Journal of Statistical Planning and Inference*, vol. 149, Jun. 2014, pp. 1–15.
[16] D. Wang, Y. Tian, and C. F. J. Wu, "Comprehensive Comparisons Of Major Sequential Design Procedures For Sensitivity Testing," *Journal of Technical Quality*, vol. 52, no. 2, Apr. 2020, pp. 155–167.
[17] P. Roediger, "GONOGO: An R Implementation of Test Methods to Perform, Analyze and Simulate Sensitivity Experiments," U.S. Army Armament Research, Development and Engineering Center, Picatinny Arsenal, New Jersey, Technical Report AREIS-CR-18007, 2018.
[18] W. Dumouchel and B. Jones, "A Simple Bayesian Modification of D-Optimal Designs to Reduce Dependence on an Assumed Model," vol. 36, no. 1, 1994.
[19] H. A. Dror and D. M. Steinberg, "Sequential Experimental Designs for Generalized Linear Models," *Journal of the American Statistical Association*, vol. 103, no. 481, Mar. 2008, pp. 288–298.
[20] A. Sarma and M. Kay, "Prior Setting in Practice: Strategies and Rationales Used in Choosing Prior Distributions for Bayesian Analysis," in *Proceedings of the 2020 CHI Conference on Human Factors in Computing Systems*, Honolulu HI USA: ACM, Apr. 2020, pp. 1–12.
[21] M. Zondervan-Zwijnenburg, M. Peeters, S. Depaoli, and R. Van De Schoot, "Where Do Priors Come From? Applying Guidelines to Construct Informative Priors in Small Sample Research," *Research in Human Development*, vol. 14, no. 4, Oct. 2017, pp. 305–320.
[22] S. W. Hyun and W. K. Wong, "Multiple-Objective Optimal Designs for Studying the Dose Response Function and Interesting Dose Levels," *The International Journal of Biostatistics*, vol. 11, no. 2, Jan. 2015.
[23] T. Zhou and Y. Ji, "On Bayesian Sequential Clinical Trial Designs," *The New England Journal of Statistics in Data Science*, 2024, pp. 136–151.
[24] A. D. Martin, K. M. Quinn, and J. H. Park, "MCMCpack: Markov Chain Monte Carlo (MCMC) Package." Comprehensive R Archive Network, Feb. 21, 2003, pp. 1.7–1.
[25] R. C. Team, "R: A Language and Environment for Statistical Computing." R Foundation for Statistical Computing, Vienna, Austria, 2021.

A Novel Repairable Wireless Blast Anthropomorphic Test Device

J. Clark[1], L. Reider[1], G. Holt[1], R. Seery[1], J. Hrivnak[1], and J. Hopping[2]

[1]The Johns Hopkins University Applied Physics Laboratory (JHU/APL), 11100 Johns Hopkins Road, Laurel, MD 20723, john.clark@jhuapl.edu
[2]Program Executive Office Soldier, 10170 Beach Road, Fort Belvoir, VA, USA, 22060

Abstract. The Wireless Blast Anthropomorphic Test Device (WBATD) is a novel untethered, field-maintainable, and highly repeatable anthropomorphic test device capable of collecting meaningful data enabling evaluation of personal protective equipment (PPE) blast attenuating performance. The Johns Hopkins University Applied Physics Laboratory (JHU/APL) designed and developed the WBATD, which includes blast-specific components (i.e. WBATD Headform, WBATD Torso, and WBATD Neck) that are integrated with arms, legs, pelvis, and neck of the Humanetics Hybrid III 50th Male Pedestrian dummy. The blast specific components contain sensors and wireless data acquisition systems to record blast relevant metrics, while the extremities allow the system to be postured in operationally realistic positions, PPE to be donned fully and functionally, motion resulting from blast overpressure to be realistic (enabling meaningful kinematic data collection), and modular parts to be easily accessible and replaceable without specialized tools or fabrication methods. The WBATD was developed with sufficient anatomical fidelity and instrumentation to enable comparison with other blast injury surrogates and to support ongoing and future blast injury research and modeling efforts. Since development of this novel surrogate, JHU/APL has executed laboratory testing of the blast-specific components, as well as live-fire field testing of the full system. Laboratory testing of the WBATD was conducted under varying blast exposure levels, ATD orientations with respect to the shockwave, and PPE configurations. This testing showed the system to have satisfactory sensitivity between test configurations and high repeatability when exposed to the same test condition. Live-fire testing of the WBATD full system under varying postures, orientations, and PPE configurations, at a number of bomb-suit-relevant blast doses demonstrated the sensitivity, durability, and usability of the system. Future plans include biomechanics-matched field testing of the WBATD in order to develop injury risk curves relevant to the test device.

1. INTRODUCTION

To better protect warfighters, there is an interest in assessing the performance of PPE such as explosive ordnance disposal (EOD) suits in comparison to previous designs, and across suit configurations. The primary function of an EOD suit is to protect the wearer from the possibility of blast-induced injury. Of the four components of blast injury, primary blast injury (resulting from the blast overpressure of an explosion) is potentially the most dangerous. Unfortunately, this injury mechanism is also the least well understood, and first article testing of warfighter PPE does not include a standard for blast overpressure attenuation performance.

A validated surrogate system for predicting risk of primary blast injury does not currently exist. ATDs which have been used for blast testing have several limitations that restrict their widespread adoption. These limitations include cumbersome and delicate external cables and data acquisition systems that restrict mobility, lack of sufficient sensors to capture important metrics in the context of armored blast exposure, and lack of anthropometry to allow for realistic postures, donning of PPE, and blast-induced motion. Other, more biofidelic surrogates are inappropriate for first article testing because their complexity means repair or maintenance can have a significant impact on system repeatability or delay test efforts. Existing ATDs that have been used to study the effects of blast include the Blast Test Device (BTD) and updated Anthropomorphic Blast Test Device (ABTD) [1], a Hybrid III-50M (HII) crash test dummy modified to house pressure transducers [2], the recently developed Blast Overpressure Mannequin (BOPMAN) [3] and Brain Injury Protection Evaluation Device (BIPED) [4], as well as the Human Surrogate Head Model (HSHM) [5] and Human Surrogate Torso Model (HSTM) [6]. While these systems differ greatly in terms of biofidelity, repairability, usability, etc., they all suffer from one or more of the previously listed limitations, and would not be appropriate systems for the first article testing of bomb suits.

To address these limitations, the aim of this work was to design, fabricate, and verify operation of an untethered, field-maintainable, and highly repeatable ATD capable of collecting meaningful data for the evaluation of the blast attenuating performance of personal protective equipment. Through inspiration from existing blast ATDs, collaboration with external institutions with expertise in blast injury research, and leveraging advanced manufacturing and low-size, -weight, and -power (SWaP) electronics, JHU/APL developed a novel, fully integrated WBATD that provides vastly improved usability, operational costs, performance, and applicability to injury risk assessment over existing blast ATDs.

https://doi.org/10.52202/081621-0040

2. DESIGN AND DEVELOPMENT

2.1 System Requirements

Meetings with other institutions with expertise in blast injury research (DEVCOM Soldier Center, Walter Reed Army Institute of Research, Army Research Laboratory, Aberdeen Test Center) were held in order to draft a set of requirements for the surrogate with regard to use cases, system performance, usability, and repairability.

Modularity and flexibility of surrogate configurations (e.g. component vs integrated testing), environments (laboratory, live-fire (LF)) and tethered and untethered testing modes were identified as critical requirements for the system, as well as the ability to test in multiple postures, orientations, and PPE configurations. It was determined that, in order for data from the WBATD to properly capture the full blast attenuating properties of PPE, baseline bare tests should be completed at whichever blast doses were used to evaluate said PPE, meaning the system must be robust enough to survive relevant blast doses unarmored.

In order to support future metrics of blast injury prevention, it was determined that the system must capture data with high enough fidelity to align with ongoing efforts to develop injury risk models, while having high repeatability and reproducibility.

Usability and repairability criteria were outlined to enable operational deployment for first article testing, including ease of use and operational cost. Additionally, repairability, and part replacement requirements were set, with the goal that high-risk repairs can happen on-site without specialized tools or manufacturing facilities.

2.2 System Overview

Based on these requirements, a robust and modular design was developed for the WBATD. A priority was placed on using COTS parts so that any part that might need repair could be accessed and replaced without any specialized tools or fabrication methods. The three blast-specific components (headform, neck, and torso) were designed to integrate with the HIII-50M Pedestrian (PED) arms, legs, pelvis, and neck. This allows the system to be postured in operationally realistic positions, PPE to be donned fully, and motion to be simulated during a blast test, enabling meaningful kinematic data collection. As the HIII-50M PED is a widely used ATD, many biomechanics test facilities already have spare HIII-50M PED parts at their disposal, allowing rapid replacement of damaged components. A fully on-board data acquisition system (DAS) and battery were incorporated into the design, eschewing the need for external cabling, amplifiers, etc., which can become damaged during the course of a test and interfere with blast-induced motion. Special consideration for the usability of the surrogate was also given. For example, reducing the number of disassembly steps required to reach critical components while maintaining their protection during blast events. Additionally, *armed* status and *power* status of the DAS can be determined externally after the system has PPE installed and postured for a blast event. Finally, the system was made to be modular and repairable: head and torso systems can be tested together or separately, the components in each system can be replaced with spare parts without the need for specialized tools. The three blast-specific assemblies, headform, neck, and torso, as well as the instrumentation and data acquisition system, are detailed in the following sections.

The fully assembled WBATD system is shown in **Figure 1.** Overview of the WBATD system showing blast-specific assemblies (head, neck, and torso) integrated with the HIII-50M PED model (cut away to show the WBATD Torso) This image shows the headform, neck, and torso as well as the integration with the remaining HIII-50M PED parts. As shown in this image, the WBATD incorporates an external geometry similar to the HIII-50M PED, allowing for both accurate incorporation of blast protective equipment as well as comparison to previously tested data using the HIII-50M PED.

The torso neck plate can interface with either the WBATD neck natively or the HIII-50M PED neck with the inclusion of an adapter plate. The torso design interfaces with the HIII-50M PED shoulder and arm assembly at a custom reinforced collar joint. The HIII-50M PED chest jacket fits over the internal torso assembly (**Figure 1.** Overview of the WBATD system showing blast-specific assemblies (head, neck, and torso) integrated with the HIII-50M PED model (cut away to show the WBATD Torso)similar to that of the HIII-50M torso, and acts as an easily replaceable protective covering. The electronics base plate is located at the base of the torso and interfaces with the HIII-50M PED lumbar spine, which can act as a mount for shocktube testing, or be installed onto the HIII-50M PED pelvis and legs, so it can be positioned and postured for LF blast testing. The battery and DAS are mounted to the electronics bracket, the access point for these electronics are located at the break of the torso and the pelvis.

https://doi.org/10.52202/081621-0040

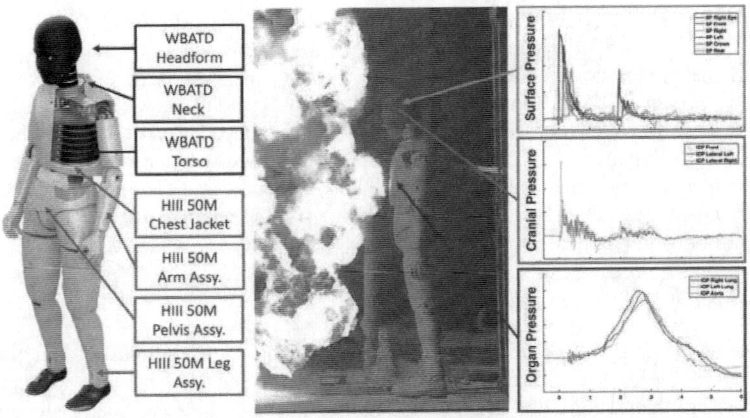

Figure 1. Overview of the WBATD system showing blast-specific assemblies (head, neck, and torso) integrated with the HIII-50M PED model (cut away to show the WBATD Torso)

2.3 WBATD Headform

The WBATD headform was designed with sufficient anatomical fidelity and instrumentation to enable comparisons with other blast injury head surrogates, as well as support continued development of future injury models. Additionally, the headform was designed to be modular and facilitate repair, consisting of multiple replaceable sub-assemblies; the crown, brain surrogate, midface, DAS, jaw, and chin insert, and kinematic DAS. The headform can be integrated into the full system, as shown in **Figure 1.** Overview of the WBATD system showing blast-specific assemblies (head, neck, and torso) integrated with the HIII-50M PED model (cut away to show the WBATD Torso)or used independently to evaluate headborne PPE.

The WBATD headform combines two existing external headform geometries. The first, which defines the crown of the WBATD headform, matches that of the large Multi-Sized Headform (LG-MSH). The remainder of the headform external geometry was derived from the Visible Human geometry. Facial features, such as ear geometries, were slightly modified to enable the donning of ear and eye PPE. While outer cranial geometry was derived from the LG-MSH, the interior cranial geometry (and therefore brain surrogate geometry) were driven by skull thickness, which was set to match literature values for the average male cranium, and ranges from 5mm to 9mm thick. The interior of the crown incorporated channels for routing sensor cables into the midface, where the DAS is located.

The midface of the WBATD headform interfaces with the crown and jaw components, and houses the headform DTS SLICE NANO (DTS, Seal Beach, CA) DAS and associated heat sink, the DAS status LED and controller, a 2200 mAh battery, sensors, as well as a shielded data connection and DAS cable ingress for wired connection during laboratory or fully integrated tests. These internal components can be accessed by removing the jaw, without removing the headform from the surrogate neck. The chin insert houses the DTS Dynamic Data Recorder (DDR) kinematics sensor and is molded into the chin insert with a silicone material to keep its placement consistent and secure during testing, and can be quickly removed from the jaw for charging, data download, and re-arming. The headform was designed to have similar mass properties as the HIII-50M PED, with a total mass of 4.44 kg, and principal moment of inertia in the mid-sagittal plane of 249 kg*cm2. The headform neck plate shares the interface of the HIII-50M system, which enables the WBATD headform to be used with a number of surrogate necks, including the Hybrid III and all previous JHU/APL surrogate necks.

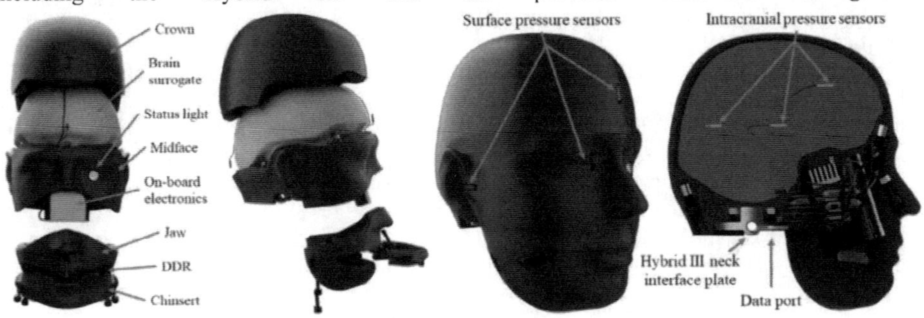

Figure 2. WBATD components

https://doi.org/10.52202/081621-0040

The crown, midface, jaw, and chin insert components are fabricated through Selective Laser Sintering (SLS) of glass-filled nylon, which has historically been used as a bone surrogate [10], and then surface-dyed black to resist discoloration during LF testing. The brain surrogate is molded from Sylgard 527 gel (Dow, Midland, MI) and is molded separately from the cranium. A thin, tougher silicone (P-656, Silicones Inc., High Point, NC) layer surrounds the brain surrogate. This silicone layer has two functions; first, it contains the Sylgard gel in the event of temperature-based volumetric and viscosity changes, and Second, it enables the brain surrogate to be stored, handled, and assembled into the WBATD headform. In the event of damage to the brain surrogate, or during routine maintenance, the brain can be removed from the headform and replaced, rather than de-molded.

2.4 WBATD Torso

JHU/APL developed the WBATD torso surrogate system to assess the blast attenuating properties of PPE worn on the torso. The torso is instrumented with sensors that enable it to take measurements relevant to primary blast injury and align with ongoing blast injury research and modeling efforts. The torso was designed to be modular and facilitate repair, and consists of multiple replaceable sub-assemblies: spine, collar, ribs, thorax assembly, and electronics bracket, which houses the SLICE NANO DAS and on-board battery. The WBATD torso is blast-specific, but interfaces with the HIII-50M PED lumbar spine, pelvis, legs, and arms to enable fully integrated, full-body blast testing.

Within the HIII-50M PED chest jacket, the WBATD torso is comprised of the spine, neck interface plate, collar assembly, thorax assembly, and base plate (**Figure 3**). The neck interface plate installs onto the spine and interfaces with the WBATD neck, or, with the addition of an adapter plate, the HIII-50M PED neck. The collar component interfaces the HIII-50M PED arm assemblies with the torso. The thorax assembly is composed of six rib pairs that surround the torso organ surrogates. Rib geometry was based heavily on HIII-50M PED rib geometry to facilitate use of the COTS HIII-50M PED chest jacket as the exterior torso covering. To balance biofidelity with repairability, the torso has three rib sizes for the six rib levels. The ribs are fabricated by SLS of Nylon 11, which was selected for its excellent toughness. The remaining hard parts were machined from glass-filled Nylon billet for strength.

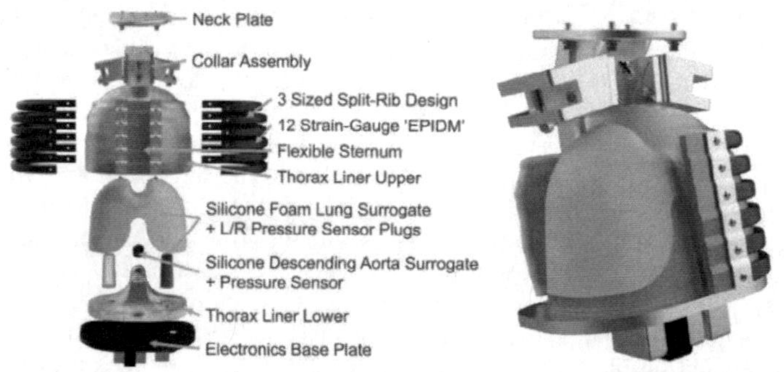

Figure 3. Exploded WBATD torso components (left) and cutaway view of the torso (right)

The ribs surround the thorax components. A silicone (P-656) back-fill component acts as a surrogate for the musculature of the back. The thorax assembly is composed of the silicone (P-656) thorax liner and liner cap, and contains the silicone foam (Soma-Foama 25, Smooth-On, Inc Macungie, PA) lung surrogate, and three sensor plugs, one for each of the left lung, right lung, and descending aorta. These intra-organ pressure (IOP) sensor locations were chosen for comparison with ongoing blast injury risk studies. The sensor plugs in which the sensors are embedded are molded from a silicone gel (Silicones, Inc., XP-429) that has historically been used for organ surrogates [8], and fit into the foam lung surrogate, which fits into the thorax liner, and is sealed and supported by the thorax liner cap. The HIII 50M PED abdominal insert, which sits in the cavity of the HIII 50M PED pelvis, protects the electronics bracket, which attaches to the base of the HIII 50M PED lumbar spine, and houses the DAS and a 4400 mAh battery.

The WBATD torso components were designed for easy repair. For example, if a rib fractures during a test, the rib design allows for easy removal of just the broken rib to be replaced with a new rib. For internal repair, the base plate and thorax cap can be removed to access the lung assembly and aorta sensor plug. The sensor plugs were designed such that they may be removed individually for calibration, repair, or replacement without the need of remolding the entire lung surrogate.

https://doi.org/10.52202/081621-0040

2.5 WBATD Neck

In previous work, JHU/APL designed and fabricated a surrogate neck to attach the HSHM to the HSTM. The objective of this previous effort was to develop a simplified neck surrogate with (1) improved reproducibility of fabrication and repeatability over previous custom neck surrogates, and (2) improved head kinematic response more similar to postmortem human subjects under relevant blast loading conditions than the HIII neck, while remaining robust enough for LF blast testing. More details on the development of this surrogate neck are available in Vignos (2022) [8].

In this current work, JHU/APL adopted and updated this design to include two additional functionalities: (1) the ability for neck rotation for proper posturing in low-crawl, or prone, position, and (2) the provision of a protected pass-through for the SLICE NANO daisy-chain cable to connect headform and torso DASs for integrated tests (Figure 4). Additionally, a steel tension cable was added to improve robustness. To achieve rotational functionality, the neck was bonded to a base plate (Figure 4) with a captured pin that can be pulled to rotate 25 degrees in both left and right direction for low-crawl posturing.

The WBATD neck consists of three layers of molded Shore 60A silicone (Silicones Inc., P-60) separated by and bonded to aluminum (6061) discs. Sitting atop the neck is the head interface plate and Shore 50A nodding blocks (Silicones Inc., P-50). This plate matches the geometry and mounting modality of the HIII neck—a pin connection between head and neck. The nodding blocks were molded and bonded to this plate, and are fit such that there is a slight pre-compression of the nodding blocks when the WBATD headform is mounted to the surrogate neck.

Figure 4. WBATD Neck with daisy-chain cable pass-through circled in orange

2.6 Data Acquisition & Electronics

The midface of the headform, and the lumbar electronics bracket of the torso each house an identical DTS SLICE NANO DAS. The DTS SLICE NANO is a modular, low-SWaP (26×31 mm footprint) DAS that has the capability of operating as a standalone system, collecting and storing high-frequency measurements in an untethered configuration, making it ideal as a DAS for blast ATDs [11]. Both SLICE NANO systems consist of one Base+ SLICE, a battery SLICE, and three bridge SLICEs, configured with a 100-kHz anti-aliasing hardware filter. The sampling rate of the SLICE NANO system scales with the number of channels collected, and can collect data at 300 kHz when the maximum nine sensors are connected (three per bridge SLICE), and up to 500 kHz if fewer sensors are connected. In the wired "lab-mode," the SLICE NANO receives external power and remains connected to a computer during testing. Lab-mode facilitates external triggering of the SLICE NANO for syncing with another DAS, as well as faster data downloading and system re-arming. For LF tests, the DAS operates in "wireless-mode," powered by an on-board rechargeable lithium-ion battery (a smaller, 2200-mAh battery housed in the midface for headform-only testing, or a larger, 4400-mAh battery housed in the torso for integrated testing) (Tenergy Power, Fremont, CA). In wireless-mode, the system operates on a circular buffer, with a signal-based trigger that is typically set to one of the surface-pressure (SP) sensors, enabling the recording of the full blast event, including pre-trigger data. A MagSafe-style quick-disconnect data connection port located under the right ear allows the user to connect a computer to the DAS via USB between tests. Using the DTS SLICEWare software, test data from the WBATD can be downloaded, system diagnostics can be run, and the system can be re-armed for the next test, before disconnecting this external cable.

For integrated headform and torso testing, a daisy-chain cable runs from the headform DAS, internally through the WBATD neck and the WBATD torso spine, and connects to the torso DAS. This allows the two systems to act in a parent-child configuration, where the headform DAS can trigger the

https://doi.org/10.52202/081621-0040

torso DAS, and data can be downloaded from both systems simultaneously via the headform data connection port. Additionally, in this configuration, the single larger battery housed in the torso powers both DASs.

The WBATD Left Eye houses an LED indicator that communicates DAS power and armed status, visible even when the WBATD is equipped with EOD PPE. The LED indicator is blue when the DAS is powered on, and green once the system is armed and ready to collect data. A custom LED controller was designed at JHU/APL to interface with the DAS and is installed on top of the DTS SLICE NANO stack.

The torso DAS and battery are housed in protective aluminum covers mounted to the base of the lumbar spine (**Figure 3**). The electronics are further protected by the HIII-50M PED abdominal insert, which can be easily removed during testing for electronics access. The torso DAS operates in the same way as the head system, allowing testing to be done independently on each of the components, or in conjunction with one another in the case of integrated testing.

2.7 Instrumentation

The WBATD headform is outfitted with a number of sensors to enable repeatable dynamic measurements of surface pressure (SP), intracranial pressure (ICP), and linear and angular acceleration (Figure 5). The DTS DDR (DTS, Seal Beach, CA) is a self-contained, battery-powered, low-SwaP kinematic sensor equipped with a 3-axis linear accelerometer, and 3-axis angular accelerometer. It is capable of recording these six sensors at a sample rate of 5.5 kHz. This system operates independently from the SLICE NANO DAS. Since the SLICE NANO sampling rate is limited by channel count, the DDR was selected as an alternative sensor package to collect the six kinematic channels without sacrificing the number of pressure sensor channels or their maximum sampling rate. Similar to the way in which the SLICE NANO system is triggered by a surface-pressure threshold, the DDR recorder is triggered by head motion and captures and stores head kinematic data. DDR data download and arming is separate from the SLICE NANO system; however, these processes can be completed simultaneously.

The headform is instrumented with 10 total pressure sensors; 6 SP sensors (Kulite LPS-LHT-145-1000A, Kulite, Leonia, New Jersey), bonded to the external surface of the crown and midface components of the headform, and 4 ICP sensors (Kulite XCL-100-500A) imbedded in the brain surrogate. The DAS can collect nine channels per test, and the selected channel configuration can be switched (by way of connectors accessible in the midface) for a given test or test series. SP sensors were placed across the surface of the head and face in areas of interest related to blast injury (e.g., eyes and ears), as well as regions that are commonly under head-borne PPE. The SP locations are described as Right Eye, Forehead, Left Ear, Right Ear, Crown, and Nape. SP sensors are bonded with a semi-permanent adhesive, and can pass through the midface without severing any wires, allowing the sensors to be removed for calibration, repair, or replacement. The locations for the four intracranial pressure sensors were selected to match ongoing blast injury risk studies; and are described as Anterior-Superior, Lateral Left, Lateral Right, and Posterior. Like the SP sensors, the ICP sensor cable routing is removable and non-permanent. All SP sensors are oriented normal to the surface at the location they are bonded. All intracranial pressure sensors are oriented in the anterior direction.

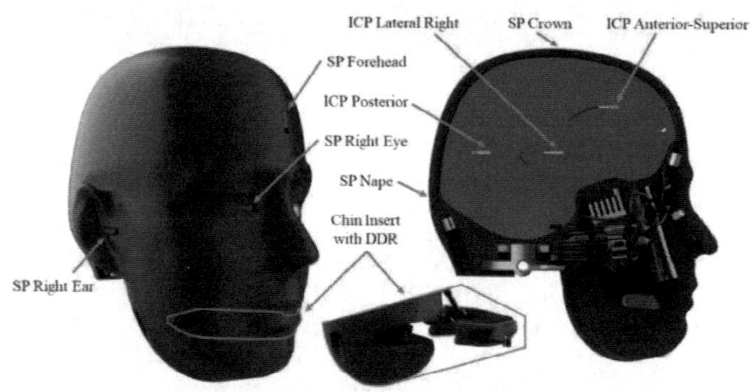

Figure 5. Location of SP sensors, ICP sensors, and the DDR kinematics sensor (SP Left Ear and ICP Left Lateral not pictured)

The WBATD torso is outfitted with sensors to enable repeatable dynamic measurements of intra-organ pressure (IOP) and thorax deformation. The locations for the three IOP sensors (Kulite XCL-100-500A) were selected to match ongoing blast injury risk studies: Left Lung, Right Lung, and Descending Aorta. Additionally, the torso can be equipped with 12 strain gauges (KFWB-2-350-C1-11 M609, Kyowa,

Tokyo, Japan) arranged and wired as 6 half-bridge pairs at symmetric locations around the left and right ribs at the 4th rib level. This rib level was selected because it is at a similar height to the HIII and HSTM chest deflection measurement systems. By wiring the strain gauges in a half-bridge configuration, the strain measurement is isolated to the bending strain experienced by the rib. An algorithm to synthesize the data from these six strain gauges into a measurement of thorax deflection, analogous to an External Peripheral Instrument for Deformation Measurement (EPIDM) [9] was developed (Figure 6). This strain-based thorax deformation measurement system can be swapped for a pair of kinematics sensor packages located on the sternum and spine of the torso (2x Endevco 7264C-2k accelerometers, 1x DTS ARS PRO-18K angular rate sensor, oriented orthogonal).

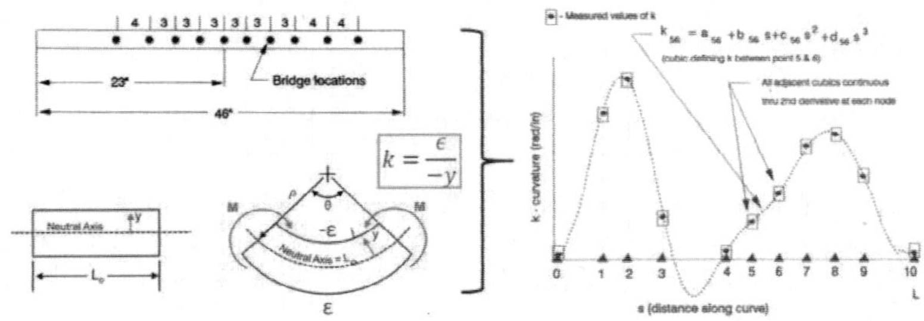

Figure 6. Overview of strain-based thorax deflection measurement

3. OPERATION & PERFORMANCE

Following fabrication and shock tube checkout testing, LF tests of the integrated WBATD system (head, neck, torso, and HIII-50M PED components) were performed at the ATC within Aberdeen Proving Ground. The primary objective of this test series was to determine the durability and usability of the WBATD when exposed to EOD-suit-relevant blast doses in standing and prone postures in a forward-facing orientation, with and without PPE. The test matrix was designed such that the blast dose would increase over the course of the test week with the goal of finding at what charge weights and standoff distances electrical or mechanical failure might occur. Testing the WBATD in bare and armored conditions would also inform the usability of the WBATD, namely the ability to quickly and easily doff and don PPE and re-posture the surrogate, as well as connect to, download data from, and re-arm the on-board DAS. The secondary objective of this test series was to determine usability of the surrogate in the side-facing, standing configuration, as well as the forward-facing, kneeling configuration. Both of these objectives were achieved through the execution and analysis of the tests performed at ATC.

3.1 Test Setup

The WBATD was positioned on the blast pad, along with a HIII-50M torso and headform modified for blast testing, at equal distances from the charge at a roughly 135° angle, to minimize reflections between the test articles (**Figure 7**). The torso and head of the HIII-50M were both equipped with a triaxial accelerometer (Endevco 7264D). The headform of the HIII-50M was additionally equipped with two pressure gauges (Endevco 8530C-100), at the crown and ear locations. Next to each test article was a pencil probe (PCB 137A24), positioned 71cm from the center of the test article, oriented directly toward the charge line, and with the center of the pressure sensor at the same standoff distance as the test article.

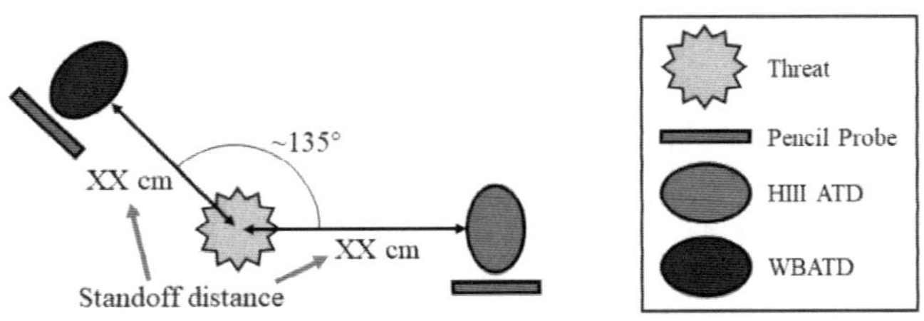

Figure 7. Live-fire blast pad configuration

Six different charge weight and standoff configurations were used for this test series (**Table 1.** Live-fire charge weights and standoffs). These conditions were selected in order to reflect a wide range of

https://doi.org/10.52202/081621-0040

relevant blast doses. ConWep was used to estimate the peak incident pressure for planning purposes. Each test used a spherical charge of C4 hung 137 and 76 cm off the ground, for standing and prone/kneeling postures, respectively, in order to simulate a free-field air blast. To achieve the proper standoff for the HIII, a measurement was taken between the charge line and the most anterior point of the surrogate chest. Because of the differences in WBATD postures, standoff was measured as the distance from the charge line to the nearest midline sensor (Table 1).

Table 1. Live-fire charge weights and standoffs

Method	Charge [kg]	Standoff Distance [cm]	ConWep Estimated Peak [kPa]
4 lb – 68 in	1.81	173	482.6
4 lb – 60 in	1.81	152	648.1
4 lb – 55 in	1.81	140	792.9
NGABS	4.55	183	896.3
4 lb – 50 in	1.81	127	979.1
Euro 1	1	100	1119.7

For the standing and kneeling postures, the National Institute of Justice (NIJ) Integrity Test Stand was used to support the surrogate in position (Figure 8). By biasing the surrogate slightly forward when standing or kneeling, and tethering it to the NIJ Integrity Test Stand, the surrogate successfully remained in the set posture until charge detonation. The stand served a secondary purpose as well, preventing the surrogate from falling completely to the steel blast pad, which could damage the headform or associated SP sensors in the bare configuration. Use of the stand does not interfere with capturing the initial surrogate acceleration kinematics directly caused by the blast.

Two PPE conditions were tested. For bare tests, the surrogate was virtually unarmored, with only boots worn for a more stable standing base. For armored tests, the surrogate was donned in a Generation 1 Advanced Bomb Suit (ABS Gen 1), including trousers, groin panel, jacket, back panel, and helmet. Figure 8 shows the various postures and PPE conditions.

Figure 8. Various orientations, postures and PPE conditions (not shown) were tested. The WBATD is free-standing prior to the shot. The NIJ Integrity Test Stand prevents it from falling to the steel pad.

4. DISCUSSION

Check-out tests in APL's Blast Overpressure Simulation System (BOSS) [11] and a small series of LF tests at ATC were completed to do an initial evaluation of the repeatability, usability, durability, and repairability of the WBATD. In general, the system showed good repeatability, with the variation in measured pressure metrics from BOSS testing similar to the variation in the generated shockwave. At a high level, the majority of the objectives set during the development process regarding usability, durability, and repairability of the system were met. Following this first LF test series a number of improvements were incorporated into the design, which are discussed in the subsections below.

https://doi.org/10.52202/081621-0040

4.1 Usability

The testing team was pleased with the usability of the system in the field based on verbal feedback from the APL and ATC testing teams given during the test week. The battery life for the system was regarded as impressive, often lasting over half of a test day, and it is estimated based on per-test voltage drop that the battery could last a full test day if the time the system was armed between tests was minimized. Additionally, there was easy access to the battery, even with the jacket donned. With or without PPE, the battery was able to be swapped in under 5 minutes by an experienced operator.

The LED light designed to alert the team to the armed and power status of the DAS before the test was visible even from under the EOD helmet, and was generally visible from ATC's live remote camera of the range. The original communication port and DDR were accessible with just the removal of the head-borne PPE, and the panel cover successfully protected the communication port from damage, even in the high-dose, unarmored test. It was found, however, that the panel cover could be challenging to fasten, because of the AM threads, and that the EOD suit itself can get in the way of plugging in the communication cable and removing the chin insert to access the DDR. For these reasons, the communications port was redesigned to the MagSafe-style connector now in use (including a magnetic panel cover), and the chin insert was angled up to its current location. It was found the original nape design was too short to fully catch the posterior helmet strap, and so this was extended down 5mm in the updated design. Additionally, after this test series, the original rotating neck design was updated to be much simpler, with the locking mechanism moved to the front, as it is more accessible here with current PPE designs.

Post-test, it generally took the team under 10 minutes to connect to the SLICE NANO, remove and connect to the DDR, download the test data from each DAS, run diagnostics on and rearm both systems, reinstall the DDR, and disconnect from the surrogate. The majority of time spent preparing the WBATD between tests was in doffing and/or donning the full EOD suit and posturing the surrogate for the next test. Repeatable posturing methods were made more efficient over the course of the test week, and were recorded in the WBATD User Guide. Generally, the doff-don-posturing process took 10 minutes for the simplest case (re-posturing the bare surrogate) and up to 40 minutes for the most complex case (doffing damaged PPE, donning new PPE, re-posturing the surrogate). This process was made difficult by both the cumbersome nature of EOD suits and the high number of degrees of freedom of the HIII-50M PED pelvis and legs, which sometimes required multiple operators working in tandem to achieve a targeted position. Further development of these components could make this process quicker and more repeatable.

4.2 Durability and Repairability

As expected, blast exposure limits of various WBATD components were found over the course of the test week, and updates to the design were made to address these limits. Generally, given the violence and potential for overmatch of the blast event the WBATD was exposed to, it was assessed that the durability of the system was sufficient. Specifically, the system survived the 'NGABS' dose unarmored without being damaged. Additionally, design features that enabled the system to remain in use in the case of damage worked as intended, as when, for example, the rear intracranial pressure sensor stopped functioning, and the channel was made use of by switching it to the left ear SP (this issue also led to the use of more robust internal connectors in the current design).

The original WBATD neck design failed in tension during a bare prone test, which led to the addition of the internal tension cable to the current design. In this case, the ability of the WBATD design to integrate with the HIII neck allowed the test week to continue without delay, by swapping in that part.

The highest dose charge configuration, in the unarmored configuration, was chosen as the last test, in order to better understand any mechanical weaknesses of the original design. During this test (1.8 kg – 127 cm), the four HeliCoils connecting the headform to the neck interface plate failed, causing the headform to fall from the surrogate, break the DAS communication cable, and impact the steel blast pad. While this caused internal rupture of the brain surrogate at the location where the ICP sensor cables exit, the GF Nylon exterior of the WBATD headform did not crack, and none of the electronics in the headform failed. It is likely that if this occurred in the middle of a test week, and a backup sensorized brain surrogate were on hand, the headform could be repaired within a few hours. As a result of this failure, the original neck plate design (and jaw-midface attachment design) was changed to use captured nuts rather than Helicoils. Additionally, the ICP cable egress from the brain surrogate was moved to the inferior surface rather than anterior surface to make that surrogate more robust, and the NANO SLICE backup battery was incorporated into the DAS stack.

While no major issues were found with Torso durability during the testing, the original hard support plate design was changed to the current strap design to allow for more realistic torso compression with a

https://doi.org/10.52202/081621-0040

hard plate, and the DAS and battery were moved from their previous location on this support plate to the current location lower on the lumbar spine base, where those components are out of the way and better protected by the HIII abdominal insert.

5. CONCLUSIONS

The Wireless Blast Anthropomorphic Test Device (WBATD) is a modular and repairable system that enables measurement of internal and external pressure, kinematics, and thorax deflection during blast events. By incorporating recent advancements in low-SWaP data acquisition systems and integrating with existing COTS ATD parts, it is the first blast-specific ATD that can be postured realistically and tested fully wirelessly, capturing data relevant to both primary and tertiary blast injury. Additionally, by taking advantage of state-of-the-art additive manufacturing of durable thermoplastics, this system is robust enough to survive EOD-suit relevant blast doses unarmored, allowing for baseline 'bare' testing and therefore full evaluation of PPE blast-attenuating performance. Usability and repairability considerations were taken into account through the design process, resulting in a system that can be reset quickly between tests, and repaired within hours, reducing risk to crucial live-fire test series. Initial laboratory testing has indicated excellent repeatability; and initial live-fire testing showed sufficient durability of the original design, which has been updated to the current iteration to address components that failed during overmatch. Future match-pair testing, and solicitation of feedback from other blast test experts is planned, in order to develop an injury-linked dataset and further optimize the design.

Acknowledgements

The authors would like to thank the PEO Soldier Product Manager Soldier Protective Equipment for sponsoring this effort, and Aberdeen Test Center for participating in the study. Any opinions, findings and conclusions or recommendations expressed in this material are those of the author(s) and do not necessarily reflect the views of the NAVAL SEA SYSTEMS COMMAND (NAVSEA).

The authors would also like to acknowledge the Walter Reed Army Institute of Research, the US Army Research Laboratory, and DEVCOM Soldier Center for giving feedback on the WBATD during the development process.

References

[1] Hsu, Y., Ho, K., and Chan, P., Anthropomorphic Blast Test Device for Primary Blast Injury Risk Assessment. *Mil. Med.*, 2020; 185(Suppl 1); pp. 227–233.

[2] Dionne, JP., Levine, J., Makris, A., Investigating Bomb Suit Blast Overpressure Test Methodologies. *Symposium on Homeland Security and Public Safety: Research, Applications and Standards,* San Diego, CA, USA, 2019.

[3] Boutillier, J., et al., A New Anthropomorphic Mannequin for Efficacy Evaluation of Thoracic Protective Equipment Against Blast Threats. *Front. Bioeng. Biotechnol.*, 2022; 9; 786881.

[4] Li, Y., Ouellet, S., Vette, A. H., Raboud, D., Martin, A., & Dennison, C. R., Evaluation of the Kinematic Biofidelity and Inter-Test Repeatability of Global Accelerations and Brain Parenchyma Pressure for a Head-Brain Physical Model, *J. Biomech. Eng.*, 2021; 143(9); 091006.

[5] Merkle, A.C., et al., Development of a Human Head Physical Surrogate Model for Investigating Blast Injury, *ASME International Mechanical Engineering Congress and Exposition*, Lake Buena Vista, Florida, USA, 2009; pp. 91-93.

[6] Wickwire, A., Carneal, C., et al., Effect of Torso Armor on Surface and Internal Pressure Response of a Human Surrogate, *Personal Armour Systems Symposium*, Cambridge, United Kingdom, 2014.

[7] Committee on Review of Test Protocols Used by the DoD to Test Combat Helmets, Review of Department of Defense Test Protocols for Combat Helmets, National Academies Press, 2014.

[8] Vignos, M., Luong, Q., Clark, J., et al. Comparison of Pressure Attenuation Performance of Bomb Suits during Free-Field Blasts using an Advanced Human Surrogate, *Personal Armour Systems Symposium*, Copenhagen, Denmark, 2021.

[9] Eppinger, Rolf H. "On the Development of a Deformation Measurement System and Its Application Toward Developing Mechanically Based Injury Indices." *SAE Transactions*, vol. 98, 1989, pp. 1626–33. *JSTOR*, http://www.jstor.org/stable/44472407.

[10] Bevan, M. Clark, J., et al. Common Helmet Test System for Blast, Blunt, and Ballistic Testing, *Personal Armour Systems Symposium*, Copenhagen, Denmark, 2021.

[11] Carneal, C., et al. Development of a Laboratory Shock Tube System for Helmet Blast Overpressure Performance Assessment, *Personal Armour Systems Symposium*, Amsterdam, NL, 2016.

https://doi.org/10.52202/081621-0040

Development of a Computational Tool for Assessing the Behind Helmet Blunt Trauma Using 3D Scanning and Impact Analysis

C. Rufino[1], M. Domingos[1], R. Aragonês[1], Ç. Bittencourt[2], R. Rodrigues[1] and I. Pereira[1]

[1]*Brazilian Army Technological Center, Américas Ave 28705, Rio de Janeiro, 23.020-470 RJ, Brazil*

clarisse.rufino@eb.mil.br

[2]*Brazilian Army Evaluation Center, Burle Marx Rd 9140, Rio de Janeiro, 23.020-240 RJ, Brazil*

Abstract. Ballistic helmets are essential defense equipment designed to improve survivability in combat situations. A key factor in evaluating these helmets is back-face deformation, which contributes to behind helmet blunt trauma (BHBT). The larger the BHBT, the greater the risk of traumatic brain injury, brain displacement, and increased intracranial pressure. Currently, standardized parameters and testing methods for accurately assessing this type of damage are lacking. This study developed a computational tool using MATLAB to quantify BHBT in thermoplastic-based ballistic helmets during a ballistic event. The BHBT assessment was performed measuring the indentation caused by the transient deformation of the helmet shell on a metal headform filled with plastiline in the coronal and occipital channels. A 3D laser scanner generated detailed point clouds of the plastiline surface before and after impact. Utilizing the alphaShape function, the tool overlays these point clouds to create a mesh from points within the analyzed region, delineating the deformed area as a closed polyhedron. This provided precise measurements of BHBT depth, area, and volume. The helmets were constructed with para-aramid plain-woven fabric and four different matrices: high-density polyethylene (HDPE), HDPE modified with exfoliated montmorillonite (MMT), HDPE modified with graphene nanoplatelets (GRF), and a hybrid stacking combining GRF and MMT matrixes (GRF/MMT). The ballistic tests were conducted using 9 mm ammunition at a 5 m distance, with a speed range of 426 ± 15 m/s. The results derived from this tool offered a more reliable and precise evaluation of the potential risks associated with BHBT to the equipment user. This, in turn, significantly enhances the assessment of helmet performance and augments user safety.

1. INTRODUCTION

Ballistic helmets are critical protective equipment designed to mitigate the risks associated with high-velocity projectile impacts in combat scenarios. Ballistic helmet performance is fundamentally dictated by the properties of their constituent materials, which must exhibit an optimal balance of strength, energy absorption, and lightweight characteristics. Para-aramid plain-woven fabrics, such as Kevlar®, are widely used in ballistic applications due to their exceptional tensile strength, impact resistance, and ability to dissipate kinetic energy through fiber deformation and fracture mechanisms [1-3]. However, to further enhance the protective capabilities of these helmets, polymeric matrices are often incorporated to improve structural integrity and energy dissipation. HDPE is a popular choice due to its high impact resistance, low density, and ability to absorb significant amounts of energy without brittle failure. Modifications to HDPE with exfoliated montmorillonite (MMT) nanoparticles enhance its mechanical properties by increasing stiffness and thermal stability while improving adhesion between fibers and the matrix [4-6]. Studies have indicated that the integration of graphene nanoplatelets (GRF) into HDPE may lead to enhanced toughness and improved load transfer efficiency, likely due to graphene's exceptional mechanical strength and high aspect ratio [7, 8]. It is also suggested that the hybrid stacking of GRF- and MMT-modified HDPE matrices could combine the advantages of both nanofillers, potentially offering a synergistic improvement in stiffness, toughness, and impact resistance [4].

A critical aspect of ballistic helmet performance evaluation is back-face deformation (BFD), which refers to the extent of deformation occurring on the helmet's inner surface after an impact. Even if the projectile does not penetrate the helmet, back-face deformation can still cause severe injuries to the user [9, 10]. However, its assessment remains challenging due to uncertainties in quantifying energy transfer mechanisms and the complex deformation dynamics involved [11-13]. Additionally, the evaluation of back-face deformation (BFD) and back-face signature (BFS) is crucial, as both contribute directly to behind-helmet blunt trauma (BHBT). BFS, in particular, represents the physical manifestation of BFD caused by ballistic impact, serving as a direct indicator of BHBT risk. It quantifies the deformation by measuring the indentation formed in a backing material, such as clay, which simulates the interface

https://doi.org/10.52202/081621-0041

between the helmet and the user's head. BHBT occurs when impact energy is transmitted through the helmet, generating internal shell deformation that can lead to cranial injuries even without projectile penetration. The larger the BHBT, the greater the risk of traumatic brain injury, brain displacement, and increased intracranial pressure [14]. Traditionally, the measurement of BFS is performed using calipers. In some cases, the diameter of the impact is measured to estimate an approximate volume for the indentation. However, this method has significant limitations. It is highly imprecise because it generalizes the shape of the indentation, often approximating it as a deformation cone [14], and fails to account for the unique details of each indentation type. This oversimplification can lead to inaccurate assessments of the energy dissipation and the associated risks of BHBT. In this study, BFS was captured using high-resolution 3D scanning, enabling a precise volumetric characterization of the deformation in clay post-impact. This approach provides an accurate assessment of the energy dissipated by the helmet, allowing for a direct correlation between the volume and depth of BFS with the severity of BHBT. By leveraging advanced computational tools and detailed point cloud analysis, this method significantly enhances the reliability and precision of BFS quantification, ultimately improving the evaluation of helmet performance and user safety.

By adopting BFS as a primary metric, this study advances the characterization of BHBT, replacing heuristic measurement methods with precise computational modeling and 3D reconstruction. This approach also overcomes the limitations of traditional methods, including those relying on Finite Element Analysis (FEA). While FEA has been utilized to simulate helmet-head interactions and address some of these challenges [5, 11, 15, 16], its predictive accuracy remains constrained by inherent issues such as mesh dependency and oversimplified material constitutive models. These limitations can lead to discrepancies between simulated and real-world results, reducing the reliability of FEA in accurately predicting BFD and BFS. Furthermore, inconsistencies in ballistic testing parameters, such as fluctuations in projectile velocity, introduce additional variability into BFD and BFS measurements, further impacting the reproducibility and reliability of results. In contrast, this study leverages high-resolution 3D scanning and computational tools to provide a more accurate and detailed assessment of BFS. By capturing the exact geometry of the deformation through point cloud data and volumetric analysis, this method eliminates the approximations inherent in traditional techniques and FEA simulations.

Given these limitations of traditional methods for quantifying BFD, BFS and BHBT, there is a critical need for more advanced assessment tools capable of capturing volumetric deformation with high spatial resolution. One promising alternative is the use of MATLAB's alphaShape function [17], which leverages computational geometry to construct a closed polyhedron mesh from high-density 3D point clouds of plastiline indentations before and after ballistic impacts. This approach provides a more accurate representation of deformation by enabling precise measurements of indentation depth, area, and volume, offering a more reliable metric for BFS and, consequently, potential neurotrauma risk. The alphaShape algorithm, with its ability to detect boundaries more naturally and tolerate noise, mitigates the overestimation and underestimation challenges commonly encountered in boundary detection, even with complex, noisy data [18, 19]. By utilizing the undeformed plastiline surface as a reference, this method correlates deformation volume with impact energy dissipation more effectively than traditional unidimensional measurements. Furthermore, the flexibility of the alphaShape function allows for the evaluation of critical parameters, such as the optimal alpha radius, which directly influences the accuracy of volume calculations [20]. This makes it a powerful tool for enhancing the precision of BHBT assessments, addressing the inherent uncertainties in traditional BFS measurements, and ultimately contributing to improved ballistic helmet design and performance evaluation.

2. EXPERIMENTAL

2.1 Materials

The helmets used in this study were manufactured by hot pressing using 28 layers of aramid (type) and thermoplastic matrices with a thickness of 60 μm. The processing parameters and layup are described in detail in the literature [21]. In this study, the following variations of thermoplastic matrices were tested: HDPE, HDPE/MMT, HDPE/GR, and HDPE/GR/MMT.

2.2 Ballistic Tests

The ballistic tests were conducted at Firing Line number 4 of the Brazilian Army Evaluation Center, LTIV – CAEx. A HPI 9 x 19 mm test barrel, in the premises of LTIV's Fire Line number 4, was used to fire 9mm ammunition against the target helmets. The distance between the end of the test barrel and the target was 5 meters. NIJ 0106.01 specifies that the ballistic helmet being tested should be positioned five meters from the muzzle of the firearm (test barrel) and must be supported on a head prosthesis (NIJ, 1981). Figure 1 illustrates the complete test setup with the instruments properly identified. It shows the position of the panels connected to the chronograph, referred to as optical barriers HPI 4272, which are responsible for measuring the impact velocity of the test. The distance between the barriers is 1 meter, and it is set up 2 meters away from the target. An aluminum head prosthesis, with 4 entrances, is filled with plastiline (Acrilex 07001 V) and supports the ballistic helmets.

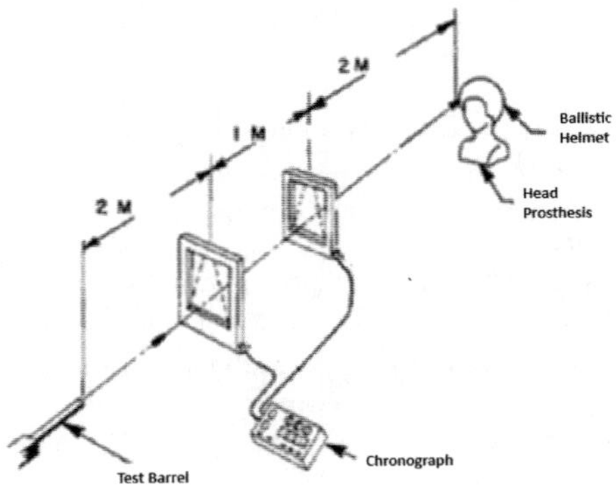

Figure 1. Schematic of the ballistic test setup for a helmet (adapted from NIJ 0106.01)

The tests were conducted using 9 mm FMJ ammunition at a velocity of 426 ± 15 m/s. Impact positions on the helmets were determined in accordance with the AR-PD 10-02 standard [22]. In the coronal plane, impacts on the right and left sides were positioned 50 mm above the ear protector. In the mid-sagittal plane, the frontal impact was placed 85 mm from the helmet's lower edge, while the rear impact was positioned 75 mm from the lower edge.

2.3 3D Scanning and Point Cloud generation

3D scanning has become a standard tool in various fields due to its ability to create highly accurate digital representations of real-world objects. Recent advancements have significantly improved its accuracy, speed, and overall results, making it indispensable for modern product development and analysis [23, 24]. In this study, the FARO Freestyle 2 3D scanning technology was employed to capture the deformation of a ballistic plastiline-filled headform before and after a ballistic impact. By utilizing this technology, the study ensures precise measurement of the headform's deformation, which is critical for evaluating the helmet's protective performance. The resulting point clouds were saved in .xyz format, a widely used file format for storing 3D point data, which ensures compatibility with various analysis and modeling software tools.

2.3 Data Processing with MATLAB

2.3.1 Preprocessing

The initial step in the analysis of the back-face signature (BFS) involved processing the raw point clouds to remove noise and irrelevant data points outside the region of interest (ROI). A bounding box was defined in the x, y, and z coordinates to isolate the points corresponding to the plastiline-filled headform, ensuring that only the relevant area was retained for further analysis. This step is crucial to focus the computational effort on the region of indentation depth and avoid distortions caused by extraneous data.

The alignment of the undeformed and deformed point clouds was performed using the Iterative Closest Point (ICP) algorithm, a widely used method for fine registration of 3D point clouds. The ICP algorithm iteratively minimizes the Root Mean Squared Error (RMSE) between corresponding points in the two clouds. The stopping criteria for ICP were set to an RMSE threshold of 10^{-5}, ensuring precise

https://doi.org/10.52202/081621-0041

alignment of the point clouds. Accurate alignment is essential for reliable analysis of the indentation depth, as even minor misalignments can lead to significant errors in volume calculations.

2.3.1 Volume Calculation using traditional method

To quantify the back-face signature (BFS) in the plastiline, a mesh variation analysis was conducted. This method calculates the point-to-mesh distance between the deformed point cloud and the reference (undeformed) point cloud. By measuring the deviation of each point in the deformed cloud relative to the reference mesh, the maximum indentation depth was determined. This indentation depth corresponds to the BFS, which is a critical parameter for evaluating helmet performance under ballistic impact. The BFS represents the height of the deformation cone used in the traditional volume calculation method, where the deformation is approximated as a cone. The base of the cone was estimated directly from the point cloud analysis, ensuring that the traditional method provides the closest possible approximation to the actual deformation geometry.

The traditional method, which approximates the deformation volume as a cone, relies on the BFS as one of its key parameters. While this approach provides a representation of the deformation, it inherently lacks the ability to capture the complex geometry and fine details of the deformation distribution. In this study, the BFS was calculated not only to enable the traditional volume calculation but also to establish a baseline for comparison with the more advanced alphaShape method. By comparing the results obtained from both methods, we can highlight the limitations of the traditional approach and demonstrate the advantages of using alphaShape for precise analysis of the indentation depth and deformation geometry.

The entire process, including noise removal, alignment, and BFS calculation, is illustrated in Figure 2. This figure provides a visual representation of the steps involved in preparing the point clouds for further analysis, ensuring that the data is accurate and reliable for both traditional and advanced methods.

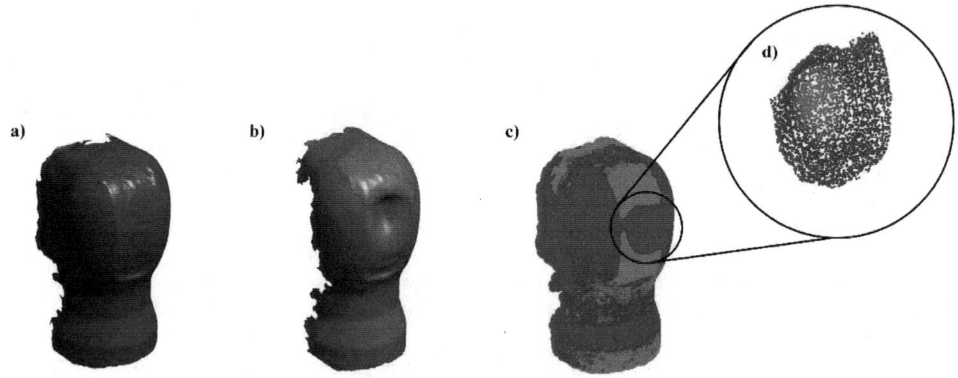

Figure 2. STL meshes generated from the point cloud: a) Without deformation, b) With deformation. c) Point clouds with deformation (red) and without deformation (blue) after ICP alignment. d) In blue, non-deformed regions and in red, deformed regions

2.3.1 Volume Calculation using alphaShape

The computational tool for calculating deformation volume was developed using MATLAB. The tool leverages MATLAB's alphaShape function, which constructs a 3D mesh from the point clouds of the plastiline before and after ballistic impact. The alphaShape function works by connecting points within a specified radius, known as the alpha radius, to create a polyhedral surface. The alpha radius controls the level of detail and smoothness of the resulting mesh, making it a critical parameter for accurate deformation analysis.

The alphaShape method is particularly useful for reconstructing surfaces from 3D point clouds, such as those captured by scanners. It works by connecting points to form triangles or lines, creating a surface that approximates the shape of the object. However, the choice of the alpha radius is crucial. If the alpha radius is too small, the resulting mesh may contain gaps or disconnected regions, failing to fully enclose the point cloud. Conversely, if the alpha radius is too large, the mesh may oversimplify the shape, losing fine details and merging separate features. A good visualization of how the alpha radius works is shown in Figure 3.

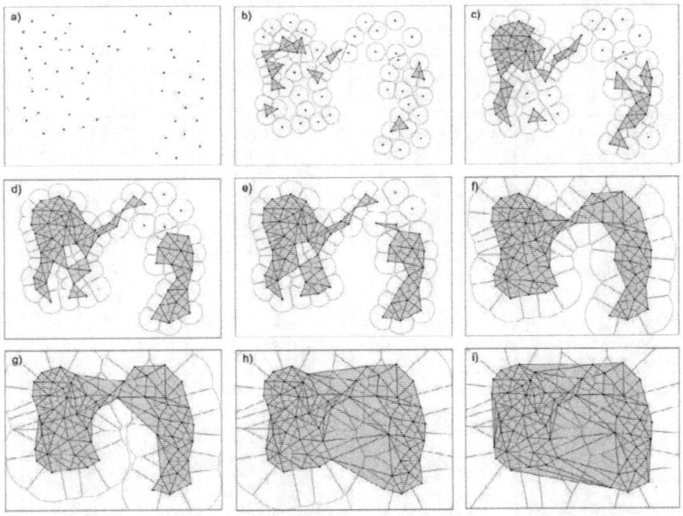

Figure 3. (a) a zero Alpha Radius producing an empty Alpha Shape; (b–h) an increase in Alpha Radius from left to right, and top to bottom, respectively; (i) an infinite Alpha Radius producing the Convex Hull [25]

Therefore, the critical alpha α_c is the smallest alpha radius that creates a fully enclosed shape without gaps or disconnected regions. It represents the point at which the mesh transitions from fragmented patches to a cohesive surface. Determining the critical alpha is essential because it ensures that the mesh accurately captures the deformation geometry while preserving fine details. In this study, the critical alpha was determined iteratively by testing a range of alpha values and selecting the smallest value that produced a fully enclosed shape.

3. RESULTS AND DISCUSSION

3.1 *AlphaShape* (AS) on Volume Calculations

To better understand the effects of the alpha radius on the reconstruction of meshes, Figure 4 presents a side-by-side comparison of the deformed point cloud region (shown in red) and the reference region (shown in blue) for the HDPE, MMT, GRF, and GRF/MMT helmets. Each row in the figure corresponds to a specific helmet material, while the sequence of images from left to right illustrates the reconstructed surface for different alpha radius values: the critical alpha α_c, 3 α_c, 100 α_c and the convex hull (CH). This sequence clearly demonstrates how the alpha radius influences the level of detail and smoothness of the reconstructed surface.

https://doi.org/10.52202/081621-0041

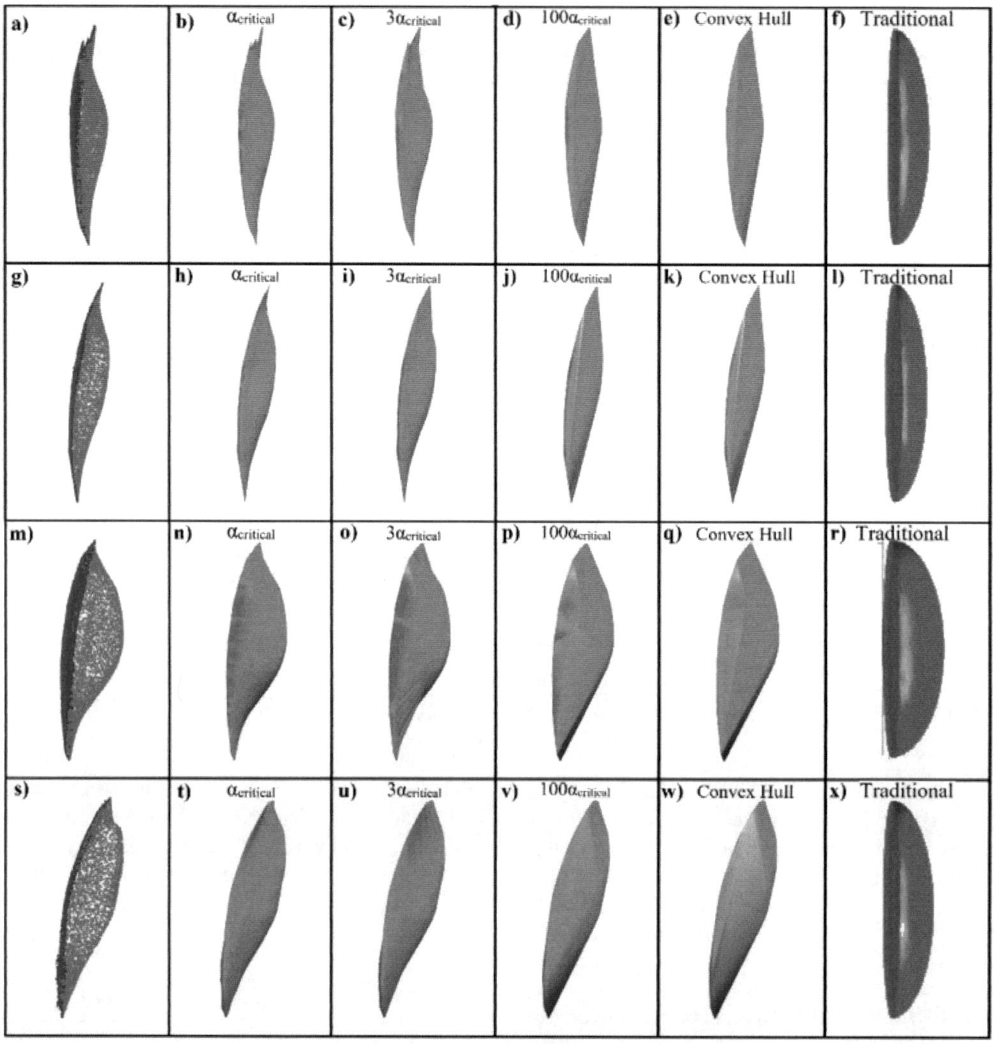

Figure 4. Side view of the deformed point cloud region (red) compared to the reference region (blue) for the HDPE, MMT, GRF, and GRF/MMT helmets (a, f, k, p, respectively). (f, l, r, x) are the respective solids if the calculation method used was the traditional method from the literature

Beyond the critical alpha, further increases in the radius lead to progressive smoothing of the surface. At $\alpha = 3\ \alpha_c$, the surface begins to lose some of the finer details, and at $\alpha = 100\ \alpha_c$, the surface approaches the convex hull (CH), which represents the outermost boundary of the point cloud. The convex hull provides a robust but oversimplified representation of the deformation, as it ignores concave features and local variations.

For comparison, closed semi-ellipsoidal surfaces were also generated for each point cloud, representing the best approximation of the deformation using the traditional method. The semi-ellipsoid method, employed for the traditional method, provides a simplified representation, it fails to capture the intricate details of the deformation distribution, such as localized depressions, ridges, or asymmetries. Besides, this method produces a much larger deformation area than what occurs in reality. In contrast, the alphaShape method, particularly at the critical alpha value, preserves these fine details, offering a more accurate and realistic representation of the indented geometry. Figure 4(f, l, r, x) highlights the limitations of traditional methods and underscores the importance of using advanced computational tools like alphaShape for precise deformation analysis.

The relationship between the alpha radius (α) and the calculated deformation volume was analyzed across four impact positions (frontal, rear, left lateral, and right lateral) for the four helmet materials (HDPE, HDPE/MMT, HDPE/GRF, and GRF/MMT). The resulting curves, shown in Figure 5, exhibit a characteristic sigmoidal profile. Initially, for small alpha radius ($\alpha < \alpha_c$), the deformation volume increases sharply, reflecting the transition from fragmented surface patches to a cohesive polyhedron. This phase is marked by a steep slope, indicating rapid volumetric growth as the alpha shape begins to enclose the point cloud.

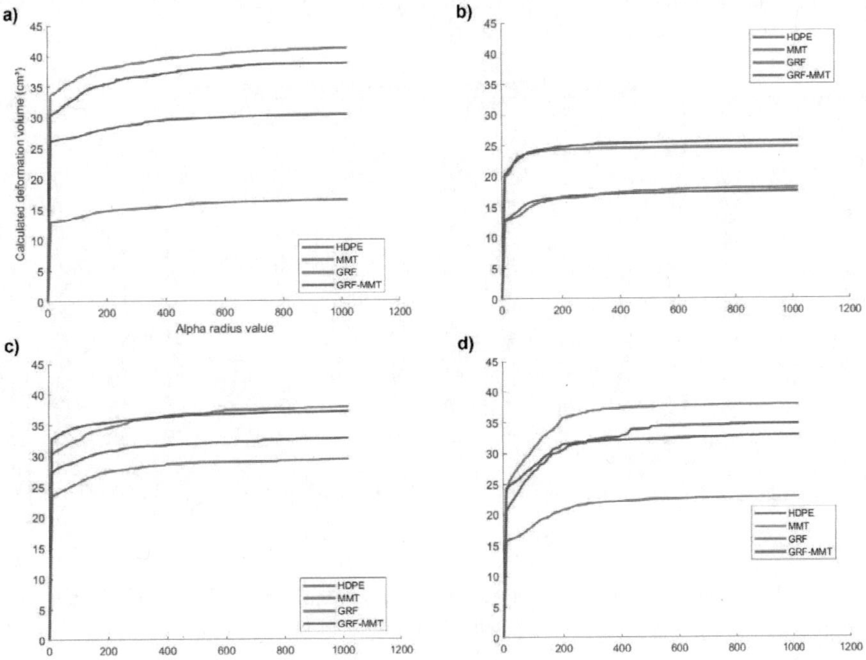

Figure 4. Deformation volume as a function of alpha radius: a) Frontal impact, b) Rear impact, c) Left lateral impact, d) Right lateral impact

Beyond the critical alpha value (α_c), the growth rate of calculated deformation volume decays exponentially, with the volume asymptotically approaching that of the convex hull (CH). At this stage, further increases in the alpha radius primarily smooth finer details of the deformation, leading to minimal residual variation in volume beyond $\alpha = 100\ \alpha_c$. This behavior aligns with theoretical predictions from Edelsbrunner's alpha-shape formalism [26], where $\alpha < \alpha_c$ represents the percolation threshold at which disconnected components merge into a single enclosed volume.

The α_c was determined through the intersection of linear regression lines fitted to the ascending and plateau phases of the α-V curve. This method improves upon traditional "elbow point" detection by incorporating material-specific deformation mechanics, ensuring that α_c corresponds to the smallest alpha radius that fully encloses the point cloud while preserving local curvature features [17]. This approach was validated through cross-correlation with micro-CT reference volumes, demonstrating high accuracy and reliability. The critical alpha value provides an optimal compromise, ensuring that the deformation geometry is accurately represented while preserving the necessary level of detail. For all helmets analyzed, the critical alpha values obtained were approximately $\alpha_c \cong 10$, with minor variations depending on the material and impact position.

Selecting an inappropriate alpha radius can lead to significant inaccuracies in volume calculations. If the alpha radius is too small, the surface may fail to fully enclose the point cloud, resulting in underfitting and underestimation of the deformation volume. Conversely, an excessively large alpha radius can over-smooth the deformation, leading to the loss of fine details and overestimation of the volume. The critical alpha value ensures that the mesh accurately represents the deformation geometry while preserving the necessary level of detail, providing a robust foundation for further analysis of deformation patterns and energy absorption in ballistic helmets.

These findings align with prior studies, such as those by Wilson et al. (2009), who utilized the critical alpha value for precise molecular volume calculations [27], and Liang et al. (1998), who applied alpha shape analysis to compute molecular surface areas and volumes, demonstrating its effectiveness in capturing the complex geometries of macromolecules [28]. By leveraging the critical alpha value, our

https://doi.org/10.52202/081621-0041

volume calculations are both precise and reliable, enabling a detailed evaluation of helmet performance and material effectiveness.

The analysis of four different helmet materials across multiple impact positions reveals substantial discrepancies between the volumes calculated using the alphaShape method and those obtained through traditional techniques, as shown in Figure 5. The relative errors range from as low as 0.11% to as high as 107.25%, with an average error of approximately 47%.

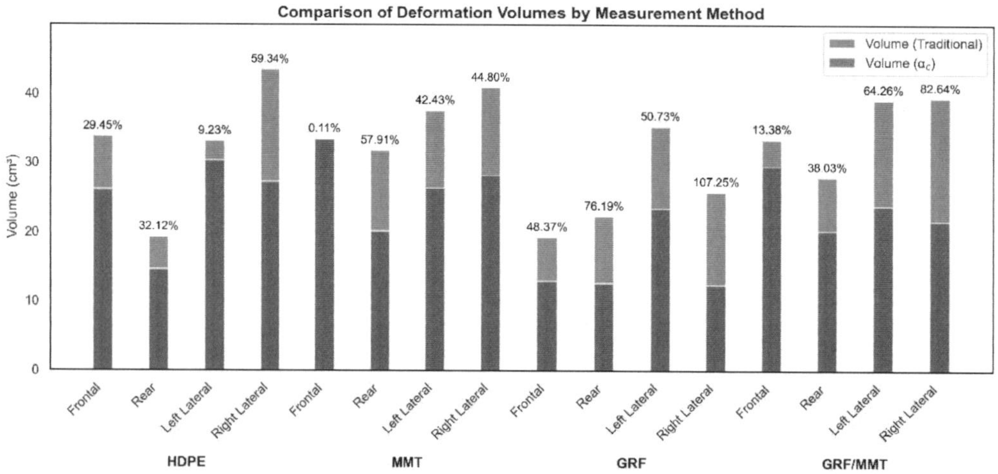

Figure 5. Comparison between volume values obtained with alphaShape and the most accurate traditional method in the literature

These findings highlight several critical aspects:

1. **Precision of Measurement:** The alphaShape method consistently produces lower volume measurements compared to traditional approaches. This suggests that conventional techniques may overestimate deformation volume, potentially due to simplifying assumptions regarding the deformation shape.
2. **Variability Across Impact Locations:** The discrepancies between methods vary considerably depending on the impact location. Lateral impacts, for instance, exhibit higher relative errors compared to frontal impacts. This variability underscores the importance of accounting for the complex geometry of helmet shells in deformation analysis.
3. **Material-Dependent Effects:** The degree of discrepancy between methods appears to be influenced by the helmet material. The HDPE/GRF composite, for example, exhibits smaller relative errors on average compared to other materials, indicating that material properties may affect the accuracy of traditional measurement techniques.
4. **Implications for BHBT Assessment:** The significant differences in volume calculations have direct implications for BHBT evaluation. Overestimation of deformation volume by traditional methods could lead to excessively conservative assessments of helmet performance, potentially resulting in unnecessary design modifications or the rejection of effective protective solutions.

The ability of the alphaShape method to capture the nuanced geometry of deformation is particularly relevant in the context of modern helmet designs. Previous studies, such as Tan et al. (2012), have indicated that traditional cone approximations can underestimate actual deformation volumes by 15-30%, particularly for non-symmetric deformations [29]. The present findings confirm that in some cases, this underestimation may be even more pronounced.

4. CONCLUSION

The alphaShape method has proven to be a critical tool for achieving precise and reliable measurements of back-face signature (BFS) and, consequently, behind-helmet blunt trauma (BHBT). Traditional methods, such as the semi-ellipsoidal approximation, fail to capture the intricate details of deformation geometry, leading to significant inaccuracies in volume calculations, with relative errors reaching up to 107.25% compared to the alphaShape method. These discrepancies directly impact the assessment of helmet performance and BHBT risks, as overestimated volumes may result in overly conservative design decisions, while underestimated volumes may understate the severity of potential injuries.

By employing the critical alpha (α_c) value, the alphaShape method strikes an optimal balance between preserving fine details and ensuring a cohesive representation of the deformation. It accurately reconstructs the surface geometry, capturing localized depressions, ridges, and asymmetries that traditional techniques often overlook. The sigmoidal relationship between the alpha radius and deformation volume further validates the method's robustness, with the critical alpha ensuring the mesh fully encloses the point cloud while maintaining essential detail. This precision is vital for evaluating the energy absorption capabilities of different helmet materials and their effectiveness in mitigating BHBT.

The study also highlights the limitations of traditional methods, which rely on generalized assumptions about deformation shape and fail to account for material-dependent and impact-location-specific variations. In contrast, the alphaShape method adapts to the complex geometry of helmet shells and the unique deformation patterns resulting from different impact scenarios. This capability is particularly relevant for modern helmet designs, where non-symmetric deformations are common and require detailed analysis for accurate performance assessment.

The findings align with prior research in fields such as molecular volume calculations, where the alphaShape method has proven effective in capturing complex geometries. By applying this method to ballistic helmet evaluation, we have established a more reliable framework for quantifying BFS and BHBT, enhancing the assessment of helmet performance and user safety. The precision of the alphaShape method not only improves deformation volume calculations but also provides valuable insights into energy dissipation mechanisms, guiding the development of more effective protective solutions.

In conclusion, the alphaShape method represents a significant advancement in ballistic helmet evaluation. Its ability to accurately reconstruct and analyze deformation geometry offers a superior alternative to traditional techniques, ensuring a more precise assessment of BHBT risks and helmet performance. This approach enhances the reliability of BFS measurements and contributes to ongoing efforts to improve helmet design and user safety in combat scenarios. Future studies should continue to explore advanced computational tools like alphaShape to further refine our understanding of helmet mechanics and traumatic brain injury prevention.

Acknowledgments

The authors would like to thank FINEP for the financial support.

References

[1] Tham, C., V. Tan, and H.-P. Lee, *Ballistic impact of a KEVLAR® helmet: Experiment and simulations.* International Journal of Impact Engineering, 2008. **35**(5): p. 304-318.
[2] Bandaru, A.K., et al., *Ballistic impact response of Kevlar® reinforced thermoplastic composite armors.* International Journal of Impact Engineering, 2016. **89**: p. 1-13.
[3] Kulkarni, S.G., et al., *Ballistic helmets–their design, materials, and performance against traumatic brain injury.* Composite Structures, 2013. **101**: p. 313-331.
[4] Junior, C.R., et al., *Mechanical properties of a laminated composite of aramid fabric and high-density polyethylene films reinforced with calcium carbonate, montmorillonite, and graphene.* Polymer Composites.
[5] Harmukh, A., et al., *Mechanical analysis of helmeted headforms under ballistic impact with implications in performance evaluation of ballistic helmets.* Frontiers in Mechanical Engineering, 2023. **9**: p. 1270905.
[6] Dias, R.R., et al., *Thermoplastic-Based Ballistic Helmets: Processing, Ballistic Resistance and Damage Characterization.* Journal of Composites Science, 2024. **8**(10): p. 385.
[7] de Tomasi Tessari, B., et al., *Influence of the addition of graphene nanoplatelets on the ballistic properties of HDPE/aramid multi-laminar composites.* Polymer-Plastics Technology and Materials, 2022. **61**(4): p. 363-373.
[8] Costa, U.O., et al., *Characterization and ballistic performance of hybrid jute and aramid reinforcing graphite nanoplatelets in high-density polyethylene nanocomposites.* Journal of Materials Research and Technology, 2024. **28**: p. 1570-1583.
[9] Zaera, R., *Ballistic impacts on polymer matrix composites, composite armor, personal armor.* 2011: Springer.
[10] Allanson-Bailey, L.S., *Improved cranial fracture metrics for assessing the protective benefits of helmet systems.* 2016: Open University (United Kingdom).

 https://doi.org/10.52202/081621-0041

[11] Cai, Z., et al., *A study on protective performance of bullet-proof helmet under impact loading.* 2016.
[12] Miranda-Vicario, A., P. Bravo, and F. Coghe, *Experimental study of the deformation of a ballistic helmet impacted with pistol ammunition.* Composite Structures, 2018. **203**: p. 233-241.
[13] Hisley, D.M., J.C. Gurganus, and A.W. Drysdale, *Experimental methodology using digital image correlation to assess ballistic helmet blunt trauma.* 2011.
[14] Magrini, M.A., *Fiber Reinforced Thermoplastics for Ballistic Impact.* 2010.
[15] Palomar, M., et al., *Relevant factors in the design of composite ballistic helmets.* Composite Structures, 2018. **201**: p. 49-61.
[16] Yang, J. and J. Dai, *Simulation-based assessment of rear effect to ballistic helmet impact.* Computer-Aided Design and Applications, 2010. **7**(1): p. 59-73.
[17] Xhimitiku, I., *Development of non-contact measurement techniques for 3d shape analysis.* 2023.
[18] Gruson, H., *Estimation of colour volumes as concave hypervolumes using α-shapes.* Methods in Ecology and Evolution, 2020. **11**(8): p. 955-963.
[19] Akkiraju, N., *Approximating spheres and sphere patches.* Computer aided geometric design, 1998. **15**(7): p. 739-756.
[20] Edelsbrunner, H., D. Kirkpatrick, and R. Seidel, *On the shape of a set of points in the plane.* IEEE Transactions on information theory, 1983. **29**(4): p. 551-559.
[21] Martin, I., et al., *Advanced thermoplastic composite manufacturing by in-situ consolidation: a review.* Journal of Composites Science, 2020. **4**(4): p. 149.
[22] Army, U.S.D.o.t., *Helmet, advanced combat (ACH) (AR/PD 10-02).*
[23] Lumme, J., et al., *Terrestrial laser scanning of agricultural crops.* Int. Arch. Photogramm. Remote Sens. Spat. Inf. Sci, 2008. **37**: p. 563-566.
[24] Stathopoulou, E.K., M. Welponer, and F. Remondino, *Open-source image-based 3D reconstruction pipelines: Review, comparison and evaluation.* The International Archives of the Photogrammetry, Remote Sensing and Spatial Information Sciences, Volume XLII-2/W17, 2019: p. 331-338.
[25] DiFrancesco, P.-M., D.A. Bonneau, and D.J. Hutchinson, *Computational geometry-based surface reconstruction for volume estimation: A case study on magnitude-frequency relations for a LiDAR-derived rockfall inventory.* ISPRS International Journal of Geo-Information, 2021. **10**(3): p. 157.
[26] Wilson, J.A., et al., *Alpha shapes applied to molecular shape characterization exhibit novel properties compared to established shape descriptors.* Journal of chemical information and modeling, 2009. **49**(10): p. 2231-2241.
[27] Edelsbrunner, H. and E.P. Mücke, *Three-dimensional alpha shapes.* ACM Transactions On Graphics (TOG), 1994. **13**(1): p. 43-72.
[28] Mittal, V., *Polymer layered silicate nanocomposites: a review.* Materials, 2009. **2**(3): p. 992-1057.
[29] Gilchrist, A. and N. Mills. *Deformation analysis for motorcycle helmets.* in *Proceedings of the International Research Council on the Biomechanics of Injury conference.* 1993. International Research Council on Biomechanics of Injury.

E_{50} testing for stab-resistant body armour evaluation

S. Jenkin[1,2], B. Cazzolato[1], C.F. Jones[1] and D. Thompson[2]

[1]*School of Electrical and Mechanical Engineering, The University of Adelaide, Adelaide, 5005 South Australia, Australia,*
scott.jenkin@adelaide.edu.au
[2]*Armor Australia Pty Ltd, 126 Frederick Street, Welland, 5007 South Australia, Australia*

Abstract. Current stab-resistant body armour certification standards, such as National Institute of Justice (NIJ) Standard-0115.00 and Home Office Body Armour Standard (CAST 2017), rely on fixed-energy impact testing, providing limited insight into armour performance across a range of impact energies. In contrast, bullet-resistant armour evaluation typically employs Penetration-Backface Signature (P-BFS) testing at reference velocities and Ballistic Limit (V_{50}) testing, which uses statistical modelling to estimate penetration probability as a function of velocity. Despite the advantages of V_{50} testing – enabling quantile-based performance assessment and greater statistical confidence – statistical penetration modelling has not been applied to stab-resistant armour. This paper introduces E_{50} testing as a framework for evaluating stab-resistant body armour, adapting statistical methodologies applied in ballistic V_{50} testing. Experimental depth of penetration (DOP) data was collected using a precision, instrumented, stab testing apparatus compliant with NIJ Standard-0115.00, across impact energies ranging from 25 J to 65 J. Testing was performed using modern, multi-threat armour, with 78 tests for the P1/A Edged Blade threat and 68 tests for the Spike threat. Binary regression techniques characterised penetration probability as a function of impact energy, and Monte Carlo simulations assessed the efficiency and accuracy of various sensitivity test designs – including Bruceton (up-and-down), Langlie, Neyer and 3pod2.0 methods – in estimating E_{50} and E_{05} quantiles. Simulations examined how sample size (n=12, 24, 36, 48) influenced test accuracy and efficiency within each test design, where each was simulated across different initial energy levels and energy increments. The results provide insight into the expected precision of E_{50} testing, highlighting the trade-offs between test design, sample size, and statistical confidence in penetration probability estimates. By adapting statistical methods from ballistic testing, this study presents a framework for a more robust, data-driven approach to stab-resistant armour evaluation, offering greater insight into protective performance across a continuous range of impact energies.

1. INTRODUCTION

Current stab-resistant body armour certification standards, including the widely adopted National Institute of Justice (NIJ) Standard-0115.00 and the Home Office Body Armour Standard (CAST 2017), rely on fixed-energy impact testing at defined protection levels. These levels were derived from studies estimating the distribution of impact energies deliverable by a male population during an attack, based on impact force and acceleration measurements from human volunteers [1]. Protection Levels 1, 2 and 3 correspond to the 85th, 90th, and 96th percentile of delivered energies, respectively [2]. Each protection level involves testing at two fixed energies: a primary test energy (E1), which assesses protective performance at the defined attack energy percentile, and a higher 'overtest' energy (E2). The maximum allowable depth of penetration (DOP) - defined as the maximum length of the threat protruding from the rear surface of the armour panel - at E1 is based on anatomical studies evaluating minimum skin-to-organ distances [3, 4]. At E2, the allowable DOP is larger, as the objective is to ensure the armour does not fail catastrophically at 50% higher energy, rather than to assess injury risk.

NIJ Standard-0115.00 requires 12 impacts per threat type, distributed across three test conditions defined by energy level and impact angle of incidence: four at E1 (0°), four at E2 (0°), and four at E1 (45°). Certification requires all impacts to not exceed the maximum allowable DOP. Since armour designs are often limited by a single test condition, performance in this one condition may determine pass or fail. The small sample size per condition reduces statistical confidence, resulting in a wide confidence interval. If all four impacts pass, the estimated 90% confidence interval (Wilson score [5]) for the true probability of protection is between 0.6 and 1. CAST 2017 and Draft NIJ Standard-0115.01 employ larger sample sizes, improving statistical confidence, yet they remain restricted to testing only at E1 and E2. Consequently, they provide limited insight into protection probabilities across a broad range of impact energies, making it difficult to assess performance beyond these two energy levels.

In contrast, bullet-resistant armour certification combines Penetration-Backface Signature (P-BFS) testing at a reference velocity with Ballistic Limit (V_{50}) testing, which applies statistical evaluation methodologies. V_{50} represents the velocity at which there is a 50% probability of complete penetration,

 https://doi.org/10.52202/081621-0042

and is estimated using generalised linear models (GLMs) with link functions such as probit and logit regression. This statistical approach enables performance assessment across a continuous range of velocities, improving confidence in protective performance evaluation and allowing estimation of critical quantiles such as V_{05}. While stab-resistant armour testing follows a fixed-energy approach to demonstrate protective performance, similar to P-BFS testing, statistical methods such as those used in V_{50} methodology have not been applied.

This paper introduces E_{50} testing as a framework for evaluating stab-resistant body armour, adapting statistical methodologies applied in ballistic V_{50} testing. It has two main objectives: first, to determine whether binary regression models can effectively characterise penetration probability as a function of impact energy; and second, to evaluate the accuracy and efficiency of various sensitivity test designs in generating data for regression-based estimation of E_{50} and E_{05} quantiles using Monte Carlo simulations. This framework provides a more statistically robust approach to assessing armour performance, enabling probabilistic evaluation across a continuous range of impact energies and improving confidence in protection levels beyond current fixed-energy certification methods.

1.1 Sensitivity test designs

Sensitivity test designs provide a structured approach to dynamically selecting stress levels, which in this context refers to the applied test variable – such as velocity in ballistic testing or impact energy in stab testing – based on observed binary outcomes. Several established test designs were considered for stab-resistant body armour evaluation, including those commonly used in ballistic V_{50} testing such as Bruceton (up-and-down) (UD) [6], modified up-and-down (MUD), and Langlie (LM) [7] methods, as well as advanced designs such as Neyer (NM) [9] and Three-Phase Optimal Design (3pod) [10] methods.

UD is widely used due to its simplicity, following a staircase approach where stress levels are sequentially adjusted based on previous outcomes. Testing begins at an initial estimate of mean response, μ_g, where 50% of tests are expected to result in a pass, and a constant step size, d, is set to an estimated standard deviation, σ_g. Subsequent stress levels are determined by $x_{i+1} = x_i - d(2y_i - 1)$, where y_i represents the binary outcome of the ith test. While UD effectively converges on the true mean, it performs poorly when μ_g or σ_g are inaccurate. Additionally, since it concentrates test points around the mean, it is not suitable for estimating extreme quantiles. A modified version, MUD, improves efficiency by employing a fixed number of variable step sizes that decrease after pass-fail reversals. This refinement has been widely adopted in ballistic V_{50} testing, including MIL-STD-662F, NIJ Standard-0101.06, and NATO AEP-2920. Further modifications are seen in NIJ Standard-0101.06, where the initial test velocity is not set at the expected mean but at a lower, reference velocity. This adjustment increases the amount of data collected at lower stress levels, improving estimation in the lower quantile range.

LM is employed in NIJ Standard-0101.04 for V_{50} estimation and in MIL-STD-331D for evaluating weapon initiation sensitivity. It begins with two initial estimates: a stress that will always result in a failure, μ_{max}, and one that will always result in a pass, μ_{min}. The first stress level is conducted at the midpoint, $x_1 = (\mu_{min} + \mu_{max})/2$. If it fails, the next test is placed midway between x_1 and μ_{min}; otherwise, it is set midway between x_1 and μ_{max}. This process continues until a pass-fail reversal is observed. Subsequent levels are determined by averaging the previous stress with the earliest prior stress where the number of passes and failures were balanced. If no such point exists, the next level is set by averaging the previous stress with μ_{max} if the last test passed, or μ_{min} if it failed. LM converges on the mean and is generally more efficient than UD, but it can perform poorly if the initial μ_{max} and μ_{min} do not bound the true transition region between pass and fail results. To improve this, a US Department of Defense modification [8] continually adjusts μ_{min} and μ_{max} by $2\sigma_g$ after the third test until a pass-fail reversal is observed.

NM, also employed in MIL-STD-331D, follows a structured three-phase approach. Phase I begins with initial estimates for μ_{min}, μ_{max} and σ_g, using a binary search algorithm [9] to efficiently locate the pass-fail transition region. This search adapts dynamically, expanding its range if initial estimates are inaccurate. Phase II starts once the difference between the lowest failure and highest success is within σ_g. At this stage, temporary parameter estimates $(\hat{\mu}_T, \hat{\sigma}_T)$ are computed, and stress levels are selected using D-optimal design, which maximises the determinant of the Fisher Information matrix. This ensures that stress levels are placed to maximise information gain, which allows for relatively efficient determination of all quantiles. Phase III is triggered once overlap occurs, meaning the lowest failure stress is below the highest passing stress. Here, maximum likelihood estimates (MLE) of $\hat{\mu}$ and $\hat{\sigma}$ replace the temporary ones from phase II, and D-optimal stress placement continues to refine quantile estimates. 3pod also follows a three-phase approach but differs from NM in phase I and phase III. Instead of the binary search used in NM, 3pod applies an alternative algorithm in phase I to locate the transition region, continuing until overlap is achieved. Phase II is identical to NM, applying D-optimal placement using MLE estimates to refine estimation of all quantiles. In phase III, 3pod introduces the Robbins-Monro-Joseph (RMJ) method [11], which focusses test points near a specific quantile of interest, improving the accuracy of extreme quantile estimation.

Monte Carlo simulations showed that 3pod estimated extreme quantiles more accurately than NM due to the targeted quantile placement method in phase III [10]. However, Neyer [12] found that NM achieved better accuracy for extreme quantiles when c-optimal design was used in phase II and III, as it placed stress levels around the estimated quantile of interest. NM also outperformed 3pod in phase I when initial estimates were inaccurate, as NM expanded its search range exponentially, while 3pod expanded linearly. In response, Wang [13] revised 3pod to 3pod2.0, modifying the phase I search logic to expand more rapidly if repeated pass (or fail) sequences were observed. This study used 3pod2.0 due to its improved phase I performance.

1.2 Binary regression models

The binary outcomes from sensitivity tests are commonly analysed using binary regression models to estimate parameters $\hat{\mu}$ and $\hat{\sigma}$, which describe the probability of response as a function of stress. The probability of response at stress x is modelled as $\pi(x) = G(z)$, where $G(z)$ is a known cumulative probability distribution function, and z is the linear predictor given by $z = (x - \hat{\mu})/\hat{\sigma}$. The choice of $G(z)$ determines how the probability of response transitions between 0 and 1 as stress increases. The corresponding link function, which maps probabilities back to the linear predictor, is the inverse of $G(z)$, given by $G^{-1}(\pi(x)) = (x - \hat{\mu})/\hat{\sigma}$.

Common link functions include the logit, probit, complementary log-log (cloglog) and log-log functions (Table 1). The logit function, based on the logistic cumulative distribution, is recommended in NIJ Standard-0101.06 for ballistic V_{50} analysis. The probit function, based on the normal cumulative distribution $\Phi(z)$, is used in NATO Standard AEP-2920 for estimating the standard deviation in V_{50} testing. Both assume a symmetric probability transition between pass and failure. The cloglog function, based on the Gumbel cumulative distribution, models asymmetric transitions, where failure probability increases gradually at low stress levels but sharply at higher levels, making it useful in survival analysis and reliability engineering. Conversely, the log-log function models the opposite asymmetry, where failure probability increases rapidly at low stress levels but transitions more gradually at higher levels. This makes it suitable for scenarios where failures are frequent at low stress but stabilise at higher stress.

Table 1. Link functions and their corresponding cumulative distributions.

Link name	Link function, $G^{-1}(\pi)$	Distribution function, $\pi(x)$
Logit	$\ln\left(\frac{\pi}{1-\pi}\right)$	$\left(1 + e^{-\frac{x-\hat{\mu}}{\hat{\sigma}}}\right)^{-1}$
Probit	$\Phi^{-1}(\pi)$	$\Phi\left(\frac{x-\hat{\mu}}{\hat{\sigma}}\right)$
Complementary loglog	$\ln\left(-\ln(1-\pi)\right)$	$1 - e^{-e^{\frac{x-\hat{\mu}}{\hat{\sigma}}}}$
Log-log	$-\ln\left(-\ln(\pi)\right)$	$e^{-e^{-\frac{x-\hat{\mu}}{\hat{\sigma}}}}$

2. BINARY REGRESSION FOR MODELLING STAB PENETRATION PROBABILITY

The suitability of binary regression methods was evaluated for modelling penetration probability as a function of impact energy in stab-resistant body armour assessment. Experimental DOP data was used in this study, which was converted to binary outcomes using thresholds of 0 mm (onset of penetration) and 7 mm (maximum allowable penetration based on an injury criterion). The experimental dataset included P1/A Edged Blade and Spike threats, capturing differences in penetration mechanics between the two.

https://doi.org/10.52202/081621-0042

2.1 Experimental depth of penetration testing

Testing followed NIJ Standard-0115.00 but included intermediate impact energies ranging from 25 J to 65 J using a rail-guided stab testing apparatus (Figure 1a). The soft armour panels featured a hybrid construction of para-aramid and UHMWPE fabric layers, measuring 400 x 400 mm with a total areal density of 6.7 kg/m^2. The stab-resistant materials were sealed within a thermoplastic polyurethane-backed 70D nylon fabric cover. This construction meets NIJ Standard-0115.00 requirements for Edged-Blade Protection Class at Protection Level 2 and Spike Protection Class at Protection Level 1. Furthermore, the panels comply with NIJ Standard-0101.06 Level II for ballistic protection, making them representative of modern, flexible, multi-threat armour solutions. Impacts were positioned with a minimum spacing of 94 mm from the armour panel edge and 60 mm from the backing material edge. Additional impacts with a 60 mm impact-to-edge spacing from the armour panel edge were also conducted during initial testing but were excluded due to observed differences in penetration behaviour, further discussed in Section 5.1.

For the P1/A Edged Blade threat, DOP was measured using a digital microscope with image processing software (ImageJ) and a calibrated scale (Figure 1b), in accordance with Home Office Body Armour Standard (2017) Section 7.7.4.1. The cut length of the rear 70D nylon fabric cover was converted to DOP using NIJ Standard-0115.00 Appendix C. For the Spike threat, DOP was recorded using a digital depth calliper (Mitutoyo, JPN) by measuring the spike length protruding from the armour rear surface. For impacts that did not fully penetrate the rear of the armour, the panel was disassembled, and an LED backlight source was used to identify the last ply with cut or displaced fibres. The number of intact plies from the rear was counted, and a negative DOP was estimated using measured ply thickness.

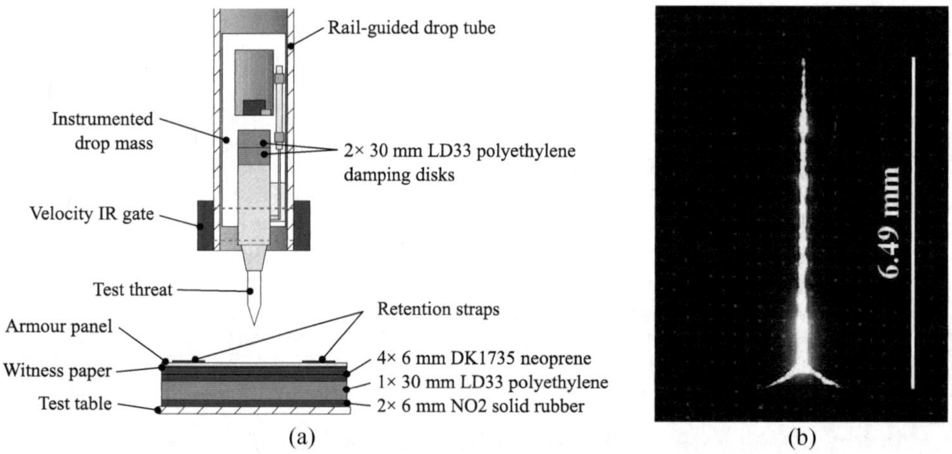

Figure 1. (a) illustration of the stab testing setup, and (b) digital measurement of cut length on 70D nylon fabric cover for P1/A Edged Blade threat testing, corresponding to a DOP of 15.1 mm.

2.2 Binary regression evaluation

The penetration probability (DOP > threshold) for both the P1/A Edged Blade and Spike threat DOP datasets was modelled using MLE with the four link functions in Table 1. Model fit was assessed using Akaike Information Criterion (AIC) [14] and leave-one-out cross-validation (LOO-CV) [15]. AIC, computed as $AIC = -2\ln(L_{max}) + 2d$, provides a relative measure of model fit, where L_{max} is the maximised likelihood, and d is the number of estimated parameters. Since all four models had the same number of parameters, rankings were determined solely by likelihood, with lower AIC values indicating better fit. LOO-CV estimated model misclassification error by systematically removing each observation, fitting the model to the remaining data, and predicting the excluded case. The cross-validation error rate, calculated as the average proportion of incorrect classifications, quantified model generalisation, with lower values indicating better predictive accuracy. Estimated $\hat{E}_{50}$ and $\hat{E}_{05}$ quantiles, AIC scores, and CV error rates for each link function and DOP threshold are summarised in Table 2.

For the P1/A Edged Blade at the 0 mm threshold, the log-log function (Figure 2b) provided the best fit, capturing the asymmetric response with a sharp increase in penetration probability at 25 J. While above the threshold, penetrations at this energy were shallow ($\leq$2 mm), indicating a sudden transition from non-penetration to minor penetration (Figure 2a). At the 7 mm threshold, the response was more symmetric, making the probit function the best fit (Figure 2c).

For the Spike threat, the cloglog function provided the best fit for both 0 mm (Figure 3b) and 7 mm (Figure 3c) thresholds, indicating penetration probability increased gradually at lower energies but transitioned sharply to near-certain penetration above 45 J. This behaviour differed from the P1/A Edged Blade, which progressively cuts fibres, whereas the Spike primarily displaced them, allowing penetration to continue with minimal additional resistance once initiated.

Table 2. Estimated $\hat{E}_{50}$ and $\hat{E}_{05}$ quantiles, AIC scores, and cross-validation error rates for four link functions applied to P1/A Edged Blade and Spike data, with DOP thresholds of 0 mm and 7 mm. The best-fitting model for each model and DOP threshold in highlighted in bold.

Threat	Description	DOP ≤ 0 mm				DOP ≤ 7 mm			
		Logit	Probit	Cloglog	Loglog	Logit	Probit	Cloglog	Loglog
P1/A	$\hat{E}_{50}$ (J)	35.0	35.2	35.6	34.3	46.3	46.2	47.0	45.4
	$\hat{E}_{05}$ (J)	22.7	23.2	15.2	26.6	38.6	38.8	37.4	39.4
	AIC	52.1	51.9	53.3	**51.0**	43.3	**42.9**	43.1	44.0
	CV Error (%)	**14.5**	**14.5**	21.1	**14.5**	**11.8**	**11.8**	**11.8**	14.5
Spike	$\hat{E}_{50}$ (J)	41.6	41.6	42.4	40.7	43.3	43.2	43.9	42.4
	$\hat{E}_{05}$ (J)	31.8	32.3	30.2	33.3	36.3	36.2	35.3	36.3
	AIC	52.6	52.0	**51.4**	53.3	41.2	41.0	**39.8**	43.4
	CV Error (%)	**17.6**	**17.6**	**17.6**	22.1	**11.8**	**11.8**	**11.8**	**11.8**

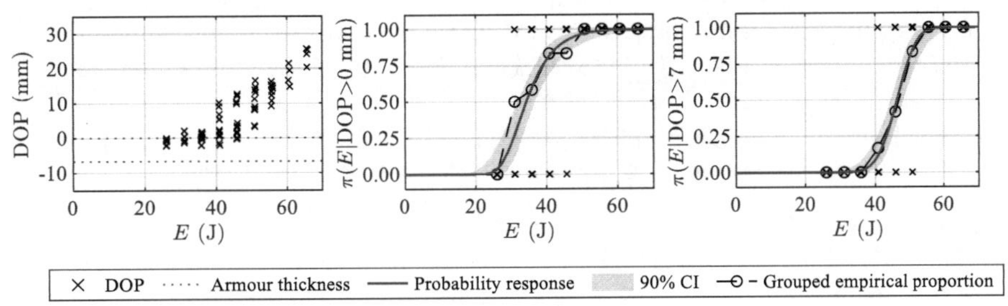

(a) (b) (c)

Figure 2. P1/A Edged Blade threat (n=76) depth of penetration results and best-fit probability response curves: (a) depth of penetration vs impact energy, (b) penetration probability response curves using loglog link for 0 mm threshold, and (c) penetration probability response curves using probit link for 7 mm threshold.

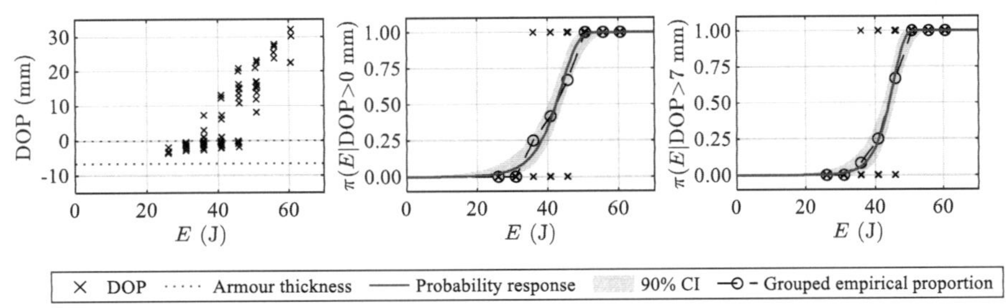

(a) (b) (c)

Figure 3. Spike threat (n=68) depth of penetration results and best-fit probability response curves: (a) depth of penetration vs impact energy, (b) penetration probability response curves using cloglog link for 0 mm threshold, and (c) penetration probability response curves using cloglog link for 7 mm threshold.

The difference in $\hat{E}_{50}$ estimates between 0 mm and 7 mm thresholds further highlighted distinct penetration mechanics between the threat types. For P1/A Edged Blade threat, the 11 J difference indicated that deeper penetration required additional energy, likely due to progressive fibre cutting. In

https://doi.org/10.52202/081621-0042

contrast, for the Spike threat, this difference was only 2 J, highlighting that once material fibre displacement occurred, further penetration required minimal added energy.

For the remainder of this study, 0 mm was selected as the DOP threshold for the Spike threat, given the small energy difference between shallow and deep penetration, it aligned with current CAST (2017) acceptance criteria for the Spike threat, and the potential biohazard risk associated with any penetration in correctional environments [16]. For the P1/A Edged Blade threat, the 7 mm threshold was used as a practical threshold, as the cutting penetration mechanism ensured shallow penetrations remain stable, aligning with NIJ Standard-0115.00 acceptance criteria.

3. MONTE CARLO SIMULATIONS

Monte Carlo simulations were conducted to evaluate the accuracy and efficiency of nine sensitivity test designs for E_{50} and E_{05} estimation. These included UD, LM, MUD, and three variations each of NM and 3pod2.0. NM variants shared the same phase I design but differed in subsequent test placement: NMD used D-optimal design, NMC applied c-optimal design targeting $\hat{E}_{05}$, and NMDC employed a hybrid approach, switching from D-optimal to c-optimal after the first 24 impacts. For the 3pod2.0 variants, 3pod2A used only phases I and II, while 3pod2B and 3pod2C targeted $\hat{E}_{50}$ and $\hat{E}_{05}$, respectively, in phase III, which initiated after 24 impacts.

The simulations were repeated for two underlying true probability distributions, based on the best-fitting experimental models, to provide penetration probabilities at each simulated test energy: a normal (probit) model for the P1/A Edged Blade ($\mu = 46.2$ J, $\sigma = 4.52$ J, 7 mm threshold) and a Gumbel (cloglog) model for the Spike threat ($\mu = 44.2$ J, $\sigma = 4.71$ J, 0 mm threshold). To reflect real-world uncertainty in μ and σ, each test design was evaluated across 81 initial parameter combinations (μ_g = 20-60 J, σ_g = 1-9 J). The MUD method followed an adapted NIJ Standard-0101.06 ballistic V_{50} testing approach, starting at 24 J and adjusting step sizes ($2\sigma_g$, $1.5\sigma_g$, or σ_g) based on pass-fail reversals, resetting every 12 impacts. For LM, NM and 3pod2.0 test designs, initial search bounds were set to $\mu_G \pm 4\sigma_G$, as recommended by Langlie [7] and previously validated as suitable for NM and 3pod2.0 methods [9, 10].

Each test design, parameter combination and assumed true distribution were simulated for 1000 test series with a sample size of $n = 48$ impacts. To assess sample size effects, results were also analysed after 12, 24, and 36 impacts. Simulated impact energy was sampled from a normal distribution centred at the target energy, E_{target}, with heteroscedastic error $\sigma = aE_{target}{}^b$ (where a = 7.12e-4 and b = 1.35, based on experimental variability), and was constrained between 5 J and 70 J to maintain realistic test conditions. A single simulated UD test series with sample size n = 12 is illustrated in Figure 4a. At the conclusion of each test series, parameter estimates $\hat{\mu}$ and $\hat{\sigma}$ were obtained using MLE, and $\hat{E}_{50}$ and $\hat{E}_{05}$ estimates were compared against their true values, E_{50} and E_{05}, respectively (Figure 4b).

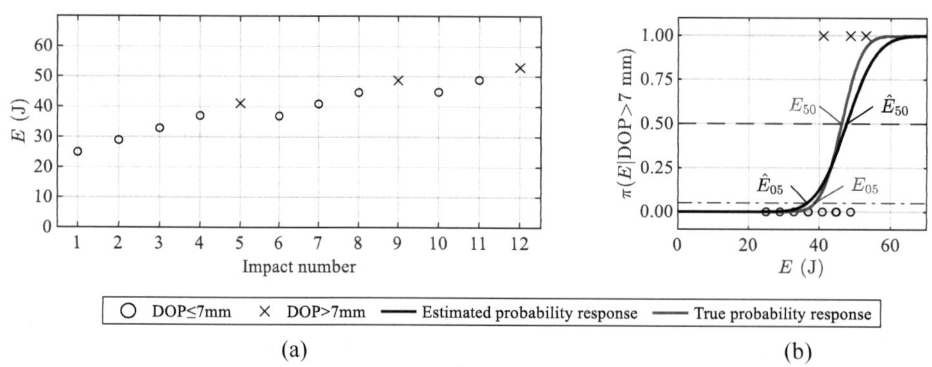

Figure 4. Simulated UD test series with μ_g=25 J and σ_g=4 J: (a) binary outcomes and energies showing test series progression, and (b) estimated probability response from the simulated data (blue) and the underlying true probability response (black).

4. RESULTS OF SIMULATED TEST DESIGNS

The accuracy of each test design was evaluated by comparing $\hat{E}_{50}$ and $\hat{E}_{05}$ estimates to their true values using root-mean-square error (RMSE), as given in Equation (1). To ensure meaningful estimates, RMSE was calculated only for simulations that met validity criteria. MLE required overlap between pass and fail results, but additional criteria were imposed: the lowest simulated energy had to result in a pass, the highest had to result in a failure, the slope of the probability response curve had to be positive, and $\hat{E}_{05}$ had to be greater than zero. The proportion of simulations that met validity criteria, P_{valid}, represents the efficiency of a test design in generating data suitable for parameter estimation (Figure 5a), where white shading indicates greater efficiency.

To compare the accuracy across test designs, RMSE values were normalised by their respective true quantiles to account for differences in energy scale. The mean normalised RMSE, $\overline{NRMSE}$, was computed across both quantiles, as given in Equation (2), and is presented in Figure 5b as an overall accuracy score, where white shading indicates greater accuracy. Although $\overline{NRMSE}$ equally weighted both quantiles, rankings were more influenced by $\hat{E}_{05}$ accuracy due to its typically larger error. This was intentional, as precise estimation of E_{05} was more critical from a protective equipment perspective, whereas E_{50} accuracy primarily reflected overall model fit.

$$RMSE_p = \sqrt{\sum_{j=1}^{n_{valid}} \frac{1}{n_{valid}} \left(\hat{E}_{p,j} - E_p\right)^2} \tag{1}$$

$$\overline{NRMSE} = \sum_{p\in\{5,50\}} \frac{1}{2}\left(\frac{RMSE_{p,P1/A}}{E_{p,P1/A}} + \frac{RMSE_{p,SPIKE}}{E_{p,SPIKE}}\right) \tag{2}$$

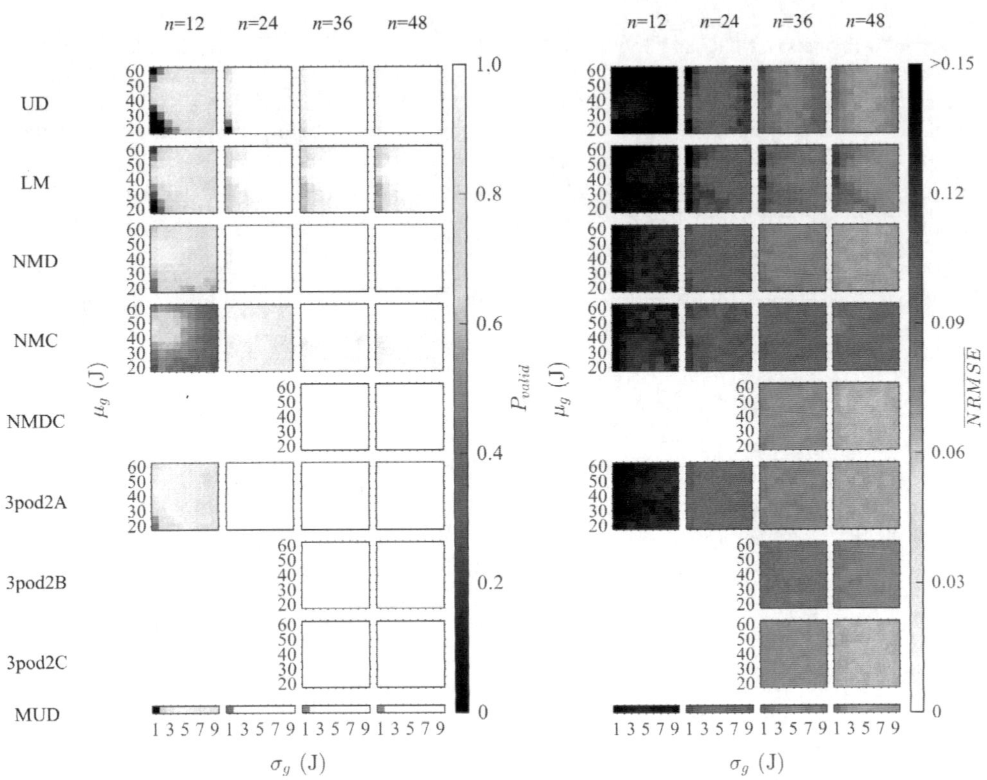

Figure 5. Efficiency and accuracy of simulated test designs for n=12, 24, 36, and 48: (a) proportion of simulated tests that achieved validity criteria, where white shading indicates greater efficiency, and (b) mean normalised root-mean-square error, where white shading indicates greater accuracy.

https://doi.org/10.52202/081621-0042

5. DISCUSSION OF SIMULATED TEST DESIGNS

The accuracy of each test design aligned with its intended purpose – methods focused on E_{50} estimation performed best for E_{50}, while those designed for E_{05} estimation excelled in that metric. UD and LM yielded the lowest $RMSE_{50}$, with typical errors of 1.8 J (4.4% from true values), 1.4 J (3.0%), 1.0 J (2.5%), and 0.9 J (2.1%) for n = 12, 24, 36 and 48, respectively, when initial parameters of μ_g (40-50 J) and σ_g (1-5 J) were close to true values. However, outside this range, P_{valid} was low at smaller sample sizes (Figure 5a). UD improved efficiency with additional sample size, whereas LM remained constrained by its poor initial bounding and was slow to adapt. Despite low $RMSE_{50}$, these methods had higher $\overline{NRMSE}$ (Figure 5b), as improved E_{50} accuracy came at the expense of E_{05} estimation, with errors typically 6.2 J (18%), 4.8 J (14%), 3.8 J (11%), and 3.5 J (10%) for each sample size. MUD improved $\hat{E}_{05}$ estimates compared to UD by incorporating additional data in the lower energy range. MUD performed best with initial parameter σ_g = 5 J, with $RMSE_{50}$ values of 2.2 J (5%), 1.6 J (3.5%), 1.2 J (2.8%) and 2.5%, while $RMSE_{05}$ was 5.8 J (17%), 4.5 J (13%), 3.5 J (10%), and 3.1 J (9%). Further work could investigate the effect of balancing step size increments and the number of impacts before resetting the energy to 24 J.

The advanced methods NMD and 3pod2.0, using D-optimal design, generally performed well across all sample sizes, offering a better balance between E_{50} and E_{05} estimation. Both yielded similar $RMSE_{50}$, typically 2.1 J (4.8%) and 1.5 J (3.4%) for n = 12 and 24, respectively, while $RMSE_{05}$ was typically 5.7 J (16%) and 4.6 J (13%). Both methods were robust across a range of initial parameter estimates, except for very small σ_g, highlighted by the shading uniformity across the combinations of initial parameters (Figure 5b). NMC provided the most accurate E_{05} estimates, however at the expense of E_{50} estimation. NMC also struggled to achieve the overlap condition at low sample sizes, potentially due to its early transition to c-optimal design. A modification where c-optimal placement is triggered immediately after overlap, rather than when the highest pass and lowest failure differ by less than σ_g, may improve its performance.

After 24 impacts, NMDC and 3pod2C were the best-performing designs, with similar results that were robust across all initial parameters. Typical $RMSE_{50}$ values were 1.4 J (3.1%) and 1.3 J (3.0%) for n = 36 and 48, respectively, while $RMSE_{05}$ was typically 3.3 J (9.7%) and 2.7 J (7.9%). These methods highlighted the effectiveness of first refining parameter estimates using D-optimal design before switching to c-optimal, concentrating energies near $\hat{E}_{05}$. Further work may assess the effect of switching the design approach earlier for smaller sample sizes. The distribution of the estimated response curves for the best-performing test designs at each sample size is illustrated in Figure 6.

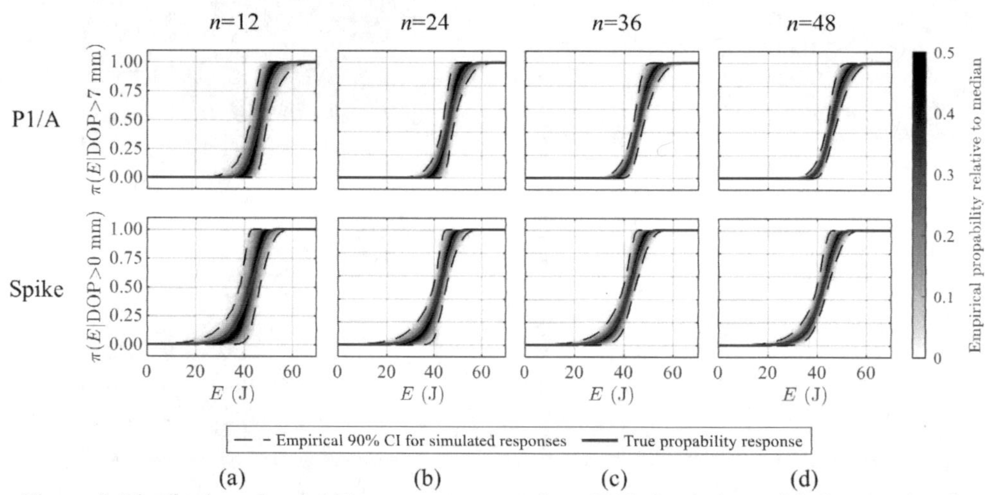

Figure 6. Distribution of probability response curves from 1000 simulations of the best-performing test designs for each sample size: (a) NMD, μ_g=55 J, σ_g=5 J, (b) MUD, σ_g=5 J, (c) NMDC, μ_g=40 J, σ_g=4 J, and (d) NMDC, μ_g=50 J, σ_g=2 J. Simulations not meeting validity criteria were excluded.

The findings in this study were based on a single armour construction, which may limit their applicability to other material systems. Armour types such as ring mesh, metal plate, or rigid epoxy-fibre composites, may exhibit different penetration mechanics that were not captured in this study. Further investigation is required to determine whether these materials follow similar probability distributions. However, provided penetration behaviour can be described by a monotonically increasing cumulative distribution function, then E_{50} methodology should remain applicable. Additionally, while 1000 simulations were conducted per test design, some statistical variation in the results is expected.

5.1 Effect of impact location on depth of penetration

Impacts placed 60 mm from the armour edge, while valid under NIJ Standard-0115.00, exhibited significantly lower DOP than equivalent tests performed closer to the centre. Figure 7 shows DOP results at an intermediate stage in testing, before edge impacts were discontinued. This discrepancy was attributed to lower effective stiffness of the armour and backing material due to lack of lateral constraint. As a result, greater material displacement occurred during impact, reducing peak impact forces and decreasing DOP. For Spike threats, asymmetric deformation of the armour and backing near the edge promoted plastic buckling of the slender spike (Figure 7b), particularly at higher energies. While edge effects can influence all armour types, flexible, loose-ply, fabric systems may be particularly susceptible.

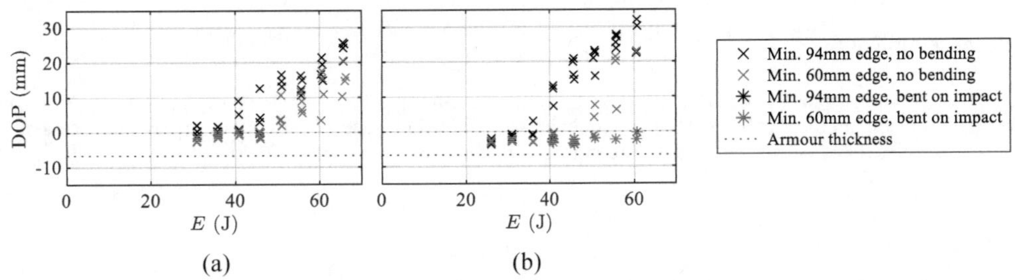

Figure 7. Depth of penetration results for minimum 94 mm and 60 mm impact-to-edge distance on armour panel, (a) P1/A Edged Blade threat, and (b) Spike threat.

If edge impact locations had been retained in the experimental dataset, the estimated standard deviation of the penetration probability response would have been artificially inflated. In practical terms, this would increase the likelihood of achieving a passing result when testing near an edge, leading to an overestimation of $\hat{E}_{50}$ and other quantiles such as $\hat{E}_{05}$, making the assessment less conservative.

6. CONCLUSIONS

This study demonstrated that binary regression models effectively described penetration probability as a function of energy in stab-resistant body armour evaluation. From the experimental DOP data, the best-fitting models captured the distinct mechanics of each threat, with the probit model providing the best fit for the P1/A Edged Blade and the cloglog model for the Spike threat. These models were applied using DOP thresholds of 7 mm for the P1/A Edged Blade and 0 mm for the Spike, providing practical performance evaluation limits while aligning with current acceptance criteria in accordance with existing standards. Among the simulated test designs, NMDC, NMD, and 3podC provided the most accurate $\hat{E}_{50}$ and $\hat{E}_{05}$ estimates, even when initial parameters deviated significantly from true values. These methods not only provided superior precision in quantile estimation but also demonstrated robustness across a range of conditions, making them strong candidates for implementation in stab-resistant armour evaluation.

E_{50} testing has the potential to enhance stab-resistant body armour evaluation by enabling probabilistic performance characterisation across a continuous energy spectrum. If adopted within standardisation and conformity assessment models, E_{50} could complement existing fixed-energy testing, similar to the way ballistic standards combine P-BFS and V_{50} methods. Within such a framework, quantile-based acceptance criteria, such as a minimum allowable E_{05} value for a given Protection Level, could be analogous to the V_{05} criterion used in NIJ Standard-0101.06 for ballistic certification. Any decision to adopt E_{50} testing rests with the appropriate standardisation bodies, which would need to define suitable DOP thresholds, acceptable quantile estimation error, and sample size requirements based on regulatory and operational considerations. The framework developed in this study may assist in informing those decisions by illustrating how such parameters could be evaluated and applied.

https://doi.org/10.52202/081621-0042

If implemented, E_{50} testing would likely require increased sample sizes compared to current certification methods, leading to a modest increase in certification cost. However, by aligning more closely with ballistic testing practices, it offers significant value to armour material developers, manufacturers, program managers, and end-users. It enables direct, threat-specific, performance comparison across armour systems, supports more informed procurement decisions, and improves confidence in meeting minimum performance requirements during batch, lot acceptance, and follow-up inspection testing. It also facilitates targeted design optimisation by identifying design-limiting threats within protection classes, reducing overdesign and supporting more efficient material use.

Beyond development and initial certification, E_{50} testing could be applied to Surveillance Testing Programs, commonly adopted for bullet-resistant armour but not yet leveraged for stab-resistant armour. Establishing an initial E_{50} dataset enables the calculation of an allowable degradation margin, as defined in NIJ STD 0101.06 Explanatory Material for Section 7.9.5. Periodic penetration probability assessments can then monitor any performance degradation as the armour ages throughout its life-of-type, critical for forecasting armour service life. This alignment with ballistic armour evaluation represents a significant step forward in the standardisation and long-term assessment of stab-resistant body armour performance.

Acknowledgments

This research was supported by the Australian Government through the Department of Education's National Industry PhD Program (project 36179). The views expressed herein are those of the authors and are not necessarily those of the Australian Government or the Department of Education.

References

[1] Horsfall I., Prosser P., Watson C. and Champion S., An assessment of human performance in stabbing', Forensic Science International, vol. 102(2-3) 1999, 79-89.
[2] NIJ Standard 0115.00, Stab Resistance of Personal Body Armor, National Institute of Justice, 2000.
[3] Connor S.E.J., Bleetman A. and Duddy M.J., Safety standards for stab-resistant body armour: a computer tomographic assessment of organ to skin distances, Injury, vol. 29(4) 1998, 297-299.
[4] Bleetman A. and Dyer J., Ultrasound assessment of the vulnerability of the internal organs to stabbing: determining safety standards for stab-resistant body armour, Injury, vol. 31(8) 2000, 609-612.
[5] Wilson E., Probable inference, the law of succession, and statistical inference, Journal of the American Statistical Association, vol. 22(158) 1927, 209-212.
[6] Dixon J.W. and Mood A., A method for obtaining and analyzing sensitivity data, Journal of the American Statistical Association, vol. 43(241) 1948, 109-126.
[7] Langlie H.J., A reliability test method for 'one-shot' items, Technical report U-1792, Aeronutronic Division of Ford Motor Company, Newport Beach, California, 1965.
[8] Johnson T.H., Freeman L., Hester J. and Bell J.L., A comparison of ballistic resistant testing techniques in the Department of Defense, IEEE Access, vol. 2 2014, 1442-1455.
[9] Neyer B., A D-optimality-based sensitivity test, Technometrics, vol. 36(1) 1994, 61-70.
[10] Wu C.F.J. and Tian, Y., Three-phase optimal design of sensitivity experiments, Journal of Statistical Planning and Inference, vol. 149 2014, 1-15.
[11] Joseph V.R., Efficient Robbins-Monro procedure for binary data, Biometrika, vol. 91(2) 2004, 461-470.
[12] Neyer B., Comparison of the Neyer D-optimal and 3pod sensitivity test designs, JSM 2014, Boston, Massachusetts, US.
[13] Wang D., Tian Y. and Wu C.F.J., Comprehensive comparisons of major sequential design procedures for sensitivity testing, Journal of Quality Technology, vol 52(2) 2020, 155-167
[14] Akaike H., Information theory and an extension of maximum likelihood principle, Selected papers of Hirotugu Akaike, Springer New York, 1998, 199-213
[15] Stone M., Cross-validatory choice and assessment of statistical predictions, Journal of the Royal Statistical Society: Series B (Methodological), vol. 36(2) 1974, 111-113.
[16] Greene M.E., Horlick J., Longhurst D.A., Miller L.A., Robinson C. and Sundstrom R.A., NIJ standards for ballistic resistance of body armor and stab resistance of body armor: new developments, Personal Armour Symposium Proceedings 2023, Dresden GER.

Moving beyond the V50 for armour performance evaluation

G. James[1], J. Keirl[1], A. Hepper[1]

[1] *Defence Science and Technology Laboratory (Dstl), Bldg 352, Porton Down, Salisbury, Wiltshire, SP4 0JQ, UK (grjames@dstl.gov.uk)*

Abstract. Evaluating the V_{50} (velocity at which 50 percent of projectiles perforate) performance of an armour system is a standard test technique to assess a protective system, but provides limited indication on the resulting injury outcome. In some situations, the threat may not be completely stopped. This paper describes a process used to measure armour overmatch performance and put it into context of the reduction of injury. This process allows a wider set of performance metrics to be generated, using 20 or more shots. An appropriate backing was required, allowing 'as worn' measured performance, along with residual velocity estimates from shots that defeat the armour. Ballistic testing was designed to cover the range of velocities from just below the V_{50} to approximately 1.5 times the V_{50}. Equations for the probability of armour perforation and residual velocity were generated and used to feed a vulnerability model. The body armour was positioned in the body model and a grid of shotlines were run at a range of velocities to establish the average vulnerability/injury risk to a given area (both with and without the armour). By taking the average injury across the body region of interest for the different velocities, a reduction in injury risk was calculated compared to an unprotected case. The UK Tier 1 Pelvic Protection was evaluated to demonstrate this process. With 30 shots, a V_{50} could be calculated and coupled with the additional shotline model analysis demonstrated the protection gave reduction in the number of injuries at velocities over the V_{50} and V_{100}. This approach combines a coverage assessment, along with the armour's ballistic performance, allowing the development of optimised protection over the widest velocity range, with a small change to the practical testing.

1. INTRODUCTION

1.1. Overview

Traditionally, Personal Protective Equipment (PPE) ballistic performance is measured using either a V_{50} or a V_{proof} assessment. The V_{50} is the velocity at which 50 percent of projectiles perforate the PPE. V_{proof} is the maximum velocity below which there is a specified confidence that projectiles will not perforate the armour. Statistical models, in this case probit or logit equations can be fitted to the (dichotomous) data to determine the risk (probability) of perforation across a range of velocities to give a broader description of the PPE perforation response. In this paper, consideration is only given to penetrating injuries and assumes that any threat stopped by the PPE does not cause a Behind Armour Blunt Trauma (BABT) risk. Where required, BABT risks can be assessed separately.

V_{50} and V_{proof} ballistic assessment methods do not articulate the protective performance, e.g. in terms of injury reduction to the user. If a shot to the PPE is above the V_{50}, it may be expected to perforate the PPE and injure the person. However, the extent of that injury (which may be reduced due to energy absorption from the PPE system) is not easily determined.

In terms of injury mitigation, a more meaningful evaluation would be to compare the overall change in injury risk, or expected injury outcome, to an unprotected case or to a baseline PPE system. This paper outlines the methodology used to generate injury reduction-based measures of PPE performance. This used practical experiments to generate the required PPE perforation response data to feed into a computer based human vulnerability Shot Line Model (SLM). The SLM is a 3D representation of a person and their tissues, along with the PPE for analysis. To demonstrate this methodology and the types of outcomes produced, the UK Tier 1 Pelvic Protection (PP) is used as the PPE and compared to an unprotected case.

Dstl has different injury and vulnerability models for different purposes. The Human Injury Modelling and Analysis (HIMA) project generated the SLM used within this paper, and so the model is referred to as the HIMA SLM. The injury risk-based PPE evaluation has a number of steps:

1. Experimental determination of probit or logit model for the PPE perforation response, and the identification of V_{50} with shots that spanned the V_{50}.
2. Experimental determination of impact versus residual velocity of the PPE for impacts up to approximately 1.5 times the V_{50}.
3. Fitting of the experimental data to required equations, to derive estimates for residual velocity.
4. Running HIMA SLM with required PPE geometry.
5. Analysis to generate injury reduction based outputs.

 https://doi.org/10.52202/081621-0043

1.2. Experimental method (steps 1&2)

For this demonstration UK Tier 1 PP [1] was chosen as it has been well characterised by Dstl, and the area of the body is comparatively simple (in terms of geometry and limited variety of tissue types) with the required tissue parameters well understood within the SLM.

A 6 mm glass sphere (nominal 0.28 g, density 2.50 ± 0.05 g cm^{-3} and a diameter tolerance of ± 0.05 mm), as defined in Reference [2]) was chosen to represent the threat to the UK Tier 1 PP[1], being similar density to stones and soil generated from buried Improvised Explosive Device (IED) events against which the PPE was designed to mitigate.

A Sabre Ballistics A1G+ gas gun was used with compressed air and compressed helium with pressures up to 30 MPa (300 Bar). A 6.05 mm smooth bore barrel was used to fire the glass spheres.

It is known that the method used to back an armour material during ballistic assessments can affect its measured perforation performance [3] and an unrepresentative backing could therefore lead to misleading results. This could lead to the acceptance of PPE or materials that do not meet the ballistic requirements or impose an unnecessary burden. The TP5 pack was chosen as it was specifically developed and validated to provide a realistic backing for soft armour assessment, as well as a representative skin perforation response [4].

The TP5 pack was critical to this methodology as:

- It enabled direct assessment of the skin perforation outcome for each shot.
- It provided validated biofidelic compliance – ensuring the measured V_{50} or outcome of the PPE matched 'as worn' performance as close as possible.
- For any shots perforating the PPE, the depth of penetration (DoP) into the pack could be measured, which allowed the DoP to be converted into residual velocity with data from some calibration shots against a bare pack.

A system response for the skin/PPE ballistic outcome was used: in cases where the PPE was not perforated, but the skin of the TP5 pack was perforated (due to pencilling of the soft armour), it was classed as a system perforation. This means it was the onset of potential injury that was used to determine the outcome of the shot. Assessing the outcome via the system performance method allowed subsequent alignment of analysis when it came to the comparison to an unprotected case within the HIMA SLM.

Projectile velocities were measured using a Photron NOVA High Speed Video (HSV) camera at 20,000 frames per second and 896x512 pixels resolution. The velocities calculated from the HSV were initially validated against 5 shots with two Oehler model 57 infra-red gates with 500 mm separation, connected to an Oehler model 36 chronograph. Validation showed velocities from each system were consistent to within 0.1 m s^{-1}, when accounting for air drag due to the measurement from the Oehler system conducted 500 mm closer to the weapon. The Oehler model 57 infra-red gates were then removed which allowed a shorter barrel to target distance.

A minimum shot-spacing of 50 mm was used, avoiding shots within 50 mm from the edge of the packs. 20 shots were conducted to generate a system perforation response relationship. A probit model within the statistical program R [5; 6] to generate this relationship. This probit determination was used to ensure the PPE response had been properly characterized over the velocities of interest. Once complete, an additional 10 shots were conducted between the V_{50} and approximately 1.5 times the V_{50}. For every shot, the system outcome was recorded. For every shot that had a system perforation, the DoP into the TP5 pack was measured.

Corresponding impacts from below the TP5 pack skin V_{50} (resulting in DoP=0) up to velocities corresponding to the maximum velocity conducted for the Tier 1 PP assessment were also conducted against the TP5 pack, but bare, i.e. no PPE covering. This enabled an estimate of residual velocity to be determined from the DoP for the shots that perforated PPE.

1.3. Analysis of the experimental data (step 3)

Although the experimental method used a probit analysis to determine the system perforation response, the HIMA SLM used a logistic regression (logit) model. This was because the bias reduction in binary-response general linear models [6] implemented within R, could not be implemented within the SLM code. The experimental outcomes were used to generate the parameters for a logit model, in the form of Equation (1):

[1] NATO Stock Number (NSN) 8420-99-873-0158 (medium).

https://doi.org/10.52202/081621-0043

$$Probability\ of\ perforation = \frac{1}{(1+e^{-(aV_S+b)})} \quad (1)$$

Where, for the Tier 1 PP with a 6 mm glass sphere; Vs= Strike velocity (or Impact Velocity) in m s^{-1}, a=0.342 and b=-88.03.

For the HIMA SLM, stochastic perforation outcomes used a random number generator with Equation (1) to predict the PPE outcome separately for each individual shotline. If the random number was lower than or equal to the probability of perforation, the projectile defeated the PPE/skin system. If the random number was higher than the probability of perforation, the projectile was stopped by the PPE and there was no skin perforation and residual velocity set to zero. If the PPE system was defeated for a given shotline, the residual velocity was required as an input to the SLM. The method used to generate the residual velocity input was based on the DoP overmatch data into the TP5 pack.

A linear fit applied to the data for the DoP into the bare pack (excluding any non-perforation data) was used to convert the DoP following overmatch of the Tier 1 PP into a residual velocity. A residual velocity equation (based on conservation of energy, assuming a constant projectile mass) from Reference [7] was used. It is a non-linear least squares regression model, given by Equation (2).

$$V_R = \begin{cases} 0, V_s < V_{50} \\ \alpha\left(V_s^{\ P} - V_{50}^{\ P}\right)^{1/P}, V_s \geq V_{50} \end{cases} \quad (2)$$

Where V_R is the residual velocity (m s^{-1}), α and P are parameters bounded by: $0 \leq \alpha \leq 1$ and $P>1$ [7].

In Reference [7], the 'limit velocity' V_L is used rather than V_{50} in Equation (2). However, for any material exhibiting a zone of mixed results in the perforation data (as occurred within this dataset), where the regression equation is allowed to solve for the V_{50} (or V_L), then there will be perforation data below this value. For clarity the V_{50} has been used, as the meaning of V_L may be open to interpretation but is not synonymous with V_0 for Equation (2).

Whilst parameters α, P and V_{50} can all be calculated through Equation (2), in this case for the Tier 1 PP; α was forced to be 1, which ensured that $V_R \rightarrow V_S$ as $V_S \rightarrow \infty$ (i.e. residual velocity will approach strike velocity as strike velocity approaches infinity), and the V_{50} was fixed based on the value previously calculated from the probit analysis. The fixing of α=1 was only necessary due to the transform from DoP into V_R from the TP5 pack. The parameter P could then be calculated by Equation (2), run in Microsoft Excel (2016).

This left an issue, as the logit fit to the PPE system perforation response meant that perforations could be predicted for velocities below the V_{50}, but these would always result in V_R=0 based on Equation (2), which is not realistic. Therefore, a modification was applied: a straight line fit between (V_S=V_{01}, V_R=0) and (V_S=1.01V_{50}, V_R as calculated from Equation (2)). The straight-line fit was applied to end at 1% over the V_{50}, to ensure that residual velocity predictions would not equal 0 when it switched from one model to the other. The final equation is given by Equation 3.

$$V_R = \begin{cases} 0, V_s \leq V_{01} \\ mV_s + c, V_{01} > V_s \geq 1.01V_{50} \\ \alpha\left(V_s^{\ P} - V_{50}^{\ P}\right)^{1/P}, V_s > 1.01V_{50} \end{cases} \quad (3)$$

For the Tier 1 PP on the TP5 pack with the 6 mm glass sphere, Equation 3 was fitted to the data with the following parameters: V_{01}=248.6 m s^{-1}, V_{50}=258.4 m s^{-1}, (1.01V_{50}=261.0 m s^{-1}), m=12.63, c=-3139.1, α=1, P=5.62.

Equation 3 fitted to the experimental data is shown in Figure 1 (red curve). Also shown in Figure 1 is the part of the non-linear regression model (Equation 2) that was removed and substituted with the linear fit between the V_{01} and 1.01V_{50} (dashed grey line). The black line is where residual velocity equals strike velocity, or a bare pack response.

https://doi.org/10.52202/081621-0043

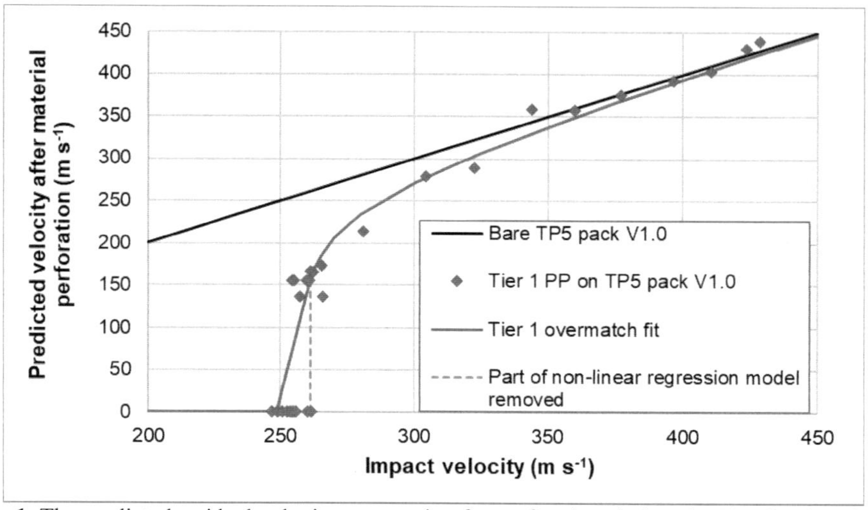

Figure 1. The predicted residual velocity, accounting for perforations below the V_{50}, for the Tier 1 PP covering the TP5 pack with 6 mm glass sphere, plotted with the raw data.

The residual velocity model and V_{50} are based on the system (combined TP5 pack skin and PPE) response. This is required to accurately model the residual velocity in the body in the HIMA SLM.

1.4. Shotline model analysis (step 4)

The HIMA SLM uses the (50$^{\%ile}$ male) Zygote model [9] for its 3D avatar representation. The current implementation of the Zygote body model within the HIMA SLM has 3D representations of 94 organs and soft tissues, 263 bones, and skin, with a fixed geometry and posture. Figure 2 shows the Zygote skin model, Tier 1 PP (area of protection in blue), alongside some of the organs and skeleton models.

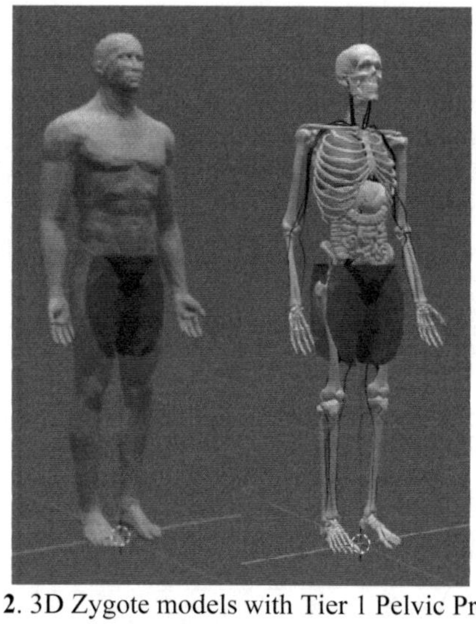

Figure 2. 3D Zygote models with Tier 1 Pelvic Protection

The SLM works by imposing user defined shotlines (rays) onto the avatar and returning the impact coordinates for each individual 3D component within the model. From these coordinates, the model indicates which PPE or tissues the projectile transects at each millimetre step of the shotline. If the shotline intersected the PPE before hitting the body, Equation (1) with a random number generator, combined with Equation 3 allowed the occurrence of /PPE/skin perforation and residual velocity (post skin as the PPE and skin were treated as a system) to be calculated. This was then used as the velocity for the next step. Equations within Reference [8] allowed the probability of skin perforation and residual velocity after perforating the skin to be estimated for impacts directly to the body (not covered by PPE). Deterministic skin perforation calculations were used.

At each millimetre step following PPE and skin or skin perforation; tissue specific retardation was applied [8], based on the residual velocity from the previous millimetre step, until the projectile came to a stop or exited the body. The retardation profile at each millimetre step was used to calculate an injured volume within each tissue type, which in turn was used to calculate the percentage volume of each tissue that was damaged. An injury score based on the percentage damage to each tissue was applied to all penetrated tissues, in a similar approach to that described in Reference [10]. The Abbreviated Injury Scale (AIS) scoring descriptions (update 2015) [11] were used as a guide with a trained and experienced AIS coder to relate each severity score allowed for that tissue from a penetrating injury to a damage volume threshold. The severities predicted through the SLM are not AIS scores or severities, nor are they directly comparable. The severity scores in the SLM serve to provide an indication of the relative injury severity outcome. For validity, the SLM is used as a comparator: either to compare multiple PPE solutions, or to compare the effect of a single PPE solution against bare skin.

The injury for each shotline could then be expressed as the maximum injury for that shotline (injury scores from 0-6, with 0 being no injury), or the overall injury which was calculated as the summed square of the three worst injuries (injury scores from 0 to 75, where any single level 6 organ injury resulted in a combined score of 75 regardless of any other injuries). This mirrors the calculation method of the New Injury Severity Score (NISS), although the scores calculated for the SLM are not NISS severities, nor are they directly comparable. Shotlines that missed the body were excluded from further analysis. Shotlines stopped by PPE were assigned an injury score of 0.

In this paper, comparison of injuries between the unprotected case and that with Tier 1 PP protection was performed by comparison of histograms of the frequency of the overall injury scores. This was because a direct comparison through averaging of scores generated from an 'AIS-like severity' system, where scores are non-linear in terms of severity and/or mortality, would lead to illogical outputs (i.e. an injury severity score of 2 is not twice as severe as a score of 1, and two injuries with a score of 1 are not equal to a single injury with a score of 2).

Future development of the HIMA SLM is to generate an injury scoring system that allows summation and averaging of scores (for example probability based), to allow a single value output from a grid for comparison between scenarios.

A grid over the entire thigh, with 10 mm spacing between shotlines was employed, at 0° incidence in the coronal plane (straight on from the front). Although, grids and/or shotlines can be specified from any impact direction and elevation, this angle was chosen for demonstration purposes and is not a representation of the expected threat to the Tier 1 PP. For simplicity and to reduce run-time the grid covered one thigh, although the outcomes (at a resolution of 10 mm spacing between shotlines) were considered symmetrical.

The grid was run at 10 impact velocities, all with the 6 mm glass sphere, from 200 m s^{-1} (0.77V_{50}) to 400 m s^{-1} (1.55V_{50}) both with and without PPE. These grids resulted in >30,000 shotlines, of which around 22,000 impacted the body. All velocities were above the TP5 pack skin perforation velocity and the skin perforation velocity calculated within the HIMA SLM for this projectile.

From the overall injury score for each shotline, a scoring heatmap was created to illustrate the predicted injury. An example of this scoring map is shown in Figure 3, showing the outcomes for a protected and unprotected thigh.

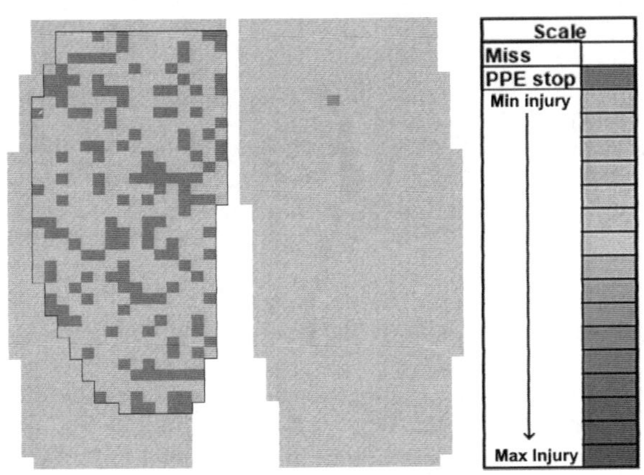

Figure 3. Injury Scoring Map based on the combined injury score for a given shotline. The leg shown on the left has the Tier 1 PP (outlined); the leg on the right has no covering. Impact velocity: 260 m s^{-1}

https://doi.org/10.52202/081621-0043

The colours in Figure 3 correspond to the overall injury score predicted by the HIMA SLM for a fragment impacting at a fixed velocity (close to the Tier 1 PP V_{50}). Each pixel in the scoring map corresponds to the 10x10 mm grid spacing. Although the impact velocity was the same for both protected and unprotected, the defeated armour still reduced the predicted injury.

To assess the effect of the PPE, the number of shotlines corresponding to the overall injury severity was compared between the unprotected and Tier 1 PP. For both the unprotected and Tier 1 PP cases, all injuries were calculated as one of six scores (plus 0 for PPE stops). For simplicity in displaying the data, the six different scores were combined into 3 categories: low (scores of 1 and 2), medium (scores of 17 and 18) and high (scores of 33 and 48). Figure 5 shows the proportion of injuries changed at the different score levels. A positive change meant an increase in the number of shotlines with that overall injury severity score with the Tier 1 PP compared to the unprotected case, and a negative change meant a decrease in the number of shotlines with those injury scores with the Tier 1 PP.

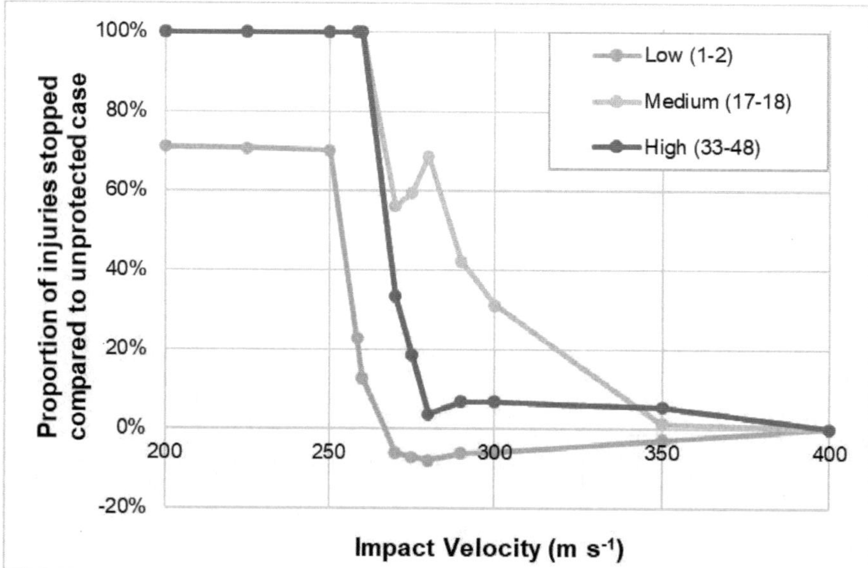

Figure 4. The proportion of injuries stopped by the Tier 1 PP at each overall injury score.

Figure 5 shows the proportional change at each overall score level as the impact velocity is varied. The Tier 1 PP shifted the injury scores to lower score levels and to PPE stops (0 injury score, not shown due to no PPE stops in the unprotected case for a proportional comparison). The increase in the proportion of injuries with a score of 1 and 2, shown by negative values at velocities ≥270 m s^{-1} with the Tier 1 PP is due to the medium and higher score injuries (17-48) being reduced to scores of 1 or 2, but not being stopped completely. At higher velocities over 280 m s^{-1}, the Tier 1 PP has limited mitigation of the higher scored injuries; <10% of 'high' (33 and 48) scored injuries are reduced to lower score levels. Figure 5 appears to show the Tier 1 PP is more effective against the higher scored injuries, but this is due to low numbers of those injuries sustained and the fact they are displayed as a proportion of those in the unprotected case.

At 400 m s^{-1}, no change in jury was observed, as this corresponds to only a 6 m s^{-1} reduction in velocity due to perforating the Tier 1 PP (Figure 1). The limit of effectiveness of the Tier 1 PP in reducing injuries determined through this SLM method is approximately 350 m s^{-1}, which corresponds to 1.35 times the V_{50}.

1.5. Comparison to other methods

To contrast this overmatch injury reduction approach to other assessment techniques, comparisons were made to the following methods (also assessed through the HIMA SLM):

- A deterministic V_{50}. This uses an experimentally determined V_{50} and assumes that any impacts below the V_{50} are stopped, those above are assumed to perforate with V_R=V_S. This method essentially equates the V_0, the V_{50} and the V_{100} as the same point. Therefore, there is no benefit from PPE when the impact velocity is greater than the V_{50}.

https://doi.org/10.52202/081621-0043

- A stochastic perforation method, based on a logit probability of perforation. Impacts that are calculated to perforate the PPE have $V_R=V_S$. This method multiplies the probability of perforation of the PPE system with the expected injury score. The benefit is inversely proportional to the probability of perforation, therefore the benefit is zero as soon as the impact velocity increases over the V_{100}. As it is stochastic, impacts within $V_0<V_S<V_{100}$ can create a mix of perforation outcomes across the PPE.
- Pure coverage assessment. Data for the PPE performance is not used. All shotlines that impact the PPE are assumed to be stopped at all velocities. Those that miss PPE have an injury calculated.

The outcomes of the four techniques in relation to the unprotected SLM case are shown in Figure 6. For simplifying the comparison, the injury scores were grouped: for the overmatch and coverage approach, injury scores were grouped into low (1-2), medium (17-18) and high (33-48). These low/medium/high groups are for comparison purposes, noting the scoring scale goes to a maximum of 75. For the deterministic and stochastic approach, all injuries were affected to the same degree, so were combined into a single group labelled 'all' (scores 1-48).

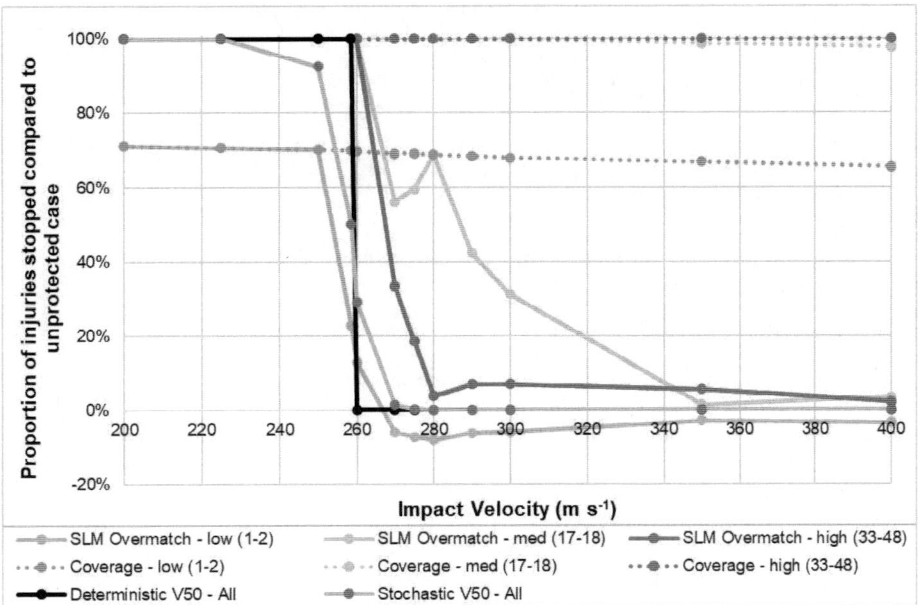

Figure 5. Proportion of injuries stopped compared to unprotected case for each model type.

Figure 6 shows that the overmatch analysis with the HIMA SLM overmatch technique indicates the Tier 1 PP is more effective (in that it gives a larger reduction in injury scores, over a wider range of impact velocities) than is indicated by the deterministic or stochastic perforation methods. The exception is that more shotlines are observed with low injury scores for the SLM overmatch method, as the higher injury scores are reduced into the low band.

Figure 6 shows large differences in the percentage injury score reduction for velocities over 260 m s^{-1} between the simple coverage and full assessment using the Tier 1 PP overmatch data. The differences highlight where the full assessment using the Tier 1 PP overmatch data may provide additional insight.

1.6. Non SLM based comparison of PPE

Noting that the development or acquisition of a SLM and underpinning resource to exploit the model is not trivial, a comparative assessment solution may be possible that utilises much of the same approach described previously but provides a simplified PPE performance comparison (rather than injury reduction prediction). This proposed, but as yet unvalidated approach, uses the same experimental procedure is followed (Section 1.2) with parameters fitted for Equation 3 to the PPE of interest to generate V_S-V_R curves (Section 1.3). If the TP5 pack skin V_{50} is then subtracted from the calculated residual velocity data, this accounts for the fact that even with no material covering, the skin layer on the pack itself has a threshold below which it will not be penetrated (i.e. no injury would be predicted to

https://doi.org/10.52202/081621-0043

have occurred). The V_S-V_R curves (with subtracted skin V_{50}) are then integrated to determine the area under each curve. The integral can be performed on V_S between the TP5 skin V_{50} and a suitable upper velocity limit.

To highlight this different approach and how/why it could lead to selection of different materials than standard test methods, data was collected for some low areal density protective materials using the same experimental approach as outlined in Section 1.2, but using a 3 mm glass sphere. The V_S-V_R curves generated for 5 materials are shown in Figure 7.

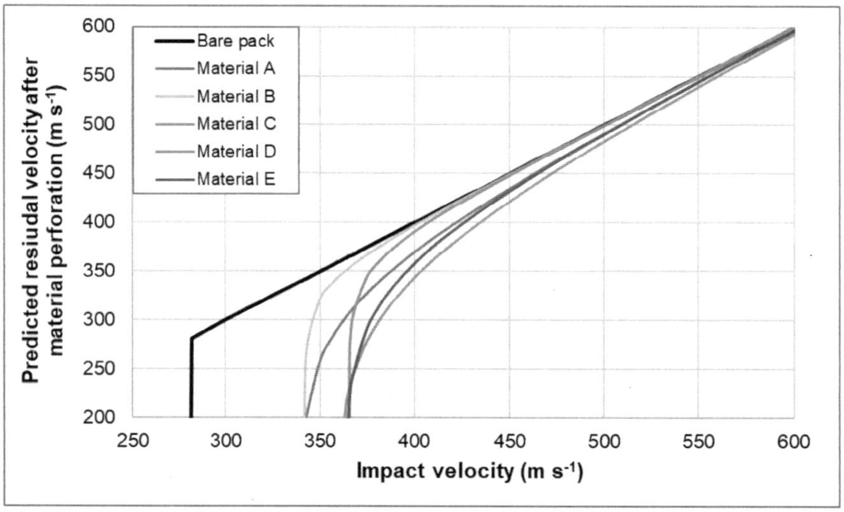

Figure 6: Residual velocity curves (Equation 3) for different materials.

Note that the data collected were not optimised for V_{01} calculations. This is a limitation of this approach and needs to be considered carefully when analysing the data to ensure sensible V_{01} velocities are used in the residual velocity predictions.

The area under the curve, the integral for V_S between the TP5 pack V_{50} (281 m s^{-1}) and 1.6 times the V_{50} of the best performing material (585 m s^{-1}) was conducted, with the TP5 pack skin V_{50} first subtracted from the calculated residual velocity. This maximum velocity of 585 m s^{-1} corresponded to the limit of the data collected. The outcomes of these integral, relative rankings and comparison to V_{50} performance is shown in Table 1.

Table 1. Performance and relative rankings of materials based on the residual velocity overmatch integral and V_{50} testing with a 3 mm glass sphere.

Material	Overmatch integral value	Overmatch % reduction to bare pack	Overmatch integral rank	V_{50} (m s^{-1})	V_{50} % increase to bare pack	V_{50} rank
Material C	4908	28.9	1	360	27.9	3
Material E	5173	25.0	2	364	29.3	2
Material A	5462	20.9	3	340	20.8	5
Material D	5656	18.1	4	365	29.9	1
Material B	5462	8.8	5	342	21.4	4
Bare pack	6901	n/a	n/a	281	n/a	n/a

Table 1 shows that the performance rankings of the materials were different when considered in terms of their overmatch or V_{50} performance. Material D had the highest V_{50}, but comparatively poor overmatch performance. In the context of these materials, they are not designed to prevent all injuries, but reduce the overall injury burden that may occur by covering a comparatively large area, but with low (mass) burden. V_{50} testing alone may result in selection of a sub-optimal protection system. Based on ballistic performance alone, Table 1 shows the optimal performing material was Material C. It also shows that considering this integral value, fabric Material A outperformed Material D, despite having a lower V_{50}.

Whilst this approach in general and these data specifically have yet to be analysed through the HIMA SLM to confirm the rankings of the integral method, Figure 7 and Table 1 are provided to demonstrate how this methodology could be utilised and the types of differences in V_{50} and residual velocities that may be encountered, potentially leading to selection of different materials. It is a similar approach to that suggested within Reference [7], mapping the region under the V_S-V_R curve to a finite area.

This integral method on the overmatch data aids in providing the additional insight to rank the PPE performance, but without the requirement for a shotline model. The main benefit of the SLM additional analysis is that it puts the performance in context, in terms of injury reduction, so that any benefits can be more confidently judged as to the required trade-offs (burden, mobility, comfort, etc.), as well as the ability to directly compare the effectiveness of solutions with different coverage areas (in addition to different ballistic performance), and articulate those differences in terms of injury reduction.

2. DISCUSSION

This study has examined the development and benefits of using an overmatch approach, coupled with vulnerability analysis for the comparison or selection of PPE. The SLM overmatch methodology could allow:

- Optimised PPE to be selected, where the threat cannot be completely protected against.
- Benefits of PPE to be quantified in terms of reduction of the number of injuries of given scores.
- Direct comparison of PPE with different areas of coverage.
- Benefits of enhancing PPE to be quantified (i.e. does adding additional PPE make a significant difference to injury outcome).
- Assessment of the relative vulnerability/protection offered by the PPE from different threats.
- Specific threat-based analysis. For example, vulnerability analysis can be run against any type of shot line grid/pattern. This could be generated via blast trial or modelled weapon fragment distributions and velocities to feed in realistic threat information. However, the key advantage is that such analysis will be based on PPE injury reduction metrics rather than simply on perforation measures.

It is important to note that being a proof of concept study, this work comes with caveats and is not intended to be used as a definitive representation of the Tier 1 PP performance, but demonstrates that using the HIMA SLM with overmatch data could lead to a better understanding of the performance of protection systems; improving operational models. This approach also allows a more complete assessment process for procurement or comparison of PPE in terms of injury reduction rather than PPE perforation.

Further development is also required with the HIMA SLM in terms of injury scoring and expansion of the available parameters within some of the algorithms used (such as appropriate tissue retardation and damage algorithms).

3. CONCLUSIONS

This study has demonstrated the benefit PPE provides across a variety of velocities, including those (significantly) over the PPE V_{50} which are not normally considered or quantified. It has demonstrated that PPE is more effective than would be understood from evaluation by traditional V_{50} type test methods. If threats cannot be stopped within the mass/burden budget of PPE, to maximize effectiveness, PPE could be optimised to provide the maximum reduction in injury over the (widest) range of expected impact conditions using this approach. This is a different approach, which may favour a different solution to selecting PPE with the highest V_{50} performance. It also allows an option for the assessment of the residual risk after mitigation.

Using the Tier 1 PP as an example, the HIMA SLM with overmatch data has been able to show a reduction in injury well above V_{50} and the V_{100}. This showed decreases in injury outcome are expected up until approximately 1.35 times the V_{50} for the Tier 1 PP (i.e. relating to a real, tangible benefit).

Using a simpler methodology, by ranking performance via the area under the V_S-V_R curve, rather than shotline vulnerability analysis, may provide a lower resource approach to comparison of PPE performance. It indicated that materials with the highest V_{50} may not provide optimised protection.

https://doi.org/10.52202/081621-0043

REFERENCES

[1] Lewis, E.A., Pigott, M., Randall, A., and Hepper, A. The development and introduction of ballistic protection of the external genitalia and perineum. *Journal of the Royal Army Medical Corps*, 2013, Vol 159(suppl 1), pp i15-i17.

[2] James, G. and Hepper, A. Ballistic simulation of fragmentation from buried improvised explosive devices. *Personal Armour Systems Symposium, Robinson College, Cambridge, UK,* 2014.

[3] James, G. and Keirl, J. The effect of backing methods on the measured ballistic performance of armour materials *Personal Armour Systems Symposium, Dresden, Germany,* 2023.

[4] Ouellet, S., Pageau, G., and James, G. A new biofidelic backing for the evaluation of the ballistic performance of soft armour and lightweight protective fabrics. *Personal Armour Systems Symposium, Copenhagen, Denmark,* 2021.

[5] R Core Team. R: A language and environment for statistical computing. V2.15.2. Vienna, Austria. R Foundation for Statistical Computing, 2012.

[6] Kosmidis, I. brglm: Bias reduction in binary-response GLMs. (R package). 2007.

[7] Lambert, J. and Jonas, G. Towards standardization in terminal ballistics testing: velocity representation, Aberdeen Proving Ground, Maryland, *Ballistics Research Laboritory, Report no. 1852,* 1976.

[8] James, G. Development of Models to assess penetrating injury from ballistic projectiles (PhD). Cranfield University, Cranfield Defence and Security. DSTL/PUB118063 V1.0. 2020.

[9] *Zygote model, https://www.zygote.com/.* [Online]

[10] Fryer, R., Breeze, J., and James, G. Development of a personnel vulnerability numerical model. *Personal Armour Systems Symposium (PASS), Amsterdam, Netherlands,* 2016.

[11] Association for the Advancement of Automotive Medicine (AAAM). Abbreviated Injury Scale Update 2015. 2016.

Implementation of NIJ Standard 0101.07 to Test and Evaluate Ballistic Resistant Body Armor

Mark E. Greene[1], Jared M. Gardner[1], Daniel A. Longhurst[2], Lance L. Miller[2], and Richard A. Sundstrom[2]

[1]*National Institute of Justice, 999 North Capital Street NE, Washington, DC, 20002, USA, (mark.greene2@usdoj.gov)*
[2]*RTI International, 3040 Cornwallis Road, Research Triangle Park, NC 27709, USA*

Abstract. The U.S. National Institute of Justice (NIJ) has revised its standard for ballistic-resistant body armor that protects law enforcement personnel against gunshots. NIJ is actively evaluating armor models submitted to test to NIJ Standard 0101.07, *Ballistic Resistance of Body Armor*, using the threats enumerated in NIJ Standard 0123.00, *Specification for NIJ Ballistic Protection Levels and Associated Test Threats*. This paper focuses on the activities that have occurred during the first year and a half of implementation of the new standard and provides a comprehensive update on the NIJ Compliance Testing Program (CTP) based on a decade and a half of experience operating with NIJ Standard 0101.06. Law enforcement purchasers can now find armors that have been successfully tested to NIJ Standard 0101.07—the seventh revision of the NIJ standard for torso-worn ballistic resistant of body armor—on the NIJ Compliant Products List (CPL). As previously presented at the Personal Armour Systems Symposium, the new standard is built with ASTM standards as components, with improvements to account for the changing landscape of policing in the U.S. Specifically, NIJ Standard 0101.07 aims to improve the testing of armor designed for women and to explore potential vulnerabilities more thoroughly in all armor designs certified by NIJ. Some of the improvements include the use of new test methods for nonplanar armor, three new rifle threats, reconfigured shot placements, testing conditioned armor at the reference velocity rather than a reduced velocity, and nonzero angles of incidence for shots in the proximity of edges. After the publication of NIJ Standard 0101.07 in November 2023, NIJ has continued to refine the standard through engagement with the body armor community and with the NIJ-approved test laboratories, currently numbered at five. Those engagements have produced a number of interpretations and clarifications to the standard which have been communicated through updates to the original publication. A potential ammunition manufacturing change to the Remington 9mm 124 grain FMJ test threat also necessitated that NIJ develop methods to reevaluate the threat specifications in NIJ Standard 0123.00. This paper will give a complete overview of NIJ's ballistic body armor activities, describing the ongoing improvements to NIJ Standard 0101.07, NIJ Standard 0123.00, the NIJ CTP, and provide guidance for agencies, purchasers, and end users.

1. NIJ STANDARD 0101.07: BALLISTIC RESISTANCE OF BODY ARMOR – IN PRACTICE

NIJ is actively evaluating armor models to NIJ Standard 0101.07, *Ballistic Resistance of Body Armor* [1], using the updated body armor test threats and protection levels enumerated in NIJ Standard 0123.00, *Specification for NIJ Ballistic Protection Levels and Associated Test Threats* [2]. Law enforcement in the United States continues to be a highly dangerous profession. Body armor worn by U.S. law enforcement, to protect the torso against handgun and rifle ammunition, remains a critical component of the safety equipment that officers rely upon. Since 1972, when NIJ published its first performance standard for ballistic-resistant police body armor [3], the minimum performance requirements and test methods for body armor have continued to evolve. Accordingly, the NIJ body armor standard's current revision, the seventh, aims to implement methodological and procedural improvements to keep pace with the evolving landscape, workforce, and threats that law enforcement officers face. Recent updates incorporate many of the lessons learned during the past decade and a half of testing using NIJ Standard 0101.06, published in 2008 [4].

In addition to updated body armor test threats and protection levels, many of the test methods and laboratory practices in NIJ Standard 0101.07 have been improved and it now includes new test methods for female body armor, as discussed previously [5, 6]. This revision also aims to explore potential vulnerabilities more thoroughly in all armor designs certified by NIJ through reconfigured shot placements, testing conditioned armor at the reference velocity rather than a reduced velocity, and nonzero angles of incidence for shots in the proximity of edges. Additionally, NIJ Standard 0101.07 is the first revision of the body armor standard for which the NIJ Mark®, a registered trademark with the United States Patent and Trademark Office (USPTO) in 2019 [7], is available at the time of certification of all armor models listed on a new Compliant Products List (CPL) for body armor.

Considering the broad scope of updates that have been incorporated in the new standard, NIJ has continued to refine the laboratory test methods and procedures after its publication in November 2023.

https://doi.org/10.52202/081621-0044

This has been accomplished through engagements with end user communities, government partners including the U.S. Army and the National Institute of Standards and Technology (NIST), body armor manufacturers, material suppliers, and with the five current NIJ-approved test laboratories. These engagements have produced substantive feedback and have been important opportunities for consultation with the body armor community during the implementation of NIJ Standard 0101.07 in the NIJ Compliance Testing Program (CTP). Interpretations and clarifications to the standard that have been produced are being communicated through updates and addenda to the original publications. These engagements have also helped make possible the successful incorporation, by reference, of ten ASTM test methods and laboratory practices in NIJ Standard 0101.07. NIJ continues to work through ASTM's Committee E54 on Homeland Security Applications, Subcommittee E54.04 on Public Safety Equipment, with its large group of end users and technical experts, to produce and maintain a suite of standardized test methods and laboratory practices for incorporation in NIJ standards [8].

Recently, a potential ammunition manufacturing variation necessitated that NIJ, with the support of all the NIJ-approved laboratories and several material manufacturers, develop methods to evaluate threat specifications in NIJ Standard 0123.00. For this purpose, an evaluation of the Remington 9mm 124 grain FMJ (full metal jacket) test threat was conducted. This was necessary due to an unexpected change in the jacket material of the ammunition, but the evaluation also serves as an initial study of potential methods for conducting similar evaluations of NIJ test threats in the future. This paper will provide a complete overview of the NIJ's body armor activities that have occurred during the first year and a half of implementation of the new standard. It includes information on some of the latest clarifications to the standard, a review of the testing that was conducted to reevaluate the 9mm test threat, updates on the NIJ CTP, and provides guidance for commercial participants, agencies, purchasers, and end users.

2. CLARIFICATIONS OF LABORATORY TEST METHODS AND PRACTICES

NIJ published Addendum 1 to NIJ Standard 0101.07, *Ballistic Resistance of Body Armor* in October 2024, which includes revisions to the standard [9]. The revisions are the product of the previously discussed ongoing collaborative engagements with stakeholders in the body armor community and the five currently NIJ-approved laboratories. Revisions are necessary for most test standards to clarify their content and methods and to provide interpretations of the standard's intent as they are being put into practice. These recent changes affect the application of NIJ Standard 0101.07 for assessing the minimum performance requirements of body amor during Initial Type Testing (ITT) and for Follow-up Inspection and Testing (FIT). Regular consultation between NIJ and the NIJ-approved test laboratories through the NIJ CTP in 2024 and 2025 has functioned as a forum to discuss the test methods and laboratory practices as they are being implemented by numerous ballistics testing experts from different testing organizations at different locations for testing body armor to NIJ's standard. This forum offers the laboratories a means to pose questions and raise concerns to NIJ and NIJ CTP staff among a ready community of experts to work through technical issues. It also offers a means for NIJ and the NIJ CTP to canvass the laboratories to gather a variety of viewpoints on matters impacting the body armor program.

NIJ is continuing to review the original publication of the standard during the implementation phase of NIJ Standard 0101.07. Additional clarifications to those published in Addendum 1 have been made during implementation of the standard. Revisions may be incorporated into the standard in several ways. They may replace or supplement the information in the original publication, or in some cases, the revisions may replace *and* supplement information in the standard. Like the updates published in Addendum 1, ongoing updates also affect the implementation of the standard for soft and hard body armor, planar and nonplanar armor designs, as well as the testing of armor designed for women. NIJ anticipates publishing these additional interpretations and clarifications on a periodic basis going forward as either new addenda or revised versions of an existing addendum to the standard.

Table 1 provides an overview of the revisions in Addendum 1 and some further revisions that have subsequently been made after its publication. While it is not possible in the scope of this paper to reproduce verbatim the complete list of all revisions that have been made and fully explore each revision within the context of the standard as discussed by NIJ, the NIJ CTP, and the NIJ-approved laboratories during the implementation phase of the new standard, this overview table summarizes a majority of the revisions. The goal here is to provide a succinct update on many of the changes that are of the greatest interest to the body armor community. For example, the first entry in Table 1 clarifies certain additional conditions that would be characterized as a complete penetration (CP) of soft or hard armor to occur during testing. The specific changes cited are not the only conditions in NIJ Standard 0101.07 that if met, may constitute a CP during testing.

Table 1. Select revisions to NIJ Standard 0101.07.

NIJ Standard 0101.07 Revision and Description
Conditions to be met for a *complete penetration* (CP) 1. (NEW) Adds for nonzero angles of obliquity on soft armor: (i) any portion of a test threat or a fragment of a test threat that is visible through the panel cover; (ii) if the shot exits the panel cover from the edge and is not retained by the ballistic panel; (iii) any portion of a test threat or a fragment of a test threat deflecting toward an edge between layers of ballistic material that meets *all* the following three conditions: (a) breaks the ballistic panel cover; (b) is visible when facing the wear face (i.e., line of sight is normal to the wear face); (c) extends beyond the edge contour of the armor panel. 2. (NEW) Adds for hard armor, a fragment of any hard component (i.e. ceramics and metals) of a test item that penetrates the back cover of the test item. Spall embedded in the clay surface beyond the perimeter of the test item shall not be considered a CP.
Definition of *crown* on a hard armor plate Modifies a crown location to be at the intersection of multiple curvatures on the strike face of a multi-curve plate.
Equation for estimated probability of complete penetration at V_{ref} Corrects the sign of the inequality to: V_{ref} shall be less than 5% (i.e., $\widehat{V}_{05} \geq V_{ref}$)
Shot locations for P-BFD testing on soft armor (NEW) Adds schematics for shot locations and order on the smallest are largest test items.
Shot pattern for ballistic limit on soft armor (NEW) Adds a schematic for shot locations and shot order.
Positioning of ballistic panels to accommodate shot 7 on soft armor (NEW) Adds that ballistic panels shall not be folded and that the test item be positioned so that the entire test item shall be supported coplanar with the backing material.
Tolerances on the location of shot 7 on soft armor (NEW) Adds tolerances at the center of the neck strip to be not more than 19 mm (0.75 in.) from the centerline.
Use of a witness panel and clay surface observations during edge shots (NEW) Adds that for any shot at a nonzero angle of obliquity on soft armor, including shot 7, a witness panel and the clay surface between the test item and witness panel shall be used to determine if any portion of a test threat or a fragment of a test threat exited the side of the armor.
Adjustments to clay block requirements Modifies that a clay block that meets the acceptance criteria may be used for up to 45 minutes after the final drop in the verification procedure prior to testing as well as removes the requirement for clay block verification between test items and after the final test item.
Clarifications regarding use of test items on a clay block (NEW) Clarifies that multiple soft armor test items can be used on a clay block, how they can be rotated, how they should be affixed with strapping and supported with extesions, and how many shots of one type of threat may be used on a clay block before reconditioning.
Exploiting vulnerabilities in conditioned soft armor (NEW) Clarifies that shot locations that overlap with features introduced by conditioning by tumbling, such as creases, shall not be avoided during testing and should be preferentially selected within the fair impact areas to strike features to exploit potential weaknesses or vulnerabilities in the armor.

https://doi.org/10.52202/081621-0044

NIJ Standard 0101.07 Revision and Description
Hard armor test item dimensions Clarifies that hard armor test items shall be no larger than 254 mm x 305 mm (10 in. x 12 in.) and no smaller than 203 mm x 254 mm (8 in. x 10 in.).
Adjustment of velocity tolerances on the first shot of the ballistic limit test series Modifies the actual velocity of the first shot to be the reference velocity for the test threat, including its tolerances, –9/+30 m/s (–30/+100ft/sec).
Supplier declares number of shots per test item for RF1 and RF2 P-BFD testing (NEW) Adds that the supplier may declare the number of shots (3 or 6) on a single test item *based on threat round*, rather than protection level alone.
Number of hard armor test items required (NEW) Clarifies the number of test items needed depending on how many shots per test item per threat round declared by the supplier.
Hard armor test item conditioning (NEW) Clarifies the requirements for the apparatus used for hard armor test item conditioning.
Shot placement for 3-shot hard armor test items for RF1, RF2, and RF3 P-BFD testing 1. (NEW) Adds that for a 3-shot plate with 0-degree shots, BFD is taken for shots 1 and 2. 2. Modifies multi-curve plate schematic so that for shot angles for a 3-shot plate with 30-degree shots: (i) the crown shot and next closest shot shall be angle parallel to each other; (ii) shots in the shaded band (around the edges) shall be angled toward the centerline.
Shot 6 on a curved hard armor test item Modifies that shot 6 shall be placed on the crown.
Shot placement for 6-shot hard armor test items for RF1, RF2, and RF3 P-BFD testing 1. Modifies that BFD measurements shall be taken for shots 1 and 2 at 0-degrees. 2. Modifies multi-curve plate schematic so that shot 5 in the cross-hatched area and one shot in the shaded band (around the edges) shall be angled at 30-degrees toward the centerline.
Shot patterns for ballistic limit on hard armor (NEW) Adds schematics for shot locations and order on 1-shot, 3-shot, and 6-shot test items.
Label test method Modifies the label test to improve the practicality of the test.

NIJ also published Addendum 1 to NIJ Standard 0123.01, *Specification for NIJ Ballistic Protection Levels and Associated Test Threats* in October 2024, which includes revisions to that standard as well [10]. The review followed the same process as with NIJ Standard 0101.07 with regular consultation among NIJ, the NIJ CTP, and the NIJ-approved laboratories. While this document is much shorter than NIJ Standard 0101.07 and did not incorporate as many revisions, several notable adjustments were made. These include correcting the listed mass of the 7.62x51mm M80 Ball NATO FMJ Steel Jacket RF1 and RF2 test threat, clarifying the ammunition identifier for the RF3 test threat, and clarifying the requirements regarding ammunition measurements. Another modification was to clarify the use of the terms "lot" and "can" for purposes of auditing 7.62x39mm mild steel core (MSC) test ammunition. More information about the complete list of revisions to NIJ Standard 0101.07 and NIJ Standard 0123.00 is available on the NIJ website [11, 12].

3. A METHOD TO EVALUATE AMMUNITION SPECIFICATIONS IN NIJ STANDARD 0123.00 - EVALUATION OF THE REMINGTON 9MM HANDGUN TEST THREAT

The Remington 9mm Luger 124 grain FMJ ammunition has a longstanding history of use by the body armor industry as a standard ammunition for the assessment of handgun ballistic protection levels for law enforcement. Recently, Remington made changes to the projectile jacket material for certain lots of this ammunition. At the time of this paper, two variants of the ammunition are being sold under Remington product number #23558. Due to the potential significance of this type of manufacturing change on the terminal ballistic response of the projectile, NIJ determined that it was necessary to develop a method to reevaluate the specification of the 9mm test threat in NIJ Standard 0123.00. With the participation of five NIJ-approved laboratories and four material manufacturers, NIJ conducted an evaluation of the two projectile variants and compared their ballistic performance against planar-soft armor panels of two common material types used in ballistic body armor, para-aramid and ultra-high-molecular-weight polyethylene (UHMWPE), which involved collecting over 1100 shots across 104 test panels.

Figure 1 shows the projectile variants, both brass in composition. To simplify the descriptions of the different the jacket types, the projectile variants are here referred to as "copper", the original projectile jacket, and "brass", the new projectile jacket. Material composition of the jackets was determined via metalurgical analysis by a partnering government agency to be aproximately 89:11 (Cu:Zn) for the "copper" jacket, and approximately 69:31 (Cu:Zn) for the "brass" jacket. The thickness and hardness (Rockwell B) of the projectile jackets was also analyzed and determined to be nominally equivalent. Sample sizes used to conduct the physical analysis were n = 30, and each sample came from a single lot of ammunition for each jacket type. Figure 1 also shows two examples of projectile core fragments for each jacket type that were recovered after testing at HG2 velocity and 30° angle of incidence with a UHMWPE soft armor panel. Minor, but recognizable differences were observed post-test in the deformation and fragmention of the projectile cores that were tested at high angles of incidence against UHMWPE panel types. However, most of the core fragments recovered after testing exhibited similar characteristics when the jacket types were compared.

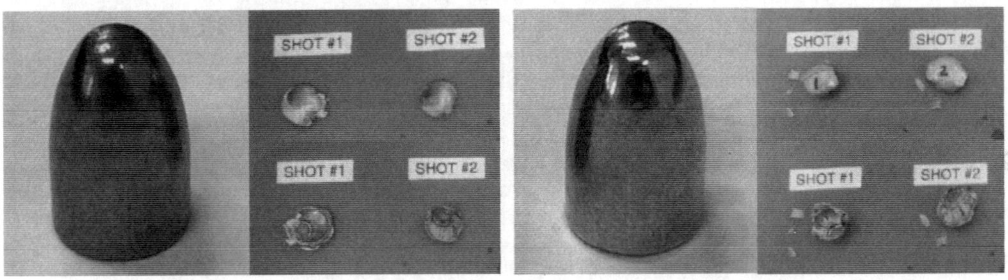

Figure. 1. Copper (left) and brass (right) projectile variants of the Remington 9mm Luger 124 grain FMJ ammunition and examples of tested projectile core fragments. The projectile fragments are shown from a perspective of the projectile's nose (top row) and base (bottom row).

Ballistic performance of the copper and brass projectile jackets was compared in several ways using two types of testing and four armor panel designs. In the first test series, one plain-woven para-aramid panel design, referred to as Aramid 1, and one cross-ply unidirectional UHMWPE panel design, referred to as PE 1, were tested at 0° angle of incidence over velocities ranging from below the HG1 protection level reference velocity, to above the predicted V_{50} ballistic limit of each panel. A total of 640 shots were conducted during these tests using forty test panels, and ten test panels were used for each of four possible projectile and panel type combinations. Test results were then analyzed by statistical regression methods, yielding full regression curves and ballistic limit values for both projectiles against a single panel design of each material type. Table 2 provides a comparison of the V_{50} values calculated from the regression analysis and the sample size "n" for each test. The absolute difference and percentage change of the V_{50} for the brass projectile compared to the V_{50} for the copper projectile are also provided for each panel type as a simple comparison of projectile performance. Absolute difference and percentage change are calculated using the V_{50} values for the copper projectile as the reference values.

https://doi.org/10.52202/081621-0044

Table 2. V_{50} ballistic limit values determined by statistical regression analysis for copper and brass variants of the Remington 9mm Luger 124 grain FMJ.

Panel Type	V_{50}			
	Copper	Brass	Absolute Difference and Percentage Change in V_{50} of Brass Relative to Copper	
Aramid 1	491 m/s (1610 ft/s) n = 160	490 m/s (1608 ft/s) n = 160	–0.6 m/s (–2 ft/s)	–0.1%
PE 1	510 m/s (1676 ft/s) n = 160	524 m/s (1719 ft/s) n = 160	+13.1 m/s (+43 ft/s)	+2.6%

Backface deformation (BFD) in the clay backing substrate was also measured and collected for the small number of shots that were conducted in the first test series at velocities in the HG1 and HG2 reference velocity ranges. Average BFD values for these shots are presented alongside the results from the second test series in Tables 3 and 4. All of the soft armor panels used during this evaluation had dimensions of 38.1 cm x 38.1 cm (15 in. x 15 in.), with 10.2 cm (4 in.) corner stitches, and ballistic nylon covers. A more complete review of the results of the first test series and of the regression analysis performed, was submitted to the conference in a separate paper.

In the second test series, two panel designs for each material type, para-aramid, and UHMWPE were used to compare projectile performance for 30° and 45° angles of incidence at HG1 and HG2 protection level reference velocities using NIJ P-BFD laboratory methods. Aramid 1 and PE 1 panel designs, used during V_{50} testing, were utilized again to compare projectile performance at HG1 protection level reference velocity. Two new panel designs were also introduced, one cross-ply unidirectional para-aramid, referred to as Aramid 2, and a different type of cross-ply unidirectional UHMWPE, referred to as PE 2, to compare projectile performance at HG2 protection level reference velocity. A total of 512 shots were conducted during these tests using sixty-four test panels. In most of these tests, four test panels were used to test one panel design of each material type at 30° and 45° angles of incidence for each reference velocity. For the Aramid 2 panel type only, two test panels were used for each projectile and shot angle combination at both reference velocities.

BFD values were measured and collected for all shots in the second test series. Tables 3 provides a comparison of the average BFD measurements and the sample size for each test. Table 4 compares the average BFD measurements for the brass projectile with the copper projectile using the absolute difference and percentage change between them for each panel type. Absolute difference and percentage change are calculated using the BFD values for the copper projectile as the reference values. Data trends observed here showed differences of less than 9% in measured BFD when comparing projectile types for the 0° test conditions. BFD percentage differences observed between the projectiles types for the 30° and 45° test conditions were comparatively smaller, and were obtained using larger sample sizes.

Table 3. Measured BFD values from P-BFD testing of copper and brass variants of the Remington 9mm Luger 124 grain FMJ.

NIJ Ballistic Protection Level	Reference Velocity	Panel Type	BFD (mm)					
			0°		30°		45°	
			Copper	Brass	Copper	Brass	Copper	Brass
HG1	398 m/s (1305 ft/s)	Aramid 1	35.3 n = 11	32.9 n = 11	25.0 n = 32	24.5 n = 32	21.6 n = 32	20.9 n = 32
		Aramid 2	*n/a*	*n/a*	25.0 n = 16	25.2 n = 16	22.9 n = 16	23.3 n = 16
		PE 1	32.8 n = 8	29.8 n = 9	26.3 n = 32	25.6 n = 32	24.9 n = 32	24.7 n = 32
HG2	448 m/s (1470 ft/s)	Aramid 1	40.5 n = 7	37.6 n = 9	*n/a*	*n/a*	*n/a*	*n/a*
		Aramid 2	*n/a*	*n/a*	26.3 n = 16	26.5 n = 16	24.2 n = 16	24.8 n = 16
		PE 1	33.3 n = 7	34.1 n = 11	*n/a*	*n/a*	*n/a*	*n/a*
		PE 2	*n/a*	*n/a*	31.7 n = 32	30.7 n = 32	31.1 n = 32	29.2 n = 32

n/a = data was not acquired for this test condition.

Table 4. Difference comparison from P-BFD testing of copper and brass variants of the Remington 9mm Luger 124 grain FMJ.

NIJ Ballistic Protection Level	Reference Velocity	Panel Type	Absolute Difference and Percentage Change in BFD of Brass Relative to Copper		
			0°	**30°**	**45°**
HG1	398 m/s (1305 ft/s)	Aramid 1	–2.5 mm, –7.0%	–0.5 mm, –2.0%	–0.8 mm, –3.5%
		Aramid 2	*n/a*	+0.2 mm, +0.7%	+0.4 mm, +1.9%
		PE 1	–2.9 mm, –8.9%	–0.7 mm, –2.8%	–0.2 mm, –0.9%
HG2	448 m/s (1470 ft/s)	Aramid 1	–2.9 mm, –7.1%	*n/a*	*n/a*
		Aramid 2	*n/a*	+0.2 mm, +0.8%	+0.6 mm, +2.6%
		PE 1	+0.8 mm, +2.3%	*n/a*	*n/a*
		PE 2	*n/a*	–1.1 mm, –3.4%	–1.9 mm, –6.0%

n/a = data was not acquired for this test condition.

Coefficients of variation of the average BFD values, calculated by dividing the standard deviation by the average value for the sample, were similar between the projectile types. Coefficient of variation values ranged between 5%–13% for para-aramid panels and 4%–8% for UHMWPE panels tested at 0° angle of incidence, and between 5%–15% for para-aramid panels and 4%–9% for UHMWPE panels tested at 30° and 45° angles of incidence. No complete penetrations of either panel type were witnessed during testing at HG1 or HG2 reference velocities. Post-test examinations of the test panels showed similar trends in the number of layers penetrated by both projectile types for all directly comparable test conditions.

4. UPDATE ON THE NIJ COMPLIANCE TESTING PROGRAM

Use of NIJ-certified body armor remains ubiquitous among U.S. law enforcement and corrections agencies, and many agencies outside the U.S. make use of NIJ standards and NIJ-certified armor. Since the NIJ CTP began accepting armor submissions to meet the requirements of NIJ Standard 0101.06 in 2009, over 2,000 unique models of ballistic-resistant body armor have been submitted to the CTP for compliance testing through the end of 2024, with an overall failure rate of approximately 36% of models submitted for initial testing which do not meet the requirements for listing on the CPL. Currently there are over 440 models of ballistic-resistant body armor listed on NIJ's Compliant Products List [13, 14].

Industry participation in the program by manufacturers is voluntary; however, it is recognized by manufacturers as the standard in body armor quality assurance. At the end of 2024, the NIJ CTP had 83 participants worldwide producing armor compliant with NIJ Standard 0101.06 at 144 manufacturing locations, with 75 locations inside the U.S. and 69 outside the U.S. The NIJ CTP also recognizes Body Armor Quality Management System Requirements called BA9000, which are optional quality assurance requirements to which body armor manufacturers can choose to have their manufacturing locations certified. Body armor manufactured at locations certified to BA9000 are required to undergo FIT less frequently than those manufactured at those that are not due to the enhanced quality assurance in place, which can have the added benefit of reduced overall costs to NIJ CTP participants to maintain their listed armor models. There are 20 manufacturing locations certified to BA 9000 operated by 19 of the NIJ CTP participants, representing 218 models, or 49%, of the ballistic-resistant body armor currently listed on the CPL.

The NIJ CTP is incorporating many improvements to the program in addition to the updated test methods and laboratory practices used to certify body armor to NIJ Standard 0101.07. The program continues to provide U.S. law enforcement and correctional agencies confidence that the body armor they purchase and use meets the NIJ's minimum performance requirements against common handgun and rifle threats, as well as stab threats. Other improvements in the NIJ CTP for NIJ 0101.07 include a new web application for the Testing Information Management System (TIMS) [15], an updated Compliance Testing Report (CTR) that helps test laboratories to generate and report test results, and reorganized process documents to assist participants in navigating program policies and procedures.

While NIJ is certifying body armor to the new NIJ Standard 0101.07, and creating the new CPL for those armor models, it continues to recognize armor certified to NIJ Standard 0101.06. Given the large number of ballistic-resistant equipment models currently worn by law enforcement officers, NIJ will continue to maintain its CPL for armor models compliant with NIJ Standard 0101.06 through at least year-end 2027 and will continue to require FIT on these models. This will allow manufacturers and

https://doi.org/10.52202/081621-0044

law enforcement agencies to transition their equipment to armor that meets the new standard as smoothly as possible over a reasonable amount of time.

5. CONCLUSION

NIJ is actively implementing NIJ Standard 0101.07, *Ballistic Resistance of Body Armor*, to evaluate the performance of ballistic resistant body armor for commercial participants, law enforcement agencies, purchasers, and users. Body armor is currently being certified to the new standard and NIJ is in the process of publishing a new Compliant Products List for those armor models. The revised standard includes updated body armor test threats and protection levels, new test methods for female body armor, and improved test methods and laboratory practices. Many functions of the NIJ CTP have also been improved through the new TIMS web application and updated program policies and procedures. With the introduction of NIJ Standard 0101.07, for the first time the NIJ Mark® is available at the time of certification of all the newly listed armor models. NIJ also continues to recognize armor certified to NIJ Standard 0101.06 and will maintain its CPL for NIJ Standard 0101.06 compliant armor models through at least year-end 2027.

Continuous collaborative engagements with the body armor community and with the five NIJ-approved test laboratories during the first year and a half of implementation of the new standard has produced numerous refinements to the standard. Insights gained from these collaborations for the implementation of reconfigured shot placements, shots at nonzero angles of incidence in the proximity of edges, new procedures for testing conditioned armor, and multiple other aspects of laboratory testing procedures, has been critical to successful implementation. New interpretations and clarifications to the standard are being communicated with the body armor community through addenda to the original publication and will continue to be published on a periodic basis to the NIJ website.

With the support of five NIJ-approved laboratories and four material manufacturers, the NIJ also developed potential methods for conducting comparative evaluations of current and future threat specifications in NIJ Standard 0123.00, *Specification for NIJ Ballistic Protection Levels and Associated Test Threats*. In the study discussed here, an unexpected manufacturing change in the jacket material of the Remington 9mm Luger 124 grain FMJ made it necessary to evaluate the effects of the change on the performance of common types of armor materials. The ballistic performance of the original copper jacketed and new brass jacketed projectiles was evaluated in V_{50} testing, and P-BFD testing, against planar-soft armor panels of para-aramid and UHMWPE materials. Over 1100 shots were conducted in total using 104 test panels, comparing the projectiles at HG1 and HG2 velocities for 0°, 30°, and 45° angles of incidence. For the materials, panel designs, and test conditions that were evaluated, similar armor performance was observed between the variants of this ammunition. While only an overview of the results of the evaluation was discussed in this paper, it provides the foundation of a potential framework for NIJ to perform similar threat comparisons in the future. Information from this initial study also prepared NIJ to adapt the threat specification for the 9mm threat if a permanent change in the manufacturing of this ammunition occurs.

Acknowledgments

The authors would like to acknowledge the ASTM staff and the participants in the ASTM E54 working groups who contributed to the development of the ASTM standards referenced in this paper. The authors would like to acknowledge the participation of the NIJ Special Technical Committee on Body Armor for contributing to the development of the two NIJ documents discussed in this paper. The authors would like to acknowledge the individual participating members of the user communities, our U.S. government partners for their assistance, and the body armor manufacturers for their feedback on the new standard as well as the new threat specification. The material suppliers are gratefully acknowledged for providing the materials that made reevaluation of the 9mm ammunition specification possible. Finally, the laboratory personnel and managers are acknowledged for their expertise on ballistic testing and evaluation of body armor during the implementation process and for their significant contributions to the reevaluation of the 9mm ammunition.

References

[1] NIJ Standard 0101.07, *Ballistic Resistance of Body Armor*, National Institute of Justice, November 2023, https://www.ojp.gov/pdffiles1/nij/307346.pdf.

[2] NIJ Standard 0123.00, *Specification for NIJ Ballistic Protection Levels and Associated Test Threats*, National Institute of Justice, November 2023, https://www.ojp.gov/pdffiles1/nij/307347.pdf.
[3] NILECJ Standard 0101.00, *NILECJ Standard on the Ballistic Resistance of Police Body Armor*, National Institute of Law Enforcement and Criminal Justice, March 1972, https://www.ncjrs.gov/pdffiles1/Digitization/7037NCJRS.pdf.
[4] NIJ Standard 0101.06, *Ballistic Resistance of Body Armor*, National Institute of Justice, July 2008, https://www.ncjrs.gov/pdffiles1/nij/223054.pdf.
[5] M.E. Greene, J. Horlick, D.A. Longhurst, L.L. Miller, C. Robinson, and R.A. Sundstrom, "NIJ Standards for Ballistic Resistance of Body Armor and Stab Resistance: New Developments," *16th International Personal Armour Systems Symposium*, Dresden, Germany, September 11-15, 2023.
[6] M.E. Greene, J. Horlick, D.A. Longhurst, L.L. Miller, C. Robinson, and R.A. Sundstrom, "NIJ Standard 0101.07, Ballistic Resistance of Body Armor: Changes and Challenges," *15th International Personal Armour Systems Symposium*, Copenhagen, Denmark, October 11-15, 2021.
[7] USTPO trademark #5906126, "NIJ Listed Model", U.S. Departartment of Justice Office of Justice Programs, 2019.
[8] "Subcommittee E54.04 on Public Safety Equipment: Matching Standards Under the Jurisdiction of E54.04 by Status," ASTM website, accessed March 7, 2023, https://www.astm.org/get-involved/technical-committees/committee-e54/subcommittee-e54/jurisdiction-e5404.
[9] "NIJ Standard 0101.07, *Ballistic Resistance of Body Armor*, Addendum 1" National Institute of Justice, October 2024 https://nij.ojp.gov/sites/g/files/xyckuh171/files/media/document/addendum-1-nij-standard-0101-07.pdf.
[10] "NIJ Standard 0123.00, *Specification for NIJ Ballistic Protection Levels and Associated Test Threats*, Addendum 1" National Institute of Justice, October 2024 https://nij.ojp.gov/sites/g/files/xyckuh171/files/media/document/addendum-1-nij-standard-0123-00.pdf.
[11] "Ballistic Resistance of Body Armor, NIJ Standard 0101.07," National Institute of Justice website, accessed February 4, 2025, https://nij.ojp.gov/topics/equipment-and-technology/ballistic-resistance-body-armor-nij-standard-010107#1-0.
[12] "Specification for NIJ Ballistic Protection Levels and Associated Test Threats, NIJ Standard 0123.00," National Institute of Justice website, accessed June 11, 2025, https://nij.ojp.gov/topics/equipment-and-technology/specification-nij-ballistic-protection-levels-and-associated-test-threats-nij-standard-012300.
[13] "Compliant Products List: Ballistic Resistant Body Armor," National Institute of Justice website, accessed February 26, 2025, https://nij.ojp.gov/topics/equipment-and-technology/body-armor/ballistic-resistant-armor.
[14] "Compliant Products Lists," Criminal Justice Testing and Evaluation Consortium website, accessed February 26, 2025, https://cjtec.org/compliance-testing-program/compliant-product-lists/.
[15] Testing Information Management System website, accessed March 12, 2025, https://nijctp.org/Identity/Account/Login.

https://doi.org/10.52202/081621-0044

Initial Studies on Roma Plastilina no. 1.2; Improvements to the Original Ballistic Testing Clay

I. Peng[1], J. Domic[2] and A. Williamson[3]

[1]*Chavant, Inc., 5600 Lower Macungie Rd, Macungie, PA. 18062, USA*
isaac@chavant.com
[2]*Smooth-On, Inc., 5600 Lower Macungie Rd, Macungie, PA. 18062, USA*
joe@smooth-on.com
[3]*Contra Threat Sciences LLP, 14 Heather Court, Elkton, MD. 21921, USA*
adam@contrathreat.com

Abstract. Roma Plastilina no. 1 (RP1) was developed as a sculpting clay and adopted as a ballistics testing clay. Several changes have been made to improve performance as a ballistic clay without changing some key attributes of the original formula. This new version, Roma Plastilina no. 1.2 (RP1.2), yields similar results as RP1 in side-by-side live fire tests using 0.44 magnum handgun ammunition on soft armour. However, RP1.2 is in calibration at room temperature and shows less shot-to-shot variability than RP1. Small scale laboratory tests suggest that RP1.2 is less thixotropic than RP1 and changes less over time than RP1.

1. INTRODUCTION

Roma Plastilina no. 1 (RP1) was chosen as a ballistic witness material in 1977 [1,2] because it was the best option among commonly available materials at the time. Its use has been problematic [3] and the need to improve the clay has been cited [4]. The US Army had awarded a research grant to Chavant, Inc. to make an improved clay [5] and a new room temperature calibrated clay was developed that successfully mimicked the original RP1 in live fire tests on soft armour, hard armour, and helmets [6,7]. However, the improved clay was not adopted, for several reasons. One problem was that the clay still hardened with age, as does RP1. Unfortunately, a year or so after manufacture, the new clay hardened out of calibration at room temperature.

Using lessons learned from developing other clays in Chavant's product line, the aging of a new ballistics testing clay (RP1.2) has been minimized. Extrapolating the data collected, RP1.2 is expected to remain in calibration for more than 10 years after manufacture. This report describes the initial characterisation of RP1.2.

2. MATERIALS AND METHODS

RP1 and RP1.2 were made in the laboratory using a 2.8 L. Readco laboratory mixer. The ingredients of both clays are proprietary but include wax, oil, kaolin clay and sulphur. Larger scale samples to pack ballistics boxes were made in the factory with industrial scale equipment.

For laboratory hardness measurements, an Ametek-Brookfield CT3 Texture Analyzer was used with a TA39 2 x 20 mm steel cylindrical probe. Using software provided by the manufacturer, the test speed was set at 0.5 mm/sec with a target value (probe travel distance) of 4 mm.

For temperature-controlled experiments, samples were incubated in a Thermo-Fisher Scientific Heratherm Incubator.

For the thixotropy experiments, a Pasta Queen® pasta maker (Himark Enterprises, Inc.) was purchased from a local retail store.

https://doi.org/10.52202/081621-0045

Live fire tests were conducted at Element US Space and Defense in Belcamp MD. All aspects of the test were in accordance with the National Institute of Justice (NIJ) standard 0101.06 [2] including drop tests, range setup, shot pattern, etc. Ballistic boxes with RP1 were incubated at 37.4 °C prior to calibration and use.

Soft armour (Pathfinder shoot pack with and 8 layers of SP71) was strapped to test boxes and fired upon with 0.44 magnum ammunition at a velocity of 430-490 m/sec.

3. RESULTS

3.1 Correlation between NIJ cylinder drop results with measurements from a Brookfield Engineering CT3 Texture Analyzer

To predict whether experimental clays made on a small scale in the laboratory would be in calibration if made on a larger scale, hardness measured by a Brookfield CT3 Texture Analyzer were correlated with the cylinder drop calibration test data [7]. Clay was sent to Aberdeen Test Center (ATC) for cylinder drop test measurements and compared with CT3 measurements on the same clay. Figure 1 shows a linear correlation between the two methodologies.

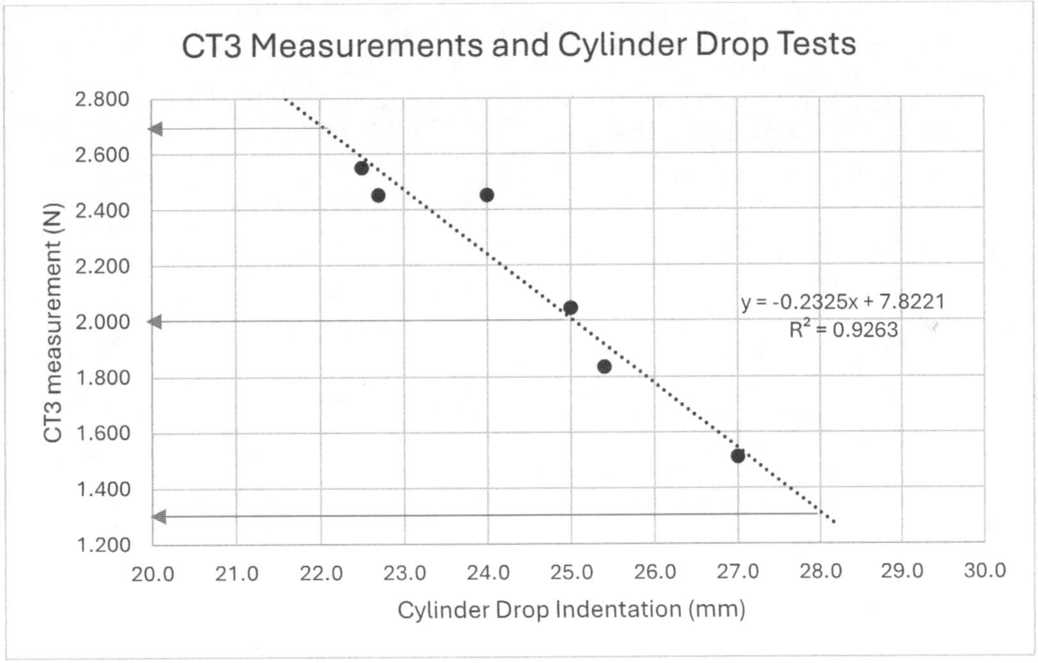

Figure 1. Correlation between CT3 measurements and cylinder drop tests done at ATC. The top arrow indicates the 22 mm drop "hard limit" (approximately 2.7N). The middle arrow indicates the 25 mm drop target hardness (approximately 2.0N). The bottom arrow indicates the 28 mm drop "soft limit" (approximately 1.3N).

The formula for the linear regression model was used to calculate pass/fail values for the CT3 that would predict pass/fail cylinder drop calibration values (Table 1). The 25 ± 3 mm cylinder drop correlates with predicted CT3 measurements of 2.707N for 22 mm lower limit, 2.010N for 25 mm, and 1.312N for 28 mm upper limit.

https://doi.org/10.52202/081621-0045

Table 1 Extrapolating the data from figure 1, calibration limits are calculated for the CT3 Texture Analyzer.

	Drop Depth (mm)	Predicted CT3 (N)
Hard Limit	22	2.707
Pass	25	2.010
Soft Limit	28	1.312

3.2 Aging

RP1 was made side-by-side with RP1.2. The hardness of each clay was measured over a period of 70 days (Figure 2). RP1 was out of the calibration range (green shaded area) the day after it was made and continued to harden. In contrast, RP1.2 remained well in calibration during the entire course of the study. Using the formula from the non-linear regression model $y = 0.0571 \ln(x) + 1.601$, the predicted hardness as measured by the CT3 would be 2.07 when x, or the age in days, is equal to 3650. In other words, RP1.2 can be predicted to remain in calibration 10 years after manufacture.

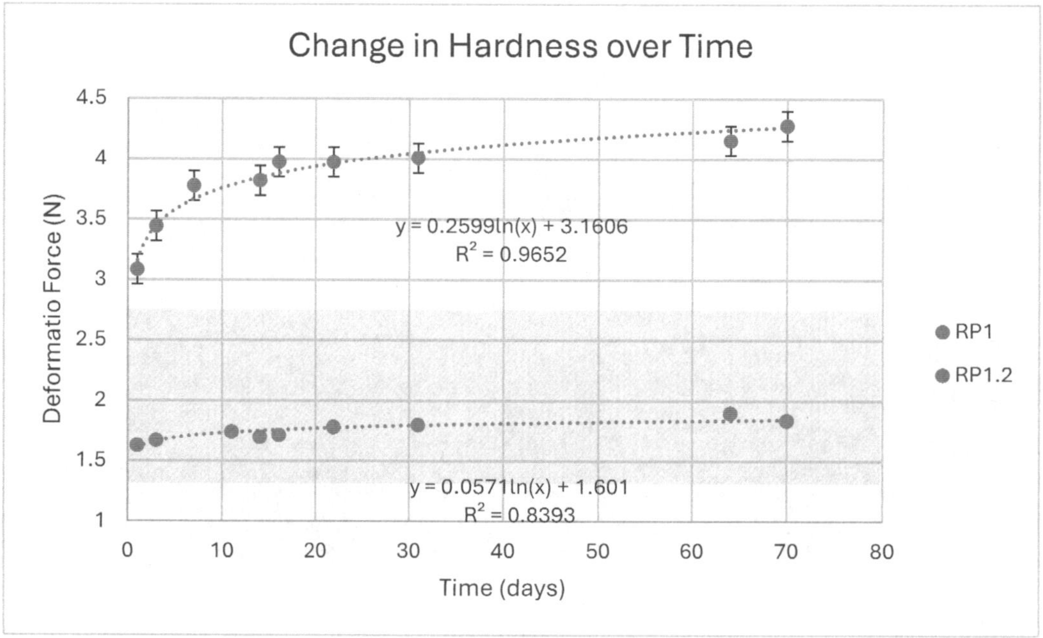

Figure 2. Time dependent changes in hardness comparison between RP1 and RP1.2. The area shaded in green represents the predicted calibration range.

3.3 Thermal Stability

Blocks of RP1 and RP1.2 were cut in blocks roughly 7 x 5 x 3 cm weighing approximately 175 gm. Between day 70 and day 72, CT3 measurements were made at temperatures between 17-27 °C in two-degree increments. For changes of two degrees C, clay was incubated for about 7 hours. For changes of six degrees, clays were incubated overnight (Table 2).

Table 2 Schedule for temperature adjustments on clay for the thermal stability tests.

Temp (°C)	Age (d)
17	71
19	71
21	70
23	70
25	72
27	72

Within the time period tested, age related changes in hardness should have been minimal. As expected, RP1 is out of calibration in this temperature range (Figure 3). In contrast, RP1.2 is in calibration in the entire temperature range tested.

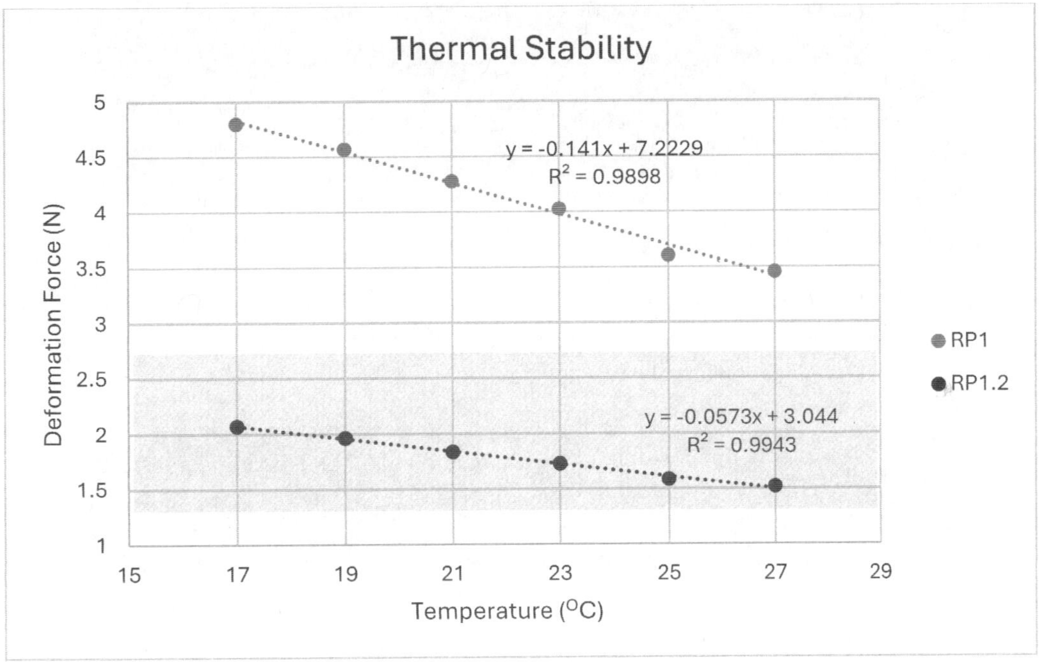

Figure 3. Temperature dependent hardness change comparison between RP1 and RP1.2. The area shaded in green represents the predicted calibration range.

Given the small number of data points, it is worth mentioning that, at the 95% confidence level, the slope for RP1 is -0.141 ± 0.020 and the slope for RP1.2 is -0.057 ± 0.006. These data indicate that RP1.2 is more thermal stable than RP1.

3.4 Thixotropic Recovery

A pasta maker was used to "cold work" the clay to varying degrees. The thixotropic impact and recovery over time were measured. Six 20-gram samples were prepared from the same 0.9 kg. block of RP1 or RP1.2. Five samples were passed through a Pasta Queen® pasta maker (Figure 4) at setting 6 according to the following schedule:

https://doi.org/10.52202/081621-0045

Sample 1 - 0 passes
Sample 2 - 2 passes
Sample 3 - 4 passes
Sample 4 - 7 passes
Sample 5 - 10 passes

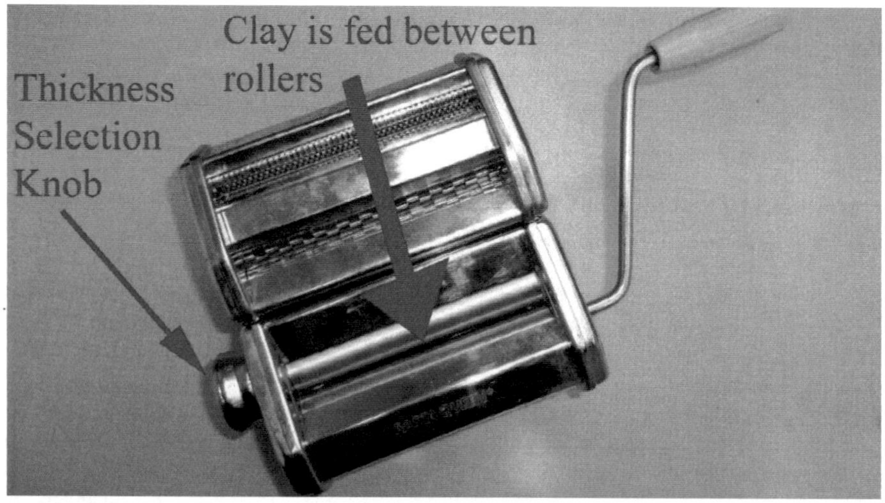

Figure 4. Pasta Queen pasta maker

Passing the clay through the pasta maker formed clay ribbons (Figure 5), which were folded and gently pressed into blocks approximating the original size and shape of the clay block (figure 6). These blocks were either passed through the pasta maker again or set aside for CT3 analysis.

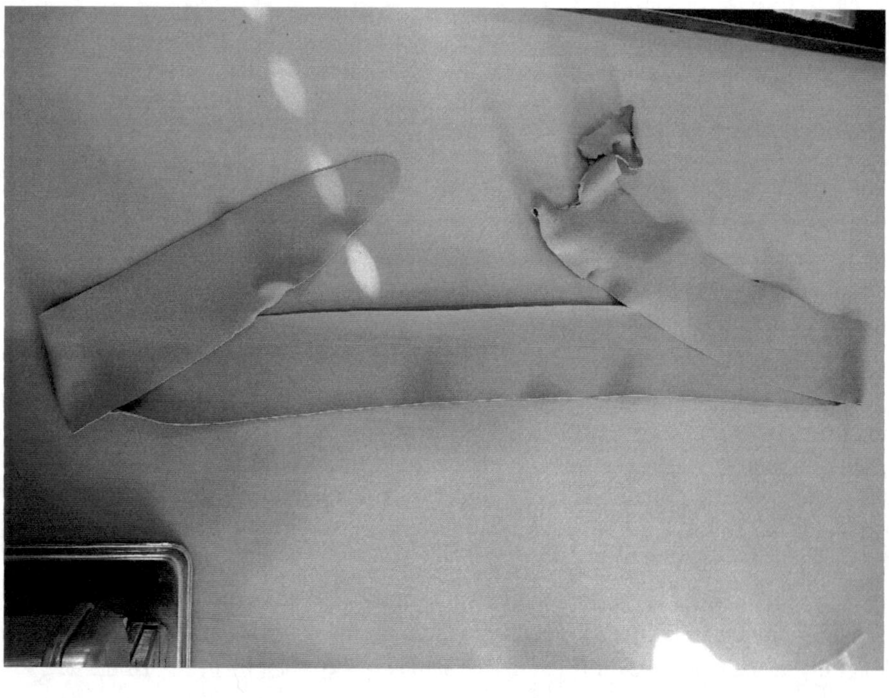

https://doi.org/10.52202/081621-0045

Figure 5. Clay ribbon formed after passing through the pasta maker.

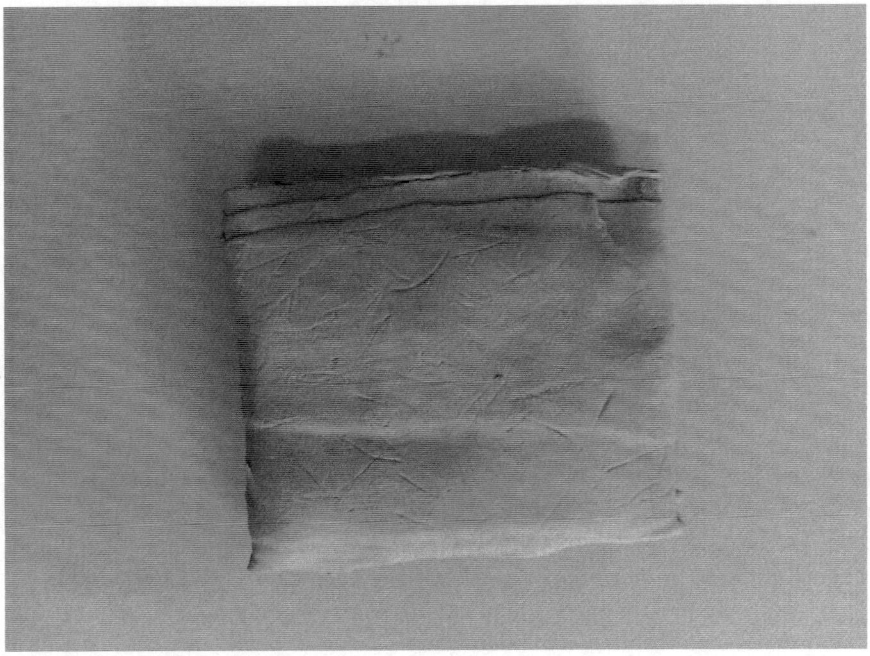

Figure 6. Folded ribbon of clay after gently pressing the layers together.

Figure 7 shows the softening of RP1 and RP1.2 from successive passes through the pasta maker. The change as a percent of the unaltered control is given in Table 3, clearly indicating that RP1.2 shows a smaller change in hardness as a result of the kinetic energy input from the pasta maker as compared to RP1.

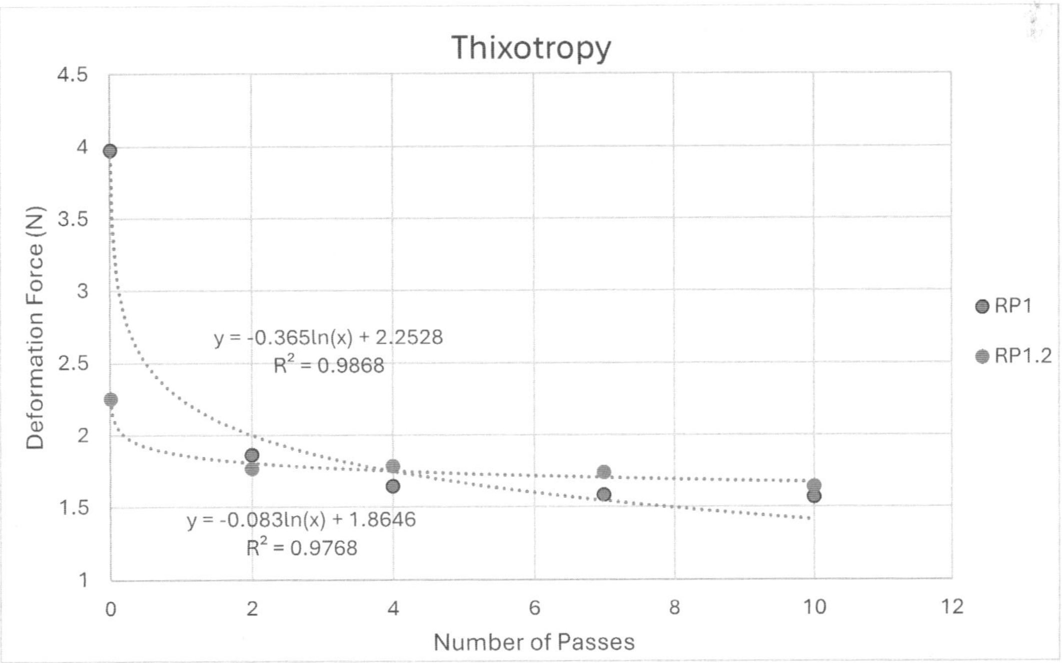

https://doi.org/10.52202/081621-0045

Figure 7. Thixotropic response of RP1 (green) and RP1.2 (orange) after successive passes through a pasta maker.

Table 3 Percent change in hardness after successive passes through the pasta maker.

	Percent Change from Control	
Passes	RP1	RP1.2
2	53.13	21.63
4	58.62	20.74
7	60.15	22.87
10	60.53	27.18

3.5 Live Fire Test

Soft armour vests were shot on two NIJ ballistic boxes of RP1 and one box of RP1.2. Heated RP1 (34.4-34.5 °C) and room temperature (21.4-21.5 °C) RP1.2 boxes passed calibration as per NJ 0101-06 immediately before live fire tests. Six shots were fired on each test article. Two boxes of RP1 were used to limit the thixotropic effect of the clay being "cold worked" by repeated firing on the same box. There was no evidence that previous shots softened the clay and increased the backface deformations (BFDs) of subsequent shots when only a single box of RP1.2 was used (Table 4).

Each of the shot locations on Roma 1.2 is similar in depth to the corresponding shot locations on Roma 1. Except for Shot 5, the standard deviation of shots on Roma 1.2 were smaller than shots on Roma 1 (Table 4).

Table 4 The backface deformations (BFDs) in mm for each shot on RP1 and RP1.2 and the mean and standard deviation of three shots at each location.

	Shot 1	Shot 2	Shot 3	Shot 4	Shot 5	Shot 6
Roma 1 Block 28	36.4	39.2	35.3	28.1	29.1	33.3
Roma 1 Block 28	34.3	33.9	36.7	26.5	27.6	35.5
Roma 1 Block 11	39.2	38.5	41.9	33.8	26.7	33.9
Mean	36.6	37.2	38.0	29.5	27.8	34.2
SD	2.46	2.88	3.48	3.84	1.21	1.14
Roma 1.2 Block 1	35.9	37.3	36.8	30.0	25.6	34.6
Roma 1.2 Block 1	36.9	35.6	36.7	28.7	29.7	34.2
Roma 1.2 Block 1	35.4	35.7	33.9	27.1	21.2	34.1
Mean	36.1	36.2	35.8	28.6	25.5	34.3
SD	0.76	0.95	1.65	1.45	4.25	0.26

4. DISCUSSION

This study is a revival of earlier work to improve RP1 as a ballistic backing material [6]. The earlier revised clay (prototype 8, also referred to as BT18c in other literature) was not adopted for a variety of reasons, including the desire for government ownership of the formula. As a result, Army Research Labs have developed ARTIC (ARL Temperature Insensitive Clay, ref. 8). While Chavant's parent company, Smooth-On, has been chosen to manufacture ARTIC under license, the problems with RP1 still need to be addressed.

RP1.2 has many advantages over the existing RP1. Being in calibration at room temperature, the clay can be used without pre-heating the clay to 35-37 °C provided that the clay is not stored at extreme temperatures. The clay will not cool down during use, extending the range time of each box. Also, a small standard deviation of mean BFD measurements can result in less shots needing to be fired for the same statistical level of confidence in an outcome. A lower level of thixotropy would allow more shots to be fired on the same block of clay, reducing the overall number of clay boxes that would be needed to pack, less clay to buy and less time needed to switch out boxes between tests.

5. Conclusions

The improved clay described in this report, RP1.2, has overcome the aging problem of both RP1 and the original prototype 8 clay. While RP1.2 has been shown to record similar BFDs as calibrated RP1 on soft armour, more live-fire testing needs to be done on soft armour, hard armour, and helmets.

Acknowledgements

The authors would like to acknowledge the financial support provided by PEO Soldier, the past owner of Chavant, John North, and the current owners of Chavant, Trey Bianco and Clay Western. We would also like to acknowledge Shane Esola, now at the Picatinny Arsenal, and Julie Ivancek, now at DOT&E for their supportive collaboration on the earlier, prototype 8 (BT18c).

References

[1] Prather, R.N., Swann, C.L., Hawkins, C.E. (1977) Backface signatures of soft body armors and the associated trauma effects. Technical report, Army Armament Research and Development Command, Aberdeen Proving Ground.
[2] Ballistic Resistance of Body Armor (2008) NIJ Standard 0101.06
[3] Tobin, L., Feeling Depressed, Personal Armour Systems Symposium, The Hague, The Netherlands, 6-10 September 2004.
[4] Lehowicz, L.G., Bass, C.R., Budinger, T.F., Denn, M.M., Fahrenholtz, W.G., Fricker, J., Gupta, Y.M., Killinger, D.K., Markov, V.B., McGuffin-Cawley, J.D., Prather, R.N., Wiederhorn, S.M., Wilson, A.G. (2012) Testing of Body Armor Materials: phase III. The National Academic Press.
[5] US Army RDECOM contract no. W91CRB-10-R-0040.
[6] Esola, S. The Development of a Room Temperature Ballistic Clay As A Replacement For Roma Plastilina #1 For Use In Body Armor Testing. Personal Armour Systems Symposium. 17-21 September 2012. Nuremberg, Germany.
[7] Peng, I. (Chavant, Inc.) and Esola, S. (Aberdeen Test Center) unpublished observations.
[8] Edwards, T.D., Bai, E.D., Cole, S.T., Freeney, R.M., Halls, V.A., Ivancik, J., Lenhart, J.L., Napadensky, E, Yu, J.H, Zheng, J.Q. and Mrozek, R.A. Mechanical Properties of Silicone Based Composite as a Temperature Insensitive ballistic Backing Material for Quantifying Back Face Deformation. Forensic Sci Intl 285 (2018), 1-12.

 https://doi.org/10.52202/081621-0045

A Novel Helmet Testing Apparatus for Assessing Rotational Head Kinematics and Its Applications in Military Helmet Design

Shuting Kou, Jaden Romero, Aravind Sundaramurthy, Ryan Doris, **Ron Szalkowski**

Team Wendy, LLC, 17000 St. Clair Avenue, Cleveland, OH 44110, USA
rszalkowski@teamwendy.com

Abstract. Traumatic brain injuries (TBIs) are a significant concern among military personnel, often leading to lasting cognitive and physical impairments. Traditionally, linear kinematics has been emphasized as a key predictor of head injuries. However, recent research increasingly points to rotational kinematics as a primary contributor. The current Department of Transportation (DOT) linear drop test, which utilizes a monorail drop tower with rigid neck and headforms, primarily captures linear kinematics, offering limited insights into rotational dynamics. To address this limitation, we developed the Device for Rotational Evaluation of Wearables (DREW), a novel testing platform designed to simulate realistic head-to-ground impacts and measure both linear and rotational kinematics. DREW incorporates a 50th percentile Hybrid-III head, neck, and torso, equipped with linear accelerometers, angular rate sensors, and neck load cells, enabling comprehensive helmet evaluations across various impact locations (front, rear, and sides) at velocities up to 6.1 m/s. Unlike monorail systems, DREW can also simulate the whiplash effects often experienced during backward falls. To determine if DREW can differentiate between the performance of different helmet designs and technologies, three bicycle helmets employing a variety of rotational mitigation technologies were evaluated. Furthermore, testing Team Wendy's ballistic helmet liners on DREW revealed that a high-density liner (optimized for impacts of 4.3–5.2 m/s), which outperformed a baseline liner (designed for 3.0 m/s impacts) under the current monorail drop test method, exhibited higher peak linear acceleration (20%) and peak angular acceleration (27%). These results suggest that the current monorail drop test methods may not accurately represent realistic impacts and could encourage the development of overly stiff helmet liners.

1. INTRODUCTION

Service members face frequent head impacts in both training and combat, with blunt force trauma posing a significant risk. Notably, Wojcik et al. documented a 15% hospitalization rate due to blunt head trauma during conflicts in Iraq and Afghanistan [1]. Although modern ballistic helmets have effectively reduced severe head injuries, such as penetrating trauma and skull fractures, they may not provide adequate protection against mild traumatic brain injuries (mTBIs). Indeed, between 2000 and the second quarter of 2024, 509,477 Service members sustained at least one traumatic brain injury (TBI), with 81.9% classified as mTBI (e.g., concussions) [2]. Recent studies have revealed that rotational kinematics, which generate substantial shear forces in brain tissue, are widely recognized as the predominant mechanism underlying concussive injuries [3, 4].

Rotational head kinematics, a common phenomenon during impact or impulsive head loading, plays a critical role in brain trauma. Due to the biomechanical properties of the brain, brain tissue deforms more readily in response to shear forces than to other forms of mechanical loading [5]. Rapid head rotations induce extensive shear forces throughout the brain, thereby elevating the risk of shear-induced tissue damage. Multiple studies across different laboratories have confirmed that shear deformation caused by rotational kinematics is the principal mechanism of concussion [4, 6–9]. This evidence underscores the critical need for next-generation helmet designs that specifically target the mitigation of rotational kinematics to reduce mTBI risk and enhance the overall protection.

Despite these insights, current ballistic helmet testing standards, such as AR/PD 10-02, predominantly focus on mitigating linear kinematics. These protocols typically utilize a Department of Transportation (DOT) headform dropped onto an anvil via a monorail drop system, a configuration that does not accurately replicate the complex rotational dynamics of the head and neck encountered in real-world impact scenarios. Alternative approaches, such as modified drop tests using a headform with flexible neck dropped onto an angled anvil to allow oblique impacts, offer improvements in simulating rotational kinematics during impacts [10]. However, these methods remain limited in replicating realistic pre-impact kinematics, such as the initial angular velocity generated by head whiplash following a fall. To better simulate a fall with accurate pre-impact kinematics, Niece et al. developed a backfall test platform that mimics the whiplash of the head during a backward fall [11]. However, this method is

https://doi.org/10.52202/081621-0046

restricted to rear impacts only, highlighting the need for a testing system capable of simulating impacts at various locations (front, sides, and rear) for a comprehensive evaluation of helmet performance.

In this study, we developed a novel test rig that simulates head-to-ground impacts while incorporating whiplash-induced head rotation, enabling a comprehensive evaluation of helmet performance. This system is capable of testing impacts at various locations, including the front, rear, and sides, thus providing a more realistic representation of the complex impact scenarios. Using this platform, we assessed the rotational kinematics mitigation capabilities of currently available bicycle helmets. Furthermore, we evaluated a high-density liner system, which had previously demonstrated improved linear acceleration mitigation in standard monorail drop tests at impact velocities of 4.3-5.2 m/s, by testing it on DREW at an impact velocity of 6.1 m/s and comparing its performance with that of a baseline liner.

2. MATERIALS AND METHODS

2.1 Device for rotational evaluation of wearables

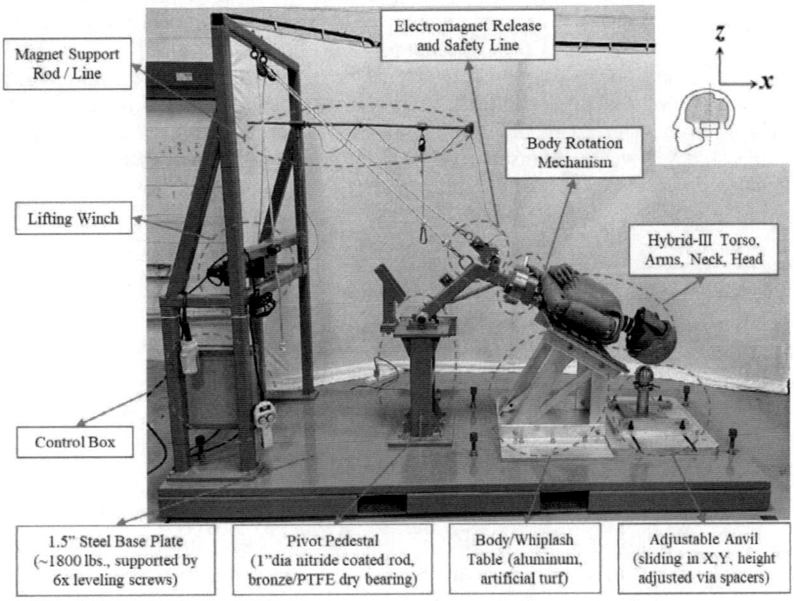

Figure 1. Experimental setup of Device for Rotational Evaluation of Wearables (DREW).

The Device for Rotational Evaluation of Wearables (DREW) is designed to simulate head-to-ground impacts while capturing both linear and rotational kinematics (Figure 1). It employs a 50th percentile Hybrid III (H3) Anthropomorphic Test Device (ATD) head, neck, and torso (56 kg; Humanetics, MI, USA) to mimic human upper-body dynamics. The ATD is mounted to a pivot leg via two adapters, one attached to the pivot leg and the other to the spine of the ATD. These adapters can rotate relative to one another and a locking mechanism secures them, permitting controlled rotation about the pivot leg's axis to simulate impacts at various helmet locations (e.g., front, rear, and side).

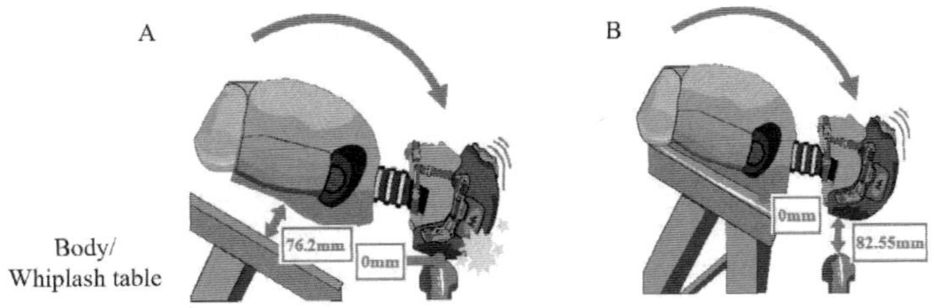

Figure 2. Description of two modes of impact: (A) headfirst and (B) whiplash test configurations.

https://doi.org/10.52202/081621-0046

The complete pivoting assembly weighs 63.5 kg, and the pivot leg is mounted on a pedestal via a shaft-and-bearing system, allowing the upper body to pivot about the y-axis (Figure 1). The pivot pedestal is rigidly fixed to an 820.0 kg steel base and supported by six leveling screws to ensure proper alignment of the test platform. An adjustable anvil system, capable of positioning along all three translational degrees of freedom, is secured to the base. In addition, a body/whiplash table is mounted on the steel base to induce two head impact modes on the ATD (Figure 2). In the first, the head contacts the anvil, followed by the torso striking the whiplash table to protect the instrumentation and ATD. In the second mode, the torso impacts the whiplash table first, inducing a whiplash motion on the head before it impacts the anvil. Finally, to perform the test, an electrically powered winch elevates the ATD to the desired height, and an electromagnetic release mechanism ensures consistent, repeatable drops by holding and then releasing the ATD into freefall. For safety, a pin through the magnet connection as well as a secondary safety line are attached until an impact test is run. The impact angle, used to characterize drop intensity, is defined as the angle between the lower pivot leg (connected to the bearing) and the horizontal plane. At the maximum impact angle of 104°, an impact velocity of 6.1 m/s is achieved.

To capture head kinematics during impact, the ATD head was instrumented with a Diversified Technical Systems (DTS, CA, USA) 6DX PRO sensor located at its center of mass. A Slice Nano modular data acquisition system, including a battery stack, Base+, and bridges, was employed to collect the sensor data. The DTS 6DX PRO sensor was selected for its ability to record head kinematics using three linear accelerometers (±2,000 g) and three angular velocity sensors (±18,000°/s), with all channels sampled at a data acquisition frequency of 30.3 kHz. Data for linear acceleration and angular velocity were recorded during impacts, and angular acceleration was subsequently estimated by differentiating the angular velocity time histories. In addition, high-speed video of the impacts was captured at 7,500 frames per second using an i-SPEED 509 camera (iX Cameras, MA, USA) to further examine the dynamics of head impact.

2.2 Testing

2.2.1 Direct vs. whiplash-induced impacts

To assess the differences in kinematics between direct head impact and whiplash-induced impact, experiments were conducted at an impact angle of 76°. For these tests, the ATD was fitted with an EXFIL ballistic helmet (Team Wendy, Cleveland, OH) and subjected to rear impacts, with five repetitions for each condition.

2.2.2 Assessment of rotational mitigation

To demonstrate the capabilities of the DREW system, we examined the performance of rotational kinematics mitigation technologies in bicycle helmets. Three helmet designs were evaluated: a baseline helmet featuring a traditional EPS (expanded polystyrene) liner (G1), a variant of the baseline helmet equipped with a Multi-Directional Impact Protection System (MIPS [12]; G2), and a helmet incorporating WaveCel [10] technology (G3). The ATD head was fitted with each helmet and subjected to frontal direct impacts on a hemispherical anvil, with each test condition repeated three times. Impacts were performed at an angle of 104°, and both linear and angular kinematics were measured and compared between the baseline helmet and those incorporating mitigation technologies.

2.2.3 Evaluation of ballistics helmet liner

Finally, we evaluated a high-density liner that was previously developed and optimized for high velocity impacts using a monorail drop test with a DOT headform. The goal for this liner was to maintain linear acceleration below 150 g during the first impact at impact velocities of 4.3 and 5.2 m/s. To assess the performance of the new liner under more realistic impact conditions of DREW, tests were conducted at an impact velocity of 6.1 m/s, and its kinematics were compared with those of a baseline liner that did not meet the 150 g criterion at 4.3 m/s. Each condition was repeated three times, and the impact angle for these tests was maintained at 104°.

2.3 Injury metrics

Universal Brain Injury Criteria (UBrIC) [13] was used to estimate the 95th percentile maximum principal strain (MPS) in the brain. UBrIC, as defined in Equation 1, computes MPS based on measurements of angular velocity and angular acceleration. It is formulated to be broadly applicable across a range of impact conditions, including automotive accidents and sports-related impacts.

$$UBrIC = \left\{ \sum_i \left[\omega_i^* + (\alpha_i^* - \omega_i^*) e^{-\frac{\alpha_i^*}{\omega_i^*}} \right]^r \right\}^{\frac{1}{r}} \quad where\ i = x, y, and\ z \qquad (1)$$

Here, ω_i^* represents the difference between the maximum and minimum angular velocities (i.e., positive and negative peaks) normalized by a critical value ($\omega_i^* = \omega_i/\omega_{icr}$) along the x, y, and z axes. Similarly, α_i^* represents the peak angular acceleration normalized by a critical value ($\alpha_i^* = \alpha_i/\alpha_{icr}$) along the x, y, and z axes. The exponent r, used in the evaluation of UBrIC, was set to 2, as recommended by the authors [13]. Although UBrIC provides MPS for various rotational kinematics parameters, it does not yield a direct injury probability value. Instead, it informs helmet and countermeasure design by capturing a broader spectrum of brain injury mechanisms associated with high brain tissue strains.

2.4 Data processing and statistical analyses

Raw experimental signals are processed following the Society of Automotive Engineers (SAE) J211 standards using a fourth-order Butterworth low-pass filter. Linear acceleration signals were filtered using a channel frequency class (CFC) of 1000 (cut-off frequency: 1650 Hz), while angular velocity signals were filtered using CFC 180 (cut-off frequency: 300 Hz). Angular velocity data, recorded in degrees per second (°/s), were converted to radians per second (rad/s) before filtering and then differentiated to compute angular acceleration. Data processing was performed using a MATLAB script (MathWorks, MA, USA).

Statistical analyses were performed using Prism (GraphPad Software, MA, USA). Differences in parameters among the three bicycle helmets were evaluated using one-way analysis of variance (ANOVA) with Tukey's post-hoc test for pairwise comparisons. An unpaired Student's t-test was used to assess the mean differences in kinematic behavior between the high-density liner and the baseline liner. Statistical significance was set at $p < 0.05$, and all peak kinematic results are reported as mean ± one standard deviation.

3. REPEATABILITY ASSESSMENT

Table 1. Percentage differences in peak linear acceleration (PLA) and peak angular velocity (PAV) across four repeatability test setups. Results are presented as mean [standard deviation (SD)], with the percentage difference (% diff) calculated as the difference between the maximum and minimum values across five tests for each condition.

Impact angle (deg)	*PLA (g) mean (SD)*	*PLA (min/max % diff)*	*PAV (rad/s) mean (SD)*	*PAV (min/max % diff)*
51° (Front)	101 (1.5)	3.5%	26.0 (0.3)	2.6%
78° (Front)	223 (1.2)	1.5%	35.8 (0.1)	0.6%
40° (Rear)	86 (1.2)	3.6%	17.2 (0.2)	3.0%
61° (Rear)	212 (1.3)	1.5%	26.7 (3.8)	1.8%

The repeatability of the DREW system was evaluated through controlled drop tests using a bare Hybrid III headform (without a helmet) onto a CADEX modular elastomer programmer (MEP) pad (hardness: 60 ± 2 Shore A; thickness: 0.025 m; diameter: 0.152 m). Four impact conditions were tested: two front-facing impacts at impact angles of 51° (low) and 78° (high), and two rear-facing impacts at 40° (low) and 61° (high), as outlined in Table 1. Each condition was repeated five times. For front impacts, the 51° impact angle produced a mean linear acceleration of 101 g [standard deviation (SD) = 1.5] and a mean angular velocity of 26.0 rad/s (SD = 0.3), with 3.5% and 2.6% variability, respectively. At the higher 78° angle, linear acceleration increased to 223 g (SD = 1.2) with a 1.5% variation, and angular velocity reached 35.8 rad/s (SD = 0.1) with only 0.6% variation.

For rear-facing impacts, the 40° drop resulted in a mean linear acceleration of 86 g (SD = 1.2) and an angular velocity of 17.2 rad/s (SD = 0.2), with respective variability of 3.6% and 3.0%. At the higher 61° angle, linear acceleration rose to 212 g (SD = 1.3) with a 1.5% difference between maximum and minimum values, while angular velocity increased to 26.7 rad/s (SD = 3.8) with a 1.8% difference. These

https://doi.org/10.52202/081621-0046

results demonstrate the DREW system's good repeatability across a range of impact angles and directions, with less than 3.6% variation in linear acceleration and less than 3.0% in angular velocity.

4. RESULTS

4.1 Direct head impact vs. whiplash-induced impact

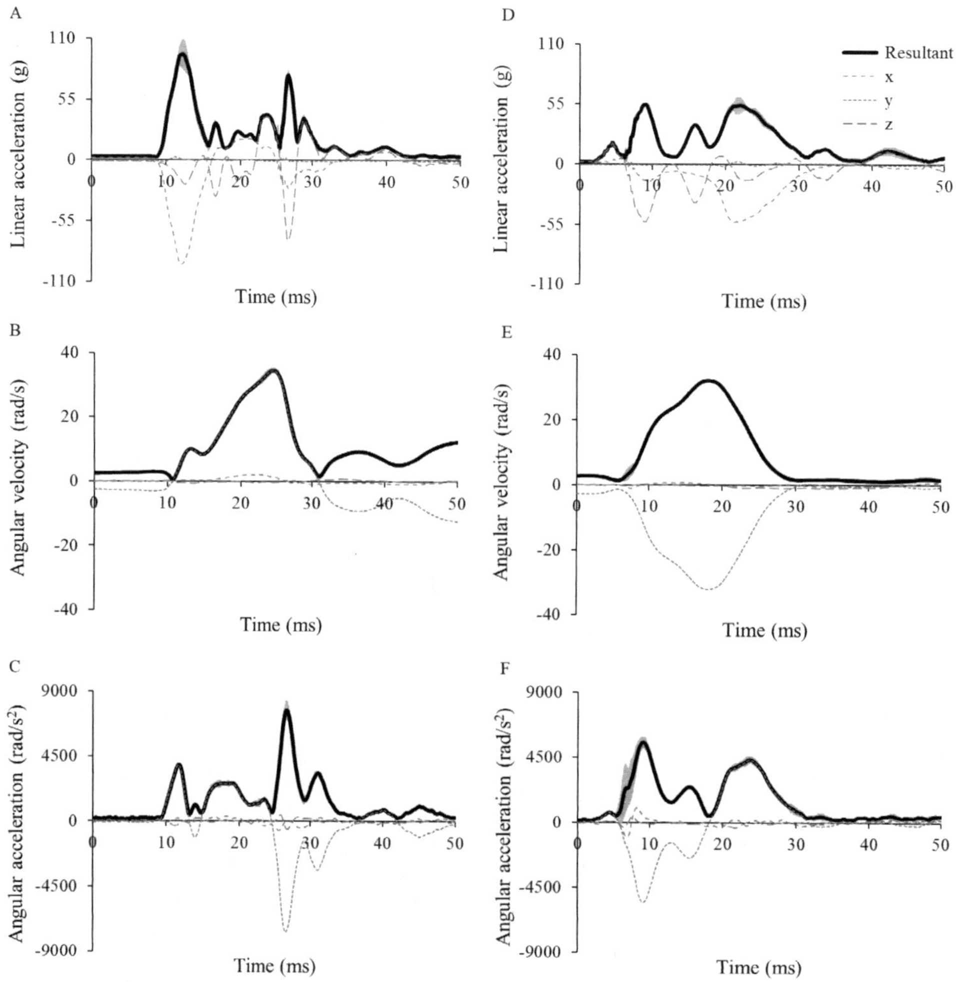

Figure 3. Mean traces of linear acceleration, angular velocity, and angular acceleration for head-first impact (Panels A, B, and C) and whiplash-induced impact (Panels D, E, and F). The solid line represents the mean resultant (n = 5), with the shaded area indicating the standard deviation. Mean component traces (along x, y, and z) are shown as dotted lines.

During direct impact, the linear acceleration peaked after the headform contacted the anvil, reaching a peak resultant value of 96 ± 13 g, with a secondary peak observed after 25 ms. Analysis of the component accelerations indicated that the first peak was due to acceleration along the x-axis, while the second peak arose from acceleration along the z-axis toward the pivot point. The angular velocity did not increase monotonically; instead, it exhibited a transient dip at approximately 12 ms, followed by a rise to a peak value of 34.6 ± 1.6 rad/s. Similar to linear acceleration, the angular acceleration displayed two distinct peaks, with the second peak (7727 ± 1079 rad/s^2) exceeding the first.

During whiplash-induced impacts, the head's linear acceleration exhibited multiple peaks: the first occurred before 10 ms (55 ± 1 g), followed by a slightly reduced peak after 15 ms, and finally a third peak, with a magnitude comparable to the first (54 ± 6 g), appeared beyond 20 ms. Component analysis revealed that the first two peaks were caused by acceleration along the z-axis toward the pivot point, while the third peak resulted from acceleration along the x-axis. Unlike direct impact, the angular velocity for whiplash-induced impacts increased monotonically, reaching a peak value of 32.1 ± 0.6 rad/s. Additionally, analysis of the acceleration traces revealed that the first peak (5566 ± 193 rad/s^2) was higher than the second.

4.2 Bicycle helmet

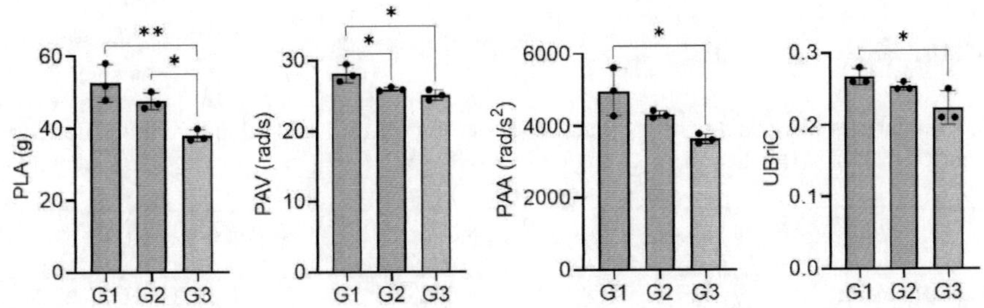

Figure 4. Kinematic evaluation of bicycle helmets under frontal head-first impacts: comparison of peak linear acceleration (PLA), peak angular velocity (PAV), peak angular acceleration (PAA), and Universal Brain Injury Criteria (UBrIC) in three helmet designs (G1: Baseline, G2: Multi-Directional Impact Protection System, G3: WaveCel). Data are presented as mean ± standard deviation (n = 3); *p ≤ 0.05, **p ≤ 0.01.

Using frontal impacts without the whiplash mode, we assessed the efficacy of rotational kinematics mitigation systems available in bicycle helmets. Three helmet conditions were tested: a baseline helmet (G1), a MIPs-equipped helmet (G2), and a WaveCel-equipped helmet (G3). The baseline helmet produced a mean peak linear acceleration (PLA) of 53 g (SD = 5), a mean peak angular velocity (PAV) of 28.1 rad/s (SD = 1.2), and a mean peak angular acceleration (PAA) of 4954 rad/s^2 (SD = 664). Compared to the baseline helmet, the WaveCel helmet demonstrated significant reductions across multiple kinematic metrics.

For PLA, the WaveCel helmet achieved a mean of 38 g (SD = 2), representing a significant 28% reduction relative to the baseline. Similarly, for PAV, the WaveCel helmet yielded a mean of 25.1 rad/s (SD = 0.7), corresponding to a 11% decrease from the baseline. For PAA, the WaveCel helmet recorded a mean of 3634 rad/s^2 (SD = 138), a 27% reduction relative to the baseline. Additionally, the UBrIC, which estimates brain's 95th percentile MPS, was significantly reduced by 16% compared to the baseline. In contrast, the MIPS-equipped helmet only demonstrated a significant reduction in PAV, with a mean of 25.9 rad/s (SD = 0.3), representing an 8% decrease relative to baseline. Notably, the WaveCel helmet outperformed the MIPs helmet in mitigating linear acceleration, however, there was no significant difference between MIPs and WaveCel among angular kinematics and UBrIC estimate.

4.3 Ballistic helmet liners

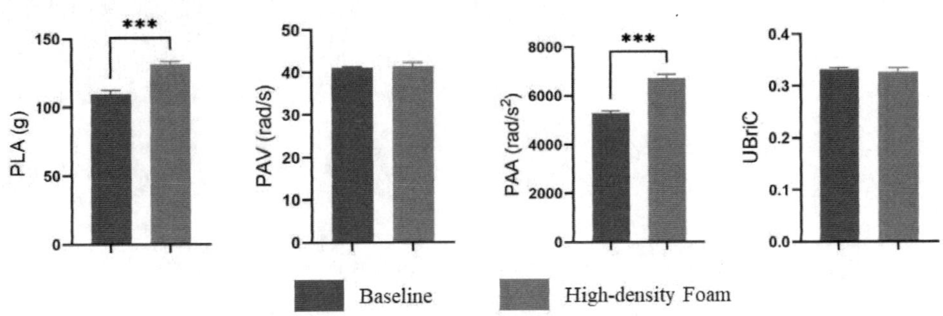

Figure 5. Kinematic evaluation of ballistic helmet liners under rear whiplash-induced impacts: comparison of peak linear acceleration (PLA), peak angular velocity (PAV), peak angular acceleration (PAA), and Universal Brain Injury Criteria (UBrIC) for two liners, Baseline (blue) and high-density liner (red). Data are presented as mean ± standard deviation (n = 3); ***p ≤ 0.001.

Whiplash-induced rear impact tests were conducted at a calculated impact velocity of 6.1 m/s to compare the kinematic performance of a high-density liner with a baseline liner. In contradiction to the standard

 https://doi.org/10.52202/081621-0046

monorail test, the high-density liner produced a significantly higher PLA, with a mean value of 131 g (SD = 2) compared to 109 g (3 g) for the baseline (Figure 5). Similarly, PAA was significantly elevated with the high-density liner (6708 rad/s², SD = 153) relative to the baseline (5286 rad/s², SD = 95). In contrast, there was no significant difference in PAV and UBrIC between the high-density liner (41.4 rad/s, SD = 0.8) and baseline (41.04 rad/s, SD = 0.2). These results indicate that the high-density liner did not perform as intended when tested in a more realistic fall simulation.

5. DISCUSSIONS

Table 3. Comparison of impact testing capabilities: monorail drop tests (department of transportation [DOT] and hybrid III [H3] headforms), backfall system, and the device for rotational evaluation of wearables (DREW) Platform.

Capabilities	*Monorail with DOT head*	*Monorail with H3 head*	*Backfall*	*DREW*
Linear motion	O	O	O	O
Rotational motion	X	O	O	O
Various impact locations	O	O	X	O
Various impact speed	O	O	X	O
Whipping motion	X	X	O	O
Biofidelic dummy	X	X	O	O

We developed a novel test rig, DREW, to simulate head-to-ground impacts for helmet evaluation. DREW is capable of generating two distinct impact scenarios: (1) head-first impacts, where the head directly strikes an anvil, and (2) whiplash-induced impacts, where the torso of the ATD first contacts a table, inducing a whiplash motion in the head before it impacts the anvil. To assess the rig's capability in characterizing rotational mitigation technologies, we tested bicycle helmets equipped with MIPs and WaveCel systems and compared their performance to that of a baseline helmet. Finally, we evaluated a ballistic helmet liner, previously optimized for high-velocity performance at impact velocities of 4.3–5.2 m/s, against a baseline liner to determine whether its enhanced performance translates to improved protection under more realistic impact conditions simulated through DREW.

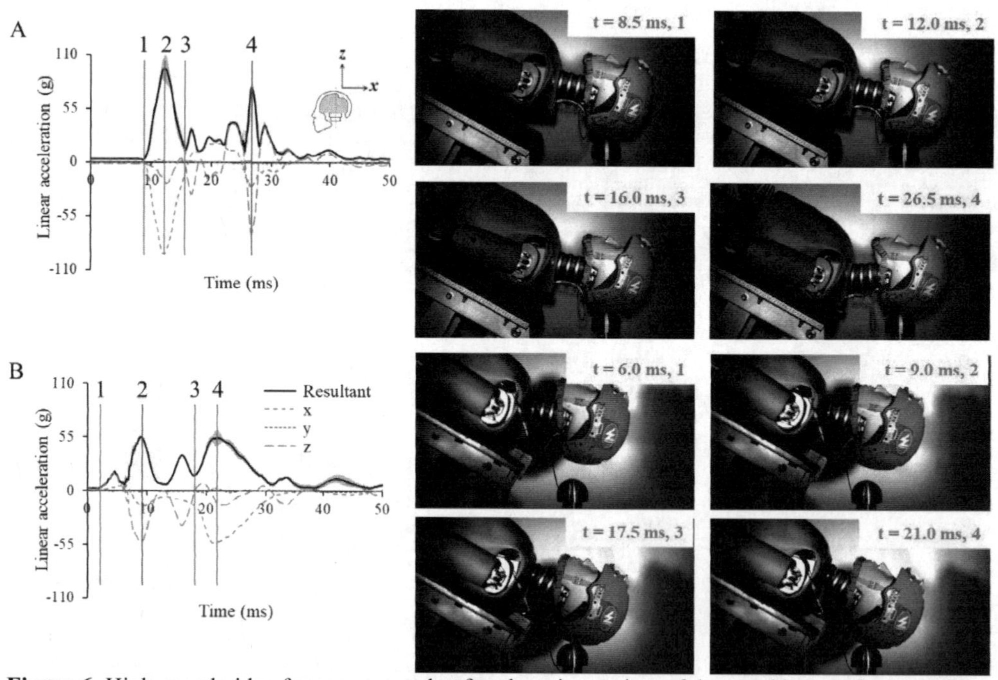

Figure 6. High-speed video frames captured at four key time points of drop and impact for head-first (A) and whiplash-induced (B) impacts are presented. The corresponding time points are numbered and marked (red line) on the linear acceleration trace to elucidate the impact dynamics.

Table 3 provides a comparison of the capabilities of conventional helmet evaluation techniques and the DREW system. Traditional methods, such as the monorail drop test with a DOT headform, reliably simulate linear motion and can accommodate various impact locations and speeds; however, they lack the ability to simulate rotational motion, whipping motion, and do not utilize a biofidelic dummy. In response, researchers have modified these methods by employing a H3 ATD head, which permits neck bending and better captures rotational dynamics, though this approach still falls short in replicating whipping motion, i.e., pre-impact kinematics. Backfall rigs that induce whiplash impacts provide an improvement by simulating the head's whipping motion; nevertheless, they are limited in that they cannot simulate impacts at multiple locations or various impact speeds. In contrast, the DREW system integrates all these critical capabilities, i.e., it can simulate and capture both the linear and rotational kinematics associated with head impacts, accommodates impacts from multiple locations at various speeds, simulates whipping motion, and employs a biofidelic dummy. This comprehensive functionality makes the DREW system a robust platform for evaluating helmet performance, particularly in the context of mitigating the rotational kinematics experienced in real-world impact scenarios.

High-speed video recordings were used to interpret the mechanics of both head-first and whiplash-induced impacts. Snapshots from the footage were synchronized with four key time points on the linear acceleration trace. For head-first impacts (Figure 6A), the video shows that the headform contacts the anvil at t = 8.5 ms, initiating an increase in acceleration that peaks at 12.0 ms. At t = 16.0 ms, the ATD torso impacts the table, resulting in a marked reduction in the head's acceleration. Component analysis indicates that the initial acceleration peak is primarily due to motion in the x-direction, corresponding to the head's impact. A secondary acceleration peak is observed at t = 26.5 ms, predominantly arising from acceleration in the z-direction. This z-direction acceleration is attributed to the centripetal acceleration generated as the torso rebounds toward the table, causing the head, while in contact with the anvil, to undergo angular motion, as evidenced by neck flexion at t = 26.5 ms. For whiplash-induced impacts (Figure 6B), the torso first contacts the table at t = 6.0 ms, inducing a whiplash motion in the head that reaches its peak acceleration at 9.0 ms. This whiplash effect generates significant angular motion, resulting in centripetal acceleration along the z-axis. Notably, oscillations in the z-axis acceleration are observed, likely reflecting dynamic changes in head motion as the torso rebounds off and back onto the table. Subsequently, the head contacts the anvil around t = 17.5 ms, with the x-direction acceleration peaking at t = 21.0 ms.

Analysis of the angular kinematics traces revealed distinct behaviors between head-first and whiplash-induced impacts. In head-first impacts, angular velocity increases non-monotonically, with a slight dip observed around t = 12.0 ms, likely due to the torso impacting the table, before rising to a peak and then decreasing. Notably, the second peak in angular acceleration exceeds the first, indicating that the rate of velocity reduction is greater than the initial rate of increase. During whiplash-induced impacts, the headform exhibits a pre-impact angular velocity, with the peak occurring before the head contacts the anvil (approximately 18 ms). Similar observations have been reported in mouthguard data from an American football player [11]. Moreover, the angular acceleration during the whiplash phase is higher than that observed at impact, likely due to helmet offering protection during impact. These findings underscore the importance of simulating pre-impact, rather than focusing solely on post-impact kinematics. Because standard monorail drop tests fail to capture these dynamics, alternative testing methods such as the DREW system may provide a more comprehensive assessment, thereby informing strategies to better prevent mTBI.

Assessment of rotational kinematics mitigation technology among bicycle helmets in non-whiplash events revealed differences in linear and rotational kinematics among different helmet designs, with the WaveCel helmet consistently exhibiting the lowest values across all metrics. Statistical analyses revealed that the PLA was significantly lower for the WaveCel helmet compared to both the baseline and MIPs helmets. For PAV, significant differences were observed between the baseline and both the MIPs and WaveCel conditions. In terms of PAA and UBriC, a statistically significant reduction was found only between the baseline and the WaveCel helmet. These findings are consistent with previous research by Bliven et al. and Abayazid et al., which reported substantial decreases in both rotational acceleration and velocity, as well as improvements in injury metrics such as the brain injury criterion [10,14]. Overall, these results underscore the potential benefits of optimizing helmets to improve protection against the rotational kinematics associated with traumatic brain injury.

Finally, the high-density liner, optimized via conventional monorail drop tests using a DOT headform, previously demonstrated promising performance by reducing PLA by 15% and 57% at impact velocities of 4.3 and 5.2 m/s, respectively, compared to the baseline liner. However, when evaluated using the DREW system with an H3 ATD, although both liners maintained the PLA below 150 g, the high-density liner produced a statistically significant increase in PLA of up to 20% relative to the baseline. In addition, while standard testing protocols do not account for rotational kinematics, emerging

https://doi.org/10.52202/081621-0046

research highlight their importance in mTBI. In our study, the high-density liner increased PAA by 27% compared to the baseline, with this difference reaching statistical significance. No significant difference was observed in PAV and UBrIC, likely because whiplash-induced impacts generate peak PAV during the whiplash phase before the liner is engaged. These results suggest that the current DOT-based monorail drop tests, with their reliance on rigid neck configurations, may favor stiffer foam designs that do not necessarily translate to improved performance under more realistic impact conditions.

One limitation of our study is that the H3 ATD system is not specifically designed for simulating backward falls. While it is traditionally used in automotive impact testing, backward falls may involve more complex kinematic patterns than those for which the surrogate was originally intended. Nonetheless, our primary objective was to assess the DREW system's ability to differentiate helmet performance in mitigating rotational kinematics. The Hybrid III provided a repeatable and standardized platform for this comparative evaluation.

6. CONCLUSION

In conclusion, the DREW test rig provides a robust and comprehensive platform for evaluating helmet performance under realistic impact conditions. DREW simulates both head-first and whiplash-induced impacts at various locations on the helmet, thereby capturing both translational and rotational dynamics. The system exhibited high repeatability, with variability in kinematic measurements remaining below 4%. Using the DREW system, our analysis of a selection of bicycle helmets revealed that those incorporating rotational mitigation technology offer improved performance in reducing rotational kinematics compared to the baseline. Furthermore, we assessed a high-density liner, developed and optimized through DOT-based monorail drop tests, against a baseline liner. In monorail tests, the high-density liner demonstrated superior performance relative to the baseline at impact velocities of 4.3 and 5.2 m/s, achieving the required standard of maintaining PLA below 150 g during. However, when tested using the DREW system at a higher impact velocity, the baseline liner achieved a significantly lower peak linear acceleration than the high-density liner. Additionally, our evaluation of rotational kinematics revealed that the baseline liner exhibited considerably lower peak angular acceleration compared to the high-density liner.

These findings indicate that a realistic simulation of impact dynamics is essential for helmet evaluation. Achieving optimal linear performance alone may not be sufficient to reduce the risk of brain injury, as rotational kinematics also play a crucial role. Therefore, incorporating the evaluation of rotational kinematics into helmet testing protocols is critical. Our results underscore that assessment methods accounting for both linear and rotational dynamics offer a comprehensive evaluation of helmet performance and are vital for developing next-generation helmets that deliver superior head protection in real-world scenarios.

Acknowledgments

This research was supported by the U.S. Office of Naval Research under PANTHER award number N00014-21-1-2916 through Dr. Timothy Bentley.

Disclaimer

References to trade names are provided solely for clarity and do not constitute an endorsement of any commercial products. It is important to note that the bicycle helmet testing described herein was conducted under specific conditions and impact velocities optimized for evaluating ballistic helmet performance; these results are not intended to serve as a direct comparison or commentary on the performance of different helmet technologies.

References

[1] Wojcik B. E., Stein C. R., Bagg K., Humphrey R. J., and Orosco J., "Traumatic brain injury hospitalizations of US army soldiers deployed to Afghanistan and Iraq." *Am J Prev Med*, 38, no. 1 (2010): S108-S116.

[2] U.S. Military Health System. DOD Numbers for Traumatic Brain Injury Worldwide 2000 – 2024 Q2. (2024). https://health.mil/Reference-Center/Reports/2024/12/30/2000-2024-Q2-DOD-Worldwide-Numbers-for-TBI. Accessed 13 March 2025.

[3] Tierney G., "Concussion biomechanics, head acceleration exposure and brain injury criteria in sport: a review." *Sports Biomech*, 23, no. 11 (2024): 1888-1916.
[4] Bian K., and Mao H., "Mechanisms and variances of rotation-induced brain injury: a parametric investigation between head kinematics and brain strain." *Biomech Model Mechanobiol*, 19, no. 6 (2020): 2323-2341.
[5] Holbourn, A. H. S., "Mechanics of head injuries." *The Lancet* 242, no. 6267 (1943): 438-441.
[6] Kimpara H., and Iwamoto M., "Mild traumatic brain injury predictors based on angular accelerations during impacts." *Ann Biomed Eng*, 40 (2012): 114-126.
[7] Rowson S., Duma S.M., Beckwith J.G., Chu J.J., Greenwald R.M., Crisco J.J., Brolinson P.G., Duhaime A.C., McAllister T.W. and Maerlender A.C., "Rotational head kinematics in football impacts: an injury risk function for concussion." *Ann Biomed Eng*, 40 (2012): 1-13.
[8] Kleiven S., "Why most traumatic brain injuries are not caused by linear acceleration but skull fractures are." *Front bioeng biotechnol*, 1 (2013): 15.
[9] Fernandes, F.A. and De Sousa, R.A., "Finite element analysis of helmeted oblique impacts and head injury evaluation with a commercial road helmet." *Struct Eng Mech*, 48, no. 5 (2013): 661-679.
[10] Bliven E., Rouhier A., Tsai S., Willinger R., Bourdet N., Deck C., Madey S.M. and Bottlang M., "Evaluation of a novel bicycle helmet concept in oblique impact testing." *Accid Anal Prev*, 124 (2019): 58-65.
[11] Neice R.J., Lurski A.J., Bartsch A.J., Plaisted T.A., Lowry D.S. and Wetzel E.D., "An experimental platform generating simulated blunt impacts to the head due to rearward falls." *Ann Biomed Eng,* 49 (2021): 2886-2900.
[12] Halldin P., Gilchrist A., Mills N. J., "A new oblique impact test for motorcycle helmets." *Int J Crashworthiness* 6, no. 1 (2001): 53-64.
[13] Gabler L.F., Crandall J.R. and Panzer M.B., "Development of a metric for predicting brain strain responses using head kinematics". *Ann Biomed Eng*, 46 (2018), 972-985
[14] Abayazid F., Ding, K., Zimmerman K., Stigson H., & Ghajari M. (2021). "A new assessment of bicycle helmets: the brain injury mitigation effects of new technologies in oblique impacts." *Ann Biomed Eng*, 49, no. 10 (2021): 2716-2733.

https://doi.org/10.52202/081621-0046

Instrumented drop mass for impact response measurement in stab-resistant body armour testing

S. Jenkin[1,2], B. Cazzolato[1], C.F. Jones[1] and D. Thompson[2]

[1]*School of Electrical and Mechanical Engineering, The University of Adelaide, Adelaide, 5005 South Australia, Australia,*
scott.jenkin@adelaide.edu.au
[2]*Armor Australia Pty Ltd, 126 Frederick Street, Welland, 5007 South Australia, Australia*

Abstract. The release of the Police Scientific Development Branch (PSDB) 1999 Stab Resistance Standard marked a pivotal development in stab-resistant body armour testing, introducing a rail-guided drop mass with a dual-mass design and a foam composite torso simulant. These innovations significantly improved test outcome consistency and continue to serve as the foundation for the apparatus employed in both National Institute of Justice (NIJ) Standard-0115.00 and Home Office Body Armour Standard (CAST 2017). However, factors such as the impact location on the armour panel and backing material, repeated use of damping disks, and test apparatus rigidity, are hypothesised to affect outcomes by altering the force-time and force-displacement response – collectively termed impact response. Typically, these sensitivities are inferred from penetration depth measurements, which usually require large sample sizes to identify statistically significant effects. An alternative approach that directly measures impact response using an instrumented drop mass could provide more precise assessment and a means to address these sensitivities. This paper details the development and validation of an instrumented drop mass, compliant with NIJ Standard-0115.00, designed to measure impact response in stab-resistant armour testing. Since it is not possible to directly measure the force at the threat tip, an analytical inertial compensation method was developed and experimentally validated to improve the accuracy of the estimated force transmitted to the armour. The system was used to evaluate impact responses from P1/A Edged Blade and Spike threat types on armour panels tested in accordance with NIJ Standard-0115.00. The results revealed distinct response characteristics for low- and high-penetration impacts, reflecting differences in threat geometry and penetration mechanics. This instrumented drop mass enables detailed investigation of variables influencing stab test outcomes, providing critical insights to guide future improvements in standardised apparatus and protocols, enhancing test outcome consistency within the validation of protective equipment.

1. INTRODUCTION

Stab-resistant body armour provides crucial protection against hand-delivered sharp-edged and pointed weapon attacks, for law enforcement, military and correctional personnel. Classified as life protection equipment, these armour systems must meet or exceed minimum performance requirements through standardised testing of representative samples. Modern stab testing standards, such as US National Institute of Justice (NIJ) Standard-0115.00 [1] and the UK Home Office Centre for Applied Science and Technology (CAST) Body Armour Standard, 2017 [2], evolved from foundational work outlined in UK Police Scientific Development Branch (PSDB) Stab Resistance Standard for Body Armour 1999 [3]. That earlier research led to the replacement of manual attack and air cannon launches with a rail-guided drop mass, a dual-mass configuration, and a composite foam torso simulant (backing material) to yield more consistent test outcomes that better replicate the dynamics of a hand-delivered stab impact. Although these improvements in apparatus design and protocol have enhanced repeatability compared to pre-1999 methods, several factors are still hypothesised to significantly influence test outcomes, including impact-to-edge spacing on the armour and backing material, the recovery time and number of uses of the damping disks, armour restraint, and test apparatus rigidity.

Much of the instrumented impact response testing during the PSDB 1999 development used drop towers with a single-component drop mass and backing material such as clay or gelatine, often with rigid composite or metallic armour samples [4-6]. More recently, Forster et al. [7] employed an instrumented drop mass, compliant with NIJ Standard-0115.00, to measure the acceleration response of various foam-based backing materials impacted by a circular punch, recommending stricter tolerances on density, hardness, and compression-deflection for future standards. However, published data on impact responses of modern, flexible stab-resistant armour with dual-mass drop masses and foam composite backing remains limited.

A detailed understanding of the dynamic impact response during stab testing, including forces and displacements, can provide valuable insight into the interactions between the armour panel, backing

https://doi.org/10.52202/081621-0047

material, and damping disks. This study presents the development and validation of an instrumented drop mass engineered for compliance with NIJ Standard-0115.00 and CAST 2017. Additionally, it provides typical impact response curves for both P1/A Edged Blade and Spike threat types, examining characteristic impact stages and the potential value of force-time and force-displacement data in evaluating stab-resistant armour performance.

2. EQUIPMENT

2.1 Instrumented drop mass

The instrumented drop mass (Figure 1a) is a dual-mass system with dimensions, masses, and materials compliant with both NIJ Standard-0115.00 and CAST 2017, ensuring equivalent impact energy, momentum, and configuration to conventional non-instrumented systems. To enable impact response measurement, the drop mass was instrumented with three sensors to capture uniaxial force, damping disk displacement, and acceleration.

The uniaxial force transmitted into the armour panel was measured using a custom strain gauge-based load cell, integrated with the threat holder and positioned behind the test threat along the axial load path. The load cell body, fabricated from high-tensile steel, employed a four-column design (Figure 1b). Dual-element biaxial (0°/90°) strain gauges (FCAB-5-350-11-1LJB-F, Tokyo Measuring Instruments Lab Co. Ltd., JPN) were bonded to the midspan of each column and connected in a full Wheatstone bridge configuration for single-channel force measurement (Figure 1c). The load cell was enclosed in an aluminium housing with castellation features to minimise potential off-axis bending or torsional errors in the uniaxial measurement. The load cell was calibrated by applying quasi-static compression-relaxation cycles using a universal testing machine (8874, Instron, USA) with a Dynacell (±25kN), achieving a maximum non-linearity error of 1.48% at full-scale output of 6.5 kN.

The damping disk displacement was measured using a linear potentiometer (PZ12-S-0075-L, Gefran, ITA) mounted between the threat holder and sabot. While not central to the present study, this measurement supports ongoing sensitivity analyses investigating how repeated damping disk use, including the number of impacts and recovery time, affects force-displacement behaviour and, in turn, the measured impact response at the threat-armour interface. Uniaxial acceleration was measured using an accelerometer (2264A, Endevco, USA) mounted to the sabot (Figure 1d). This was used to estimate the velocity and displacement of the sabot during impact, and, when combined with the displacement measurements from the linear potentiometer, to estimate the motion of the threat holder. The acceleration measurement also enabled inertial compensation of the measured force, as discussed in Section 2.3.

Sensor signals were recorded using an on-board data acquisition system (DAS) (SLICE NANO, Diversified Technical Systems, USA) mounted to the sabot (Figure 1e). The recorded signals were sampled at 500 kS/s, and were passed through zero-phase, 4th-order Butterworth lowpass filters with cutoff frequency adjusted by impact velocity. All instrumentation and power sources, including the sensors, DAS, and Li-ion battery, were rigidly coupled to the drop mass, forming a self-contained system without external connections or added dynamic compliance.

2.2 Stab test apparatus

Dynamic impact testing with the instrumented drop mass was conducted using an NIJ Standard-0115.00 compliant rail-guided drop apparatus (Adelaide T&E Systems, AUS) at Armor Australia's Research, Development and Engineering Facility in Adelaide, Australia (Figure 2a). The apparatus allowed precise programmatic control of impact energy up to 70 J and included an adjustable height test table, and hook-and-loop retention straps to secure the armour panel to the backing material (Figure 2b). Drop mass velocity, v_m, was determined by counting a 1 MHz clock as the sabot passed between two infrared (IR) emitter-detector pairs, spaced 50 mm apart. The table height was adjusted so that when the lowest edge of the sabot crossed the lower IR beam, the threat tip was 25 mm above the target surface, resulting in a 50 mm standoff distance, h_s. The adjusted impact energy, E_{adj}, was calculated as the sum of kinetic and potential energy associated with h_s, as shown in Equation (1), where m is the drop mass (1.902 kg) and g is gravitational acceleration. The adjusted impact velocity, v_{adj}, was then determined using Equation (2).

https://doi.org/10.52202/081621-0047

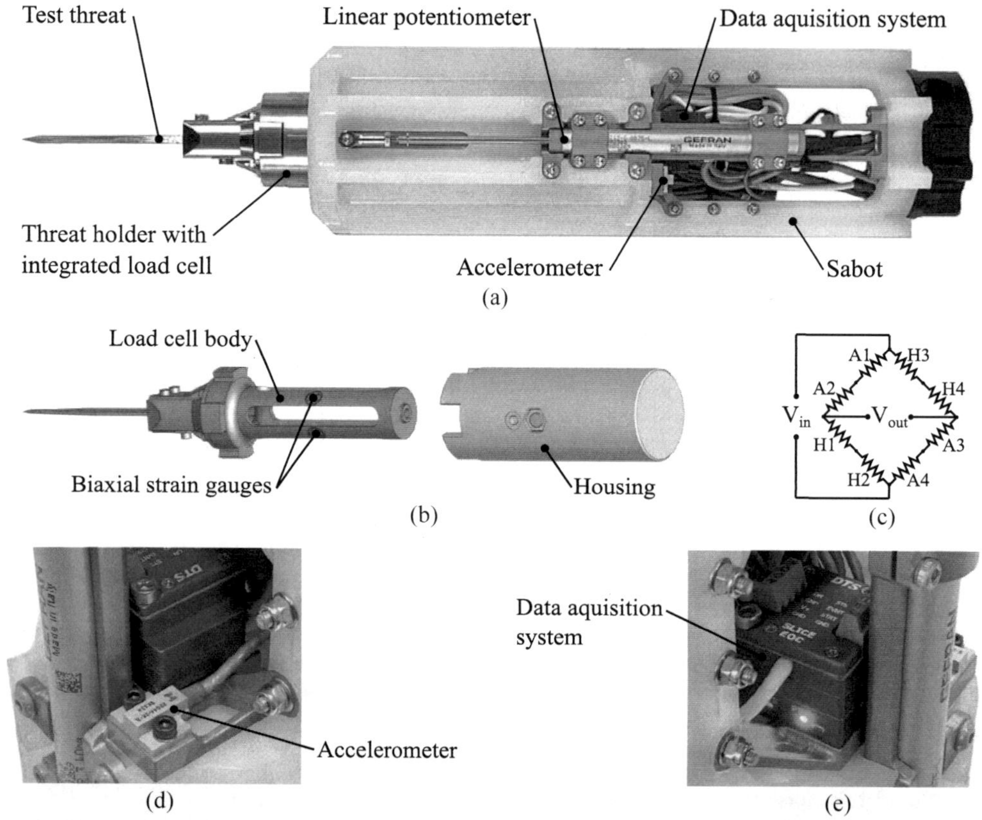

Figure 1. Instrumented drop mass: (a) front view of sabot and threat holder, (b) rendered exploded view of the threat holder with integrated load cell, (c) strain gauge wiring diagram with strain gauges A1 to A4 aligned in the axial direction and H1 to H4 aligned in the hoop direction, (d) accelerometer mounted to the sabot, and (e) data acquisition system mounted to the sabot.

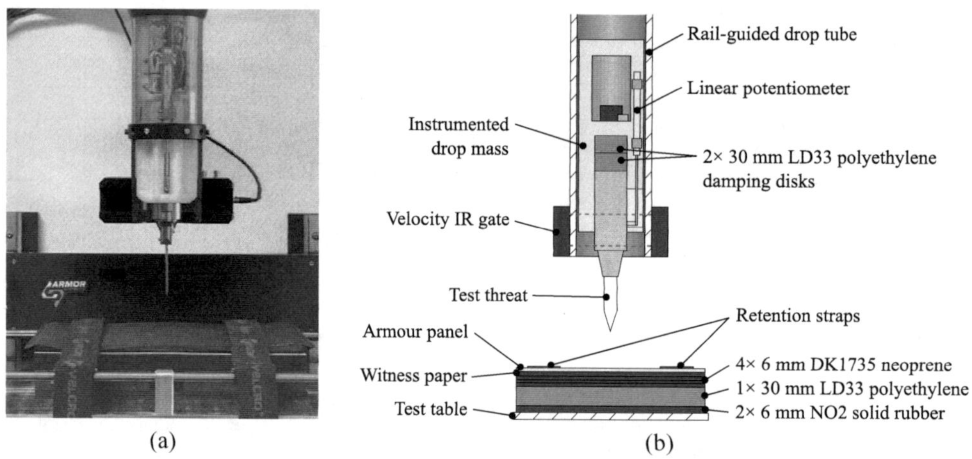

Figure 2. NIJ Standard-0115.00-compliant stab testing apparatus and experimental setup: (a) photograph of apparatus, and (b) cross-sectional schematic. The linear potentiometer in (b) is rotated 90° relative to its actual orientation in (a), to better illustrate its assembly configuration.

$$E_{adj} = \frac{1}{2} m {v_m}^2 + mgh_s \tag{1}$$

$$v_{adj} = \sqrt{\frac{2E_{adj}}{m}} \tag{2}$$

2.2.1 Test table rigidity

Test table rigidity was assessed separately from the main experimental work to facilitate comparison of apparatus used in other laboratories. Testing was conducted with NIJ-compliant backing materials, a stab-resistant armour panel, and a blunt hemispherical impactor (11 mm diameter) to replicate NIJ Standard-0115.00 conditions. A laser displacement sensor (CP24MHT80, Wenglor Sensoric, GER) was tripod mounted and physically isolated from the stab test apparatus to record deflection directly beneath the impact location. Three impacts at 24 J each resulted in maximum displacements of 0.33mm, while at 65 J, the mean maximum displacement was 0.96 mm (*SD* = 0.02 mm).

2.2.2 Backing materials and damping disks

The stab testing setup employed a multilayered backing material comprising foam and rubber layers, as specified in NIJ Standard-0115.00. The standard permits two grades each of expanded neoprene and low-density polyethylene (LDPE), which vary in mechanical properties and can significantly influence the impact response [7]. This study used DK1735 expanded neoprene and LD33 Plastazote® (Zotefoams, UK), supplied in 400 x 333 mm sheets, due to their mechanical properties closely aligning with specifications proposed by Forster et al. [7] and Draft NIJ Standard-0115.01 [8], ensuring impact responses remain relevant to both current and future standardised testing. Additionally, two 6 mm thick NO2 solid rubber sheets were placed beneath the foam backing materials. All backing materials were subjected to restitution testing in accordance with NIJ Standard-0115.00 Section 5.5.7 [1]. The rebound height of the steel sphere was determined by measuring the time of flight between its first and second bounces using an IR gate. All backing material packs satisfied the rebound height requirements of NIJ Standard-0115.00 and Draft NIJ Standard-0115.01, with a mean height of 528 mm (*SD* = 4 mm, *n* = 36).

The LD33 foam was also used for the damping disks, positioned between the threat holder and sabot (Figure 2b). The disks were precision cut to 50 mm diameter using a CNC milling machine and a custom hole saw. Table 1 details the relevant properties of all materials.

Table 1. Material properties of the backing materials.

Property	Expanded neoprene DK1735 [9]	Polyethylene foam LD33 [10]	Solid rubber NO2 [11]
Thickness (mm)	5.8-6.2	29.0-30.4 (no skin)	5.8-6.1
Density (kg/m^3)	150-200	33	1500
25% Compression Deflection (kPa)	60-90 (ASTM D1056)	66 (ISO 7214.2012)	-
50% Compression Deflection (kPa)	100-130 (ASTM D1056)	133 (ISO 7214.2012)	-
Shore Hardness	50-60 (OO)	58 (OO)	60-70 (A)

2.3 Inertial compensation of force measurement

Impact force was measured at the strain gauge location, positioned in a region of high strain uniformity but offset from the test threat tip. As a result, the measured force, F_{cell}, includes inertial contributions from the effective mass between the impact point and measurement location, and does not directly represent the forces at the threat tip, F_{tip}, or through the damping disks, F_{disk}. To estimate these quantities more accurately, an analytical inertial compensation method was developed, following an approach similar to Casem [12] for the split Hopkinson pressure bar.

2.3.1 Analytical inertial compensation method

The threat holder and load cell were analysed as a system sectioned at the strain gauge location, allowing the force components to be resolved (Figure 3a). Free-body diagrams illustrate the force interactions between the load measurement location and the threat tip (Figure 3b) and between the load measurement location and the damping disk interface (Figure 3c). The force measured by the load cell, F_{cell} is expressed in Equation (3), where $F_{inertia,th,b}$ represents the inertia force from the lower effective mass of the threat holder, $m_{th,b}$, and its acceleration, a_{th}. By incorporating the measured acceleration of the

https://doi.org/10.52202/081621-0047

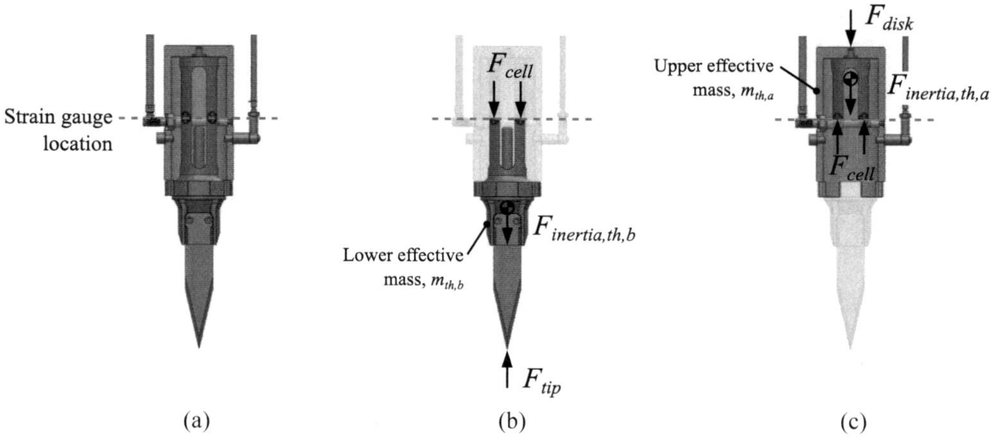

Figure 3. Rendered section view of threat holder and load cell (a) strain gauge location, (b) free-body diagram between load measurement location and threat tip, and (c) free body diagram between load measurement location and damping disk interface.

sabot, a_{sabot}, and relative displacement of the sabot and threat holder, s_{rel}, the inertia-compensated force acting at the threat tip is given by Equation (4). The sabot acceleration was first double-integrated to estimate its global displacement, which was then summed with the relative displacement from the potentiometer. The result was passed through a zero-phase, 4th-order Butterworth lowpass filter with 700 Hz cutoff frequency, before double-differentiation to remove high-frequency noise in the compensated force estimate. Similarly, the inertial compensation method was applied to estimate F_{disk}. From the free-body diagram (Figure 3c), the force balance yields Equation (5), where $m_{th,a}$ represents the upper effective mass between the strain gauge and damping disks.

$$F_{cell} = F_{tip} - F_{inertia,th,b} \tag{3}$$

$$F_{tip} = F_{cell} + m_{th,b} \frac{d^2}{dt^2}\left(\int_0^t \int_0^\tau a_{sabot}(\tau)d\tau dt + s_{rel}\right) \tag{4}$$

$$F_{disk} = F_{cell} - m_{th,a} \frac{d^2}{dt^2}\left(\int_0^t \int_0^\tau a_{sabot}(\tau)d\tau dt + s_{rel}\right) \tag{5}$$

2.3.2 Experimental validation of inertial compensation

Experimental validation of the inertial compensation method was conducted using a blunt hemispherical indenter (11 mm diameter) impacting an ultra-high molecular weight polyethylene (UHMWPE) soft armour panel with an areal density of 5.9 kg/m^2, restrained on an NIJ Standard-0115.00-compliant backing material (Figure 4). A six-axis load cell (MC5-2500, Advanced Mechanical Technology, Inc., USA) was positioned beneath the backing material to measure the transmitted force, $F_{backing}$, for comparison against the inertially compensated F_{tip}. Time alignment was performed by minimising the root-mean-square error (RMSE) between measured force signals, from initial contact to the occurrence of peak F_{tip}.

Force-time responses from five impacts at 20 J (Figure 5a) and 70 J (Figure 5b) showed that the compensated F_{tip} closely corresponded to $F_{backing}$, with differences small in magnitude and consistent in shape. The average RMSE up to peak force was 60 N for the 20 J series and 81 N for the 70 J series. These differences were primarily attributed to uncompensated inertial effects from the accelerated armour, backing material, and aluminium backplate. The consistency of these differences across all five impacts at 20 J (Figure 5c) and 70 J (Figure 5d) demonstrates that the inertial compensation method provides a reliable and accurate estimate of the force at the threat tip.

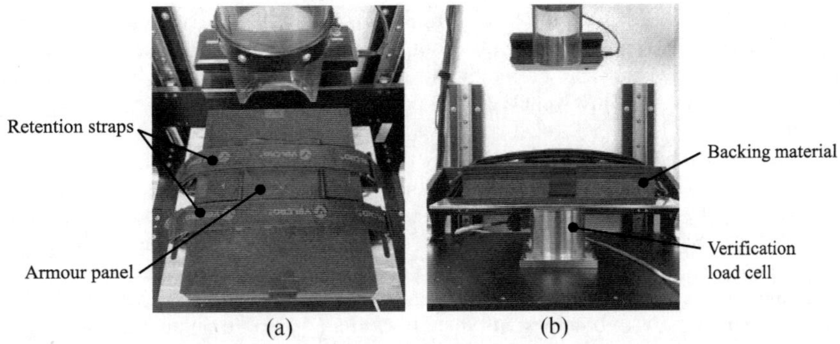

Figure 4. Experimental setup for inertial compensation validation: (a) armour panel restrained on backing material, and (b) side view of load cell mounted below the backing material.

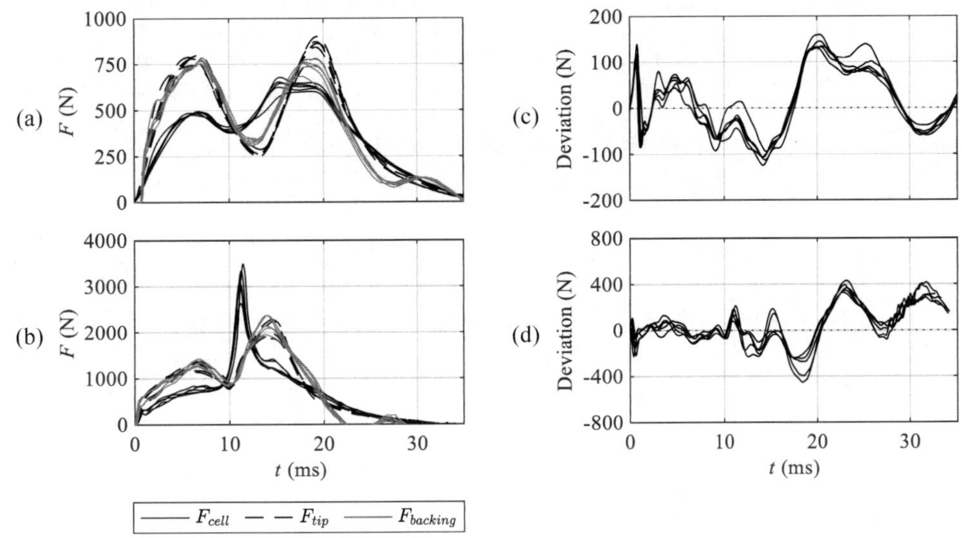

Figure 5. Experimental validation of inertial compensation method: overlay of F_{cell}, F_{tip} and $F_{backing}$ force-time responses for (a) 20 J and (b) 70 J impacts, and deviation between F_{tip} and $F_{backing}$ for (c) 20 J and (d) 70 J impacts.

3. IMPACT RESPONSE MEASUREMENT

The impact responses presented in this section were selected from a larger test series using the P1/A Edged Blade threat (*n*=76) and the Spike threat (*n*=68). While the primary objective of the larger test series was to evaluate regression methods for DOP outcomes within a novel E_{50} test methodology for stab-resistant armour [13], force-time and force-displacement responses were also recorded to characterise impact behaviour. Testing was conducted in accordance with NIJ-Standard 0115.00 on soft armour panels with a hybrid construction of para-aramid and UHMWPE fabric layers, measuring 400 x 400 mm with a total areal density of 6.7 kg/m^2. This construction meets NIJ Standard-0115.00 requirements for Edged-Blade Protection Class at Protection Level 2 and Spike Protection Class at Protection Level 1. Furthermore, the panels comply with NIJ Standard-0101.06 Level II for ballistic protection, making them representative of modern, flexible, multi-threat armour solutions used in law enforcement and security applications. Two distinct impact behaviours were observed: low-penetration impacts, where the threat did not fully penetrate the armour, and high-penetration impacts, where greater penetrative displacement between the threat and armour panel occurred.

3.1 Impact response for low-penetration impacts

Low-penetration impacts exhibited highly repeatable response curves with similar behaviour across both threat types. Force-time (Figure 6a) and force-displacement responses (Figure 6b) for a representative

https://doi.org/10.52202/081621-0047

30 J impact using the P1/A Edged Blade threat, resulting in 0 mm DOP, are presented. Corresponding acceleration, velocity, and displacement for the sabot and the threat holder are presented in Figure 6c. A consistent sequence of interactions between the test components was observed, which can be generalised into four characteristic stages of low-penetration impacts, discussed below and illustrated in Figure 6d.

Stage 0 began at the initial contact between the threat tip and the armour panel, causing rapid deceleration of the threat holder. Meanwhile, the sabot continued largely unopposed due to the low initial stiffness of the damping disks. The force at the threat tip increased linearly and reached a local peak before transitioning to Stage 1, at approximately 7 ms, when the threat holder momentarily came to rest and began to reverse direction as the sabot continued decelerating and compressing the damping disks. As the disks compressed further, they exerted increasing force until reaching a peak at Stage 2, around 14 ms. The peak damping disk compression typically exceeded 80% strain at 25 J and surpassed 90% strain at 65 J, placing them in the densification region of closed-cell foams [14]. In this region, the collapsing cell walls, elevated internal gas pressure, and compression of the bulk polymer led to a rapid rise in stress with additional strain. Consequently, the threat holder experiences peak downward acceleration into the armour, while the sabot simultaneously reached peak deceleration. Stage 3 occurred at peak threat tip force, at about 16 ms. For low-penetration impacts, the relationship between impact energy and peak threat tip force was approximately linear. From this point, the threat holder continued to displace into the backing material, while the sabot had already displaced to a global peak and had begun to reverse direction. Stage 4 is marked by the threat holder achieving its global peak displacement into the backing material, at approximately 20 ms, after which it started to recoil.

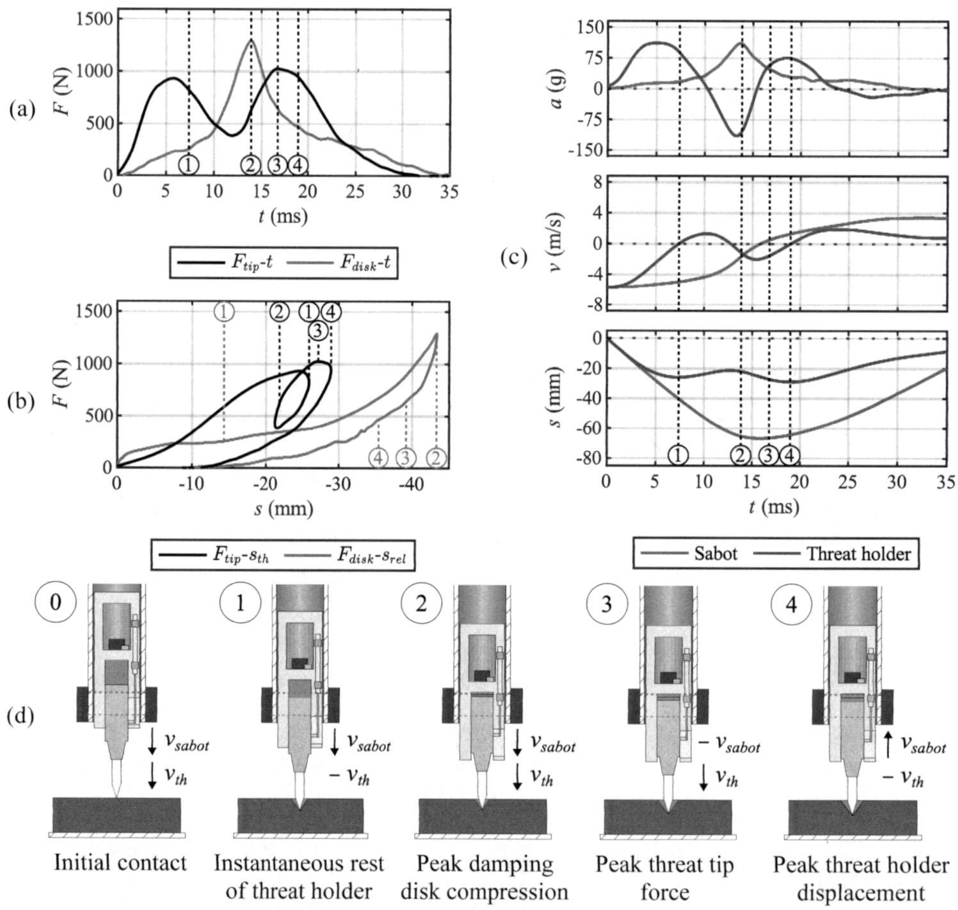

Figure 6. Impact response of a P1/A Edged Blade threat impact at 30 J (0 mm DOP) showing the characteristic stages of low-penetration impacts: (a) force-time response curves, (b) force-displacement response curves, (c) time-domain plots of acceleration (top), velocity (middle) and displacement (bottom) for both the sabot and threat holder, and (d) illustration of impact stages, indicated by encircled numbers, depicting instantaneous velocity directions of the sabot and threat holder.

3.2 Impact response for high-penetration impacts

To examine the transition from low- to high-penetration impacts, single representative impacts are presented at four energies, ranging from 30 J to 60 J for the P1/A Edged Blade threat (Figure 7) and 25 J to 55 J for the Spike threat (Figure 8). All impacts were conducted at a consistent impact location, maintaining a 94 mm impact-to-edge distance on the armour panel and 98 mm on the backing material. DOP was measured per CAST 2017 Section 7.74 [2]. For non-penetration impacts, the number of intact armour layers was counted to estimate a negative DOP using measured ply thickness. Table 2 summarises the impact response maxima and test outcomes, including the threat tip sharpness measured per NIJ Standard-0115.00 Section 5.5.5 [1], and whether the drop mass bounced out of the armour panel after impact.

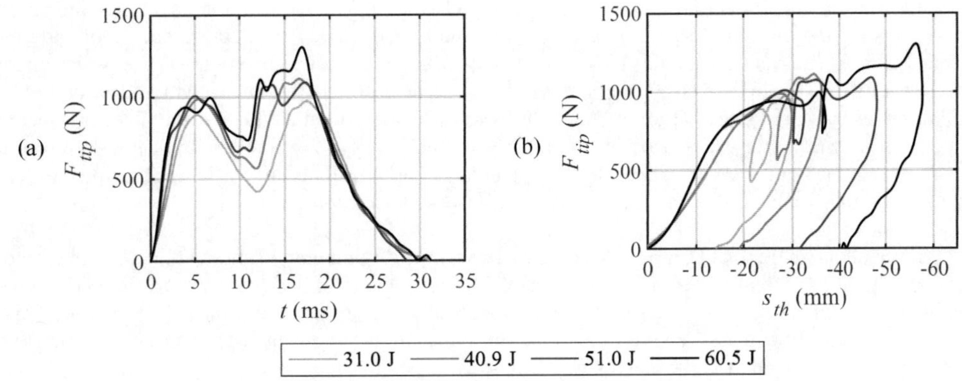

Figure 7. Impact response curves for selected single impacts using the P1/A Edged Blade threat over 31.0-60.5 J energies: (a) force-time response curves, and (b) force-displacement response curves.

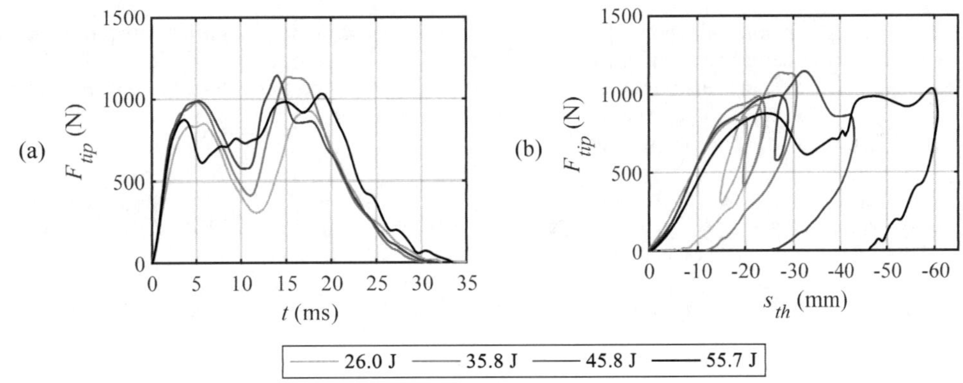

Figure 8. Impact response curves for selected single impacts using the Spike threat over 26.0-55.7 J energies: (a) force-time response curves, and (b) force-displacement response curves.

Table 2. Impact response maxima and test outcomes for selected single impacts. $F_{tip,peak}$ is the peak force at the threat tip, and $s_{th,peak}$ is the peak downward displacement of the threat holder.

E_{adj} (J)	Threat	Sharpness (HRC)	Bounce	$F_{tip,peak}$ (N)	$s_{th,peak}$ (mm)	DOP (mm)
31.0	P1/A	-139	Yes	977	-30.1	-1.3
40.9	P1/A	-142	Yes	1112	-36.1	1.8
51.0	P1/A	-123	No	1091	-48.2	14.7
60.5	P1/A	-123	Yes	1305	-57.7	21.5
26.0	Spike	-137	Yes	928	-23.5	-3.3
35.8	Spike	-132	Yes	1138	-30.7	-1.3
45.8	Spike	-141	No	1146	-43.2	10.8
55.7	Spike	-128	No	1033	-60.7	25.5

https://doi.org/10.52202/081621-0047

Impacts up to nominal 40 J for the P1/A Edged Blade, and 35 J for the Spike (shown as light grey lines in Figures 7 and 8, respectively) resulted in minimal or no penetration. Within these ranges, peak force at the threat tip increased with energy. However, once the penetration-resistance capacity of the armour was reached, further increases in energy resulted in greater threat holder displacement rather than a higher peak force. For impacts resulting in a DOP exceeding approximately 3 mm, the impact responses became more complex and exhibited greater variability compared to low-penetration impacts. The increased complexity of impact responses at higher DOP suggested that additional factors, including significant fibre cutting or parting, threat tip deformation, and friction or slippage between the threat and armour layers, introduced further variability as the threat penetrates deeper.

For the P1/A Edged Blade threat, the force-displacement curve typically exhibited a linear loading region up to approximately 20 mm displacement, followed by a penetration region where the force either plateaued or exhibited a slight increase at higher energies (Figure 7b). At 60 J, the peak force increased slightly above that observed at 50 J, suggesting that as the Edged Blade threat cuts deeper, more armour material engages, thereby increasing its penetration-resistance capacity. In contrast, the Spike threat displayed a more complex penetration phase. At 55 J, the peak force was lower than at 35 J and 45 J, particularly in the early stages (Figure 8b). This behaviour may be attributed to the slender, tapered geometry of the Spike, for which cross-sectional area increases only marginally with depth. As a result, once fibre parting occurred, the engaged armour area remained relatively unchanged as penetration progressed, limiting any further increase in penetration-resistance capacity.

Another notable trend in Figure 7a and Figure 8a was the diminishing bi-modal feature of the force-time response at the threat tip as energy increased. Deeper penetration during the early impact stages caused the threat holder to decelerate more gradually, delaying damping disk compression. Consequently, the drop mass behaved more like a single-mass system at higher DOP, reducing the distinct second force peak usually seen using a dual-mass system.

4. CONCLUSIONS

The development and validation of an instrumented drop mass marks a significant step forward in the study of stab-resistant body armour. Unlike conventional testing, which relies solely on DOP as a performance measure, the instrumented drop mass captures force-time and force-displacement responses, offering a deeper understanding of how armour materials and constructions interact with different threats under dynamic impact loading.

To improve the accuracy of force measurements at the threat tip, an analytical inertial compensation method was developed and experimentally validated. Testing with P1/A Edged Blade and Spike threats demonstrated that when DOP was small – indicating the penetration-resistance capacity of the armour was not exceeded – the impact responses were similar across both threat types and typically exhibited four characteristic stages: instantaneous rest of threat holder, peak damping disk compression, peak threat tip force, and peak threat holder displacement. However, at DOP values exceeding approximately 3 mm, impact responses became more complex, with distinct differences in force profiles and displacement behaviour observed between the threat types. For these higher DOP impacts, peak force did not directly correlate with increased DOP; instead, once the penetration-resistance capacity of the armour was reached, remaining energy was converted into further penetrative displacement.

While the instrumented drop mass is not intended to replace the drop mass currently specified in NIJ Standard-0115.00 or CAST 2017, nor is impact response intended as a certification criterion, it provides an accurate and robust method for investigating impact response within the NIJ Standard-0115.00 or CAST 2017 frameworks. The ability to measure and compare impact responses has practical applications for stab-resistant armour research and development, enabling assessments of penetration-resistance capacity, energy dissipation mechanisms, material interactions and the influence of layering strategies, all under normal test conditions. These insights can contribute to the development of more advanced armour constructions, ultimately enhancing personal protection against Edged Blade and Spike threats.

Beyond armour material and state-of-the-art stab-resistant armour solution development efforts, the instrumented drop mass is currently being used in sensitivity analyses to investigate factors hypothesised to contribute to test outcome variability. These include impact location on both the armour and backing material, the recovery time and repeated use of damping disks, armour restraint, test apparatus rigidity and variations in backing material properties. By systematically evaluating these variables, identifying those with greatest influence, and implementing appropriate controls, the instrumented drop mass has the potential to improve test outcome consistency, strengthening the foundation of standardised evaluation methods for stab-resistant armour systems.

Acknowledgments

This research was supported by the Australian Government through the Department of Education's National Industry PhD Program (project 36179). The views expressed herein are those of the authors and are not necessarily those of the Australian Government or the Department of Education.

References

[1] NIJ Standard 0115.00, Stab Resistance of Personal Body Armor, National Institute of Justice, September 2000.
[2] Payne T., O'Rourke S. and Malbon C., Home Office Body Armour Standard (2017), CAST publication number 012/17, July 2017.
[3] Pettit M. and Croft J., PSDB stab resistance standard for body armour, PSDB publication number 6/99, 1999.
[4] Horsfall I., Stab resistant body armour, PhD thesis, Cranfield University, March 2000.
[5] Chadwick E., Nicol A., Lane J. and Gray T., Biomechanics of knife stab attacks, Forensic Science International, 105 (1999), 35-44.
[6] Nicol A., Chadwick E. and Gray T., Investigation of knife stab characteristics – development of body tissue simulant, Technical report, University of Strathclyde, Scottish Executive Central Research Unit, 1998.
[7] Forster A.M., Chan-Ou-Teung A. Guigues E. Forster A.L., Davis J. and Rice K., Foam composite specifications and qualification in support of NIJ 0115.00, Personal Armour Symposium Proceedings 2014, Cambridge UK.
[8] Draft NIJ Standard 0115.01, Stab resistance of body armour, National Institute of Justice, March 2020.
[9] Alanto Ltd, Specification Sheet, DK1735 Expanded Neoprene Sponge Rubber, Issue 2, 2015.
[10] Alanto Ltd, Specification Sheet, LD33 – Low Density Polyethylene Foam, Issue 1, 2018.
[11] Alanto Ltd, Specification Sheet, NO2 Commercial Economy Solid, Issue 1, 2016.
[12] Casem D., Weerasooriya T. and Moy P., Inertial effects of quartz force transducers embedded in a plot Hopkinson pressure bar, Experimental Mechanics, 45, 368-376.
[13] Jenkin S., Cazzolato B., Jones C.F. and Thompson D., E50 testing for stab-resistant body armour evaluation, Personal Armour Symposium Proceedings 2025, Bruges BEL.
[14] Gibson L., Modelling the mechanical behaviour of cellular materials, Materials Science and Engineering: A, 110 (1989), 1-36.

 https://doi.org/10.52202/081621-0047

Ballistic testing of armour grade ceramics

E. P. Carton

TNO Group Platform Protection and Advanced Materials, Ypenburgse Boslaan 1, 2496ZA, The Hague, The Netherlands, erik.carton@tno.nl

Armour grade ceramic materials are used in ceramic based armour systems in vehicles and body armour. Ceramic based armour provides light-weight solutions against a range of threats, including armour piercing bullets. Generally, the ceramic is supported by a backing material that typically is a fibre composite (like aramid or UHMW-PE) or a metal. The function of the ceramic is to erode the core of the AP round and to distribute the dynamic loading over a wider area of the backing material. The function of the backing material is to prevent fragments (ceramic or projectile) to penetrate and absorb the (remaining) kinetic energy of the system.
The ballistic efficiency of a ceramic based armour highly depends on the performance of its ceramic component. Using the Residual Energy Method (REM) on bare ceramic tiles, the efficiency of the ceramic to erode and decelerate the AP core can be determined. Using a bare tile allows a pure ceramic-projectile interaction, which can be very reproducibly tested. In the test the mass and velocity of the residual AP core are measured, hence the loss in kinetic energy of the core is known. Also, the diameter of the first axially responding part of the rear of the tile is determined (from high-speed video recording), which provides information about the cone-crack angle.
Using several ceramic tiles that have the same areal density ($AD = \rho * t$) their ballistic efficiency can be determined and used to rank them. Examples are provided of REM results obtained with (armour grade) SiC tiles of various suppliers.

1. INTRODUCTION

The lightest personal and vehicle armour systems make use of (armour grade) ceramic materials, like Al_2O_3 (alumina), SiC (silicon carbide) and B_4C (Boron carbide). Production of armour grade ceramic materials like SiC and B4C has largely been extracted from Europe. This leads to a dependency of other countries which cab suffer from long lead times and high prices.
To explore alternative sources for these important armour components, TNO was asked to perform ballistics tests on SiC tiles of various suppliers. This provides an insight in the ballistic protection capability of such ceramic components.
Many efforts have been made to rank armour ceramic tiles using the Depth-of-Penetration (DoP) method [1]. Here, the tile is adhesively bonded to a semi-infinite backing material (like an aluminium alloy block) and shot using an armour piercing bullet. The depth of penetration in the backing material is used by many researchers to rank various ceramic tiles. However, this DoP method has serious drawbacks such as the semi-infinite backing which is not representative of realistic armour concepts and the large variation in test results [2]. The latter can be compensated by performing a large number of repeated experiments, however this is seldom done.
Using the Residual Energy Method (REM) the reduction in kinetic energy of the core of an armour piercing (AP) bullet can determined. This method allows the ballistic efficiency (BE) to be determined for any target (material or armour system), including the bare SiC tiles used here. The testing of bare ceramic tiles provides a very reproducible target-bullet interaction, as the influence of a backing material, adhesive layer and internal stresses as well as confinement are excluded.
The ballistic efficiency expresses the relative reduction in kinetic energy of the AP core (as a percentage). In this paper the REM is used to compare SiC tiles of various suppliers and explore if there are any correlations with their material properties.

https://doi.org/10.52202/081621-0048

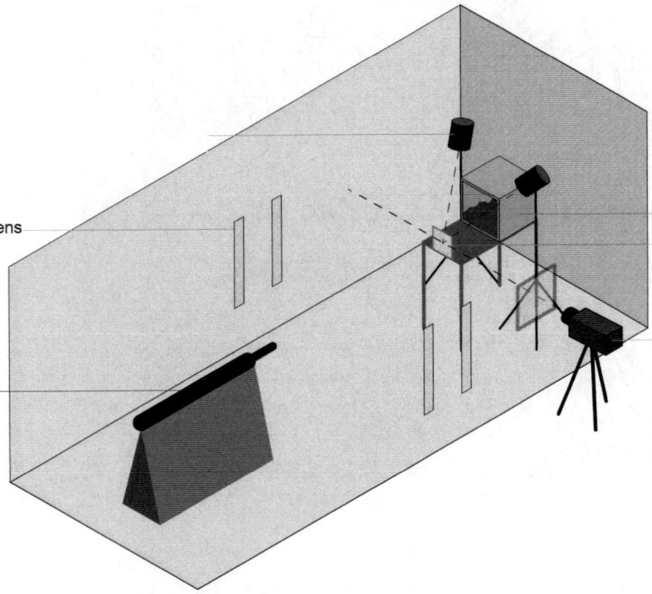

Figure 1. Test setup for the combined test method REM.

2. RESIDUAL ENERGY METHOD

Figure 1 shows a typical test setup for the REM. The projectile is fired using the appropriate gun and propellant. The infrared velocity screens allow the determination of the speed of the incoming projectile. The ceramic tile is positioned perpendicular to the line of fire on a steel platform, a small amount of ballistic clay is used in between in order to avoid the tile from dropping before the shot is fired. A high-speed video camera captures the interaction of the projectile with the target. By analysing the high-speed images, both the impact velocity of the projectile and the residual velocity of the projectile/fragment cloud can be determined. It is assumed that the velocity of the residual AP core is equal to that of the tip of the fragment cloud. Experimental support for this assumption was been obtained from flash X-ray images, as well as high speed video recordings of a fragment cloud after passing a gap between two steel plates (positioned 10 cm behind the ceramic tile) that showed both the bullet and the (now 2-dimensional) fragment cloud. Both recording methods, as well as simulations [3] showed that residual projectile parts have a similar velocity (at that position) as the tip of the fragment cloud.

The soft catch (residual projectile catching device) is able to decelerate the residual projectile or fragments without any plastic deformation, allowing the retrieval of the eroded residual projectile or its core fragments.

The REM ballistic testing method more specifically determines the degree of erosion and deceleration of the projectile after target interaction.

A scenario where an AP bullet interacts with an armour grade ceramic tile is schematically shown in Figure 3. Using a very high frame rate (order of 10^5 fps) with the high-speed camera, behaviour of the material can be observed such as bullet deceleration and cone formation in the tile. The dimension of the cone base (2R) is observable, from which the cone angle (α) can be calculated using the tile thickness and projectile core diameter [2].

https://doi.org/10.52202/081621-0048

Figure 2. Photograph of a ceramic tile on the platform in front of the soft catcher.

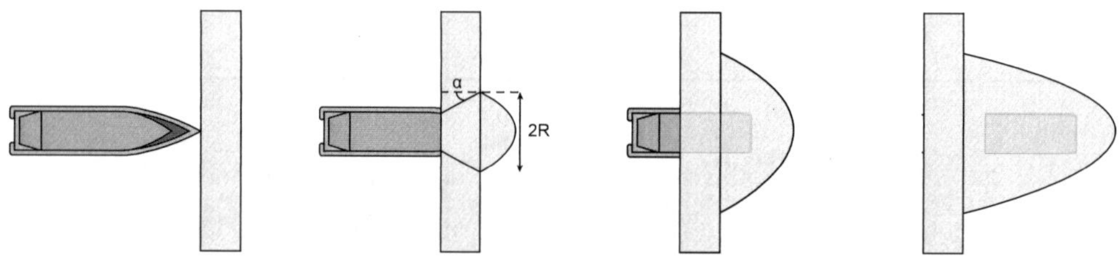

Figure 3. Schematic view of high-speed video images at various interaction times.

Based on the masses and velocities of the AP core before and after impact on the ceramic tiles the kinetic energy loss ΔE_{kin} of the AP core can be calculated:

$$\Delta E_{kin} = 0.5 * m_{core} * v_0^2 - 0.5 * m_{core,residual} * v_{cloud}^2 \quad (1)$$

Where m_{core} is the initial AP core mass and v_0^2 the square of the impact velocity.
The material or armour system can be ranked using the fraction of energy loss of the AP core they provide.

$$Ballistic\ efficiency = \frac{\Delta E_{kin}}{E_{kin,0}} \quad (2)$$

Where $E_{kin,0}$ is the kinetic energy of the AP core before impact.

As the REM uses mass and velocity of the bullet/core before and after penetration, it can be applied to any armour system (a single plate or a complete armour system), as long as the target is overmatched by the projectile. In this work the tiles have been shot using an armour piercing rifle bullet (7.62 mm AP) that strikes the bare SiC tile at a velocity of 825 m/s in the center of the tile at normal impact condition.

Table 1. Overview of SiC samples and some of their properties

Sample	Density [g/ccm]	Thickness [mm]	Areal Density [kg/m2]	Longitudinal Sound velocity [km/s]	Transversal Sound velocity [km/s]	Youngs Modulus [GPa]	Shear Modulus [GPa]	Compression Modulus [GPa]	Poisson's ratio [-]	Vickers HV1 [GPa]
A	3,154	8,05	25,4	12,00	7,80	435	192	198	0,13	25,1
B	3,134	8,03	25,2	11,80	7,53	411	178	199	0,16	23,6
C	3,156	7,89	24,9	11,24	7,56	392	180	158	0,09	23,6
D	3,155	8,10	25,6	11,70	7,20	391	164	214	0,20	21,7
E	3,134	9,08	28,5	11,83	7,56	414	179	200	0,15	22,9
F	3,144	7,26	22,8	11,65	7,50	405	177	191	0,15	22,8
G	3,097	7,70	23,8	11,23	7,43	380	171	163	0,11	22,2
H	3,152	7,95	25,1	11,99	7,54	420	179	214	0,17	22,7

3. MATERIALS

Several suppliers of armour grade siliconcarbide (SiC) material have been requested to send samples free of charge. We have received samples from several countries, including China, USA, Italy, and South Korea. The squared SiC tiles that were supplied should ideally have dimensions of 100x100x8 mm.
Table 1 shows an overview of the samples and some of their mechanical properties. The actual tile thicknesses vary between 7.26 and 9.1 mm. The density of the samples was measured using He-pycnometry, while the longitudinal and transversal sound velocity were measured using ultrasonic equipment (at 5 MHz). From the density and sound velocities, the elastic moduli (E, G and K as well as the Poisson's ratio) have been calculated. The Vickers hardness was obtained using a micro-Vickers hardness indenter with an indentation load of 10N.
The hardness values of all SiC samples ranged between 21.7 and 25.1 GPa. Such high hardnesses are considered sufficient for the erosion of AP core materials. The material properties provided in Table 1 allow to explore possible correlations. Although the range in moduli seems limited, in Figure 4 a linear correlation between the elastic moduli (E, and especially G) and the Vickers hardness of the samples is shown. This can have practical applications, as the sound velocity and density of a tile can be measured before a ballistic experiment, and it does not change/damage the material. Hence, it is rather easy to determine the elastic modulus of a material and estimate its hardness through the linear correlation in Figure 4. To measure the hardness of a ceramic material directly, a very small piece/fragment needs to be embedded, grinded and polished before the hardness indentation can be made.
Sample G has a relatively low density of 3.097 g/cm^3 and it also shows the lowest elastic modulus and hardness of all samples. Although not investigated in this study, this sample may have some porosity which is generally considered to be detrimental to the mechanical properties of a ceramic.

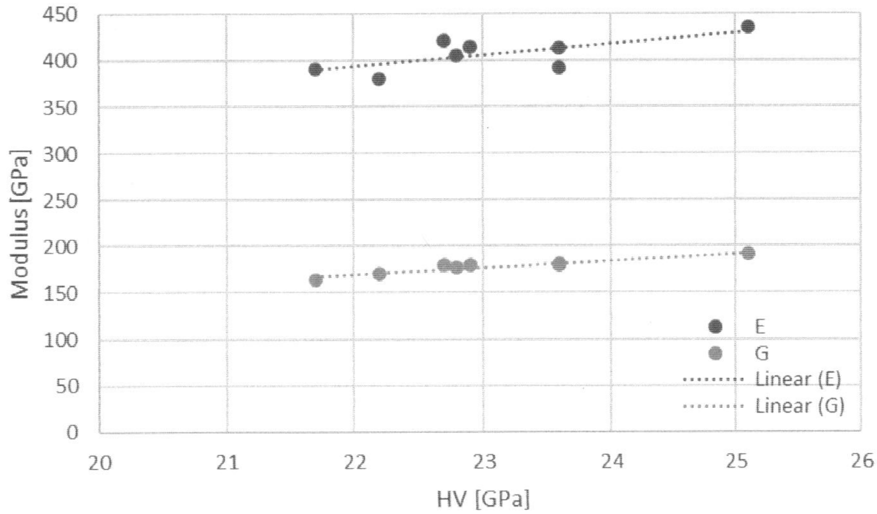

Figure 4. Calculated elastic moduli (E, G) versus the Vickers hardness of the SiC samples

https://doi.org/10.52202/081621-0048

4. TEST RESULTS

Table 2 shows the average of 3-10 (typically 4) repeated shots on the SiC samples. The column '**BE**' shows the (average) ballistic efficiency (BE) of the tiles. There are some differences in tile thickness, hence areal densities (AD) between the samples. When the AD differs too much, this prevents a direct comparison between these materials. To compensate for these differences, the BE-values have been divided by their AD and multiplied by 25 kg/m^2. This results in a linearly AD-corrected BE, which allows the samples to be compared and ranked more effectively. The results of this AD-correction of BE ('BE eff') is shown in the last column of Table 2.

Figure 5. High speed video stills showing bullet just before and after impacting a bare ceramic tile

Table 2. Overview of REM test results for bare B_4C tiles

Sample	AD [kg/m2]	Vo [m/s]	Vr [m/s]	mr [gram]	Ekin,0 [J]	Eres. [J]	ΔE [J]	BE	BE eff
A	25,4	827	591	2,33	1279	406	873	68%	67%
B	25,2	826	613	2,63	1276	494	782	61%	61%
C	24,9	825	616	2,79	1273	529	743	58%	59%
D	25,6	825	597	2,60	1273	463	809	64%	62%
E	28,5	825	578	2,41	1273	403	870	68%	60%
F	22,8	825	658	2,95	1273	639	634	50%	55%
G	23,8	823	623	2,83	1267	549	717	57%	59%
H	25,1	823	579	2,52	1267	422	844	67%	67%

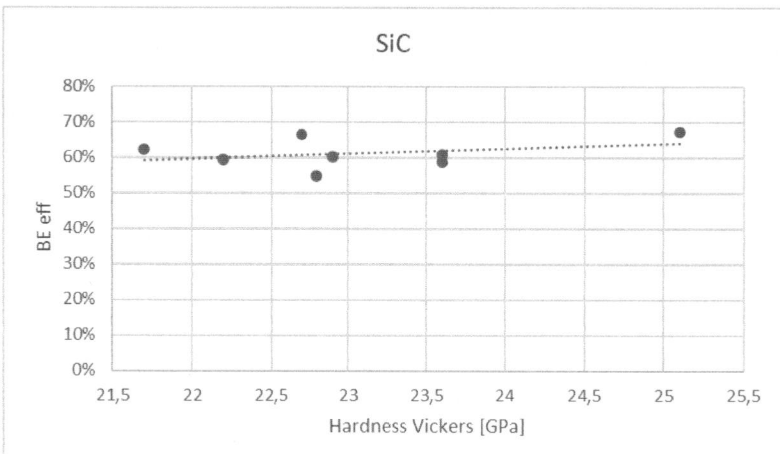

Figure 6. Graph showing the effective ballistic efficiency of the SiC tiles.

5. DISCUSSION

It may feel strange to test ceramic tiles in their bare state. However, the fracturing processes in ceramic (with extreme high sound velocities) are not different when the tile is backed by a material or not. The local bullet impact results in a cone-crack (strike-face to the rear of the tile) under an semi-cone angle that is defined by both intrinsic and extrinsic tile parameters (like strength and thickness). The impulse of the local impact causes the tile to bend bi-axially, which leads to high tensile stress at the rear of the tile (due to its thickness and stiffness). The tensile stress leads to radial cracks and later circumferential cracks, which due to interaction of cracks leads to fragmentation of the tile (and a fragment cloud due to impulse conservation). It is mainly inertia of the ceramic cone that controls the interaction; deceleration and erosion of the AP core as it cannot penetrate the ceramic material. Therefore, it is of high importance to compare ceramic tiles based on an equal areal density. And (second best) to correct for differences in areal density (AD). As was done in this work using AD-compensated effective BE.

The mass of the cone-fragment that is formed needs to be accelerated by the impacting/eroding bullet/core. For a ceramic with a backing material, the same occurs to the ceramic tile (same crack formations), however now the inertia of the cone fragment has increased due to the mass of the backing material (supporting the cone). Thus, the inertia of the ceramic cone fragment has increased a lot, and the bullet will be more eroded and decelerated. The residual kinetic energy of the projectile, cone and backing (part that is in direct contact with the cone fragment) can be absorbed by the strength/toughness of the backing material (plastic deformation, visco-elastic strain, fiber strain, delamination, etc.). This is why a ceramic tile needs a backing material in order to stop a threat. Yet, for the REM we need a residual penetrator, hence the projectile should overmatch the target. This can be done for thin backing materials, and certainly for a bare ceramic tile. The REM test method is energy based and as such can be applied to any target material or armour system. This includes that of otherwise difficult to test armour materials such as ceramic tiles. It can also be used to quantify the effect of damage or aging of complete armour systems. In all cases the REM is used with impact conditions that overmatch the target. The perforation allows the energy of the residual projectile to be determined (together with the status of the residual penetrator; intact, broken, deformed or eroded/ shattered). For a fair comparison between different targets, one needs to make sure that all samples involved have similar areal density (hence density times thickness).

Figure 6 shows the BE eff. values of the SiC samples versus their Vickers hardness. The limited range in hardness of these samples does not allow us to draw conclusions about any correlation or trend. The rather close ballistic efficiencies of these samples show that a pure armour grade ceramic material has much less extrinsic properties (like residual porosity) that can influence the mechanical behaviour. Also, the influence of lateral tile dimensions was excluded, as all tiles had the same shape and size (100 x100 mm squares). Performing ballistics tests using the ceramic tile in bare tile state avoids any influence of the backing material and its adhesive bonding. This results in a rather reproducible test and a good average mass and velocity of the residual projectile can be obtained in 3-5 repeated shots. The SiC samples can now be ranked according to their effective BE values; samples A and H show the highest ballistic efficiency (67%), while sample F has the lowest BE eff value (55%). This order is reflected in the Youngs modulus as samples A and H also have the highest values (420 GPa or above).

6. CONCLUSION

Ballistics tests using 7.62 mm AP bullets have been performed on bare SiC tiles of various suppliers. The density, hardness and sound velocities have been determined from which the elastic moduli have been calculated. The Vickers hardness of the samples show a linear correlation with elastic moduli (E and G). This may be used to predict the hardness of a SiC tile, when only its sound velocity is known or measured. The ballistic efficiency (normalized energy loss of the AP core) has been determined for the SiC samples obtained from various suppliers. After correcting for differences in areal density (tile thickness), the samples can be ranked using the effective ballistic efficiency (eff BE). The range in BE efficiency for the tested SiC samples lies between 55% and 67%. The highest eff BE was obtained for the SiC samples that also have the highest Youngs modulus (>420 GPa).

https://doi.org/10.52202/081621-0048

Acknowledgments

The help in contacting various material suppliers by Rob de Buijzer is acknowledged. Hein Jager is acknowledged for financing this research and his continuous interest in and stimulation of this work.

References

[1] Carton E.P., et al., Defence Technology, 2019; Volume; 15 (6) 829-836

[2] Creusen F., Carton E., PASS proceedings 2023 (Dresden)

[3] M. Burkins et al., 30th International symposium on ballistics, 2017.

Development of a drop-mass test method for assessing the back face signature reduction potential of trauma attenuation backings

G. Pageau[1], T. Bayne[2], N. Shewchenko[2], S. Ouellet[3]

[1]*Promaxis Systems Inc., 2385 St. Laurent Blvd, Ottawa, Ontario K1G 6C3, Canada.*
gilles.pageau@ecn.forces.gc.ca
[2]*Biokinetics and Associates Ltd, 2470 Don Reid Dr, Ottawa, Ontario K1H 1E1, Canada.*
[3]*Defence R&D Canada (DRDC) – Valcartier, Department of National Defence,*
2459 de la Bravoure Rd., Québec City, Québec G3J 1X5, Canada.

Abstract. Larger-calibre sniper rifles have become a threat on the battlefield for hard armour designed to protect against less severe threats. Trauma Attenuating Backings (TABs) have been developed to reduce the Back Face Signature (BFS) generated by higher Kinetic Energies (KE) rifle threats; however, these TABs may not be used with the same armour for which they were designed. It is difficult to compare the relative BFS reduction potential of TAB systems, as there is no dedicated test method or associated performance levels. To partially address this gap, a guided-fall drop-mass test method was developed, in which a heavy steel impactor is dropped onto a TAB sample in the standalone condition, supported on a conditioned clay block to record the BFS. The striking KE and impactor diameter were adjusted to match the transmitted KE encountered during ballistics tests of hard armour inserts. The clay deformations observed from the drop impacts on a new rigid TAB model closely resembled the damage profile obtained from ballistics tests on a reference armour plate with the same TAB placed behind. The resulting BFS were characterized by measuring Back Face Deformation (BFD) and clay cavity volumes during the ballistics tests with the four TAB models. The novel TAB model was also tested with a second reference hard armour plate against a larger-calibre soft-core round, and the BFD obtained was substantially less than the threshold of 44 mm. However, the resulting clay cavity volumes were very large, likely leading to an underestimation of the actual injury severity. The relative performance ranking of the TAB models vs. the no-TAB case was then calculated using cavity volume reduction normalized by the TAB thickness. DRDC's VD^2 injury risk function was used to estimate the probability of AIS-3 injury, taking into account volume and BFD. Although the loading mechanisms involved differ between the drop-mass and ballistics tests, the same overall performance ranking was obtained for the four TAB models tested, with the two rigid models outperforming the flexible and semi-rigid models. The larger striking impulse during drop testing resulted in significantly larger cavities in the clay backing, highlighting one limitation of this method, which was primarily designed for quality control. The need remains for ballistic testing of TABs in conjunction with body armour, for which a reduction of BFS may be necessary for emerging higher-energy threats.

1. BACKGROUND: BFD OVERMATCHING FROM LARGER-CALIBER THREATS

The protection levels of soft and hard body armour are classified based on reference handgun and rifle threats, classified into protection levels. For each level, the armour must successfully withstand the specified projectiles at muzzle velocity (point-blank) while being supported on a clay backing to evaluate resistance to perforation (RTP) and back face deformation (BFD). The clay backing serves as a medium for recording BFD and representing the as-worn condition of the armour. Each protection level encompasses multiple projectiles, with one often chosen for its greater perforation ability and another selected for its greater capacity to deform the armour. RTP is typically assessed at a constant velocity (Vproof) using a series of non-perforating shots, with the BFD pass/fail criterion in clay set at 44 mm, as per the National Institute of Justice (NIJ) standard [1], which corresponds to an acceptable risk of injury [2]. The UK Home Office standard [3] is more stringent, with the BFD limit set to 30 mm. The VPAM German standard [4] also incorporates a performance criterion based on the maximum cavity volume in clay. In Canada, DRDC has investigated the use of an injury predictor that combines BFD and cavity volume (the VD2 model) [5], from which injury risk functions were developed. To address the limited medical foundation and correlation between clay-based BFD measurements and Behind Armour Blunt Trauma (BABT), numerous studies [6-8] have been initiated in recent years to enhance the understanding of the injury mechanisms involved and establish relevant injury metrics and tolerance limits. Improved laboratory methods, including physical models [9] and injury functions (e.g., viscous criterion and peak pressures), are also being considered and may eventually replace clay-based testing for body armour certification once performance thresholds are established. For soft armour, there are many cases of serious to severe BABT documented by the law enforcement community. On the military side, there are fewer cases, but BABT is still a concern, as reported during the conflict in Ukraine [10].

https://doi.org/10.52202/081621-0049

The highest threat levels included in most test standards for hard armour are the 7.62x51mm and 7.62x54mm calibres associated with machine guns and sniper rifles. Larger-calibre threats, such as the 12.7mm and 14.5mm calibres, which are primarily used for anti-material defeat, are included in the VPAM standard. However, the newly introduced calibres (see Table 1), which bridge the gap between 7.62 mm and 12.7 mm, are not yet included in any body armour standard. Since the war in Iraq, there has also been a marked increase in the deployment of sniper rifles [11], with the newer, larger calibre rifles being able to kill unprotected soldiers at distances of up to 2 km. Projectiles for sniper rifles are typically designed with higher masses to lower velocity retardation and make them more incapacitating and lethal than lighter bullets which are designed to be more effective at shorter ranges. Even though armour-piercing (AP) bullets have been developed for larger-calibre sniper rifles, soft-core ball match rounds are more frequently used over the AP variants due to their greater accuracy and lower cost. Commonly used NIJ Level RF2/3 [12] hard armour inserts are capable of stopping these heavier and faster soft-core projectiles even at close range; however, the resulting backface deformation (BFD) will likely exceed the 44 mm BFD threshold, leading to potentially more serious injury. Figure 1 compares the Kinetic Energy (KE) of the projectiles of Table 1 as a function of range, where KE serves as a proxy for estimating the likelihood of BFS. For NIJ RF2/3 hard armour inserts, BFS overmatch could occur at distances up to 300 m when impacted by a 178-grain bullet fired from a .300 Win Mag rifle. For the .375 calibre projectiles, which possess nearly three times the kinetic energy of the 7.62mm M80 bullet at point-blank range, BFD overmatch could occur up to 800 meters.

Table 1. Comparison of larger calibre sniper rifles.

Calibre	Cartridge case capacity (mL)	Bullet weight g (gr)	Nominal diameter (mm)	Muzzle velocity (m/s)	Muzzle kinetic energy (J)	Muzzle kinetic energy density (J/mm^2)
0.223 NATO (5.56x45mm)	1.85	4.0 (62)	5.6 (4.6 core)	930	1730	71 (105 core)
0.308 NATO (7.62x51mm)	3.60	9.7 (150)	7.82	850	3504	73
0.300 Win Mag (7.62x67 mm)	6.10	11.5 (178)	7.82	945	5135	107
0.300 Norma Mag (7.62x63 mm)	6.75	14.9 (230)	7.82	915	6237	130
0.338 Norma Mag (8.59x63.3 mm)	6.95	19.4 (300)	8.6	830	6696	116
0.338 Lapua Mag (8.58x70 mm)	7.53	18.5 (285)	8.6	837	6372	111
0.375 Swiss P (9.5x70 mm)	NA	22.7 (350)	9.55	865	8492	119
0.375 Chey Tac (9.5x77 mm)	10.71	20.7 (320)	9.53	930	8952	126

The Kinetic Energy Density (KED) vs. range curves (Figure 1) follow the same trends as the KE curves but with a tighter grouping between the various calibres, still exhibiting a significant offset relative to the baseline 7.62 mm M80 round. KED is used as a proxy to assess the relative perforation capability of the projectiles. However, the diameter of the soft-core rounds expands significantly upon impact, making them much less penetrative (i.e., lower effective KED) than non-deforming hard-core bullets. At ranges below 800 meters, the potential for the BFD overmatch is likely much more substantial than the potential for the hard armour to be perforated by the heavier soft-core projectiles, since current RF2/3 hard armours are only constructed to comply with the 44 mm BFD limit from the projectiles' KE they are rated for. The resulting cavity volume in clay, which is related to the transmitted energy, typically increases more than the cavity depth when firing larger-calibre soft-core bullets on hard armour, which produce wider cavities. For spherical-cap-shaped cavities, the volume increases with depth raised to the third power and cavity diameter squared. There is then a risk of underestimating the resulting injury severity when using only BFD compared to measuring both cavity volume and depth. For coping with more severe injury risks from larger-calibre sniper rifles, one option would be to use current RF2/3-rated armour inserts with Trauma Attenuating Backings (TAB) engineered to prevent BFD overmatch.

Sondén [13] demonstrated that TABs can significantly reduce the size of lung contusions and peak thoracic pressures during tests conducted on live animals protected with hard armour impacted by 7.62 mm rifle projectiles. Additionally, the mean BFD in clay was decreased from 28 mm to 19 mm.

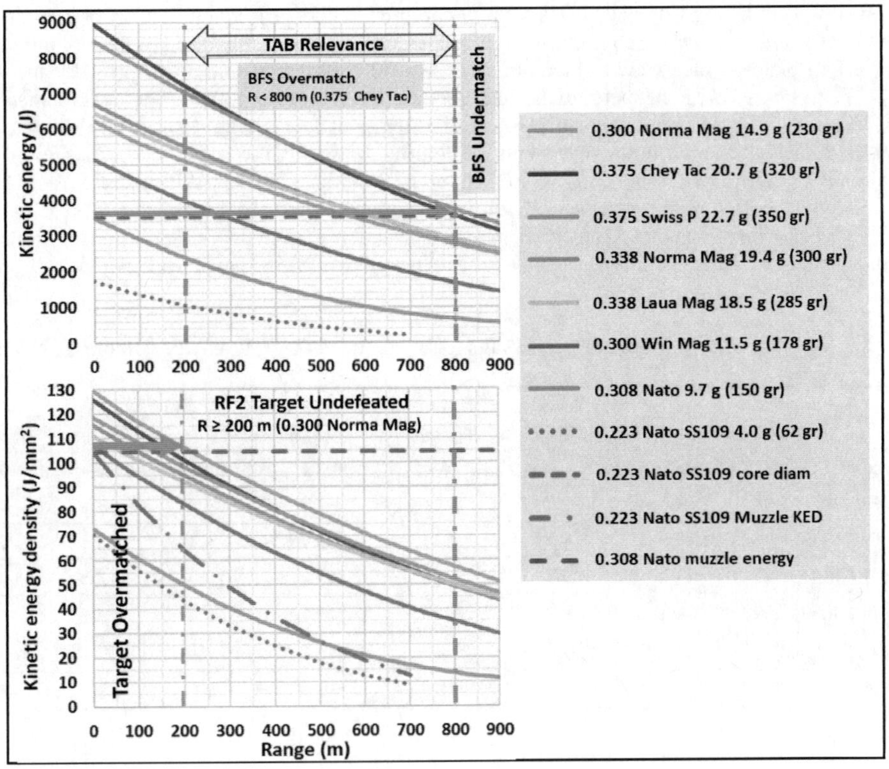

Figure 1: KE (upper graph) and KED (lower graph) of Table 1 projectiles vs range.

2. TAB SYSTEM REVIEW, SELECTION AND BFS MITIGATION ASSESSMENT

A detailed market survey of Commercial Off-the-Shelf (COTS) TAB systems was conducted to identify promising systems in the flexible, semi-rigid and rigid categories. TABs, often called anti-trauma liners, pads, or plates, utilize a combination of materials to absorb, dissipate, and redistribute the energy transferred from the armour. This process effectively reduces the deflection and velocity of the thoracic wall, thereby minimizing the risk of injury. Over twenty COTS models of non-ballistic TABs have been identified as being available online. Attractive marketing strategies are used to promote TAB products, such as claiming they would eliminate thoracic blunt trauma. Innovative materials for TAB, such as auxetic foams with negative Poisson's ratios and non-Newtonian materials (e.g., shear-thickening fluids and viscoelastic compounds), have been explored [14-16] to enhance impact energy absorption. The absence of a standardized BFS mitigation testing method for TABs, coupled with the lack of operationally relevant higher-calibre protection levels in current body armour standards, enables the promotion of TAB products with uncertain and potentially overstated performance claims, which may mislead end-users and explain their limited use. Also, since TABs are typically designed for a specific armour configuration and a designated threat level, there is no guarantee that an adequate level of BFS mitigation will be achieved for other body armour constructions and threat levels, leading to a risk of overestimating the actual protection provided. To address these deficiencies, it was decided to develop a repeatable laboratory test method that simulates the loading from ballistics tests and assesses the TAB relative BFS mitigation performance in the standalone condition, i.e., not in conjunction with a plate.
A test method that does not involve ballistics tests on hard armour plates would facilitate quality controls during the production of TABS and help end-users identify more promising TAB technologies.

 https://doi.org/10.52202/081621-0049

Based on the survey results, the three most promising TAB systems available (models A, B, and C) were selected, and their properties (mass and areal density) are presented in Table 2. The desired performance criteria included a thickness of ≤ 6.5 mm, a weight of ≤ 250 g, full coverage, and curvature matching of the primary reference hard armour plate (RP1). The maximum TAB thickness and the plate curvature requirements for the TAB rigid models are essential to ensure that the plate inserts and the TAB fit well within the existing plate carrier while minimizing bulk. Although the flexibility feature of TAB Model A was appealing for conforming to the plate inner shape, it only covered 80% of the surface area of the RP1 plates, which were of medium size (25 x 30 cm) with a shooter cut profile. Since none of the COTS systems reviewed met the end-user requirements, DRDC initiated the development of a custom TAB model. After many concept and material iterations, the design successfully converged to TAB Model D, which met all the requirements while providing a high level of BFS mitigation for the principal reference hard armour plate insert (RP1) rated RF2+ [12]. The four TAB models were tested in conjunction with the RP1 armour insert (RP1) supported on a calibrated clay block as per ASTM standard E3004 [17]. The performance of TAB model D was also tested in conjunction with a secondary reference hard armour plate (RP2), also rated RF2+, using a single shot of a 0.338 Lapua Mag calibre ball bullet.

Table 2. TAB models, properties, and test results (AD: areal density, CV: cavity volume, n: repeats).

Target	TAB type	Thickness t (mm)	TAB AD (kg/m²)	Mean CV (ml)	Mean BFD (cm)	ΔCV/t (%/mm)	Prob AIS3 (%)
Ref. plate RP1 (n=10)	None	----	----	313.4	5.2	----	99.9
TAB A + RP1 (n=3)	Flexible	9.7	3.1	198.3	4.0	3.8	46.0
TAB B + RP1 (n=3)	Semi-rigid	6.8	2.6	178.6	4.1	6.3	39.8
TAB C + RP1 (n=3)	Rigid	18	5.0	127.6	3.3	3.3	5.5
TAB D + RP1 (n=8)	Rigid	6.5	3.7	146.1	3.6	8.2	11.1
Ref. plate RP2 (n=2)	None	----	----	240.5	3.5	----	37.8
TAB D + RP2 (n=2)	Rigid	6.5	3.7	156.4	3.0	5.4	5.7

Each plate and plate/TAB combination was impacted three times by the 7.62 mm rifle projectile, with a first left corner shot at 42 mm from the bottom left and the left edge, a second shot at 42 mm from the right edge and 75 mm from the plate bottom, and with the third shot near the crown area. The velocity of the 7.62 mm projectile was adjusted to ensure the BFD exceeded the 44 mm limit (i.e., 52 mm without TAB) for generating BFD overmatch conditions, thereby enabling the assessment of the relative BFS mitigation capability of the TAB models. The measurement of the three BFD depths in the clay backing was performed according to the procedures outlined in ASTM Standard E3068 [18], using a precision digital caliper, after completing the three shots on each plate. The cavity volumes were measured only for the crown shot using the water-filling method (Figure 2). Once the cavity was filled with the clay block levelled, the water was extracted with a syringe and transferred to a beaker, which was weighed using a precision scale to determine the water mass. The cavity volume was then computed by assuming a specific gravity of one for water. The test results are provided in Table 2 and Figure 3.

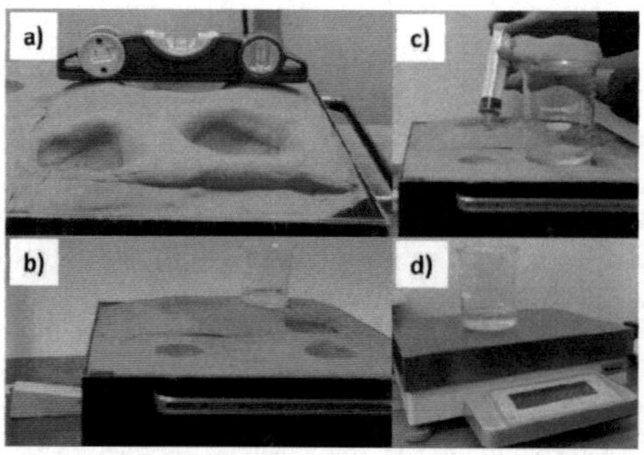

Figure 2: Water method for cavity volume measurement (right).

The BFD values reported in Table 2 are the mean values for the crown shots. The probability of AIS-3 injuries is calculated using the Injury Risk Function (IRF) given by Equation 1. This IRF [5] is referred to as the VD^2 model since it uses cavity volume (V) times BFD squared as the injury predictor. The IRF was obtained from the logistic regression of the Cooper test results [19] on live anaesthetized animals impacted at the anterior mid-sternum with rigid projectiles (ϕ37 mm, masses:140-380 g, velocities: up to 72 m/s) for cases leading to AIS-3 injuries. The risk of 50% AIS-3 injuries is estimated to occur for a ***Vol·BFD²*** value of 3280 cm^5, as shown by the solid purple line in Figure 3, where the red lines correspond to the VPAM criteria (harder and softer clays).

$$\boldsymbol{Prob(AIS3) = 1/(1 + e^{(4.919 - 0.0015 * Vol \cdot BFD^2)})} \qquad \textbf{Eq. (1)}$$

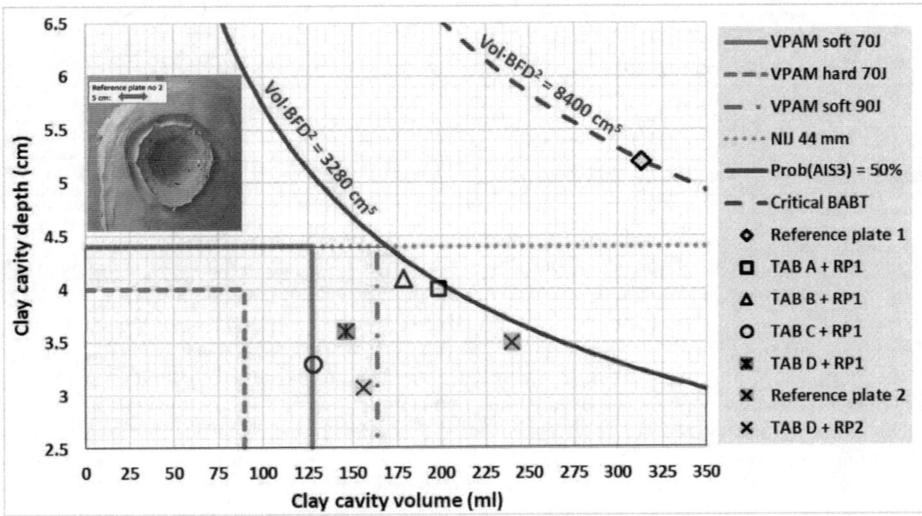

Figure 3: Clay cavity depths and volumes results compared to VPAM and NIJ requirements.

The Table 2 results indicate that all the TABs tested succeeded in significantly reducing BFD and cavity volume. However, TAB models A and B, flexible and semi-rigid types, respectively, were not as efficient for the corner shots. TAB model C was the one that provided the smallest cavity volume and the most severe VPAM requirement, but it exceeded the weight and thickness requirements. When normalizing the reduction in cavity volume over the TAB thickness, TAB model D performed the best with a cavity volume reduction of 8.5% per millimetre of plate thickness. As shown in the picture included in Figure 3, the post-impact clay cavity for the RP2 insert impacted by the 0.338 ball round is enormous, with a volume of 240 ml, despite having a BFD value of 35 mm, which is well below the 44 mm threshold. Such a cavity would likely lead to serious thoracic injuries and would be underestimated by the NIJ criteria. In comparison, the VD2 model predicts a notable risk of AIS-3 injury at 37.8%, which is reduced to 5.7% when wearing TAB model D. This highlights the benefits of being able to scale up the BFS mitigation level of in-service armour when high-intensity threat environments are anticipated.

3. DROP-MASS TEST METHOD DEVELOPMENT

A literature review was conducted on test methods and standards [20-26] for assessing the impact performance of chest protectors, which may also mitigate thoracic injuries and blunt trauma, as with the TAB. The relevant standards are summarized in Table 3. The review included older research work and more recent studies [6-8] where large-mass (100 to 200 grams) rigid impactors are launched with gas guns at low velocities (≈55 m/s) to reproduce the deformation profile of body armour impacted at high velocity (≈850 m/s) by low-mass (≈ 8 grams) small-calibre bullets. Gas-gun testing with rigid impactors is primarily used for performing tests with Post Mortem Human Subjects (PMHS) and Live Animal Subjects (LAS), as it does not require shooting bullets on actual armour, thus providing better repeatability. In a recent study, McMahon [7] observed that rib fractures occurred for impact energies exceeding 60 Joules with a 50% risk for impulses as low as 2.54 Newton-seconds (Ns).

https://doi.org/10.52202/081621-0049

Table 3. Laboratory methods for BFS studies and PPE tests (*maximum energies and impulses).

Study/standard	Launch method	Impactor shape	Mass (kg)	*Impact energy (J)	*Impulse (N·s)
Op't Eynde-Yogonandan: BFS	**Gas gun**	**Spherical cap**	**0.214**	**306**	**11.4**
Prather: BFS soft armour	**Gas gun**	**Hemispherical**	**0.200**	**303**	**11.1**
Arborelius: BFS hard armour	**Gas/powder gun**	**Spherical cap**	**0.058**	**317**	**6.1**
Bowen: Non-Lethal Weapons	**Gas gun**	**Flat**	**0.196**	**350**	**11.7**
Bir-Cooper-Whilhem-NATO AEP-99	**Gas gun**	**Flat**	**0.140**	**112**	**5.6**
CSA Z617-06 Torso PPE	Linear impactor	Hemispherical	1.460	75	14.8
HOSDB 20/07 2007 BT PPE	Guided drop	Bar & wedge	1.60	40	11.3
VPAM KDIW 2004 Stab/impact PPE	Free fall drop	Pyramidal	5.00	100	31.6
NOCSAE ND200-22 Chest protector	Gas gun	Lacrosse ball	0.15	36.4	3.2
EN 356 Security glazing	Free fall drop	Steel ball	4.10	370.0	55.0
EN 1621 Motorcycle armour	Free fall drop	Hemispherical	5.00	50.0	22.4

The gas-gun option was considered, but since it is more difficult to achieve reproducible impact conditions with no yaw and since not as widely available in independent test laboratories, it was decided to explore the drop-mass testing methodology currently used for assessing the performance of thoracic Personal Protective Equipment (PPE) and chest protectors. The approach consisted of adjusting the impactor's shape, mass, and energy levels to replicate loading conditions similar to those observed during the ballistics tests [6-8], but with the TAB tested standalone rather than in conjunction with a hard armour insert. Although numerous physical surrogates have been developed specifically for testing non-penetrating blunt ballistic impacts [27], it was decided to support the TAB samples on back-filled clay backings, as this approach was more straightforward and allowed for direct comparison with the ballistic results obtained. Due to limitations in drop height, the impact velocities achievable through drop testing are significantly lower (≈ 8 m/s) compared to those obtained with gas guns. Consequently, increasing the impactor's mass was necessary to achieve the required impact energy levels (≈300 J) that produce injuries comparable to those observed in gas gun studies using rigid indenters. Matching the impact energy with a higher mass and lower velocity results in more significant impact momentum, as illustrated in Table 3, which usually leads to much larger target deformation. The first drop-mass configuration implemented was a free-fall drop test (i.e. no guide tube or rail) inspired by the EN-356 standard, where a 4.1 kg (100 mm diameter) steel ball generates impact energies of up to 370 Joules. Due to ceiling height limitations, a larger and heavier steel ball (127 mm, 8.37 kg) was used instead, as shown in Figure 4.

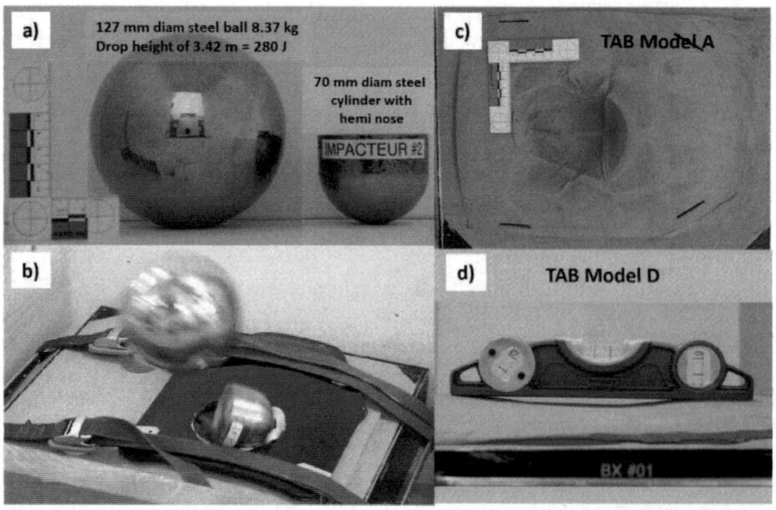

Figure 4: Drop tests with large steel ball and resulting clay cavities for TAB models A and D.

The tests were performed on TAB Model D samples with a drop height of 3.42 m, resulting in an impact energy of 280 joules. TAB Model D was selected as the reference for comparing the post-impact residual TAB damage between the drop and ballistic test conditions. The BFDs in clay obtained with the TAB Model D samples were much smaller than those with the ballistics tests, and they did not sustain any permanent damage or residual deformation. The test configuration was then changed to make a ball-on-anvil drop where the 8.37 kg ball hit a hemi-nose cylinder of 70 mm diameter fixed at the impact point on the TAB sample, as shown in Figure 4b. This configuration succeeded in generating more localized deformations than the direct ball impact method, especially for the flexible and semi-rigid TAB Models (A and B), as illustrated in Figure 4c. However, it was quite challenging to make the large steel ball consistently hit the centre of the fixed anvil cylinder, leading to poor repeatability and unreliable outcomes. To ensure repeatable results, it was decided to adopt a guided free fall configuration (Figure 5) using the drop apparatus specified in the VPAM stab/impact [23] and NIJ stab standards [28]. The guiding assembly offers the benefit of maintaining vertical free-fall conditions with minimal friction, providing effective control over the impact point and angle, as well as excellent velocity reproducibility. Instead of using the large steel ball, the drop impactor was replaced with a hemisphere steel cylinder (30HRC) of 9 kg mass, enclosed within a plastic sabot, as in stab testing, to minimize friction. The impactor was also designed to allow the nose piece to be interchanged with different diameters (i.e., 50.8 mm and 63.5 mm) while keeping the same total mass (i.e., 9 kg). The drop tests were performed at energy levels ranging from 280 to 330 Joules. It was observed that the impactor with a 50.8 mm diameter nose, dropped from a height of 3.8 m, resulting in an impact energy of 330 Joules, produced sufficient localized blunt impact loading for the residual shape of the TAB Model D to closely match the post-impact denting profiles obtained during the ballistics tests as shown in Figure 6. For the 330 J drop impacts, the striking impulses are about 6.4 times larger than those achieved with gas gun tests (≈12 N-s). A much lighter impactor (i.e., 1.4 kg) would be used to match this impulse level with the same drop height, but with generating impact energies 6.4 times lower than those with ballistic impacts using a rigid indenter.

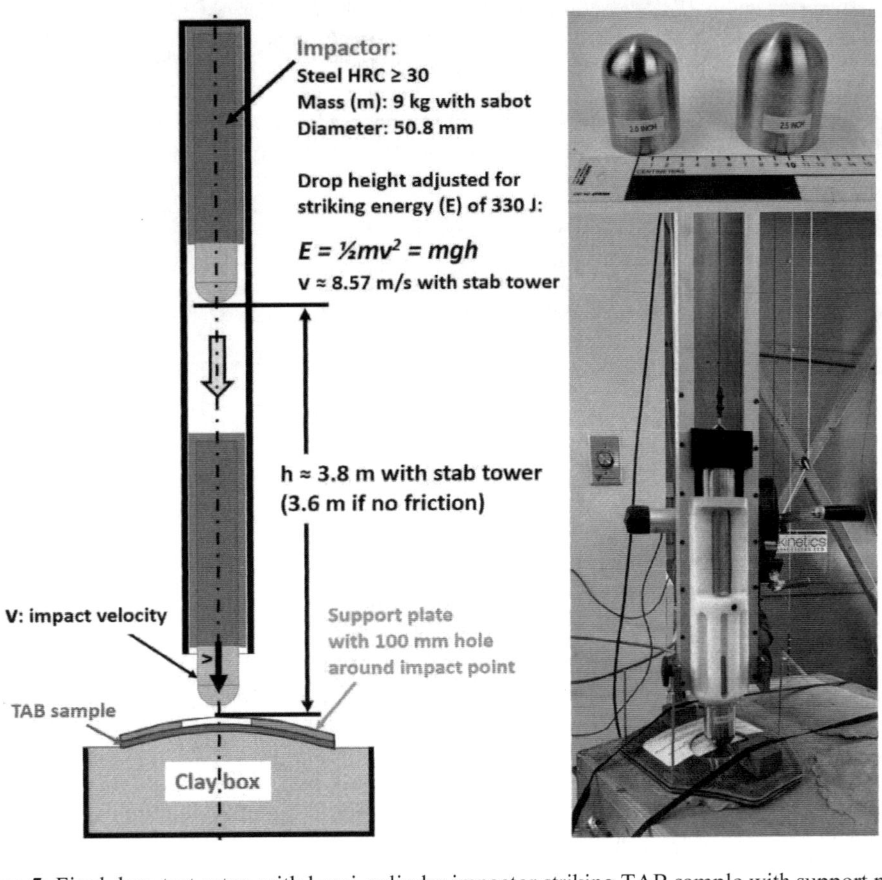

Figure 5: Final drop test setup with hemi-cylinder impactor striking TAB sample with support plate.

https://doi.org/10.52202/081621-0049

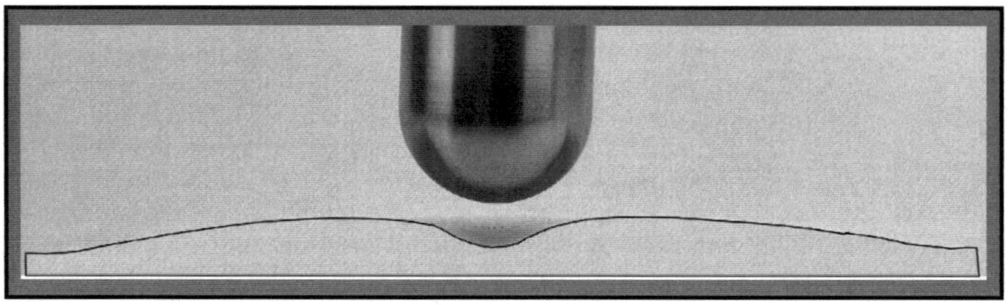

Figure 6: Denting profile of TAB Model D from 330 J impacts with the 50.8 mm diameter impactor.

For drop tests, the energy (E) to impulse (I) ratio (E/I) is independent of the mass, i.e., $E/I = 2.2 \cdot h^{1/2}$, while for gas-gun tests, the ratio is proportional only to impact velocity, i.e., $E/I = v/2$. Therefore, to achieve a match for 330 J impacts, a drop height of 160 meters would be required using the same impactor. Since the drop test method was intended for quality control of rigid TABs, such as Models C and D, replicating the post-test residual profile was used as the deciding factor for adopting a relevant impact energy level rather than matching the impulse level, given the limitations on drop height.

As shown in Figure 5, the TAB samples were supported by calibrated clay blocks shaped to follow the inner curvature of the reference plates used (RP1 and RP2), thereby reproducing the ballistic test conditions. Each TAB sample was impacted only once at the crown location. To more accurately replicate the TAB boundary conditions during ballistics tests, which are restricted by a hard armour plate, a support plate that corresponds to the shape of the hard armour plate was positioned over the TAB samples during the drop tests. A 100 mm diameter hole was incorporated into the support plate at the designated impact point (i.e., the crown location) to permit the impactor to pass through without obstruction. The support plate and the TAB samples were firmly secured to the backing material using two 25 mm wide webbing straps, one positioned near the top of the support plate and the other at the bottom. This configuration effectively prevented the edges and corners of the TAB sample from bending upwards during impact, particularly for the flexible and semi-rigid variants (TAB models A and B), while ensuring more realistic loading conditions. The support plate also incorporated four holes in each corner, allowing for the insertion of spike blades to mitigate any potential slippage of the support plate and TAB sample during impact by the heavy drop mass. After each impact, the TAB sample was removed from the backing material fixture and checked for penetration by the impactor. The resulting BFD in the clay backing was then measured relative to the original surface of the clay backing using a precision digital depth caliper, as illustrated in Figure 7. After each impact, the clay backing was smoothed out and reshaped to the original reference plate (RP1) curvature. The drop tests were repeated only once for COTS TAB Models A, B and C, with 5 repeats for TAB Model D.

4. DROP-MASS TEST RESULTS AND DISCUSSION

For TAB Model D, an average BFD in clay of 33 mm was obtained with the final drop test configuration described in the previous section, with the damage and residual denting profile (Figure 6) of the TAB samples being quite similar for both the drop impacts and ballistics tests since the drop test was designed this way. However, the resulting clay cavity shapes obtained for the four TABS tests were quite different between the drop and ballistics tests. When the TAB models were ballistically tested in conjunction with the hard armour plates, the profiles of the cavities formed in the clay backings followed a spherical cap shape, with the cavity diameters typically twice the cavity depths. In contrast, the resulting cavities were significantly larger for the drop tests on the four TAB models, especially for the flexible and semi-rigid TAB models A and B ($n = 2$), with mean cavity depths exceeding 100 mm, as shown in Figure 7. For the TAB Model D, although the cavity depths were similar to those of the ballistics tests (≈33 mm), the cavity shapes obtained were different, being larger and non-axisymmetric. TAB D is a rigid model, and the increased impact momentum from the drop tests may have caused the observed cross-folding, making the cavity volume measurement with the water method impossible. The maximum clay cavity depth for TAB Model C ($n = 2$) was larger (≈ 50 mm) and more localized than for TAB D but much smaller than those obtained with TAB Models A and B. TAB Model C sustained extensive damage, including a through-hole at the impact point.

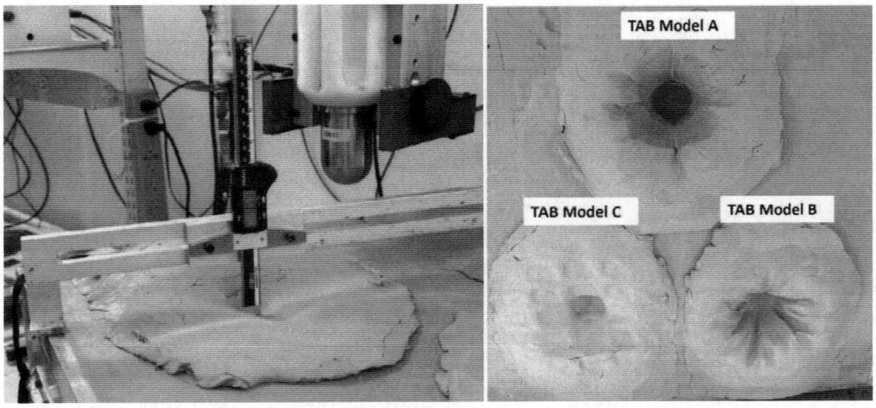

Figure 7: Cavity depth measurement of TAB D (left); clay deformations for TABs A, B and C (right).

Although the post-impact shapes and damages obtained differed, a similar overall qualitative performance ranking was observed between the drop-mass and ballistics tests. The results indicate that the rigid TAB models (C and D) provide increased BFS mitigation compared to the flexible and semi-rigid models tested (A and B) for both the ballistic (in-conjunction) and the drop test cases (standalone). The drop test method was shown to produce repeatable results with the TAB model D samples.

5. SUMMARY AND CONCLUSIONS

Larger-calibre sniper rifles (e.g., 338 Lapua Mag) are becoming more prevalent in military operations capable of producing BFD overmatch conditions at typical engagement ranges for hard armour plates engineered to provide ballistic protection against standard 7.62 mm NATO small arms. BFS mitigation can be achieved by using COTS TABs; however, the lack of standardized testing protocols and performance metrics limits progress in developing TABs optimized for BFS mitigation with reduced weight and bulk. The ballistics tests conducted with the four TAB models underscored the limitations of relying solely on BFD reduction in clay backings as the exclusive metric for BFS mitigation. Applying the VD2 injury risk function, which accounts for both clay cavity volume and depth, enabled more effective discrimination of the relative performance of the tested TABs. The drop-mass test method implemented as a complementary approach to ballistics tests, involving a 9 kg impactor striking with 330 J of energy, succeeded in replicating the localized deformation obtained with the TAB model D, used as a reference. For the three commercial TAB models tested (A, B, C), the damage incurred was more significant. The post-impact clay cavities were also notably larger, especially for the flexible and semi-rigid TAB models (A, B), which may be attributed to the greater impulse experienced during drop impacts and the extended duration of momentum transfer relative to ballistics tests. The challenges of simultaneously achieving both impact energy and impulse while under similar boundary conditions represent limitations of the drop-mass test method. Nonetheless, the proposed method can be used for initial performance assessments and primarily for quality control during production. There will always be a need to conduct ballistic testing with hard armour and the larger-calibre rifle threats expected on the battlefield, highlighting the need to update body armour standards to include larger-calibre sniper threats. One of the industry's challenges in assessing the performance of TAB systems through ballistic testing is the high cost of obtaining multiple hard armour samples, particularly when various models, sizes, and threat projectiles are required. Therefore, the capability to conduct some performance screening without the need for armour would be advantageous. Future research may investigate the application of novel physical torso surrogates recently developed [9] in conjunction with more relevant performance metrics for evaluating the effectiveness of TABs. Rigid impactors launched by gas guns could be explored in future studies; however, this approach also has its limitations, as it does not necessarily replicate the complex BABT loading mechanisms involved.

Acknowledgements

The authors wish to thank the technical staff who conducted the ballistics and drop tests, particularly Mr B. Gauthier and Mr P. Grenier from DRDC, as well as Mr D. Richard from the Munitions Experimental Test Centre (METC).

https://doi.org/10.52202/081621-0049

References

[1] National Institute of Justice NIJ, Ballistic Resistance of Body Armor, Standard 0101.07, U.S. Department of Justice, Washington, DC, 2023.

[2] N. Yoganandan, et al, A Novel Paradigm to Develop Regional Thoracoabdominal Criteria for Behind Armor Blunt Trauma Based on Original Data, Military Medicine, Nov./Dec. 2023.

[3] T. Payne, et al, UK Home Office, Body Armour Standard, CAST report number 012/17, July 2017.

[4] VPAM BSW 2006, Ballistic Protective Vests - Requirements, Classifications and Testing Methods, May 2009, https://www.vpam.eu/wp-content/uploads/2022/12/20090514_II_BSW2006_ENG.pdf

[5] G. Pageau, S. Ouellet and A. Bouamoul, Injury risk functions for behind armour blunt trauma based on clay backing cavity volume and depth, Proc. PASS-2023, Dresden, Germany.

[6a] J. Op 't Eynde, et al, Behind armor blunt trauma indenter simulating high velocity impacts from rifle rounds on hard body armor, Proc. PASS 2020 Conf., Copenhagen, Denmark, 2021.

[6b] J. Op 't Eynde, et al, The fundamental limitations of clay for assessing human response for behind armor blunt trauma, Proc. PASS 2023 Conf., Dresden, Germany, 2023.

[7] J.A. Mc Mahon, et al, Development of Impulse-Based Rib Fracture Injury Criterion for Behind Armor Blunt Trauma, Proceedings of IRCOBI conference, 2023, Cambridge, UK.

[8] N. Yoganandan, et al, Matched-pair hybrid test paradigm for behind armor blunt trauma using an experimental animal model: Trauma Surgery & Acute Care Open, 2024.

[9] N. Shewchenko, et al, The Development of the f-BTTR and its Use for Hard Armour Testing", Proc. Personal Armour Safety Symposium 2021, Copenhagen, Denmark.

[10] T. Talmy, et al, Close-Range Fire Inflicting Behind Armor Blunt Trauma: Case-Series and Implications for Battlefield Care, Military Medicine, Vol. 189, Jan./Feb. 2024.

[11] J.L. Plaster, Sniping in Ukraine by John L. Plaster, July 2022, https://www.americanrifleman.org/content/sniping-in-ukraine/

[12] National Institute of Justice NIJ, Specification for NIJ Ballistic Protection Levels and Associated Test Threats, Standard 0101.07, U.S. Department of Justice, Washington, DC, 2024

[13] A. Sondén, et al, Trauma Attenuating Backing Improves Protection Against Behind Armor Blunt Trauma, Journal of Trauma: Injury, Infection, and Critical Care 67(6), December 2009.

[14] S. Ouellet, et al, Parametric Study of An Anti-Trauma Layer to Reduce BFS, Proc. Personal Armor Systems Symposium 2004, The Hague, The Netherlands.

[15] D. Pacek, et al, Anti-trauma pads based on non-Newtonian materials for flexible bulletproof inserts, Proc. 29th Int. Symp. on Ballistics, Edinburgh, Scotland, UK, May 2016.

[16] R. Critchley, The preparation and characterization of auxetic foams for the application of trauma attenuation backings, PhD thesis, University of Southampton, UK, Feb. 2015.

[17] ASTM E3068-20, Standard Test Method for Contact Measurement of Backface Deformation in Clay Backing During Body Armor Testing, ASTM International, 2020.

[18] ASTM E3004-22, Standard Specification for Preparation and Verification of Clay Blocks Used in Ballistic-resistance Testing of Torso Body Armor, ASTM International, 2022.

[19] G. Cooper, et al, Prediction of chest wall displacement and heart injury from impact characteristics of a non-penetrating projectile, Proc. Int. IRCOBI Conference, Salon de Provence, France, 1981.

[20] NATO STANREC 4744 AEP 99, Thorax Injury risk assessment of non-lethal projectiles, 2021.

[21] CSA Group, standard CSA-Z617-06, Personal protective equipment for blunt trauma, 2011.

[22] C. Malbon, UK Home Office, HOSDB Blunt Trauma Protector Standard for UK Police: Limb and Torso Protectors, Publication No. 20/07, 2007.

[23] VPAM KDIW 2004, Test Standard: Stab and Impact Resistance, May 2011, https://www.vpam.eu/wp-content/uploads/2022/12/20110518_II_KDIW2004_ENG.pdf

[24] NOCSAE ND200-22, Standard test method and performance specification for evaluating the performance characteristics of protectors for commotio cordis, Feb. 2023.

[25] European Standard EN 356, Security glazing testing and classification of resistance against manual attack, 1999.

[26] European Standard EN 1621-3, Motorcyclists' protective clothing against mechanical impact – Part 3: Chest protectors - requirements and test methods, Dec. 2018.

[27] M. Chaufer, et al, Review of non-penetrating ballistic testing techniques for protection assessment, Proc. Inst. of Mech. Engineers, Part H. 2024;238(4):383-402.

[28] National Institute of Justice NIJ, Stab resistance of personal body armor, Standard-0115.00, U.S. Department of Justice, Washington, DC, 2020.

Combined Blast & Fragment Test for Textiles: Effect of the stochastic response of electrical detonators on the time interval between consecutive loads

T. Gerlache[1], G. Kechagiadakis[1] and D. Lecompte[1]

[1] *Royal Military Academy, Avenue de la Renaissance 30, 1000, Brussels, Belgium*

Abstract. Protective textiles for military applications must withstand both fragment and blast impacts. However, current testing protocols solely assess ballistic resistance, overlooking the complex dynamic interactions between blast waves and high-velocity fragments. This results in an inadequate evaluation of protective performance for real-world explosive scenarios. To address this, an experimental setup capable of generating a synchronized planar blast wave and high velocity fragment is used to evaluate the explosion resistance of various textile solutions. A key challenge of the setup lies in precisely controlling the time interval between blast and fragment impacts, as the resistance of the textiles varies depending on the impact phase. At critical time intervals, textiles exhibit reduced impact resistance. The primary source of uncertainty in this timing control stems from variability in the detonator's response time after receiving the initiation signal. To quantify and refine control over this delay, two electrical circuits are designed: one employing a voltage divider to measure the response time and one integrating a custom-made shunt to monitor the current flowing through the detonator prior to detonation. Tests are conducted under various laboratory power supply conditions (20V/2.5A, 30V/2.5A, 60V/2.5A) and utilizing a blast machine to characterize the detonator's response across different operating configurations. By accurately characterizing the detonator's initiation parameters, including timing precision and energy input, a predictive rough estimation of the detonation time delay can be developed in function of the voltage input. The results will guide the development of the next generation of protective textiles.

1. INTRODUCTION

Protective textiles used in military applications have to withstand both fragment and blast impacts. In real-world scenarios, such as encounters with grenades or improvised explosive devices, these materials are subjected to a combination of high-velocity fragments and intense pressure waves. The interaction between these two loading conditions is complex and can significantly alter the textile's protective performance. A key factor influencing this response is the timing between the blast wave and the fragment impact. Depending on several factors such as the phase of the textile's oscillation at the moment of impact, the interaction can either enhance energy absorption or reduce penetration resistance 0. In certain cases, this coupled effect results in greater damage than the sum of both loads applied independently. Despite this known phenomenon and the ability of fabrics to reflect and absorb blast waves depending on material and structural properties such as internal damping, layer sequencing, and density, protective materials are still solely evaluated based on ballistic resistance, neglecting the potential vulnerabilities introduced by combined loading conditions.

As found in 0, the time interval between the two loading events is a critical factor influencing textile performance. The blast wave induces oscillations that alter key parameters such as relative impact velocity and the fabric's stress state, affecting its resistance to penetration. This interaction occurs within a narrow timeframe of approximately a few hundred microseconds. Therefore, to systematically investigate different loading configurations—including blast-first versus fragment-first impacts and varying time delays—microsecond control over the time interval between the two loads impacting the textile is essential.

The primary source of uncertainty in timing control arises from variations in the electrical detonator's response time after receiving the initiation signal. This response is inherently inconsistent, exhibiting fluctuations of more than 125 microseconds compared to the intended detonation time for the used 30V/2.5A laboratory power supply, making precise microsecond-level control in a combined loading setup unfeasible.

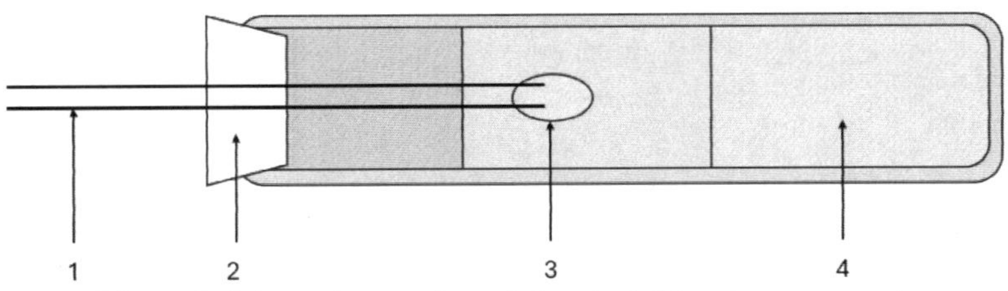

Figure 1. Schematic overview of electrical detonator components

https://doi.org/10.52202/081621-0050

As shown in Figure 1, the electrical detonator consists of four main components. When an electrical voltage is applied across the detonator wires (1), energy is supplied to ignite the primary explosive (3), which in turn initiates the detonation of the secondary explosive (4). A closing tip (2) is used to shield the internal components from external influences.

Variations in detonation time primarily arise from two factors: the electrical power supplied to the detonator and inconsistencies in the masses of the primary and secondary explosives. This paper focuses on investigating the influence of the first factor. While manufacturers specify the minimum energy required for detonation through a recommended input voltage, this information provides little insight into how variations in input voltage influence the speed of the detonation process. A higher input power is expected to shorten detonation time, yet its influence on timing variability has not been stochastically described.

This study aims to characterize the detonation time and variations to changes in input power. This by examining the stochastic behaviour of in-field used electrical detonators under various power supply configurations. In total, 45 detonators are tested experimentally using the same power supply available in the laboratory for the following configurations: 20V/2.5A, 30V/2.5A, and 60V/2.5A. This provides an overview of the influence of lowering and raising the supplied voltage compared to the currently used 30V/2.5A. The current is limited to 2.5A, which is maximal for the used power supply. Furthermore, 13 experiments are conducted using a blast machine, as shown in Figure 2. This device is employed in military applications for initiating electrical detonators and is able to generate high voltages compared to the available laboratory power supplies.

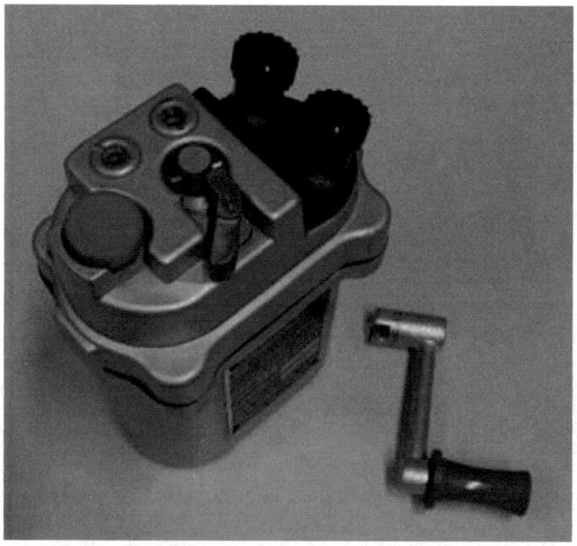

Figure 2. Blast machine used in military applications

The results of the experiments highlight the current level of control and offer insight into how further improvements could enhance synchronization and reliability in combined loading scenarios.

https://doi.org/10.52202/081621-0050

2. METHODS

To characterize detonation time and its variations in function of the power supply, a series of experiments was designed. Each test measures the detonation time delay (Δt), defined as the time interval between the activation signal received by the detonator ($= t_{begin}$) and the detonation of the primary explosive ($= t_{end}$). A custom-built electrical circuit, illustrated in Figure 3, was implemented to capture both moments in time and is used for each experiment for the four power supply configurations.

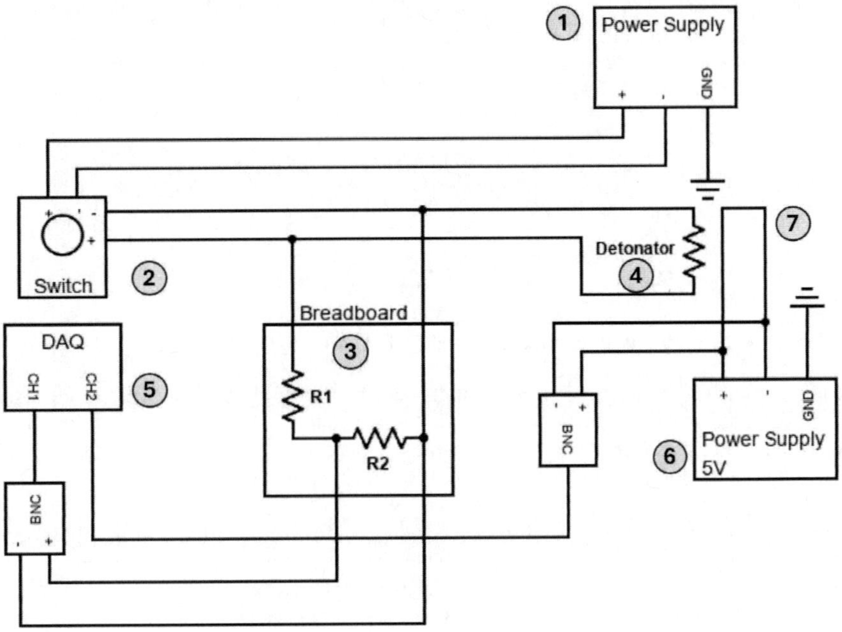

Figure 3. Custom-built electrical circuit for activation- and detonation signal detection

The switch (2) serves as the activation signal by connecting the power supply (1) to the detonator (4), initiating current flow and defining the measurement start point ($= t_{begin}$). A voltage divider, connected in parallel with the detonator via a breadboard (3) is used to capture this moment in time. This by measuring the voltage across the second resistor (R_2), referred to as the 'R_2 voltage'. The voltage divider was made using laboratory-available resistors, selected to keep the measuring voltages below the 10V limit of the data acquisition system (DAQ). Given the known ratio between R_1 and R_2, the actual power supply voltage over time can be deducted using Equation 1.

For all power supply configurations, except for the blast machine, the voltage divider is implemented with resistor values $R_1 = 99.7k\Omega$ and $R_2 = 4.661k\Omega$. This results in an output voltage below 3V for a 60V power supply input. Since the exact voltage output of the blast machine is unknown, the voltage divider is configured to remain below the DAQ safety limit for kilovolt-level inputs with $R_1 = 1.005M\Omega$ and $R_2 = 330.3\Omega$.

$$V_{source} = V_{R_2} \cdot \frac{R_2}{R_1} \quad (1)$$

To detect the explosion of the primary explosion, a separate wire (7) connected to a 5V power supply (6) is taped onto the detonator, with its signal recorded by the DAQ (5) as the 'Taped wire voltage'. Upon detonation, the explosive force severs the wire, causing a change in the measured electrical signal, which is registered by the DAQ. This break in continuity marks the detonation event of the primary explosive ($= t_{end}$), and in combination with the trigger voltage, allows for the measurement of the detonation time delay (Δt).

An example of a detonation time delay measurement for the 20V/2.5A configuration is presented in Figure 4. The start- and end times are indicated with black crosses, with the duration in between corresponding to the detonation time delay. At the beginning of the measurement, a voltage peak is observed, followed by a non-smooth fluctuation before stabilizing. This behaviour is attributed to internal impedance of the power supply and ends with the activation of the current limit, resulting in a peak. Next, the current stabilizes until detonation of the primary explosive occurs, with an intermediate peak reflecting the moment where the current limit is reached.

https://doi.org/10.52202/081621-0050

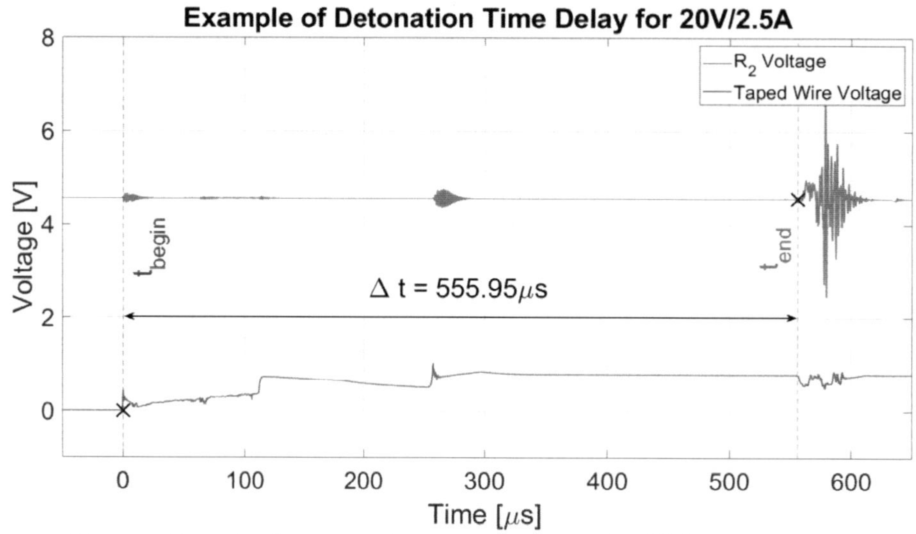

Figure 4. Example of detonation time delay deduction for 20V/2.5A power supply

The energy dissipation with the laboratory power supply is discussed in section 4.2, where a shunt resistor is placed in series with the detonator to measure the current flowing through the detonator. Additionally, interference between the 'R_2 voltage' and the 'Taped wire voltage' is observed, originating from the DAQ itself. The influence of this interference does not interfere with the accuracy of the measurements.

In total, 45 detonators are tested experimentally for each of the three laboratory power supply configurations: 20V/2.5A, 30V/2.5A, and 60V/2.5A. Additionally, 13 experiments are conducted using a blast machine. For each test, the applied power supply voltage is recorded, and the detonator resistance is measured. Additionally, it is noted if a detonator is taken from a different pack. This to have an overview on the influence of the batch on the detonation time. Data acquisition is performed at a 20MHz sample rate, and the detonation time delay is determined from the resulting measurement plots.

3. RESULTS

This section gives an overview of the results, given in Table 1. These outcomes together with the corresponding sample mean $\bar{x}$ and sample standard deviation s are shown in Figure 5.

Table 1. Experimental results for 20V/2.5A, 30V/2.5A, 60V/2.5A, and blast machine power supplies

20V/2.5A — Total of 15 experiments

Voltage [V]	$R_{detonator}$ [Ω]	Δt [μs]	Pack
20.3	2.8	555.95	1
20.4	2.8	513.95	1
20.4	2.8	475.05	1
20.4	2.7	475.40	1
20.3	2.7	456.50	1
20.2	2.7	597.70	2
20.2	2.7	532.00	2
20.2	2.7	551.40	2
20.2	2.8	526.90	2
20.1	2.7	481.70	2
20.2	2.7	446.85	2
20.1	2.7	465.05	2
20.1	2.7	568.70	2
20.2	2.7	521.60	2
20.1	2.7	514.45	2

30V/2.5A — Total of 14 experiments

Voltage [V]	$R_{detonator}$ [Ω]	Δt [μs]	Pack
30.2	2.7	554.30	1
30.4	2.7	364.45	1
30.4	2.7	408.90	1
30.4	2.6	361.40	1
30.4	2.7	352.45	1
30.5	2.9	326.75	1
30.4	2.6	410.00	1
30.4	2.6	401.20	1
30.4	2.6	441.80	1
30.4	2.6	544.75	1
30.3	2.7	447.80	2
30.4	2.7	329.50	2
30.4	2.6	414.05	2
30.4	2.6	382.70	2

60V/2.5A — Total of 16 experiments

Voltage [V]	$R_{detonator}$ [Ω]	Δt [μs]	Pack
60.5	2.7	261.05	1
60.7	2.8	231.60	1
60.7	2.7	275.70	1
60.7	2.6	350.70	1
60.5	2.5	251.70	1
60.6	2.8	272.95	1
59.9	2.8	270.85	2
59.9	2.7	289.00	2
60.0	2.6	324.05	2
60.0	2.6	268.80	2
60.0	2.8	336.15	2
60.0	2.6	384.15	3
60.0	2.6	313.70	3
60.0	2.7	260.00	3
60.0	2.7	301.70	3
60.0	2.7	265.20	3

Blast Machine — Total of 13 experiments

Voltage [kV]	$R_{detonator}$ [Ω]	Δt [μs]	Pack
27.4	2.6	106.50	1
27.2	2.6	98.65	1
25.7	2.6	108.00	1
25.3	2.7	102.30	1
27.1	2.6	94.25	1
25.9	2.5	82.75	1
26.7	2.5	104.90	1
27.8	2.6	92.70	1
28.9	2.6	72.55	1
25.0	2.9	117.45	2
26.1	2.6	119.65	2
25.1	2.7	86.45	2
24.7	2.7	89.35	2

https://doi.org/10.52202/081621-0050

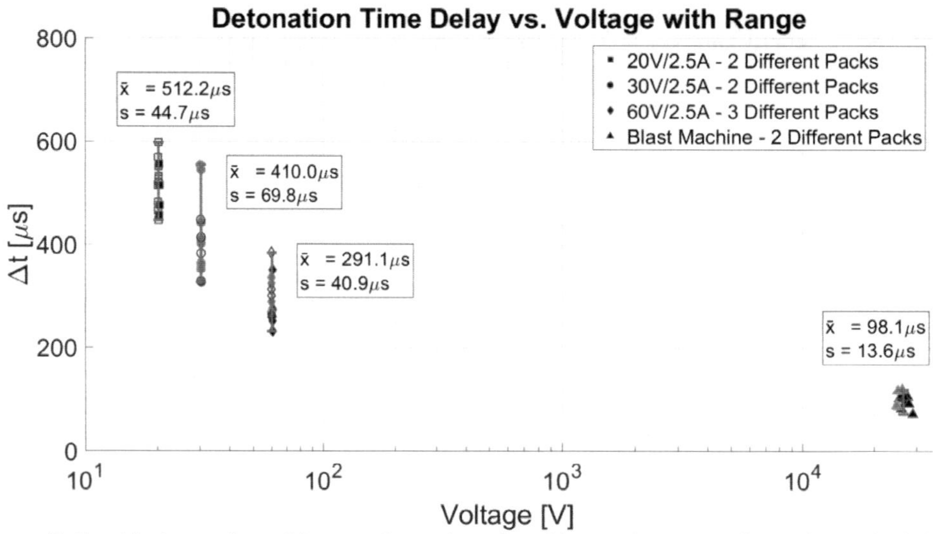

Figure 5. Graphical overview of the experimental results with sample mean and sample standard deviation indicated

4. DISCUSSION

This section discusses the results by describing the statistical distribution of the detonators as a population, after which the energy dissipation using the laboratory power supplies is detailed. Finally, an overview of the blast machine is discussed and a predictive rough estimation graph can be setup.

4.1 Normal distribution

Given the small sample sizes, it is necessary to describe the statistical distribution of the detonator population as a whole. The probability plot shows for each dataset that the population follows a normal distribution. This conclusion is further supported by p-values well above 0.05 for each power supply configuration. Consequently, t-statistics can be applied to estimate the confidence intervals for each configuration.

The confidence interval of the population mean (μ) can be calculated using t-statistics using Equation 2. For a 95% confidence interval, α is set to 0.05. With corresponding t-statistic, $t_{\alpha/2}$ is obtained from standard statistical tables, with n representing the sample size and s the sample variance, which determines the standard error of the mean $\sigma_{\bar{x}}$. These values, along with $\bar{x}$, are illustrated in Figure 5.

$$\mu = \bar{x} \pm \left(t_{\alpha/2}\right) \sigma_{\bar{x}} = \bar{x} \pm \left(t_{\alpha/2}\right) \frac{\sigma}{\sqrt{n}} \approx \bar{x} \pm \left(t_{\alpha/2}\right) \frac{s}{\sqrt{n}} \tag{2}$$

Similarly for the population variance, confidence interval is calculated using Formula, where the χ^2 obtained from standard statistical tables as well using Equation 3.

$$\frac{(n-1)s^2}{\chi^2_{\alpha/2}} \leq \sigma^2 \leq \frac{(n-1)s^2}{\chi^2_{1-\alpha/2}} \tag{3}$$

The results for the four different power supplies are given in Table 2 and the best- and worst-case scenarios for the standard deviation control for a 95% (= 2σ) certainty are listed in Table 3.

Table 2. 95% certainty intervals of the population mean and standard deviation for each power supply

Population 95% Certainty Intervals	μ [μs]	σ [μs]
20V	512.2 ± 24.7	$32.7 \leq \sigma \leq 70.4$
30V	410.0 ± 40.3	$50.6 \leq \sigma \leq 112.5$
60V	291.1 ± 21.8	$30.2 \leq \sigma \leq 63.3$
Blast Machine	98.1 ± 8.2	$9.7 \leq \sigma \leq 22.4$

Table 3. Best- and worst-case 2σ intervals for each power supply

95% certainty (= 2σ)	Best-case [μs]	Worse-case [μs]
20V	65.4	140.8
30V	101.3	225.1
60V	60.4	126.5
Blast Machine	19.4	44.9

4.2 Laboratory Power Supply

A decrease in the confidence interval for the population mean and standard deviation is present between 20V and 60V. The results of 30V do not follow this same trend as two measurements leading to a noticeable deviation, as shown in Figure 5. One of the causes of this discrepancy is thought to be a difference in the amount of primary explosive in the detonator itself, needing more energy to initiate the explosion. These two extremes come from detonators of the same pack. Hence, an argument can be made to investigate the differences in detonation time caused by the discrepancies in amount of explosives present in the detonator. However, too few experiments were conducted to make a conclusion regarding the batch influence. Nonetheless, the overall trend of the population statistics getting more controllable going from 20V/2.5A to 60V/2.5A is still present. While the improvement of the 95% certainty interval with a 40V is present, it remains relatively limited as listed in Table 3. Nonetheless, it shows the influence of the input voltage which can be understood by analysing the energy dissipation of the detonator using the power supply.

The energy distribution is determined from the current flowing through the detonator, measured using a shunt. A custom shunt was created using a simple wire, which was relatively long, introducing inductance into the measurement. This electrical component was placed in series with the detonator and the voltage divider was removed to minimize interference of other components. As a consequence of the increased inductance due to the shunt, the detonation time delay is slightly extended, making it an imperfect representation of the detonation time itself. However, it still provides an overview of the energy dissipation relative to the detonation time and is shown in Figure 6. The power supplied to the detonator is calculated using Equation 4, and integrating it over time yields the dissipated energy.

$$P = I^2 \cdot R_{detonator} \tag{4}$$

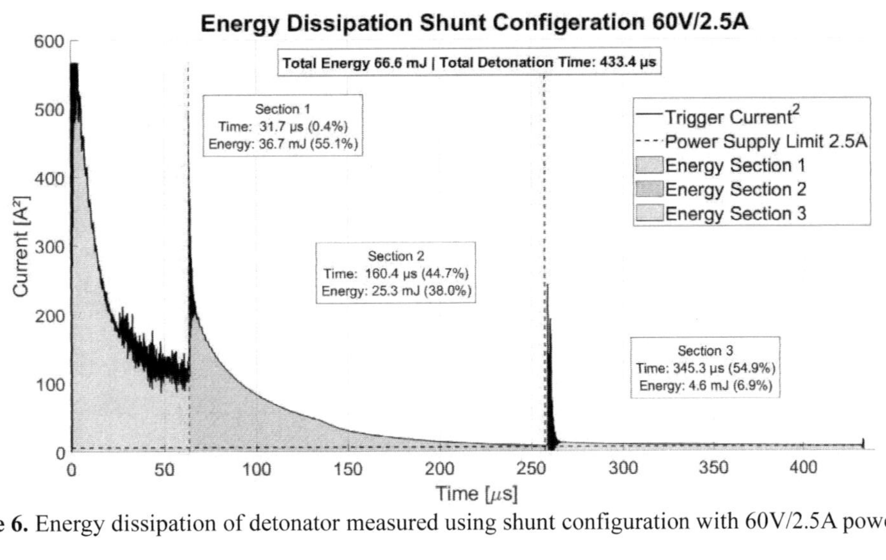

Figure 6. Energy dissipation of detonator measured using shunt configuration with 60V/2.5A power supply

https://doi.org/10.52202/081621-0050

The energy dissipation process can be categorized into three distinct sections. The first phase corresponds to the period before the activation of the power supply's current limit control loop. During this phase, the current reaches the theoretical short-circuit value, leading to a high power output dissipated by the detonator. As shown in Figure 6, over 55% of the total power is delivered to the detonator in less than 0.5% of the total detonation time delay. A gradual decline in power is observed in this phase due to the internal impedance of the power supply. Section 1 concludes with the activation of the current limiting circuit, resulting in a characteristic peak. The transition between the first- and second phases is not consistent, exhibiting variations of up to 100 μs and will differ from power supply to power supply.

The second phase is characterized by a gradual exponential decrease in current, regulated by the current limiting circuit, which reduces the voltage to maintain the current within the specified limit. During this phase, approximately 38% of the total power is dissipated over a proportion of about 44% of the detonation time delay. The end of this phase is marked by the power supply's control loop recognizing that the current has fallen below the maximum threshold of 2.5A. Consequently, the voltage is no longer reduced, leading to the deactivation of the voltage-limiting circuit. This transition may result in a secondary peak, as observed in Figure 6.

The third section is characterized by a relatively small energy output over a big proportion of the detonation time delay. It would be expected that a higher maximum current would result in a higher energy output in the second but mostly in the third section, leading to a lower detonation time delay. To investigate this effect, additional experiments were conducted using 40V/30A and 30V/5A configurations. However, the results remained within the same order of magnitude, showing no significant improvement.

It is concluded that the first phase is the most crucial in energy delivery, as it corresponds to the short-circuit current condition, where a substantial amount of energy is transferred within a very short time interval. This first section is constrained by the capabilities of laboratory-grade power supplies, indicating that the power source required for precise microsecond control of the detonation time delay is external, such as the use of a blast machine.

4.3 Blast Machine

Unlike laboratory power supplies, the blast machine is not restricted by current-limiting mechanisms and is capable of delivering a high energy peak over a short duration. As a result, it is particularly effective in the initial phase of energy delivery, as shown in Figure 6.

Figure 7 depicts a simplified version of the electrical circuit of a blast machine. These devices use a small single-phase alternator (1), which is mechanically driven by a handle-actuated plunger. The electrical output from the alternator is rectified (2) and subsequently stored in one or more capacitors (3). Once sufficient energy has been accumulated, the Zener diode allows current to flow and an LED indicator (4) lights up, signalling that the system is ready to discharge. By pressing the switch (5), the circuit releases the stored energy to the output terminals, where the detonator (6) is connected. Additional resistors are incorporated to safeguard circuit components. In the absence of a connected blasting circuit such as a detonator, the stored energy is dissipated through an internal bypass circuit, which is not included in the schematic representation in Figure 7.

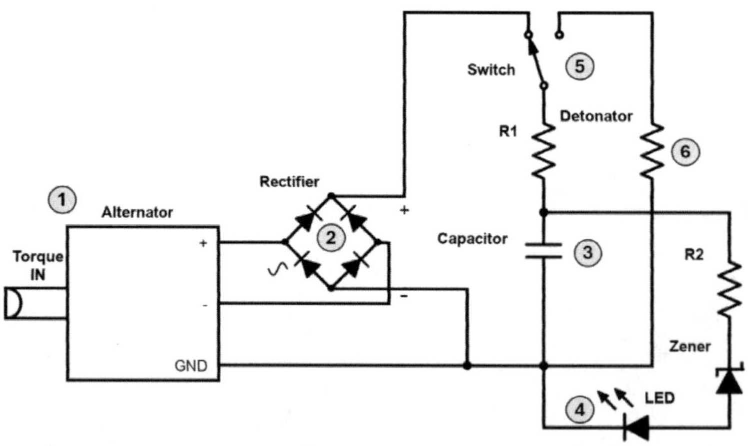

Figure 7. Simplified schematic overview of blast machine electrical circuit

The voltage output of the blast machine reaches a peak of approximately 25 kV. Even after the initial peaks start to settle, the voltage remains at approximately 350V between 20 μs and 40 μs, as illustrated in Figure 8.

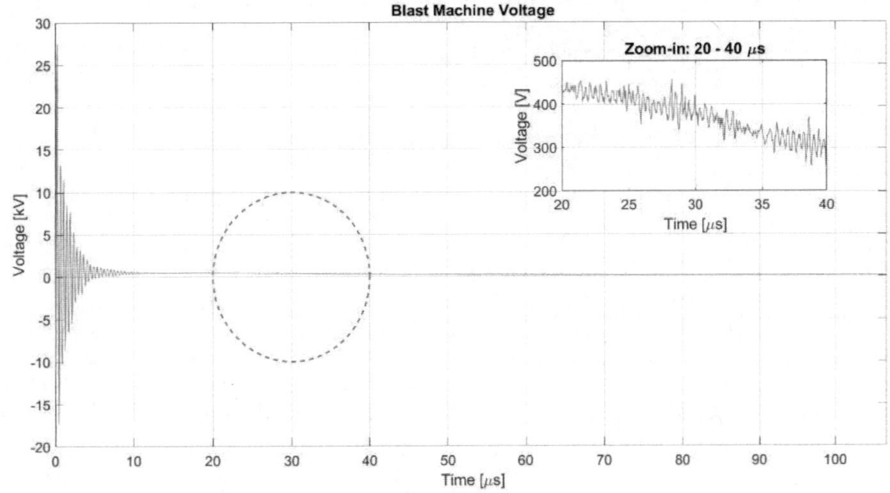

Figure 8. Blast machine voltage with 20-40μs zoom

The capacitor discharge principle significantly increases the energy supplied to the detonator upon the moment of pressing the trigger switch, reducing the mean detonation time delay by a factor of more than five and its standard deviation by a factor of more than three as listed in Table 3. Consequently, it is concluded that a kilovolt-range peak voltage is needed for microsecond level control between consecutive blasts and fragment impacts of which a blast machine is an effective example.

4.4 Predictive Rough Estimation

With increasing voltages, the control in detonation time delay increases. This trend is visible in Figure 5 and allows to fit a logarithmic predictive curve through the obtained datapoints and estimate the change in control with further increasing the supplied voltage. Given the normal distribution of the data, the sample standard deviation can be estimated using Equation 5. This enables the formulation of predictive rough estimates for the expected outcome of increasing the peak voltage, as illustrated in Figure 9.

$$\sigma_{\bar{x}} = \frac{Range}{4} \tag{5}$$

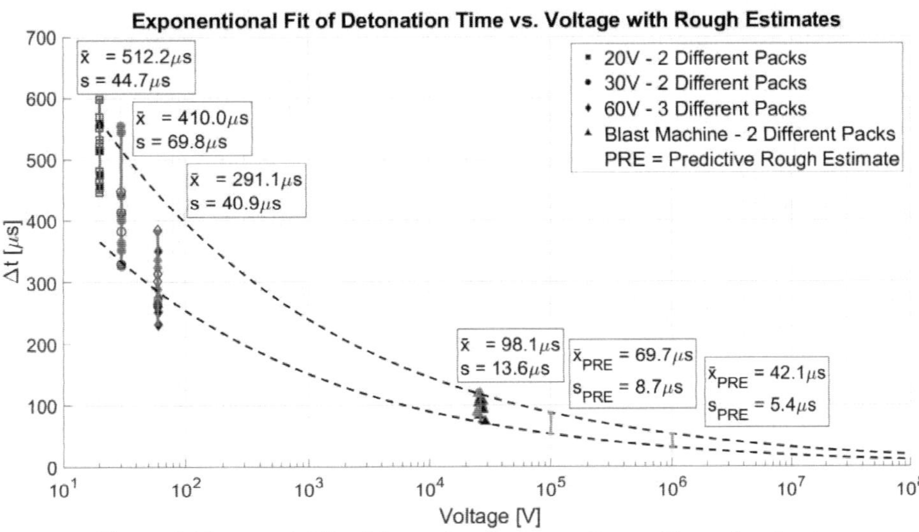

Figure 9. Exponential fit of detonation time vs. voltage with rough estimates

https://doi.org/10.52202/081621-0050

The exponential fit predicts that increasing the input voltage from 25 kV to 1000 kV results in a reduction of both the sample mean and sample standard deviation by more than a factor of two. While this remains a rough estimate, it demonstrates that the level of control over detonation timing can be improved for the tested detonators.

However, practical implementation of capacitor discharge circuits of the range of 1000 kV is challenging. Additionally, the achievable control remains within the microsecond range. Reducing this further to the scale of a few hundred nanoseconds is not considered realistic with the tested in-field used detonators, as these are not designed for such high precision. Therefore, the use of specialized detonators tailored for this level of accuracy should be explored.

5. CONCLUSION

The influence of input voltage on the variability of the detonator response time was experimentally investigated by characterizing the detonation time delay across different configurations using a laboratory power supply. Results showed that increasing the voltage generally improved timing control, with the mean detonation time decreasing from 512.2μs at 20V/2.5A to 291.1μs at 60V/2.5A. However, the reduction in variability was limited, as the worst-case standard deviation (with 95% certainty) only improved from 140.8μs to 126.5μs. These advancements remain constrained by the inherent limitations in laboratory power supplies. To overcome these limits, a capacitor discharge system from a blast machine was tested, achieving peak input voltages of 25 kV. A reduction of the mean detonation time delay to 98.1μs and the worst-case standard deviation to 44.9μs was observed, showing noticeable improvements. This enabled the development of a predictive rough estimation graph of the performance at even higher voltages, showing further increase in controllability with increasing input voltage.

However, achieving microsecond-level precision, which is desired for controlling the synchronized loading of textiles, remains complex. This is due to the difficulties of integrating high-voltage capacitor discharge systems into existing setups and the inherent batch-to-batch variability of in-field-used detonators. Addressing these challenges may require a combination of an optimized capacitor discharge firing system together with dedicated high-precision detonators.

Acknowledgements

I would like to thank **Maxym Desauvage** for his support during the experiments and for his valuable contributions to the discussion of this study.

References

[1] Kechagiadakis, G., Lecompte, D., Van Paepegem, W., Coghe, F., and Pirlot, M.. International Journal of Impact Engineering, 185 (2024), 104830.

Comparative Analysis on the Ballistic Performance of Planar Soft Armours Against a Common Projectile With Different Jacket Alloys Through Binary Regression Analysis

E.Gavelda[1], R. Burris[1], M. Denhart[1], J. Gardner[2], A. Gibson[1], and M. Greene[2]

[1]*Energetic Test Labs, 5055 S. 33rd Street, Phoenix, 85040 AZ, USA, eric@energetictestlabs.com*
[2]*National Institute of Justice, 710 7th Street, Washington DC, 20531, USA*

Abstract. The Remington 9mm Luger 124 grain FMJ ammunition has a longstanding and widespread history of use by the body armour industry and is engrained as a standard ammunition for the assessment of handgun ballistic protection levels for law enforcement. Specifically, in the National Institute of Justice (NIJ) Compliance Testing Program (CTP) this applies to NIJ armour type II and IIIA in NIJ 0101.06 and the future HG1 and HG2 protection levels in NIJ 0101.07 and NIJ 0123.00. To date a copper jacket projectile has been the only type produced by Remington and utilized by the NIJ, but recently Remington has made a production change in the jacket alloy from Copper to Brass for part number 23558. To better understand the change in the jacket alloy versus terminal ballistics, a test plan has been created to evaluate the performance of copper jacketed and brass jacketed variants of the Remington 9mm Luger 124 grain FMJ ammunition against planar-soft armour panels of two commonly used ballistic materials. The approach uses established methods to assess and compare the performance of projectiles at 0° angle of incidence using statistical regression analysis and confidence intervals to evaluate potential changes in the bullets penetration power against soft body armour systems. This paper details a test plan for evaluation of changes of bullet jacket alloys, results of a binary regression analysis of V_{50}, V_{20}, and V_0 testing of the jacket alloys against standard soft armour shoot packs of aramid and UHMWPE, and provides ballistic results compared in graphical form using S-Curves followed by suggested further refinement to the test method and analysis for future evaluation of changes in testing ammunition.

1. REMINGTON 9MM LUGER AMMUNITION

The Remington 9mm Luger 124 grain full metal jacket (FMJ) cartridge has long served as a benchmark round within the body armour industry, widely adopted for assessing handgun ballistic protection levels in law enforcement applications. Its role is particularly established within the National Institute of Justice (NIJ) Compliance Testing Program (CTP), where it is designated for evaluating NIJ armour types II and IIIA under NIJ Standard 0101.06, as well as the forthcoming HG1 and HG2 protection levels under NIJ 0101.07 and NIJ 0123.00. Historically, this ammunition has featured a copper-jacketed projectile exclusively. However, a recent manufacturing change by Remington—specifically for part number 23558—replaces the traditional copper jacket with a brass alloy. To assess the potential impact of this material substitution on ballistic performance, a comparative test plan was developed to evaluate both copper- and brass-jacketed variants against soft body armour panels composed of two widely used materials. Using standardized test methods, the study focuses on 0° angle-of-incidence impacts and employs statistical regression analysis with confidence intervals to determine whether the change in jacket composition affects the projectile's penetration capability.

This test plan provides the procedures and requirements to evaluate the terminal ballistic performance of copper jacketed and brass jacketed variants of the Remington 9mm Luger 124 grain FMJ (Full Metal Jacket) ammunition against planar-soft armour panels of two commonly used ballistic materials. The approach used establishes a method to assess and compare the performance of projectiles at 0° angle of incidence using statistical regression analysis techniques over a broad range of velocities, relative to the threat protection level test requirements in the National Institute of Justice (NIJ) NIJ 0101.07 [1] and NIJ 0123.00 [2].

The Remington 9mm Luger 124 grain FMJ ammunition has a longstanding and widespread history of use by the body armour industry and is engrained as a standard ammunition for the assessment of handgun ballistic protection levels for law enforcement. Specifically, in the NIJ standard this applies to NIJ II and IIIA (in NIJ 0101.06) and the future HG1 and HG2 (in NIJ 0101.07 and NIJ 0123.00). The copper variant of the projectile has been the only type utilized by the NIJ. Figure 1 provides a visual comparison of the copper and brass projectiles. and Table 1 compares the physical properties of the two jacket variants.

https://doi.org/10.52202/081621-0051

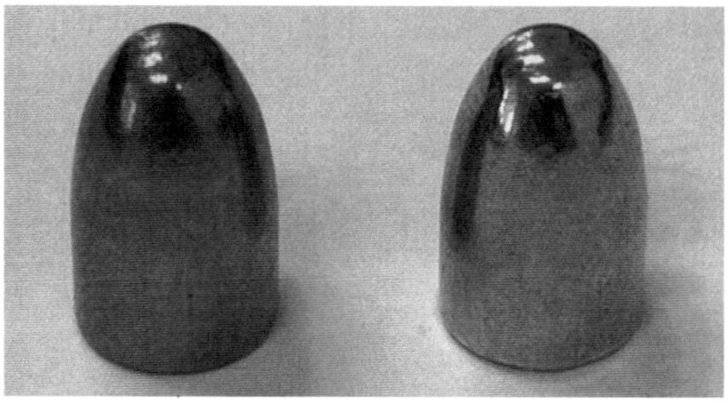

Figure 1. Sample projectiles taken from lots used in this evaluation of the copper (left), and brass (right) variants of the Remington 9mm Luger 124 grain FMJ ammunition

From a sampling of 30 projectiles for each variant, physical characteristics of length, diameter and weight were measured and documented. Table 1 summarizes those measurements. The Brass variant showed a small average decrease in length and corresponding weight. No difference was measured in diameter.

Table 1. Physical measurements and weights from 30 samples of each ammunition projectile jacket type used in this evaluation

Sample	Copper Jacket			Brass Jacket		
	Length (mm)	Diameter (mm)	Weight (grain)	Length (mm)	Diameter (mm)	Weight (grain)
Min.	14.98	8.9	124.96	14.55	8.93	123.52
Max.	15.28	8.98	126.15	14.92	8.99	124.24
Average	15.12	8.96	125.34	14.73	8.96	123.85
Delta	-	-	-	-0.39	0.00	-1.48
St. Dev.	0.085	0.017	0.371	0.086	0.019	0.205

2. TEST PROTOCOL AND ARMOuR SAMPLES

During ballistic limit testing, test articles are repeatedly subjected to projectiles across a range of impact velocities. This range is carefully balanced to ensure the occurrence of both complete and partial penetrations. The primary objective is to determine the ballistic limit, specifically the V_{50} value, which represents the impact velocity at which there is a 50% probability of complete penetration. Several established methodologies exist for conducting such tests and calculating V_{50}. This metric is particularly useful for determining the specific energy absorption of materials, facilitating comparisons of ballistic efficiency across different armour systems. Additionally, V_{05}, defined as the impact velocity at which the probability of complete penetration is only 5%, serves as a more conservative measure for assessing the safety margin of body armour. To estimate V_{50} and V_{05}, a logistic regression method will be used as detailed in work performed by C. Boettger [3], which involves applying linear regression to the logarithm of the odds of complete penetration.

2.1 Regression Analysis Test Protocol

Three subtests shall be performed for each projectile and test panel combination (i.e. "test pair") at 0° angle of incidence and a fourth subset is planned at 30° and 45° angles of incidence. Data from all testing performed at 0° angle of incidence for each test pair will be combined for statistical regression analyses and comparison, utilizing a total of 160 shots per test pair.

- V_{50} Ballistic Limit: This test produces point estimates of the V_{50} ballistic limit for each test panel at 0° angle of incidence using the 3x3 up-down method. Four panels, with 16 shots per panel, will be tested for 64 data points.
- Velocity Sweep Across V_{20}: This test utilizes data from the V_{50} ballistic limit subtest to calculate the V_{20} ballistic limit for each test pair. Testing shall be performed at 0° angle of

incidence from 60 meter per second (m/s) below, to 60 m/s above the V_{20}. Four panels, with 16 shots per panel, will be tested for 64 data points.

- V_0 HG1 and HG2 Ranges: Testing shall be performed at 0° angle of incidence at velocities below, and in accordance with the NIJ HG1 and NIJ HG2 protection levels in NIJ Standard 0123.00. Two panels, with 16 shots per panel, will be tested for 32 data points.
- Angled Incident Shots: Future testing planned to be performed at 30° and 45° angles of incidence at velocities in accordance with the NIJ HG1 and NIJ HG2 protection levels in NIJ Standard 0123.00.

The test panels used in this evaluation have an estimated V_{50} ballistic limit of approximately 488 m/s fps, which is 40 m/s above the NIJ HG2 reference velocity in NIJ Standard 0123.00 for the Remington 9mm Luger 124 grain FMJ. Figure 2 provides representation of the velocity ranges, or estimated velocity ranges, for each subtest in relation to the test type and NIJ soft armour protection levels.

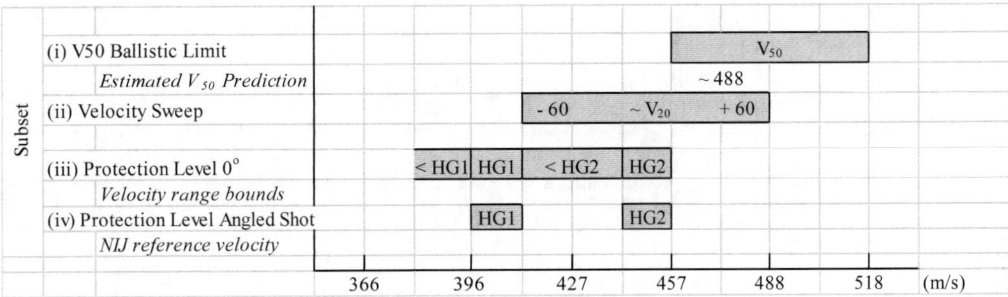

Figure 2. Test protocol and velocity range for each test subset

2.2 Armour Samples

Copper and Brass projectile jacket variants of the Remington 9mm Luger 124 Grain FMJ ammunition shall be evaluated against two soft armour panel types: woven aramid, and unidirectional (UD) cross-ply ultra-high-molecular-weight polyethylene (UHMWPE). All soft armour samples were built in planar construction, 380 mm by 380 mm dimensions with 100 mm corner diagonal stitches and a ballistic nylon cover Test panel construction and the four test pairs, i.e. combinations of projectile type and test panel type, are documented in Table 2.

Table 2. Combination of projectile and armour sample type in this evaluation

Test Panel Type	Test Panel Construction	Projectile Jacket Type	Test Pair
Aramid (A)	34 Plies 600d KM2 Kevlar 34 x 34 Plain Weave 0.48 g/ cm^2	Copper (C)	Aramid vs. Copper (AC)
		Brass (B)	Aramid vs. Brass (AB)
UHMWPE (P)	28 Plies Dyneema SB31 Unidirectional Cross Ply 0.37 g/ cm^2	Copper (C)	UHMWPE vs. Copper (PC)
		Brass (B)	UHMWPE vs. Brass (PB)

2.3 Shot Pattern

Each soft armour panel was shot 16 times evenly placed on the 380 mm by 380 mm panel staggering the shots by at least a diameter off line from prior horizontal and vertical shots. Figure 3 below displays the shot pattern for the test plan.

https://doi.org/10.52202/081621-0051

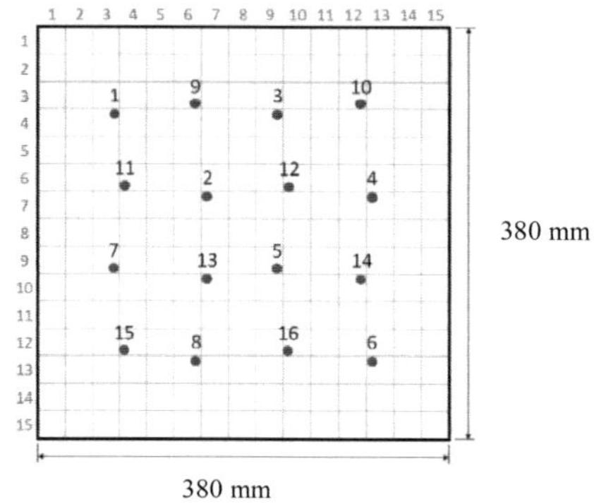

Figure 3. Shot pattern and order

3. BALLISTIC RESULTS

For each test pair the logistic regression analysis was run after the first test subset (V_{50} Ballistic Limit) to estimate the V_{20} to set up the second test subset velocity sweep for each test pair. After the three subset tests were completed, the logistic regression was performed again, and each test panel type was compared against each projectile test type using a 95% confidence level over the curve.

3.1 Aramid

The logistic regression analysis and s-curves, Figure 4, are the results of 160 shots taken on aramid test panels against the copper jacket 9 mm round and brass jacket 9 mm round. From these curves the V_{50} probability of penetration is estimated to be 490.7 m/s for the copper jacket and 490.2 m/s for the brass jacket. The new brass jacket has a decrease in V_{50} of only 0.10% than the original copper jacket.

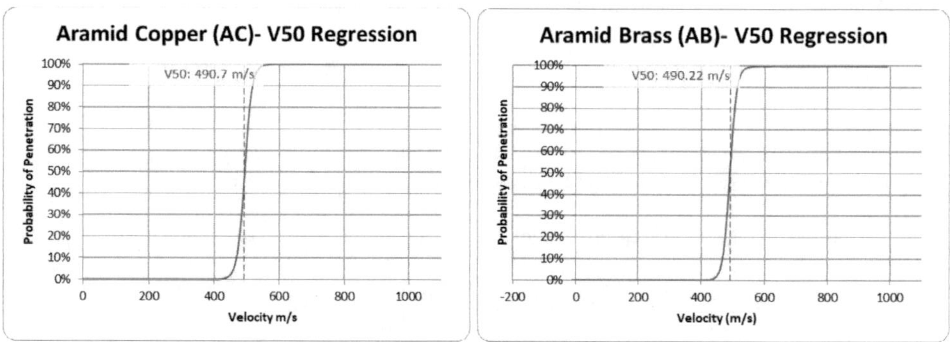

Figure 4. Logistic regression S-Curve for Aramid test panels versus copper and brass jacket 9 mm

Ballistic armour with steeper S-curves is most desirable since the confidence bounds at different V_{xx} will be narrower, increasing confidence in the measured probability. At lower probabilities like V_{05}, an increase in shot sampling is needed to also help decrease these confidence bounds. The three step test protocol covering the V_{50} range, sweeping across the V_{20} range and testing across the NIJ 0101.07 V_0 range was specifically designed to increase shots at these velocity ranges. Figure 5 compares the aramid test panels and includes a 95% confidence bound in response to each type of jacket to help calculate these confidence bounds.

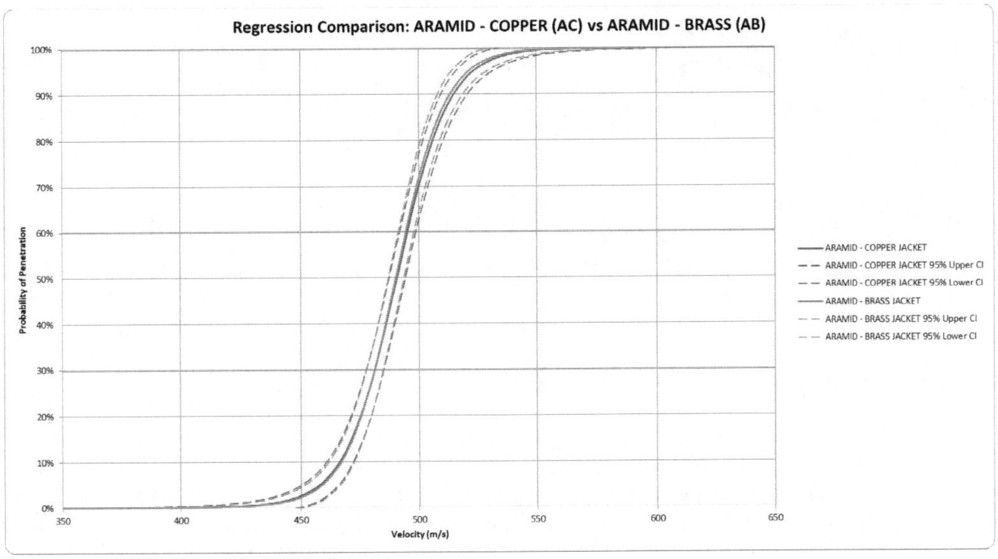

Figure 5. S-curve with 95% confidence interval for aramid versus copper and brass jackets

From the S-curves with 95% confidence intervals it can be seen both curves generated for each jacket fall within the 95% confidence interval of each other at the V_{50} and V_{05}. The range of confidence intervals did increase at the lower V_{05} versus V_{50} to 14.7 m/s from 6.52 m/s as seen in table 3. These results suggest the jacket type does not influence ballistic penetration against standard aramid test panels.

Table 3. Velocity at various probability of penetration percentages for aramid test panels versus copper and brass jackets and 95% confidence bound of V_{50} and V_{05}

	Aramid Copper (AC)	95% Lower Limit	95% Upper Limit	Range	Aramid Brass (AB)	95% Lower Limit	95% Upper Limit	Range	% Increase
V_{60}, m/s	495.2				494.5				-0.14%
V_{50}, m/s	490.7	487.3	494.1	6.8	490.2	487.0	493.5	6.5	-0.10%
V_{40}, m/s	486.2				485.9				-0.06%
V_{20}, m/s	475.3				475.6				0.06%
V_{05}, m/s	458.1	450.3	465.7	15.4	459.2	451.8	466.5	14.7	0.23%

3.2 Ultra High Molecular Weight Polyethylene

The logistic regression analysis and s-curves, Figure 6, are the results of 160 shots taken on UHMWPE test panels against the copper jacket 9 mm round and brass jacket 9 mm round. From these curves the V_{50} probability of penetration is estimated to be 510.5 m/s for the copper jacket and 523.9m/s for the brass jacket. The new brass jacket has an increase in V_{50} of 2.58% than the original copper jacket.

https://doi.org/10.52202/081621-0051

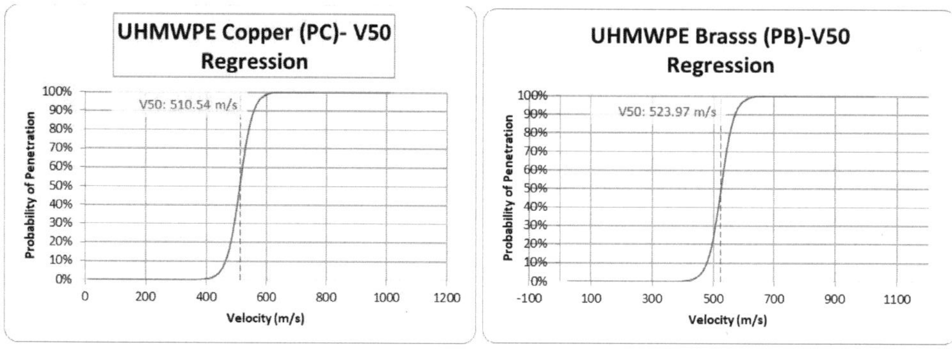

Figure 6. Logistic regression S-Curve for UHMWPE test panels versus copper and brass jacket 9 mm

As mentioned earlier, ballistic armour with steeper S-curves is most desirable since the confidence bounds at different V_{xx} will be narrower, increasing confidence in the measured probability. From figure 7 comparing the UHMWPE test panels and the 95% confidence bound in response to each type of jacket shows a less steep S-curve than the prior aramid test panel results but also a higher V_{50} of 18- 30 m/s for both types of jackets. The higher performing UHMWPE fibers even at 25% less areal density than aramid still had a higher V_{50} than the aramid packs leading to two major variables that could cause an excess in variability: lower areal density and higher V_{50} starting point compared to the NIJ 0101.07 V_0.

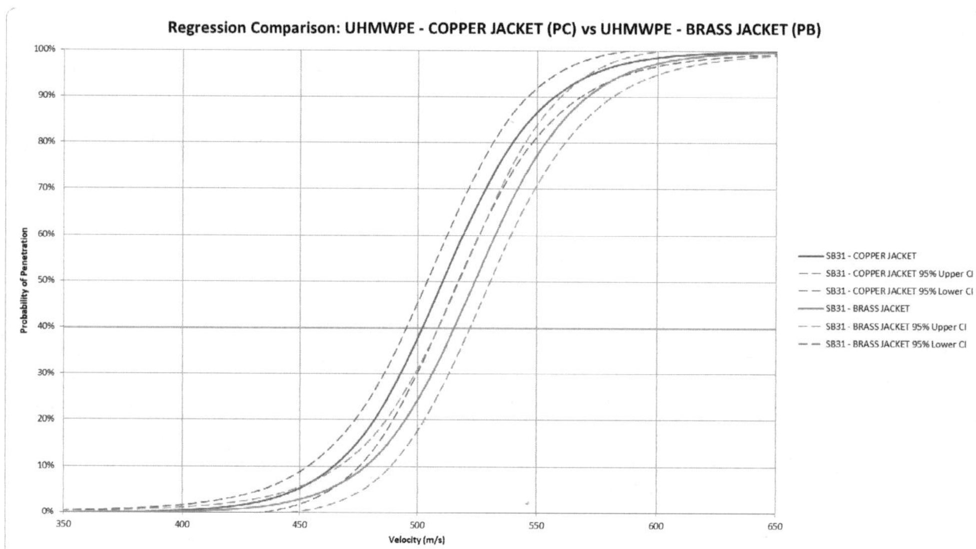

Figure 7. S-curve with 95% confidence interval for UHMWPE versus copper and brass jackets

From the S-curves with 95% confidence intervals it can be seen the curves generated for each jacket do not fall within the 95% confidence interval of each other at the V_{50} and V_{05}. The V_{50} of the brass jacket is 18 fps above the upper limit for the brass jacket and just 1 fps above the upper limit for the V_{05}. The range of confidence intervals did again increase at the lower V_{05} versus V_{50} to 29.4 m/s from 13.0 m/s as seen in table 3, with both ranges being higher than seen in the heavier aramid test pack. These results suggest the jacket type is slightly influencing the ballistic penetration against standard UHMWPE test panels. At the lower areal density of the UHMWPE test packs (0.37 g/cm^2 versus 0.48 g/cm^2 of the aramid) the manufacturer expected the variability of the V_{50} testing to be higher with that factor alone. This variable could be investigated further to better understand if and to what extent the relationship of armor areal density has on the confidence interval spread.

Table 4. Velocity at various probability of penetration percentages for UHMWPE test panels versus copper and brass jackets and brass jackets and 95% confidence bound of V_{50} and V_{05}

	UHMWPE Copper (PC)	95% Lower Limit	95% Upper Limit	Range	UHMWPE Brass (PB)	95% Lower Limit	95% Upper Limit	Range	% Increase
V_{60}, m/s	519.8				532.5				2.45%
V_{50}, m/s	510.8	503.5	518.7	15.2	524.0	517.5	530.5	13.0	2.58%
V_{40}, m/s	501.7				515.4				2.73%
V_{20}, m/s	479.9				494.7				3.10%
V_{05}, m/s	445.1	425.4	461.4	36.0	461.9	447.1	476.6	29.4	3.77%

4. CONCLUSIONS

Logistic regression analysis has been shown as a useful tool to be utilized for evaluation of ballistic penetration performance when a material or design change occurs within a specific ammunition type. With an adequate sample size and range of velocities spreading from the V_0 range through the V_{50} estimate of the armour sample, the S-curve can be analysed at the 95% confidence interval to understand if the ballistic penetrating power of ammunition changes will have a more or less severe effect on standard armour samples.

Future work should include investigating effect on areal density of the armour test pack versus change in 95% confidence bound at V_{50} and V_{05} using the same armour material to better define standard materials and areal densities that are most repeatable for evaluating changes in ammunition materials and design.

Acknowledgments

Energetic Test Labs and the National Institute of Justice would like to acknowledge the support from DuPont Specialty Products USA, LLC and Avient Protective Materials for manufacturing standard soft armour samples for the test evaluation.

References

[1] Ballistic Resistance of Body Armor, NIJ0101.07, NJC 307346, October 2023.

[2] Specification for NIJ Ballistic Protection Levels and Associated Test Threats, NIJ 0123.000, NJC 307347, November 2023.

[3] Boettger. C., Impact of Mechanical Stress on Ballistic Performance of Body Armour Materials, Personal Armour Systems Symposium 2023, Dresden, Germany, September 11- 15th, 2023, pp. 436-443.

https://doi.org/10.52202/081621-0051

Testing vs. Reality: A Comparative Analysis of Body Armour Performance Under Lab Conditions

A. Azevedo[1] and F. Coghe[1]

[1] *Department of Weapon Systems & Ballistics, Royal Military Academy, Brussels, Belgium, ana.ferreira@dymasec.be*

Abstract. Body armour is critical and essential for the protection of military and law enforcement personnel, yet the gap between laboratory testing and real-world performance remains a significant concern. Standardized testing protocols, such as those measuring resistance to ballistic impact, often fail to capture the complexity of real-world scenarios where body armour is subjected to a wider range of environmental and situational variables. These include factors like user movement, varying impact angles, multi-hit scenarios, and the influence of everyday objects under the body armour vests.

This study explores some of the discrepancies between the controlled conditions of body armour testing in a laboratory environment and its actual use in operational environments. During a series of controlled experiments, coins and mobile phones were placed beneath various types of body armour, including polyethylene and polyaramid-based vests. Ballistic impacts were then applied to measure the extent of backface deformation and to assess whether these objects altered the armour's protective capacity, potentially compromising the safety of the wearer.

The findings of this research reveal significant differences in armour performance when such objects are present, highlighting the possible need for updated testing protocols that consider the real-world conditions in which body armour is used. This study aims to contribute to a more accurate understanding of body armour effectiveness, ensuring that protective equipment can address both ballistic and non-ballistic threats posed by carried objects.

1. INTRODUCTION

Body armour serves as a critical component of personal protection for military and law enforcement personnel, designed to mitigate ballistic threats and reduce fatal injuries in combat and high-risk operations. Modern body armour systems primarily consist of soft armour (made from high-performance fibres such as polyaramid and ultra-high-molecular-weight polyethylene) and hard armour (ceramic, steel or composite plates). Standardized testing procedures, such as those defined by the National Institute of Justice (NIJ) [1] and the STANAG 2920 standard [2], provide a controlled framework for evaluating the ballistic resistance of protective vests. These protocols typically measure factors like penetration resistance and backface deformation (BFD) using specific ammunition types and impact conditions.

While these tests ensure a baseline level of protection, they often do not fully replicate the dynamic conditions experienced in operational environments. Real-world use of body armour introduces numerous variables, including environmental exposure, multi-hit scenarios, user movement, and the presence of everyday objects carried under the vest. These factors may significantly alter the actual performance of armour in ways that are not currently captured by standardized laboratory assessments.

Several studies have highlighted that body armour performance is influenced by factors such as temperature, humidity, and mechanical wear over time [3], [4], [5]. However, a less-explored aspect is the effect of foreign objects carried under the armour.

While specific studies on items like pens or keychains are limited, there is evidence suggesting that carrying rigid objects under body armour can alter its performance. Discussions on backface deformation in body armour suggest that the presence of hard objects can change the way ballistic energy is absorbed and dissipated, potentially increasing the risk of injury. One overlooked factor is the influence of items such as mobile phones and coins positioned benhind the armour. These objects may alter the armour's ability to dissipate energy, potentially increasing backface deformation (BFD) and even leading to localized penetration. This study investigates how such objects affect body armour performance under ballistic impact, with a specific focus on soft polyethylene and polyaramid-based vests. Regarding the influence of clothing and/or objects present under or over the body armour, no scientific literature was retrieved, except for a study regarding the possible increased risk of injury for underwire bras [6]. Only indirect reference to the risk of objects carried under body armour have been found in documents related to how body armour should be used, but without any supporting references backing these claims.

This study aims to assess the risks posed by everyday objects worn beneath body armour, evaluating how common foreign objects, such as coins, mobile phones, and identification tags (also known as 'dog tags'), influence the ballistic performance of soft body armour systems, particularly in terms of backface deformation and potential penetration risks.

https://doi.org/10.52202/081621-0052

2. EXPERIMENTAL SETUP

The research described here, considers two different ballistic personal armour systems, respectively aramid- and polyethylene- based vests. Three different objects, namely mobile phones, coins and military ID positioned benhind armour were studied (see setup in Figure 1). The selected objects were chosen based on their common presence in the shirt pockets of personnel wearing body armour. These items differ in composition, thickness, and structural integrity, making them intersting candidates for evaluating their effect on armour performance.

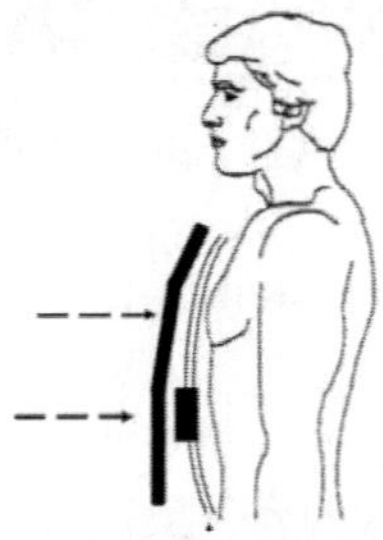

Figure 1. Example of equipment worn under the torso body armour

All the experimental ballistic tests were carried out in the firing range of the Department of Weapon Systems and Ballistics at the Royal Military Academy (Belgium). . A series of tests followed the NIJ0101.04 standard, and in particular the threat level IIIa was considered for this research which among other things evaluates the backface signature of the body armour when impacted by specific handgun rounds (9 mm FMJ RN and .44 Mag JHP). For the witness material, a gelatine block (NATO composition) of 15 x 15 x 45 cm and Roma Plastilina No. 1 modelling clay were used. Table 1 shows the overview of the different ballistic tests that were performed.

Table 1. Ballistic tests performed over the different equipment

	Polyethylene vest	Aramid vest
50 cents coin	9 mm FMJ RN	9 mm FMJ RN
	.44 Mag JHP	.44 Mag JHP
Smartphone	9 mm FMJ RN	9 mm FMJ RN
	.44 Mag JHP	.44 Mag JHP
Military ID tag	.44 Mag JHP	----

For the experimental tests, a universal receiver with interchangeable barrel was used and the targets were placed at 15 m from the muzzle. A double velocity base was used to measure the projectile velocity and a high speed camera to capture correctly the movement of the projectile and/or target inside the gelatine block.

3. RESULTS AND DISCUSSION

3.1 50 cents coin test

3.1.1 Roma plastilina No. 1

The experimental tests for the 50 cents euro coins are presented in Figure 2. Figure 3 presents the maximum backface signature (BFS) measured in Roma Plastilina No.1 witness material for two types of body armour—polyethylene and aramid—when impacted by different projectiles (9 mm FMJ and .44 Magnum JHP), with a 50 euro coin placed beneath the vest. Additionally, baseline tests were conducted

https://doi.org/10.52202/081621-0052

for the polyethylene vest without the coin, providing a direct comparison of how the presence of a rigid object affects armour performance.

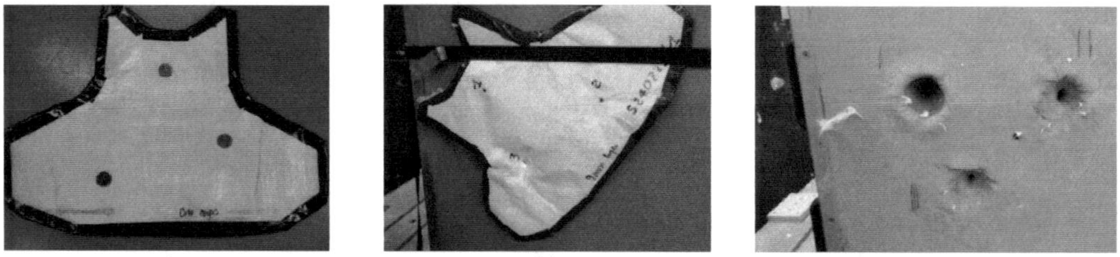

a) b) c)

Figure 2. Experimental tests for the 50 cents euro coin: a) body armour vest with the coins (backface shown); b) body armour vest in place against the witness material; c) BFS results after the test

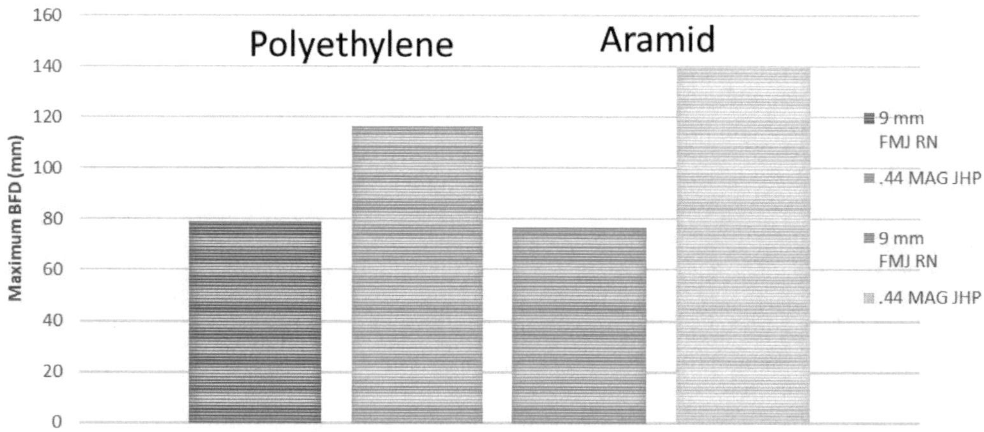

Figure 3. Maximum BFS as a function of the considered threats for two different body armour protection systems for 50 cents coin experimental tests

For polyethylene-based body armour, the baseline BFS values (without a coin) were 25.5 mm and 42.2 mm for 9 mm FMJ and .44 Magnum JHP respectively. Compared to the baseline tests for the 9 mm FMJ and for the .44 Magnum JHP an increase of over 200% and 160% in BFS occurs. These results indicate that the presence of a rigid object (the coin in this case) under the armour possibly significantly amplifies backface deformation, which could lead to a much higher risk of behind armour blunt trauma (BABT) for the wearer, or that the coin is also accelerated at the moment of impact to become a secondary projectile. The coin likely acted as a stress concentrator, altering the way the armour absorbs and disperses the impact energy. Both types of vests materials performed similarly for 9 mm FMJ, with BFS values around 75–80 mm when the coin was present. The aramid vest showed significantly greater BFS (140 mm) for the .44 Magnum JHP, compared to polyethylene (110 mm). This suggests that polyethylene armour distributes the energy more effectively than aramid when a rigid object is present.

The 9 mm FMJ round produced significantly higher BFS values in both vests compared to the baseline, emphasizing that even a low-energy projectile can lead to critical increases in BFS when foreign objects are present. The .44 Magnum JHP caused extreme BFS increases, particularly for the aramid vest, reinforcing concerns about how soft body armour interacts with large-calibre, expanding bullets in real-world conditions, especially during operational use.

In several tests involving the 50 cents coin positioned beneath the body armour, full perforations of the vest were observed when impacted by .44 Magnum JHP projectiles. These perforations occurred in both polyethylene and aramid-based soft armour vests, despite their certification for Level IIIA threats under NIJ 0101.04. The coin itself was not perforated (Figure 4); rather, it acted as a rigid interface that disrupted the normal deformation behaviour of the armour. The projectile, upon impact, was effectively supported by the coin, which appears to have focused and intensified the energy transfer onto a smaller area of the vest, leading to localized failure.

Figure 4. 50 cents euro coin after an impact (full vest perfotation)

3.1.2 Gelatine

In the initial tests, Roma Plastilina No.1 was used as a backing material to measure BFS when a 50 cents coin was placed beneath the polyethylene vest. However, it was observed that the maximum depth of indentation was quite large (see Figure 2c) making it challenging to measure accurately due to the material's non-uniform response and the difficulty in precisely determining the deepest point.

To improve measurement accuracy and obtain a clearer visualization of the deformation, a ballistic gelatine block was used in subsequent tests. Next to this, a skin simulant (based on NATO AEP-94) was added to the gelatine block in order to assess whether additional injury risk would be mostly due to a change in blunt trauma or rather penetrating trauma, while maintaining compliance with recognized ballistic testing methodologies. Figure 5 shows the schematic representation of the experimental tests using the .44 Magnum JHP projectile against the polyethylene vest, with the 50 cents coin, the skin simulant and the gelatine block. Ballistic gelatine is transparent, allowing for direct observation and precise quantification of the amount of deformation.

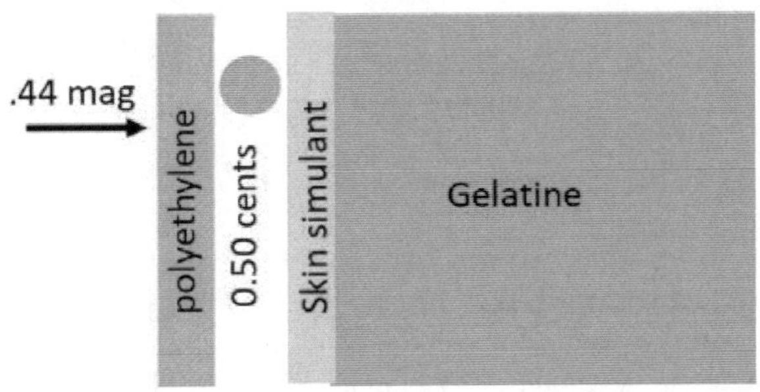

Figure 5. Schematic representation of the experimental tests using the .44 Magnum JHP projectile against the polyethylene vest with, 50 cents coin, the skin simulant and the gelatine block

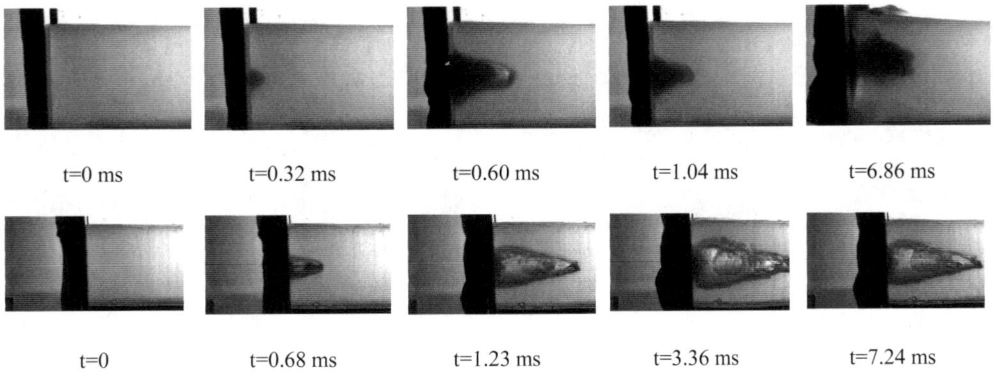

Figure 6. High-speed image sequences for the experimental tests of the .44 Magnum JHP projectile against polyethylene vest with 50 cents coin. Top sequence: no perforation of the vest; bottom sequence: full perforation of the vest

https://doi.org/10.52202/081621-0052

Figure 6 shows the experimental results of the .44 Magnum JHP projectile against the polyethylene vest with the 50 cents coin. No perforation of the body armour vest was observed for the top sequence while a full perforation of the vest occurred for the bottom sequence. The test clearly shows that the high levels of BFS as measured during the previous testing are not due to an actual increase in BFS of the armour systems, but rather due to the acceleration of the coin, turning it into a secondary projectile. Although the AEP-94 skin simulant was not intended to be used in this type of testing, the tests indicate that in reality there would also be a high risk of having full skin perforation of the coin and a penetration into the tissue behind it.
Next to these non-perforating tests (i.e. for the vest) multiple instances of full perforation of the body armour vest were also observed. Despite the high-energy impact and the resulting failure of the armour, the coin itself was not penetrated but rather sustained a significant plastic deformation (Figure 7). This suggests that the coin acted as a "rigid" intermediary, concentrating the impact energy and disrupting the vest's ability to absorb and effectively dissipate the projectile's energy. Significant differences were observed in terms of maximum penetration depth. When no perforation of the vest occurred (Figure 6 top), the coin penetrated up to 74 mm into the gelatine block. However, in cases of full perforation of the vest (Figure 6 bottom), the maximum penetration increased substantially, reaching up to 112 mm. These results highlight the critical role of the armour in mitigating projectile energy and demonstrate how the presence of a rigid object like a coin can contribute to deeper wounding potential, even if the object itself is not fully perforated. The use of transparent gelatine allowed for precise depth measurements and offered a more detailed understanding of the internal damage mechanisms associated with such interactions.

a) b)

Figure 7. a) and b) Example of a 50 cents euro coin and .44 Magnum JHP ammunition after an impact (full vest penetration)

3.2 Smartphone test

In this part, a series of experiments investigated how the presence of a smartphone placed underneath two different types of body armour vests — polyethylene and aramid aramid — affects the maximum BFS when impacted by 9 mm FMJ and .44 Magnum JHP ammunition (see Figure 8). The BFS was measured using Roma Plastilina No.1 as a witness material. Figure 9 shows the maximum BFS as a function of the different calibres for two different body armour protections in the smartphone tests. As the screens readily splintered upon impact, this may lead to additional injuries, but this effect was not evaluated during these tests.

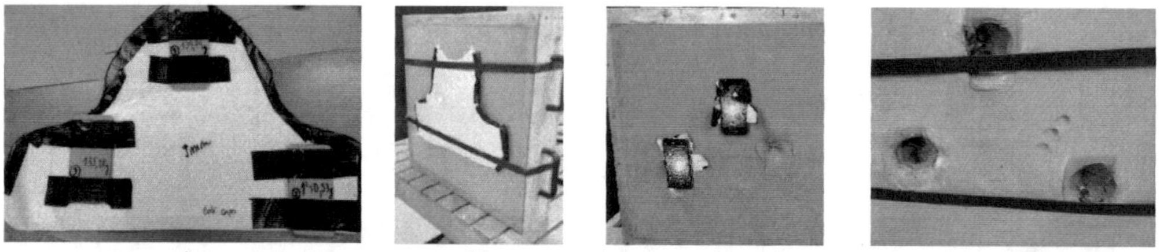

a) b) c) d)

Figure 8. Experimental setup for the smartphones tests: a) body armour vest with the smartphones (backface); b) body armour vest in place against the witness material; c) experimental results after the test with smartphone; d) results after removing the smartphone

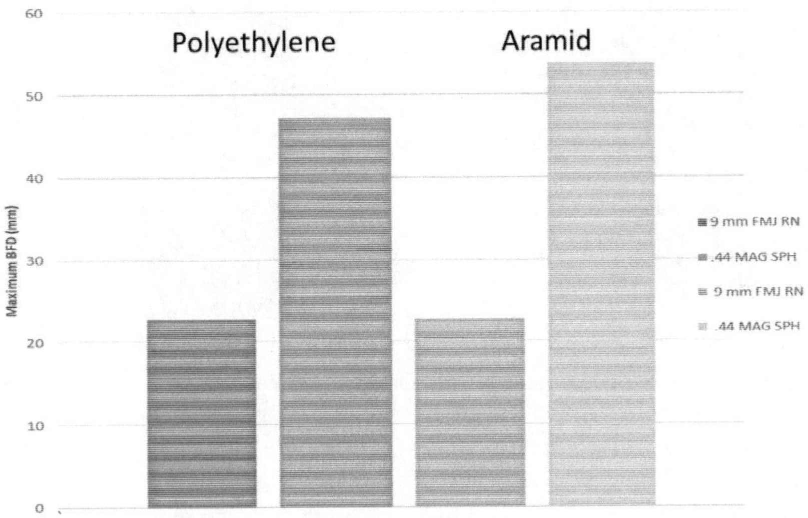

Figure 9. Maximum BFS in function of the different calibres for two different body armour protection in the smartphone experimental tests

The presence of the smartphone did not significantly affect the BFS for both vests under 9mm FMJ (BFS values around 23 mm), likely due to the reduced energy transfer and relatively high stiffness of the phone's internal structure, as the smartphone presents a larger impact area for energy distribution compared to the coin!. The .44 Magnum JHP round resulted in substantially higher BFSs for both armour types, indicating that the higher energy of the .44 Magnum round, combined with the localized resistance and (stiffness of the smartphone, created stress concentration zones, leading to increased deformation. While the polyethylene vest showed better performance for both types of ammunition (especially with the .44 Magnum JHP) the aramid vest demonstrated greater deformation, suggesting that it may be less efficient at redistributing energy around the smartphone. Compared to the baseline test, the presence of a smartphone can reduce the BFS for lower-energy impacts but increase it for high-energy rounds, especially in materials like aramid. An additional and critical observation from the smartphone tests was the occurrence of battery ignition following ballistic impact. In several cases—particularly when the smartphone was impacted by a .44 Magnum JHP projectile—thermal events were observed almost immediately after impact. The lithium-ion battery within the device ignited, producing visible flames and smoke. This reaction is attributed to the rapid mechanical failure of the battery's internal structure, which can lead to a short circuit and thermal runaway when subjected to high-velocity impact. Such ignition events pose a severe secondary risk to the wearer, potentially causing burns or contributing to further tissue damage. Last but not least, in one of the tests involving the polyethylene vest and the .44 Magnum projectile with a smartphone placed beneath, a complete perforation of the vest and the smartphone was observed, as can be seen in Figure 10. This result is particularly concerning, as it again demonstrates that the presence of rigid objects like smartphones can not only increase backface deformation, but in extreme cases, may entirely compromise the structural integrity of the armour. The rigid body likely altered the local stress distribution, creating a focal point for penetration that would not have occurred under standard test conditions.

https://doi.org/10.52202/081621-0052

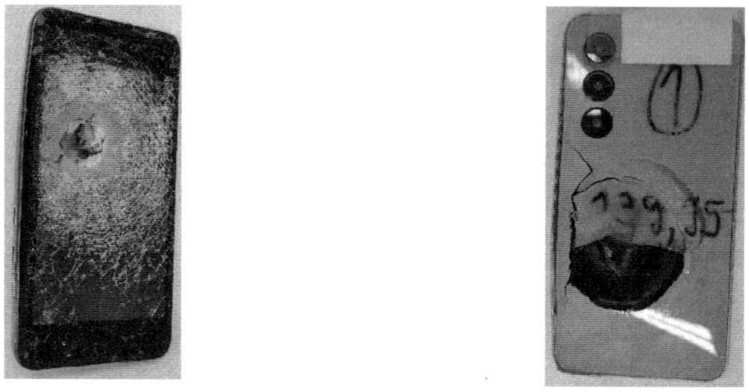

a) b)

Figure 10. Example of a full penetrating smartphone test (polyethylene vest and the .44 Magnum projectile): a) front side; b) back side

3.3 Military ID tag test

To further evaluate the potential effect of rigid objects commonly worn under ballistic vests, a series of four tests were performed using a polyethylene soft armour vest and a metal military ID tag (see Figure 11) as the underlying object against the .44 Magnum JHP projectile. A skin simulant was positioned in front of the ballistic gelatine block to replicate more realistic tissue interaction. The recorded indentation depths into the gelatine varied significantly, ranging from 29.74 mm to 197.0 mm, as shown in Table 2. The deepest penetration (Test 1) suggests a case where the ID tag likely contributed to either localized failure or redirection of the projectile energy, possibly due to fragmentation or secondary projectile behaviour. On the other hand, the shallower indentations (Tests 2 and 3) indicate scenarios where the armour was able to better distribute the impact, even in the presence of the object. Figure 12 shows the high speed video sequence of test number 2.

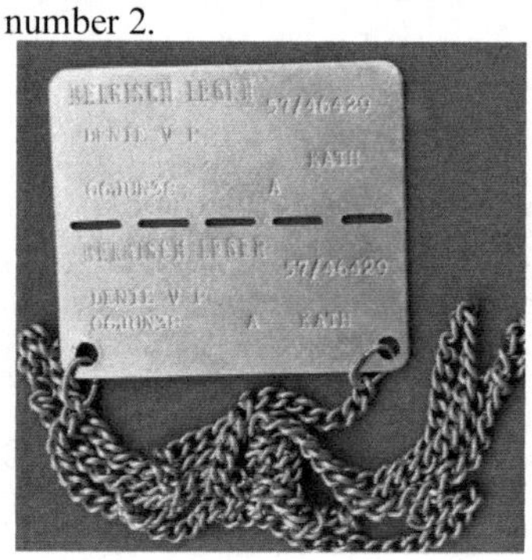

Figure 11. Example of the used military ID tag

Table 2. Indentation result on gelatine after the impact.

Test number	Indentation on gelatine (mm)
1	197,0
2	73,2
3	29,74
4	102,15

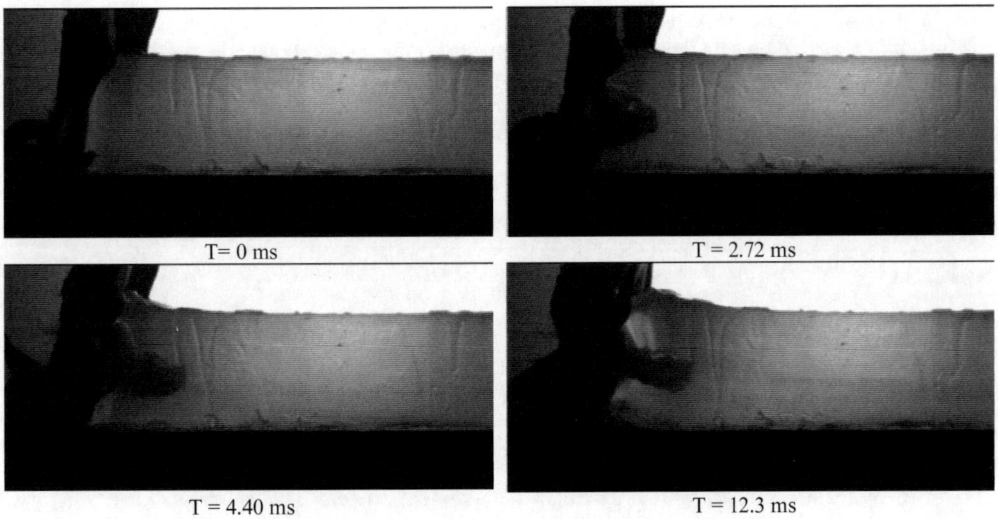

Figure 12. Sequence of high speed video frames for Test number 2

4. CONCLUSIONS

This study investigated the influence of common personal items—specifically a 50 cents euro coin, a smartphone, and a military ID tag—placed beneath soft body armour, using both a modelling clay and ballistic gelatine as witness materials. Three separate experimental setups were used to assess backface signature (BFS), armour performance, and potential wound profiles when these objects interact with ballistic threats.

The presence of a 50 cents coin beneath the body armour significantly altered the vest's performance. In tests with Roma Plastilina No.1, both polyethylene and aramid vests showed increased backface signature, and in multiple cases with .44 Magnum JHP projectiles, full perforation of the vest was observed. Gelatine-based tests further revealed that when no perforation occurred, the coin acted as a secondary projectile and penetrated up to 74 mm into the gelatine. However, in cases of vest perforation, this depth increased to a maximum of 112 mm. The coin itself was deformed but not penetrated, suggesting it played a critical role in concentrating energy and altering the armour's performance.

Smartphones placed beneath body armour also had a significant effect on backface deformation. In comparison with baseline tests, both 9 mm (for aramid) and .44 Magnum (for aramid and polyethylene) impacts showed increased BFS values when a smartphone was present. Notably, for .44 Magnum impacts, one of the polyethylene vest tests resulted in full perforation of the armour. These findings underline the hazardous interaction between electronic devices and ballistic protection, as the "rigid" structure of smartphones—especially components like batteries and metal frames can amplify the localized stress on the vest and increase the likelihood of trauma or penetration. This can potentially lead to higher risks of BABT or even armour failure depending on the energy of the projectile and armour composition.

Finally, in a series of tests, a standard military ID tag was placed behind a polyethylene vest and in front of a ballistic gelatine block, with a skin simulant to replicate more realistic conditions. When struck by a .44 Magnum JHP projectile, the armour was perforated, though the ID tag itself was only deformed. These tests highlight that even relatively thin metal objects can interfere with the ballistic performance of soft armour and may contribute to secondary injuries due to increased penetration depth or fragmentation effects.

Overall, these findings demonstrate that common items worn under body armour can significantly influence its protective performance. They introduce new risks not currently accounted for in standard testing protocols. The study advocates for a reevaluation of certification procedures to incorporate realistic wearing conditions and improve end-user safety. Standardized body armour testing (without objects beneath the vest) does not fully capture real-world threats, as personal items carried by

https://doi.org/10.52202/081621-0052

users can dramatically alter armour performance. New testing methodologies could incorporate scenarios with commonly carried objects, as current testing standards may underestimate the risk of BABT and/or perforation when foreign objects are present. Next to that it is important to stress the risks introduced by foreign objects and advise users not to place hard items between the armour and the body. Dog tags may be the only item that could not be place outside the armour.

References

[1] National Institute of Justice, "NIJ Standard-0101.04 Ballistic Resistance of Body Armor," *Law Enforcement Standards Laboratory of the National Bureau of Standards*, Sep. 2000, Accessed: Sep. 20, 2017. [Online]. Available: http://www.aashield.com/nij-standard-0101-06-ballistic-resistance-of-body-armor/

[2] NATO Standardization Agency, "STANAG 2920 - Ballistic test method for personal armour materials and combat clothing," North Atlantic Treaty Organization (NATO), Jul. 2003.

[3] M. Wilhelm and C. Bir, "Injuries to law enforcement officers: The backface signature injury," *Forensic Science International*, vol. 174, no. 1, pp. 6–11, Jan. 2008, doi: 10.1016/j.forsciint.2007.02.028.

[4] K. A. Rafaels *et al.*, "Injuries of the head from backface deformation of ballistic protective helmets under ballistic impact," *Journal of Forensic Sciences*, vol. 60, no. 1, pp. 219–225, Jan. 2015, doi: 10.1111/1556-4029.12570.

[5] Y. K. Wen, H. Zheng, J. B. Zhang, W. M. Yan, G. Y. Cui, and C. Xu, "Analysis of dynamic back face deformation of a body armor impact by a rifle bullet using 3D-DIC," *J. Phys.: Conf. Ser.*, vol. 1507, no. 3, p. 032051, Mar. 2020, doi: 10.1088/1742-6596/1507/3/032051.

[6] G. Burrell *et al.*, "Do underwire bras affect wounding potential during non- perforating ballistic impacts onto police body armour?," in *Personal Armour Systems Symposium (PASS)*, Cambridge (UK), Sep. 2014.

A framework for a multi-dimensional composite combat helmet scoring system

C. Bradfield[1], D. Midgett[1], L. Voo[1], A. Lurski[1], S. Swetz[1], G. Holt[1], J. Leon[1], M. Tumperi[1], Q. Luong[1], and J. Hopping[2]

[1] *The Johns Hopkins University Applied Physics Laboratory, 11100 Johns Hopkins Road, Laurel, 20723 MD, USA, connor.bradfield@jhuapl.edu*
[2] *Program Executive Office Soldier, 10170 Beach Road, Fort Belvoir, VA, USA, 22060*

Abstract. The current study presents a framework that scores combat helmet blunt impact performance for different exposure environments. Selecting the best helmet for the operation of interest (e.g. mounted, dismounted, or airborne) depends on several competing factors, such as likelihood of impact locations, impact severities, temperatures, and helmet sizes. These competing factors make it difficult to select the best helmet using the current combat helmet blunt impact evaluation standards that rely on a pass/fail outcome for all test conditions and lack head injury probability estimates. This study hypothesizes that combining field exposure data with biomechanically based head injury risk criteria can result in an improved combat helmet evaluation system. To accomplish this, helmet-mounted acceleration data collected from theatre were analysed to estimate frequency of impacts at different helmet locations, severities, and temperatures. These data were used to inform exposure weights for laboratory blunt impact test results. The risk of injury for seven different combat helmet systems were evaluated using the Head Injury Criterion, calculated from monorail drop tower impacts. The laboratory impact data were combined with the theatre-based exposure weights to estimate the number of injuries that would result from 100 head impacts with the helmet system of interest being worn. For exposure conditions corresponding to a mix of mounted and dismounted exposure conditions, the Enhanced Combat Helmet with Team Wendy pads performed the best (1.45 injuries per 100 impacts) while the Advanced Combat Helmet with custom prototype pads performed the worst (3.39 injuries per 100 impacts). While more helmet sensor and biomechanical research is needed to ensure all types of head injuries are represented in helmet evaluation, this study illustrates how a helmet scoring framework can be an improvement over the current blunt impact evaluation standards.

1. INTRODUCTION

Combat helmets provide protection in a range of operational environments and from a range of operational insults, including blunt and ballistic impacts. Test and evaluation of combat helmets is critical to ensuring they are suitable for head protection. Blunt impact performance is evaluated in a variety of conditions, including various temperatures, impact velocities, and impact locations meant to represent a range of relevant operational exposures.

The current United States (U.S.) Army combat helmet blunt impact standards are based on the drop tower test protocol detailed by McEntire and Whitley [1]. This report focused on the accelerative response of a headform outfitted with an Advanced Combat Helmet (ACH) with a foam pad suspension system and a Personnel Armor System for Ground Troops (PASGT) helmet with a sling suspension system. The approach was based on motorcycle helmet test methods from U.S. Federal Motor Vehicles Safety Standard 218 [2]. For this standard, the U.S. Department of Transportation headform with a modified foam chin was outfitted with either an ACH or PASGT helmet and subjected to either a 3.05 m/s (10.00 ft/s) or 4.31 m/s (14.14 ft/s) impact on a monorail drop tower system. These velocities were chosen based on estimated descent velocities of paratroopers during parachute landing fall, with 3.05 m/s simulating the lower value of the estimated range while executing a poor landing manoeuvre. The 3.05 m/s case is often referred to as the baseline impact energy case, while the 4.31 m/s impacts represented twice the kinetic energy of a baseline impact. McEntire and Whitley found that the ACH with foam pads resulted in more favourable head accelerations compared to the older PASGT helmet and sling suspension system [1]. This assessment was based on the average headform acceleration not exceeding 150 g across all impact conditions. The 150 g acceleration threshold was based on results from Slobodnik's accident reconstruction study with U.S. Army Aviator helmets [3].

Currently, combat helmets must meet a 150 g peak acceleration threshold requirement under all blunt impact test conditions. This includes seven different impact locations (front, crown, left nape, right nape, left side, right side, and rear), three different temperatures (cold, ambient, hot) and four different helmet shell sizes (small, medium, large, x-large). Some helmets are only evaluated at 3.05 m/s while others are tested at 3.05 m/s and 4.31 m/s. While there are benefits to the simplicity of the uniform 150

 https://doi.org/10.52202/081621-0053

g pass-fail threshold, there are several drawbacks. First, it can be difficult or impossible to determine if one helmet or suspension system is more protective than another system. One system will often perform better than another system for some test conditions, but worse at other test conditions. With a simple pass-fail criterion, it is difficult to understand if one system is more protective than another if both meet the 150 g criteria for all conditions or if both fail the 150 g criteria for at least one test condition. Secondly, the pass-fail threshold does not accurately reflect the relative risk of injury, which is better represented by a sigmoid shaped curve rather than a step function [4][5]. Additionally, the step function does not incentivize helmet systems to reduce peak accelerations even lower than 150 g and does not penalize helmet systems more so for having even higher peak accelerations. Thirdly, the 150 g requirement uniformly applied to all test conditions does not reflect operational conditions and does not allow for a tailored assessment for different mission profiles. Operationally, there are certain impact conditions (i.e. temperatures, impact locations, impact velocities) that more prevalent that others, and thus a helmet system that provides better protection for the more common impact conditions would likely result in overall lower prevalence of injuries compared to a different helmet system, even if both helmet systems meet the 150 g criteria for all impact conditions. The uniform pass-fail criterion also makes it difficult to determine the most protective helmet system for different end-users or missions that have different exposure profile (e.g. artic paratrooper versus dismounted soldier in the tropics).

From a helmet and suspension system development perspective, the pass/fail requirement incentivizes manufacturers to simply meet the 150 g acceleration threshold. There is no incentive to further improve the energy absorption properties of the pads if they already meet the current pass/fail requirements. This limits the ability of new pad designs to further mitigate head injury risk than what is currently available. Finally, the binary outcome (injury/no injury) in relation to the 150 g acceleration thresholds prohibits graded injury risk, which would be beneficial for comparing different helmet pad responses for a variety of impact conditions.

The issues with the current uniform 150 g pass/fail standards can be addressed by a helmet safety scoring framework. Researchers at Virginia Polytechnic Institute and State University (Virginia Tech) and Wake Forest University developed a scoring system for American football helmets, which is referred to as the Summation of Tests for the Analysis of Risk (STAR) system [6]. STAR consisted of each helmet given a single score, which was based on the number of injuries that would occur in a single season if player wore that helmet. Helmets with lower injuries scores are considered safer than those with higher injury scores. The STAR system also incorporated impact exposure data into its framework, with frequent impact locations and velocities being weighed more heavily than rare impact conditions in a composite helmet score. Injury probability was assessed with a continuous injury risk function based on peak translational head acceleration during impact, which ranged from 0-100% risk of concussion.

An approach similar to the STAR rating system would addresses the shortcomings of the current combat helmet standards by considering an objective performance metric that expands beyond a simple/pass fail outcome. Having a system that assess performance on continuous injury risk and field exposure data would ensure the selected helmet system minimizes injuries for operational conditions of interest. This paper discusses an effort to develop a multidimensional composite helmet scoring system than can be integrated with the current U.S. Army drop tower evaluation method while leveraging prior blunt impact exposure monitoring data from theatre to inform operational exposure frequency. Not discussed in this paper, is the implementation of this multidimensional composite helmet scoring system into a user-friendly software tool.

2. METHODS

2.1 Combat Helmet Scoring Framework

The proposed combat helmet scoring framework relies on two types of information: (1) exposure weighting factors and (2) injury risk data. The exposure weighting factors represents the frequency of impact conditions a Soldier experiences for a given threat operational environment. The injury risk data represents the risk of head injury for a given impact condition. Multiplying the frequency of the exposure condition (out of 100 theoretical impacts) and the associated probability of the injury yields the number of estimated injuries for that specific condition. Summing the number of injuries for all impact conditions produces a single multi-dimensional composite helmet score for each helmet type. Rowson and Duma found it useful to renormalize the amount impacts to a predetermined value person football season, such as 1000 impacts per player per season [6]. The current study assumes the composite score for combat helmet simply refers to 100 impacts for the operational environment considered. Adapting Rowson's and Duma's STAR rating equation [6] for combat helmets is as follows:

https://doi.org/10.52202/081621-0053

$$\text{Injuries Per 100 Impacts} = \sum_{L=1}^{7}\sum_{V=1}^{5}\sum_{T=1}^{3}\sum_{S=1}^{4}\mathbf{W}(L,V,T,S)\cdot\mathbf{R}(L,V,T,S) \quad (1)$$

Where **W**() refers to the exposure weight and **R**() refers to the injury risk. The independent variables L, V, T, S refer to the current U.S. Army drop tower impact conditions for helmet impact location, velocity, temperature, and shell size, respectively. The exposure weight **W** will change depending on the combination of location, velocity, temperature, and shell size. There are a total number of 420 exposure weights for each combination of helmet impact location (7), impact energy (5), temperature (3), and size (4). Similarly, the injury risk **R** will also change based on the combination of these four different variables. Equation 1 demonstrates that the weighting factor is multiplied by the injury risk for each condition, then summed across all four dimensions for a single composite score. The following sections describe the basis for **W** and **R** in detail.

2.2 Blunt Impact Exposure Weighting Analysis

Equation 1's exposure weight **W** refers to the number of estimated impacts for a single combination of helmet impact location, velocity, temperature and shell size, per 100 impacts. Different operational environments will have different weighting factors. For example, a paratrooper may be more likely to impact the front of their helmet during landing compared to the crown. Therefore, all front impact locations would receive higher weightings than crown locations, which places more importance on the helmet/suspension system blunt impact response for front impacts. Determining the exposure weights for each operational environment is challenging. In their approach for American football, Rowson and Duma relied on instrumented helmet data that recorded helmet accelerations for multiple players for several games [6]. This allowed the authors determine the number of impacts each helmet locations. Rowson and Duma also conducted a series of drop tower tests with the instrumented helmet to determine a relationship between helmet acceleration and drop tower velocity, which was incorporated into their STAR framework.

The current study leveraged a similar approach with instrumented combat helmet data collected in theatre from 2013 and 2014. The second generation of the Headborne Energy Analysis and Diagnostic System (HEADSII) is a helmet mounted sensor package that measured triaxial acceleration, overpressure, temperature, and date and time. Thousands of HEADSII systems rigidly mounted to the crown location of the ACH were fielded by U.S. Soldiers, and helmet acceleration data for triggered impact events was recorded and reported in a limited distribution report [7]. In the current effort, recorded acceleration data were analysed to estimate the location of impact on the helmet, and the severity of the impact. The original dataset contained 31,103 triggered events, but the current analysis evaluated a subset of these events due to concerns about false triggered recordings and potential bias to recording impacts on the helmet's crown location. This resulted in 4,371 triggered events being selected for the exposure weighting analysis.

Impact location on the ACH helmet shell was determined by finding the x, y, and z components of the peak acceleration, and constructing an impact vector that was the oppositive direction of the measured acceleration vector. The intersection of the impact vector with the ACH helmet shell geometry determined the impact location on the helmet, and triggered events where the impact vector did not intersect the helmet shell were excluded from the analysis. Impact locations coincided with the U.S. Army's drop tower impact locations over the shell's front, crown, left nape, right nape, left side, right side, and rear, as shown in Figure 1.

The HEADSII dataset was also analysed to estimate impact velocity. The ideal approach to determining the relationship between impact velocity and helmet acceleration would be to conduct a series of drop tower experiments measuring both quantities. While prior studies have measured headform accelerations during drop tower impacts, those data cannot be directly used for this study since head and helmet accelerations differ during impact. Performing such tests was not possible for the current study since the HEADSII sensors are no longer manufactured. The current study instead relied on a scaling approach that (1) assumed helmet and head acceleration followed the same relative trend with impact velocity, and (2) helmet acceleration will always be higher than headform acceleration for a given impact velocity. First, prior U.S. Army drop tower test data with an ACH were analysed to determine a relationship between peak headform acceleration and impact velocity, ranging from 2.16 m/s to 6.10 m/s, reflecting an impact energy change of 1/2x to 4x the baseline impact energy. This was done to ensure a range of impact severities from theatre were accounted for. Headform accelerations bin limits were established for each impact severity (1/2x, 1x, 2x, 3x, and 4x of the baseline impact energy). Next, the

https://doi.org/10.52202/081621-0053

HEADSII peak acceleration data were analysed and the majority of the triggered events from theatre were assumed to occur at a 1x impact energy on the drop tower (i.e. the majority of the triggered events were assumed to fall within the scaled headform acceleration bin at 1x impact energy). The bin limits for the HEADSII data were scaled upwards from the headform acceleration bin limits, since the helmet was assumed to experience higher accelerations than the padded headform for a given impact. The remaining HEADSII events were scaled to follow the previously defined headform acceleration-impact velocity relationship at 1/2x, 2x, 3x, and 4x the baseline impact energy bins. This resulted in each of the 4,371 recorded impacts HEADSII events coinciding with one of the potential five impact energies tested with the standard drop tower.

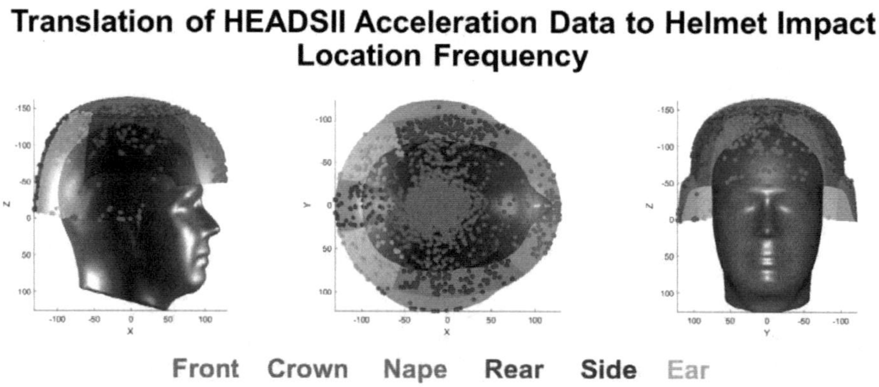

Figure 1. Examples of estimated impact location on the helmet shell taken from fielded helmet accelerometers.

Equation 1 specifies that exposure weights for different temperatures. This is important since the mechanical response of the suspension system changes with temperature, ultimately influencing probability of head injury if the impact is severe. Temperature data recorded by the HEADSII sensors for each of the 4,371 recorded impacts were recorded, and binned accorded to the current U.S. Army specifications (cold: 14±5 °F, ambient: 68±5 °F, hot 130±5 °F). If a recorded temperature was between either of these three ranges, it was considered to be classified at the closest temperature range. As a result, each recorded HEADSII event was labelled as a cold, ambient, or hot impact event.

Equation 1 calls for the weighting factors to depend on helmet size. This is critical since injury risk can be influenced by helmet size. The current study assumes the distribution of helmet sizes follows that of a recent U.S. Army helmet order. This results in the weighting factor consisting of small, medium, large, and x-large helmet shells.

2.3 Blunt Impact Injury Risk Analysis

The multi-dimensional composite score equation relies on calculating the risk of head injury for each combination of impact location, velocity, temperature, and size (see **R** in equation 1). Injury probability ranges from 0% - 100% risk of injury. There are several head injury risk functions proposed in the literature, and many rely on either linear (i.e. translational) or rotational head acceleration as a basis for injury risk. Since the current U.S. Army test standards only allow for the measurement of linear acceleration, the Head Injury Criterion (HIC) was proposed as the metric for evaluating injury probability [4][5]. HIC accounts for the magnitude and duration of the headform's accelerative response. The head injury criterion is calculated from the linear acceleration time history as:

$$\text{HIC} = \max\left\{(t_2 - t_1)\left[\frac{1}{t_2 - t_1}\int_{t_1}^{t_2} a(t)\,dt\right]^{2.5}\right\} \tag{2}$$

where t_1 and t_2 correspond to the start and end times of integration (limited to 15 ms) that maximizes the HIC value, and a(t) represents the linear headform acceleration measured from impacts according to the current U.S. Army combat helmet test standards. Different helmet and pad designs will produce different HIC values for the same impact conditions. HIC values have previously been correlated to injury using injury risk curves according the Abbreviated Injury Severity (AIS) scale [8]. The equation for an AIS 3 injury is defined as:

https://doi.org/10.52202/081621-0053

$$\text{Risk of AIS 3 Injury} = \frac{1}{1 + \exp\left[\left(3.39 + \frac{200}{\text{HIC}}\right) - 0.00372 \cdot \text{HIC}\right]} \tag{3}$$

Where HIC refers to the calculated HIC score from equation 2. In this instance, the risk of AIS 3 injury is considered a serious head injury, but does not predict specific injury patterns.

The current study leverages prior U.S. Army drop tower data from seven different combinations of helmet shells and suspension systems. Test data were obtained from the U.S. Army Combat Capabilities Development Command Soldier Center and the U.S. Army Aberdeen Test Center, which rely on consistent helmet positioning guidelines, such as the Head Positioning Index and retention strap specifications, to improve repeatability and reproducibility. The first two helmet systems consisted of the standard Team Wendy seven pad suspension system in an Advanced Combat Helmet (ACH) and an Enhanced Combat Helmet (ECH). The remaining five systems consisted of the ACH with five different prototype pads, all from different pad vendors. These pad systems are referred to as company A, B, C, D and E pads in the current analysis, and additional information on their design and material is found in Table 1 [9][10]. Risk of AIS 3 injury was calculated for all seven helmet/pad systems, and the resulting values were used for **R** in equation 1. The multi-dimensional composite score for each of the seven helmet/pad systems were calculated using the exposure weights (**W**) and injury risk (**R**) using equation 1. The headform acceleration data from the seven helmet/pad systems were used to calculate the HIC values for each impact location, velocity, and temperature. Since some of the helmet systems had incomplete data for size small, medium, and x-large, only size large data were used for the current analysis. This was achieved by reducing **W** to zero in equation 1 for small, medium, and x-large helmets. Additionally, HIC values were not calculated for impacts at 2.16 m/s and 6.10 m/s due to lack of data. The exposure coefficient, **W,** for these velocities were also set to zero in equation 1. The non-zero values of **W** were renormalized to sum to 100 impacts to account for the missing helmet data current analysis.

Table 1. Material description of the suspension systems analyzed in current study, taken from [9], [10]

	Suspension System Type	Number of Main Foam Layers	Materials of Main Foam Layers
Team Wendy	Individual Pads	2	Polyurethane
Company A	Hard Foamed Shell with Interior Pads	3	Polypropylene, Polyether Urethane, Polyurethane
Company B	Individual Pads	2	Polyether Urethane
Company C	Individual Pads	3	Polyurethane, Filled Silicone
Company D	Individual Pads	2	Polyether Urethane
Company E	Helmet Liner with Pads	1-3 Depending on Pad	Alkyd, Polyurethane Ethylene-vinyl Acetate, Polypropylene

3. RESULTS

3.1 Blunt Impact Exposure Weighting Analysis

The analysis of the 4,371 HEADSII events resulted in a range of impact locations, peak accelerations, and temperatures. Normalizing the 4,371 impacts to 100 theoretical impacts resulted in the exposure weights shown in Figure 2. The impact location analysis suggested that left and right sides were impacted the most frequently, with 44.4 out of 100 impacts occurring at these two locations. The crown and front were the next regions that were impacted the most frequently, with exposure weights of 14.3 and 13.5 impacts, respectively. Finally, the rear and nape had a combined exposure weight of 27.7 impacts.

The majority of HEADSII impact severities corresponded to the baseline impact energy on the monorail drop tower, with 86.6 out of 100 theoretical impacts. This result was expected since the analysis assumed the majority of the HEADSII events were comparable to a 3.05 m/s impact (i.e. the baseline impact energy). The next highest impact energy was 2x the baseline impact energy, with 10.4 out of 100 theoretical impacts. While several assumptions were made in translating helmet accelerations from

https://doi.org/10.52202/081621-0053

theatre to monorail drop tower impact energies, these results suggest the majority of triggered events in theatre can be represented by 1/2x, 1x, or 2x, the baseline impact energy on the monorail drop tower.

Analysis of the HEADSII temperature readings determined that the majority (92.4 impacts) corresponded to ambient temperature conditions, while the next highest temperature range reflected hot temperature conditions (7.3 impacts). This suggests that the helmet performance for ambient and hot locations should be weighed higher than performance for cold impact conditions.

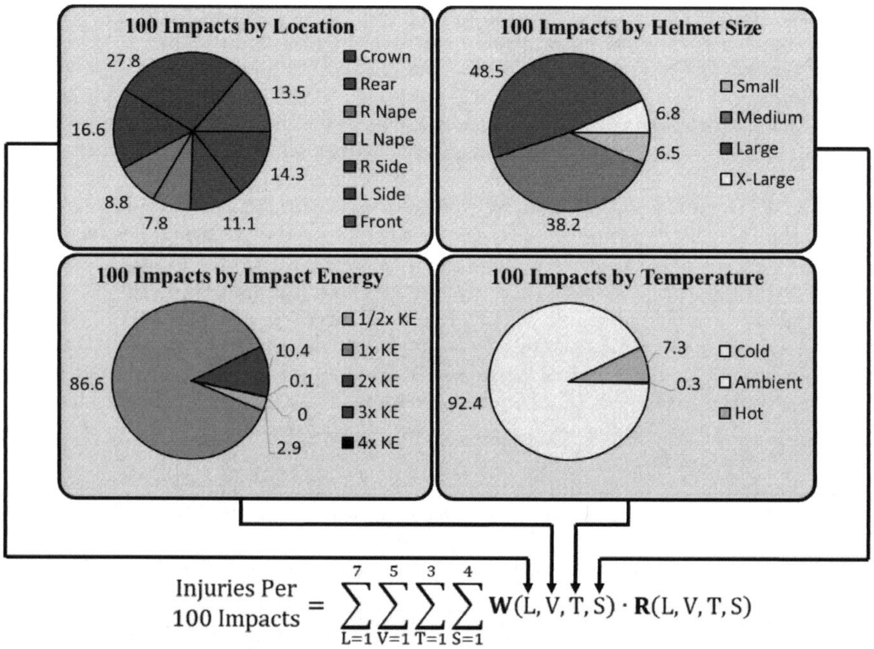

Figure 2. Weighting factors obtained through analysis of HEADSII helmet acceleration data collected in 2013 and 2014. Numbers next to each pie wedge correspond to the number of impacts (out of 100 impacts) for each condition.

There were 420 exposure coefficients (**W**) in equation 1. These were based on the individual combinations of location, size, impact energy, and temperature. As an example, the largest exposure coefficient value was 0.108. This represented the number of impacts occurring on the size large (48.5/100) helmet's left side (27.8/100), of a 1x KE impact (86.6/100) during ambient temperature conditions (92.4/100). The second highest exposure coefficient was 0.064, and represented the same impact conditions on the helmet's right side.

3.2 Example Injury Risk Values for Helmet Drop Tower Testing

HIC scores varied with impact location, energy, temperature and helmet size. Figure 3 illustrates the range of measured HIC values for all seven helmet systems of interest for size large helmets in ambient conditions. None of the helmet suspension systems exceeded a 50% probability of AIS 3 (HIC score = 967) for 3.05 m/s and 4.31 m/s monorail impacts. The ECH with the Team Wendy pads had the lowest average HIC score for all seven helmet locations, with average values of 125 and 306 for 3.05 m/s and 4.31 m/s impacts, respectively. The worst-performing helmet varied with impact severity. At 3.05 m/s, the ACH with company C pads had the highest average HIC score (219), but the ACH with company B pads had the highest average HIC score (496) at 4.31 m/s impact velocities. Company A resulted in the highest average HIC scores at 5.28 m/s impact, with one of the seven impact locations exceeding 99% risk of AIS 3 injury. These results indicate that the relative performance of helmet systems can vary with impact location and severity.

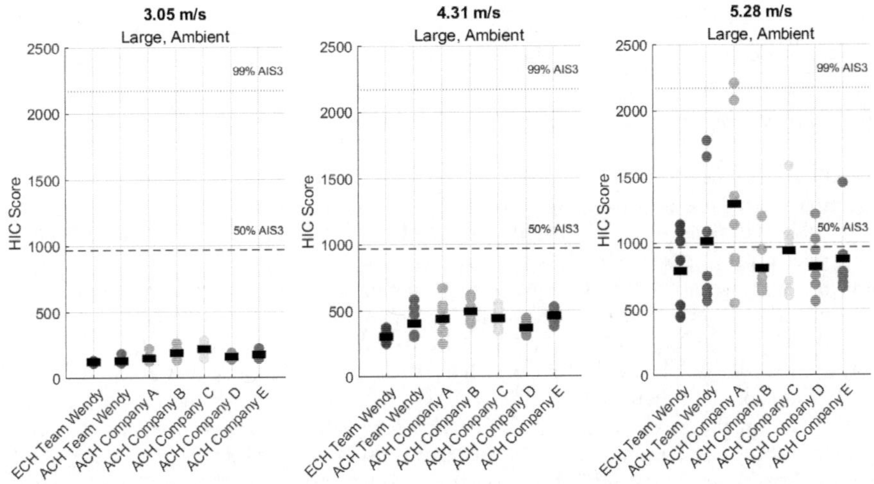

Figure 3. Range of HIC scores calculated from prior monorail drop tower test data for ambient temperature conditions. Size large helmet data are shown. Each coloured dot illustrates the HIC score for each of the seven impact locations at the specified impact energy, and the single black dash represents the average of all seven different locations for each helmet system. The two horizontal lines represent the HIC score associated with a 50% and 99% risk of AIS 3 injury, according to equation 3.

3.3 Example Composite Scores for Helmet Systems of Interest

The composite helmet scores for each of the seven helmet systems are presented in Figure 4. These helmet scores represent the number of AIS 3 injuries that would be expected for 100 theoretical impacts with the exposure weights presented in Figure 2. Using these exposure weights, all seven helmet systems had a low number of predicted AIS 3 injuries per 100 impacts. The ECH with the Team Wendy Pads resulted in the least amount of AIS 3 injuries (1.45 injuries per 100 impacts), followed by the ACH with Team Wendy Pads (1.90 injuries per 100 impacts). The helmet/pad combination that resulted in the worst score was the ACH with company C pads (3.39 injuries per 100 impacts). It should be noted that the scores presented here are based only on the size large helmet response since the current study did not have complete test data for the other helmet sizes. These results indicate that the ACH and ECH with Team Wendy pads offer the best protection against AIS 3 injuries for the current blunt impact exposure weights used.

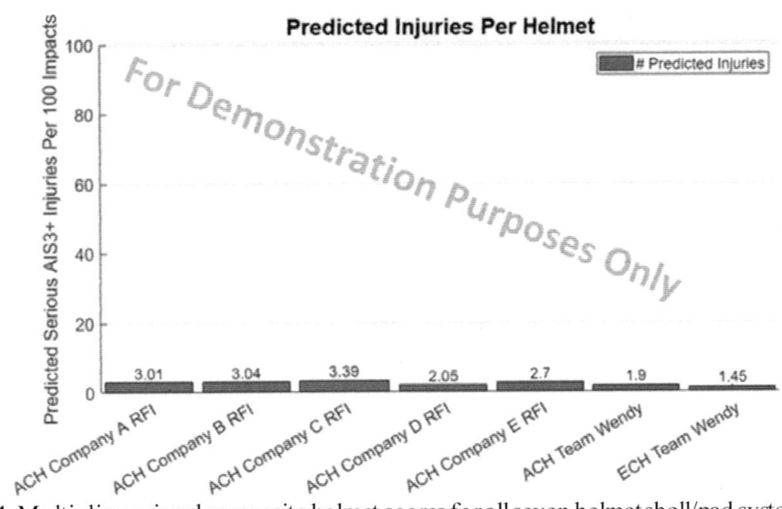

Figure 4. Multi-dimensional composite helmet scores for all seven helmet shell/pad systems. Each value represents the number of expected injuries for 100 impacts if the given helmet system is worn. A lower number of injuries (i.e. a better score) is indicative of a more protective helmet system

https://doi.org/10.52202/081621-0053

4. DISCUSSION

4.1 Rating Framework

The current study outlines a performance scoring framework for combat helmets based on Rowson's and Duma's approach for American football helmets [6]. The scoring framework has been modified to account for four test dimensions in accordance with the U.S. Army drop tower test methods for combat helmets: impact location, impact velocity, temperature and helmet size. The relative importance of each test condition is weighted according to combat exposure environments. Impact conditions that occur more frequently in operations are weighted more heavily in the composite score in equation 1. As an example, the ECH with the Team Wendy Pads had the lowest average HIC scores at ambient conditions for impacts at 3.05 m/s and 4.31 m/s (figure 3). This same pad system resulted in the second highest average HIC score for hot impacts at 5.28 m/s. If one were to were to solely examine the 5.28 m/s hot temperature condition, they may assume the ECH with Team Wendy pads performs poorly compared to the other suspension systems. However, the ECH had the best composite helmet score since the 5.28 m/s hot temperature conditions for size large helmets across all impact locations had a small weighting factor (3.5e-05) compared to 3.05 m/s and 4.31 m/s at ambient conditions (weighting factors of 0.388 and 0.047, respectively). The small weighting factor for hot impacts at 5.28 m/s resulted in this test condition have a negligible effect on the overall helmet score. This example highlights (1) the importance on selecting the appropriate set of weighting factors that represent the exposure environment of interest and (2) that solely analyzing HIC scores for single conditions may not represent the helmet system response for all conditions.

A benefit of having a single composite score for each helmet is that multiple helmets can be ranked for an exposure environment of interest, such as a mounted or dismounted operation. This is proposed as an improvement over the current standard that simply passes or fails a helmet system if it results in peak accelerations less than 150 g for all conditions. With the current standards, selecting the helmet out of a group that all 'pass' is challenging since the relative performance changes drastically depending on the test condition considered. The single composite score, which could potentially be used in addition to or in lieu of the current standard, makes this selection relatively straightforward.

The scoring system's metric of performance is the number of serious injuries (AIS 3) that occur for 100 theoretical blunt impacts in the given combat environment. The analysis of the seven different helmet/pad systems demonstrates that the number of expected serious head injuries are small for the given operational environment (<4 injuries per 100 impacts with the current exposure weights). While the current analysis only examines serious injury (AIS 3), the number of expected injuries would increase if mild (AIS 1) or moderate (AIS 2) were considered. Additionally, the number of injuries would change depending on the assumed relationship between the HEADSII acceleration values and impact velocity. The current analysis assumes the vast majority of impacts in theatre are comparable to 3.05 m/s on the monorail drop tower, however this assumption has not been validated. If future exposure studies determine that the majority of impacts are comparable to a different impact energy, then the predicted number of injuries for each combat helmet would likely change.

One potential improvement to the injury prediction framework in equation 1 would be to evaluate specific types of head injuries based on the head's mechanical response. Early biomechanics research found a correlation between linear head acceleration and skull fracture [11], [12], while others determined that rotational head acceleration and velocity are likely responsible for brain injuries in the absence of skull fracture [13][14]. There is growing narrative that linear accelerations are correlated with skull stress (and by extension skull fracture) and intracranial pressure response, while head rotation is correlated with brain deformation (and by extension injuries such as intracerebral hematoma and diffuse axonal injury) [15][16]. The current combat helmet monorail test procedure restricts the headform's rotation during impact, so the current analysis was limited to injury risk functions that are based on translational acceleration, such as the head injury criterion. This suggests that the injuries predicted by the current scoring framework are more indicative of skull fracture rather than diffuse brain injuries. Future efforts can also measure the effect of different helmet systems on the headform's rotational response to determine how helmet and pad design potentially influence risk of brain injury, similar to the approach taken by Neice and Plaisted [17]. This would be an improvement over the current study's use of AIS specific HIC injury risk curves (equation 3), which does not distinguish between different mechanisms and types of injuries.

4.2 Exposure Weights Limitations

The relative weighting of each test condition was based on the HEADSII combat helmet acceleration and temperature data, which captured similar data as the Head Impact Telemetry (HIT) system leveraged by Rowson and Duma [6]. Prior analysis of HEADSII recordings during known operational events determined the majority of recorded events occurred in either a mounted or dismounted environment [7]. This suggests that exposure weights presented in figure 2 reflect a mixture of these threat environments. However, it remains unclear how each of the 4,371 HEADSII data recordings used to develop the exposure weights related to specific threats, such as accelerations due to blast, falls, or automotive accidents. Future data collection under different combat scenarios may lead to different exposure weights, ultimately affecting the overall helmet scores.

Preliminary analysis of the HEADSII sensor results revealed a number of challenges in data interpretations, including sensor reliability, potential recording sensitivity to crown impacts, empty helmet recordings, pairing of acceleration data to known events, and sensitivity to different helmet impact locations and impact severities. One of the biggest assumptions of the current analysis was relating the severity of impacts in theatre, via the HEADSII accelerations, to specific impact velocities during helmet drop tower testing. The current study assumed the majority of HEADSII acceleration events corresponded to impacts at 3.05 m/s, but additional work is needed to determine if this is accurate. In their scoring framework for American football helmets, Rowson and Duma conducted a series of drop tower impact tests with the HIT system, and developed a relationship between helmet acceleration and impact velocity [6]. This allowed the authors to relate in-game impacts with the HIT system to drop tower test conditions in the laboratory. A similar approach could be implemented for combat helmets, but this was not implemented for this study mainly because the HEADSII system is no longer manufactured and available to procure. Instead, future work can rely on computational models, such as finite element models, to simulate helmeted impacts with the HEADSII sensor to determine the relationship between helmet acceleration and impact velocities for different locations on the helmet shell. This relationship could be leveraged to develop improved exposure weights for the helmet scoring analysis.

Another limitation of the exposure weight analysis assumed the temperature measurement from the HEADSII sensor was valid for all pads within the helmet. The HEADSII sensor is located inside the helmet shell, underneath the crown pad. The temperature measurement is likely different than the external environment due to the body temperature of the Soldier's head. To the authors' knowledge, this difference is not well characterized. Future work can seek to characterize how temperature measurements taken inside the helmet relate to body temperature and the external environment.

4.3 Conclusion

In order to address some of the limitations of current combat helmet blunt impact test methods and requirements, the current study developed a multidimensional composite helmet scoring system for combat helmets that leverages the framework of the STAR American football helmet scoring system. Current U.S. Army combat helmet blunt impact test standards require helmet drop tower test results to not exceed 150 g for all test conditions, which include 7 helmet locations, 1 to 2 impact velocities, 3 temperatures, and 4 helmet sizes. The multidimensional composite helmet scoring framework leveraged an analysis of helmet accelerometer data recordings of impact events from theatre to determine the relative importance of each test parameter. Composite scores were calculated for seven different helmet systems by combining the exposure weights of each combination of test parameters and along with the associated the probability of head injury as determined by prior drop tower testing. The resulting composite helmet scores represented the number of estimated injuries that would result from 100 impacts with the same exposure frequency from theatre data. Data from seven different helmet/pad systems were assessed using the current combat helmet blunt impact requirement as well as using the new scoring system with improved correlation to injury, weighting based on exposure frequency, and a composite score representing the estimated number of injuries for 100 impacts. While it was difficult to determine the relative performance ranking of each helmet/pad system using the current combat helmet blunt impact requirement due to different helmets outperforming others in different test conditions, the new scoring system provided a composite score enabling a relative ranking of performance. Of the helmet/pad systems evaluated with the new scoring method, the ECH with the standard Team Wendy pads performed the best. While the current study has several assumptions behind the exposure weighting and injury risk predictions, it provides a useful framework to assess the relative performance of combat helmet systems which could be used to spur innovation and improvements in helmet systems.

https://doi.org/10.52202/081621-0053

Acknowledgments

The authors would like to thank Program Executive Office Soldier Product Manager Soldier Protective Equipment for sponsoring this effort. The authors would also like to thank the U.S. Army Combat Capabilities Development Command Soldier Center and the U.S. Army Aberdeen Test Center for supplying the monorail drop tower test data in the current study. Any opinions, findings and conclusions or recommendations expressed in this material are those of the author(s) and do not necessarily reflect the views of PEO Soldier or Naval Sea Systems Command (NAVSEA).

References

[1] J. McEntire and P. Whitley, "Blunt Impact Performance Characteristics of the Advanced Combat Helmet and the Paratrooper and Infantry Personnel Armor System for Ground Troops Helmet.," 2005.

[2] U.S. Department of Transportation, "Federal Motor Vehicles Safety Standard 218." p. Sec. 571.218, Standard No. 218, FMVSS 218-49 CFR.

[3] B. A. Slobodnik, "SPH-4 helmet damage and head injury correlation," *Aviat. Sp. Environ. Med.*, vol. 50, no. 2, pp. 139–146, 1979.

[4] P. Prasad and H. J. Mertz, "The position of the United States delegation to the ISO Working Group 6 on the use of HIC in the automotive environment," *SAE Tech. Pap.*, no. June 1985, 1985, doi: 10.4271/851246.

[5] W. C. Hayes, M. S. Erickson, and E. D. Power, "Forensic injury biomechanics," *Annu. Rev. Biomed. Eng.*, vol. 9, no. May, pp. 55–86, 2007, doi: 10.1146/annurev.bioeng.9.060906.151946.

[6] S. Rowson and S. M. Duma, "Development of the STAR evaluation system for football helmets: Integrating player head impact exposure and risk of concussion," *Ann. Biomed. Eng.*, vol. 39, no. 8, pp. 2130–2140, 2011, doi: 10.1007/s10439-011-0322-5.

[7] J. Jessup *et al.*, "Vehicle and Individual-Worn Sensor Data Analysis in Support of the JTAPIC Program Final Report," 2016.

[8] Natl. Highw. Traffic Saf. Adm. (NHTSA)., "Final economic assessment, upper interior head protection," 1995.

[9] K. Masters *et al.*, "Joint Live Fire Advanced Combat Helmet Report," 2008.

[10] "Team Wendy Science and Technology." [Online]. Available: https://www.teamwendy.com/about/science-and-technology

[11] V. R. Hodgson, L. M. Thomas, and S. W. Greenberg, "Bone Against Cylindrical Surfaces1," 1970.

[12] V. R. Hodgson and L. M. Thomas, *Breaking Strength of the Human Skull Vs. Impact*. 1971.

[13] A. . Ommaya, A. E. Hirsch, P. Yarnell, and E. H. Harris, "Scaling of Experimental Data on Cerebral Concussion in Sub-Human Primates to Concussion Threshold for Man," 1967.

[14] A. . Ommaya and A. E. Hirsch, "Tolerances for Cerebral Concussion from Head Impact and Whiplash in Primates," *J. Biomech.*, vol. 4, pp. 13–21, 1971, [Online]. Available: http://www.sciencedirect.com/science/article/pii/002192907190011X

[15] A. I. King, K. H. Yang, L. Zhang, W. N. Hardy, and D. C. Viano, "Is head injury caused by linear or angular acceleration," in *IRCOBI Conference*, Lisbon, 2003, pp. 1–12.

[16] S. Kleiven, "Why Most Traumatic Brain Injuries are Not Caused by Linear Acceleration but Skull Fractures are," *Front. Bioeng. Biotechnol.*, vol. 1, no. November, pp. 1–5, 2013, doi: 10.3389/fbioe.2013.00015.

[17] R. Neice and T. Plaisted, "Evaluation of a Combat Helmet Under Combined Translational and Rotational Impact Loading," 2020.

Residual Helmet Deformation from Non-Penetrating Ballistic Impacts on Various Head Surrogates

E. Butler[1], C. Howes[1], N. Hahne[1], Q. Luong[1], M. Clark[2], **A. Iwaskiw[1]**

[1] *The Johns Hopkins University Applied Physics Laboratory (JHU/APL), 11100 Johns Hopkins Road, Laurel, MD 20723, alexander.iwaskiw@jhuapl.edu*
[2] *U.S. Special Operations Command, 10 General Greene Avenue, Natick, MA 01760*

Abstract. Ballistic performance of helmets is affected by boundary conditions including the backing material (e.g., clay, air, skull) behind the helmet. This is a critical factor when developing and employing test methods and test equipment such as head surrogates. By comparing residual armour backface deformation (BFD) measurements from different backing-condition ballistic tests, the hope is to understand how particular helmet testing methods may influence helmet ballistic performance and material behaviour. This study assessed residual armour BFD measurements of helmets with different backing conditions: a clay headform, the Ballistic Load Sensing Headform (BLSH), the Adaptable Testing and Load Assessment System (ATLAS) Headform, and postmortem human surrogates (PMHS). PMHS were assumed to most accurately simulate the boundary conditions of live humans. Helmets were impacted with ballistic rounds at a range of velocities at four locations where the residual armour BFD was characterised. Results indicate comparable helmet BFDs between the PMHS-backed helmets and load cell headform-backed helmets. The clay-backed helmets had a larger range of peak residual BFDs, with the peak residual BFDs being nearly 15 mm greater than BFDs seen with PMHS-backed and load cell headform-backed helmets. Additionally, it was found that shot velocity was not a strong predictor of peak deformations across all backing types. This study demonstrates that backing materials and boundary conditions have an effect on helmet deformation behaviour and thus head surrogate materials and designs should be carefully considered. The measurements from this study are not meant to predict injury, but rather to understand the correlation between test conditions and the operational environment. These results indicate a potential need for test methodologies and test equipment to simulate the boundary conditions of armour in operational use as it will influence helmet performance during laboratory testing.

1. INTRODUCTION

Helmets play a crucial role in providing protection for the warfighter against a wide range of threats. Recent material advancements have led to the development of helmets capable of withstanding ballistic impacts from higher energy, small arms ammunition [1]. Modern helmets mitigate threats by dissipating kinetic energy and preventing armour perforation, thereby reducing the risk of penetrating ballistic injuries to the wearer [2]. Consequently, during impact the round causes rapid material deformation of the helmet. This deformation causes the backface of the helmet to strike the wearer potentially resulting in behind helmet blunt trauma (BHBT). BHBT can lead to severe injuries such as skull fracture, brain contusion, hematomas, and axonal injuries [3].

Given the critical need to protect warfighters, the ballistic performance criteria of helmets must be examined. This includes developing standardised measurement techniques to inform manufacturing processes. Ideally, a single test methodology would be used for both penetrating and non-penetrating ballistic performance assessments due to the cost associated with destructive testing. Previous studies have indicated that penetrating ballistic performance is influenced by backing material [4], [5]. This indicates the need for operationally relevant ballistic performance qualification testing including the use of biofidelic backing materials. While experiments to determine penetration levels associated with different backing materials are possible, they are costly and pose a risk of damaging expensive testing components such as load cells in headforms. An alternative approach is to analyse residual helmet backface deformation (Figure 1) to provide some insight into ballistic performance and allow for comparison across backing conditions [6].

 https://doi.org/10.52202/081621-0054

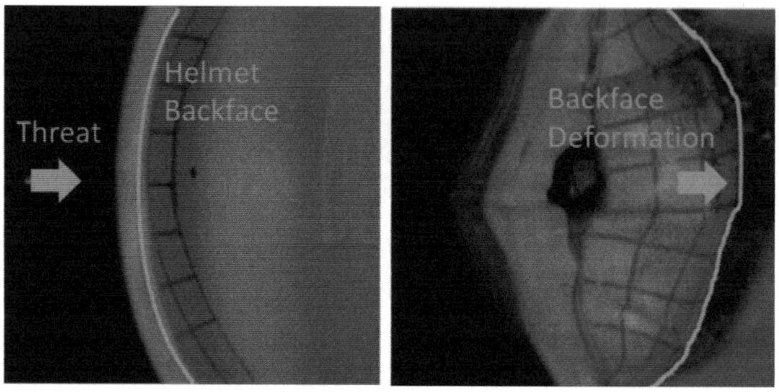

Figure 1. Undeformed helmet just prior to impact (left) with threat direction indicated and backface deformation of helmet during impact (right). Note the helmet used in this visual demonstration is not the helmet used in this study.

It is hypothesised that biofidelic test methods that simulate operational conditions are essential for obtaining representative ballistic performance measurements. Various platforms have been developed to evaluate helmet ballistic performance characteristics using different backing materials. Current ballistic performance measurements rely on a clay headform which provides a penetration outcome, and in cases where there is no complete penetration, a static measurement of the maximum deformation of a helmet in the impacted clay from a BHBT-induced event. However, clay testing does not capture dynamic load transfer during impact from BHBT. Surrogate headforms equipped with load cells were developed to understand this dynamic loading, but the effect these load cell backed headforms have on ballistic performance dynamics remains unclear. Additionally, ballistic performance testing has also been performed on postmortem human surrogates (PMHS) to understand BHBT related injuries, providing a realistic response comparable to the warfighter. Data collected across these backing conditions varies significantly which limits the ability to make direct comparisons of their effects on ballistic performance.

Residual BFD is defined as the backface deformation remaining in a deformed helmet after a non-penetrating ballistic impact. This residual deformation is typically less than the peak dynamic deformation of the backface due to the material rebound. Capturing maximum dynamic deformation is challenging due to the rate of the deformation and the obscured backface on a headform. Conversely, residual BFD measurements are taken after a ballistic event, allowing for the same collection technique to be used across backing conditions. This enables comparisons between experimental conditions to determine if the backing material affects the degree of helmet deformation during non-penetrating ballistic impact. If significant differences are observed, this suggests the need for helmet ballistic performance qualification testing to be performed on a backing material aligned with operational conditions to ensure accurate assessment of field performance. This study presents the residual backface deformation of a light weight helmet impacted with the same ballistic threat on various backing materials commonly used to assess helmet ballistic performance.

2. METHODS

2.1 General experimental methods

Each experimental set up utilised a modern lightweight helmet primarily composed of ultra-high molecular weight polyethylene that was fit to the head surrogate according to manufacturer instructions. Proper pads and suspension systems to achieve a 23mm±0.5 mm standoff[1] regardless of backing material were in place during testing. All ballistic testing was performed at an NIJ certified range (Element U.S. Space and Defense, Belcamp, Maryland, USA). The helmets were impacted with an 8.0 g (124 grain), 9 mm full metal jacket threat. The same armour-threat paring was used for each experimental setup. Velocities were varied to achieve a range of backface deformation without helmet penetration. The velocity range was chosen to align with previous PMHS tests where both injurious and non-injurious fracture outcomes were recorded. Projectile velocity measurements were obtained using infrared screen-based velocity gates (Oehler Research Model N0. 57) with universal counter chronographs (Hewlett-Packard model No. 53131A). Velocity was measured 1.5 m from the surface of the helmet, then velocity

[1] Airgap between backface of the helmet and underlying surface

loss equations were applied specific to the projectile to calculate striking velocity. The distance between the barrel and surface of the helmet was held consistent across test setups at 4.6 m. The crown, front, side, and rear impact locations (Figure 2) were chosen to remain consistent with current clay standard testing protocols used to define allowable clay deformation for helmets. The NVG mount was removed for all tests, while the ARC rails were left on but not directly impacted during the tests. All tests were conducted with the same 0° obliquity at each shot location where the ballistic striking vector was normal to the impact location.

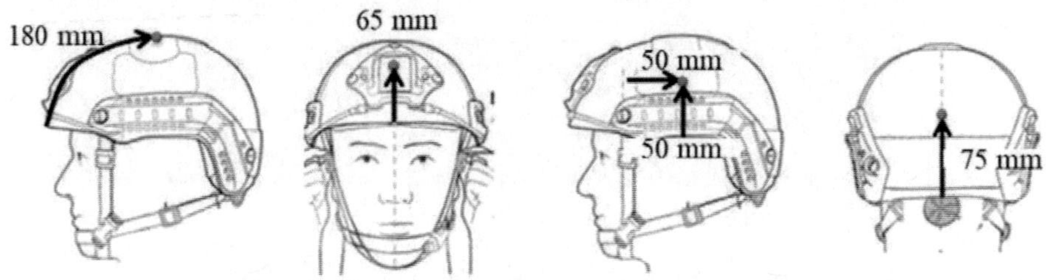

Figure 2. Helmet impact locations (red circle) for (left to right) crown, front, left side, and rear. Arrows indicate direction of measured distance across the surface of the helmet.

2.2 Experimental setup

In this study, four backing materials were characterised for their effects on residual backface deformation: the multi-sized clay headform, the Adaptable Testing and Load Assessment System (ATLAS) Headform, the Ballistic Load Sensing Headform (BLSH), and PMHS. This study collected ballistic impact velocity and residual BFD measurements for each of the four experimental conditions. Identical ballistic test methods were used across the four setups to allow for the comparison of each type of backing material.

2.2.1 Clay-headform backed experimental setup

The clay backface deformation measurement is a key metric in helmet ballistic performance qualification [7]. In this standardised test protocol, the multi-sized clay headform (size Medium/Large) was packed and moulded with temperature-calibrated Roma Plastilina No. 1 clay and the helmet was fit as previously described. The external geometry of this headform was standardised to uniformly produce the 23 mm standoff distance for each impact location when outfitted with the tested helmet. The test range was controlled within test standards to an ambient temperature of 20°C ±12°C and 50 ±20% relative humidity. The headform was rigidly mounted on a fixture that allowed for vertical, horizontal, and rotational adjustments to achieve the desired obliquity.

The helmet was then impacted with a non-penetrating ballistic round. After testing, the helmet was removed and the maximum depth of the indented clay was measured to define the clay backface signature for a specific armour-threat pairing. As this is a widely accepted standard for helmet performance testing, impact locations, velocities, and general experimental set up for remaining residual BFD test protocols was modelled off the clay standard test protocol.

A total of 56 clay backed tests for this study were performed according to the standardised helmet ballistic performance qualification tests. Figure 3 shows the helmeted clay headform prior to ballistic impact (top row) and indented clay headform after the helmet was removed post-impact (bottom row).

https://doi.org/10.52202/081621-0054

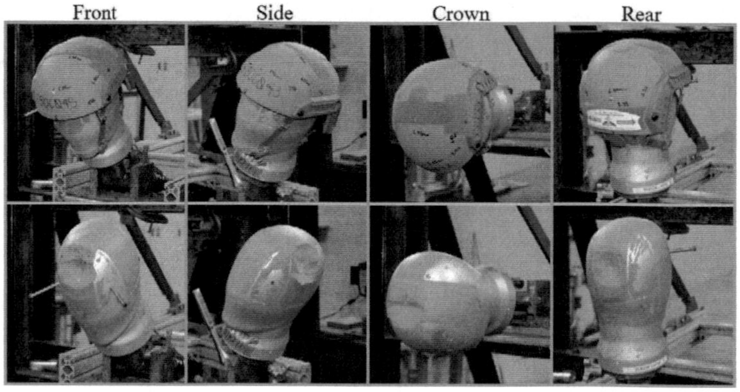

Figure 3. For each impact location, an intact helmet (top) fit onto the clay headform and the deformed headform (bottom) after a non-penetrating ballistic impact at the point indicated by the laser.

2.2.2 The Adaptable Testing and Load Assessment System (ATLAS) headform experimental setup

The ATLAS headform (JHU/APL, Laurel, Maryland, USA) is a load-cell based surrogate headform developed to evaluate helmet performance and the potential for BHBT [8]. The system consists of modular headforms used to test each of the impact locations described in Table 1, where the centre of the load cell is positioned directly under the impact location. Figure 4 shows a schematic of the headform configured for a front impact test. The ATLAS headform is additively manufactured from acrylonitrile butadiene styrene (ABS-M30) and polyurethane elastomer (Huntsman RenCast 6443) and equipped with a stainless-steel impact cap (304/304L) with an outer radius that conformed to the curvature of the headform at each impact location. A neoprene impact pad with a thickness of 12.7 mm covered the impact cap and contoured with the headform. The impact cap was magnetically attached to a piezoelectric load cell (111 kN, 224C/FCS-DI IC, PCB Piezotronics) and attached to a post and base assembly. The headform base was then fixed to a ballistic alignment fixture. The ATLAS headform was used to collect non-penetrating ballistic impact data across four shot locations for a total of 45 shots.

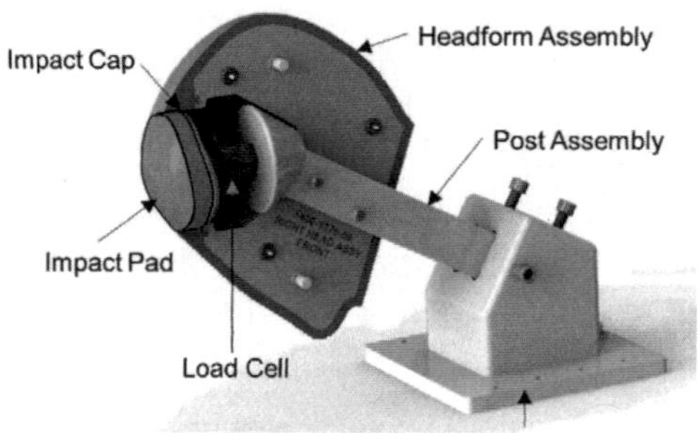

Figure 4. Schematic of ATLAS Headform system components. Schematic shown is for the front impact configuration.

2.2.3 The Ballistic Load Sensing Headform (BLSH) experimental setup

The Ballistic Load Sensing Headform (BLSH) (Biokinetics and Associates Ltd., Ottawa, Canada) is used to assess BHBT from non-penetrating ballistic impacts [9]. The headform was mounted on a flexible neck and was equipped with an array of 7 load cells (Figure 5) at each impact location to measure contact force, distribution, and impulse during ballistic impact. A metal impact cap was mounted to each load cell, and a silicone rubber skin pad covered each array of 7 impact caps. The BLSH experimental setup was similar to that of the ATLAS Headform where the centre of the loadcell array was aligned with the intended impact location. A total of 7 non-penetrating shots were conducted.

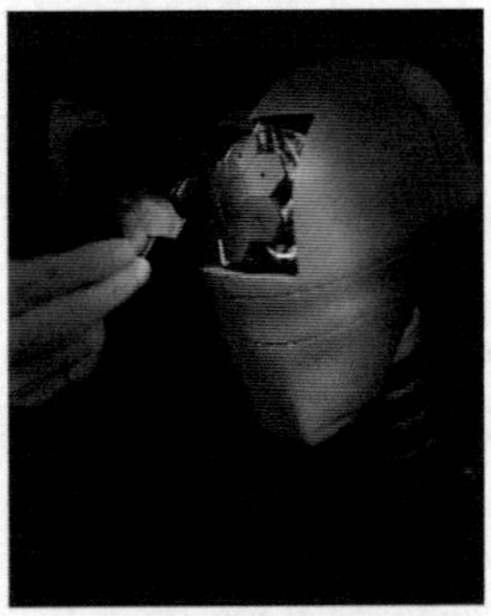

Figure 5. Biokinetics Ballistic Load Sensing Headform with transparent helmet to highlight use of front location - silicone skin pad not shown (left) and BLSH front impact location with impact pad held to side (right) [10].

2.2.4 Postmortem human surrogates (PMHS) experimental setup

Experimental testing with PMHS offers insight into the biomechanical response to ballistic impacts warfighters may face, providing what is likely the most biofidelic backing condition. This study uses data collected by JHU/APL, detailed in Iwaskiw, et al. [11]. Exclusion criteria related to bone health, head size, and age were implemented to select fresh frozen specimens to reduce variability in the experiment. The specimens were affixed to a rigid suspension fixture to allow for alignment of target anatomy with the path of the projectile for all shot locations (Figure 6). A total of 58 non-penetrating shots were analysed for this study.

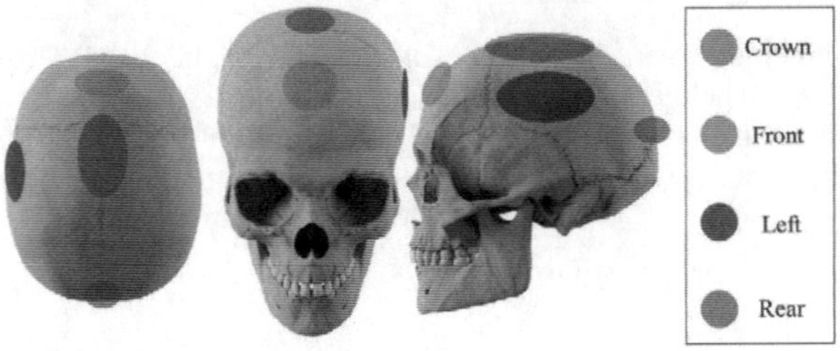

Figure 6. Approximate anatomical impact locations for the crown (red), front (green), left Side (blue), and rear (purple) for the PMHS experimental setup

2.2.5 Test matrix and experimental summary

Table 1 summarises the four experimental conditions used for data collection in this study. It includes a breakdown of impacts at each location for all backing conditions.

Table 1. Summary of experimental setup for various backing materials.

	Clay-Headform	**ATLAS**	**BLSH**	**PMHS**
Backing Material	Clay	Neoprene pad, metal impact cap, and load cell	Silicone pad, metal impact caps, and load cells	Skin/skull
Helmet	Modern lightweight ultra-high molecular weight polyethylene			
Standoff	23 mm ± 0.5 mm			
Threat	8.0 g (124 grain), 9 mm full metal jacket			
Velocity Range (m/s)	~280 – ~450			
Distance to barrel (m)	4.6			
Obliquity	Held consistent across experimental setups			
Crown Impacts	13	10	0	11
Front Impacts	18	12	1	16
Side Impacts	12	8	6	16
Rear Impacts	13	15	0	13
Total Impacts	56	45	7	58

https://doi.org/10.52202/081621-0054

2.3 Residual BFD measurement technique

Residual BFD is measured by comparing the surface scan of an intact helmet to the surface scan of a tested helmet (Figure 1). A coordinate measurement machine (ROMER Absolute Arm, 7535SEI-3401-UC, Hexagon Manufacturing Intelligence, North Kingstown, RI) was used to scan each helmet's backface surface post-test, creating a point cloud representation of the residual BFD. The intact helmet was also scanned for a reference surface. All helmets were scanned using a standard fixture to assist with co-registration in 3D space. Metrological inspection software (Geomagic Control X, 3D Systems, Rock Hill, South Carolina, USA) was used to process the data scans to generate a digitised version of each helmet surface. A tested helmet scan and the intact helmet scan were loaded into the software and coregistered. The deviation analysis method used involved comparing the measured data from a tested helmet to the intact helmet surface by calculating the shortest distance between paired points. The measured points are projected onto the reference data using the shortest distance calculation.[2] This method generates a colour deviation map, visually representing positive and negative deviations across the entire part seen in Figure 7.

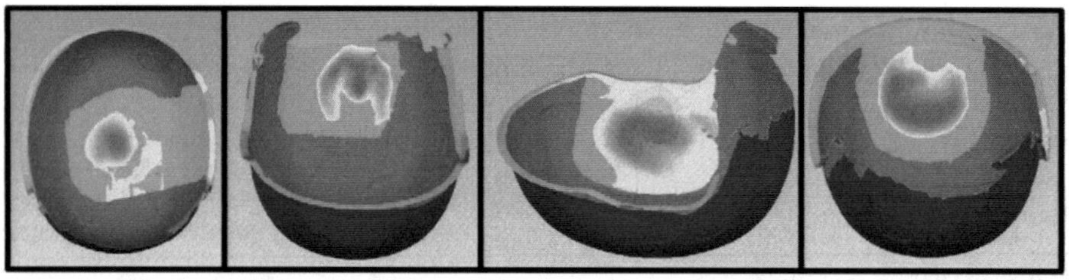

Figure 7. Point clouds scans of shot helmets at (left to right) crown, front, side, and rear impact locations. Each helmet is oriented to have a straight on view of the deformation surface (i.e. inside surface). Colour map (scale not shown) represents varying levels of deformation from the intact surface (grey). Light blue/green indicates minimal deformation, orange/red indicates maximum deformation

3. RESULTS

Maximum residual BFD was recorded with respect to impact velocity for each of the helmets tested (Figure 8). The range of residual BFD results across all backing conditions was 5.88 mm to 39.56 mm. The crown impact location consistently had the lowest residual BFD measurements, regardless of backing material (Figure 9). The front impact location had the highest residual BFD for all but the PMHS backing condition. In clay and ATLAS backed tests, deformation was highest at the front impact location, followed by rear, average over all locations, side, then crown (Figure 9). Although deformation magnitude varied by backing condition, this trend of differences in magnitude for a given backing condition across impact locations was a notable finding of the study. While the BLSH backed results were included in this study, there are significantly less impacts for this test condition. These results were included in the current study but were originally collected for a different purpose.

[2] This was executed through the "3D Compare/Shortest" method

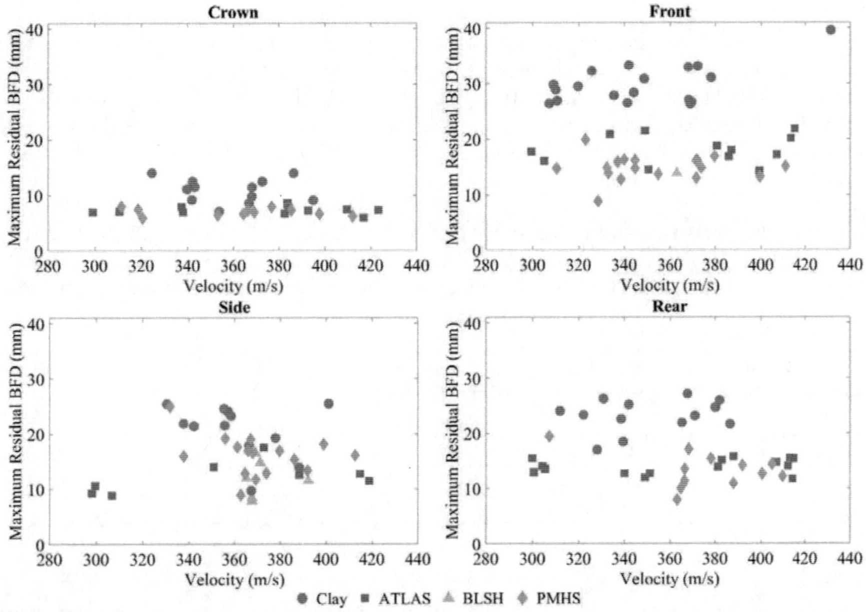

Figure 8. Point of maximum residual BFD results by impact location and backing condition.

Helmet tests on the clay headform resulted in the highest residual BFD measurements across all shot locations (Figure 9), while the three other backing materials had more similar residual BFDs. An initial single factor Analysis of Variance (ANOVA) was conducted to compare the means of residual BFD for all backing materials. The resulting p-value was 0.0000, indicating that at least one of the groups was significantly different from the others. In order to determine which groups were statistically different, a Tukey's honestly significant difference (HSD) test was conducted to do pairwise comparisons of each of the backing materials. Clay was statistically different from each of the other backing conditions with p-values less than 0.0001. The ATLAS Headform and BLSH were not statistically different with a p-value of 0.9640. The BLSH and PMHS were not statistically different with a p-value of 0.9601. The ATLAS Headform and PMHS were not statistically different with a p-value of 1.000. This initial analysis considered all shot locations in the groups. The full analysis results are in Table 2.

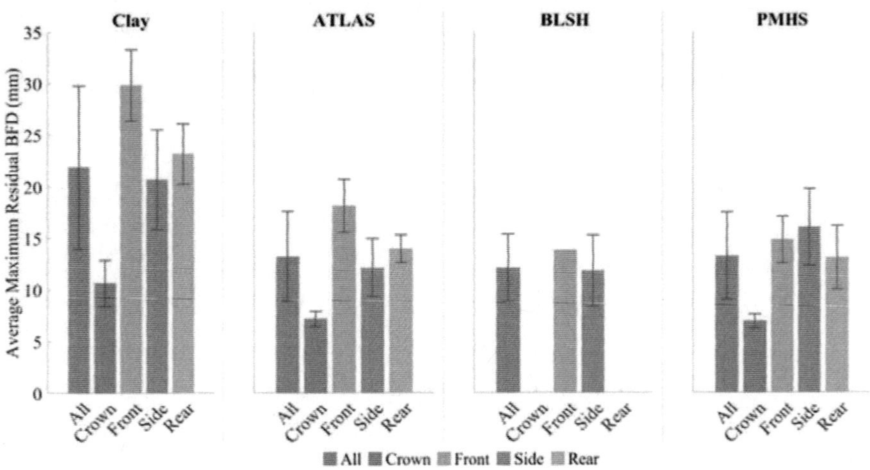

Figure 9. Average maximum residual BFD results by impact location and backing condition with standard deviations included

The same analysis process was conducted to compare the mean residual BFD for each backing condition across shot locations. For all impact locations, the clay residual BFD measurements were

https://doi.org/10.52202/081621-0054

significantly different from all other backing materials. For the crown and rear impact locations, the ATLAS Headform and PMHS backing conditions were not significantly different. The BLSH was not considered for crown or rear impact location analysis. For the side impact location, the ATLAS Headform was not significantly different from the BLSH or the PMHS. The BLSH and PMHS were also not significantly different at this location. For the front impact location, the ATLAS Headform and the BLSH were not significantly different. The BLSH and PMHS were not significantly different. The ATLAS and PMHS were significantly different.

Table 2. Tukey's honestly significant difference (HSD) results for determining significant differences in mean residual BFD on various backing materials across shot locations. Residual BFD mean differences are significant at the 0.05 level. Groups with means that were statistically significant are indicated with (*).

Impact Location	Group 1	Group 2	Mean Difference (mm)	95% Confidence Interval	P-value
All	Clay	ATLAS	8.63	(5.67, 11.6)	0.0000
		BLSH	9.74	(3.82, 15.7)	0.0001
		PMHS	8.60	(5.84, 11.4)	0.0000
	ATLAS	BLSH	1.12	(-4.89, 7.12)	0.964*
		PMHS	-0.02	(-2.96, 2.91)	1.000*
	BLSH	PMHS	-1.14	(-7.05, 4.77)	0.960*
Crown	Clay	ATLAS	3.45	(1.94, 5.04)	0.000
		PMHS	3.74	(2.22, 5.25)	0.000
	ATLAS	PMHS	0.25	(-1.36, 1.86)	0.924*
Front	Clay	ATLAS	11.7	(8.87, 14.5)	0.000
		BLSH	16.0	(8.18, 23.8)	0.000
		PMHS	15.0	(12.5, 17.5)	0.000
	ATLAS	BLSH	4.28	(-3.62, 122)	0.478*
		PMHS	3.31	(0.484, 6.14)	0.0159
	BLSH	PMHS	-0.969	(-8.77, 6.83)	0.987*
Side	Clay	ATLAS	8.54	(3.76, 13.3)	0.0001
		BLSH	8.84	(3.60, 14.0)	0.0003
		PMHS	4.64	(0.640, 8.64)	0.0175
	ATLAS	BLSH	0.295	(-5.36, 5.95)	0.999*
		PMHS	-3.90	(-8.44, 0.631)	0.113*
	BLSH	PMHS	-4.20	(-9.21, 0.814)	0.128*
Rear	Clay	ATLAS	9.20	(6.86, 11.5)	0.000
		PMHS	10.1	(7.67, 12.5)	0.000
	BLSH	PMHS	0.894	(-1.45, 3.23)	0.624*

4. DISCUSSION

It is hypothesised that ballistic helmet performance is influenced by multiple factors and residual backface deformation could be an indicator of this performance. This gives insight into the helmet's ability to stop a round from penetrating and to prevent BHBT injuries from non-penetrating events. This post-test measurement reveals evidence of different deformation patterns for less biofidelic backing materials.

Regional variations in maximum residual BFD calls for additional exploration. The front impact location had the highest maximum residual BFD for all but the PMHS backing condition. This may indicate there are geometric effects caused by the unique interaction between the 3D curvature of the helmet and the backing material. This may be driving the differences when compared to PMHS. The front impact location also had HSD results that did not agree with trends for the other impact locations across backing conditions. Further analysis should be done to confirm these potential sources of region-specific residual deformation effects.

From Figure 9, the standard deviation for some backing conditions is higher when considering the total group compared to each individual impact location. This can be explained by the differences in mean residual BFD between impact locations. The PMHS mean residual BFD was 0.02 mm greater than the ATLAS Headform mean and 1.14 mm greater than the BLSH mean. The clay mean residual BFD was at least 8.60 mm greater than the other surrogate headforms. Considering the standard deviations reported, it is plausible the differences in mean of the PMHS, ATLAS, and BLSH could be within the noise of the measurement systems used to collect residual BFD.

The clay backing material was the only one that resulted in mean helmet deformations at various impact locations greater than the airgap reported earlier. This is evidence of a rebound phenomenon occurring for the load cell and PMHS backed conditions that is much less significant or not present in clay backed tests. This could highlight the role in the stiffness of the backing material to residual deformation, where the typical Young's modulus values in literature of steel (approximately 100-200 GPa) [12] and skull (18 GPa) [13] are larger than clay (150 MPa) [14].

Examining Figure 8, differences can be seen in residual BFD across backing materials, but there is no clear correlation between residual BFD and velocity across the velocity range tested. The expectation would be for a positive linear relationship, but this is not seen across any backing material. The reason for this non-intuitive trend is unclear. This study did not cover a wide range of ballistic velocities. A larger velocity range, especially at lower velocities, may reveal a more intuitive trend between velocity and residual BFD. One hypothesis for the observed behaviour is the stiffness of the backing material has a stronger influence on residual BFD than the threat at this higher range. A study using soft armour and a 9 mm threat saw that backing material stiffness had a significant impact on the deformation response [15]. Another hypothesis is that the volume of the residual backface deformation may have a stronger correlation to velocity than the peak deformation. A previous study found if the backing material is stiff enough, it could arrest deformation in the shot direction where transverse deformation (radial to the shot direction) is unconstrained by backing material stiffness [16]. Further analysis of the residual backface volume would be needed to confirm this hypothesis.

Multiple factors constrain this study and the analysis done on the results. Residual BFD was collected as a one-dimensional metric. Other techniques could have been used in place of this measurement such as volume, surface area, or BFD shape. The one-dimensional metric was chosen for consistency with the clay standard test metric for max deformation. The other metrics have not yet been fully explored for this study, however additional analysis of the results may be suitable for comparison with results of previous studies [16]. Lastly, this study explores the relationship between residual BFD and backing material for one armour-threat pairing, and it is unknown if these findings can be extended to other armour or threats.

5. CONCLUSION

The results of this study show that helmet backing condition affects at least one aspect of an armour performance correlate, residual helmet BFD, and thus it is important to consider helmet backing conditions when performing helmet ballistic testing. Of the four backing conditions evaluated in this study, it is assumed that PMHS best simulates the backing condition of a living individual. For the same range of input velocities, ballistic helmet testing on the clay headform resulted in significantly higher residual BFD measurements than helmets tested on PMHS, whereas helmets tested on the BLSH and ATLAS Headform had residual BFD results that closely aligned to results from PMHS, suggesting these load cell backed headforms are more biofidelic than clay platforms.

To ensure warfighter protection when helmets are fielded, helmets should be tested in a manner that sufficiently simulates the operational environment to accurately assess helmet performance. While the relationship between residual helmet deformation measurements and penetrating and non-penetrating ballistic response of the helmet remains to be determined, the results of this study indicate there could be differences in helmet performance that are not captured accurately when testing helmets on backing materials that are significantly different than PMHS.

Acknowledgements

The authors would like to express their appreciation to the U.S. Special Operations Command for funding this work under Contract No. H92236-10-D-001, Task Order 0521.

 https://doi.org/10.52202/081621-0054

References

[1] Y. Li, H. Fan, X.-L. Gao, "Ballistic Helmets: Recent advances in materials, protection mechanisms, performance, and head injury mitigation," *Compites Part B: Engineering* , p. 109890, 2022.

[2] K. Loftis, E. Matheis and K. Rafaels, "Assessment of Head Injuries: Blunt versus Penetrating," in *Proceedings of the Personal Armour Systems Symposium*, Cophenhagen, Denmark, 2020.

[3] V. Nair, et al., "Review of Department of Defense Test Protocols for Combat Helmets," National Research Council, Washington DC, 2014.

[4] T.-T. Nguyen, et al., "The Critical Role of a Backing Material in Assessing the Performance of Soft Ballistic Protection," *Journal of Transformative Technologies and Sustainable Development,* 2022.

[5] G. James, et al., "The effect of backing methods on the measure ballistic performance of armour materials," Platform Systems Division, Wiltshire.

[6] M. Bevan, et al., "Effect of Backing Material on Residual Armor Deformation," in *Proceedings of the Personal Armour Systems Symposium*, Copenhagen, Denmark, 2020.

[7] Y. Li, et al., "Review of mechanisms and research methods for blunt ballistic head injury," *Journal of Biomechanical Engineering,* 2023.

[8] M. Trexler, et al., "Development of a Modular Load-Sensing Headform," in *Personal Armour Systems Symposium*, Washington DC, 2018.

[9] M. Philippens, et al., "Results of a Round Robin Ballistic Load Sensing Headform," in *Personal Armour Systems Symposium*, Cambridge, 2014.

[10] Biokinetics, "Ballistic Load Sensing Headform," Biokinetics, 24 February 2025. [Online]. Available: https://biokinetics.com/prouct/ballistic-load-sensing-headform/. [Accessed 24 February 2025].

[11] A. Iwaskiw, et al., "Loads Assoicated with behind helmet blunt trauma: Matched-pair load sensing headform tests correlated with skull fracture severity," in *Personal Armour Systems Symposium*, Dresden, 2023.

[12] Z. Chen, "Variation and consistency of Young's modulus in steel," *Journal of Materials Processing Technology,* vol. 227, pp. 227-243, 2016.

[13] A. D. Brown, "Shear behavior of human skull bones," *Journal of Mechanical Behavior of Biomedical Materials,* vol. 116, p. 104343, 2021.

[14] E. Carton, "Characterization of dynamic properties of ballistic clay," in *Personal Armour Systems Symposium*, Cambridge, 2014.

[15] T. Goode, "Soft body armor time-dependent back face deformation (BFD) with ballistics gel backing," *Composite Structures,* vol. 220, pp. 687-698, 2019.

[16] Y. Zhu, "Ballistic impact wave and bulge profile propagation characteristics and blunt injury assessment of UHMWPE laminate composite," *Journal of Materials Research and Technology,* vol. 31, pp. 3760-3780, 2024.

https://doi.org/10.52202/081621-0054

Experimental Assessment of Aluminum Witness Sheets as Human Skin Surrogates for Vulnerability and Lethality Studies

G. Pageau[1], and A. Bouamoul[2]

[1]*Promaxis Systems, 2385 St. Laurent, Ottawa, ON, K1G 6C3, Canada,*
gilles.pageau@ecn.forces.gc.ca
[2]*Defence Research and Development Canada – Valcartier, Department of National Defence, 2459 de la Bravoure Rd., Québec City, Québec G3J 1X5, Canada.*

Abstract. Many ballistic test standards for vulnerability and lethality studies of armour and anti-armour systems specify the use of aluminum witness sheets to assess the potential for body injury or incapacitation from fragment impacts. The aluminium grades, tempers, and thicknesses vary across different standards. Al 2024-T3 sheets, typically 0.5 mm in thickness, are primarily used when performing V_{50} ballistic limit tests to assess the complete perforation of armours as per the Protection criterion. Following a literature review, only limited evidence was found to correlate to some degree the relevance of aluminum witness sheets as a suitable surrogate for assessing human skin and soft tissue injuries during ballistic and explosive munitions tests. The test and analyses conducted at Defence Research and Development Canada (DRDC) as a part of this study showed that none of the three aluminum alloy sheets tested (i.e., 0.5 mm 2024-T3, 0.4 mm 1060-O and 0.5 mm 1000-O) provided a good match to the human skin lower bound perforation threshold over the broad range of relevant projectile sectional densities. From the review and analysis of previous work studies, a 0.55 mm Al 1100-H12 showed better agreement, but further experimental characterisation would be required, especially for the smaller and lower-density fragments. To account for the effect of skin thickness and density, the regression analyses of the V_{50} data were performed as a function of the dimensionless ratio of the skin areal density over the projectile sectional density. The analysis results for the polystyrene foam used in the body armour mannequin demonstrated that the predicted V_{50} response for 50 mm-thick polystyrene foam would fit within the skin perforation targeted response corridor, making such a witness material worth further characterising experimentally.

1. BACKGROUND: THE USE OF WITNESS SHEETS AND PACKS IN BALLISTIC TESTS

The first reported uses of witness sheets in ballistic test applications were documented in a 1948 study by the Battelle Memorial Institute [1] on laminated composite armour tests, followed by a study of Victory Plastics in 1953 [2] on experimental personal armour materials. The witness material specified was Duralumin 24ST, now referred to as Al 2024-T3, with a thickness of 0.51 mm. In 1959, Mascianica [3] published at the Watertown Arsenal Laboratory (WAL) his monograph on ballistic testing methodologies, where he was the first to introduce the concept of Army, Navy and Protection ballistic limits (V_{50}), which is the velocity for which there is a 50% probability for complete penetration of the armour, calculated from the statistical analysis of the binary outcome obtained, i.e., partial or complete penetration. The specific degree of target damage required to consider the projectile as having successfully perforated the armour sample differs under the Army, Navy, and Protection criteria. Under the Protection Criterion (PC), an impact is defined as a complete penetration if the projectile or back spall fragments from the target have sufficient energy to perforate the witness plate mounted securely and parallel to at a given distance (typically 15 cm) behind the target with the presence of a crack or a hole permitting the passage of light. Any impact rebounding from the armour plate, remaining embedded in the target, or passing through the target with insufficient energy to create a crack without visible light passing through is a partial penetration. For the Army Criterion (AC), a complete penetration occurs when a projectile or fragment has penetrated or caused enough damage to permit the passage of light through a hole or a crack in the armour, including cases when the projectile remains embedded in the armour with the front of the fragment or nose of the projectile being visible from the rear of the armour sample. Impacts that create bulges without cracks, allowing light to pass through, are considered partial penetrations. The effects of potentially lethal behind-armour debris from back spalling or scabbing of the armour are not considered with the Army criterion, which is then less severe than the Protection Criterion. For the Navy Criterion (NC), which is the most severe of the three, a complete penetration occurs when the entire projectile or a significant portion passes entirely through the armour, with all other impacts classified as partial penetrations. No witness plates are employed for tests using the Army and Navy criterion, as a significant portion of the projectile must pass through the armour to obtain a complete penetration.

https://doi.org/10.52202/081621-0055

Figure 1 illustrates a case where the projectile became embedded in the armour target, ejecting the plug and piercing the aluminium sheet. The use of aluminium witness sheets has been adopted in many standards (Table 1, Refs. 4-12), including the assessment of potential injuries from the front spall of hard body armour [11].

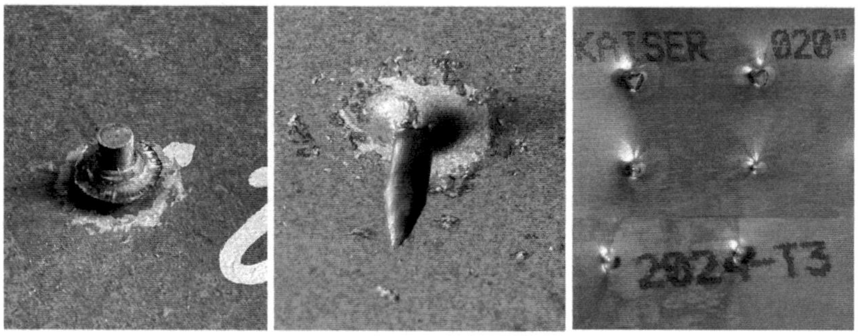

Figure 1: Embedded projectile in an armour plate: front (left), rear (centre), and witness sheet (right).

NATO AEP-2920 [13] is not listed in Table 1 because it does not specify the material type and thickness of the witness sheet. Since the aluminum material grades, temper, and thickness differ between standards, this will likely lead to diverging injury risk assessment and lethality/vulnerability outcomes for the same armour and ammunition conditions which would prevent making reliable performance comparisons, especially for lower areal density armour targets with V_{50} ballistic resistance of the same order as the aluminum witness sheet. The 0.5 mm-thick aluminum witness plate has long been assumed to be a surrogate or simulant for human skin in assessing the potential of soft tissue penetrating injuries. TOP-2-2-710 [5] refers to the study by F. James [14] in 1969 on standardising witness materials for ballistic testing, but no copy of the study report was found. From the literature review, only four relevant studies [15-18] were found that correlated the relative ballistic resistance to projectile penetration between thin aluminum witness sheets and human skin. Lee [15] and Hsiao [16] in Taiwan have explored the validity of using aluminum 1100-H14 in thicknesses ranging from 0.5 to 0.65 mm for assessing the safety and wounding capability of air guns. Jakobsen [17] in Norway conducted tests with a 4.5 mm steel sphere and found that the 0.5 mm aluminum 1050-H4 witness sheet was suitable as a skin simulant for the abdomen and back areas but not adequately simulating other body areas (thorax, thigh, buttocks), raising the need to explore aluminum sheets with lower ballistic resistance for making conservative injury assessments.

Metallic Witness Packs (WPs) also utilise aluminum sheets in a spaced array configuration for behind-armour debris assessment and live-fire arena testing of fragmenting munitions. The WP sandwich structure, with aluminum plates and low-density expanded polystyrene foam layers [19], enables the recovery of fragments between the plates, making them also more rigid when scaled to large dimensions. They have been used extensively during the development of Personal Defence Weapons (calibre 4.6mm and 5.7mm) for assessing the relative incapacitation overmatch capability when defeating the NATO's CRISAT target (20 aramid layers + 1.6-mm titanium plate, [20]), simulating a typical body armour at a range of 100 m. WPs have the advantage of sustaining multiple fragment impacts and maintaining their performance regardless of temperature, which is important for outdoor trials. According to NATO STANAG 4164 [12], the first layer of the standard WP consists of a 1 mm thick 1050-H14 aluminum. G. James [18] found that such a WP configuration was not suitable for assessing soft tissue injuries from secondary fragments (e.g., soil ejecta) during buried Improvised Explosive Device (IED) tests, as only denting of the first aluminum layer occurred without any perforation. By comparing the relative ballistic perforation resistance of the 1-mm aluminum sheet to the one of human skin, he demonstrated that for lower sectional density fragments (i.e., small diameter and low specific gravity), the soft tissue injury severity gets significantly underestimated, such that using WPs for collateral damage and safety assessments should be made with caution. Expanded polystyrene foam with a mass density of 15 kg/m^3 was found to be too fragile and incapable of resisting blast loading. NATO AEP-97 [10] also specifies the use of witness plate target arrays with the first layer made of 0.5-mm Al 2024-T4 for the standard armour V_{50} tests used in the development and acceptance testing of armour systems.

Table 1. Aluminium witness sheet grades and thicknesses used in test standards.

Organisations & Standards:	VPAM PM [4]	FFI [17] TNO Witness Pack [32]	TOP 2-2-710 [5]	MIL-STD-662F MIL-STD-3038 NIJ 0108.01 NATO AEP-55&97 US DOD ESAPI	STANAG 4164 [12] DRDC WP	Taiwan [15, 16]	DRDC Current Study	
Alloys / Properties	2017-T4	1050-H4	5052-H36	2024-T3/T4 (24ST Dural)	1100-H14	1100-H12	1060-O	1000-O
Thickness (mm)	0.5	0.5 FFI 1.0 TNO	0.4	0.5	1.0	0.55	0.4	0.5
Density (kg/m³)	2790	2710	2680	2780	2710	2710	2780	2710
YS (MPa)	276	110	240	270-280	117	92	24	26
TS (MPa)	427	103	280	400-430	124	110	70	90
Elongation (%)	20	8	9	10 to 15	5	10	30	32

YS: Yield Strength; TS: Tensile Strength; FFI: Swedish Defence Agency; TNO: Dutch Defence Agency

2. BALLISTIC TESTS: MATERIALS AND METHODS

The test series to assess the ballistic limit (V_{50}) were conducted on three aluminum grades: the standard 0.5 mm 2024-T3, a 0.4 mm (26 gauge) 1060-O (annealed/soft) flashing roll, and a 0.5 mm (24 gauge) Al-1000-O cold-rolled (see Table 1). The aluminum 1100 and 1060 compositions are identical to the 1050 alloy, as they are part of the commercially pure family (1xxx series). A total of 168 shots were performed for the nine (9) test series. The Al 2024 and Al 1100 witness plates measured 30.5 cm x 30.5 cm, while the Al 1060-O samples had a width of 35.6 cm and were cut into lengths of 45.7 cm and flattened from their roll form. The experimental setup for the test firings is shown in Figure 2. The aluminum sheets were tested at room temperature in the air-backed condition and were secured to a rigid frame with four clamps placed around their edges. The Al 1060-O and Al 1000-O were also tested (series B4 and C2 in Table 2) in conjunction with a 50 mm-thick extruded (XPS) foam backing (Owens Corning Foamular NGX-600), similar to the expanded polystyrene foam (EPS) used in US Army body armour foam mannequins [21]. Aluminum 2024-T3 and 1060-O sheets were tested against 2.5 mm (0.065 g) and 6.35 mm (1.045 g) chrome alloy steel spheres (grade G20, 60 HRC) and the 5.4 mm fragment simulating projectile (un-skirted FSP, 1.105 g [13]). The projectiles were selected to cover a range of shapes and sectional densities (1.3 to 4.8 g/cm²). Aluminum 1000-O was only tested against the 6.35 mm ball. The projectiles were launched with a single-stage gas gun, designed to have interchangeable smooth-bore barrels of various inner diameters that matched the projectile's dimensions, eliminating the need for using sabots. The pressure tank can be filled with helium, nitrogen, or dry air at pressures up to 41 MPa from high-pressure cylinders, allowing velocities of up to 750 m/s ± 5 m/s with the 5.4 mm FSP. The V_{50} tests were conducted under standard conditions of AEP-2920 [13], using the up-and-down firing method as outlined in Annex H to determine the number of shots required to obtain reliable V_{50} estimates. The required velocities were attained by adjusting the gas pressure.

Figure 2: DRDC gas-gun test setup with the high-speed camera (a), mirrors (b), and launch tube (c).

https://doi.org/10.52202/081621-0055

The plate perforation assessment was performed after each impact using the Army criterion. A shot spacing of 10 calibres was used; however, for the softer Aluminium alloys (1060-O and 1000-O), the spacing was increased to prevent overlapping of damage, as shown in Figure 3. For the tests with the 2.5- and 6.35-mm spheres, the target was positioned at a distance of 30.5 cm from the gun barrel, while for the tests with the 5.4 mm FSP, the muzzle-to-target distance was increased to 45.7 cm to allow for using a double mirror system [22] to measure the projectile yaw. The impact velocity of all three projectiles and the yaw angle for the 5.4 mm FSPs were measured using a high-speed video camera (Photron FASTCAM SA-Z) with a rate of 15,000 frames per second. The Yaw HSV image analysis software [23] was used to compute the FSP total yaw angle at 15.2 cm from the witness sheet from the two orthogonal views of the FSP in flight. For the FSP tests, a gun extension was used to help limit the initial yaw, which remained highly variable, ranging from 2° to 10°. More shots were fired to obtain a sufficient number of partial penetrations with yaw angles lower than 5°. The instability of the 5.4 mm FSP is likely caused by firing from a smooth-bore gas gun, which is necessary for achieving low-impact velocities not achievable with powder guns having rifled barrels.

Table 2. Projectile types, aluminium grades, and thicknesses tested with V_{50} ballistic limit (±95% CI) results. (S: series, #: number of shots, T: thickness, AD: areal density, SD: sectional density, AC/NC: Army/Navy criteria).

S #	Projectile	Mass (g)	Aluminum grade & temper	T (mm)	Density (g/cm³)	AD (g/cm²)	SD (g/cm²)	$V_{50\text{-}AC}$ $V_{50\text{-}NC}$ (m/s)	ΔV_{50}* (%)
A1 18	2.5mm ball	0.065	2024-T3	0.5	2.78	0.139	1.32	178.7±3.4 209.6±14.7	+28.0 +50.1
A2 22	6.35mm ball	1.045	2024-T3	0.5	2.78	0.139	3.30	117.4±1.6	+16.5
A3 14	5.4mm FSP	1.105	2024-T3	0.5	2.78	0.139	4.83	77.9 (0° yaw)	-11.6
B1 17	2.5mm ball	0.065	1060-O	0.4	2.78	0.111	1.32	138.2±0.1 148.8±7.7	-0.9 +6.7
B2 17	6.35mm ball	1.045	1060-O	0.4	2.78	0.111	3.30	67.1±0.1	-33.4
B3 22	5.4mm FSP	1.105	1060-O	0.4	2.78	0.111	4.83	52.9 (0° yaw)	-40.0
B4 17	6.35mm ball	1.045	1060-O + XPS foam	0.4	2.78	0.111	3.30	72.2±0.4	-28.3
C1 25	6.35mm ball	1.045	1000-O	0.5	2.71	0.136	3.30	110.5±2.1	+9.7
C2 16	6.35mm ball	1.045	1000-O + XPS foam	0.5	2.71	0.136	3.30	112.4±4.4	+11.5

* Relative to human skin Probit V_{50} equation from James [18] for adult thigh spherical/pointed fragments

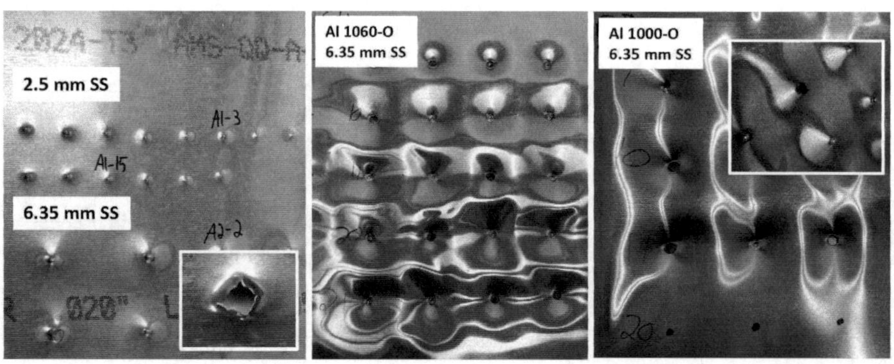

Figure 3: Post-impact damages on witness sheets: 2024-T3 (left), 1060-O (centre), 1000-O (right).

https://doi.org/10.52202/081621-0055

3. BALLISTIC TESTS: RESULTS AND ANALYSIS

The V_{50} results were obtained by performing a logistic regression analysis of the fair shots for each test series using the V_{50} Assist software [23] and the Probit link function with 95% confidence intervals (Table 2). For test series A3 and B3, logistic regressions were performed using two independent variables (velocity and yaw) and with the penetration status as the dependent variable. The probability plot obtained using the Test Science application [24] is shown in Figure 4 (left), with similar results presented in a different format using the V_{50} Assist Yawgit function [23]. For the tests with 2.5 mm spheres (A1, B1), gradual cracking failures of the aluminium sheet occurred at velocities near V_{50}, leading to different interpretations of the penetration outcome (partial or complete) depending on whether the Army or Navy criterion is applied. The analyses were then performed by applying both the Army and the Navy criteria (red colour in Table 2) to allow comparison with previous studies. The Navy criterion is often used since it is less prone to misinterpretations. For the tests with the 6.35 mm sphere and the 5.4 mm FSP, failure by cracking did not occur; instead, the plates failed due to a combination of petalling and plugging (Figure 3), with no perforation issue observed between the Army and Navy criteria. For the V_{50} values computed using the Army criterion, the width of the 95% confidence intervals is less than 4 m/s due to the very narrow zones of mixed results, making the V_{50} values reliable even though only one repeat was performed for each test series. The human skin probit V_{50}s were computed using James' [18] perforation equation, with coefficients for intact adult thigh skin versus spherical and pointed fragments. The difference between the human skin V_{50} values (139.6 m/s for the 2.5 mm sphere, 100.8 m/s for the 6.35 mm sphere, and 88.1 m/s for the 5.4 mm FSP) and those obtained for the tested configurations is provided in the last column of Table 2. It can be observed that the perforation resistance of 0.5 mm Al 2024-T3 is 28% higher for the 2.5 mm sphere and 16.5% for the 6.35 mm sphere, while being 11.6% lower for the 5.4 mm FSP. A reverse trend is observed, with the 0.4 mm Al 1060-O being 25% easier to perforate on average than human skin. The 0.5 mm Al 1000-O alloy provides a closer match to human skin, with an 11.5% higher V_{50} for the 6.35 mm sphere. The foam backing used in test series B4 and C2 did not significantly increase the V_{50s} obtained compared to the no-foam condition.

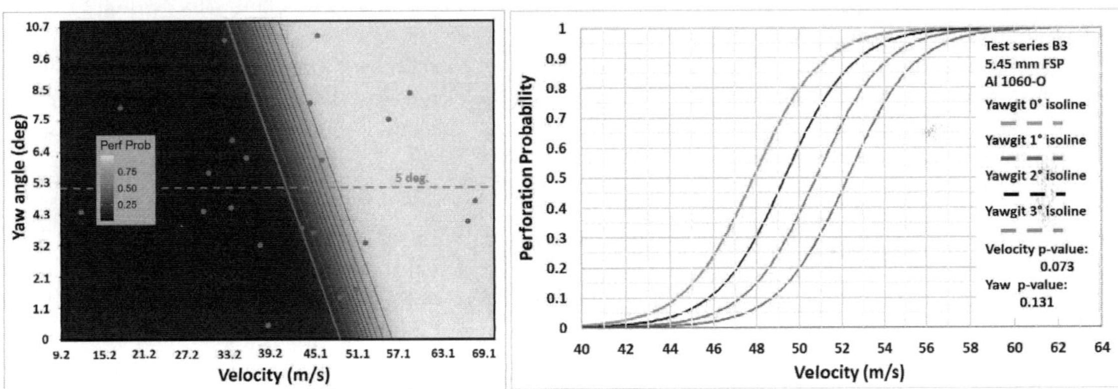

Figure 4: Logistic regression probability plots for the B3 test series: Test-Science (left), and Yawgit (right).

4. COMPARISON WITH PREVIOUS STUDIES AND TARGETED RESPONSE SELECTED

The logistic regression curves (Probit) obtained for the 2.5 mm sphere versus the 0.5 mm Al 2024-T3 witness sheet are compared in Figure 5 with the experimental one provided by Hudgins [25] for goat skin impacted by 2.38 mm (0.055 g) steel spheres. The skin perforation equations by James [18] are presented in Figure 5, along with the predicted V_{50} value for goat skin against the 2.38 mm sphere. The Probit curves obtained for the 0.5 mm Al 2024-T3 are shown to provide significantly higher ballistic resistance for such a low sectional density (SD) projectile compared to the experimental data, highlighting the risk of underestimating skin perforation injury risks from these smaller fragments when using such a witness sheet specified in most standards.

https://doi.org/10.52202/081621-0055

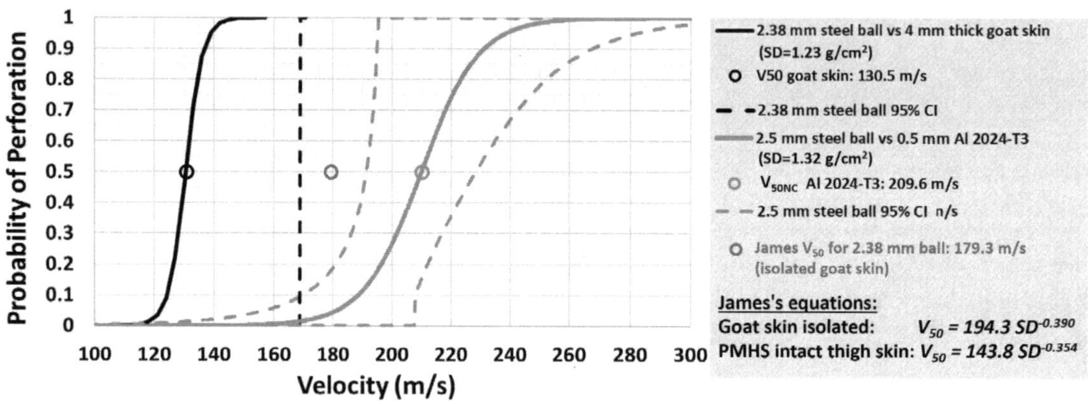

Figure 5: Probit of 2024-T3 vs. 2.5 mm ball (green curves) and goat skin vs. 2.38 mm ball (black curves, [29]) compared to James's equation prediction (red dot) for goat skin vs. 2.38 mm ball.

For better assessing the suitability of the various aluminum witness materials for estimating the risk of skin injury, relevant data from previous studies with 1-mm thick aluminum witness sheets [26-28] and recently published data on Post Mortem Human Subjects (PMHS) [29] and porcine skins [30] were analysed taking into account the effect of target thickness (T) of the aluminum sheet and PMHS/animal skins by following the approach of Cunniff [31] where the V_{50} of armour materials is correlated to the dimensionless areal density ratio: $\boldsymbol{CR = AD/SD}$, and where $\boldsymbol{AD = \rho \cdot T}$ is the target Areal Density and $\boldsymbol{SD = Mp/Ap}$ is the projectile Sectional Density with **Mp**: projectile mass, **Ap**: cross-sectional area, **T**: target thickness, and **ρ**: mass density. The skin perforation model of Hudgins [29] employs a similar parameter (**P**), **where** $\boldsymbol{P = 2/CR}$. In addition to accounting for target thickness, this correlation is more intuitive, as a target thickness of zero (**CR=0**) results in a null ballistic limit (**V_{50}=0**), thereby preventing a vertical asymptote at low values of **SD** and **P**. The results are displayed in Figure 6 using a logarithmic scale on the horizontal axis to better visualise the V_{50} response for smaller fragments. When skin thickness data were unavailable, a value of 2 mm was used based on the mean thickness reported in the literature.

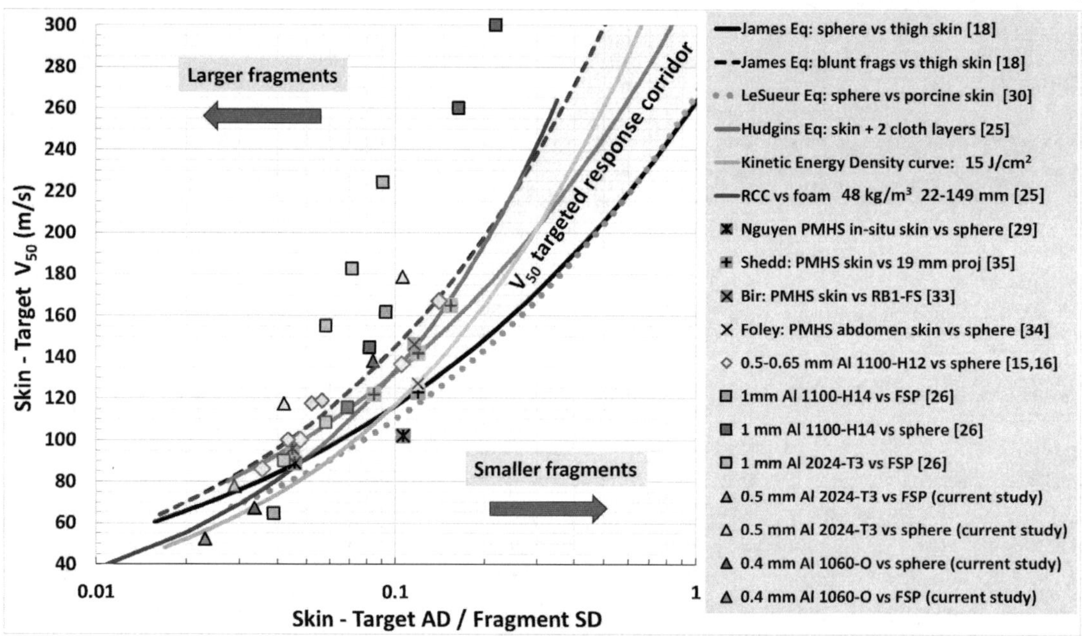

Figure 6: V_{50} vs. AD/SD for aluminum, PMHS/animal data and polystyrene foam.

In Figure 6, the skin perforation equation from James [18] for the thigh area, with the coefficients for spheres used to provide the lower bound for the smaller and lower-density fragments. The upper bound of the proposed targeted V_{50} response corridor was computed using James' equation with the coefficients for blunt fragments. When considering skin thickness, the perforation resistance curves from Hudgins [25] and LeSueur [30] follow the lower bound from James's model for spheres. Kinetic Energy Density (KED) has also been proposed as a threshold for skin perforation, with reported values varying considerably between 5 and 25 J/cm^2 [32]. A KED value of 15 J/cm^2 provides a curve fitting midway between the lower and upper bounds defined by James's equation; however, it underpredicts V_{50}s for the larger and higher-density fragments. Using larger KED threshold values (e.g., 20 J/cm^2) would shift the curve upward, making human vulnerability assessments less conservative. The data points from Bir (anterior on rib, [33]), Foley [34], and Shedd [35] for non-lethal projectiles on PMHS and animal skins are also shown to be consistent, as they lie within the targeted response corridor. It can be observed that the ballistic resistance of the 1-mm 2024-T3 and 1-mm 1050-H14 aluminum sheets used as the first layer in witness packs falls well outside the corridor with much higher V_{50}s, making them unsuitable for fragment hazard studies, as the injury risks to personnel would be underestimated. The Hudgins perforation equation [25], with coefficients for skin covered by two cloth layers, is shown to give a good average fit between the proposed targeted response corridor, which can be used to evaluate the suitability of candidate surrogate skin materials.

Regression analyses of the data points in Figure 6 for the various aluminum grades and thicknesses, as well as the data of LeSueur for porcine skin, were performed using Curve Expert Professional software [36] and a two-parameter power function. The data of the body armour mannequin [21], designed for live-fire testing, were also curved fitted (light blue curve in Fig. 6) for the three thicknesses (22 mm, 73 mm, 149 mm) of extruded polystyrene (EPS, 48 kg/m^3 density) foam tested which was tested against the four standard US Right Circular Cylinder (RCC) fragments. The EPS foam response can be seen to fall mainly within the targeted response biofidelity corridor. The curve fit equations, which were performed as a function of AD/SD, were then re-computed as a function of the reciprocal of the projectile sectional density, referred to as the form factor **F = 1/SD**. The calculations were performed using the original AD values for each case except for the polystyrene foam, where the predicted curve was computed for the more standard panel thickness (50 mm) and density (35 kg/m^3). The curves obtained are shown in Figure 7, along with the 95% confidence and 95% prediction bands from the regression analysis of LeSueur's data, which encompass the previously defined response corridor (red curves).

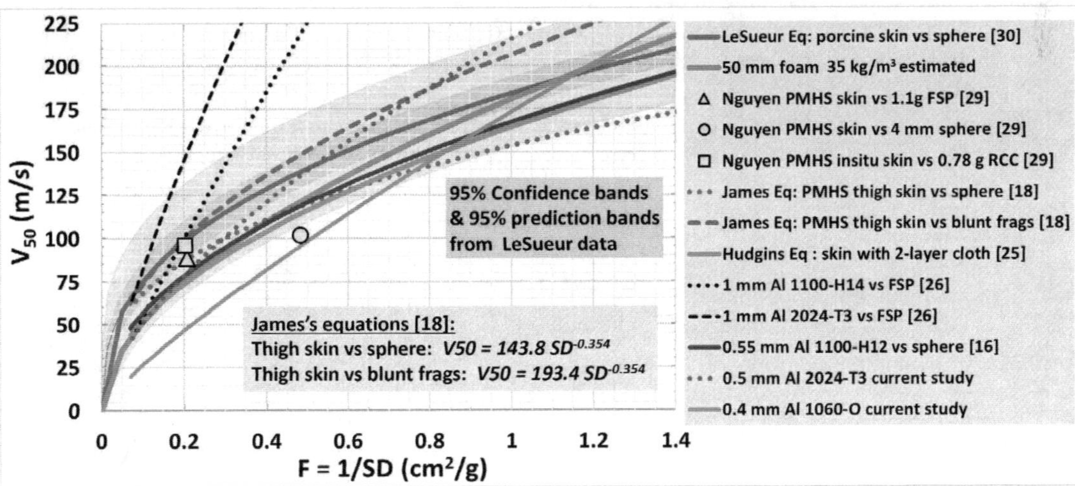

Figure 7: V_{50} versus 1/SD for aluminum, PMHS/animal data, and 50 mm foam (35 kg/m^3).

In Table 3, the areal density of XPS foam panels (50 mm thick, 35 kg/m^3 density), more easily available than EPS foam, is shown to lie between the AD for the 0.5 mm aluminum sheet and the one for human skin (2.5 mm thick).

https://doi.org/10.52202/081621-0055

Table 3. Aluminium and EPS/XPS foam material properties compared to human skin.

Material	Thickness (cm)	Density (g/cm³)	AD (g/cm²)	Strength
Aluminum alloys	0.050	2.7	0.135	24-280 MPa (tensile)
XPS Foam 35 kg/m³	5	0.035	0.175	0.41 MPa (compressive)
EPS Foam 14 kg/m³	5	0.014	0.070	0.07 MPa (compressive)
Human skin	0.25	1.1	0.275	17-28 MPa (tensile)

The overall trends in Figure 7 are similar to those observed in Figure 6, with the 1 mm thick aluminum sheets in both grades exhibiting significantly greater perforation resistance than animal and human skin for form factor (F) values beyond 0.2 cm^2/g. The linear horizontal scale used in Figure 7 explains the changes in the shape of the V_{50} response curves relative to those presented in Figure 6. Some minor curve shifting also occurred, as not all data points in Figure 6 from previous studies were for the same skin thicknesses. The 0.5mm Al 2024-T3 response (dotted blue line) remains above the lower band based on James's equation (dotted red line) for the full range of form factor while exceeding the James upper band for F values beyond 0.8 cm^2/g. The 0.4 mm Al 1060-O is shown to have insufficient ballistic resistance for F values under 0.8 cm^2/g. The regression line from Hsiao's data [16] for the 0.55 mm 1100-H12 aluminum aligns with the lower corridor and could be considered more suitable than the standard 0.5 mm Al 2024-T3. It would be worthwhile to explore the response of thinner sheets (e.g., 0.45 and 0.5 mm) of the same alloy over a broader range of projectile sectional densities and shapes (i.e., RCC, FSP) to validate whether a better match could be obtained for F values beyond 0.45 cm^2/g. The recent data for PMHS skin from Nguyen [32] fall within the response corridor, except for the 4 mm sphere, which falls below it.

The predicted response (green line) for the 50 mm-thick polystyrene foam is situated in the central area of the proposed corridor, indicating that such a foam could be used as a replacement for aluminum sheets for ballistic limit tests of armour materials using the Protection criterion. The significantly higher strength of the 35 kg/m^3 XPS foam compared to the 14 kg/m^3 EPS foam tested by James [18] suggests that this material could be suitable for live fire explosive trials involving combined blast and fragment loadings. Its performance is anticipated to be comparable to that of the ARL body armour mannequin, which is constructed with a slightly denser foam grade (48 kg/m^3). The foam panel would provide the advantage of being able to recover the embedded fragments from overmatch shots and measure the Depth of Penetration (DoP) of the fragments, which is impossible with witness sheets, where only a binary outcome is obtained. Figure 8 compares the DoP response of PMHS (red curve, [18]) to that of EPS foam (black curve, 48 kg/m^3 density [24]) for impacts with 16-grain RCC (1.037 g, 5.5 mm) surrogate fragments. The DOP response curves of polystyrene foam and PMHS soft tissues are observed to differ less than anticipated. The foam material exhibits a smoother response, characterised by deeper cavities, particularly at low velocities, with no identifiable threshold velocity required to initiate penetration. The foam's greater compliance would allow for improved velocity discrimination from the DoP measurements.

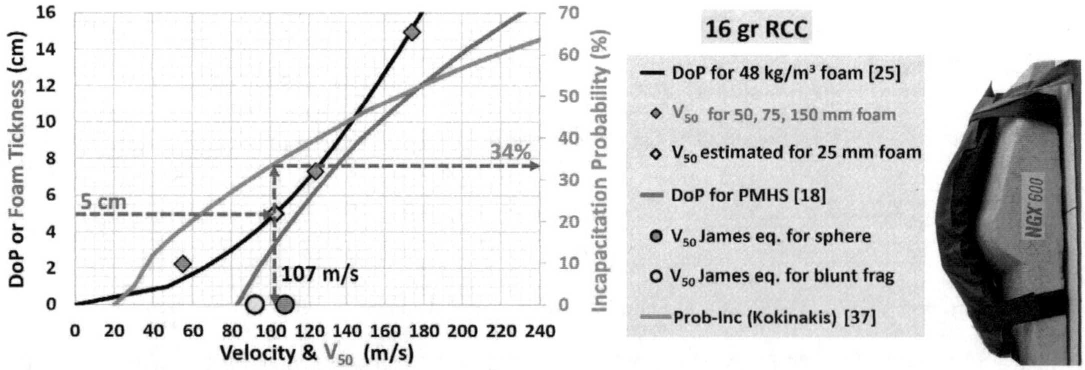

Figure 8: DoP response of XPS foam vs PMHS with V_{50} values and incapacitation probability for the 16-gr RCC (left); an example of a shaped XPS foam backing for armour testing (right).

The corresponding incapacitation probabilities were also computed using the Kokinakis-Sperrazza incapacitation model [37], employing coefficients for the entire body in a defensive role. In Figure 8, a DoP of 5 cm within the 48 kg/m^3 polystyrene foam is shown to correspond to a probability of incapacitation of 34% using the impact velocity (107 m/s) derived from the foam's DoP response curve (black line).

5. SUMMARY AND CONCLUSIONS

The ballistic resistance of standard 2024-T3 aluminum alloy, along with two potential substitute aluminum witness sheets (0.4 mm Al 1060-O and 0.5 mm Al 1000-O), with lower mechanical strength due to their annealed condition, was evaluated against three projectiles with varying sectional densities and shapes (2.5 mm and 6.35 mm steel spheres, and 5.4 mm FSP). Ballistic data from relevant studies on aluminum alloys used in metallic witness packs (1 mm Al 2024-T3 and Al 1100-H14), as well as those from the ARL body armour foam mannequin, were also included for comparison. The V_{50} data for the various skin and material configurations were correlated against the Cunniff dimensionless areal density parameter. The data from previous studies using non-lethal projectiles demonstrated greater consistency when analysed using this approach, which is physically more relevant since it accounts for skin thickness and for which V50 = 0 when AD = 0. The data analyses conducted also confirmed that the ballistic resistance of a 1-mm-thick aluminium used in witness packs is significantly too high, leading to non-conservative injury assessments during vulnerability-lethality trials. Although the V_{50} of 0.5 mm 2024-T3 aluminum is substantially lower than that of 1 mm thick sheets, it remains notably above the lower bound of the targeted response corridor derived from James' equation [18] for PMHS adult skin subjected to spherical or pointed fragments, particularly for the higher 1/SD form factors corresponding to lower SDs. The 0.55 mm 1100-H12 aluminum witness sheet configuration more closely matched the targeted response corridor of the PMHS thigh skin, making it a more suitable alternative to the 2024-T3 alloy as a skin surrogate. However, thin 1100-H12 aluminum sheets are less readily available compared to 2024-T3, which could potentially restrict wider adoption. The analysis of the foam mannequin data indicated that the ballistic resistance of 50 mm-thick panels with a density of 35 kg/m^3 would fall within the response corridor, where the upper bound was defined using James's equation for PMHS adult thigh skin with the coefficient for blunt fragments. EPS and XPS foam materials, being rigid in contrast to soft tissues and gelatin, which are hyperelastic, would not be suitable as a backing material for replicating the as-worn condition. However, a foam panel could replace the 2024-T3 witness sheet for testing fragmentation-resistant body armour material in air-backed conditions when secured onto rigid frames. Polystyrene foam panels would also allow for the soft recovery of target fragments when applying the ballistic limit Protection criterion, as the effect of backing spalling/plugging of the target must then be assessed. Future experimental tests are needed to validate which combination of polystyrene foam grades (EPS or XPS), densities and thicknesses would provide the best response compared to the proposed targeted V_{50} corridor. The ballistic resistance of Al 1100-H12 with lower thicknesses (0.45 and 0.5 mm) should also be characterised over a broad range of sectional densities against projectiles of various shapes (sphere, RCC, FSP) to identify which thickness would better fit the skin perforation targeted corridor. One issue requiring further assessment is the effect on V_{50} of the edge sharpness of non-spherical fragments (FSP, RCC, cubes) [38, 39], which could explain some of the discrepancies in the reported literature. Future research involving PMHS and animal skin samples should evaluate the V_{50} response for lower densities, non-metallic, and smaller fragments, for which the available data are limited. The thickness, body location, and condition (in situ or isolated) of the skin samples used should be well-documented, as they significantly impact perforation resistance. The skin perforation criterion should also be standardised to enhance the repeatability of the results needed for developing improved skin surrogate materials.

References

[1] H. W. Kuhlmann, et al., Interim report on the development of laminated armor, Battelle Memorial Institute, Contract W33-019, Sept 1948, https://apps.dtic.mil/sti/tr/pdf/ADA954855.pdf.

[2] Victory Plastics Co., Ballistic test panels of experimental and standard personal armor materials, Watertown Arsenal Laboratory, File No. WAL 710/1025-3, Dec. 1953, https://apps.dtic.mil/sti/tr/pdf/AD0026368.pdf

[3] Mascianica, F.S., Ballistic Concepts Employed in Testing Lightweight Armor, Watertown Arsenal Laboratory, Monograph Series, Oct 1959, https://apps.dtic.mil/sti/pdfs/AD0609690.pdf

[4] VPAM-PM, Bullet resistant plate materials Requirements, classification, and test procedures, V3-E15, 2021.

 https://doi.org/10.52202/081621-0055

[5] US Army, Ballistic tests of armor materials, TOP 2–2–710, Aberdeen Proving Grounds, 1984.
[6] MIL-STD-662F, V50 Ballistic Test for Armor, US Department of Defense standard, Dec 1997.
[7] MIL-STD-3038, Test Methods for Ballistic Defeat Materials, US Department of Defense, May 2011.
[8] National Institute of Justice NIJ, Ballistic Resistant Protective Materials, Standard 0108.01, Sept 1985.
[9] NATO AEP-55, STANAG 4569, Procedures for evaluating the protection level of armoured vehicles -Vol 1: kinetic and artillery threat, NATO Standardization Office (NSO), May 2014.
[10] NATO AEP-97, Multi-Calibre Manual of Proof and Inspection (M-CMOPI) for NATO Small Arms Ammunition, NATO Standardization Office (NSO), October 2020.
[11] US DOD, Purchase Description Enhanced Small Arms Protective Insert (ESAPI), CO/PD 04-19H, 2013.
[12] NATO STANAG 4164, Test Procedures for Armour Perforation Tests of Anti-Armour Ammunition, NATO Standardization Office (NSO), AC/225 D/1375, Nov. 1998.
[13] AEP 2920, (Ed. A, V2), Procedures for the evaluation and classification of personal armour, bullet and fragmentation threats, NATO Standardization Office, Brussels, Sept. 2016.
[14] Frederick, James C., Final Report of Technique and Study of. Witness Materials for Tests Standardization (Ballistic test) (U), Report APG-MT-3187, Tecom Project 9-9-0008-73, Jan. 1969.
[15] Lee, H.C, Meng. H.H., The development of witness plate method for the determination of wounding capability of illegal firearms, January 2011.
[16] Hsiao, Y. T., Meng, H. H., Evaluation of wounding potential of airguns using aluminium witness plates. Australian Journal of Forensic Sciences, 52(4), 2018.
[17] Jakobsen, M.F., An analytical review of perforation statistics – using aluminium as witness plates in perforation experiments, Norwegian Defence Establishment (FFI), FFI Report 23/02387, Dec. 2023.
[18] James G., Development of Models to Assess Penetrating Injury from Ballistic Projectiles, PhD Thesis, Cranfield University, UK, 2020.
[19] McIntosh, G., The Effect of Styrofoam Interlayers in Fragment Witness Packs, DREV-TM-1999-110.
[20] Saw, D., https://euro-sd.com/2025/02/articles/42600/pdws-a-revolution-that-never-quite-happened/
[21] N. Moholkar, Characterising the Ballistic Performance of Polystyrene Foam for the Development of Body Armor Mannequin, DEVCOM, Army Research Lab, Report ARL-CR-0835, Oct 2018.
[22] https://biokinetics.com/product/openbox/
[23] Magnan, S., Pageau, G., et al., "Beyond V50, A More Comprehensive and Efficient Methodology for Assessing Armour Performance", Proc. PASS-2023, Dresden, Germany, 2023,
[24] IDA, Binary data logistic regression, https://test-science.shinyapps.io/LogisticRegression/
[25] Hudgins, H., 2005. Estimating ballistic limits of skin and clothing for projectiles, 40th Annual Armament Systems: Guns - Ammunition - Rockets - Missiles (GARM).
[26] Y. Baillargeon, C. Lalanne, Hole Growth Equation Related to Al-1100 Witness Plates, Defence R&D Canada, Report DRDC-TM-2003-122, August 2004.
[27] D. Bourget, Y. Baillargeon, S. Northrop, Witness Pack Calibration for Human Vulnerability Assessment, Proc. 20th Int. Symposium on Ballistics, Orlando, Florida, US, Sept. 2002.
[28] Verolme, J.L., M. Szymczak, and J. Broos, Metallic witness packs for behind-armour debris characterisation, International Journal of Impact Engineering, Volume 22, Issue 7, August 1999.
[29] Nguyen TN, Tsukada H, James GR, et al, Experimental model and simulant for studying blast penetrating injury to the skin, J Mech Behav Biomed Mater, May 2025.
[30] LeSueur J, Koser J, Hampton C, Kleinberger M, Pintar FA. Penetration Thresholds of Porcine Limbs for Low Sectional Density Projectiles in High-Rate Impact. Mil Med. 2024 Aug 19;189(Suppl 3).
[31] Cunniff P., Variability in ballistic impact performance due to projectile physical properties and dimensions, Proc. 24th Int. Symp. on Ballistics, New Orleans, LA, Sept 2008.
[32] Pandey, P.K., et al, Review of Interaction of Bullets and Fragments With Skin-Bone-Muscle Parenchyma, ASME J of Medical Diagnostic, Paper No: JESMDT-24-1044, Nov 2025.
[33] Bir, C. A., Stewart, S. J., & Wilhelm, M., Skin Penetration Assessment of Less Lethal Kinetic Energy Munitions, Journal of Forensic Sciences, 50(6), 2005.
[34] Foley, S. et al., Evaluation of Skin Penetration from Less Lethal Impact Munitions, 33rd Int. Symp. on Ballistics, 2023, Bruges, Belgium, https://ibrc.osu.edu/wp-content/uploads/2023/05/Foley-IBS-Paper.pdf
[35] Shedd, D., et al., The risk of skin injury caused by high-rate blunt impacts to the human thorax. Human Factors and Mechanical Engineering for Defense and Safety, Vol. 6, No. 5, 2022.
[36] Hyams, D.G., 2021, https://curveexpert-professional.software.informer.com/1.6/
[37] Kokinakis W, Sperrazza J., Criteria for incapacitating soldiers with fragments and flechettes, Ballistic Res Lab 1965, Available at https://apps.dtic.mil/sti/citations/AD0359774.
[38] Cline J., et al., The ballistic response of woven Kevlar fabric as a function of projectile sharpness, US Army Research Laboratory Report ARL-TR-8694, May 2019.
[39] F. Barbillon, et al., review of fragment simulating projectiles definition and associated quality controls needed, Proc. PASS-2020, Copenhagen, Denmark, October 2021.

Review of Bladed Weapon Threats

J. Barnes-Warden[1], P. Mahoney[2]

[1]*Metropolitan Police Service, Physical Protection Group, 60 Albany Street, London, NW1 4EE*
jane.barnes-warden@met.police.uk
[2]*Cranfield University, Shrivenham, Swindon, SN6 8LA*

Abstract. Knife crime is prevalent in the UK with the Office of National Statistics recording 50,510 bladed weapon offences in the year ending March 2024. Hence, bladed weapon attacks are a significant threat to the public and policing, with background knowledge of these types of attacks being highly relevant to law enforcement. There is a considerable body of published work ranging from academic studies investigating the mechanics of knife attacks and real-world injury profiles, as well as open-source reporting of incidents. This study summarises relevant publications, whilst offering a perspective from the Metropolitan Police Service of the current threat. Finally, it proposes areas of future development to improve the preparedness of frontline responders to bladed weapon threats.

1. INTRODUCTION

This paper draws on government funded, national statistics to quantify the motive and frequency of knife attacks within the UK. To contextualise the blade weapon threat, these statistics will be compared with data for known threats, such as firearm attacks. Furthermore, national statistics and a case study from a major trauma centre emphasise the clinical consequences of bladed weapon attacks on the public and the emergency services.

There is a considerable body of published work ranging from academic studies investigating the mechanics of knife attacks and real-world injury profiles, as well as open-source reporting of incidents. The principal aim of this paper is to outline key issues relating to reported knife attacks and the supporting academic literature. This study summarises relevant publications, whilst offering a perspective from the Metropolitan Police Service of the current threat. Finally, it proposes areas of future development to improve the protection of frontline responders to bladed weapon threats.

2. POLICE STATISTICS FOR BLADED WEAPON ATTACKS

According to police data recorded in England and Wales, published by the Office of National Statistics (ONS), knife-enabled crime in the year ending March 2024 saw a four percent increase to 50,510 offences, from 48,409 in the previous recording year.

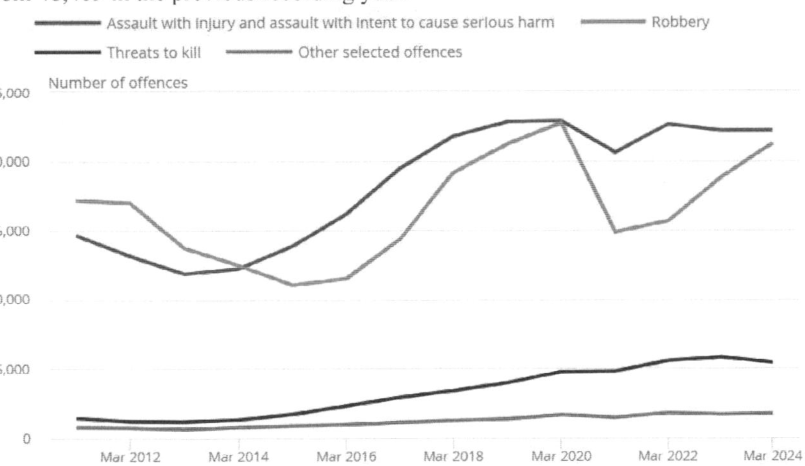

Figure 1. Graph reproduced from the ONS website detailing knife-enabled crime [1]

The majority of knife crimes, in year ending March 2024, were categorised as "assault with injury and assault with intent to cause serious harm" (n=22,617), this is closely followed by "robbery"

https://doi.org/10.52202/081621-0056

(n=21,226) which has steeply risen by 13 percent, compared with the previous year. Finally, reported "threats to kill" (n=5,411) were 7 percent lower, compared with March 2023 (n=5,801) [1].

2.1 Offences involving firearms

The ONS also publish police data relating to recorded offences involving firearms. These statistics include various weapon types including imitation firearms, handguns and shotguns. In the year ending March 2024, in England and Wales, there were 6,268 reported offences involving firearms [2]. Hence, when comparing UK recorded data for offences involving firearms with offences involving knife-enabled crime, firearms offences are approximately one eighth in frequency.

3. CLINICAL STATISTICS FOR BLADED WEAPON ATTACKS

In the UK, the National Health Service (NHS) has published counts of inpatient finished admission episodes (FAEs) with a cause code of 'assault with a sharp object' [3]. Figure 2 below illustrates data, between years ending 31st March 2013 to 31st March 2022, namely the total number of "assaults with sharp object" and the proportion of these assaults on persons under 25. In this dataset, patients under 25 represent 38.2 to 41.8 percent of the total FAEs [3].

Furthermore, it is interesting to note that that, although the police and clinical data sets are from different sources, the typical frequency of "assaults with sharp object" is circa one tenth of the total data recorded by police for knife-enabled crime.

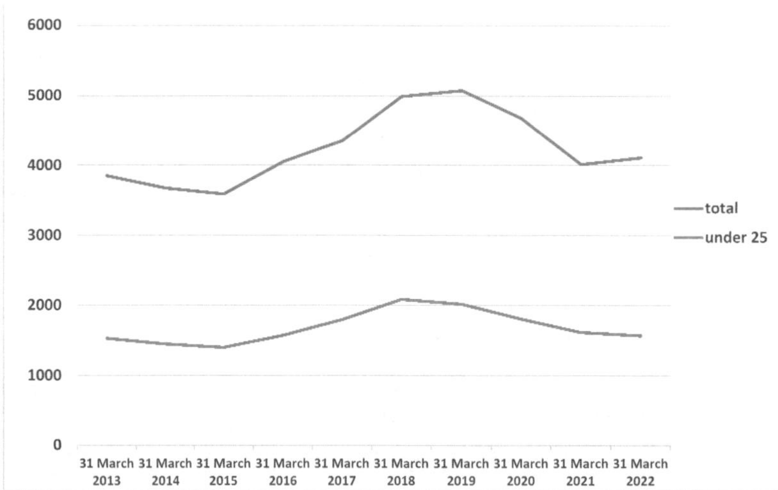

Figure 2. Graph reproduced from NHS data for 'assault with a sharp object' [3]

3.1 Case study of bladed weapon attacks from a UK trauma centre

A research paper by Malik et al, published in EClinicalMedicine in 2020, concludes that knife injuries constitute nearly 13 percent of a UK trauma centre workload (Birmingham, UK) [4]. Reoccurring trends include recidivism, violent, intoxication in males, (who constitute 93 percent of the patients). Domestic violence was predominated in females, constituting the remaining 7 percent of patients. This paper also includes data relating to the age of the patient. For the three years considered under 25s were 39 percent, 38 percent and 51 percent, respectively, of the reported population.

Malik et al plot the time of arrival of patients with knife injuries to the Trauma Centre and there were higher frequencies of patients on Friday evening from 1930 to 0030, on Saturday evening from 2130 to 0330 and Sunday evening from 2230 to 0530 [4].

The description of the type of bladed weapon in patient records was often ambiguous, as in 63.7 percent of cases the bladed weapon was described as a "knife". Household knives were specifically mentioned in 9 percent of patient records and small folding blades were cited as a causative weapon in less than 3 percent of patients. Machetes were named as the cause of knife injuries in 14 percent of UK major trauma centre patients [4].

Finally, the estimated financial cost of treatment for a stab victim is £4375 (£2574 to £8490; circa 2020) [4].

4. UK LAWS ON CARRIAGE AND POSSESSION OF KNIVES

In the UK, with the exception of non-locking folding pocket knives no longer than 3 inches, it is illegal to carry most knives in public without a good reason [5].

Acceptable reasons include carrying knives for work (such as a chef or deer manager) or for religious reasons (for instance a Sikh carrying a kirpan). However, some knives in the UK law are deemed prohibited to possess, sell or hire out, lend or give, or bring into to the UK. Examples of such knives include a:

- belt buckle knife
- butterfly knife or balisong
- cyclone or spiral knife
- disguised knife
- flick knife / gravity knife
- push dagger
- stealth knife
- zombie knife [5].

4.1 Metropolitan Police Service Bladed Weapon Data

In the Metropolitan Police Service (MPS), Physical Protection Group (PPG) conduct physical counts of confiscated bladed weapons within Criminal Exhibits. This work is to determine the types of bladed weapons seized within the Metropolitan Police District, ie nominally Greater London. Over the last 15 years, the majority of knives impounded are domestic, ie kitchen knives as illustrated in figure 3 [6].

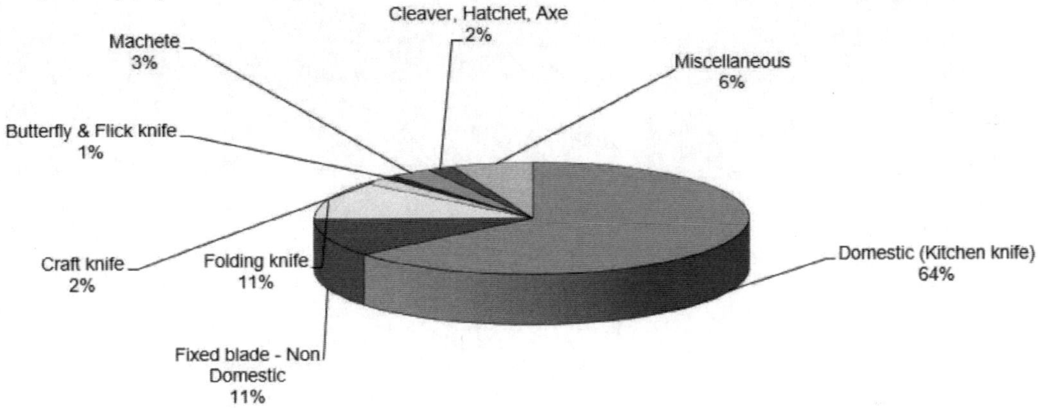

Figure 3. Pie chart illustrating MPS bladed weapon analysis 2022 reproduced from reference [6]

In the latest blade weapon analysis, the length of each knife was also measured. A summary of the data showed that 56 percent of knives were less than 150mm (6") in length [6] and, therefore, easy to conceal.

5. CONCEALED CARRIAGE AND GRIP OF KNIVES

5.1 Concealment of knives

A study by Mahoney et al [7] investigating the use of concealable and disguised knives, reports that a concealed knife, or knives, are typically carried at various points on a person's body including in:

- a belt
- a boot
- a trouser leg
- around the neck
- up a sleeve
- across the back [7].

However, it is the clothing worn over the concealed knife that determines the ease of accessing and drawing the knife. Loosely worn garments, ie a shirt or t-shirt, which are not tucked into a waistband are far easier to "clear" than tight fitting garments, which may be hindered by the interaction with other garments. This enables the person carrying the knife to minimise the time taken to move layers of clothing, and prevent snagging, thus enabling the bladed weapon to be accessed and drawn most efficiently [7].

https://doi.org/10.52202/081621-0056

5.2 Perspective from frontline policing

During conversations with a serving frontline police officer, the locations reported by Mahoney et al [7] were confirmed as accurate and applicable to frontline policing in London. Furthermore, most concealed knives are thought to be carried around the belt / waistband, or hip pockets of an assailant [8].

5.3 Effect of grip

Consideration of the concealment position is important, as this influences the type of grip used to draw and hold the knife, thereby influencing the angle of attack used by the assailant (see figures 4, 5, 6) [9]. Research into the motion of knife attacks by Miller et al [10] concluded that the maximum speed generated during stabbing is influenced by **the way the knife is held**.

Figure 4. Capped grip to stop hand slipping from the weapon handle to the blade. Image reproduced from reference [9]

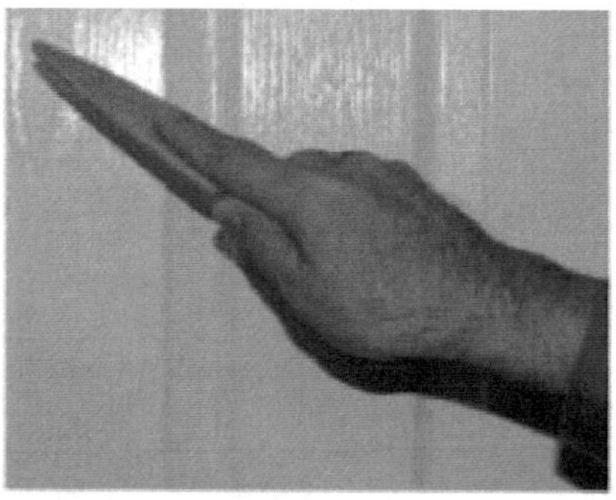

Figure 5. Finger point grip increases accuracy of bladed weapon attack. Image reproduced from reference [9]

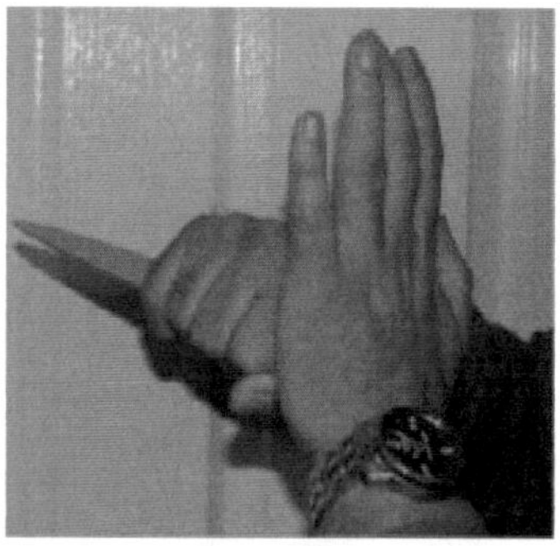

Figure 6. Hand thrust, with full body weight behind, increases the force applied behind the bladed weapon. Image reproduced from reference [9]

6. MOTION OF BLADED WEAPON ATTACKS

6.1 Stabbing with a bladed weapon

6.1.1 University of Wales Institute

In 1996 Miller et al [10], at the University of Wales Institute, studied the kinematics (motion) of ten right-handed subjects (9 males, 1 female) using four methods of stabbing, namely:

- **Long Over** – where the subject's feet are immediately behind a line 1.25 m from the target, using an overarm motion in which, when gripped, the blade of the knife emanates from the ulnar aspect of the hand. Subjects aim for an upper target.
- **Long Under** – same setup as 'Long Over', but using an underarm movement, whereby the blade emanates from the radial aspect of the hand. Subjects aim for a lower target.
- **Short Over** – same setup as 'Long Over', ie using an overarm motion, but where the subject's feet are behind a line 0.50 m from the target.
- **Short Under** – same setup as 'Long Under', ie using an underarm motion, but where the subject's feet are behind a line 0.50 m from the target.

The upper and lower targets were mounted on a punch bag. Each subject was instructed to stab as hard and fast as possible, and were allowed to take one step forward (with the opposite foot to their knife holding hand) for the 'long over' and 'long under' methods of stabbing [10].

Data was captured using a video recorder and analysed using proprietary software. The bladed weapon tip speeds and standard deviation are summarized in table 1 below [10].

Unsurprisingly 'long over' and 'long under' motions produced greater entry speeds, than the corresponding 'short over' and 'short under', due to these stabbing methods enabling extra speed to be built up over the longer acceleration path of the knife.

From the results of this study Miller et al suggest that there are two independent kinematic (motion) strategies employed by subjects, one for overarm, and the other for underarm patterns [10].

Table 1. Bladed weapon tip speeds, extract of data from reference [10]

	Long over		Long under		Short over		Short under	
	Entry speed (m/s)	Peak speed (m/s)	Entry speed (m/s)	Peak speed (m/s)	Entry speed (m/s)	Peak speed (m/s)	Entry speed (m/s)	Peak speed (m/s)
Bladed weapon tip	9.2	11.4	6.3	8.2	8.5	10.4	5.8	7.9
Std dev	2.03	1.04	0.89	0.74	1.72	1.89	1.34	1.57

Finally, Miller et al conclude that the maximum speed generated during stabbing is influenced by the way the knife is held [10].

6.1.2 Strathclyde University

In 1999 the Scottish Executive Central Research Unit published a comprehensive research programme (led by Strathclyde University), entitled "investigation of knife stab characteristics". In part I the "biomechanics of stab attacks" are examined [11].

Twenty subjects (17 males, 3 females) performed three types of stabs against a target comprising of a 10mm thick body armour (28 layers of Kevlar) mounted onto 30mm thick foam (density 33kg/m^2) on a box of plasticine. (Preliminary trials mounted the body armour directly onto the plasticine and the authors considered the plasticine did not sufficiently mimic the response of flesh, hence foam was added between the armour and the plasticine) [11].

The three stabs consisted of a:

- short thrust forward (abbreviated to **thrust** in tables 2 and 3)
- horizontal sweep around the body (abbreviated to **sweep** in tables 2 and 3)
- overhand stab (abbreviated to **overhand** in tables 2 and 3) [11].

Each subject repeated the three types of stabs, five times, using an instrumented knife.

Table 2. Energy of system before impact (joules (J)), extract of data from reference [11]

Target →	Body armour + FOAM + plasticine			Body armour + plasticine		
Type of stab ↓	Max (J)	Min (J)	Mean (J)	Max (J)	Min (J)	Mean (J)
Thrust	61	14	39	85	9	40
Sweep	90	17	45	103	15	73
Overhand	78	12	38	72	7	35

Table 3. Knife velocity prior to impact (m/s), extract of data from reference [11]

Target →	Body armour + FOAM + plasticine			Body armour + plasticine		
Type of stab ↓	Max (m/s)	Min (m/s)	Mean (m/s)	Max (m/s)	Min (m/s)	Mean (m/s)
Thrust	7.6	3.4	5.2	8.1	3.2	5.2
Sweep	8.8	4.2	6.5	9.2	3.8	6.6
Overhand	8.3	3.6	6.0	8.6	2.6	5.9

It is interesting to note that the mean velocities, particularly for the overhand stab, are lower than those reported by Miller et al [10] and the range of velocities increases with a larger sample size.

6.1.3 Horsfall

A PhD study undertaken by Horsfall [12] includes stabbing performance data for a large number of subjects, as summarised in table 4 below. Furthermore, peak forces of 800 newtons (N) are cited.

An instrumented knife, measuring force and velocity, was used to impact a 12 layer Aeroflex sheet (plain woven aramid encapsulated in a thromplastic binder) mounted onto Plastalina modelling clay. Photographs of the test set up show a vertical target for underarm stabs (nominally at chest height), whereas the overarm stabs are onto a horizontally positioned target (nominally at waist height).

It is considered by the authors that the horizontal orientation of the target has an influence on the human stabbing performance for overarm impacts and, when compared with the maximum energies

https://doi.org/10.52202/081621-0056

and velocities recorded by Strathclyde University [11], the data recorded by Horsfall is considerably higher [12].

Table 4. Summary of human stabbing performance data, extract of data from reference [12]

Stab type	Number	Mean energy (J)	Maximum energy (J)	Mean velocity (m/s)	Maximum velocity (m/s)
Underarm	310	27.5	83.5	6.1	11.5
Overarm	200	37.1	114.9	9.0	12.3

Finally, Horsfall observes that there is a significant difference in performance between male and female subjects, as the mean energy (detailed in table 5) for underarm stabs is 29.1J and 10.6J respectively.

Table 5. Human stabbing performance data for underarm stabs from male and female subjects, extract of data from reference [12]

	Number	Mean energy (J)	Maximum energy (J)	Mean velocity (m/s)	Maximum velocity (m/s)
Male underarm	283	29.1	83.5	6.2	11.5
Female underarm	27	10.6	30.9	4.8	7.9

6.1.4 University of Greenwich

During a scoping study in May 2024, unpublished data using a male subject (trained in boxing) to undertake an underarm stabbing motion (a thrust) resulted in a mean velocity of 6.47m/s (SD 0.47m/s) for eighteen stabs [13]. This velocity is 0.27m/s higher than the mean velocity data reported by Horsfall [12] for males (6.2m/s). During this study a high consistency of stabbing velocity was observed. Furthermore, the time to draw and stab a target at 1.25m from the subject was determined to be 0.7 seconds.

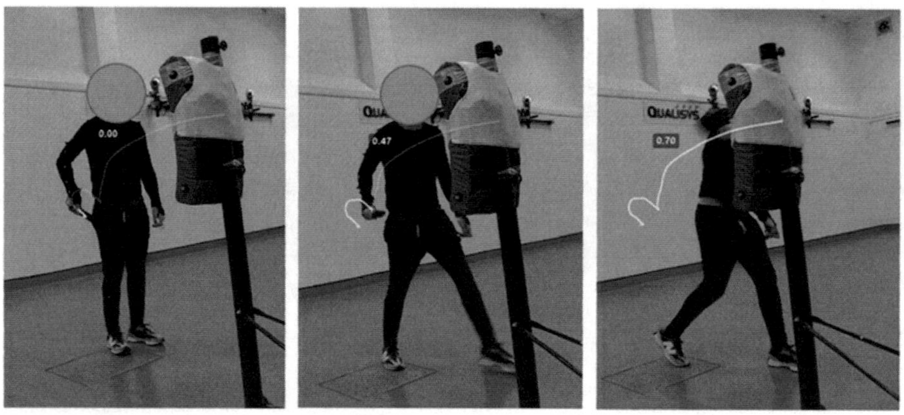

Figure 7. Timeframe images from drawing a knife to stabbing a target at a distance of 1.25m [13]

Distance from target	1.25m (approximately)
Time of event – draw to impact	0.7 seconds
Velocity of stab	6.4 m/s

6.2 Slashing with a bladed weapon

6.2.1 Cranfield University

In a research paper by Bleetman et al in 2003 87 (79 male; 8 female) untrained subjects attacked a white cartridge paper mounted on a Kevlar/elastomer, vertical target. This paper categorises slash events into two groups, namely i) **chop and drag** and ii) **sweep** motions in which the forces and velocities are detailed table 6 below [14].

Table 6. Human stabbing performance data for slashes, extract of data from reference [14]

Type of slash	% of sample population	Mean peak force (N)	Maximum peak force (N)	Mean velocity (m/s)	Maximum velocity (m/s)
Chop and drag	26	124	212	6.44	13.30
Sweep	74	97	178	5.88	14.88

Bleetman et al noted [14] that the mean and maximum peak forces are 124N and 212N newtons, respectively, which is far lower than the 800N recorded in earlier trials by Horsfall into stabbing performance [12]. This can be attributed to the very different mechanisms in stabbings and slashes.

7. LOCATION & FREQUENCY OF INJURIES CAUSED BY BLADED WEAPON ATTACKS

7.1 Non-terrorist related blade weapon attacks

7.1.1 Bleetman

In 2000, Bleetman published his PhD in which the injuries of 431 patients [405 male; 26 female] with open wounds were analysed [15]. All patients in this study had been admitted to the Accident and Emergency department of Glasgow Royal Infirmary, between 1993 and 1996, with penetrating trauma. Each patient record was analysed to determine the location of all penetrating injuries, and these were defined and marked onto anatomical figures. The wounds were categorised in accordance with the following definition and in accordance with the abbreviated injury scale (AIS):

i) minor - not a threat to life or limb (average AIS score of 1);
ii) major - wounds that resulted in major blood loss, threat to life, tendon damage, bone involvement, major neurovascular injury or permanent major cosmetic deformity (average AIS score of 3);
iii) devastating - fatal or near-fatal (patients in the near-fatal category arrived at hospital in cardiac arrest or in a moribund state) (average AIS score of 5) [15].

Table 7. Distribution of wounds by region and severity, reproduced from reference [15]

Region	Minor wounds	Major wounds	Devastating wounds	Total wounds
head (face and scalp)	203	29	0	232
neck	45	10	12	67
shoulders	25	3	0	28
left chest	29	92	29	150
right chest	31	37	14	82
right abdomen	27	21	18	66
left abdomen	21	39	2	62
right groin	0	3	2	5
left groin	0	3	2	5
thighs	61	21	0	82
buttocks	47	4	0	51
right arm	80	21	1	102
left arm	87	18	0	105
TOTAL (approx %)	**656** (63%)	**301** (29%)	**80** (8%)	**1037**

Of particular note is the predominant distribution of injury to the patients' left side from attacks by right-handed assailants.

7.1.2 Barnes-Warden and Mahoney

This study (by the authors) reviewed a dataset of 75 postmortem reports of deaths due to bladed weapon assaults in 2019, within the Metropolitan Police District (Greater London). It included anatomically correct body maps to illustrate the location of the injuries and identified the principal location of fatal injury, from a bladed weapon assault, to be the torso. Furthermore, it attributed the main causes of death from a knife attack to be due to catastrophic damage of an organ, great vessel, or both, within the torso [16].

https://doi.org/10.52202/081621-0056

Statistics are presented with regards to the depth of the fatal stab wound and blade type. In Bleetman's [15] and the author's study [16] over double the number of wounds are located on the front of the victim's body, compared with their back.

7.2 Terrorist blade weapon attacks

7.2.1 Merin

Merin et al [17] compares injuries from terrorist and non-terrorist related knife attacks in Israel, and précised research by Mayo et al, to conclude that there are significantly more body regions injured in terrorist attacks compared to non-terrorist ones. Merin refers to research by Kluger et al (2004, cited in [17]) which determines that victims of terrorist knife attacks have a higher injury severity score compared with non-terrorist attacks. In addition, Merin also refers to Heldenberg et al (2014, cited in [17]) who observed that vascular injuries are more than 9 times more frequent in terror attack victims (almost 10%), than among non-terrorist trauma victims (1.1%), and that the injuries tend to be more complex [17]. Merin [17] states that "exceptional force is used in terror incidents" and that "traditional stab wounds (…) tend to have a lower lethality":

7.2.2 Rozenfeld

Rozenfeld et al [18] compare the injuries sustained from interpersonal stabbings (IPS) and terror related stabbings (TRS) in Israel. These attacks were predominantly on Jews during the Knife Intifada circa 2015-16 [18].

It concludes that TRS are more severe attacks, on a different demographic to IPS, with differing anatomical locations of the injuries (such as more injuries to head and neck during TRS). [18]
There are some interesting statements about stabbing technique and power as the TRS victims tended to exhibit wounds to chest, and much higher proportion of head and neck injuries, compared with IPR victims. Furthermore, terrorists predominantly attacked using the more powerful ''overhand'' grip, thrusting downwards with the knife. Whereas IPS attacks often use the less powerful ''underhand'' grip.

7.2.3 London Bridge attacks

On the 3rd June 2017, during the London Bridge attacks, a van was deliberately driven at pedestrians on London Bridge, which resulted in the death of three victims. The three terrorists inside the van then proceeded on foot to Borough Market with knives strapped to their wrists. In total, the terrorists murdered eight people and injured forty eight. It is reported that a witness to the attack, Gerard Vowls, told the BBC [19] "…they stabbed this girl maybe 10 times, 15 times" describing the ferocity of the bladed weapon attack.

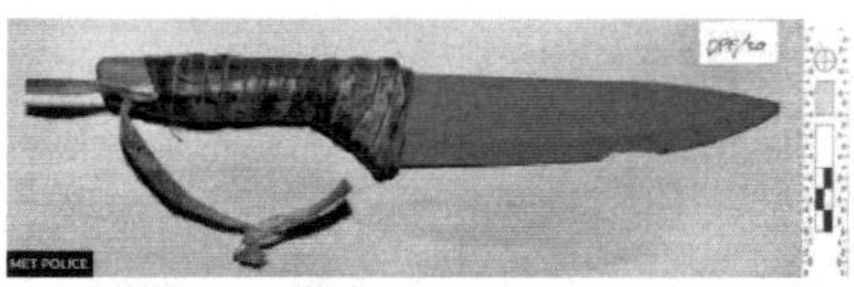

Figure 8. image of a knife strapped to a terrorist's wrist in the London bridge attacks, reproduced from the BBC news online, reference 20

7.2.3 Fishmongers Hall attacks

At the Fishmongers Hall attacks, on the 29th November 2019, Usman Khan murdered two people and injured three others in a terrorist knife attack. Khan used multiple knives and tape (see figure 8) to secure the knives to his wrists. He also was wearing a fake suicide vest [20].

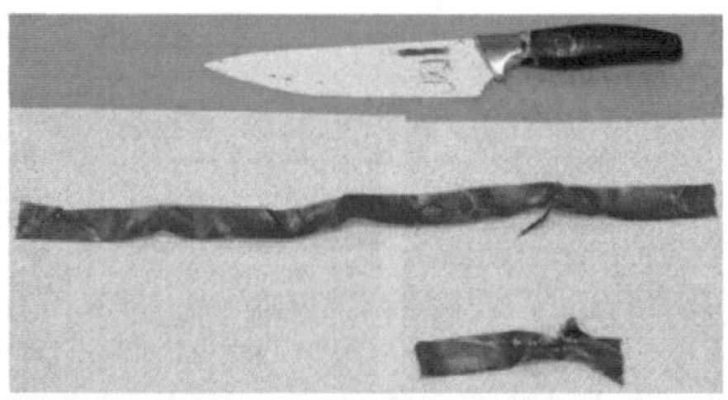

Figure 9. image of a knife strapped to the terrorist's wrist in the Fishmongers Hall attacks, reproduced from Sky news website, reference 20

Thus, in both of the above UK attacks, bladed weapons were strapped/taped to an assailant's hand, thereby making him more difficult to disarm, and influencing the angles of attack that can be used.

8. DISTANCE FROM A BLADED WEAPON ATTACK

The "21-foot rule" (created by Lieutenant John TUELLER in the US in the 1980s) was declared a safe distance to deal with potentially dangerous suspects armed with knives. However, modern research by Sandel et al (2020) [21] addressed i) how fast can a suspect run 21 feet (approx. 7m)? ii) is an officer able to respond (draw and fire) before the suspect can move 21 feet? iii) is there anything that an officer can do to provide more reaction time?

On average individuals charging at an officer were able to cover a distance of **21 feet in 1.5 seconds**, and run times ranged from 1.24 seconds to 1.83 seconds [21]. US officers trained in firearms, under sterile (no stress) conditions, **on average took 1.8 seconds** to fire a single shot at a silhouette target, using a conventional firearm [21]. Furthermore, when evaluating the effect of tactical movement by an officer being charged at, the most effective tactical movement is to sidestep [21].

Finally, the conclusion from this study is that 21-feet is an ineffective distance for officers to draw and fire their weapon at a charging assailant [21].

9. CURRENT THREAT & FUTURE PROTECTION SCHEMES

Bladed weapon attacks remain a principal threat for MPS frontline officers and staff.

From the statistical data, academic research and the open-source reporting of incidents summarised within this paper, we have shown that bladed weapon attacks are typically fast-time events and, in non-terrorist related attacks, domestic (kitchen) knives are commonly used. Furthermore, in untrained subjects the speed of a bladed weapon attack is highly variable. Increasing the area of armour protection is likely to be beneficial for frontline staff safety, particularly if this does not impact on mobility. In 2024 a new concept of supplementary knife resistance (SKR) was introduced into MPS service, by enhancing body armour designs in "difficult to protect areas" [16]. Furthermore, SKR has been well received by the user.

Future proposals are to seek designs for neck protection, as this is a highly vulnerable area to bladed weapon attacks. Additional tactics and/or protection should be considered for terrorist related attacks.

10. CONCLUSIONS

Knife enabled offences are prevalent within England and Wales, with annual figures of 50,510 (year ending March 2024), with the main motives being "assault with injury and assault with intent to cause serious harm", followed by "robbery". These attacks are a threat to the public as well as frontline policing and have clinic consequences for the NHS.

UK law prohibits the carriage of knives over 3 inches without a good reason. MPS data from confiscated blade weapon analysis shows that over the last 15 years, the majority of knives impounded are domestic, ie kitchen, knives. Knives are easy to obtain and conceal, and are typically carried around the belt / waistband, or hip pockets of an assailant. The location of a concealed bladed weapon can determine the grip used to draw then hold the knife, which in turn can influence the velocity of attack.

Multiple studies into the motion of knife attacks have shown that knife attacks are typically fast time events, with a scoping study measuring 0.7 seconds as the time to draw and stab a target at 1.25m. There is much variation in the reported velocities of stabs and slashes by untrained subjects. However, overarm stabs are typically reported at higher velocities than underarm stabs. Slash attacks have also been observed to be highly variable in velocity.

In non-terrorist related bladed weapon injury data, the torso and neck are highly likely to receive devastating wounds. Furthermore, the main cause of death due bladed weapon injury is catastrophic damage of an organ, great vessel, or both.

In terrorist related bladed weapon attacks there are significantly more body regions that were injured compared to non-terrorist ones and, typically, greater force is used. The overhand grip and downward thrusts are cited in terrorist attacks. In the London Bridge and Fishmongers Hall attacks there was evidence of strapping of the knife to the terrorist's wrist making the blade harder to disarm and influencing the angle of attacks.

https://doi.org/10.52202/081621-0056

Since the 1980s the "21-foot rule" has been deemed to be a safe distance from a suspect armed with knives. However, this distance is now considered to be too close for the majority of (armed) officers to react (with a conventional firearm).

In 2024 a new concept of supplementary knife resistance (SKR) was introduced into service, by enhancing body armour designs in "difficult to protect areas". Furthermore, SKR has been well received by the user. Future proposals are to seek designs for neck protection in frontline policing, as this is a highly vulnerable area to bladed weapon attacks. Additional tactics and/or protection should be considered for terrorist related attacks.

Acknowledgements

The authors would like to sincerely thank Professor Mark Goss-Sampson, Dr Mark Colpus and Callum St Romaine (University of Greenwich), for their permission to publish the biomechanics of stabbing study.

References

[1] Office for National Statistics (ons.gov.uk). Crime in England and Wales: year ending March 2024. https://www.ons.gov.uk/peoplepopulationandcommunity/crimeandjustice/bulletins/crimeinenglandandwales/yearendingmarch2024

[2] Office for National Statistics (ons.gov.uk). Other related tables: year ending March 2024. https://www.ons.gov.uk/peoplepopulationandcommunity/crimeandjustice/datasets/crimeinenglandandwalesotherrelatedtables

[3] National Health Service. Monthly hospital admissions by sharp object December 2022. https://digital.nhs.uk/supplementary-information/2023/monthly-hospital-admissions-for-assault-by-sharp-object-december-2022

[4] Malik N S. Munoz B. De Courcey C, Rizwana I, Lee K C. Chernbumroong S, Bishop J, Lord J M. Gkoutos G. Bowley D M. Foster M A. Violence-related knife injuries I a UK city; epidemiology and impact on secondary care resources. EClinicalMedicine. 2020 Mar 3; 20:100296

[5] GOV.UK. Selling, buying and carrying knives and weapons. Available at: https://www.gov.uk/buying-carrying-knives

[6] Mapp G A. PPG-RPT-22-035. 2022 Knife Blade Analysis Audit Report. Metropolitan Police Service. 2022.

[7] Mahoney P F, Godhania K, Carr D J. Investigating the use of concealable and disguised knives. Police Journal: Theory, Practice and Principles. 2017; 1-11.

[8] PC Turpin, Metropolitan Police Service. Conversation with Jane Barnes-Warden. 21 May 2024.

[9] Carr D J, Godhania K, Mahoney P F. Edged weapons awareness. International Journal of Legal Medicine. 2018; 1-11. https://doi.org/10.1007/s00414-018-1966-6

[10] Miller S A. Jones M D. Kinematics of four methods of stabbing: a preliminary study. Forensic Science International. 1996; 82: 183-190

[11] Great Britain. Scottish Executive Central Unit. Investigation into knife stab characteristics I) Biomechanics of knife stab attacks II) Development of body tissue simulant. Edinburgh. 1999.

[12] Horsfall I. Stab Resistant Body Armour. PhD thesis. Cranfield University; 2000.

[13] University of Greenwich, unpublished work. May 2024

[14] Bleetman A. Watson C H. Horsfall I. Champion S M. Wounding Patterns and Human Performance in Knife Attacks: Optimising the Protection provided by Knife-Resistant Body Armour. Journal of Clinical Forensic Medicine. 2003; 10: 243–248

[15] Bleetman A. Safety Standards for Police Body Armour. PhD thesis. The University of Birmingham; 2000

[16] Barnes-Warden J. Mahoney P F. Bladed Weapon Assaults and Human Vulnerability. Proceedings of the Personal Armour System Symposium. Dresden: IPAC; 2023

[17] Merin O. Sonkin R. Yitzhak A. Frenkel H. Leiba A. Schwarz A D. Jaffe E. Terrorist stabbings – distinctive characteristics and how to prepare for them. The Journal of Emergency Medicine. 2017; 53: 451-457

[18] Rozenfeld M. Givon A. Peleg K. Violence-related Versus Terror-related Stabbings Significant Differences in Injury Characterists. Annals of Surgery. 2018; 267:965–970

[19] BBC News. London Bridge attack: What happened. 3 May 2019. Available at: https://www.bbc.co.uk/news/uk-england-london-40147164

[20] Sky news. London Bridge attack: Harrowing accounts of Usman Khan's deadly rampage – and the people who stopped it. Available at: https://news.sky.com/story/london-bridge-attack-harrowing-accounts-of-usman-khans-deadly-rampage-and-the-people-who-stopped-it-12296765

[21] Sandel W L. Martaindale M H. Blair J P. A scientific examination of the 21-foot rule. Police Practice and Research. 2021; 22:3, 1314-1329

Implications of Lead-free Shotgun Ammunition to Police Body Armour

J. Barnes-Warden[1]

[1]*Metropolitan Police Service, Physical Protection Group, 60 Albany Street, London, NW1 4EE*
jane.barnes-warden@met.police.uk

Abstract. Within the UK it is estimated that there are in excess of 1.33 million shotguns legally owned under licence. Therefore, with such large quantities of weapons in legal possession, there is the potential for some of these to be misappropriated by criminals. With the exception of "solid slug", the majority of projectiles within shotgun ammunition consist of multiple spherical balls, known as "pellets" or "shot". Traditionally, pellets or shot has been manufactured from lead, however an official review by the UK government has suggested that lead ammunition may be phased out, due to its toxicity to wildfowl and their predators. This paper initially reviews the current calibres of this weapon and ammunition. It then considers alternative materials for shotgun ammunition when moving away from lead to lead-free ammunition. Finally, this paper considers the effect of alternative shotgun ammunition on typical materials used in police body armour.

1. INTRODUCTION

According to UK government statistics, in March 2024 there were 495,798 shotgun certificates held legally within the UK and, it is estimated that, there are in excess of 1.33 million shotguns [1, 2]. Therefore, with these large quantities of shotguns, it is possible that some of these weapons may be misappropriated and used in criminal activity.

Traditionally, shotgun ammunition contains "pellets" or "shot" which are spherical projectiles manufactured from lead [3]. However, a recent government proposal considers the phasing out of lead, due its toxicity to wildfowl and their predators [4, 5, 6].

This paper describes shotguns, lead and lead-free shotgun ammunition, and their effect on typical materials used in police body armour.

2. SHOTGUNS

The principal use of a shotgun is for hunting game, such as flying birds, ground game (eg rabbits) or larger animals, such as deer. They are also used for International Sporting Shooting Federation and Olympic disciplines such as skeet and trap clay target shooting [3].

Shotguns are smooth-bore weapons; shotgun barrels do not have rifling cut into their inner surface, unlike handgun and rifle barrels, where rifling is used to impart spin stabilisation to bullets in flight. The effective range of a shotgun is approximately 30 meters and pellets typically fall to the ground within several hundred meters [3].

2.1 Gauge of shotgun

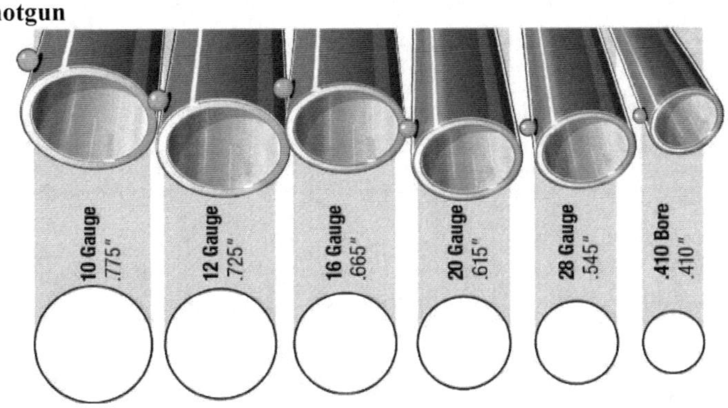

Figure 1. Common gauges, or sizes, of shotgun, reproduced from reference [7]

https://doi.org/10.52202/081621-0057

There are many gauges, or sizes, of shotgun. The nomenclature of denoting the gauge is based on an antiquated method of measuring the bore size of artillery, whereby the size of the shotgun barrel is determined by the mass of spherical lead ball that fits the bore. A shotgun that fits a lead ball with a mass of one twelfth of an imperial pound is known as a 12 bore, and this nomenclature is used for the majority of gauges of shotguns (see figure 1). The exception is the .410" gauge of shotgun, which was introduced later than other gauges, hence modern calibre notation is used [3, 8, 9].

2.2 Tapered and parallel sided barrels

Shotgun barrels are often tapered on the inside of the barrel, known as a choke, which narrows the bore toward the muzzle. This feature is designed to constrict the shot, and by restricting the width dispersion of the shot there is a higher probability of hitting the target at longer ranges. A parallel sided shotgun barrel is known as a cylinder [8].

3. SHOTGUN AMMUNITION

Ammunition for shotguns is known as cartridges, which are parallel sided cylinders. The walls of shotgun cartridges are made from cardboard or plastic, and a brass base contains the primer. The shotgun's firing pin strikes the primer, thereby igniting the propellant within the cartridge. The wad, and sometimes a cup, sits behind the pellets to apply a force to the projectiles within the cartridge. The end of the shotgun cartridge is typically sealed with a card disc or plastic cover [8].

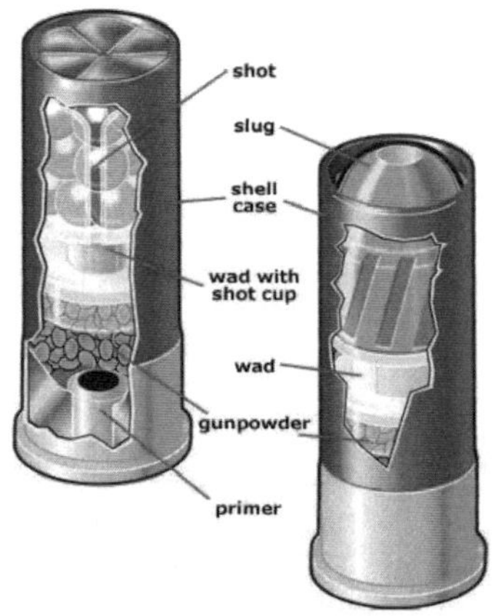

Figure 2. Diagram of shotgun cartridges [10]

containing multiple pellets (left) and a solid slug on right

3.1 Mass and size of pellets

Most shotguns fire approximately one ounce to one and an eighth of an ounce (nominally 28 to 32grams) of pellets [8]. Pellet sizes are illustrated in figure 3.

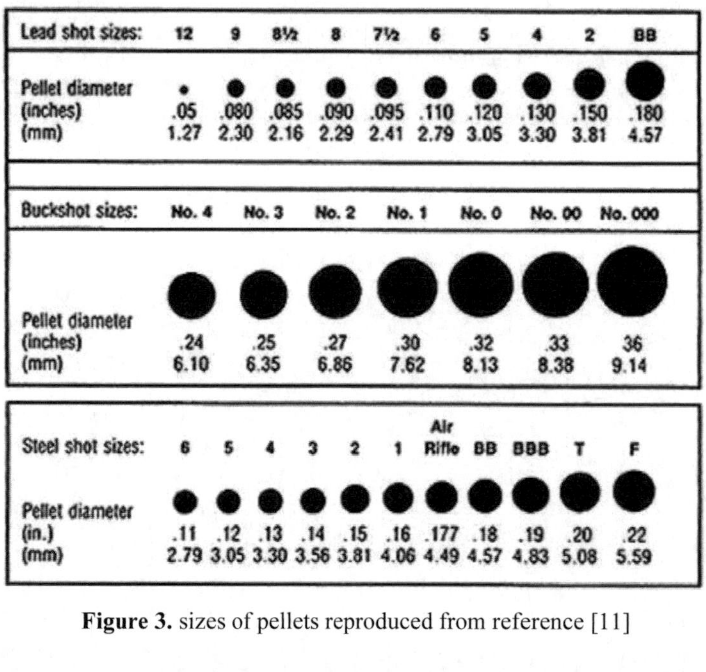

Figure 3. sizes of pellets reproduced from reference [11]

4. DISPERSION OF SHOTGUN PELLETS

When multiple pellets are fired from a shotgun, they initially emerge from the muzzle together. However, as the pellets travel away from the muzzle they spread out in three directions, in width, in height and in length. Figure 4 illustrates the dispersion of multiple pellets as the distance from the muzzle increases [12].

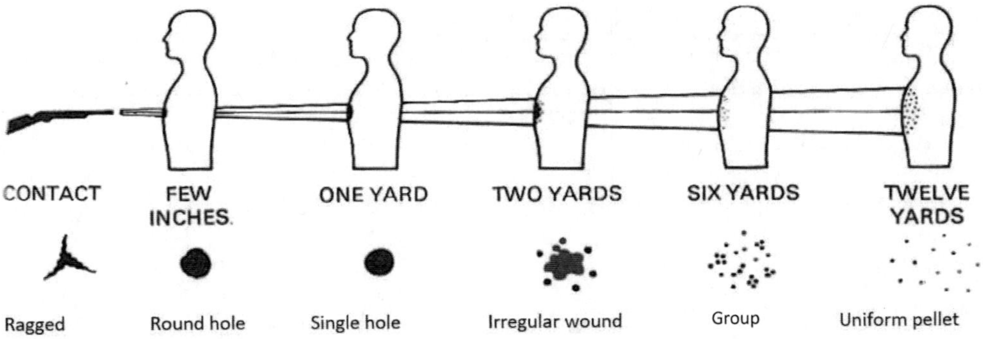

Figure 4. Characteristics of shotgun wounds at various distances, reproduced from reference [12]

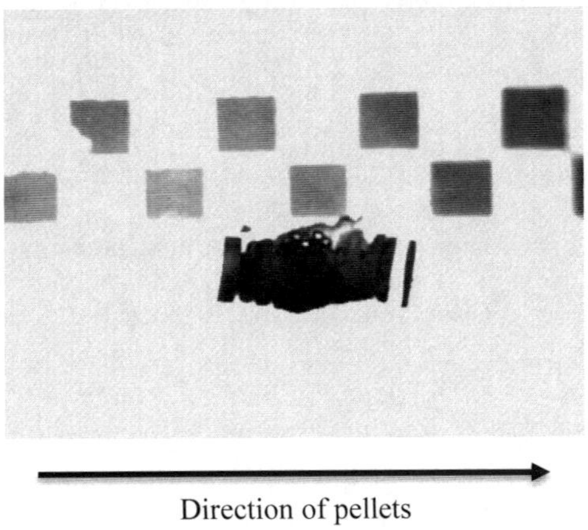

Figure 5. image taken from high-speed video illustrating the beginning of the opening of the plastic shotgun cup, and the ejection of the wads (on the right).

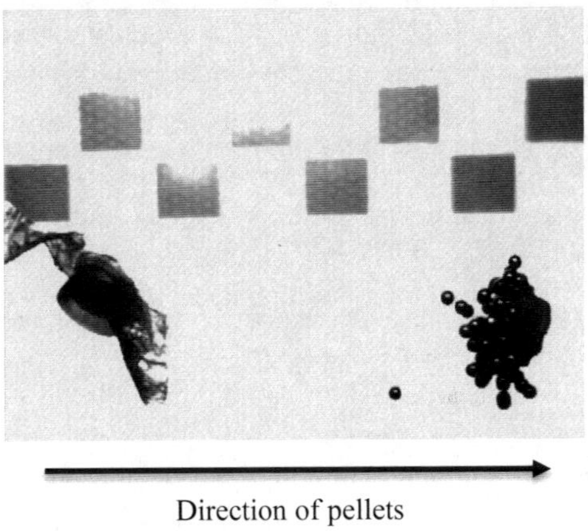

Figure 6. image taken from high-speed video illustrating the fully open plastic shotgun cup (on left) and the pellets on their flight path (on right).

5. MATERIALS USED IN SHOTGUN AMMUNITION

Traditionally shotgun cartridges have used lead pellets (or shot) however, this may be phased out due its toxicity to wildfowl and their predators. The main alternative materials to lead are steel, bismuth alloys (typically containing tin) and tungsten matrix (ie tungsten bound in a polymer) [4, 5, 6].

https://doi.org/10.52202/081621-0057

5.1 Density of lead and alternative materials

Table 1. Density of materials reproduced from reference [13]

Material	**Density (ρ) kg/m³**
Lead	11,300
Mild Steel	7,860
Tungsten	19,300
Bismuth	9,790

5.2 Hardness of alternative materials

According to the Vickers hardness test [14] using HV 10kgf the steel pellets are 101HV and tungsten matrix pellets (13g/cm³) are 297HV. Hence, the tungsten matrix tested in this study is around three times the hardness of the steel pellets. (It should be noted that higher densities of tungsten matrix are manufactured, such as 18g/cm³, however these were not available at the point of testing).

The lead and bismuth pellets were subjected to the same hardness test, however using HV 1kgf the lead pellets are 9.9HV and the bismuth 17.5HV. Hence lead and bismuth are very soft metals, particularly when compared with steel and tungsten.

5.3 Price of alternative materials

Price comparisons for lead shotgun ammunition compared with alternative materials have been published by Thomas [15]. Although prices are subject to change, typically steel is a comparable price to lead, and other materials are more expensive than lead. For instance, bismuth shotgun ammunition is approximately 5 times the price of lead ammunition, and tungsten ammunition is 10 times the price of lead ammunition [15, 16].

6. UK HOME OFFICE BODY ARMOUR STANDARD

Within the UK Home Office body armour standard [17] there is a shotgun protection level, namely SG1, for plates (ie hard armour, typically used to augment soft body armour panels). The projectile used in the Home Office body armour standard is a one-ounce, rifled lead slug (see figure 2).

Currently, there is not a protection level for soft, body armour designs (eg HO1 and HO2 protection schemes) against lead shotgun pellets. However, as the lowest ballistic protection level (HO1) specifies a proof test against 9mm full metal jacket (FMJ) and jacketed soft point (JSP) ammunition at 365m/s (+/-10m/s) lead shot is likely to be stopped within soft, body armour designs accredited to HO1 and HO2.

Protection level	Classification and calibre	Test round designation	Bullet mass	Range (min) (m)	Max. mean BFS (mm)	Single shot BFS limit (mm)	Velocity ($m.s^{-1}$)
SG1	Shotgun 12 gauge True Cylinder	Winchester 1 oz. Rifled 12RSE	28.4 g	10	25.0	30.0	435 ± 25

Figure 7. Ballistic protection levels reproduced from the UK Home Office body armour (2017) standard [17]

https://doi.org/10.52202/081621-0057

7. IS LEAD-FREE SHOTGUN AMMUNITION MORE PENETRATIVE AGAINST ARMOUR MATERIALS?

A series of trials were conducted to determine whether alternative shotgun ammunition is more penetrative against armour materials, typically used in police protection schemes, when compared with lead ammunition. As the size and mass of shotgun pellet can significantly vary (as described in section 3.1) a nominal diameter of pellet was chosen in which lead and each of the alternative materials could be sourced. During these trials 3" magnum (76mm) ammunition was fabricated with 32g of i) lead ii) steel iii) bismuth (alloy unspecified) and iv) tungsten matrix pellets ($13g/cm^3$).

7.1 Test set up

Throughout this study a 12-gauge cylinder shotgun barrel (length 725mm) was coupled with a universal receiver. The shotgun ammunition (3" magnum, 76mm) was loaded by the Ballistic Test Facility (BTF) using the same mass of pellets (32g) in a "Gualandi" plastic cup and 2 off 4mm cork wads. An appropriate mass of propellant was used to achieve pertinent velocities of 425m/s $\pm$ 25m/s [16]. The diameter of the lead, steel and tungsten pellets were nominally 4mm in diameter and, due to availability, the bismuth pellets were nominally 3.56mm.

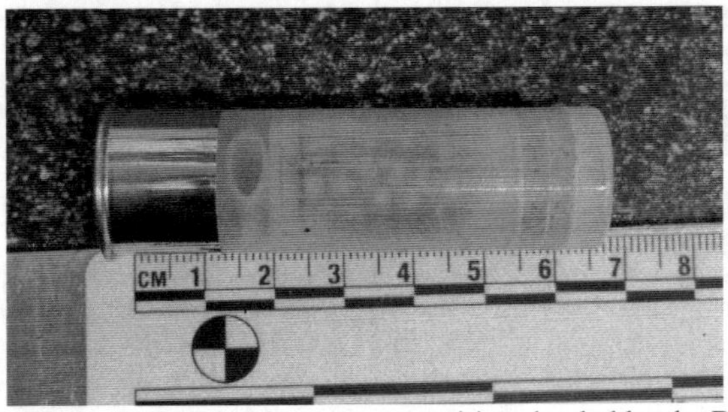

Figure 8. Photograph illustrating the shotgun ammunition, loaded by the BTF, prior to firing

When testing individual projectiles against armour designs typically the velocity of the projectile is varied and statistical techniques, such as V_{50} or Critical Perforation Analysis [18], are used to estimate the armour performance. However, in these tests the distance was varied (nominally 0.5m, 1.5m and 2.5m) so that the concentration of pellets decreases at longer distances.

A high-speed video (HSV) camera with known frame rate (30,000 fps) and a scale in the HSV footage were used to evaluate the average velocity of the pellets by measuring distance travelled over time. This method was employed as miscalculations often occur when chronographs estimate the velocity of multiple projectiles, and other objects (such as wads) can also produce erroneous results.

Each target was mounted on a microcrystalline wax block[1] (nominally 420 x 350 x 100 mm), which was affixed into a steel frame. This set up allowed for the backing to be perforated without damaging Home Office specified backing materials (ie Roma Plastilina No 1) which would be expensive to repair. Furthermore, the effect of momentum on the target, or back face signature, was not within the scope of these tests.

7.2 Targets

In this study the performance of the following generic armour materials, used within police body armour schemes, were evaluated. Firstly, a comparison of the performance of:

- 20 off layers of nominally $80g/m^2$ of ultra high molecular weight polyethylene (UHMWPE). Areal density of target $1.6kg/m^2$, and
- 8 off layers of nominally $200g/m^2$ of aramid. Areal density of target $1.6kg/m^2$.

Then a simple, generic armour target of 20 off layers of nominally $200g/m^2$ of aramid was evaluated. Areal density of target $4.0kg/m^2$.

[1] ref:3795, Poth Hille & Co Ltd, RM13 9BP

https://doi.org/10.52202/081621-0057

8. RESULTS

The following tables 2 to 8 summarise the results.

Table 2. Performance of 20 layers of UHMWPE ($80 g/m^2$) at nominally 0.5m, against lead and 3 lead-free materials for shotgun pellets

#	Mass and material of shot	Nominal diameter of pellets (mm)	Average velocity of pellets (m/s)	Distance between muzzle and target (m)	P / NP*	Notes & observations
PE-1	32g LEAD	4.00	407	0.53	NP	Considerable indentation of target into backing.
PE-2	32g STEEL	4.00	427	0.53	P	
PE-3	32g BISMUTH †	3.56	426	0.53	NP	† alloy unspecified Considerable indentation of target into backing.
PE-4	32g TUNGSTEN MATRIX ‡	4.00	403	0.53	P	‡ $13 g/cm^3$ density

*P = Perforation NP = Non-perforation

Table 3. Performance of 8 layers of aramid ($200 g/m^2$) at nominally 0.5m, against lead and 3 lead-free materials for shotgun pellets

#	Mass and material of shot	Nominal diameter of pellets (mm)	Average velocity of pellets (m/s)	Distance between muzzle and target (m)	P / NP*	Notes & observations
A-1	32g LEAD	4.00	429	0.50	NP	Considerable indentation of target into backing.
A-2	32g STEEL	4.00	450	0.50	P	
A-3	32g BISMUTH †	3.56	426	0.50	NP	† alloy unspecified Considerable indentation of target into backing.
A-4	32g TUNGSTEN MATRIX ‡	4.00	449	0.50	P	‡ $13 g/cm^3$ density

*P = Perforation NP = Non-perforation

Table 4. Performance of 20 layers of UHMWPE at nominally 1.5m, against 2 lead-free materials for shotgun pellets

#	Mass and material of shot	Nominal diameter of pellets (mm)	Average velocity of pellets (m/s)	Distance between muzzle and target (m)	P / NP*	Notes & observations
PE-5	32g STEEL	4.00	433	1.49	P	
PE-6	32g TUNGSTEN MATRIX ‡	4.00	449	1.49	P	‡ 13g/cm^3 density

*P = Perforation NP = Non-perforation

Table 5. Performance of 8 layers of aramid (200g/m^2) at nominally 1.5m, against 2 lead-free materials for shotgun pellets

#	Mass and material of shot	Nominal diameter of pellets (mm)	Average velocity of pellets (m/s)	Distance between muzzle and target (m)	P / NP*	Notes & observations
A-5	32g STEEL	4.00	428	1.50	NP	
A-6	32g TUNGSTEN MATRIX ‡	4.00	428	1.50	P	‡ 13g/cm^3 density

*P = Perforation NP = Non-perforation

Table 6. Performance of 20 layers of UHMWPE at nominally 2.5m, against 2 lead-free materials for shotgun pellets

#	Mass and material of shot	Nominal diameter of pellets (mm)	Average velocity of pellets (m/s)	Distance between muzzle and target (m)	P / NP*	Notes & observations
PE-7	32g STEEL	4.00	407	1.49	P	Borderline perforation, the majority of pellets were stopped in the armour.
PE–8	32g TUNGSTEN MATRIX ‡	4.00	431	1.49	P	‡ 13g/cm^3 density. Borderline perforation, the majority of pellets were stopped in the armour.

*P = Perforation NP = Non-perforation

https://doi.org/10.52202/081621-0057

Table 7. Performance of 8 layers of aramid (200g/m^2) at nominally 2.5m, against 1 lead-free material for shotgun pellets

#	Mass and material of shot	Nominal diameter of pellets (mm)	Average velocity of pellets (m/s)	Distance between muzzle and target (m)	P / NP*	Notes & observations
A-7	32g TUNGSTEN MATRIX ‡	4.00	448	2.50	P	‡ 13g/cm^3 density. Borderline perforation, the majority of pellets were stopped in the armour.

*P = Perforation NP = Non-perforation

Table 8. Performance of 20 layers of aramid (200g/m^2) at nominally 0.5m, against lead and 3 lead-free materials for shotgun pellets

#	Mass and material of shot	Nominal diameter of pellets (mm)	Average velocity of pellets (m/s)	Distance between muzzle and target (m)	P / NP*	Notes & observations
A-8	32g LEAD	4.00	416	0.53	NP	Considerable indentation of target into backing.
A-9	32g STEEL	4.00	421	0.53	NP	Considerable indentation of target into backing.
A-10	32g BISMUTH	3.56	426	0.53	NP	Considerable indentation of target into backing.
A-11	32g TUNGSTEN MATRIX ‡	4.00	438	0.53	NP	‡ 13g/cm^3 density Considerable indentation of target into backing.

*P = Perforation NP = Non-perforation

9. DISCUSSION

At close proximity to the shotgun muzzle, nominally 0.5m, the UHMWPE and aramid (8 layer) targets (both with an areal density of 1.6 kg/m^2) were not perforated by either the lead or bismuth shotgun pellets. However, the steel and tungsten matrix pellets, which are much harder than lead and bismuth, perforated the UHMWPE and aramid (8 layer) targets at this range.

In order to reduce the intensity of concentration of the pellets, the distance between the muzzle and the target was increased to nominally 1.5m for the lead-free pellets, which had perforated the target at 0.5m. At nominally 1.5m both the steel and tungsten matrix pellets perforated the UHMWPE target, and the steel pellets were stopped within the aramid (8 layer) target, whereas the tungsten matrix pellets perforated it.

Once again, to reduce the concentration of pellets, the distance between the muzzle and the target was increased to 2.5m for perforating lead-free shotgun ammunition. Perforations from the steel and the tungsten matrix pellets were observed within the UHMWPE target. However, as the majority of the steel and tungsten matrix pellets were retained within the armour (ie partial perforations) this performance indicates that 2.5m is close to the threshold of this armour/projectile combination. A similar performance threshold was observed with the tungsten matrix pellets impacting the aramid (8 layer) target at 2.5m.

The simple, generic armour target of 20 layers of $200g/m^2$ of aramid was able to stop all of the lead and lead-free pellets at 0.5m. Many body armour schemes used in UK frontline policing have greater quantities of ballistic resistant materials in their construction, hence all of the nominal 4mm lead-free shotgun pellets tested here are unlikely to defeat such armours.

Finally, although the effect of momentum was outside of the scope of this project, considerable indentation of the target into the backing material, particularly at short range, was observed in non-perforations in both UHMWPE and aramid targets.

10. CONCLUSIONS AND FURTHER WORK

In the UK shotguns are common weapons, with over 1.33 million shotguns legally owned under licence. Traditionally, pellets or shot has been manufactured from lead, however an official review by the UK government has suggested that lead ammunition may be phased out, due to its toxicity to wildfowl and their predators.

The main alternative materials to lead pellets are steel, bismuth alloys and tungsten matrix. Bismuth alloys are the closest match in density and hardness to lead, compared with steel and tungsten matrix. In this study typical armour materials (used in police body armour) were tested with commercially available bismuth shot, and a similar performance to lead shot was observed.

Steel shotgun ammunition is a similar price to lead and is readily available. Steel is less dense than lead and, therefore, more pellets are required to produce same total mass of pellets in the cartridge (if they are the same diameter). Due to the difference in mass, if fired at the same velocity, steel pellets will have less kinetic energy per pellet than lead. However, as steel pellets are harder than lead, steel pellets are less easily deformed.

Tungsten is approximately ten times the price of lead and is a material of significantly higher density and hardness. Although tungsten pellets are fabricated from tungsten powder within a polymer matrix, when subjected to the Vickers hardness test the pellets in this study were around three times the hardness of the steel pellets. As the tungsten matrix pellets have a higher density than the lead equivalent, for the same mass of pellets (32g), the kinetic energy per pellet is higher for tungsten matrix pellets than for lead pellets since there are fewer of them for the same total mass.

Both steel and tungsten pellets were found to be significantly more penetrative against typical armour materials than lead pellets of the same diameter.

The simple, generic armour target of 20 layers of $200g/m^2$ of aramid was able to stop all of the lead and alternative, lead-free pellets tested at 0.5m. Many body armour schemes used in UK frontline policing have greater quantities of ballistic resistant materials in their construction, hence all of the nominal 4mm lead-free shotgun pellets are unlikely to defeat such armours.

Finally, this study focuses on nominally 4mm pellets of lead and alternative, lead-free shotgun ammunition, fired from a 12-gauge shotgun barrel. As the choice of shotgun pellet size was determined by the availability of pellets, it is recommended that further work is undertaken on larger sizes of pellets since each pellet with have a higher kinetic energy than those with smaller diameters. Furthermore, it should be noted that the 4mm tungsten matrix pellets under test had a density of $13g/cm^3$, however tungsten matrix of higher density is also commercially available.

Acknowledgments

The author would like to thank Dr A H Jones, MERI, Sheffield Hallam University for evaluating the Vickers hardness of the lead and non-lead pellets used within shotgun ammunition. I would like to thank Steph Dale at the Steel Shot Company Limited for his advice on lead-free pellets. Finally, my thanks also goes to Mr P Hall at Hallrite, the Ballistic Test Facility.

 https://doi.org/10.52202/081621-0057

References

[1] Home Office. Statistics on firearm and shotgun certificates, England and Wales: April 2023 to March 2024. Published 27 June 2024. Available at: https://www.gov.uk/government/statistics/firearm-and-shotgun-certificates-april-2023-to-march-2024/statistics-on-firearm-and-shotgun-certificates-england-and-wales-april-2023-to-march-2024#summary-of-statistics

[2] Home Office. User guide to statistics on firearm and shotgun certificates, England and Wales. Updated 27 June 2024. Available at: https://www.gov.uk/government/publications/user-guide-to-firearm-and-shotgun-certificate-statistics/user-guide-to-statistics-on-firearm-and-shotgun-certificates-england-and-wales-accessible-version

[3] Chan C (ed.) How weapons work. London: Marshall Cavendish Publications Limited; 1976. pp. 27.

[4] Government Department for Environment, Food & Rural Affairs. 23 March 2021. Press release Plans announced to phase out lead ammunition in bid to protect wildlife [online].
Available at: https://www.gov.uk/government/news/plans-announced-to-phase-out-lead-ammunition-in-bid-to-protect-wildlife

[5] All-Party Parliamentary Group. Technical report Alternatives to lead shot: Assessing supply and demand. September 2022. Available at: https://leadammunitionappg.org/wp-content/uploads/2022/09/Alternatives-to-lead-shot-Assessing-supply-and-demand.pdf

[6] Health and Safety Executive. Annex 15 Restriction Report Proposal for a restriction Substance name(s): Lead IUPAC name(s): Lead, EC NUMBER(S):231-100-4, CAS NUMBER(S): 7439-92-1. V1.0. April 2022.
Available at: https://consultations.hse.gov.uk/crd-reach/restriction-proposals-004/supporting_documents/Annex%2015%20restriction%20dossier%20%20lead%20in%20ammunition.pdf

[7] A shotgun's gauge. Hunter-ed. Available at: https://www.hunter-ed.com/pennsylvania/studyGuide/A-Shotgun-s-Gauge/20103901_88442/

[8] Burrard G. Chapter 1 The barrels. The modern shotgun. 2nd edn. London: Lowe & Brydone Ltd; 1969. pp.17-36 & pp.273.

[9] Hogg I. Guns and how they work. London: Marshall Cavendish Publications Limited; 1981.

[10] Shotgun cartridge. Available at: https://qph.cf2.quoracdn.net/main-qimg-33aa0a5da341f1831b84d645dbfbff91.webp

[11] Shotgun pellet sizes. Available at: https://forum.cartridgecollectors.org

[12] Knight B. Legal Aspects of Medical Practice. 5th edn. Edinburgh: Churchill Livingstone; 1992. pp. 194-195

[13] Engineers Edge. Densities of metals and elements table. Available at: https://www.engineersedge.com/materials/densities_of_metals_and_elements_table_13976.htm

[14] Philip M, Bolton B. Technology of Engineering Materials. Oxford:Butterworth-Heinemann; 2002. pp. 55.

[15] Thomas VG, Kanstrup N, Gremse CY. Key questions and responses regarding the transition to use of lead-free ammunition. Proceedings of the Oxford Lead Symposium Lead Ammunition: understanding and minimising the risks to human and environmental health. 2014. pp. 176. Available at: http://www.oxfordleadsymposium.info/wp-content/uploads/OLS_proceedings/papers/OLS_proceedings_thomas_kanstrup_gremse.pdf

[16] Dales S. Managing Director, Steel Shot Company Limited, HU14 3RS. Conversation with Jane Barnes-Warden. 20 February 2023.

[17] Payne T, O'Rourke S, Malbon C. Home Office Body Armour Standard (2017). CAST Publication number 012/17. July 2017. pp.26.
Available at: https://assets.publishing.service.gov.uk/government/uploads/system/uploads/attachement data/file/634517/Home Office Body Armour Standard.pdf

[18] Leeming DW, Quelch LJ, Fenne PM, Gotts PL, Kelly PM. Fundamental Data and Analysis on the Penetration Resistance of a Textile Armour. PASS Conference. The Hague, Netherlands. 2002.

Viability and Performance of Improvised Ammunition

D. Cooper[1]

[1]Physical Protection Group, Threat Mitigation Technologies, Metropolitan Police Service, Albany Street Police Station, 60 Albany Street, Camden, London, NW1 4EE, United Kingdom
dean.cooper2@met.police.uk

Abstract. Within the UK, ammunition is tightly controlled by the Firearms Act (1968) and, as such, not readily available for legal purchase by the general populous. The Metropolitan Police Service assess the threat picture within the UK and have highlighted open source designs and instructional material available online to construct improvised ammunition. This paper has two aims i. to investigate the viability and performance of improvised handgun ammunition ii. To assess the ability of such ammunition to defeat typical materials used in law enforcement protection schemes. This paper will initially review open-source material available online with a view to down selecting the most promising and investigating the viability and performance of each option. Finally, this study will consider modifications and alternative materials for the fabrication of improvised ammunition and, once again, its performance will be evaluated against typical materials used in law enforcement protection schemes.

1. INTRODUCTION

Within the UK, both firearms and ammunition are tightly controlled by the Firearms Act (1968) and, as such, not readily available for legal purchase by the general populous. The UK operates a firearms licensing structure, in which citizens can apply to hold firearms or shotguns; for this they require a 'good reason' for owning the firearm, as well as an assessment by the police to ensure they do not pose risk to public safety [1]. As of 31 March 2024 147,364 people hold firearms certificates and 495,798 hold shotgun certificates in the UK. There are 510,717 individuals who hold these due to dual certification [2]. Due to this strict firearms licensing, firearms offences in the UK are considered some of the lowest in the world with 6268 offences in England and Wales reported between April 2023 and March 2024 [3]. Although globally the UK has one of the lowest rate of firearm related violence, death and crime in the world, this annual figure has shown an upward trend since 2014, and in that time has risen by over 29% from 4856 cases in 2014 [3]. One possible cause for this increase in offences over the past decade is the introduction of the 3D printed firearm.

Although 3D printed firearms are not the first kind of improvised firearms in existence with published designs and assembly instructions, they have significantly reduced the fabrication skills and equipment needed to produce a functional semi-automatic weapon. The first 3D printed firearm design was published on the internet on the 7th May 2013. The Liberator was a single shot firearm almost entirely produced from plastic using a 3D printer. Since this there have been huge advancements in the field of homemade printed firearms. Models available for download on the internet vary in calibre, complexity, and construction materials. At the time of writing most downloaded and produced firearms are hybrid 3D printed firearms. These systems use both 3D printed components in conjunction with traditionally manufactured components such as barrels and firing pins. By using a hybrid approach these firearms are durable, powerful, and accurate. Currently the most downloaded 3D printed firearm is a pistol-calibre carbine known as the 'F**k Gun Control 9 mm' (FGC 9) which is used by hobbyists, insurgents, militia members, terrorists, and drug traffickers in at least 15 countries [4]. The FGC9 free download has an extremely comprehensive guide to its production, and is also distributed with a link to producing improvised ammunition for use in tightly controlled areas such as Europe [5]. The FGC9 and many similar hybrid 3D printed weapons are chambered in 9 mm, and will still require ammunition that cannot legally be obtained in the UK without a firearms licence. Data from the Metropolitan Police Service (MPS) shows that 12 3D printed guns were recovered in London between 2020 and 2024 [6]. Data from the National Crime Agency (NCA) show that nationally 17 3D printed firearms or components were recovered in 2022 and 25 recovered in 2023 [7]. In addition to the threat of 3D printed firearms, converted blank firing weapons pose a threat in the UK. Data from police figures show that in 2023 there were 64 discharges from converted blank firing weapons compared to 42 from real equivalent weapons [8]. Recently legislative changes have made owning certain types of blank firing weapons illegal in the UK [9].

Whether the threat is an illegal i) converted blank firing firearm, ii) 3D printed firearm or iii) factory produced firearm, without a firearms licence, criminals will not be able to legally obtain

 https://doi.org/10.52202/081621-0058

ammunition for any of these weapons in the UK. This leaves the criminal two available options; illegally obtain commercial off the shelf (COTS) ammunition, or the production of improvised ammunition.

2. OPEN-SOURCE INSTRUCTIONAL MATERIAL

A specialist team within the MPS assess the firearms threat picture within the UK and have highlighted open-source designs and instructional material available online to construct improvised ammunition. The team highlighted the documentation that had received high levels of traffic and downloads. In addition using MPS registered IT equipment and as part of their employed role, the author undertook an assessment of open source material provided by a popular search engine with relevant key terms.

Instructional material highlights techniques to produce improvised ammunition in a range of materials and calibres. From the material reviewed improvised handgun ammunition fell into three broad categories; 3D printed ammunition, reactivated ammunition, and modified ammunition.

2.1 3D printed ammunition

3D printed ammunition files are widely available on many 3D model file repository sites. Files are available in a variety of calibres and designs. The files are usually accompanied by suggested material type, most of which suggest Poly Lactic Acid (PLA+) as a build material as well as printer settings and instructions from the designer on how to achieve the best results. The files that are provided are usually in a Stereolithography (STL) file format, which can be edited and amended by the user to customize or enhance the print to their specification. Files are available for both projectiles and casings. Projectiles are available which are modelled on conventional ammunition including Full Metal Jacket (FMJ) and Jacketed Hollow Point (JHP) as well as custom models designed as sabots to carry ball bearings for additional mass. It should be noted that 3D models will not behave the same as the ammunition they are geometrically modelled on due to their different material properties. Some casing files are modelled on COTS ammunition and as such are dimensionally accurate to its brass counterpart, others are specifically designed with modified internal dimensions to help withstand the firing of the bullet in a weapon, or allow the seating of an alternative sized projectile. One such design, the '9mm Deep' is well documented and designed to fire a range of different sized projectiles from a 9mm chambered weapon using a custom designed case with external dimensions of a 9mm Lugar [10]. Videos are available on host platforms such as YouTube of 3D printed ammunition being successfully discharged from both 3D printed weapons and conventional handguns into a range of targets including paper targets, watermelons and gels.

2.2 Reactivated ammunition

Within the UK and many other heavily regulated areas it is possible to obtain deactivated ammunition, as well as some component parts for the ammunition such as the projectile and casing. Various resources on the internet discuss how to re-activate this de-activated ammunition. Distributed alongside the FGC9 is the document *'But What About Ammo? V1.0'* [5] which documents using Hilti DX 6.8/11 cartridges for Hilti DX tools to provide the priming compound and propellant to re-activate them. The document is a comprehensive guide with high quality imagery for the user to follow. It provides details of all the required tooling, as well as links to 3D printed parts to assist in the process.

Following on from V1.0 *'But What About Ammo Volume 2'* [11] has also been published. This document focuses on casting and electroplating lead projectiles for use in 9mm ammunition. This document follows a similar format to V1.0 and equally has a high quality pictorial guide to the process. Since casting of lead projectiles has been a process followed by hobbyist shooters in the UK for a long time, and unjacketed lead projectiles can be purchased from some specialist ammunition suppliers and known to be less penetrative than its jacketed counterpart, this process was not explored further.

Various resources are available online for the creation of homemade primers. These typically use a spent primer and reactivate it with an alternative high explosive, making use of the existing anvil and primer cup. *'Homemade Primer Course'* [12] details three methods that can be used; Primers made from toy caps, primers made from strike-anywhere matches, and primers made with H-48 home made potassium chlorate priming compound. Strike anywhere matches are hard to obtain in the UK and have been banned from sale since 2018, the chemical components in H-48 are also restricted and hard to obtain, as such the toy cap method is the most viable.

2.3 Modified ammunition

Another method documented within the open-source literature focuses on the modification of blank ammunition. Blank ammunition can be purchased in the UK without the licensing restrictions other ammunition is subjected to. Blank ammunition is designed to generate a muzzle flash and explosive sound, for instance to start a race, rather than fire a projectile. A paper *'Modification Methods of Blank Pistols in Turkey in 2006'* [13] explores how both blank firing weapons of the time could be modified to discharge a projectile, as well as modification of blank cartridges by adding projectiles to them. The paper makes reference to how blank cartridges with buckshot projectiles can be lethal.

3. PRODUCTION OF IMPROVISED AMMUNITION

Assembly of improvised ammunition took place at a Ballistic Test Facility (BTF), under the supervision of the facilities range officer. Assembly was conducted using basic hand tools similar to those recommended in open-source material. Where appropriate some professional reloading tools were used, for example a primer seating tool and reloading press in the interest of efficiency and safety.

3.1 3D printed ammunition production

3D printing, which is otherwise known as additive manufacturing, is a process which has been available to hobbyists since 2009. Within the field of 3D printing there is a vast array of different 3D printers that are available for purchase, using different filament types and technologies. Prices for 3D printers for home use currently range from around £100 upwards. Most open source resources recommend PLA+ as a build material. For this study ammunition was produced in alternative materials. Ammunition was produced using two different 3D printers;

i. Form 3+ [1] Stereolithography (SLA) or resin 3D printer, which can produce very high quality prints using UV curing of resin. Components produced in; Black Resin V5, Durable Resin and Tough 2000 Resin. Unit price approximately £2250 [14].

ii. Bambu Lab P1S[2] Fused Deposition Modeling (FDM) 3D printer, which uses a more traditional filament spool to melt and fuse layers of material. This unit was used to produce Acrylonitrile Butadiene Styrene (ABS), PLA+ and Polyethylene terephthalate glycol (PETG) components. Unit price approximately £510 [15]

Utilizing the FDM printer ABS and PETG prints were evaluated. Compared to PLA+, PETG is a tougher but softer material, ABS is a tougher and more rigid material. Prints on the SLA printer use specialist resin. In addition to a standard resin (black Resin V5), using the Formlabs material database [16] resins design to withstand force were selected to produce prints from Durable resin and Tough 2000 resin.

3.1.1 3D printed projectiles

Figure 1 below shows a selection of 3D printed projectiles. Projectiles were produced in three different designs, and were sized to fit COTS 9mm x 19mm brass cartridge cases and loaded using COTS primers and propellant. The three designs of 9mm projectiles that were evaluated were:

i. Solid projectile,
ii. Projectile with opening for three 6mm ball bearings (BBs),
iii. Projectile with opening for single 4mm BB.

All projectile only designs were downloaded from freely available 3D printing file repositories and had a nominal diameter of 8.93mm.

---

1 Form 3+, Formlabs®, Massachusetts, USA, Catalogue #PKG-F3-WSVC-BASIC

2 P1S, Bambu Lab®, Shenzhen, China

https://doi.org/10.52202/081621-0058

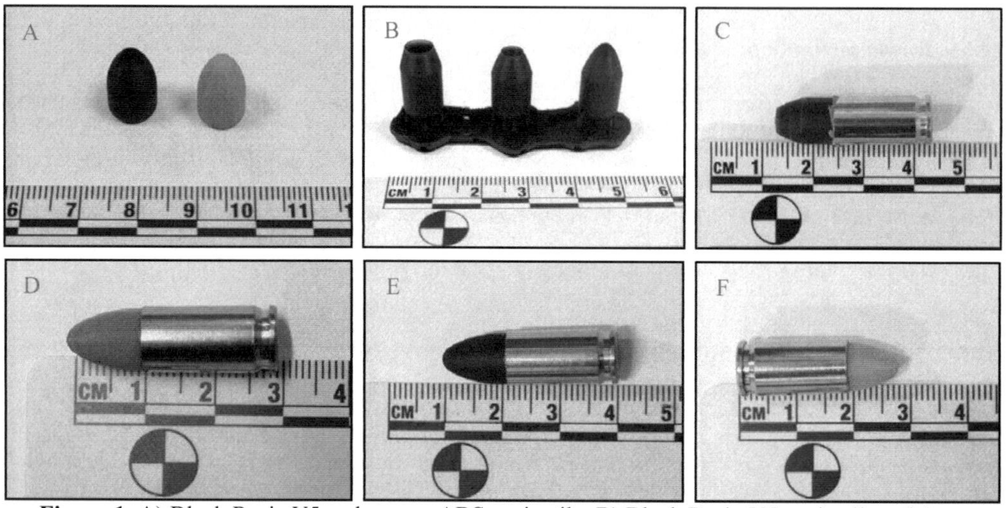

Figure 1. A) Black Resin V5 and orange ABS projectile. B) Black Resin V5 projectiles of three designs on a raft C) Black Resin V5 with 4mm BB projectile in brass case D) Red PETG projectile in brass case. E) Black Resin V5 projectile in brass case. F) Orange ABS projectile in brass case.

3.1.2 3D printed cartridge cases

Figure 2 below shows 3D printed cartridge cases, with projectiles assembled as complete rounds. The projectiles were the same as described above. Although cartridge cases were downloaded from the internet, to achieve the desired fitment custom case STL files were created. Cases were initially modeled on the 9 x 19mm brass NATO cartridge design using freely available CAD software. After initial trials it was found that 3D printed cartridge cases offered insufficient support to hold the primer in place, and it moved axially towards the neck of the case when struck with the firing hammer and pin. Hence the primer pocket was modified to offer more support to the primer in this area these cases are referred to in the results section as 'enhanced (1)' with an extra 1mm of support around the primer pocket, and 'enhanced (2)' with an extra 2mm support around the primer pocket. Cartridge cases of this design were produced on the SLA printer, capable of producing high tolerance and thin prints. A further custom design projectile and cartridge case combination was then created to be printed in PLA+ on the FDM printer. Due to the capability of the printer, and material characteristics this had a thicker walled case and projectile of only 8mm diameter. As such this projectile would not engage the rifling of the 9mm barrel.

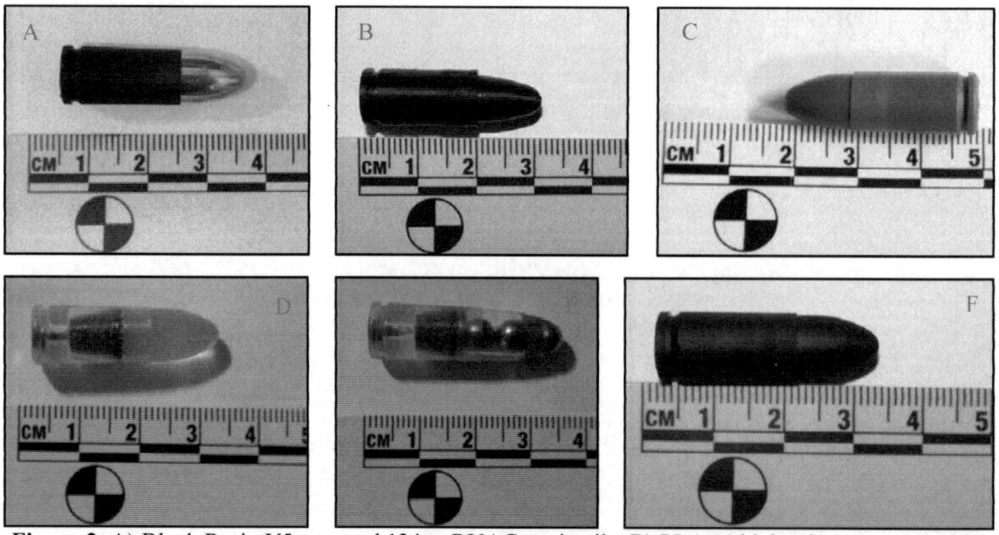

Figure 2. A) Black Resin V5 case and 124gn RUAG projectile. B) PLA+ 'thick print' case and 'thick print' projectile. C) Tough 2000 Resin case and projectile. D) Durable Resin case and projectile. E) Durable Resin case and projectile with 3 x 6mm BB. F) Black Resin V5 case and projectile.

https://doi.org/10.52202/081621-0058

3.2 Reactivated ammunition

Reactivated ammunition was produced using two different methods, one which produced both a reactivated primer and the propellant for the round, the other to produce just the primer. Following the process described in *'But what about AMMO?'* [5] A Hilti 6.8/11 cartridge was disassembled, and propellant and priming compounds were harvested. The priming compound was added to a previously fired primer and this was inserted into a new brass case with a new projectile. It should be noted that all of the above materials described in *'But what about AMMO?'* can be legally obtained in the UK with a license. Figure 3 below show the Hilti cartridge and one of the rounds produced.

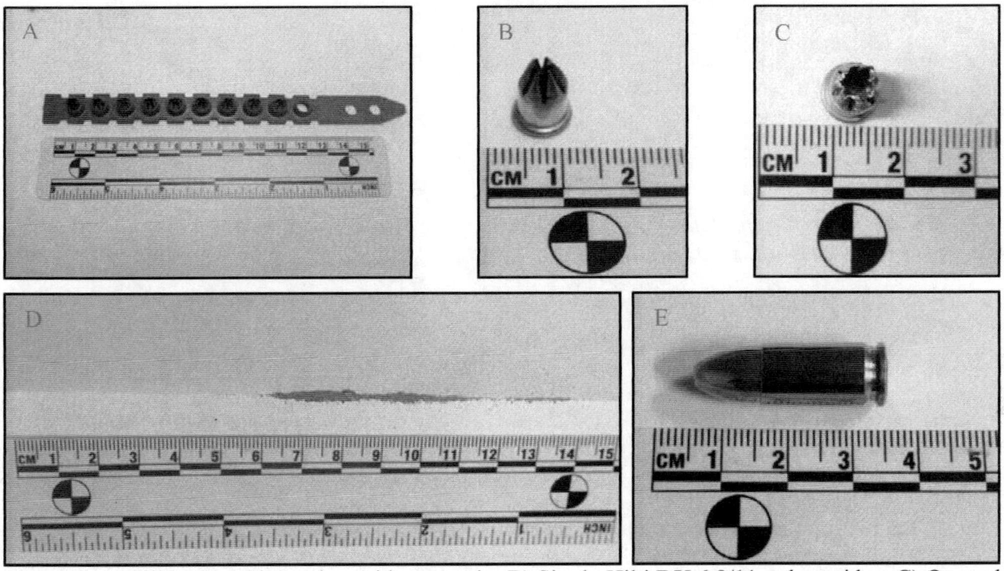

Figure 3. A) Hilti DX 6.8/11 red cartridge on strip. B) Single Hilti DX 6.8/11 red cartridge. C) Opened Hilti DX 6.8/11 with propellant and priming compound removed. D) Priming compound recovered from 2 Hilti DX 6.8/11 Cartridges. E) Assembled reactivated bullet.

Following the process described in 'Homemade Primer Course' [12] three improvised primers were produced, see Figure 4 below. By utilizing a cap gun ring, designed to create a loud sound and smoke simulating a gunshot when used in conjunction with a cap gun, the charge was removed from the ring and in conjunction with a small amount of fast pistol propellant an improvised primer created. The improvised primer was created from a previously discharged COTS primer, which can be obtained from 'deactivated ammunition' without license in the UK.

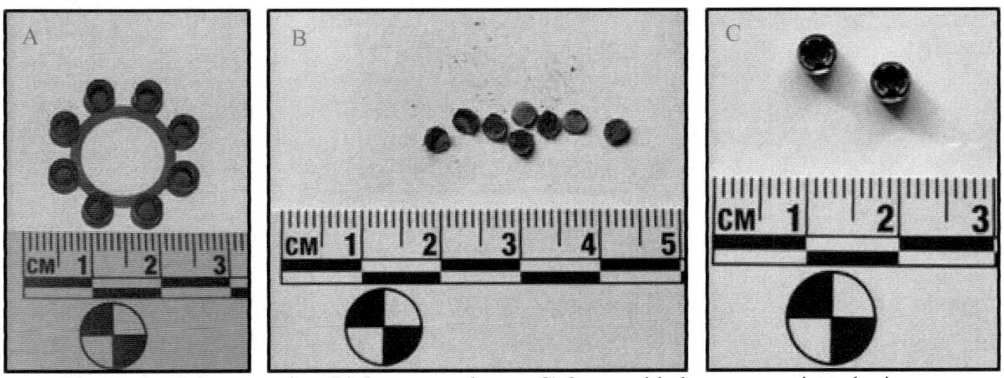

Figure 4. A) Cap gun ring. B) 8 recovered caps. C) 2 assembled cap gun activated primers.

https://doi.org/10.52202/081621-0058

3.3 Modified ammunition

Figure 5 below shows some of the modified ammunition produced during this study. Modified ammunition were created in a similar method to that as documented in *'Modification Methods of Blank Pistols in Turkey in 2006'* [13] by affixing BBs to the end of blank ammunition. BBs were affixed with either microcrystalline wax, superglue, or melted plastic to the top of the blank. In this study stainless steel BBs were used. A crimp was applied to the neck of the 9mm x 21mm blank to reduce the neck diameter and allow it to chamber into a 9mm x 19mm barrel.

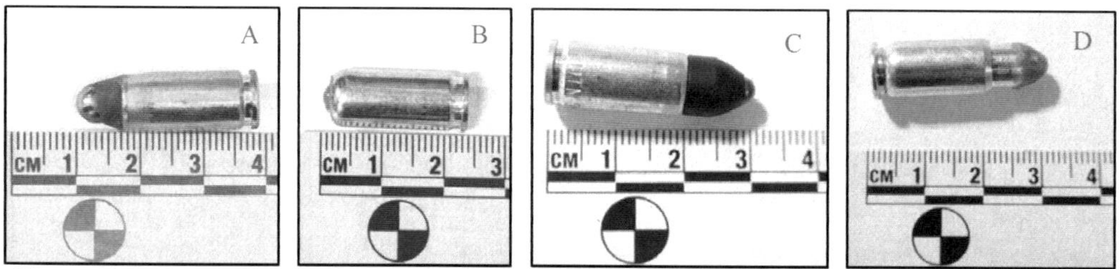

Figure 5. A) 9mm x 21mm blank with 8mm BB held in place with wax. B) 9mm x 19mm blank with 4.2mm BB melted into top. C) 9mm x19mm blank with 2 x 3.8mm BB glued in. D) Blank with 4.2mm BB held in place with wax.

4. BALLISTIC TESTING

A series of trials were conducted to determine the performance of the improvised handgun ammunition, constructed as described in section 3. In this study the performance of generic materials used within police protection schemes were evaluated. Figure 7 below shows the following materials which were evaluated;

i. 3mm Polycarbonate sheet 400mm x 400mm
ii. 0.9mm mild steel galvanized sheet 500mm x 500mm
iii. 20 layers of 200g/m^2 Aramid. Aerial density of target 4.0kg/m^2

The Rheinisch-Westfälischen Sprengstoff-Fabrike (RWS) DM11A1B2 (produced by RUAG since 2002) FMJ 124gn, as identified in the UK Home Office body armour (2007) standard [17] was used as a benchmark for testing, referred to as the RUAG DM11 in this report.

4.1 Method

Throughout this study a 9mm barrel (length 254mm) was coupled to a universal receiver. The improvised ammunition was loaded by the BTF range officer. Velocity was recorded using a chronograph system positioned 2.0 meters in front of the target. The aramid target was mounted on a microcrystalline wax block (nominally 420mm x 350mm x 100 mm), which was affixed into a steel frame. The polycarbonate and steel targets were affixed directly to the steel frame with no backing support. Figure 6 below shows a diagram of the experimental set up. Aramid and polycarbonate targets were held in place with elasticated hooked straps, and steel targets were affixed using clamps. High speed photography was captured a sample of the impacts and flight paths during the study.

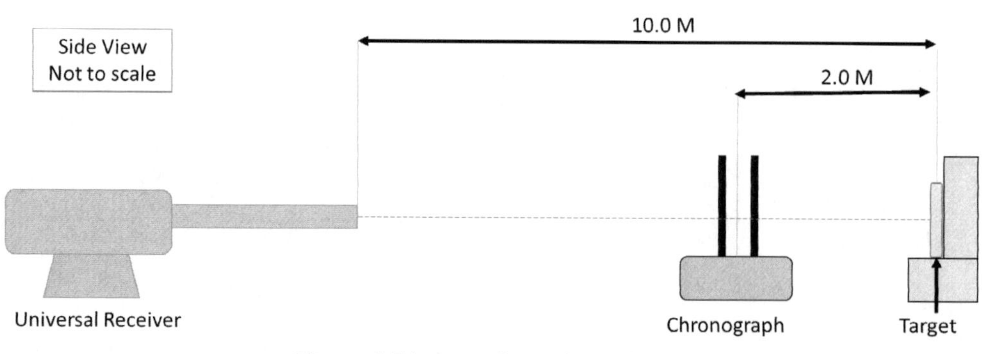

Figure 6. Diagram of experimental set up

Figure 7. A) Universal receiver fitted with 9mm barrel. B) Chronograph system. C) 3mm polycarbonate. D) 1mm galvanized steel plate. E) 20 layers of 200gsm aramid.

4.2 Results

Table 1 below summarized the results of the ballistic trial. Figure 8 plots velocity against mass of the three different types of improvised ammunition as described above, ammunition suffering a misfire and not recorded on the chronograph system has been omitted.

Table 1. Ballistic testing results

Shot reference	Type	Projectile	Mass (gn)	Cartridge case	Primer	Propellant	Propellant mass (gn)	Velocity (m/s)	Target type	Perforation
1	COTS	9mm RUAG DM11	124	9 x 19mm brass	CCI 500	APP	5.4	362	Polycarbonate	YES
2	COTS	9mm RUAG DM11	124	9 x 19mm brass	CCI 500	APP	5.4	365	Steel	YES
3	COTS	9mm RUAG DM11	124	9 x 19mm brass	CCI 500	APP	5.4	368	Aramid	NO
4	COTS	9mm RUAG DM11	124	9 x 19mm brass	CCI 500	Lovex DO36	3.5	156	Aramid	NO
5	Reactivated	9mm RUAG DM11	124	9 x 19mm brass	CCI 500	Hilti red	3.5	295	Steel	YES
6	Reactivated	9mm RUAG DM11	124	9 x 19mm brass	Improvised Hilti	Hilti red	4.1	Misfire	Steel	N/A
7	Reactivated	9mm RUAG DM11	124	9 x 19mm brass	Improvised Hilti	Hilti red	4.1	Misfire	Aramid	N/A
8	Reactivated	9mm RUAG DM11	124	9 x 19mm brass	Improvised Cap	APP	4.1	Misfire	Aramid	N/A
9	Reactivated	9mm RUAG DM11	124	9x19mm brass	Improvised Cap	APP	4.1	Misfire	Aramid	N/A
10	Reactivated	9mm RUAG DM11	124	9x19mm brass	Improvised Cap	APP	4.2	Misfire	Aramid	N/A
11	3D printed	9mm RUAG DM11	127	9 x 19mm black resin V5	CCI 500	APP	5.0	287	Aramid	NO
12	3D printed	9mm black resin V5	14.3	9 x 19mm black resin V5	CCI 500	APP	5.0	Misfire	Aramid	N/A
13	3D printed	9mm red PETG	13.7	9 x 19mm brass	CCI 500	APP	5.0	343	Aramid	NO
14	3D printed	9mm orange ABS	10.7	9 x 19mm brass	CCI 500	APP	5.0	367	Aramid	NO
15	3D printed	9mm black resin V5	14.3	9 x 19mm brass	CCI 500	APP	5.0	408	Polycarbonate	YES
16	3D printed	9mm red PETG	13.7	9 x 19mm brass	CCI 500	APP	5.0	361	Polycarbonate	NO
17	3D printed	9mm orange ABS	10.7	9 x 19mm brass	CCI 500	APP	5.0	404	Polycarbonate	NO
18	3D printed	9mm black resin V5	14.3	9 x 19mm brass	CCI 500	APP	5.0	387	Steel	NO
19	3D printed	9mm orange ABS with 1 x 4mm BB (friction)	17.5	9 x 19mm brass	CCI 500	APP	5.0	424	Steel	YES
20	3D printed	9mm orange ABS	10.7	9 x 19mm brass	CCI 500	APP	5.0	372	Steel	NO
21	3D printed	9mm red PETG HF	13.7	9 x 19mm brass	CCI 500	APP	5.0	398	Steel	NO
22	3D printed	9mm black resin V5	14.3	9 x 19mm brass	CCI 500	APP	5.0	329	Aramid	NO
23	3D printed	9mm black resin V5	14.3	9 x 19mm black resin V5	CCI 500	APP	5.0	Misfire	Aramid	N/A
24	3D printed	9mm black resin V5	14.3	9 x 19mm black resin	CCI 500	Red dot	5.0	Misfire	Aramid	N/A
25	3D printed	9mm black resin V5 with 2 x6mm BB (friction)	41.5	9 x 19mm black resin	CCI 500	Red dot	5.0	Misfire	Aramid	NO
26	3D printed	9mm tough 2000 resin	15.6	9 x 19mm tough 2000 resin	CCI 500	APP	4.1	Misfire	Polycarbonate	N/A
27	3D printed	9mm tough 2000 resin	15.6	9 x 19mm tough 2000 resin	CCI 500	APP	4.1	Misfire	Polycarbonate	N/A
28	3D printed	9mm tough 2000 resin	15.6	9 x 19mm tough 2000 resin	CCI 500	APP	4.1	125	Polycarbonate	NO
29	3D printed	9mm tough 2000 resin	15.6	9 x 19mm tough 2000 resin	CCI 500	APP	4.1	351	Steel	NO
30	3D printed	6mm BB (friction)	49.6	9 x 19mm tough 2000 resin	CCI 500	APP	4.1	N/R	Steel	NO
31	3D printed	9mm tough 2000 resin	15.6	9 x 19mm tough 2000 resin (1)	CCI 500	APP	4.2	Misfire	Aramid	N/A
32	3D printed	9mm tough 2000 resin	15.6	9 x 19mm tough 2000 resin (2)	CCI 500	APP	4.2	274	Aramid	NO
33	3D printed	9mm durable resin	14.6	9 x 19mm durable resin (2)	CCI 500	APP	4.2	313	Aramid	NO
34	3D printed	9mm durable resin	14.6	9 x 19mm durable resin (1)	CCI 500	APP	4.2	322	Aramid	NO
35	3D printed	9mm durable with 3 x 6mm BB (wax)	50.9	9 x 19mm durable resin (1)	CCI 500	APP	4.2	343	Aramid	NO
36	3D printed	9mm durable with 3 x 6mm BB (glued)	50.9	9 x 19mm durable enhanced (1)	CCI 500	APP	4.2	304	Aramid	NO
37	3D printed	9mm durable with 3 x 6mm BB (glued)	51.9	9 x 19mm tough resin	CCI 500	APP	4.2	Misfire	Aramid	NO
38	3D printed	9mm durable with 3 x 6mm BB (glued)	51.9	9 x 19mm tough resin	CCI 500	APP	4.2	N/R	Aramid	NO
39	3D printed	9mm durable resin	14.6	9 x 19mm durable enhanced (1)	CCI 500	APP	6.5	Misfire	Aramid	NO
40	3D printed	9mm durable resin	14.6	9 x 19mm durable enhanced (1)	CCI 500	APP	6.5	358	Steel	NO
41	3D printed	8mm PLA Thick Print	11.8	9 x 19mm Thick PLA 3d printed	CCI 500	APP	4.1	152	Polycarbonate	NO
42	3D printed	8mm PLA Thick Print	11.8	9 x 19mm Thick PLA 3d printed	CCI 500	APP	4.1	N/R	Steel	NO
43	Modified	8mm BB (wax)	32.3	9 x 21mm blank	Preloaded	Preloaded	Preloaded	279	Steel	YES
44	Modified	3 x 4.2mm BB (friction)	24	9 x 19mm Force n Force training blank	Preloaded	Preloaded	Preloaded	N/R	Steel	NO
45	Modified	6.34mm BB (wax)	16.2	9 x 21mm blank	Preloaded	Preloaded	Preloaded	32	Steel	NO
46	Modified	4.2mm BB (melted in)	8	9 x 19mm blank	Preloaded	Preloaded	Preloaded	N/R	Steel	NO
47	Modified	4.2mm BB (melted in)	8	9 x 19mm blank	Preloaded	Preloaded	Preloaded	N/R	Polycarbonate	NO
48	Modified	8mm BB (wax)	32.3	9 x 21mm blank	Preloaded	Preloaded	Preloaded	298	Polycarbonate	YES
49	Modified	6.34mm BB (wax)	16.2	9 x 21mm blank	Preloaded	Preloaded	Preloaded	149	Polycarbonate	NO
50	Modified	8mm BB (wax)	32.3	9 x 21mm blank	Preloaded	Preloaded	Preloaded	295	Aramid	NO
51	Modified	2 x 3.8mm BB (glued)	8	9 x 19mm battlefield blank	Preloaded	preloaded	preloaded	N/R	Aramid	NO
52	Modified	4.2mm BB (wax)	8	9 x 19mm Force n Force training blank	Preloaded	preloaded	preloaded	N/R	Aramid	NO
53	Modified	8mm BB (wax)	32.3	9x19mm brass	CCI 500	APP	4.2	166	Aramid	NO

https://doi.org/10.52202/081621-0058

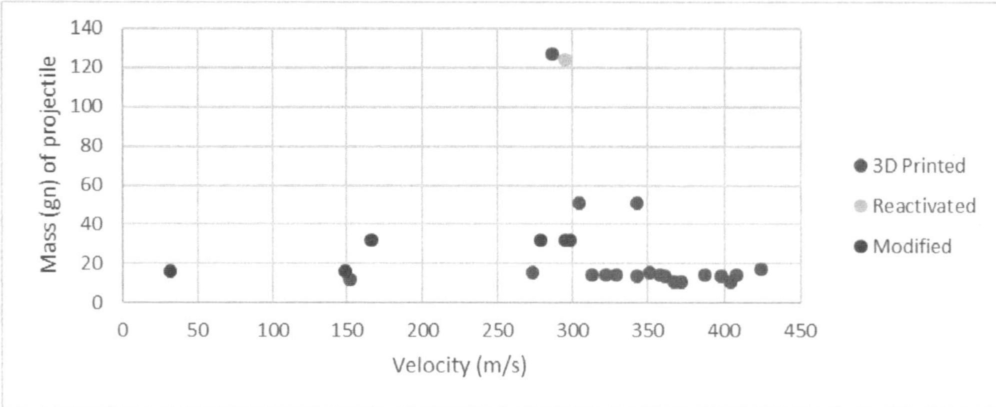

Figure 8. Chart showing velocity plotted against mass of the three categories of improvised ammunition.

4.3 3D printed ammunition discharged using a handgun

Following the ballistic trials using the universal receiver system, three 3D printed rounds were assembled and inserted into the magazine of a Glock 17 pistol mounted in a Ransom Rest, as shown in Figure 9 below. The pistol was remotely fired using a draw string. Results of this trial are displayed in Table 2 below. It should be noted that although the pistol successfully discharged the rounds, the weapon did not successfully cycle, and required the slide to be manually drawn backwards to chamber the next round.

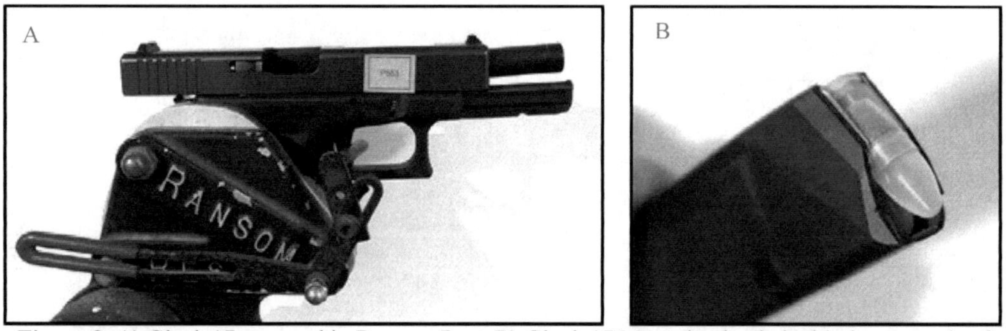

Figure 9. A) Glock 17 mounted in Ransom Rest. B) Glock 17 Magazine loaded with three rounds of 3D printed ammunition.

Table 2. Results from handgun discharge

Shot reference	Type	Projectile	Mass (gn)	Cartridge case	Primer	Propellant	Propellant mass (gn)	Velocity (m/s)	Target type	Perforation
54	3D printed	9mm durable resin	14.6	9 x 19mm durable enhanced 1	CCI 500	APP	4.2	339	Steel	NO
55	3D printed	9mm tough 2000 resin	15.6	9 x 19mm durable enhanced 1	CCI 500	APP	4	Misfire	Steel	N/A
56	3D printed	9mm durable resin	14.6	9 x 19mm durable enhanced 1	CCI 500	APP	4.2	365	Steel	NO

5. DISCUSSION

Of the 14 impacts on the aramid target, none perforated the target. Both polycarbonate and steel targets were perforated on one occasion by both the 3D printed and modified ammunition. Figure 10 below shows some impact sites from 3D printed ammunition.

In this trial all of the improvised primers lead to a misfire. The instructional material available online was followed as accurately as possible from the instructions given. As only two Hilti based primers were produced, and three cap gun primers, the process is at best difficult to successfully achieve for the layperson. It should be noted that spent primers that had previously been used had a heavy strike on their previous use with a universal receiver when compared to that normally left by a pistol. This may have impacted the primers ability to function for a second time. The propellant harvested from the Hilti DX 6.8/11 was able to be utilized in conjunction with a COTS primer, and performed in a very similar manner to the COTS Alliant Power pistol (APP) which was also used within this trial for 3D printed ammunition.

Of the 31 rounds of 3D printed ammunition that were loaded into the universal receiver system, nine resulted in a misfire. One perforation of steel and one perforation of polycarbonate were observed, both 3D printed projectiles using brass cartridge cases. On inspection of the misfired ammunition it was normally due to one of two causes; primer seated too deep into the primer pocket which lead to a light strike, or the cartridge case shattering on impact from the firing pin and again the primer not being struck as intended. Most 3D printed cartridge cases that had been successfully fired sustained some form of damage from the firing process, this was often a fracture around the primer pocket, or split to the neck of the case. Some cases split into two or more pieces in the firing process. Figure 11 below shows some of the recovered 3D printed casings and projectiles from the trial. Upon inspecting the impact on targets, as well as high speed photography, at least two of the 3D printed projectiles with BBs separated before reaching the target, leading to multiple impact locations. Accuracy of 3D printed ammunition with a 9mm projectile was very good, behaving in a similar manner to COTS ammunition. This is likely due to the ammunition engaging with barrel rifling. All fired shots impacted within 25mm of the intended location. The 'thick print' bullet was less accurate, impacting 90mm from the intended target in one instance, likely due to the projectile diameter being smaller than the barrel.

Of the 11 rounds of modified ammunition that were loaded into the universal receiver, none resulted in a misfire. Unfortunately due to the limitations of the chronograph system used it did not reliably detect projectiles less than 7mm in diameter passing through. As such no reading was attained for seven of the BBs discharged. It should be noted that accuracy for modified ammunition was poor, shots were up to 280mm from the intended impact location in one instance. This is likely due to the projectile diameter being smaller than the barrel.

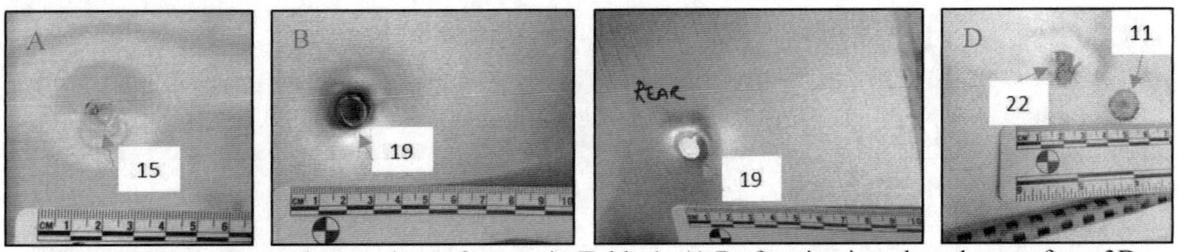

Figure 10. Numbers relate to shot reference in Table 1. A) Perforation in polycarbonate from 3D printed Black Resin V5 projectile. B) Perforation (front) in steel from 3D printed ABS projectile. C) Perforation (rear) in steel from 3D printed ABS projectile. D) Two non -perforations in aramid from 3D printed Black Resin projectile (22) and COTS projectile fired from 3D printed case (11).

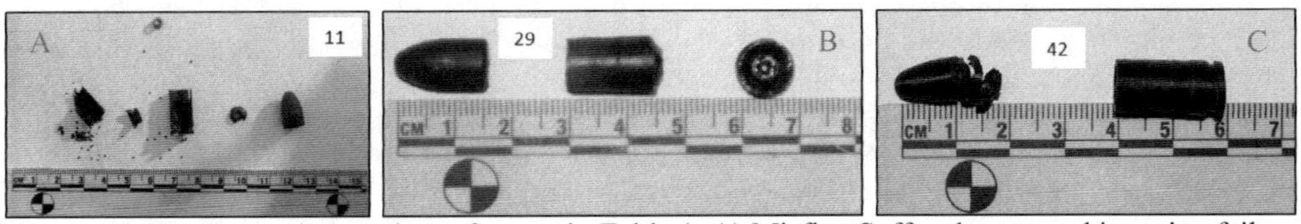

Figure 11. Numbers relate to shot reference in Table 1. A) Misfire, Suffered catastrophic casing failure upon strike from firing pin. B) Successfully fired at steel sheet, projectile recovered, case fracture at primer pocket. C) Successfully fired at steel sheet, projectile recovered showing impact damage.

6. CONCLUSION AND FURTHER WORK

The UK is subject to strict licensing laws which heavily restrict public access to firearms and ammunition. Because of this criminals looking to obtain ammunition for illegally held weapons are left with two options; illegally obtain COTS ammunition or produce improvised ammunition.

Instructional information is available open-source on the construction of improvised ammunition, which broadly falls under 3 categories;

i. Reactivated ammunition,
ii. 3D printed ammunition,
iii. Modified ammunition.

A significant amount of information is available, some techniques described require components that are not legally obtainable within the UK.

https://doi.org/10.52202/081621-0058

Of the 14 impacts made on the aramid targets, none perforated. The simple, generic armour target of 20 layers of $200g/m^2$ of aramid was able to stop all of the improvised ammunition tested at 10 meters. Many body armour schemes used in UK frontline policing have greater quantities of ballistic resistant materials in their construction, hence all improvised ammunition of these construction materials and types are unlikely to defeat such armours.

All improvised primers produced resulted in misfires. Both modified and 3D printed ammunition perforated steel and polycarbonate targets on one occasion each during the trials.

Finally the 3D printed element of this study focusses on plastic based materials, as is currently commonplace for hobbyist users due to the prohibitively expensive setup and material costs of other forms of 3D printers. As advancements are made in this area of manufacturing, alternative techniques such as metal 3D printing may become accessible to hobbyists, resulting in alternative projectile threats. Further consideration should be paid to assessing fiber reinforced materials which have recently become available for hobbyist use, and if they have merit in the improvised ammunition field.

Acknowledgments

The author would like to acknowledge Richard Pryor for his assistance in this work. I would like to convey a sincere thank Christopher Collinge for his design and production of 3D printed components. I would also like to thank Peter Hall and the team at Hallrite Engineering BTF [18] for their expertise, guidance and assistance.

References

[1] Guide on firearm licensing law, UK Home Office https://www.gov.uk/government/publications/firearms-law-guidance-to-the-police-2012/guide-on-firearms-licensing-law-accessible-version#chapter-1-an-overview--frequently-asked-questions-about-firearms-licensing accessed 05/03/2025

[2] Accredited official statistics, Statistics on firearm and shotgun certificates, England and Wales: April 2023 to March 2024 https://www.gov.uk/government/statistics/firearm-and-shotgun-certificates-april-2023-to-march-2024/statistics-on-firearm-and-shotgun-certificates-england-and-wales-april-2023-to-march-2024 accessed 05/03/2025

[3] Dataset Crime in England and Wales: Office for National Statistics Centre for Crime and Justice https://www.ons.gov.uk/peoplepopulationandcommunity/crimeandjustice/datasets/crimeinenglandandwalesotherrelatedtables accessed 05/03/2025

[4] Dearden, Lizzie; Gibbons-Neff, Thomas (10 September 2024). "He's Known as 'Ivan the Troll.' His 3D-Printed Guns Have Gone Viral". New York Times.

[5] IvanTheTroll, Deterrence Dispensed presents – but what about AMMO v1.0 (2020)

[6] Metropolitan Police Service, Freedom of Information Request reference no: 01.FOI.24.041779 available at https://www.met.police.uk/foi-ai/metropolitan-police/disclosure-2025/january-2025/3d-printed-guns/ accessed 28/02/2025

[7] National Crime Agency, National Strategic Assessment 2024 Firearms https://www.nationalcrimeagency.gov.uk/threats/nsa-firearms-2024 accessed 07/03/2025

[8] Adrian Goldberg & Paul Grant, Converted handguns fired more than 'real' weapons in UK crimes (BBC News, UK, 2024) https://www.bbc.co.uk/news/uk-67895627 accessed 07/03/2025

[9] National Police Chiefs' Council, Firearms Amnesty available at https://news.npcc.police.uk/releases/over-1-000-top-venting-blank-firers-handed-in-with-less-than-10-days-remaining-of-national-gun-amnesty accessed 08/03/2025

[10] 9mm Deep Build instructions https://3dgunbuilder.com/accessories/9mm-deep/

[11] Cathode, But What About Ammo Volume 2 (2020)

[12] W. Marshall, Homemade Primer Course (2014)

[13] Aylin YalÅın Sarıbey, Modification Methods of Blank Pistols in Turkey in 2006 (J Forensic Sci, May 2009, Vol. 54, No. 3)

[14] Bambu Labs P1S 3D printer specification https://uk.store.bambulab.com/products/p1s?id=42280371159100 Accessed 08/03/2025

[15] Formlabs 3+ 3D printer specification https://formlabs.com/store/3d-printers/form-3-basic-package-without-service/#/ accessed 08/03/2025

[16] Formlabs material database available at: https://formlabs.com/uk/materials/ accessed 10/03/2025

[17] Croft J., and Longhurst D., HOSDB Body Armour Standards for UK Police (2007) Part 2: Ballistic Resistance Publication No. 39/07/B (2007) pp 12

[18] Hallrite Precision Engineering, Markham Works, Markham Road, Chesterfield, England, S40 1SU

The threat of (high energy) laser against military combat clothing

B. Meuken[1], S.A. van Binsbergen[1], M. Geljon[2] and J.N.A. van Leeuwen[1]

[1] *TNO Defense Security and Safety, Ypenburgse, Boslaan 2, 2496 ZA, The Hague, the Netherlands (denise.meuken@tno.nl)*
[2] *TNO High Tech Industry, Stieltjesweg1, 2628 CK, Delft, the Netherlands*

Abstract. The High Energy Laser (HEL) is becoming a technologically mature weapon system. Although high energy lasers are not intended to be used as a direct threat to dismounted personnel, incidental exposure to radiation from high energy lasers will be an increasing (indirect) threat. In the past few years TNO performed research on laser-material interaction, to investigate the effects of lasers against different targets in terms of lethality. Parallel to this, work has been done on laser reflections from targets and hazard-risk analyses related to the use of high energy lasers in the field. It was shown that reflections of targets irradiated with high energy lasers show a highly dynamic and irregular intensity distribution in both temporal and spatial manner. A dismounted combat soldier may also be exposed to these laser reflections. Due to the relative novelty of this threat, the influence or effect of indirect exposure of clothing and equipment to a high energy laser was unknown, e.g., the occurrence of melting, smouldering or ignition of the fabric materials. This paper discusses the experiments performed within TNO's high energy laser facility (L30) on a range of single and multi-layered samples taken from military combat clothing and personal equipment (vests and/or helmets). The intensities used during the experiments were calculated from laser safety scenarios and reflected intensities from typical targets determined from earlier test series. These were then translated into the possible reflective intensities that could reach the dismounted combat soldier in the same scenarios. In this paper the experimental set-up and selected results in terms of effect and transmitted heat will be shown and discussed.

1. INTRODUCTION

High energy laser weapons will increasingly be used, for example as a counter drone system, by both own troops and opponents. Accidental exposure to radiation from high energy lasers is an increasing (indirect) threat for the dismounted combat soldier. It has been shown that laser reflections from targets show a highly dynamic and irregular intensity distribution in both a temporal and spatial manner [1, 2]. The reflections when engaging a potential target show a highly dynamic pattern that varies in magnitude and direction and depends on bulk material, material surface condition, phase state of the material (solid or liquid) and geometry. The dismounted combat soldier may be (inadvertently) exposed to these laser reflections. The focus of this paper is on the experimental investigation of the influence or effect when exposing military clothing and equipment to these laser reflections.

Using relevant reflection intensities from typical target engagement scenarios, experiments were performed within TNO's high energy laser facility on a range of single and multi-layered samples taken from military combat clothing and personal equipment. The data on heat flux obtained from the experiments are compared to existing thermal thresholds doses [3, 4]. The goal of the study was to gain insights into the effects of exposed military clothing to laser reflections (at typical reflection intensities), as well as the associated potential risks for the dismounted soldier.

1.1 Laser safety

The principal approach to protect personnel or bystanders from laser exposure is to define a space in which the direct, reflected and scattered radiation from the laser exceeds the Maximum Permissible Exposure (MPE). The MPE is a non-damage threshold for eye and skin damage due to laser radiation (at a given wavelength and duration). This zone is known as the Nominal Hazard Zone (NHZ) and controlling access to this NHZ promotes safety, see Figure 1, left [5]. This NHZ is translated into a NOHD (Nominal Ocular Hazard Distance), which is a linear distance. Deterministic approaches to calculate these areas exist for purely specular (worst-case) and purely diffuse (best-case) reflections, see Figure 1, right.

https://doi.org/10.52202/081621-0059

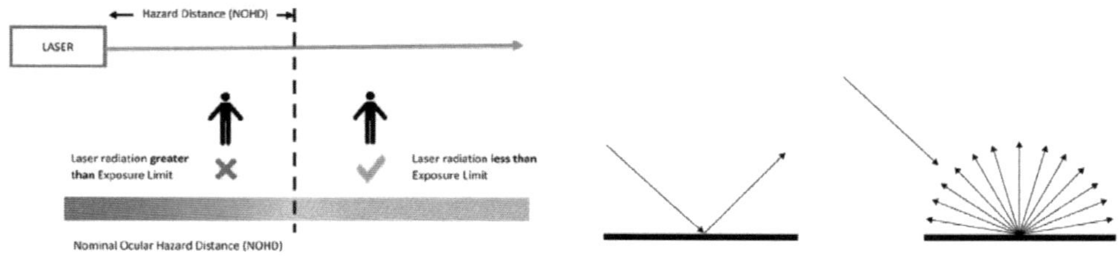

Figure 1. Nominal Ocular Hazard Distance schematically explained [5] (left). Specular and diffuse reflections (right).

Conservative approaches in respect to laser safety assume a worst-case scenario: the reflection is perfectly specular and the laser beam properties are conserved. In reality, the experimental results suggest that reflections generally are a mix of specular and diffuse components, and that the reflection process increases the beam divergence (see Figure 2 for some experimental examples) [6]. This implies that the safe stand-off distance (NOHD) for observers is somewhere between the values found for purely specular and those for diffuse reflection. This is also the approach taken to determine the intensities used during the experiments.

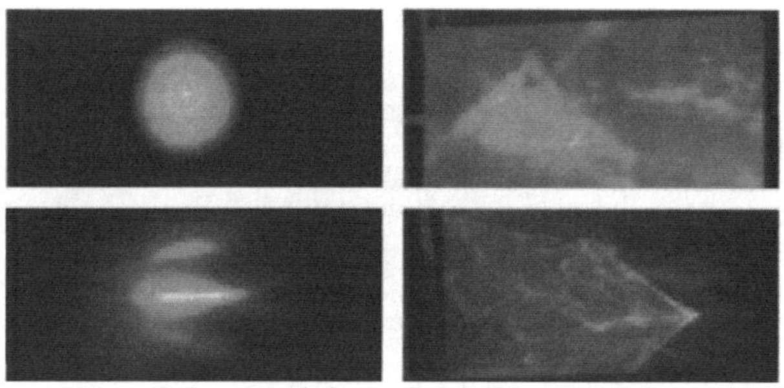

Figure 2. Reflection patterns of glass (top-left panel), machined steel (bottom-left), and curved steel (top-right and bottom-right). Scales and intensities are not comparable. The shape of the reflection screen itself is visible in the images on the right [6].

1.2 Scenario description

The focus is on reflections from targets engaged by a high power laser beam, irradiating personnel present in the vicinity as an incidental or collateral effect. The potential exposure will depend on the intended target, atmospheric conditions, laser specifications as well as on circumstantial aspects such as range, direction, orientation, doctrine and protection. Ideally, all relevant aspects are accounted for as they determine the relevant exposures applied in the experiments. To determine laser power and intensity levels to which samples will be exposed during the experiments, two scenarios and vignettes were created.

Scenario 1 ("drone") involves a HEL system trying to engage and intercept an approaching drone. A recon team crosses an open field ahead of their unit as suddenly a hostile armed drone appears at a low altitude (~7 m. The drone moves at 5 km/h towards the recon team posing a lethal threat. A support unit at the rear (~1 km) targets the drone with a laser. The drone is hit and destroyed, but reflections of the laser beam reach the recon team members. Scenario 2 ("155 mm") involves a HEL system trying to eliminate a 155 mm artillery shell before it hits the ground. A platoon attempts to cross an open field when it comes under artillery fire using 155 mm shells. The shell, aimed near the platoon, is detected on radar and engaged with a laser nearby (~1 km). The shell is hit and destroyed but reflections of the laser reach the recon team members.

Table 1 highlights the most important input parameters for the experiments discussed in this paper. The reflection intensities are estimated from previous TNO laser material experiments performed on 3 mm ABS (Acrylonitrile Butadiene Styrene) material and inert 155 mm artillery shells where the reflection data was gathered. This data was then used to extrapolate relevant reflection intensities for the laser, at target and observer positions defined in the scenarios. The performed laser hazard assessment assumes worst case numbers, e.g., the target is considered to be close to personnel and the peak reflection intensities are taken from the earlier experiments.

Table 1. Scenario data for experiments.

	Scenario 1: armed drone interception	Scenario 2: incoming artillery
Target characteristics	Material: 3 mm ABS (Acrylonitrile Butadiene Styrene) Distance of personnel to target 5 to 30 m	Material: 10 mm steel (155mm artillery) Distance of personnel to target 11.5 to 0.1 km
Laser definition	Wavelength 1.07 μm Output Power 30 kW	Wavelength 1.07 μm Output Power 300 kW
From laser hazard assessment > Required laser power setting*	Peak (@5m) > 65 Watts Peak (@30m) > 1.9 Watts	Peak (@100 m) > 913 Watts

* using an 8 cm laser spot diameter.

The laser facility is usually used for experiments using up to 30 kW of optical power. The minimum power obtainable with the setup is 155 W. Therefore, this power was used instead of the lower powers mentioned in Table 1. These experiments are intended as an initial exploration of exposure and the consequences.

1.2 Heat flux

(Reflected) laser irradiation may lead to exposure of the dismounted soldier to a high heat flux caused by interaction of the laser beam with clothing and/or personal equipment or even direct exposure if the material is (partially) transparent. Heat flux encountered in reflected laser scenarios are likely to be high intensity and very short in duration. The role of (protective) clothing is to prevent or minimize skin burn damage by reducing the heat transfer towards the skin underneath the clothing. In these situations the garment is assumed to maintain its integrity during the exposure. For the laser scenarios, the (protective) characteristics of the clothing are derived in a manner similar to what has been done in the past for heat flash [7].

LaChance [8] summarizes harm criteria (injury or lethality) associated with people exposed to flames, high air temperatures or high heat fluxes. Burn injuries may vary in degree (e.g., first, second or third degree burn), but also in the amount of body area covered. The harm is described in both flux intensity (kW/m^2) as well as exposure time (s). This leads to the expression in a thermal dose unit:

$$\text{Thermal Dose Unit} = I^{4/3}t \qquad (1)$$

Where I is the radiant flux in kW/m^2 and t the exposure time in seconds.

Table 2 lists some of the thermal doses presented in literature that may result in first, second or third degree burns, as a result of exposure to UV or IR radiation [3]. The dose considers the intensity as well as the time of exposure. The IR radiation range is of most concern for generating burns, and also of concern considering the L30 laser wavelength (1070 nm). These radiation ranges will be compared against the heat flux data retrieved during the experiments for an indication of possible burn severity.

Table 2. Radiation burn data [3].

Burn severity	Threshold dose (kW/m^2)$^{4/3}$s	Threshold dose (kW/m^2)$^{4/3}$s
	UV	IR
First degree	260-440	80-130
Second degree	670-1100	240-730
Third degree	1220-3100	870-2640

The values for the dangerous dose and the LD50 thermal dose levels of infrared radiation are 590 and 1460 (kW/m^2)$^{4/3}$s, respectively [4]. Here a dangerous dose is defined as dose resulting in death to 1% of the exposed population. The LD50 dose is the lethal dose where 50% of the exposed population would die.

https://doi.org/10.52202/081621-0059

2. EXPERIMENTAL SET UP AND MATERIALS

Experiments at the L30 facility at TNO were performed in an indoor environment (see Figure 3) with a 30 kW multimode laser (IPG, Model YLS-30000-Y18, λ: 1068-1080 nm). The laser system operates using 23 separate modules, each providing approximately 1.3 kW of power. The laser was powered by a generator which was located outside of the building and cooled using 2 chillers which were located inside the building. The laser fibre was connected to a custom-built optical platform (OP) which can focus or defocus the beam over a 2 – 40 m range from the aperture to a spot diameter of 2-15 cm at target. The beam can also be steered up to ±2 degrees in horizontal and vertical direction when necessary. Laser output power can be selected in the L30 software anywhere between 155 W and 30 kW, depending on the number of modules being operational and system settings and calibration.

The target area was located at ~20 m from the optical platform. The diagnostics used (e.g. cameras) were mostly located near the target area. Shown in Figure 3 are the laser unit itself, the OP, and chiller positions, as well as the target location set up, including target rack with sample, TREMs (Target Reflected Energy Measurement unit), back stop, heat flux sensor and multiple cameras.

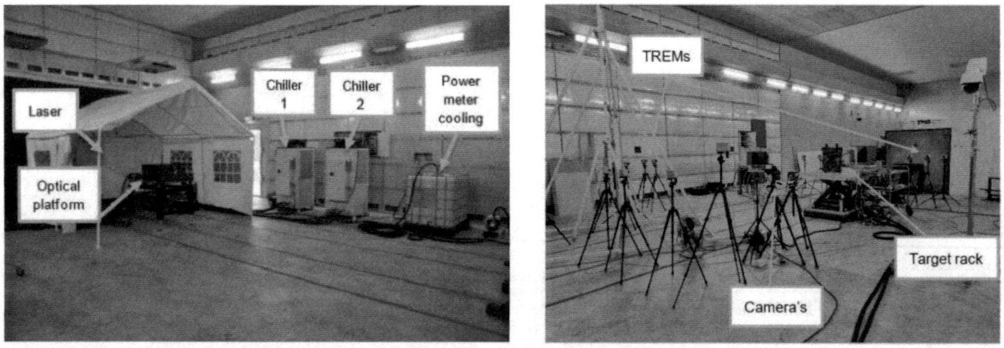

Figure 3. View of the L30 set up, including laser, optical platform and chillers (left) and the target location set up, including target rack with sample, TREMs, back stop, heat flux sensor and multiple cameras.

Figure 4 shows a detailed view of the heat flux sensor position in relation to sample. The sensor was aligned to the centre of the laser spot using the red guide laser of the L30 system. There was an airgap of approximately 1 cm between the sensor and the fabric. The HFS01 is a water cooled sensor that measures heat flux [9]. It is mainly used to test reaction to fire and fire resistance. HFS01 measures heat fluxes in the range of (0 to 800) x 10^3 W/m^2.

Figure 4. View of the sample rack and heat flux sensor position in relation to sample. The sensor was aligned to the centre of the laser spot.

Several samples of combat fabric materials were supplied for the experiments and an indication of the appropriate layering of the samples in relation to the scenarios was given. Table 3 lists the different configurations tested.

Table 3. Configurations of combat fabric materials for testing.

ID	Layering	Description
1	NFP green *(Woven Fabric for Combat Uniform)*	Single layer worn on arms and legs
2	NFP green FR *(Woven Fabric for Combat Uniform (Flame Resistant)*	Single layer worn on arms and legs
3	Cordura FR Sioen	Cover material Dutch vest and pouches
4	Combat shirt knitwear *(Knit with FR properties)*	Single layer worn on torso
5	NFP green + Underwear (moisture regulating, *Synthetic Knit*)	Double layer worn on torso and lower body
6	NFP Tan FR *(Woven Fabric for Combat Uniform (Flame Resistant)*	Single layer in lighter colour, worn on arms and legs
7	Tarp *(Lightweight Coated Fabric)*	Example behaviour of thin material
8	Snow camouflage *(Lightweight Coated Fabric)* + Rain wear + Combat shirt knitwear + Thermo underwear *(Synthetic Knit)*	Combination of layers worn in arctic circumstances
9	Rain wear *(3 layer Laminated Fabric)* + NFP green	Double layer worn on arms and legs
10	Rain wear + Combat shirt knitwear	Double layer worn on torso and lower body
11	Backpack cover 120L + Cordura Fr WebeR *(Woven Fabric with FR coating for Body Armour, Pouches, etc.,)*	Backpack combination

Where appropriate the samples were cut down to sizes that fit the target rack. The fabric was secured using clamps and tightened to form a (as much as possible) flat surface. The experiments were performed using the appropriate calculated power at target (e.g. 155 W and 913 W), an 8 cm spot diameter and irradiation times of 1 or 5 seconds, or up to a visible effect on the fabric tested (e.g. perforation, burn, melt, smoke formation). In this way a broad experience was gained towards to the effects of laser reflections on clothing.

3. RESULTS

Both power levels, 155 W based on the drone scenario and 913 W based on the 155 mm scenario, are on the high end of what is to be expected in the field under the given circumstances. Typically, samples were exposed to the laser for 1 or 5 seconds to simulate brief and sustained exposure, respectively. Additional experiments were performed where exposure was continued until the samples were visibly affected by the laser illumination or even perforated. Some experiments are discussed below for illustration.

3.1 Experiments on fabric samples

An example of an experiment on the NFP Green material is discussed below (ID1 from Table 3). This material is often worn as a single layer on arms and legs. In the drone scenario the power affecting the dismounted soldier is that of 155 W due to the system limitations as mentioned earlier. Figures 5, 6 and 7 show the experimental data acquired. The sample is exposed to the laser beam (seen as a purple haze) and after 15 seconds the material is perforated, see Figure 5. The temperature data is shown in Figure 6, obtained from the MWIR (MidWave-InfraRed) camera. To convert the IR measurements to temperatures, the emissivity of the target material is required. Because this can change (significantly) during irradiation, it was decided to operate under the assumption that the emissivity equals 1, i.e. the target is a black body. In reality, the emissivity was (slightly) lower, meaning that the produced temperature values are minimum temperatures - an overestimation is not to be expected. Also note that the camera is limited by its dynamic range at a temperature of 140°C. This is due to the available calibration profiles and ranges. In reality, the temperature will have increased to significantly higher

https://doi.org/10.52202/081621-0059

values. Higher temperatures are typically found for the darker parts of the camouflage pattern, as those absorb more laser radiation. This is also where the material starts to disintegrate.

Figure 5. Left and middle: NFP Green non FR during laser irradiation (back and front view), settings: 155W, 8 cm spot diameter. Right: results after 15 seconds (perforation).

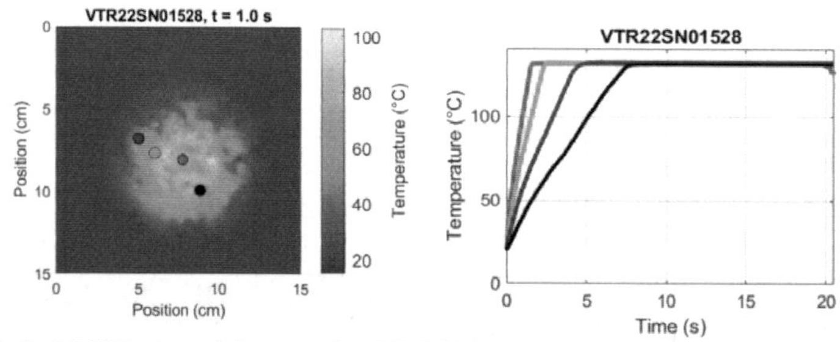

Figure 6. Left: MWIR plot of shot number VTR22SN01528, 155 W, 8 cm spot diameter on NFP Green. Highest temperatures at darkest parts of the camouflage pattern. On the right the temperature on the coloured dots as a function of exposure time is displayed. Note: the emissivity was assumed to be 1, meaning that the converted temperatures shown are minimum values.

The heat flux data is presented in Figure 7. The heat flux through the fabric seemed to increase in several steps which corresponded to effects witnessed at the materials surface; these are indicated with the orange arrows in Figure 7. After 4 seconds the front surface of the sample started to produce some light smoke, while around 8 seconds the smoke is also visibly forming at the rear side of the sample. At this time the material is being scorched (see also Figure 5). Around 15 seconds a perforation occurs in the dark part of the camouflage pattern. The laser is shut off around 17 seconds (visible from the black lines in the graph). Looking at the first few seconds (maximum 5 seconds is of interest in the drone scenario), a heat flux of ~1 W/cm^2 was obtained.

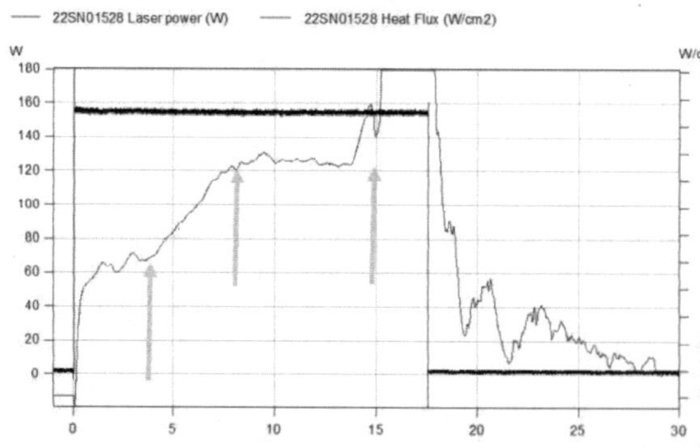

Figure 7. Heat flux data graph of VTR22SN01528, 155 W, 8 cm spot diameter on NFP Green. Different stages of effect are indicated by the orange arrows (explained in the text). Perforation occurred after about 15 seconds, the material scorched between 8 and 13 seconds while also producing smoke.

The same material was also exposed to 913 W (artillery scenario). The results are shown in Figures 8 to 10. As can be expected the effects occurred much more rapidly. Smoke formation occurred within 2 seconds, after which the material started to burn. Local temperatures reached at least 300°C while a heat flux of ~5W/cm^2 was obtained in the first 2 seconds.

Figure 8. NFP Green with a power setting of 913 W and a spot diameter of 8 cm (artillery scenario). Figures from left to right show smoke formation after 1.5 s, start of burn after 2.3 s and fire after 2.4 s.

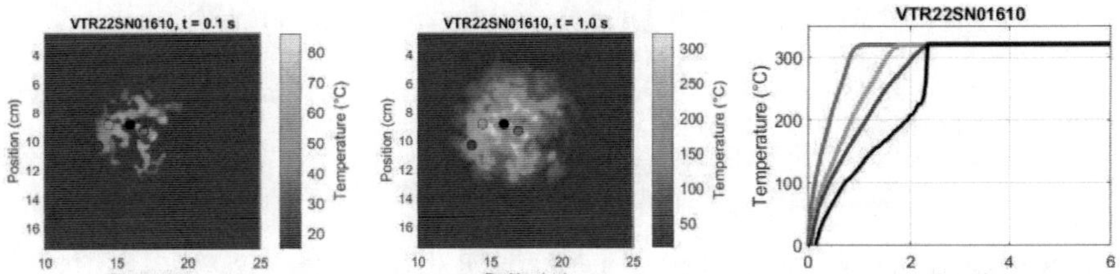

Figure 9. Left and middle: MWIR plots of shot number VTR22SN01610, 913 W, 8 cm spot diameter on NFP Green after 0.1 s and 1.0 s. Darker parts of the pattern heat up faster, as is clearly visible in the leftmost image (t = 0.1 s). On the right the temperature of the coloured dots as a function of exposure time is displayed. Note: the emissivity was assumed to be 1, meaning that the converted temperatures shown are minimum values.

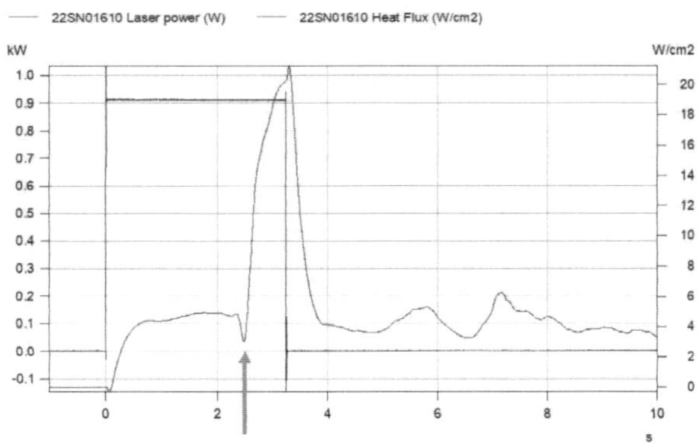

Figure 10. Heat flux data graph of VTR22SN01610, 913 W, 8 cm spot diameter on NFP Green. Indicated by the orange arrow is the fire starting at 2.3 seconds, after which the laser was shut down.

Similar results were obtained for the other materials and layer combinations presented in Table 3. The low weight polyamide materials for the rain wear, tarp and backpack cover, melted upon laser irradiation while the other fabrics started to scorch, smoke or burn (although mostly when exposed to the laser beam longer than 5 seconds as shown for the NFP Green example). A special note is to be made for the combat shirt knitwear which seemed to be transparent for the laser wavelength. Please note that for reflections, a direct 5 second exposure is considered an extreme worst case as reflections tend to be very dynamic in both time and space. The time to perforation varied from sample to sample due to the camouflage pattern in the area irradiated by the laser spot. The darker dye is affected much more rapidly compared to the lighter dyes.

In most experiments the ventilation was set to low when the laser was in operation. This is part of the protocol to reduce smoke and to prevent hazardous gases from moving towards the laser set-up. To determine whether airflow would influence the experimental outcome, some experiments were

https://doi.org/10.52202/081621-0059

repeated with ventilation off. In some cases these tests resulted in the ignition of the fabric and fire (although not necessarily within a 1 or 5 seconds exposure). In all cases the flames self-extinguished as soon as the laser was shut down.

3.2 Heat flux

Table 2 provided an indication of what the measured heat flux during the experiments could mean to the soldier while wearing an outfit consisting of the clothing layer(s) as used in the experiments. As the underlying study is a first introduction, the data retrieved will only be used as an indication to determine the heat flux caused by laser reflections on military clothing. Besides the heat flux also the effect on the material is of interest. For practical reasons the data is processed for the two scenarios separately, e.g. a 5-seconds 155 W laser exposure in the drone scenario versus the 1-second of 913 W laser exposure for the artillery scenario. Please note that a "mean" value for the heat flux was taken from the heat flux graphs over the given time period (5 or 1 seconds) as the heat flux is not always constant.

From Table 2, the following threshold doses were mentioned for IR radiation [3]:

- First degree burns: 80-130 $(kW/m^2)^{4/3}s$
- Second degree burns: 240-730 $(kW/m^2)^{4/3}s$
- Third degree burns: 870-2640 $(kW/m^2)^{4/3}s$

Figure 11 shows the calculation results for the thermal dose in $(kW/m^2)^{4/3}s$ as derived from the heat flux data during the 5-second exposures simulating the drone scenario. Indicated in red are the thresholds for first (80-130 $(kW/m^2)^{4/3}s$) and second (240-730 $(kW/m^2)^{4/3}s$) degree burns [3]. It shows that first degree burns may be expected behind the NFP Green (*light green*), NFP Green FR (*dark green*), NFP Tan FR (*light brown*), the tarp (*purple*) and the backpack + Cordura FR Weber (*brown*) layer combinations. No experiments resulted in levels exceeding the second degree burn limit, only combat knitwear (light blue) got close.

Figure 12 shows the calculation results for the thermal dose in $(kW/m^2)^{4/3}s$ as measured by the heat flux sensor obtained during the 1-second exposures simulating the artillery scenario. Indicated in red are the thresholds for first and second degree burns [3]. It shows that first degree burns may be expected for (*light green*), NFP Green FR (*dark green*), NFP Green + moisture regulated underwear (*yellow*), NFP Tan FR (*light brown*), the tarp (*purple*) and the backpack + Cordura FR Weber (*brown*) layer combinations. Second degree burns may be expected when the soldier only wears one layer of combat knitwear (*light blue*).

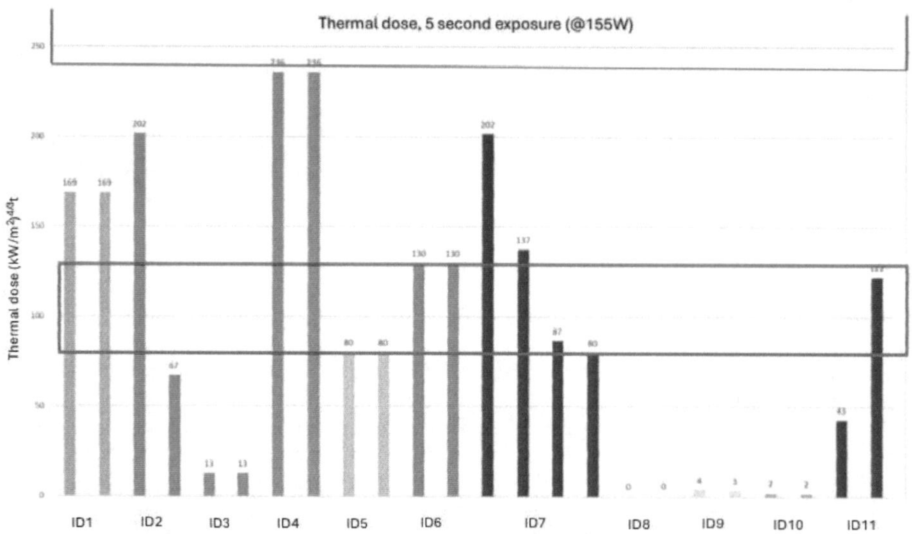

Figure 11. Thermal dose in $(kW/m^2)^{4/3}s$ as measured by the heat flux sensor behind the fabric as obtained during the 5-second exposures simulating the drone scenario. ID# as indicated in Table 3. Indicated in red are the thresholds for first (80-130 $(kW/m^2)^{4/3}s$) and second (240-730 $(kW/m^2)^{4/3}s$) degree burns [3].

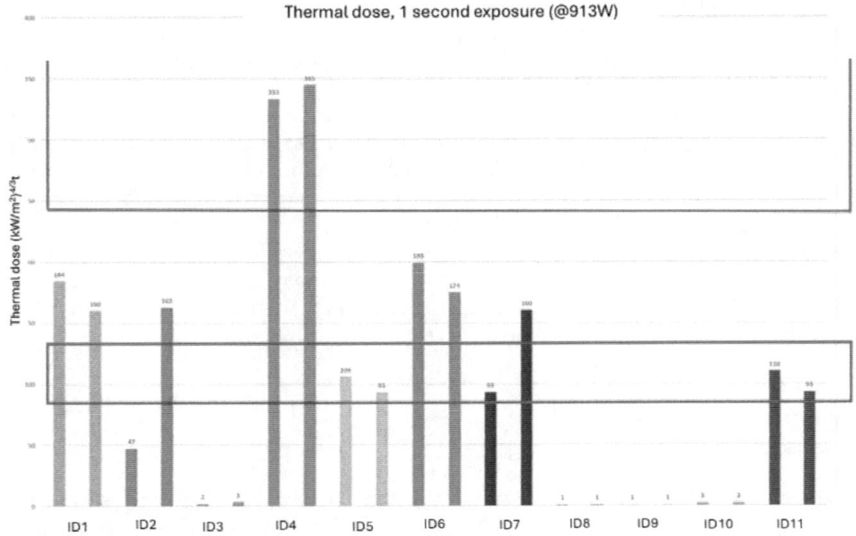

Figure 12. Thermal dose in $(kW/m^2)^{4/3}s$ as measured by the heat flux sensor behind the fabric as listed for the 155 mm reflective scenario for a 1 second exposures. ID# as indicated in Table 3. Indicated in red are the thresholds for first (80-130 $(kW/m^2)^{4/3}s$) and second (240-730 $(kW/m^2)^{4/3}s$) degree burns [3].

The dangerous dose and L50 dose stated by TNO in CPR 16E [4] (being 590 and 1460 $(kW/m^2)^{4/3}s$, respectively), were not exceeded in any experiment. Considering the thresholds defined by Pew [3], one may expect first degree burns after exposure to the reflections, even when exposed for only 1 second. Second degree burns are less likely: only for Combat shirt knitwear the measured heat flux exceeded the thermal dose in the 155 mm reflective scenario. In addition, more fabric layers provide better protection.

4. DISCUSSION AND CONCLUSIONS

The dynamic nature of the power, size and direction of the laser reflections means that even for well-defined scenarios, one can justify a wide range of typical values, often ranging from harmless to very harmful. While (future) laser safety models should be able to generate probabilistic results, this aspect is expected to remain a significant uncertainty in future work. Doses that cause pain should be considered while looking at heat flux, what are the affected surfaces of the skin (actual irradiated area from reflections) and within what time can one withdraw from the exposure. This is also part of the work performed internationally on establishing a probabilistic laser risk approach.

Slightly different conditions or targets (position, orientation, material, movement, etc) can result in significant changes to reflected intensities. Therefore, the presented results should not be extrapolated to, for example, another 'drone scenario' without critical consideration. The heat flux meter measures both (broadband) thermal radiation from heated objects such as fabric or flames, as well as (narrowband) laser radiation, as its functionality is purely based on absorption of radiation in visible and IR wavelengths. This means the measured heat flux is a combination of both aspects. As long as it is assumed that both have the same effect on skin this does not pose a problem. However, considering the difference between thermal dose limits found in heat flux literature and MPE's as set by laser safety standards, this may not be the case.

The camouflage pattern on the fabrics plays a significant role as to where the effect of the laser radiation occurred. Darker areas of the camouflage pattern absorb more of the laser energy compared to the lighter parts. Reproducibility of results therefore varies between the duplicate tests. For example, the 'black green' colour in the NFP Green material is known to have a reflectance of ~10% at the wavelength of the laser, whereas the lightest colour, 'light grey' has a reflectance of ~60% [10].

The low weight polyamide rain wear, tarp and back pack cover materials melted quickly upon laser irradiation, while the other fabrics, blends containing cellulose, started to scorch, smoke or burn

https://doi.org/10.52202/081621-0059

(although mostly when exposed to the laser beam longer than 5 seconds). Due to the extremely dynamic nature of reflections, a continuous 5-second exposure should be considered as an absolute worst case that can only be achieved for static targets and personnel.

Ultimately, thermal doses observed for single-layer experiments in some cases exceeded the thresholds for first degree burns, but never did so for second degree burns with a single exception (Combat shirt knitwear). This is the case for higher laser power (913 vs. 155 W) and longer exposure times (5 vs. 1 s). For two or more layers of clothing, the thermal doses were shown to be significantly lower in the majority of cases. Taking into account that the laser-based MPE (skin) value is just below the second degree burn heat flux threshold, and that scenarios were evaluated for worst-case assumptions, one should be able to conclude that as long as skin is covered, reflections of the simulated laser engagements should not be expected to cause skin damage.

The dangerous dose or L50 dose stated in CPR 16E [4] was not exceeded in any experiment. However, considering the thresholds defined by Pew [3], one may expect first degree burns after exposure to the reflections, even when exposed for 1 second only. Second degree burns are less likely and more layers provide better protection, unless the (top) layer perforates.

Acknowledgments

TNO would like to thank the Clothing & Personal Equipment Branch of the Defence Materiel Organisation of the Netherlands Ministry of Defence for funding and supplying fabric material for this research.

References

[1] Daigle J.-F., Pudo D., Théberge F. and Châteauneuf M., Laser safety evaluation for high-energy laser interaction with solids, Opt. Eng., Vol. 56(2), nr. 026106, 2017.

[2] Henrichsen, M., Schwarz, B., Ritt, G., Azarian, A., and Eberle, B., Laser safety assessments supported by analyses of reflections from metallic targets irradiated by high-power laser light, Applied Optics, vol. 60, nr. 22, 2021.

[3] Pew, P., LD50 equivalent for the effect of thermal radiation on humans, WS Atkins Safety & Reliability, 1997.

[4] TNO, Methods for the determination of possible damage to people and objects resulting from releases of hazardous materials, TNO 121647, CPR-16E: Committee for the Prevention of Disasters, 1992.

[5] Mikrocentrum, Laser safety instruction, Delft, 2018.

[6] Meuken, D., van Binsbergen, S.A., Scheers, L.C.W., van den Berg, P.J., and van Eijk, A.M.J., Reflection measurements in TNO's 30 kW laser facility, Proc. SPIE 12273, High-Power Lasers and Technologies for Optical Countermeasures, 122730D (2 November 2022); https://doi.org/10.1117/12.2634047

[7] Jager, H., Heat Flash Testing - Combat Clothing, Minitry of Defence - Defence Materiel Organisation, 2014.

[8] LaChance, J., Tchouvelev, A., and Engebo, A., Development of uniform harm criteria for use in quantitative risk analysis of the hydrogen infrastructure, International Journal of Hydrogen Energy, vol. 36, nr. 3, pp. 2381-2388, 2011.

[9] Hukseflux, „User Manual HFS01," [Online]. Available: https://www.hukseflux.com/uploads/product-documents/HFS01_manual_v1906.pdf. [Opened 21 04 2023].

[10] Jager, H., KPU SWIR reflection measurements (private communication)

AUTHOR INDEX